G000124744

A Million And One Random Digits

A Million And One Random Digits

Douglas Crockford

**virgule
solidus**

A Million And One Random Digits
Douglas Crockford
Copyright ©2022 Douglas Crockford

Copyright is claimed only for the Random Notes.
All random digits and their orderings are in the Public Domain.
The digits can be obtained from `https://github.com/douglascrockford/million`.
The pseudo random number generator functions `rash` and `srash` can be obtained from `https://github.com/douglascrockford/fash`.

978-1-949815-03-0 Hardcover.
978-1-949815-04-7 Paperback.
978-1-949815-05-4 EPUB.

Books By Douglas Crockford

JavaScript: The Good Parts

How JavaScript Works

A Million And One Random Digits

A Million Nines

Forthcoming

Lower Mathematics

Misty System

Capability Security For Beginners

Contents

Random Notes

Note 0:
About This Book

This book contains one million random digits. The digits are broken into lines. Each line contains a line number and one hundred digits. Each line is broken into ten blocks of ten digits. This is a preview of the first line:

0000: 5730610812 7021212098 6943131922 4255393035 4818150524 0598782178 0418153681 6371467827 2290027795 8403371492

There are fifty lines per page. There are two hundred such pages. There are ten thousand lines numbered 0000 to 9999.

Each of the million digits has a six digit address. The first four digits are the line number. The last two digits, 00 thru 99, designate a digit in that line. The digit at address 000000 is 5. The digit at address 000064 is 1.

There was a time when a book like this was useful for performing monte carlo simulations, testing statistical models, selecting sample populations, and awarding sweepstakes prizes. Nowadays, we rely on the random number generation ability of computers to perform these functions, so books like this are obsolete, except perhaps as objects of meditation on the concept of the million and the nature of randomness.

We talk a lot about millions, but as humans, we really don't understand what that means. We can grok numbers like six. A million is more abstract. But here, now, you can hold a million of something in your hands. You can try to count them. If you do, you will quickly come to appreciate what a wonderful thing multiplication is. Counting the number of digits on a line, the number of lines on a page, the number of pages, and multiplying them together is a whole lot faster than counting to a million.

Should you desire to use this book to select random numbers, you could simply open the book to any page, close your eyes, and touch a digit. That technique is easy, but there is a potential pitfall: The digits are small and your fingers are comparatively fat, so you will have to make a judgment as to which digit was selected. That could introduce an unintended source of bias.

That problem can be mitigated with indirection. Select six digits starting at the point that you touched. Use those six digits as the address from which you will harvest random digits.

As you proceed left to right, top to bottom, you are incrementing your digit address. If you ever feel that the set of random digits is getting stale, you can get an entirely new set by changing the increment. You can increment by -1, going right to left, bottom to top. You can increment by 100, advancing vertically, or by 99 or 101 to advance diagonally. You can use any function that modifies a six digit number. There are a multitude of functions, and each will give you a different random set. This book can provide a lifetime supply of randomness.

> **Warning:** Do not use this book to produce random keys or passwords. It should be assumed that any public source of randomness is also available to miscreants.

In addition to the million digits, this book also contains an extra digit, a bonus digit, if you will. I was careful and rigorous in creating the million digit set. That was not the case with the extra digit. I just picked a number out of the air.

Note 1:
Random

In the vernacular, the word random can mean weird, unordered, unfamiliar, shameful, or undesired. A dictionary might define random as lacking a definite plan, purpose, or pattern. But the sense of randomness that we are examining in the digits here is quite different.

Random systems can be examined as individual samples, or as a large collection of samples. The samples, it is said, are unpredictable, but that is not strictly true. We predict random events all the time. One of the participants in a coin toss will win. The winner might have been absolutely certain by some faulty logic, and by winning, receive confirmation of the correctness of his profound misunderstanding.

As you read each random digit, you can try to guess the next digit. With average luck, you should be able guess correctly about 10% of the time.

Luck sometimes comes in streaks. We celebrate business leaders who get lucky, and reward them with fabulous wealth. So while unpredictability seems to be a key feature of randomness, there is clearly more going on. For example, my sequel to this book is *A Million Nines*, and it is exactly what the title states: a book containing between its covers a million nines. That book is the same size as this one, but its contents are superfluous. Everything you need to know about the book can be determined from the title. It may be the most completely descriptive title in the history of literature. It is much more compact than the contents, so we can say with confidence that *A Million Nines*, which may very well be an excellent book, is not random. The title demonstrates a type of compression called run-length encoding, expressing its contents as a million instances of 9. There are other compression schemes, such as Huffman encoding, arithmetic encoding, and LZW encoding. All would be highly effective on *A Million Nines*. None of them are effective on this book.

The smallest binary representation is obtained by treating the million digits as a single integer (approximately $5.73061081270212 * 10^{999999}$), converting it to hexadecimal, and packing the result into 413,241 bytes. Further compression is not possible.

At the level of the individual samples, we say that random numbers have no pattern because, despite our delusions, we can not with certainty predict the next sample. That is because true random number generators have no memory. A coin does not remember the results of previous tosses, so after a streak of several tails, the likelihood of the next toss being tails (or heads) is still 1 in 2. Streaks happen in random systems because there is nothing to prevent them.

When analyzing a long random sequence, definite patterns emerge. The patterns are not specific enough to allow reliable prediction or compression, but the sights and sounds of a random sequence are unmistakable. If you take these million digits, feed them to a digital-to-analog converter, and send that signal to a speaker, you will hear 20 seconds of hiss or white noise. In theory, a random sequence could sound like anything, but in practice, it always sounds like white noise.

Very complex systems can be indistinguishable from random systems. God might not play dice, but sometimes it can sure feel like it.

Note 2:
Random Number
Generation

Random digits are easy to make. Here is how you can make your very own book of a million random digits. All you need is a coin. A coin is the simplest true random number generator. The coin must not be biased: it must not have any defects or deformations that could make one result more likely than the other.

This is the procedure:

0. Set the accumulator to 0.
1. Flip the coin. If it is heads, add 1 to the accumulator.
2. Flip the coin. If it is heads, add 2 to the accumulator.
3. Flip the coin. If it is heads, add 4 to the accumulator.
4. Flip the coin. If it is heads, add 8 to the accumulator.
5. If the accumulator is 10 or more, go to step 0.
6. The result is in the accumulator.

To make your own version of this book, repeat that procedure a million times.

Note 3:
The Rand Corporation

In 1955, The Rand Corporation published the first book containing a million random digits. It contained a set of digits that they constructed in 1947. They made their digits using a roulette circuit. It was basically two clocks. The fast clock incremented a five bit counter. The slow clock sampled the counter once per second. Randomness was obtained because the clocks were carefully not synchronized. Unfortunately, over time, the clocks tended to drift into sync, necessitating algorithmic fiddling. The digits were punched onto 20,000 hollerith cards.

The Rand set had 1361 more 2s than 9s. That is a bias, but it is small enough that it can be ignored in most applications.

By the 1980s, cheap computers were capable of conveniently generating random numbers, making the Rand book obsolete. It is still interesting as an historic artifact. It is a large, public fount of randomness that was created before the National Security Agency took an interest in public computer networking.

Note 4:
How These Digits
Were Produced

I had a much easier time producing my digits than Rand had. Rand did not have a computer that could hold the complete set of digits in memory at once. They did not have access to disk drives, which were being invented about the same time. They were using IBM punch card machines. Not to brag, but my laptop contains 64 gigabytes of random access memory (not including the display processors), a vastness that was inconceivable in 1947. My central processing unit is also crazily faster. This permitted me to take a software oriented approach that was not available to Rand.

I started with an array containing 100,000 copies of 0 thru 9, which I then shuffled 100,000 times using a variety of random functions that I seeded with bits from `random.org`. In theory, a single shuffle with a suitable random function should have been adequate, but this was for publication, so I went with the extra 100 kiloshuffles. The riffle card shuffle demonstrates that several passes of a poor shuffle can produce a good result, so excessive passes of a good shuffle could not hurt.

The random functions included an antique version of `lehmer` (a linear congruential generator), Microsoft's `rand` and `rand_s`, `xorshift128plus`, and my own `rash` and `srash`. The random functions and the array of a million digits were passed to a shuffling function that implemented the Fisher–Yates–Durstenfeld–Knuth Shuffle algorithm.

Note 5:
TLDR If you do not have time to closely read all two hundred pages of random digits, I offer this summary:

Divide the digits into twenty chapters of ten pages each. Count the populations of digits in the chapters.

Chapter	Digit										χ^2
	0	1	2	3	4	5	6	7	8	9	
0	5022	5021	5009	4983	5042	5011	4847	5133	4879	5053	12.345
1	4897	4942	5129	5083	4988	4985	5016	4934	4973	5053	9.204
2	4857	5018	5006	5061	5093	5052	5095	4863	4967	4988	12.982
3	5048	4949	4928	5108	4999	4944	4978	5082	5035	4929	7.672
4	4992	5017	4939	4925	4983	5069	4874	5099	5030	5072	9.302
5	4876	5040	4971	5054	5015	4899	5014	5065	4926	5140	12.131
6	5076	5005	4951	4950	4987	5073	5017	5041	4978	4922	4.947
7	5091	4975	5043	4969	4944	4932	5070	4974	4987	5015	5.089
8	5069	4921	5019	5042	4986	4964	4945	4855	5119	5080	11.846
9	4956	5111	4960	5035	5029	4995	4946	4944	5006	5018	4.872
10	4953	4989	4953	5066	4885	5025	4972	5118	5098	4941	10.107
11	5059	5081	5119	4979	4889	5019	4983	4993	4948	4930	9.053
12	4983	4913	5030	4985	5076	5024	4988	5036	5015	4950	3.900
13	4979	4924	5003	4963	5064	4964	5018	5035	5019	5031	3.171
14	4986	4973	5005	4979	4958	5037	5105	5025	5033	4899	5.492
15	5079	5017	5093	5023	5081	4957	4901	4919	4957	4973	8.611
16	5078	5057	4972	4833	4969	4972	5164	4938	5091	4926	16.849
17	4926	4986	4951	4953	4959	5002	5130	5029	5058	5006	6.621
18	5010	5060	4882	5112	5049	5069	4901	4989	4948	4980	10.071
19	5063	5001	5037	4897	5004	5007	5036	4928	4933	5094	7.163
Total	100000	100000	100000	100000	100000	100000	100000	100000	100000	100000	0.000

There are approximately 5000 instances of each digit in each chapter. There are small variances, which are expected in a random distribution. The surprising thing is that the digit totals are all exactly 100,000. This is a consequence of using a shuffling algorithm rather than taking digits directly from the random functions. This looks statistically suspicious, but it is by design. It avoids a small bias that the Rand set has.

Note 6:
Run

A run is a string of one or more of the same digit. A run of a single digit might not be considered a run at all, but it is the most common case with a total of 810205 occurrences. Notice that a run of 2 is about a tenth as likely as a run of 1, and so on. In a set of a million random digits, a run of 7 is quite rare, and longer runs are even rarer than that.

Digit	Run length						
	1	**2**	**3**	**4**	**5**	**6**	**7**
0	81034	8084	797	88	11	0	0
1	80754	8223	804	82	12	0	0
2	81116	8047	791	94	7	1	0
3	81230	7971	813	87	7	1	0
4	81116	8033	821	72	11	2	0
5	81150	8033	807	79	7	2	0
6	80991	8079	805	99	8	0	0
7	80937	8094	821	84	14	1	0
8	81154	8101	746	92	5	1	1
9	80723	8207	819	89	10	0	0
Expected	81000	8100	810	81	8.1	0.81	0.081

A run of seven 8s occurred at 105787.

Note 7:
Skip

A skip is sort of the opposite of a run. It measures a span of digits not containing a particular digit. The shortest possible skip is a skip of 0, which happens between each pair of characters in a run. There were 99932 skips of zero length in this set, the first occurring at 000020. The longest skip is at 814034: 127 digits without a 9.

The following bar chart shows the number of occurrences (on a log scale) of skips in this set.

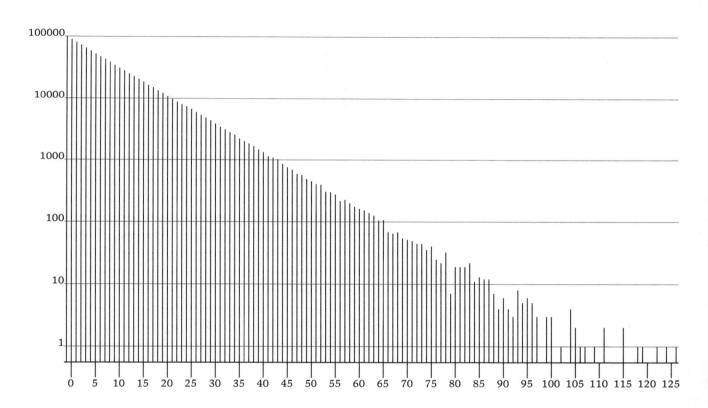

There is a strong association between random numbers and games of chance, where people make bets on their ability to predict the unpredictable. Sometimes there are winners, but always there are losers.

Let's play Five Digit Poker. We divide our million digits into 200,000 sets of five that will be evaluated as poker hands. We can then compare the scores with the expected outcome. The actual scores vary a little from the expected scores, which is not a bad thing. If they matched exactly, that might be a little bit suspicious. That is the nature of randomness. You expect them to be close to the expected values, but not too close.

Hand	Number of hands	
	Actual	Expected
No pair	60615	60480
One pair	100801	100800
Two pair	21459	21600
Three of a kind	14407	14400
Full house	1795	1800
Four of a kind	904	900
Five of a kind	19	20

In Five Digit Poker, one pair is more common than no pair, and five of a kind is possible. The kind hands are runs that are constrained to five digit boundaries.

Many games of chance are played with dice. Dice can be fine random number generators. Adding together two dice can produce a more interesting distribution than throwing a single die.

Let's roll dice. We divide our million digits into 500,000 rolls of 2 ten sided dice. The most likely roll is a 9, but the odds of rolling a 9 are only 1 in 10. Though the odds are against you, it is likely that you will roll a 9 about a tenth of the time. The odds of rolling an 18 are only 1 in 100, but if you keep rolling, it is likely that you will eventually roll the double nines.

Roll	Number of rolls	
	Actual	Expected
0	5061	5000
1	9881	10000
2	15160	15000
3	19877	20000
4	25003	25000
5	29927	30000
6	35092	35000
7	39823	40000
8	45097	45000
9	49763	50000
10	45153	45000
11	40408	40000
12	35147	35000
13	29726	30000
14	25027	25000
15	19976	20000
16	14764	15000
17	9967	10000
18	5148	5000

Below are the million random digits printed as a 1000x1000 image. The 0s are white, and the 9s are black.

The image looks gray, but it is not a clean uniform gray. It resembles an aerial view of landscape. There are ridges and valleys winding through, but nothing regular or repeating.

This was a common sight in the late twentieth century when television was distributed primarily by analog broadcasting. When a receiver was set to an empty channel, a rapid succession of images like this one were presented sixty times per second, accompanied by the sound of white noise.

Note 11:
Pi

The mathematical constant pi is the ratio of a circle's circumference to its diameter. Even though pi is a ratio, it can never be the ratio of two whole numbers. Pi is incommensurable. It can be approximated with fractions like 22/7 or 355/113. It can be approximated with decimal representations such as 3.1415926535897932. We can get sufficiently close to the true value of pi to make practical use of it, but we will never be able to build a digital computer that can know pi exactly. Because pi is irrational, its decimal expansion goes on forever without repeating, looking like a random sequence.

Can we find digits of pi in this book? Yes. There are two occurrences of 314159 at 749285 and at 881042. There is nothing special about those digits. As we will see in the next note, most six digit strings can be found within a million random digits, just like runs.

Pi	3	31	314	3141	31415	314159	3141592
Actual	89970	9050	867	97	14	2	0
Expected	90000	9000	900	90	9	0.9	0.09

Going the other way, can this book's million digit sequence be found in the decimal expansion of pi? The answer is probably yes. This book's sequence will likely start somewhere in the vicinity of the $10^{1,000,000}$ decimal place, but it is impossible to determine exactly where because we will never know pi to that precision.

Note 12:
PIN

A Personal Identification Numbers (PIN) is a numeric password that is used to secure credit cards and other accounts and locks. PIN is a misnomer because it does not identify a person.

Most PIN numbers are four digits because that is about the limit of what many humans can remember. Every one of the 10,000 possible four digit PIN numbers can be found in this book.

Only 99,995 of the 100,000 five digit PIN numbers are in this book. The five missing numbers are 00312, 12410, 16572, 63249, and 75450. It would have been nice if they had been included, but they were not lucky enough to make the cut.

Only 632,258 of the million 6 digit PIN numbers are in this book. To list the missing numbers would take another book. If you read this book backwards as well as forwards, then 864,163 of the six digit PIN numbers are in this book.

Note 13:
The Millionplex Library

The Millionplex Library contains a copy of every possible book of a million digits. Its stacks include this book, the Rand book, *A Million Nines*, *The First Million Digits of Pi*, and more.

A random process, like the one in Note 2, is potentially capable of generating every book in the library. Every book is equally likely. Most of the books will have statistical properties similar to this book. However, a truly random process can also produce *A Million Nines*. If your random process produces several 9s in a row, you would be right to feel suspicious and concerned that your coin is somehow violently biased. It is possible to randomly and fairly generate *A Million Nines*, but it is so unlikely that we like to consider it impossible. But generating *this* book is equally impossible. Without knowing the seed values and precise application of the functions, it is impossible to recreate this book by a random process. The odds of making this book were 1 in $10^{1000000}$. And yet somehow I pulled it off.

The complete, unabridged, unexpurgated Millionplex Library contains $10^{1000000}$ volumes. To understand how large that number is, read *A Million Nines* as a single number and add 1 to it.

Book titles are useless in the Millionplex Library, so books are cataloged by number. A book's catalog number is its million digit contents.

The Millionplex Library contains more matter and occupies more space than our universe. We can never hope to understand it in its entirety. We can only ever hope to see a vanishingly tiny sliver of it. You are holding a piece of that vastness in your hands.

The Millionplex Library is much smaller than Jorge Luis Borges's Library of Babel, which is well worth a visit.

A Million Random Digits

```
0000: 57306 10812  70212 12098  69431 31922  42553 93035  48181 50524    05987 82178  04181 53681  63714 67827  22900 27795  84033 71492
0001: 06954 43401  12553 47642  39595 57268  41957 52205  75814 95486    63583 13537  40655 15785  47348 52899  82600 65639  84574 66466
0002: 70298 89115  73736 68598  48225 93061  07010 50225  10417 14270    30524 79647  33601 00597  21875 58019  26755 31768  83986 62973
0003: 70165 95948  41960 39293  32041 71883  02427 78120  14169 39048    00036 23007  12869 30971  58944 95748  88176 81955  09632 60219
0004: 66962 84604  26338 44569  19685 54272  99785 59615  55245 19338    03151 61849  56900 30723  17946 24845  87490 08331  04987 21217
0005: 94917 72966  84341 04439  29362 05731  73870 72059  24958 09148    65195 83283  08983 48430  74957 65875  64853 98615  92404 53134
0006: 75555 93552  65503 13465  36808 53210  26451 07921  31437 40000    49447 34963  84864 21872  24169 31300  85627 75219  59050 44846
0007: 98125 60666  04674 72200  33705 87979  38047 76167  27276 40847    67668 40574  05496 89050  85172 51712  73959 94957  19195 15058
0008: 11482 85436  21842 57793  71955 04902  81522 77778  02256 99459    97978 68389  24265 01094  39718 37584  92365 88474  06907 00717
0009: 88017 88916  93042 99135  96879 67886  93531 08029  10627 86097    91498 65289  69123 17396  30463 35650  57925 28273  66068 49724
0010: 18724 25246  73334 98039  17430 47204  54649 21886  28987 34784    04332 17555  54318 66902  44743 09564  68866 76912  55949 13533
0011: 95253 39546  85444 92559  66670 06300  32598 92041  02933 36266    71545 20723  67136 98669  59674 50010  92017 78304  37044 93753
0012: 04631 82696  35935 34770  59682 91475  49447 67698  93587 34837    81157 04645  51821 68123  26237 23709  37586 31635  84905 00333
0013: 68821 64590  60022 24974  60121 41081  26564 75687  10977 54029    10824 50644  85802 34271  64350 34376  10005 88345  62997 87358
0014: 19376 25685  49700 84240  29517 79630  92217 18264  79975 60033    62625 41064  22310 83514  70883 73068  93308 93686  35295 63593
0015: 27121 35571  35196 83496  85614 86344  93408 80527  61894 84831    10409 77303  46495 88601  18958 60937  34498 19112  12717 25940
0016: 98092 58792  84990 07756  86625 57997  47839 21573  96371 69159    34124 94304  88597 10156  31970 17079  66502 51095  60124 40708
0017: 18378 53144  28608 03397  45773 00643  30967 43643  33215 39307    10544 65172  97345 14414  52453 59342  56362 75799  88942 74805
0018: 02782 33384  87644 52962  20163 14377  97089 26969  20376 39636    03053 72624  33484 64437  04970 14561  21052 65794  36115 16991
0019: 45388 93683  66361 27448  40345 48573  60456 53577  36844 92986    59388 97941  32069 39207  34403 84956  77730 05157  66836 17377
0020: 52445 76659  57504 69767  92657 07114  25069 08147  40037 37293    41454 21185  14900 84853  69540 54985  95520 78258  54346 93566
0021: 71485 53610  44870 47775  54408 02271  19281 85740  97124 52202    04113 62907  58553 59551  49558 15977  07846 96686  22609 92048
0022: 34003 33278  51414 21956  83293 39913  01623 81118  82116 10468    77491 27581  22884 11586  42284 36016  86384 84266  72111 26877
0023: 67405 23451  08086 23678  86099 51263  69618 57118  96360 84321    60207 89621  86307 55234  00951 13227  09096 30968  00677 98886
0024: 99940 39675  60697 68232  63978 60497  24258 60612  40824 97691    85243 04705  47170 57436  20171 98432  37056 23227  26764 66912
0025: 74617 76018  67652 84357  19357 78409  46544 51595  23720 51772    14854 11616  24400 96754  78299 03074  61738 77164  37126 07619
0026: 84162 48100  12447 96729  52299 97124  97563 07805  57342 86581    23278 17574  79650 38236  42100 68378  35896 22470  06260 06762
0027: 99763 53437  72126 26098  22108 70306  92887 18761  85066 79837    33610 98981  53431 28066  72377 64305  13228 47799  23848 72089
0028: 52276 68338  45420 05691  99242 17595  20035 24019  51708 89071    78592 25827  87702 71644  29308 44801  07357 15609  63402 27934
0029: 36310 47509  74378 14118  43447 71292  66833 68689  49425 78654    53387 52672  47546 69629  73273 14801  57795 13140  23208 55017
0030: 59386 46750  68011 90591  36618 51367  85927 34728  49738 25539    90704 86526  93373 32931  49875 64311  20955 66441  27642 90635
0031: 59088 82112  72131 27572  64832 35111  11471 91468  40128 09547    42734 77229  31805 76933  68190 16977  80464 20962  96491 82801
0032: 46137 01953  99937 79770  64724 45962  18778 04614  70103 37196    07297 30367  13083 17219  04714 24441  56493 11031  08278 86941
0033: 78382 22295  45610 43618  49502 57413  93214 50981  65087 41864    84220 87463  74287 33990  42603 38912  87634 23929  38913 31111
0034: 10204 30951  92555 52610  15041 53309  71833 36515  54294 44396    69722 70304  61629 01464  47349 93920  77352 63787  57628 03293
0035: 56307 91667  07100 53509  82289 05827  69581 11953  04228 58172    21136 23879  50432 35747  18825 24308  45597 82896  71935 51291
0036: 58032 12970  89293 89136  14755 18712  48564 14050  06969 09882    20315 71851  13465 05728  55312 52456  44658 33974  72329 25118
0037: 25036 91442  44567 21121  81356 68071  56580 45985  83775 75611    10918 74113  23735 66535  41835 74052  95524 58750  47450 93095
0038: 43281 72776  21787 86853  35354 70843  83517 07178  00615 14439    00836 24925  98833 48345  30908 80420  49137 24340  08448 99297
0039: 69759 66589  23470 36072  99384 61656  91311 81103  69091 61345    08852 26090  75890 52777  72518 39831  46114 66039  16966 78880
0040: 62056 85802  49395 36988  54780 13620  55974 43178  51608 41668    88047 43890  31087 41986  68371 14493  21408 90750  44051 56916
0041: 24706 58509  12125 27722  30259 61862  03308 88375  45441 44323    95938 69286  73764 23676  88359 53662  61694 86384  07831 50884
0042: 24511 56561  99388 31214  19597 16258  55159 30071  22472 15801    67636 77371  69211 24240  75263 57709  46628 24073  94108 40092
0043: 03930 31529  51545 06821  17507 75199  69664 30410  28813 68169    44039 55059  86660 43729  26125 87991  42174 50203  65281 32261
0044: 67959 31791  72094 52338  77049 94415  68633 48852  98943 62031    68276 35255  68814 25606  72110 05021  22527 70183  64873 25941
0045: 29884 55445  12458 32093  19893 81111  86274 45230  41464 09536    57714 32245  93196 54847  78934 08205  33459 96929  53728 57141
0046: 82117 99326  24628 57082  58201 52797  59374 99892  40148 81870    19578 04832  50078 76307  12591 60622  75237 71643  27968 88380
0047: 52143 57999  58998 15371  97215 14654  82974 35479  22801 07795    66904 48727  58280 30710  89697 12833  50426 09875  63385 65086
0048: 42104 27353  29375 92635  76119 74566  13490 73089  51198 41198    33399 55225  32474 70556  36459 29382  23516 97529  08024 28338
0049: 27652 81689  25390 90796  77655 71105  42528 47532  21219 08942    00274 03550  26390 95549  91964 25539  10438 14021  23778 04072
```

```
0050: 59681 08977  84265 42431  61553 94929  92079 54542  63192 27283    34848 54432  36080 99867  10321 55568  46834 09666  63781 65154
0051: 82335 44637  59469 82176  27432 26146  60126 75464  69260 40627    01102 43067  96175 25804  29900 71971  96426 09026  70659 01884
0052: 20639 37224  85270 50812  97778 29593  59804 24028  00015 29814    74777 36235  56086 17742  61669 83579  11670 97139  10736 75613
0053: 49214 61422  06579 95216  17360 23661  07360 25212  75099 73687    50328 60873  27085 61960  26356 69293  66536 73479  09752 88853
0054: 04301 76375  48377 06193  86401 68771  74168 64430  54682 30823    07803 81889  23100 17827  64164 59702  80404 30399  52004 96067
0055: 52702 15765  40886 91617  76196 80309  91244 46236  24730 92188    43871 16032  87794 06024  16784 42919  54694 41606  30137 64063
0056: 15606 58822  16362 73517  04857 13688  18657 33925  78176 63098    19366 27390  49139 88296  96797 15059  23517 76172  57372 24637
0057: 41871 91231  02332 52450  00990 47927  84618 36811  20218 73629    29862 86552  31041 54085  11120 05912  54591 51819  54065 06589
0058: 85559 92667  25144 01113  50039 73636  43155 27746  17460 72441    17515 08878  95505 95303  72927 11442  44116 02091  28080 57179
0059: 27147 24647  91373 44151  34444 29412  85043 57373  43882 04575    71953 33895  82260 17302  28216 22203  43306 75253  91733 46376
0060: 05282 64528  91861 62425  91852 05900  30731 13361  94098 61354    68405 15592  59526 42708  11496 86019  67812 25037  79282 07236
0061: 43768 58061  96478 40974  08602 06541  59924 29942  29368 07243    70611 88657  10279 04845  39700 52070  22384 73196  47851 81502
0062: 77711 11422  74004 06129  49804 71845  72200 56301  39039 10874    83654 13530  00963 40720  01385 34749  60541 15774  19200 53535
0063: 86009 59357  98267 03751  78027 16480  03433 65304  59154 80688    71895 69564  21748 84955  01727 86792  06471 12564  40924 76463
0064: 40752 52359  10509 49966  59817 72081  39311 01459  27603 81873    17878 25027  28853 95723  93237 77688  74480 27846  60126 11883
0065: 42021 07082  10390 19119  42070 34219  88597 50142  97732 35380    47574 87775  79091 68516  75217 85652  97459 38900  95964 57511
0066: 62436 84977  38204 07812  78289 95764  54227 26237  71826 42366    52095 33409  87128 21018  12425 46177  57378 17973  91814 17334
0067: 47753 27656  06759 67881  01666 85023  51859 13249  05440 87054    37910 14826  77680 30010  11417 09885  86843 29124  40570 65470
0068: 17988 58377  86777 08621  18597 68134  67871 81193  64844 22908    67294 90904  33688 15437  44416 02551  54267 29649  31013 80420
0069: 32300 78817  61412 67965  50904 92364  39571 13910  26385 92345    22411 73744  40566 91172  03158 54635  58071 46948  07368 32036
0070: 49621 95492  37187 76992  73588 33646  41624 86192  30101 05540    25991 01040  77876 20324  66874 97293  17001 18303  89200 60181
0071: 93395 65933  48119 53578  26976 05823  10029 81685  23883 50473    00428 33181  85038 65852  43788 20269  93883 12673  97398 94209
0072: 24293 01127  04042 59093  52675 40085  37589 83115  31522 92047    60689 55055  41440 55233  46335 90560  66087 77006  22447 57617
0073: 10972 40281  28376 25559  20683 49701  46715 57050  69842 62276    78547 30966  80876 16024  95340 40186  83964 45664  47313 29349
0074: 77751 79072  07731 77396  53315 27408  12192 73473  08684 58464    10116 12342  10151 81353  86204 51122  55946 03116  54593 75782
0075: 28757 70397  19103 91676  61139 44526  42818 19131  39849 08721    77398 71020  77148 13343  37960 29984  73368 81049  94327 48679
0076: 45207 86626  86388 89387  39753 79909  34696 90071  02504 06679    48929 03233  54846 97568  31369 09737  90008 78689  57149 04591
0077: 24547 56947  66233 29708  10099 60529  04835 83385  64537 95928    47418 22209  80199 57631  50596 32167  65068 65089  94961 24412
0078: 42995 98545  04042 73757  19198 58159  07594 00441  42580 48570    45622 43004  02012 69720  50826 14180  94803 54104  15685 73550
0079: 74199 15309  69272 32697  61768 86194  72405 09936  71534 31713    02281 64062  56040 28545  43197 00018  97703 43340  47878 97294
0080: 73880 94873  29596 34001  45547 44660  69622 06824  71555 47382    69764 42812  22739 43795  04312 15638  25040 53299  01741 34403
0081: 61853 84208  46043 15605  23013 61558  45635 71764  59739 40845    43502 27230  15160 14643  09894 37057  70268 45146  56527 74056
0082: 15476 60101  56095 54740  43571 97289  77734 49228  46591 72992    69821 55550  05499 31050  07832 45367  74501 15296  75230 08973
0083: 71875 90509  83539 28016  65641 31179  65429 21138  31892 80683    07183 48819  35573 57723  71014 41424  07234 49051  46228 71081
0084: 39837 30036  73245 88486  24790 55874  68622 45646  45747 30759    00643 90078  70422 35350  19376 58993  73064 80148  43528 95830
0085: 39477 17350  77846 17819  61443 60261  03896 11058  72288 38781    02716 47309  38886 75953  90690 92052  56641 60001  42994 29828
0086: 11706 75123  41376 19705  56133 00610  08153 66906  01518 16118    15858 78379  68270 22874  91763 41933  63062 35708  17684 08324
0087: 97193 42454  05941 45040  65424 03574  12213 10086  29791 79703    87667 67882  94959 99253  50802 78031  36028 81722  21120 08630
0088: 57359 06166  89328 64351  67795 93264  93254 09051  41869 12239    90420 90342  33042 61946  55186 78963  60814 82933  40911 56976
0089: 46394 07475  32808 45465  52073 18169  55691 84400  05692 76735    41907 34835  34787 01199  87245 17589  79796 25120  89670 90292
0090: 26975 06871  69009 87041  64671 53938  98998 02941  81843 08053    48705 20246  23973 76913  31846 37260  25974 62265  28832 71813
0091: 20987 71170  96272 61774  69090 83919  83156 86228  57091 43166    80729 35275  28519 22975  90987 82300  05353 01687  69172 92875
0092: 03428 88905  28771 61229  49082 89734  21791 94217  76353 26925    57353 82781  23315 43908  95776 86841  94809 36852  24207 27920
0093: 46359 14948  84198 25640  21733 20991  55947 14516  75693 90867    75009 23034  52766 23581  94804 35771  85003 06406  40232 61619
0094: 71899 98439  41362 24397  20202 30582  94081 60020  82445 56874    39916 48101  58941 71741  40294 67715  62566 52547  25881 88347
0095: 91188 73200  65507 88295  44890 26644  68577 68524  05182 79849    71028 82171  82349 77320  68027 20078  19388 68216  79534 35393
0096: 66284 97300  59735 01211  28572 98361  43485 92398  84563 20394    37993 71755  46572 81169  19939 58872  33448 56908  55537 65567
0097: 49638 18744  31017 39310  27390 28932  69558 08821  20245 72993    78123 50933  27915 12002  91249 06573  85166 52540  98184 68625
0098: 03173 68131  43536 27706  79973 13842  16370 51096  77093 60319    89009 22168  62870 19235  98655 97793  50702 71442  04317 08401
0099: 32373 05778  51578 78055  00030 60123  87625 49927  08210 95875    19427 86220  79415 19060  01618 88449  05743 58887  37430 07876
```

```
0100:  24042 32210   06440 24019   31141 60612   63718 29518   65104 42962      51412 71045   22342 31422   50737 18532   03157 16429   48524 63200
0101:  38954 68766   95876 98396   60732 65621   16510 73539   60758 33143      53062 48691   45015 33756   39454 99850   84117 96824   65997 01795
0102:  71226 15404   25843 66992   30342 44388   03341 54592   01617 55245      36164 42977   16188 60496   03823 10820   67579 80180   48814 90937
0103:  11427 45298   21739 85247   70033 09378   14729 88149   63947 73488      95084 28067   95544 10614   66308 71222   94704 08137   87949 92348
0104:  03765 23416   92723 06749   83237 15672   77836 78270   45442 32911      12865 90445   87703 50507   79480 30331   53697 94378   61238 87692
0105:  98171 98899   81967 39345   43313 10835   85157 94942   15729 04492      56389 46528   96967 37413   88340 49155   97065 47421   95477 13256
0106:  82803 09617   36775 10665   40653 95792   88050 62014   78992 33864      03171 87519   92720 12461   31101 00030   22554 42359   70988 50758
0107:  18748 46358   15008 79611   94256 93643   80744 97446   12884 47573      61249 00968   43129 51521   47735 88880   83468 18355   70464 36182
0108:  01851 44732   01146 16160   12473 18745   20826 80691   06511 64672      52129 49491   51734 96982   65843 85449   86561 55715   84567 56440
0109:  99059 06149   64762 86660   43575 90578   04406 93320   49729 58107      38402 17307   62205 22959   49102 51319   62349 87664   42417 23286
0110:  26780 19815   03544 43323   13178 16735   07024 59880   85775 97729      42566 60059   95252 75769   29418 40618   37584 20377   43951 45147
0111:  31312 22156   56018 23409   75233 35579   00685 21283   59466 54343      83996 13468   82685 38550   40458 54985   12147 74669   46057 16583
0112:  78262 55669   17562 36907   02157 62413   03427 73449   90412 93080      84617 92421   99904 72408   71636 05569   81876 87728   42986 34257
0113:  34407 62329   75431 34887   98440 31421   59769 88402   44891 69504      81455 15740   10573 82908   07980 42323   57663 60828   03551 27174
0114:  14482 50772   55888 75384   12013 20378   61914 92506   39414 91689      03189 88784   54040 76871   37480 76461   48019 39257   28850 69467
0115:  64770 53158   01946 11993   48229 94215   01845 61700   86442 66705      37615 23900   16235 23645   23827 48626   28853 67687   46185 66042
0116:  83645 81751   23809 92470   64135 72831   81565 74535   40349 11811      73482 53770   18061 46639   18723 76899   56072 01149   53088 25361
0117:  80358 90698   03919 13045   14528 99616   41974 86909   78016 45625      90241 30012   08277 86380   00057 00916   22408 11716   70359 29594
0118:  73545 32677   81799 58456   33725 46915   32706 50361   46924 89812      86843 85979   30456 68161   27419 01746   03236 35284   72224 75550
0119:  62204 44442   62618 66009   82792 47864   38141 51698   89415 08761      63193 31640   78729 81987   23990 34415   46535 90038   84019 69597
0120:  56924 82861   63368 01721   39662 62055   87008 44185   12827 87368      51895 29782   22199 75080   94216 97643   57208 42971   75986 70654
0121:  60555 98121   09608 33381   49413 58612   78751 64411   11481 89133      23737 98476   62451 47133   39012 06212   93086 92335   13444 69617
0122:  74386 90245   86105 97999   97305 45081   48558 43227   26536 43712      22465 09210   99141 99758   62886 66097   78156 69684   88141 76220
0123:  33093 97393   13625 72909   09326 83052   95364 30170   49520 10806      37785 53230   52824 69975   89706 33074   24034 51826   55173 20815
0124:  57930 82733   09610 59498   24231 41867   96984 77609   55082 75332      38690 03864   65743 95421   83001 34410   06152 89755   82365 51015
0125:  92294 87603   60568 92558   14085 28274   78178 80108   71986 12511      27700 94845   87487 56747   41411 50067   16107 67143   75944 00341
0126:  72064 41807   48811 76926   76139 15694   92609 85244   20320 02569      41872 49143   75559 78189   81891 20616   15190 88323   43031 71599
0127:  48751 64575   37768 96568   48517 38027   12626 84924   26298 43326      50237 42143   82051 79898   90185 44172   54276 80724   73959 43223
0128:  92753 07119   64674 01645   89727 86916   12587 12462   10772 21546      79597 45467   16323 43109   28417 68719   46530 36556   62039 36186
0129:  60410 04755   53307 61114   32629 61650   43548 45881   01084 77277      03416 07747   71841 32263   98039 60096   15190 12758   29659 51024
0130:  56581 67182   02608 59314   87877 14139   11749 04481   04652 80453      56988 59768   85929 45666   47098 96048   31256 96963   35698 79967
0131:  60361 94340   94464 03660   77777 30306   45659 44791   50123 92813      17002 94136   77222 13451   99335 77012   20247 59501   32457 89784
0132:  91549 46134   40996 41687   47482 55353   36130 16253   24076 39302      33193 92231   76433 29483   73294 88760   16854 93344   68253 33146
0133:  23098 10680   44824 05568   64815 21952   44279 29296   61556 22537      26302 50898   51420 22298   85194 26484   82923 48157   28584 26112
0134:  19317 43582   85202 34541   71231 63120   30916 10326   00420 96977      43070 85381   51677 50956   85959 28523   85714 33870   46792 94176
0135:  57218 62373   10934 89104   70635 78939   94864 57579   96823 45455      52251 54798   93163 23476   34943 22370   43583 84616   76634 23225
0136:  87776 74304   38738 17398   27045 17020   22308 95321   32144 98752      30991 15427   25206 81611   00284 01557   80780 14395   51129 28130
0137:  60153 74512   47465 86811   30575 07896   69195 17707   50093 64565      68026 98774   89356 70103   03037 91696   53493 38849   12452 77351
0138:  16175 86813   91251 29998   13022 36244   11588 85862   26143 40935      60390 75470   11956 61723   89509 04951   77124 47296   72249 95495
0139:  40004 72163   58809 63411   76002 09510   64805 10666   42407 43820      25733 11052   06698 20477   88177 98175   45667 12294   46762 85541
0140:  76495 38821   52983 43460   98915 96037   50832 68981   78451 60613      92002 17879   69315 22567   17200 95735   34145 89313   50996 65516
0141:  34445 11777   50626 68399   58170 96395   17327 28337   16718 99183      12502 40788   82040 06925   57225 10614   79480 77391   58386 50872
0142:  27604 54277   15288 00690   70889 43257   67332 94476   80182 05414      11578 90738   27638 17229   35052 03798   69987 85560   93364 90466
0143:  21571 76426   89595 75145   75693 78431   98345 82241   58833 30866      20919 02845   53888 19860   86507 67523   60219 32759   01353 82310
0144:  00834 24270   17769 97332   96174 12698   85130 17779   36149 08094      15880 48856   76098 10853   60748 14765   96874 17438   26007 19883
0145:  28280 03818   15973 86022   06438 45058   45385 74004   08462 12434      12004 03843   31617 78938   73492 51822   72989 00135   79430 14692
0146:  77246 05337   01919 39838   31484 25693   51349 52660   53854 11379      63571 61739   11789 05280   87875 39556   27753 21797   27993 35079
0147:  02833 21584   70652 71998   33375 31374   36307 59203   22654 29512      25224 21605   82247 92964   26413 52494   18563 54300   04676 84970
0148:  99436 67908   84007 59714   96912 05476   55371 86522   87193 13969      89304 37130   70308 79421   49902 10163   92437 20499   53662 82613
0149:  19157 36268   61024 65614   05755 00531   12772 64667   71228 80467      55961 04545   78515 30418   35567 11892   18011 40511   31352 10961
```

```
0150: 34799 06480  49079 03873  98146 91061  53233 92611  08593 91572    10957 79901  14026 67423  76823 48626  79071 21961  40341 57264
0151: 23229 11507  24844 29351  52673 91374  51102 63789  48266 99579    73159 96935  15864 24865  93530 89418  76917 24405  16903 10425
0152: 86312 43380  62581 55284  07827 73654  11394 14819  72813 96488    07040 15790  60267 61677  84745 13342  71405 00123  50460 97036
0153: 35849 06361  95193 84441  38596 09683  81157 14186  04752 48294    98009 54878  27631 55737  46902 83325  00635 53211  34265 65674
0154: 42139 56393  18582 43975  58949 22079  47371 90449  84306 97139    46488 98674  15795 48958  20645 19356  96637 76738  67403 17527
0155: 57567 61301  82771 17404  60694 51288  62818 66851  00570 53891    17075 51448  05333 57483  83979 22927  10009 26135  55536 01688
0156: 54797 46561  15501 68363  74937 77597  17388 88622  70588 67837    07805 01623  44545 18370  07274 78378  25807 25799  60433 29281
0157: 52244 47315  95797 42190  17968 60045  88932 46176  91401 80319    79479 74599  47444 29978  85283 02673  54940 69259  95287 20329
0158: 35302 12912  72816 18572  64650 70601  51804 19262  66036 98724    87530 18002  33176 50469  04821 40787  59581 71330  99770 74956
0159: 61760 14101  23677 14063  72152 76823  10805 32241  33289 44643    92730 80410  05415 70772  12115 27900  39956 44828  72220 89514
0160: 92019 65074  08018 27363  84944 18138  43691 24550  64847 20068    60126 14967  26700 80516  14535 24282  99649 13477  88845 67161
0161: 16428 80172  46020 18746  60828 07334  40045 45478  44110 73532    19726 14251  62091 72009  17874 02357  14996 55535  05693 60949
0162: 12091 83858  10036 11184  67251 66872  13361 48898  66782 61470    02805 51317  68323 68422  45939 32652  30370 81134  54054 67796
0163: 95072 48785  78333 63675  15705 91915  77101 84072  23981 98148    90237 34252  68103 56662  11923 53816  22494 91790  45566 76671
0164: 68401 69990  23604 86000  09256 85109  51455 47926  20249 27524    91262 96718  07016 35657  60189 09450  12959 69113  29785 87301
0165: 06243 50024  16094 58208  22295 31586  42191 58724  53011 13314    13478 68920  44900 08958  88803 06866  96297 11021  77299 78040
0166: 75727 81603  33713 22865  56417 05702  65608 95092  89862 56746    92455 23109  00021 58509  82291 96101  91123 45637  03062 53664
0167: 76010 43336  94621 44637  76098 73472  46786 01340  44378 25022    15316 22555  75572 35710  42043 09232  40206 85115  04936 11564
0168: 50159 85701  08067 25340  11709 38999  99404 70885  57948 41806    36633 69947  54465 61763  02152 92392  62616 75129  18957 05299
0169: 08719 73147  94438 99821  66086 46288  05946 64908  17939 33007    72324 04933  78273 24595  68477 10186  62465 77062  20030 55471
0170: 14804 17419  89215 64473  92116 28882  96989 14539  78876 75082    28242 44386  12851 16346  86978 62926  81427 15033  90111 10485
0171: 52885 62815  39471 26816  90928 96118  84750 52706  03905 48204    23954 48032  21936 51791  98044 91406  80779 17639  35748 29726
0172: 03863 95564  19365 67825  93437 78176  61043 18246  22364 22014    75602 62574  79832 37182  13670 02222  39128 50273  55641 40880
0173: 23022 06500  36214 56477  48963 22247  33129 12643  24365 27070    04753 10492  00418 93741  51818 20457  70138 50335  00655 83933
0174: 56319 70674  31776 09337  01172 93187  19661 07181  43180 55359    91956 69931  74720 69455  25466 20147  18275 94733  09963 33647
0175: 10465 12395  91171 54343  94123 22871  97939 32876  04055 56579    19286 72004  22985 55654  14081 26420  74722 18392  81467 89341
0176: 14514 23493  07704 74581  18542 36438  98827 14650  46135 99381    86933 49384  92765 22803  78816 78462  60895 86294  75910 58277
0177: 28382 15445  95405 44040  59974 55141  28842 65405  40079 84478    70106 60972  79124 04437  13075 01383  38364 04072  48759 33222
0178: 00277 28270  66534 77656  12178 98622  21007 71139  20654 67903    33490 43772  52766 31044  25223 43820  97420 27160  42601 39613
0179: 00641 40269  05328 99964  57523 69290  19902 57730  34822 52970    02888 75070  00790 64462  06950 75831  36236 70272  18752 32050
0180: 27649 16900  29753 41145  80596 58357  84443 94296  49324 60324    17927 55665  95266 82725  76505 38183  27514 78344  13091 25236
0181: 76684 51590  34567 54997  92656 69640  50816 15403  18164 23600    84031 28995  97837 78543  00222 31145  18545 13689  25134 58856
0182: 71129 99923  17729 49708  65263 59749  25739 88107  11947 73003    32450 33984  29732 83367  75910 19762  09268 95048  33415 48352
0183: 19183 04034  67337 12345  12287 82707  87653 60675  83645 56545    97622 80674  39900 69017  65110 04655  38778 47689  72060 85341
0184: 44521 65389  44110 42451  93825 10498  65767 07169  86226 51836    25928 21106  13841 22492  17715 52223  91684 56292  08502 88411
0185: 19308 55205  49231 71436  76933 99389  03127 63375  67199 02760    71486 68650  59096 33786  76939 64878  68649 30670  94913 92748
0186: 79094 47170  44481 78607  45047 95418  51233 02936  54029 13744    20175 97279  55845 03562  06914 09306  39577 00701  63485 93314
0187: 01530 80127  10706 89787  23711 27544  09307 05495  31791 18854    43192 99125  17653 60587  50317 08998  63671 39707  06091 54950
0188: 59764 48252  13838 76627  62005 30612  20148 51038  63987 93063    40722 63376  14811 30999  22658 46043  67364 34241  39821 64142
0189: 64022 35530  10248 41455  83292 99795  33546 91088  03026 90127    32186 87025  87500 99061  85177 68363  20956 52293  12076 62235
0190: 11788 75766  49709 76887  93221 48360  42291 85381  98203 59521    36148 95199  77288 66377  44809 28198  31006 64491  30767 81110
0191: 48052 73337  75449 38870  36381 12392  12650 81211  70608 68597    65043 74746  79932 04147  57761 23364  04813 32693  24410 09004
0192: 41872 05207  07292 92221  65916 28741  22753 02118  05520 85070    73795 11123  79734 21547  59649 47129  83192 88993  57387 76568
0193: 30734 20612  00933 88642  95627 71572  84993 50476  90133 37507    26342 66487  66229 89405  98557 65937  70117 08029  03094 95108
0194: 17068 02438  82146 40075  00258 79890  10388 36451  03284 94036    00096 30611  45918 39813  00192 12135  48212 38182  09969 87365
0195: 14356 45022  25359 65295  53456 58398  39304 74997  43100 46101    62774 69274  23031 01994  00516 68594  08151 35264  35263 79680
0196: 23305 20347  83270 28392  15594 23309  35073 77312  90957 02666    98172 36269  82138 15083  14053 95810  39977 89617  72129 18106
0197: 67543 84324  87190 02931  96942 29126  59489 38315  15021 38708    96236 13367  90577 77483  92266 40410  85509 93379  73008 72492
0198: 17598 19312  60021 85138  78100 92359  23201 76891  13536 08565    53919 99532  15496 67370  80572 17646  51749 95345  95440 02105
0199: 59653 89135  10099 50130  01934 34321  33190 12550  30110 64115    01672 77023  42894 65096  13134 18644  05066 81762  39370 80575
```

```
0200: 78206 60794   72958 13094   27475 73711   27026 94634   23917 48249      94332 03718   48432 52800   04585 37807   68221 12745   22050 48983
0201: 66663 60779   58991 08020   84024 12523   85651 11172   17823 25102      45244 19931   63922 64787   57915 49231   18200 74516   09809 73563
0202: 77590 34099   75361 31053   03973 91043   46479 14309   92794 05446      22468 36711   30576 13143   33592 98831   64008 45546   81232 07825
0203: 77644 23846   70564 44803   52131 39502   76090 03018   49579 61831      10503 23520   94620 42490   43670 28973   36545 41139   98046 85388
0204: 25298 42282   00774 91247   03249 85485   39034 48776   71545 34866      04245 14886   40935 20295   94568 22504   98482 95861   30897 18583
0205: 06753 44458   11911 80877   29594 20498   45187 38154   74705 71994      32096 49575   37789 80205   77737 61552   06391 01352   95049 25671
0206: 86821 37293   01776 13323   79267 13647   03144 66099   76994 03377      91834 62554   98160 57208   75243 46216   60525 33518   07905 62132
0207: 48076 05898   92628 29614   26645 85758   18165 80677   59693 97667      79832 77231   12047 91944   80563 94608   75365 60626   34572 77139
0208: 02457 97858   50303 13797   04411 60316   00529 67107   91629 39817      55900 16232   36924 00304   44808 02541   45162 72653   15616 06184
0209: 41539 26605   64761 79901   53396 74812   65316 49928   83126 58772      23794 37697   32628 10584   48268 32054   52822 31283   94418 54666
0210: 78948 58923   07193 53300   27191 99540   10875 54197   99516 86892      68117 80889   33615 49478   24318 78320   25091 17051   14358 06508
0211: 51490 93353   15191 88245   08066 59865   61176 45858   81720 30601      33828 93991   10836 28345   27848 57485   90508 07049   55654 43800
0212: 75540 84838   91699 17672   87884 33133   38503 88997   80358 97677      71310 89464   29498 47697   47544 11246   29666 70954   69576 25348
0213: 52193 05526   98191 86448   73284 78347   70679 75671   95373 63926      83011 54267   51209 38768   54583 20803   84086 23859   61160 32545
0214: 43482 38798   62586 20428   94797 75658   60476 32372   55222 67357      29477 07332   80556 45973   18160 97777   99749 06641   04596 64030
0215: 62753 60166   05454 75891   14370 80274   64022 07161   15551 72030      62703 89527   22729 23058   30489 10749   47459 79447   53257 04197
0216: 30121 66592   59849 71956   12343 19603   61802 43471   19742 39254      47325 45601   48282 40116   97636 66032   38401 73971   88582 62728
0217: 66181 12874   39930 88872   93267 82259   81998 23955   07510 07524      50971 96560   56344 18363   04224 97669   54237 25675   33825 81480
0218: 19417 40581   79041 52356   91690 10297   59799 31925   17266 49890      21003 05673   83225 07792   44262 75040   88623 65052   94203 55056
0219: 44269 00067   60990 91237   18624 15450   99849 47552   89593 58734      45021 23606   57271 21306   08976 43532   98824 93718   82226 13921
0220: 32571 96625   35527 85008   04094 87336   79064 35672   40372 32034      75037 27247   84954 14379   81642 98643   85740 40255   78131 14144
0221: 33735 08152   59172 68205   91937 38636   31319 50319   08161 18690      72772 90889   34637 61468   81263 17637   74195 43994   01130 32052
0222: 95291 34803   74457 00085   03238 68164   31619 77715   35504 26459      20354 09631   18236 93837   43910 31649   37207 61301   05451 91612
0223: 07793 09654   01611 31066   19802 71810   82662 62782   57605 38508      51822 96615   07355 99505   17496 41282   36042 81713   23807 75023
0224: 90538 91471   12707 90157   94553 89111   40513 98996   91228 79627      10343 74759   44741 57179   12923 55362   60109 49580   33837 79559
0225: 00316 46750   92517 24574   10956 93222   93967 87156   85535 86682      69155 49721   75149 86110   27943 10929   00741 04955   57562 71779
0226: 74438 19870   31233 53643   19947 29732   05043 81617   82297 51035      89464 50054   04146 27280   50541 32517   03977 78633   36833 13391
0227: 88188 37867   85308 62427   98512 73704   64907 89584   95228 07434      97752 99363   94311 18459   47990 02828   43489 01416   63083 87763
0228: 15455 63367   85289 97340   12652 64861   69434 89340   41248 27337      83116 56738   61487 17810   71850 17544   73996 98970   37103 50209
0229: 07384 65747   91808 81722   49871 05267   94736 19405   29999 03884      62484 00964   98355 45523   43956 30224   40603 78473   60991 57495
0230: 99634 94847   49475 43153   23934 10231   58973 32767   21136 98373      91477 31851   80912 44338   63643 40985   49076 92964   25209 79505
0231: 94368 89504   36309 47223   30690 53069   99435 05689   82683 67314      39381 80425   63289 86062   53766 72396   01253 11553   69337 18472
0232: 89119 49143   94084 54457   26739 23183   50935 62551   71864 69968      93465 49459   46475 44045   71978 35766   16189 97612   13912 11707
0233: 69972 94936   06108 17855   24100 80581   42629 84214   07091 81893      48699 66013   34033 72029   36577 35088   65409 22332   43118 89910
0234: 65931 83763   91375 93232   10622 33138   71712 98240   21152 47846      80794 24357   76153 94473   98147 12606   89277 58540   32684 14908
0235: 32868 89010   77118 80259   08171 53510   84845 64504   11900 94107      78431 17783   07184 46162   67295 21588   00271 70541   56983 54707
0236: 64872 39584   29918 21055   83924 46294   15843 76015   54669 32203      72176 97805   13255 31255   46464 75631   02114 51742   19733 20058
0237: 78918 67519   01411 45071   83057 87811   17723 00366   50505 74209      83684 99855   98493 16579   13826 86453   55302 40006   98425 03163
0238: 36102 88002   46059 02499   48568 95578   05845 25026   66359 50053      62167 18376   92828 18256   81544 49015   38382 05447   20953 72095
0239: 75715 50628   62820 54850   67077 81081   02795 69133   41638 43553      98624 36059   28431 67585   74317 53467   11435 95859   22578 90257
0240: 19160 43388   34140 44964   74570 53287   28428 11159   97508 59846      74399 69614   09207 95890   06995 59578   52890 62657   94390 62626
0241: 42067 13723   64439 24137   62876 36157   22067 36953   72312 94999      80612 90603   19562 83963   66236 77136   55723 36966   02970 24266
0242: 75772 05422   89939 39525   87933 98823   65801 02287   37825 07010      77946 80471   63426 72270   70884 43598   05202 60293   06130 56327
0243: 30975 54897   79180 66477   79727 12414   23657 42353   66095 13025      98306 77206   93212 78461   70123 03790   26000 23391   20887 94979
0244: 29666 01601   45414 23833   45854 15372   64588 51768   50983 80754      59162 76896   18508 66059   43210 77081   78999 47663   23022 07276
0245: 34015 07058   13734 29549   36577 52555   46018 10622   22413 03473      13128 80205   51119 77775   96537 23518   36810 58483   84764 24655
0246: 99350 08918   46663 17369   51454 02318   24130 04492   37915 94689      87350 56344   49425 86429   68725 47574   89887 73847   05710 33968
0247: 41495 73139   67442 05030   26145 83273   29257 62792   39154 49463      72942 62817   44246 26281   57766 48288   62829 28806   98330 11261
0248: 41391 82438   71997 28987   23913 27951   92110 25914   42994 34915      35556 42217   60752 34514   11516 47647   33951 09466   75449 65316
0249: 50202 86225   98429 66857   07247 30426   07959 78090   13925 45197      69047 44300   66583 94968   06549 30715   29719 55555   67703 55056
```

```
0250: 59364 20522  47867 55337  35359 94528  19104 21882  05241 21187   01877 24015  10749 78351  00443 97894  05327 37874  03073 55605
0251: 89032 46440  06437 94908  83934 49639  78480 30029  81834 97680   61657 50002  30358 99181  02520 39790  31103 32297  63198 12592
0252: 02432 01833  70131 01862  38489 82135  80155 22947  87724 43989   09509 64572  95200 12931  17771 27638  65402 59150  41587 91069
0253: 41955 76715  39058 82662  02608 05597  78618 43269  58770 36094   97846 14747  08755 54151  21124 65033  48925 39464  44435 76776
0254: 05844 72615  17736 39196  47860 92052  18724 64798  35854 12655   44805 28168  92411 53085  03959 61373  69197 38468  71500 37637
0255: 50316 06103  31922 52752  16378 97975  35972 57048  26576 55562   13386 47814  09749 41431  34767 34987  02790 57333  48153 55622
0256: 22172 76762  74027 17820  72897 32752  43642 62440  67142 78901   41887 76851  77059 52593  27532 39232  65485 04592  23987 89731
0257: 86764 09095  54002 31915  58139 82292  93879 28463  14983 67169   35250 04418  86176 62834  29344 18340  09187 56542  55419 12487
0258: 78812 63096  40499 00111  04843 34129  96587 64701  80750 30542   54895 88062  65279 93808  54729 07214  32502 31526  91532 01002
0259: 16185 64069  91499 04645  68903 56501  19320 15350  96196 16592   36848 99323  44184 60858  98899 09614  20134 30059  08556 37938
0260: 73193 91058  77563 15648  30593 29515  44015 49702  02104 20746   76777 30006  53002 98217  21785 98017  96477 73014  87197 09279
0261: 10297 19101  39254 99501  89310 52219  18128 47132  19813 10436   67064 59215  29942 65756  39625 36754  89066 88912  24779 07221
0262: 79880 72881  28220 94753  58816 88060  90161 76151  86361 31471   20076 53791  39039 01456  26883 09825  50184 60687  85649 71962
0263: 99880 51326  24726 29226  08283 04717  04146 74392  35162 51193   18381 20559  84605 68709  87287 94370  58559 83753  37117 41568
0264: 32973 00431  29543 44537  11147 76062  32162 82696  07593 13135   22520 96033  84778 55444  77004 74255  45920 67598  46621 79609
0265: 34551 59217  03379 48333  34783 89690  38399 70566  49564 49704   33907 79668  38953 02478  57545 34330  61828 75639  41070 70380
0266: 68064 05722  90041 49055  77085 34710  51218 82727  47238 36010   91769 70484  34420 86718  03858 49359  33001 66787  53222 28671
0267: 04110 52096  44124 60674  33561 54069  35277 85463  94510 68229   92726 57235  11070 90361  07796 69811  44916 21357  86728 34475
0268: 99198 30288  81561 24823  42265 49785  53322 21113  64597 83320   72100 72149  48607 61788  28871 98958  97883 37337  35609 65712
0269: 22760 26257  57408 79524  01394 83829  30423 81506  50225 17588   47256 66621  32591 52668  16242 91951  17270 29378  64718 27469
0270: 79961 15162  92444 54374  58333 39053  72994 99980  12827 12609   24827 79328  72291 81872  17236 91583  75837 58351  85591 86558
0271: 23645 52455  64109 31676  01720 73026  18247 92598  99788 01093   16861 01788  40589 67053  47813 75952  07934 99969  32168 56579
0272: 18638 83560  29159 15189  61845 68441  42284 87983  20040 40764   93623 60625  97686 97200  66881 40385  48601 01472  53440 07811
0273: 39449 58550  12039 58541  65353 87184  51819 80069  55674 88759   59361 21410  61693 74952  47320 07044  77111 50973  44228 53170
0274: 05552 56742  97877 31328  37530 84875  40513 65057  06151 06528   47145 36248  02813 14736  66265 04032  08257 94893  72496 60751
0275: 55463 39246  51018 84031  47761 21371  35555 82904  32841 12751   57965 35915  66138 37243  83242 44357  34699 10626  71472 78900
0276: 53074 37607  61556 99132  18420 44399  68259 12395  16507 09346   72532 85353  16036 42619  43359 17050  75300 42566  80092 86311
0277: 88158 81994  72279 50897  31982 47134  86289 13454  09699 06343   26316 15040  76143 67516  31850 85081  72490 39763  84076 81066
0278: 32980 50391  51082 25831  17171 42625  15883 22094  24157 08509   01357 10618  95414 77324  00119 04555  10677 44556  81136 54435
0279: 33971 32523  59193 10149  91130 02748  33947 36505  88188 39138   69860 55862  06481 20848  65608 69011  98705 87086  89574 61995
0280: 05809 33491  58010 19157  39235 84710  39989 76539  24401 16653   53462 21152  18487 39356  82836 22210  61592 40152  22208 70642
0281: 86182 49630  42624 18357  31807 25234  68934 62651  61420 42235   35483 64706  54723 30146  11381 51639  90634 28139  28690 45319
0282: 18428 85671  42120 51719  50681 78408  52338 17312  48832 02636   10111 98226  59760 44481  48134 24948  75398 23584  75097 76956
0283: 77389 09558  37385 55056  94870 92816  10998 37938  08330 51035   07982 33660  36741 46222  91527 50581  64988 15042  98478 02653
0284: 79068 15120  14768 46245  14285 21027  21507 23898  48178 62127   46957 95812  97440 14088  84928 86723  16596 71811  03582 93646
0285: 81163 92691  08952 09076  36731 80024  51729 38573  76057 52458   87261 34014  13734 56376  91261 69173  96373 72164  85259 92476
0286: 16846 93338  94799 04667  27383 51708  24858 20505  00921 39989   53763 72474  92021 50050  30663 82714  14842 78415  21757 15210
0287: 13279 60273  88047 60361  45831 06211  83726 89027  09095 15986   22533 29159  35547 86793  35144 76662  33761 99831  70255 84899
0288: 64773 82258  99871 01739  47846 07248  60120 27033  64004 03006   45767 90012  66391 84563  31066 48388  99665 41329  63087 43370
0289: 49613 84490  00477 42939  26498 11946  81428 65930  18395 91338   43814 02928  11845 14334  93482 75716  20929 57517  76923 79124
0290: 09101 97387  84851 23505  57982 28017  05294 32729  83977 58149   43854 77497  66939 87145  98602 28406  17236 51455  72488 16262
0291: 91367 38780  01553 51246  29364 70997  65550 58813  75949 75957   94651 13647  56400 61339  45960 35450  13776 70802  56437 60802
0292: 76749 14861  40150 78355  75596 36221  19744 67049  14660 67780   36768 38139  70164 78375  03818 47014  83736 11750  84746 84534
0293: 54037 65608  01516 22241  30271 87112  89312 54193  58271 70247   39042 07065  55183 72475  04069 72896  56141 51178  17395 29975
0294: 39128 86903  77694 29828  90482 56566  96476 07493  12470 77722   64209 50407  13683 65555  62939 34083  30661 18568  31991 81889
0295: 31470 55023  28772 66000  86169 48978  50327 12507  52176 02952   60516 89886  64467 17595  93334 62643  17689 01986  11176 31677
0296: 37115 81507  73250 23014  19294 57167  69397 46391  71595 55450   55945 40921  93141 22735  13155 27647  99537 07915  41428 74708
0297: 07744 34062  85751 57931  13630 31124  69862 09301  54088 25222   31128 30226  91454 57843  99532 56197  91140 81630  42058 06499
0298: 76371 38663  82563 71893  65780 56115  62343 29135  36084 13795   22114 31098  21743 49957  24433 29650  63911 45155  46572 55197
0299: 75040 26113  22121 60088  40549 25243  77665 56679  34741 36099   95557 12368  57172 69636  37219 22463  72866 62636  78395 20955
```

```
0300: 14107 89776   05211 44253   70098 31825   62891 68374   35936 80804      74224 32112   31417 12998   56105 39443   08387 63789   46393 57641
0301: 17899 65268   24705 40968   54954 37647   40047 90715   94127 25545      24004 64495   04857 35024   63150 72429   56605 33048   17029 42044
0302: 46566 73954   40785 85718   83267 63554   01916 90900   31699 38577      55566 46860   53910 09843   52214 32112   07559 77519   64898 09159
0303: 26975 85842   74206 12423   21394 19282   33147 45044   75881 28816      16702 81845   61882 49741   92230 02063   48108 57945   54037 26984
0304: 69034 56062   90473 20773   47270 99391   50068 97483   95072 40276      71079 27961   32720 78946   79947 80146   61424 44064   05468 58434
0305: 10561 86190   20212 25396   71780 56944   10073 29445   88129 45850      93015 38965   90386 34333   01370 43710   82056 07826   10086 68663
0306: 56833 94954   91729 56198   25428 59928   41374 75653   73769 38015      31582 20670   02256 07809   67334 17427   26510 78645   50067 17889
0307: 38516 66274   86566 28826   36894 36099   85116 17854   99361 74825      33216 02936   45550 03859   75488 26359   58812 31253   11171 85893
0308: 62550 16887   32139 63867   30996 02320   31093 94503   98503 80342      29411 06651   27443 31154   79421 95969   93728 55185   81486 26565
0309: 21424 17262   57892 98189   78719 22628   94521 85732   00901 62907      00605 35929   29758 60807   77256 14620   41312 69568   45753 03621
0310: 89917 77603   07117 93328   97274 61458   85470 84319   46125 60438      01383 81432   02339 37608   60805 97195   85598 02566   42774 27386
0311: 03931 89191   27987 40153   29263 57832   50217 90904   73933 14070      08953 90940   24302 26033   17742 20654   65594 71686   82040 45348
0312: 46946 87778   28337 46674   12976 70140   29284 44064   15011 32273      00379 06073   46584 39871   96834 91417   09988 37778   88876 94864
0313: 52665 74366   25014 66662   60982 53830   93079 23291   83294 02283      59837 22618   28555 02703   89568 77086   31468 02399   25539 43469
0314: 94677 19190   33701 90276   23354 06691   19163 94631   57285 54285      75495 33928   72377 99884   69703 48798   40260 88983   58835 32723
0315: 82683 15403   05800 29671   52422 63076   04208 60480   80910 71339      32810 40108   59487 81598   48160 08279   78315 99264   70683 33704
0316: 20995 52012   18418 54217   01119 52768   55849 42114   34216 85579      03826 00741   33062 95396   61227 04714   14834 90471   56926 65804
0317: 33180 77721   00475 70694   56819 77716   28596 42861   35209 92736      30046 29822   03540 21965   77710 47498   67677 23753   99492 84146
0318: 01454 77246   66392 58456   03745 91013   39098 96203   47977 68277      02009 15371   78012 61919   16863 69788   73724 75179   70883 48077
0319: 95218 69436   60957 94397   81573 76817   91659 88698   37842 02936      63986 45719   49094 93991   21847 89847   66379 70684   31391 32604
0320: 66065 25654   30821 66477   91929 70552   57557 95862   87825 50054      00964 28950   58853 10602   72427 78090   25539 48152   29766 28884
0321: 17056 05067   82099 73758   06712 10424   00304 91227   36825 77704      36029 89803   91549 67953   54866 42051   46017 32305   88626 83037
0322: 42909 98171   72115 23309   80567 06671   47343 40375   34918 18365      36874 42918   94371 39852   39807 51431   02795 11824   52036 67496
0323: 07979 54851   54376 58103   37306 47220   06138 92525   26976 30317      94406 99259   91125 75494   57771 68486   73406 43831   19190 64041
0324: 17471 11342   58204 71353   91341 21026   74458 58422   50442 46455      27572 94200   61148 29537   94543 80294   05703 27871   24680 89880
0325: 92253 01870   89106 59256   08516 92236   60389 81844   38202 11837      17644 61519   63144 60862   12277 58333   93897 55014   95484 67784
0326: 04904 82580   67913 35868   82833 84747   86174 39734   66307 23046      69799 64839   24164 22085   43419 24137   20002 53916   20098 71475
0327: 44477 53955   34982 47421   16132 57441   88010 08675   83081 63631      10170 23161   92024 74334   86658 08959   81296 42915   29032 72425
0328: 30670 00515   41609 29765   07977 12321   37641 75766   61744 40886      92509 98868   41021 64690   66230 11211   42448 90596   34972 47216
0329: 23954 00004   14569 92317   76906 01934   20460 73488   00574 78194      62954 73063   90017 70006   51236 44391   90704 28080   92271 72921
0330: 34550 67392   32493 36202   75270 65710   00908 38940   47629 50413      42301 51035   99086 99731   74854 88441   76396 80925   60382 82337
0331: 89524 15614   20164 75591   65529 19702   11620 04362   82461 36318      50673 62521   49299 15352   19156 94465   66486 57532   44251 84661
0332: 34606 37025   67151 78048   03730 36199   63191 25830   03838 09245      11412 26492   31741 07940   50990 82979   91112 63837   54313 54545
0333: 59063 12382   24721 48273   43668 91234   28561 59402   09043 55990      25419 09553   91341 21701   91816 13814   45168 80140   41465 38738
0334: 16046 25703   70658 02580   77145 69591   55241 25374   95598 02863      85473 45930   36874 72463   19059 67879   93222 89053   82405 01109
0335: 23367 58159   70293 21508   72880 22806   00292 87571   48180 44750      56814 58043   15575 50105   38399 83891   07376 16830   93573 34161
0336: 96943 38508   38683 48563   91415 71013   31717 88704   25662 60547      20660 98714   39139 35395   32711 48905   61361 66334   29847 98317
0337: 00848 21516   29034 31934   67382 27348   90902 22382   29082 51553      20027 37574   50857 12820   98678 92299   03771 12813   07155 54911
0338: 87759 75529   70454 66990   73840 52650   16206 99106   10276 47465      18853 77081   99182 38802   52454 89993   50728 74223   65510 40139
0339: 38102 52103   39359 17843   97032 47734   46145 46028   10950 55244      77901 80151   04200 04500   96422 71816   07962 69180   80427 52120
0340: 59985 95561   98966 47965   57435 67821   46912 40982   52320 31389      88879 51095   15571 57327   86924 41595   11171 10350   40977 15291
0341: 84772 80076   19844 92668   93105 97563   05961 19122   99670 00904      18283 32680   71103 01050   28246 87285   88160 54733   81015 44531
0342: 14495 43035   13813 27813   15581 24320   66814 50474   76447 41522      72768 54823   09498 08348   33782 72201   32823 07645   43211 18814
0343: 85654 28116   43557 27415   60239 27936   54042 78917   96858 27215      55695 33350   68455 05572   19950 25346   32582 24411   82615 88488
0344: 59096 56076   43203 55719   69821 38708   14823 77794   89109 92032      30177 28403   30377 26198   76343 23876   84192 87389   85102 08336
0345: 96835 47761   08849 09287   00791 97385   06398 81739   27384 53560      29594 20693   85993 73029   60762 21433   15770 21183   17006 78486
0346: 23478 53640   18329 15928   54235 68989   33992 73193   61237 07566      75100 69673   99166 80051   14883 44558   38799 44899   29318 31606
0347: 39472 24743   20265 44451   26956 25511   87507 78754   10020 44524      49111 65114   07832 04281   49458 21186   04283 66885   37408 95766
0348: 65187 55733   16398 59944   27604 91049   78120 69960   55675 64497      99904 66302   68285 64492   52069 89368   32511 93007   86849 88088
0349: 35617 81930   27938 76614   57820 31099   41292 44785   95745 36082      74865 05906   04271 36422   41578 84235   14469 63055   80079 81589
```

```
0350: 11128 15576   73551 74448   32241 96608   08256 16965   68568 51997     78123 17438   39064 98060   52081 90724   13884 17720   38407 69701
0351: 44735 46178   68409 92724   81975 68672   55181 74829   67471 78267     86733 93930   29286 28605   32736 14625   76814 37947   24733 78325
0352: 57624 74636   44875 56268   51202 71812   84614 88917   10361 17905     26354 66628   76604 95528   38846 68032   11967 52965   23259 58981
0353: 54724 59183   44459 29088   01120 50337   06594 30136   75545 38551     80372 19412   41580 67408   70481 69853   06633 46007   51256 27715
0354: 86727 34651   11119 40140   55750 94313   20963 88943   05718 62690     00565 83451   73559 54712   54883 79747   63732 47673   56589 96904
0355: 75923 97682   08601 89587   34758 10803   35362 93082   40269 29649     70397 19469   91124 53701   37983 53208   68890 33945   74375 54279
0356: 76981 89659   82754 46918   99622 38296   75937 95734   28626 95497     25414 40299   27059 51091   16601 78950   25718 92719   73830 84531
0357: 73184 00673   30051 53630   75544 05740   98116 14583   48763 41440     89820 48408   92403 69348   89741 98078   99842 21480   43325 03503
0358: 99474 29695   63091 86386   58284 39765   95361 77747   63192 22957     27418 04102   18829 30597   65128 25506   77817 27397   27149 53215
0359: 82781 80408   80072 16006   57329 07844   91805 91070   02761 83933     96681 67112   70605 70752   94146 77247   51756 79596   13352 09378
0360: 68462 07120   44437 07147   46328 53807   33333 19984   55150 44912     71027 55524   02292 47089   04858 77141   49752 64162   37659 30113
0361: 93070 78129   59601 48902   90285 01368   23519 27386   43355 26501     58458 73631   13591 26333   35661 56890   82757 83387   96733 94449
0362: 32345 54382   25507 84198   97669 21455   19709 58879   13697 31980     80589 39720   26664 97798   44182 77893   88960 32779   16621 84027
0363: 29644 35158   32289 39330   92207 17025   98779 66673   00504 90163     96665 44418   46091 92285   99442 35303   35543 35662   97593 77208
0364: 36668 37299   79109 22359   23152 77458   36031 43086   63703 35816     39522 29038   31231 87227   79972 77686   45944 69533   37300 73633
0365: 89071 77106   50223 87723   72473 38260   73390 44536   92040 52127     85911 08700   01970 50418   01712 37009   53105 94210   18487 08671
0366: 17354 13906   92253 48986   32907 03425   34954 19063   71031 91857     89842 25852   45963 94257   84742 59645   51117 72120   81744 08426
0367: 60554 97386   45274 74318   96986 92038   86127 65577   23328 52357     40685 41871   90989 27677   20792 45797   05544 99365   52493 93296
0368: 84681 38297   19969 38459   82120 94916   21079 93829   75689 05714     11237 33822   16775 75870   25973 20626   96307 79644   28750 16283
0369: 86500 60086   87249 95934   00356 88724   11992 41794   19677 76906     72613 31323   97692 74028   37094 80774   99501 05444   60304 64970
0370: 83574 62650   71915 20003   75430 39392   30629 47429   24042 63542     87339 45287   63867 21733   28009 82652   22965 07849   11767 62669
0371: 29233 98210   50829 57917   37509 50939   75631 56285   04005 07339     38522 95790   99931 12651   06452 85038   90502 31107   95049 16393
0372: 12187 66012   77494 41102   42653 77568   49107 30276   16709 87026     25250 79998   65223 33913   84439 49885   65347 00040   96380 41132
0373: 34957 42679   34686 86926   78789 11901   50589 70202   92384 86186     12202 48942   79290 46306   96903 63050   45039 73362   98046 90587
0374: 77571 79211   84482 23914   85526 28723   02365 17289   31362 08053     13516 21318   98032 95657   80779 40245   86465 63333   62925 61794
0375: 41722 28266   93504 62418   84573 58683   83469 78982   41990 91607     55222 95190   90675 11813   72687 12144   16920 25051   68463 98112
0376: 14269 18553   05480 77734   49523 35797   28303 19482   30630 53192     22825 01415   18680 12746   78601 51788   11114 65437   82119 83076
0377: 87798 20869   23173 69450   92017 87578   35316 17220   56766 77457     53868 16646   05720 85867   77900 75876   21786 17365   44898 70195
0378: 55231 10180   66228 66952   49142 20730   91666 97848   53113 73750     15947 67483   08225 80203   63336 99986   61322 67975   38613 59924
0379: 61453 19845   72995 76207   47256 76453   76912 88605   41082 61521     67084 89490   30479 58688   58413 22266   57514 23254   70124 13302
0380: 40851 42354   73526 14238   04701 65679   87183 78334   31307 95862     42393 41878   37207 47431   57017 76843   62785 13006   85733 35394
0381: 40232 88627   43784 15056   52547 09593   07247 25833   76401 12590     86264 10684   25512 68462   46288 22007   26050 09313   39040 71703
0382: 31922 45362   10515 39850   48458 98240   31867 12899   32789 52515     17010 18876   70250 76640   09684 91639   53097 18568   31904 46669
0383: 74690 77724   14151 47468   75939 05080   04714 93386   02323 35098     79439 15112   67072 92060   65165 21613   05973 18576   18124 38589
0384: 63613 30694   49222 27142   71104 16184   65120 82533   01711 42034     57620 02935   25563 32688   41184 07893   09804 96062   68920 87603
0385: 84126 92331   31453 01725   36414 33801   36841 29335   74069 53770     15660 55904   52091 08204   16328 11103   92715 78971   97720 02077
0386: 39369 22887   09774 86478   16013 73647   13855 37816   58034 23697     45902 85659   28644 95430   47507 22782   33750 81108   48708 35873
0387: 98474 82543   57225 75650   20633 96928   68628 93810   18431 03447     99643 16375   46262 55050   53744 83083   81800 10935   73877 02755
0388: 35190 62390   00229 74275   45137 44832   30999 21518   59163 18829     23791 25097   62378 61790   28945 84237   11126 46707   99860 60072
0389: 72941 88177   70530 61754   23442 45501   95504 50979   95445 19753     63611 64782   74769 30635   04175 82067   78304 43896   20954 10021
0390: 29183 37200   86416 47670   55959 74549   14669 76304   33770 49798     27356 67870   05006 33070   37242 79310   93501 78417   51889 42482
0391: 64734 75073   38115 26729   88112 56393   42459 11510   71682 95905     65871 17607   22448 45161   15355 86549   57822 42001   07607 01607
0392: 78218 92161   54319 85288   07642 52028   25236 21847   80502 14885     96012 44076   83013 90664   80664 18397   34420 73647   77737 98923
0393: 38706 96950   58398 13380   52668 51496   09137 28276   83322 86681     38163 01697   91603 81229   11749 69097   14499 72983   76066 85783
0394: 98701 62442   58560 27692   78919 66647   61469 75703   91622 75254     72258 35411   92378 55678   24098 60922   65318 33280   47188 89190
0395: 90413 79074   13917 65587   78068 42312   56104 20978   05369 35238     40597 53446   75264 47528   71489 89720   24912 94360   00422 35203
0396: 78551 87913   30472 08286   49150 84278   52765 55825   49483 14678     92338 34034   31092 02657   58151 29193   35954 11911   70834 44290
0397: 87932 71448   10829 06947   03559 69294   23588 43059   55119 20729     84840 59684   81646 19883   58136 52725   93200 77013   61058 30346
0398: 72270 99153   80906 14780   67050 54944   72790 64979   02843 30490     83480 13063   12647 49165   02979 72407   13144 43738   77250 74126
0399: 32064 43831   62711 02336   56335 50252   42898 29733   47391 91002     21317 22918   21695 65438   35699 54882   33233 15164   57471 17838
```

```
0400:  68856 09622  12059 13656  09012 51953  10365 80064  85369 84548    09757 01254  92683 73081  20028 83603  10186 57167  43944 10733
0401:  24764 34771  24528 95663  41447 51930  86927 93184  49242 55603    51614 21686  01505 22835  10059 73253  84800 71635  81845 86362
0402:  07345 06860  05649 94473  52699 31718  63049 18475  20455 94250    29875 76850  86290 10797  76864 05157  66982 29387  68920 31253
0403:  65848 13726  53596 78974  50459 72074  89934 90272  26492 12543    89902 31064  00229 34144  57736 91148  07273 10752  89945 80185
0404:  79670 32745  71201 51540  89444 77895  90063 75622  98120 53230    18976 71199  56057 70348  46138 26616  91078 97366  51814 99251
0405:  90551 16626  82978 04218  81493 43667  51518 43086  12330 22746    38400 42979  63150 13257  78140 03367  49204 62470  38025 19091
0406:  95022 68776  96267 38880  08529 80633  53990 34007  15475 04260    27624 72818  42646 82500  91241 34269  20116 07849  02575 47948
0407:  32630 75512  42647 52303  48044 93997  78203 64893  00020 04547    89960 35618  95232 21164  42990 03406  55193 62729  52240 61245
0408:  59221 83528  53847 01039  54482 57481  83899 76706  15303 32005    96006 77415  57471 96005  43037 07982  08269 68310  24164 44520
0409:  26875 58333  81023 60997  28740 52521  66718 78399  59973 61867    02937 79043  27678 35222  79274 99957  13061 45776  15643 87453
0410:  71622 56811  33397 34449  56570 72492  87751 48594  08676 20333    94078 70047  15002 69047  73603 61299  43607 63375  52839 22735
0411:  70662 76938  43939 07219  70124 87359  79248 89150  47251 91028    33699 77307  52871 64880  54649 24764  37490 11894  12918 62901
0412:  24683 99160  05533 51382  14735 19271  51990 64219  83581 78195    47403 54558  49464 34877  42539 46152  19021 52473  76114 66604
0413:  78051 75824  29092 15981  80993 19526  15298 45465  43223 51818    74611 94592  97019 16633  27663 40090  64744 30612  02121 44395
0414:  78466 03285  09994 08751  83897 75760  41885 47587  06629 70505    71876 43459  74933 97462  01063 73194  60564 67325  22628 38596
0415:  43954 57240  27595 89669  91725 49038  58153 80016  87752 95070    15609 37895  17131 60993  03092 13300  64985 37550  97191 79667
0416:  25941 90152  85059 97474  38907 01084  29842 30245  99923 05158    89644 62757  13995 73075  30211 97369  93467 42579  26340 93890
0417:  59551 08043  93036 80623  79773 15825  78330 84167  35663 63384    22973 74289  14312 06930  42611 78743  48324 83083  81299 39013
0418:  72985 60735  54842 15284  58069 89633  24128 87430  95031 62939    42070 24776  39033 67356  74760 00695  51108 64831  74004 73000
0419:  78654 06651  56660 37721  95828 30612  39402 46341  50715 75271    12937 48922  70975 20680  16870 69524  87685 21945  91593 85153
0420:  54781 57631  48127 94212  90990 96016  63431 11531  91459 11505    32578 39956  15791 68113  96652 49026  69903 81978  38908 64412
0421:  53222 38419  18596 23036  52835 01092  25326 84389  01460 92150    83172 04749  80305 17952  02545 02949  02739 61141  86886 57227
0422:  24773 15014  40302 78604  75576 81162  83526 04368  32258 95653    69512 34466  11850 27869  89711 09419  94831 90746  47516 34302
0423:  77764 95978  22412 84596  78768 12419  04847 14814  77197 30187    51555 55349  86183 70863  13400 18288  76832 97771  71006 30644
0424:  68403 88867  24446 32380  39507 24264  37237 54427  87908 89137    75647 73484  01904 10532  04342 02119  63361 28728  40232 37287
0425:  19254 44611  09692 79436  87183 92396  56551 24448  13050 80778    65409 47800  08909 43070  96055 24103  16138 65353  40566 39757
0426:  27427 00488  42607 89477  65357 53071  93019 49950  68675 77643    24614 66294  20195 88059  94976 44706  79637 72916  90571 37391
0427:  51235 35418  47820 16536  91099 96791  39312 95373  03405 02406    04941 66279  75875 51699  29052 56613  32555 92812  41845 52769
0428:  88012 93116  49091 35388  26746 31428  89598 31731  32086 42449    44497 58475  95744 53945  80336 68521  78387 69077  47242 58188
0429:  65296 24732  73688 86881  61713 87251  56262 32729  71453 29027    48084 48545  05695 95986  14408 93790  51557 63818  79978 07754
0430:  69075 93550  84147 00769  56833 71223  61035 75742  22096 54741    70693 87725  64424 52659  15962 54088  97192 58315  90468 18430
0431:  28439 33050  45824 52299  94662 43261  63885 21583  59068 15562    84071 53243  68436 49106  46009 59530  62655 49499  07926 97590
0432:  13127 39834  99513 82425  28926 25715  44246 43370  56373 81281    15368 12354  80273 78353  69332 45977  83238 48211  61976 61204
0433:  61019 17914  09043 19760  72636 94261  73849 45255  53387 37666    41409 30351  28736 48690  61483 45483  79603 58229  59965 14951
0434:  75381 86574  24815 77206  65086 64171  62502 17299  65327 19999    52819 69692  13556 60279  01063 87934  99780 18985  37649 38906
0435:  99028 25469  43656 41307  19496 55681  16432 69658  19453 29391    32744 61677  16645 19892  29179 31116  56177 79605  00539 20699
0436:  23177 18534  45323 61602  75674 90176  74357 91070  00755 29096    24685 85615  67067 79675  35560 69438  07409 66483  26436 33093
0437:  86119 28822  16831 56158  08723 07768  08847 26606  06349 31317    51188 32513  69731 06870  92890 89593  42283 36693  74785 92108
0438:  36989 90759  29512 09498  36825 43966  21785 91275  35572 75367    15695 27886  16664 40667  20691 48268  76740 78890  87894 84787
0439:  71956 63708  89800 48790  97548 69421  70694 28918  02737 23490    21892 20073  37776 86228  60831 27615  48158 55497  58157 24748
0440:  22733 08051  41543 42953  35666 82015  79543 24854  88307 44427    58200 27462  02540 42885  42063 87139  44114 91679  48529 34166
0441:  32666 95208  60127 58065  47346 90751  17386 84616  65690 95569    37899 43105  54217 12859  43880 90356  74869 36952  93935 37094
0442:  36794 79244  44958 97385  89825 67103  12148 35553  21775 59313    06128 99323  85140 46698  50609 90792  02129 75777  27926 29773
0443:  28974 98774  17012 15386  58579 49624  29236 03705  24963 42019    72357 07846  10612 27523  94088 89007  67303 71873  99490 36832
0444:  77260 30348  27628 86294  01560 80845  07799 62924  98007 51859    65066 47028  71752 22849  88117 30206  22542 88643  53117 05867
0445:  26982 43898  77911 06154  22826 34315  33686 86963  90272 51094    64933 60769  29620 62791  76044 18154  50139 95666  24024 84544
0446:  54477 61139  42834 19109  43278 67342  19009 07973  40474 77552    70350 01877  86325 48251  26925 34356  78863 02140  57652 18796
0447:  48120 65580  81873 19669  77088 32201  32818 61187  52583 15942    27175 50891  61979 56217  15994 58199  07145 93728  13420 41963
0448:  27557 29844  63433 95594  56018 11842  32857 01439  75217 36558    35959 51151  74597 70564  10133 55399  79701 35082  44348 82272
0449:  92058 82799  13278 20072  71982 73984  81378 49650  18168 40644    66650 92202  48590 27202  45998 43705  39695 29903  95720 12334
```

```
0450: 9321818935  6451196876  7074278648  5768110319  2036084220    4687097138  4042050797  2651340587  9875243419  9887493302
0451: 5474606923  3787071454  2258083353  3442150346  1583647573    2504451407  5418433522  9763000380  7253069303  5668799277
0452: 3241731780  6802318948  7567119789  0629438890  8961816770    2075124815  9452013715  5550264615  8976044056  8118152333
0453: 9422324020  0986132697  4945790116  2574018764  8694975572    8145942066  5962502664  1706223092  8222022316  7473061476
0454: 1724954504  5495126955  5164804919  6550838922  3442380416    1186683082  9070738412  1443572388  7913543003  7538167785
0455: 0973271781  5403012671  4230388229  1407065275  4678461274    3061133738  1089672107  9680573375  6462635748  6932101496
0456: 5769751588  9861941008  5408749604  2465843657  5619295531    7581622385  2081428465  9857392547  3438967988  1862366127
0457: 2946429753  2080531856  8705235276  3108898591  1060225030    6983996830  5259030370  6387088852  5229388784  3564223975
0458: 4709987588  6992870520  1545775878  4928543309  1405597953    1982678910  3394844607  8163753810  5613561030  9188777323
0459: 9263602109  5803112610  4915412866  5394956767  3744312941    9311728669  1266677303  5365645831  4388660634  2350996283
0460: 4701647695  4187672857  6035009032  0566127393  3397538072    8280278992  1634932616  2599463910  7435495785  6165936426
0461: 2837639967  6853761547  9033302595  7158921686  5905604263    6539859227  6420785694  1857220424  6009511743  2818148953
0462: 8354634509  6535478807  6079422640  4381990258  3228842423    8039704179  2454208940  6630661034  2368777817  6217319312
0463: 8150942805  8057105488  5642920712  2083377463  7913130039    9118138652  4351327917  2212651031  7101925421  5619669034
0464: 0430654392  5905831910  1013698874  0771586177  6017297946    1139755675  7779699459  6283864556  1013983725  0427101585
0465: 5126324581  8692031124  8981962389  1228940258  4512705265    9444745394  4178136310  2989327477  0130988492  3858998394
0466: 0283494654  8140600554  3125354674  5934010718  4491608969    0426357331  5887554254  5420847122  8991838641  4797062623
0467: 2101943211  3074217913  8229120807  3195096984  5303063732    5039980955  3514960273  3107483551  0155570850  0319378877
0468: 5161248138  7647492933  1837164153  9650998440  7026618929    3052876215  3944429629  7155810759  2198848839  1638812272
0469: 4783475498  5012962934  3649864820  7224892221  2131242430    3954992320  9742503018  1722944642  8455822079  0450353147
0470: 9139454328  2688711065  8552043665  4726281474  3919798573    8228223453  6479649359  0753109421  8027965151  6352196292
0471: 7991784255  8207575878  5244273162  1159648281  7184792583    5927356532  0363764626  6927868781  5695806222  4332298401
0472: 5706422495  5688487244  9007607471  4563053800  3486376664    1201039561  4840106677  3664363671  1276090002  7765983507
0473: 6849566166  0113049328  4110224236  2699978855  1120430659    5086896193  3102236953  1227483535  1763773489  4017879292
0474: 2113058995  3911592602  6174983461  6904309622  5320686919    1631427618  0585153344  4165807664  0169048530  3770990081
0475: 4367747542  4443959304  2553165277  4127866887  9860323549    1474788153  2859606565  2837613151  7130306982  4960027928
0476: 3574100279  9606840649  2450411490  8697731835  1379551541    2591790261  0073359456  4173592914  1695285397  2067348128
0477: 6570242625  2089190053  3943299203  0796161659  8409545794    0803508342  9050173314  2570166601  4366915412  2108603472
0478: 1465905118  8585620587  0313894465  4527358183  7661547924    9641586042  0248046910  9016536344  5631950893  8250987797
0479: 4219718434  8362419570  5838760164  2723367040  0690008158    4643012147  6660716739  9124629820  7698488737  6144057853
0480: 4041357048  6451586778  0224164807  0758661114  3201158835    4077473208  6723925712  7462097600  0564286927  5712894036
0481: 8419217947  8413820925  0013778574  2113471887  0699163738    5002355189  4714749885  6298845340  5671386894  5869602245
0482: 8749629099  8131738375  0157270637  2058734093  5927252347    0456275389  1748740271  6708607366  5104207373  6925654689
0483: 4909629561  9898711100  9361162367  3568257732  2966664847    6217985792  7175290627  0913611008  8544529823  2412826710
0484: 7430001336  9049761956  7030376018  6116887888  0937595591    3970533433  2763091394  9602073357  2894134252  2141748921
0485: 8574059247  4085763514  6885798282  9591080848  0780494685    1702355244  4360980103  4588310737  0059572947  4991093890
0486: 9724322411  9461038336  0367737690  3480119607  0073011756    7408459134  2195982868  6664576608  3230922830  5794807419
0487: 6375120507  3419524327  0911764588  9604837075  9260147829    2524460118  0585887036  5410935194  2634970751  4735963679
0488: 9290288560  8831394425  1306544026  0235199385  0695950422    1785262672  7556626279  2743082936  5492649097  6279338041
0489: 6918206775  2263440408  6455840179  1680790806  6297300670    2162923605  9407121064  2110767149  5884415372  0244076516
0490: 5009862275  1252686533  3282656690  9373475622  3870693446    3202131143  2037846429  9892384836  3084744467  6343025590
0491: 9375912391  4762011912  8541711556  5851395587  2655309033    3463203642  0911157440  4865393631  4821122551  6908093974
0492: 5076661515  6168616963  0243069570  0329467873  8254172953    3192920670  0300446286  5931582319  3386001862  1510848063
0493: 9129007505  7505758482  7036566951  7636688544  9864159057    2199946847  1755202150  6269242657  0581226454  5155798395
0494: 3308851978  0236439698  4791940111  7032279026  4934115388    9683959481  6274884687  3172456184  8094666246  7124895655
0495: 3827762883  0368633483  0516960815  5441071196  5877035366    5509137313  8509636420  1737910797  4412725351  0393224286
0496: 4174176201  9738862382  9855784136  7954488949  2492944343    8817032174  6789188026  6328402504  0311088806  5985657725
0497: 6409996700  7211521509  3639254908  9054498226  6384769287    9085369658  7042943085  3460935320  7800334835  0178791948
0498: 8476567165  3200482876  6985866606  0264222940  8318990675    1380994540  5482757929  8633587545  6102277897  5774328664
0499: 2308236084  8114027557  3617624800  1148190243  6233029321    7870076629  0324498923  0382797055  4061154046  6497796041
```

```
0500:  62393 36343   27434 98710   28129 80883   15389 33935   31605 81893      13384 46306   53256 75999   06045 00819   97403 64132   36635 37156
0501:  45937 40594   16043 10022   17347 92746   19256 69725   83828 97470      56879 83794   98496 55080   33086 50389   75429 48562   00895 90198
0502:  68635 67640   14231 44113   24363 42539   40213 13412   50846 21764      00202 67989   12901 29831   73541 40242   02478 51535   47536 60322
0503:  33159 11837   79601 74463   09636 04442   73129 44196   57959 34392      99734 88850   94833 60429   87352 13202   17378 37425   89742 87814
0504:  48063 01492   37940 23520   78784 83777   09240 46173   24490 65174      70656 36113   88673 84878   56207 67949   05916 03986   89916 53845
0505:  52002 78133   14149 88047   57365 76623   76732 10045   33337 80854      86966 46945   71774 67897   27037 92534   66478 85189   16012 61578
0506:  65607 92855   68367 02527   09157 37317   94260 89069   90597 53092      43078 16650   19742 17054   18524 65125   77797 13459   89703 29227
0507:  50826 35603   28633 26821   73172 45689   76865 37616   26399 33355      76617 97585   52877 91393   84119 37646   71591 84757   01694 41532
0508:  04305 71520   95370 36111   25354 34968   80657 74690   28956 26413      65562 37955   46254 61361   22536 30643   88804 02404   09575 23434
0509:  16774 26522   75965 74206   27652 19257   91151 80280   98933 56785      48491 16392   10263 76978   19275 16785   19564 45835   78956 16542
0510:  70096 27843   80642 00103   73957 10086   70287 69664   36203 51846      14829 59647   83282 95424   15468 52157   37824 39081   58313 69404
0511:  15315 26280   05559 17866   70033 18825   56804 30555   56757 01278      97553 26751   31161 10938   75064 41618   32794 26199   68063 45626
0512:  68548 97767   46942 39240   20222 21778   12572 17575   08643 33912      07695 50149   26315 73457   24911 15069   63046 52147   16666 41756
0513:  94157 29353   50258 64828   99231 76850   01320 90576   93711 00817      13270 44055   33846 85205   82804 04296   55690 19215   52551 38226
0514:  63949 11153   88086 56042   06274 55940   62594 94716   28249 30841      48965 14218   09287 70498   37771 26599   25112 19567   03921 63636
0515:  06212 98907   78068 29829   47615 09856   42651 49896   41430 77991      68126 81774   73601 82707   07593 47098   46666 03666   76339 28032
0516:  07270 21487   29370 94534   94521 92059   00041 69225   93939 81710      36080 73768   64512 30615   67665 12423   92155 60127   75358 84429
0517:  85695 10830   77785 88399   82793 89381   73039 87831   65661 14585      41515 82642   34573 39946   08336 65165   96635 76947   60749 69049
0518:  02643 98223   36685 87975   85871 83882   01240 68721   53565 75353      36127 90194   58702 88569   48735 91503   91766 29017   36406 20162
0519:  05227 45418   24530 81836   63729 63897   11485 66874   99232 90601      88080 66405   69619 01456   64555 94035   28295 79647   31260 16938
0520:  32062 97490   68277 47114   40402 07846   26547 07905   26465 66957      14017 24494   22048 65858   25780 87762   08393 01685   68644 83740
0521:  51797 26943   22745 42817   83115 54835   52296 77781   64583 95154      84556 14839   25979 17204   86835 86473   74639 43118   41106 55511
0522:  76860 21761   32796 11054   81878 13442   32947 63015   71694 66139      95211 52086   40417 28452   22886 94181   39734 31039   50941 36952
0523:  87546 27321   87406 00051   26780 02573   51235 16381   76970 96674      90867 03550   44375 20525   66622 40080   94179 41187   42844 73654
0524:  84233 26720   53455 99182   39272 83567   71004 87067   77237 73488      42984 38429   28050 56867   91699 44183   02024 98290   63615 28926
0525:  09836 70143   21474 43323   15225 09062   72253 49648   56655 63537      16067 32064   92663 02580   15215 37780   03434 25060   59658 28493
0526:  77923 76174   60287 31654   13056 98054   42389 38661   66117 41752      16656 97839   87574 27875   36273 60960   75223 31113   66329 09094
0527:  76549 93649   41932 17655   85638 57534   21732 18987   54570 66599      34205 22701   30387 53496   84456 85051   04535 29140   10643 04236
0528:  89072 31259   51183 15239   75685 84265   25537 14904   40973 29360      80703 66079   40066 90694   51796 69422   61207 91599   21435 45262
0529:  50282 42986   45684 30345   17716 18220   40162 62787   06228 52126      95439 38803   38611 30622   59538 94766   08416 43598   63293 36105
0530:  83885 05057   80905 84252   34961 71049   40852 98607   45881 42696      64331 50665   76023 54695   50919 71955   05679 30480   53068 37505
0531:  07399 08305   79166 29347   87602 53987   88355 11868   76087 07038      53810 17053   75321 52247   42548 72228   69255 66544   30171 25951
0532:  70979 26163   57823 90067   64205 46871   89380 12426   69580 44879      89263 96736   00861 98682   43231 41665   80994 05082   28501 81867
0533:  98306 51600   04459 82642   31863 34485   30449 51297   97023 32179      35346 71676   91730 31789   20247 98050   82358 82001   67619 00412
0534:  36556 70465   08237 67127   44816 85388   94538 64622   64510 85268      46338 06480   50080 36495   80437 81571   74455 95232   63535 77911
0535:  91840 20222   94923 01672   64444 23814   07081 72943   78368 12310      54563 20636   36549 82101   40058 59987   51797 83883   17652 53586
0536:  84485 88191   61350 78040   16711 22806   51120 17305   10436 85726      60469 41919   27237 45416   78359 24023   19551 23095   29372 70403
0537:  62378 44082   34564 10730   03641 23087   43284 22912   21949 51265      23820 11963   98656 68450   81544 35127   53152 40722   39284 32549
0538:  05413 71013   60423 71458   22787 04949   56859 04722   80124 33810      16398 41069   61872 50576   72711 53584   59759 66623   94057 29951
0539:  10364 87432   83221 15126   65359 69112   58735 94337   84324 82414      38771 19136   25559 91547   56999 70960   46523 43109   48507 51452
0540:  73110 35995   58401 26141   39102 65853   13289 84759   08984 25330      59188 04149   73642 45218   85539 08202   81379 22930   02588 15552
0541:  11962 51173   94682 13335   05390 23137   38040 88736   23414 86231      51846 14547   32638 44219   89301 77833   93244 21297   93248 22882
0542:  47143 02547   25362 74431   37465 53386   84178 64552   60574 13605      86218 05201   46899 12833   27699 98641   59911 92329   41955 37893
0543:  93212 81380   34296 05097   04259 55333   02668 95906   53152 10493      48894 40215   30916 69766   78915 87209   82210 06798   54471 85111
0544:  71509 87676   01969 63299   71632 17610   63216 02165   28617 43469      15578 57670   41524 33649   84133 57298   12569 43949   97758 46327
0545:  55610 29744   82647 77576   80450 72906   48997 39794   09561 92437      42401 88828   84726 00560   57150 12467   83903 52356   03812 20784
0546:  41290 75825   09664 17964   03641 02705   56381 44228   55068 53339      40411 50159   78003 43020   40907 67876   95280 25828   42213 69733
0547:  25998 35432   05057 30018   57590 82158   88413 12945   62918 98165      83648 38889   33328 62672   34752 14455   48669 57318   53937 15942
0548:  58288 59159   82134 76087   24822 17994   70311 73095   69833 36233      58394 92801   72193 37847   95723 13617   35000 42055   63360 03446
0549:  24149 74158   20402 14118   16699 31395   20996 52819   63786 95421      65861 06759   01376 88571   89682 24622   74409 91975   83327 18394
```

```
0550:  64094 22207   17493 49109   82581 82840   20827 62010   83206 57342      89638 27149   85778 05176   76480 62367   66191 18009   22182 56502
0551:  89774 74765   41007 85023   06630 41398   26495 06605   54753 96778      90435 85491   11617 14230   28493 81171   84967 43022   99149 56642
0552:  21173 96535   83537 80020   34930 72646   34892 48612   60463 23998      95406 66230   76044 14190   44732 27702   30238 77167   72627 00024
0553:  26091 33646   62400 37639   84360 29253   81252 72797   04568 84778      55272 47034   55301 86789   33686 88299   88696 99407   65542 74036
0554:  27952 43308   04585 85435   16028 97058   98699 56332   15876 48197      40940 02731   70046 41678   38467 13781   95751 83254   13509 82681
0555:  64201 02522   94901 65335   30581 65396   57562 08520   52078 49729      65963 53953   29154 79574   94208 61487   77347 21259   31957 31612
0556:  27718 99017   83231 99585   15870 35985   31896 40065   91961 87931      77535 42385   13642 84910   83876 18594   63324 88836   96323 61598
0557:  09594 45055   78336 82734   42257 90577   70700 92034   56407 92248      76215 12870   80507 47700   35184 82647   68728 30143   31686 74134
0558:  31397 59545   16279 86462   22281 89715   16972 99242   85899 82888      80723 53532   43178 19942   71448 60947   68435 28853   99129 34407
0559:  40430 65740   10928 64701   16871 29881   70358 85606   07515 29234      65493 65951   47214 72029   18948 20192   23036 01303   41773 04472
0560:  64127 17886   21472 46007   94728 01389   43835 10612   79450 74927      89403 16954   30046 94809   55827 14613   42337 98239   04245 56597
0561:  74644 26562   94141 87608   10889 84615   83364 10386   54760 31098      99792 32923   96528 38848   63605 96415   66026 43307   31102 21844
0562:  37984 14285   67074 26586   44102 47459   73363 03682   74204 46047      40945 40664   60967 12462   77250 83471   41172 00965   52787 91491
0563:  11024 94577   19791 86420   00265 88601   82737 72085   05747 89997      23256 91801   36844 88686   87427 52715   71113 87911   38926 29876
0564:  33207 46636   24578 43389   13518 99919   26160 91270   15176 95585      62887 19191   14247 89965   53903 30648   64698 98284   20432 90319
0565:  37251 80441   06202 62042   10275 10536   20454 97913   24732 30102      47043 22072   50261 11820   21514 73151   53902 84601   71694 35996
0566:  90530 78864   63486 41504   27673 80019   76417 25862   18171 96767      35104 91191   80135 57848   53239 77274   25380 04424   12880 16463
0567:  33019 21802   94307 75692   33920 49461   04257 16274   89520 20161      41983 05330   82509 38579   37295 13875   25509 83543   18365 27079
0568:  33991 45160   48709 16161   54638 44097   77707 09499   99440 88744      38871 12315   45176 87965   73572 31589   78768 95317   70960 58656
0569:  77469 13610   02734 73392   99178 78050   43804 70159   88268 27238      22748 13174   71921 77197   31348 05795   56165 36426   97367 89706
0570:  43343 82600   38424 63585   19154 36742   23194 29003   76728 92105      60171 42321   39243 55952   38466 69251   40671 46580   34701 77760
0571:  32541 70725   70391 87131   16446 33645   42568 63642   45912 26200      50642 36245   80627 49879   32981 24395   62760 80652   14594 68010
0572:  36678 46503   18335 06930   95078 03812   48545 58244   69482 77416      99902 45461   03111 70337   74526 32832   94331 96018   29260 01629
0573:  28310 12578   34214 80866   05582 28621   56871 51056   54927 79668      97875 69216   01758 58637   97806 73587   40776 71582   42858 64081
0574:  95518 15625   38310 69866   31967 25253   35240 87573   38422 82169      32989 07010   50260 81409   95788 21082   51409 35739   13336 18004
0575:  99015 11453   93301 62580   21827 96807   59549 53213   34025 91023      01860 62685   24951 96837   88230 76447   50963 61871   97842 68812
0576:  39938 89599   44036 68727   66195 28770   70161 91762   22666 36249      81532 12760   36136 38409   16134 00864   81655 08294   77486 15514
0577:  55564 63672   16542 85420   24001 20392   86660 98548   58191 22325      50574 49046   41471 06847   93198 25499   24894 72759   87650 66936
0578:  09556 53499   81577 90975   17469 59610   55539 95948   28085 11503      19339 65299   45262 90996   35436 93682   20460 19164   30558 78380
0579:  47997 76309   10522 76461   16241 44167   73693 19502   06885 91537      39791 64057   22858 74734   52464 40556   96741 49691   24162 37572
0580:  20114 73971   70687 40212   93302 60498   91006 71925   08804 96603      98965 15728   00482 42304   83189 82774   02967 41250   68731 28644
0581:  36653 91393   12569 15145   18397 70158   98659 02003   83802 59405      30826 47360   27480 22015   55309 25644   40985 06997   27244 30496
0582:  59274 98830   60616 12591   90209 99814   27552 98636   49755 81499      53566 23994   96774 74969   78610 72586   48382 91123   65263 98020
0583:  47989 54841   93168 51568   67944 49316   03591 05726   24013 86836      89684 00884   84792 08530   33164 10960   62722 63042   20433 48983
0584:  63629 24688   10642 33709   84210 34898   16175 39927   65805 29060      83317 68838   13610 37022   67563 50756   66615 99399   26366 53239
0585:  07826 88625   30336 26669   19420 73172   14828 26233   90664 15947      32553 34132   54922 55279   32069 73375   98456 55319   48094 15522
0586:  57804 04876   52659 46907   33219 47071   18076 73173   20274 64035      16339 07998   02822 65166   08539 83463   20073 85048   26546 08382
0587:  50390 95882   74630 33834   55998 96389   33371 49240   96092 99874      38555 80862   59114 49196   67045 01735   52392 46459   20694 99491
0588:  21314 30353   08021 13848   90874 77710   24227 88616   95460 41777      10014 07042   81655 89960   48495 31497   47257 04251   47370 71590
0589:  42530 65087   76206 62338   10483 97757   50952 40668   81792 57051      97101 00165   41184 38427   81543 36793   87271 58481   74875 46215
0590:  30988 35165   58879 16029   94682 99900   25556 41919   59445 49307      64905 49454   38644 94329   25648 77943   14150 35269   98779 28936
0591:  31431 12988   35969 26742   44356 66168   73841 05313   04747 26275      85246 90938   76697 60762   01167 92437   24817 01430   37767 15954
0592:  35223 06444   72165 61565   68288 50157   40656 96194   24665 68325      55992 15549   35502 43183   99710 72748   98302 09831   51269 67893
0593:  99984 85115   28436 19119   53398 10590   22574 95158   55077 14878      80362 56730   97910 26883   88029 80306   74216 08677   71666 78436
0594:  65370 02389   07723 52204   58945 65273   77059 61724   66049 91455      04146 97194   08120 95479   95662 89583   20962 85284   94052 65082
0595:  44231 21050   07336 24063   94588 93804   96660 47093   16474 87808      75621 00487   21044 03674   18257 30677   57312 87446   82878 10072
0596:  13599 99385   52026 87545   96069 30496   73173 99933   08274 01858      11835 73250   85020 07520   05097 48546   40856 72340   51098 91969
0597:  02500 40942   98628 16316   56579 54616   78312 27564   59957 40797      08904 39180   86717 73719   58173 47574   58388 85952   24014 02693
0598:  89878 52019   73489 23857   32737 47334   35747 57252   56009 21914      33526 10318   02041 08067   11779 73344   61899 86203   21908 93821
0599:  84015 70900   13518 17444   11101 66350   44532 75363   09494 69879      90306 91454   86129 12981   10382 71873   74792 86009   88515 39916
```

```
0600:  1427467094  7401624242  8674530913  5360543683  8618152135    7551309083  3616441503  9454861087  6658525054  9439905816
0601:  4954233117  5478008295  3488943792  8740066224  8044151430    8930874837  6699592303  1711940406  3268712173  9251936991
0602:  0639240604  3633793295  3288932670  4151067651  4608963741    8286807427  9383884513  6952969296  3182693257  9870330833
0603:  3867718538  5694244882  7705122987  7222931536  8039959942    1758319584  5340379921  1366038506  3658713711  3179791695
0604:  5249623693  3000753213  5434519058  1626557532  7525898784    4571869611  0101155113  0309071464  0402045589  2878316629
0605:  0058771904  9304731090  4159379773  1561395117  2672522163    5564465038  3786123037  8830863301  7733514165  3072353521
0606:  4532865748  9777999140  5600811365  0739168693  9696983952    8840849558  5394408551  0076899301  7314966334  8019727245
0607:  7675535208  5181632567  2333497343  3913700843  5530280821    5265509685  6215094404  9592228888  2761417076  6708644284
0608:  1483141972  9760911705  4827524998  3065020045  6466476453    3672849442  5648969573  9304856419  7071564182  2929997033
0609:  7136084010  8262847388  7322518511  1623001944  3821821747    6266036804  8536491313  9190412063  5138806913  1192698678
0610:  8005912339  6556346269  0692963932  8710870013  5066244896    2967570496  3820130902  9101349966  3702208434  3483233502
0611:  9901497316  4261404425  9442290249  1551635308  5830196306    8898372403  8009034575  8318841266  1667345796  3157537322
0612:  9343186661  7712926940  8354710485  4831384940  4633416877    5337314184  6917280103  0151539916  2581311833  6229342392
0613:  3862174822  1808663771  3805689588  1333714261  1693073933    4196341859  4385972028  3918762853  6047558459  9956458381
0614:  7325213937  7475091392  2942170501  6120728238  1086978784    8598028550  3436228984  6391449815  7846315546  4484348603
0615:  0088364680  9259196127  2366150397  2547627323  1680988658    1617348259  1802190859  3547002667  4571735085  6762210521
0616:  5979750553  8159298947  5955764624  9251977816  7838714497    4375260067  1283960434  9691034097  6565190739  1168959120
0617:  0017849173  2646290400  5167920673  6592073970  5251014187    0526451637  8938439215  0598779195  5867243760  5227375666
0618:  7427392474  2927355912  7131017480  7246521941  8601861393    8247132448  7925125598  1007201935  4443511790  1598038559
0619:  9443889053  3088956454  7155622142  7681602082  0090052218    8151270912  8201992023  3368019874  6491573262  3005237784
0620:  2700715622  5403514927  8966902945  8375766400  8327181419    1962761592  0980402525  1028327914  2340515744  6995055587
0621:  3195643373  1396016093  2696350340  0940733858  8035280951    2017474121  7680319624  5518936997  0932925717  5473258836
0622:  0387470673  6300809639  4476226883  2127207519  9623456268    2671424504  2413252408  4656694318  3089683848  4762537535
0623:  0150420341  0544809231  3201639658  9638156027  2362794891    9478588574  1041014306  3423444760  9216728968  2179453371
0624:  3392636077  0308698493  2408019127  0951518761  6339263048    3778609616  1902776175  7436934503  3870260982  1565580050
0625:  6044524362  5391971467  6713928679  8792476701  1299897243    3582651771  8104766451  3665517670  8345495896  7857938339
0626:  2757096277  5345444214  6250466182  2036245333  5515242966    3313605881  3206218539  3245370891  1870489997  4433999293
0627:  4022102035  2047785785  1608191750  6684006099  2568579839    1601380883  0021422317  7153856246  0251477704  1202125516
0628:  7219996159  2996611789  5529418241  1208471220  2181232807    1818242289  2360507846  0524738784  1527337459  3506886650
0629:  7079121831  0323229401  4934128984  8054968059  1683240022    8915543779  8301670335  0873852478  7299214293  7449854613
0630:  7075220022  5544749627  3397084360  6423930037  1579659030    2635012222  9798003610  4749853025  0462063295  1981557674
0631:  9499032344  4781608591  6818058453  4093508822  1653947659    1333296552  6393246758  5972112840  2948972246  7403344108
0632:  9210278863  6205255887  5738711364  3229556261  9215171799    7884696479  9772760762  6510805583  6991339428  6382590546
0633:  2196921749  6196912183  8701123994  6176027229  8071470560    0781071808  1328777314  0985430082  5756289589  4292913574
0634:  7727944581  2222533199  3029935783  9579010893  0564401167    6121504457  5405548190  9915953709  9827381446  1385630984
0635:  1277144086  1238068508  0238921058  2421117355  1162574519    3955348398  8081030805  4768912160  9426439159  7275669177
0636:  7965620743  1353916674  2197980143  2541460221  2544933287    6501214515  5533795483  7208232296  8073681127  2239733747
0637:  0378262394  9643676096  4339970791  7617038340  9641886343    1596184013  9604181804  9488725647  0622479297  0603381870
0638:  2599230227  4773619008  4059034564  7831472221  0553722787    0111674660  4536586187  4540363137  4583327198  8171126257
0639:  2542498461  3627777646  3234982097  6662207563  5865792005    3634604428  5855886303  6197208659  2897078001  5688210080
0640:  7653864520  0369867270  6093337317  6211669767  9103745775    9995249976  4425803704  7833911069  3955933790  4744684672
0641:  3473287757  0393034570  8329978990  8630175570  3033621900    7120056535  3209768648  8731899777  2217406177  5564238317
0642:  8855154357  3764154774  5036931066  3896877433  6333139177    6067003906  5787533279  2176188846  5940450249  0674348893
0643:  1135478366  8912619905  9647557133  6990828058  0880704979    4796192776  5161348770  5816775358  7728580678  2076748014
0644:  1158431793  4059504282  0731005266  6581704985  9304093205    4507599151  3416306008  0525569951  3021883642  3834264056
0645:  6837854500  7694587570  3658592611  0955179405  8683378491    6765101944  5218857442  7778809198  3891735613  7198141051
0646:  1958586360  6624626153  1864217942  5171150546  0060350921    0697353480  3952002644  1085412495  3643563781  9704353828
0647:  9427003299  4633544794  9712034768  2918007505  8316060753    2958308378  8389353339  1111387575  9833379254  1422862675
0648:  1855056230  2048829543  4086633209  0950529667  1578913048    3946353200  5253162160  2377115114  5059886664  8983990786
0649:  8112999507  9233167364  7630505662  0531219452  7413820878    4899922425  3490734387  8352146178  2479674033  7291843940
```

```
0650:  6927474003  9605676883  9983974335  4793442047  7217132186    9908217537  4411271994  0017324690  3940860272  5300074960
0651:  8669362884  2342369353  6253235168  4831295930  1739190143    7850715756  8571607079  4327439502  5299052209  1499995033
0652:  0071189645  1380850125  6523410263  3311799703  4568698969    5167189677  8683833277  3293082478  4290371720  6743448521
0653:  8741241394  8442505502  6957822582  1561441432  6203933341    4058375048  5463113583  6281117669  1726089184  6854759012
0654:  3891961951  5122085887  8099731078  3805903212  2320683344    3448565264  4597147393  1951046638  4419480064  5668333307
0655:  6287808852  6498158049  3475692720  9501988347  5834001122    8526570875  2877300472  1219052673  6955211907  8001236648
0656:  0832080179  5591228959  5707790130  0137903543  7836033560    4314913461  6505493244  5437664223  9112377269  5364964995
0657:  9286672727  1781992240  9344822921  1621166099  3015960198    0374137044  1574935543  6446963630  4400694676  2382576434
0658:  7658923368  4729046462  7872375154  0876346951  2740364602    1828109488  0591203485  8137921798  3395017297  1541432987
0659:  2751333925  9340418574  7363574327  9639289860  3303590999    6400159701  3920600016  1253449067  2563925007  7567621985
0660:  9146843083  0459042079  1542000771  6664581355  5952359429    2412668790  3817095108  0665477063  8769670141  3263493438
0661:  6093144339  6780277985  4986272701  8071267249  5930067398    7269671814  4823469898  4735763314  6544945025  7506131723
0662:  6937487776  5591571788  6590502240  6683996422  3051577330    8121538630  2707927044  0462328801  0748875048  4248936321
0663:  3242769087  6880692099  2377773061  6406444215  0432508146    4716494267  7473745125  0041433303  3724685085  7083082622
0664:  2228554949  4025506545  6852829777  5092791616  2511567992    0159066544  9887180048  5190749415  4441240749  5275354502
0665:  1294939124  3565686312  8079817000  1433218029  7307863113    8241509565  8411773147  0550863221  7364237981  4628749726
0666:  2700987288  9903319567  2909483263  0175204839  6945111751    0908740328  7879388369  3600683485  6378917786  6043804544
0667:  0539189744  5327548021  3465632609  9406970209  3153426703    6329943871  6139076363  1454329896  3497020838  9181591540
0668:  2663662411  6065792989  8772425825  8565213225  3281975204    8319302615  6165930935  2956800254  6977851914  7179842254
0669:  7669557487  6283797095  1829756129  2271665328  7262609287    1164078595  8843394652  0111474783  2034561211  7442394750
0670:  9962744652  1026512828  1095674302  0590925236  9934590169    2694308037  3780927564  8912425629  1859753932  8105938771
0671:  3263307546  4466020839  5117509057  3623937697  8075692933    1704979313  7072808066  8514476996  3361950516  5678876925
0672:  1544692650  5506974985  2423214602  3665157019  7549367871    3775596496  9649670297  2599964475  9303670828  9134990920
0673:  2796255946  2492984726  6531768809  8260720945  3413819841    4085768429  0526407985  0759543969  5907668797  3649427418
0674:  8132962062  5810494363  9974673776  3575466670  2330458698    1937306215  6592983433  3736011415  0253163933  3134115882
0675:  0174644196  6652881234  6803621872  0157319568  0703903858    3126548537  9851514287  0489180958  0989159654  7982094016
0676:  0254832541  9290414942  0362710529  2106005057  4003161175    1164736741  8786296522  0502144128  1755082014  6770058651
0677:  0424833197  9845528271  6023669374  9077398406  2114340216    9729925279  3830949959  0904496418  2099172082  4126620709
0678:  0016682985  8847587462  3810948206  2684046309  4393833632    9212402587  3821757778  4582148856  3287677796  6398727075
0679:  2385247508  7665835378  9818534140  3769360215  5882622644    9321891243  0292385794  9331307962  1233325657  1009284150
0680:  1444030110  5220418308  1793606806  2285665336  5233135255    7962648842  0461843059  6787450650  5940011100  0128100418
0681:  6916905036  0245587504  3561373569  5887636041  2184080957    0275187502  7625516956  0423249507  2812391437  3895137436
0682:  3286899241  3246754216  9491346298  5135096548  7009879437    0632522655  3984773460  2017325502  6989816364  0535955116
0683:  4672993772  8637223110  9700770601  6138433190  0812842572    5638435615  8529784608  0915538063  4633272547  5844322147
0684:  1938708408  9599152220  8500878794  0924835343  4019735703    7262884832  3185037976  3429153957  1184002828  8210315374
0685:  1425071656  0088257670  1370878691  1790940341  9001306174    2627266576  0084168407  6730818553  0248455467  0071585682
0686:  1576733363  7459906557  0423938102  2872469380  3886815214    8056705804  1229637847  6663490922  0628754085  8031722220
0687:  9612562760  3990646770  9128177286  9749049957  5774502581    2436702308  0003589508  7772702233  4675673098  5431938728
0688:  5380177577  9707716497  7602417151  6926312266  0521549623    5576264297  4396551666  2697812565  9225141686  7188293375
0689:  5078406263  7942055352  2504017948  7248772049  0189524024    8852401611  9477518176  3425987363  0497904143  4688071833
0690:  4040518420  7596393874  4900766581  7578498871  3308358023    0872483522  3295303405  7696754525  8592759169  3371593800
0691:  4881169020  4966953242  0934725879  9637125023  0905839612    2296260974  6302735830  6003656023  1797116819  6926178613
0692:  9410204644  5474455090  2132388496  5426716407  2813264141    6602573929  5668589134  9097232032  9494001824  2103263328
0693:  0273320315  4899752105  3943117945  9055557801  4875046111    3111999057  2915508301  1494241106  7979730143  2099207843
0694:  1494915671  7932441105  9571680727  4987176452  8508671586    1566421672  3498307255  2371026237  1911663010  3550321705
0695:  7940870552  4391143755  0647399120  7602019402  6021352533    8242919849  8479308107  0036299614  5917525021  0032111578
0696:  2744856423  3342606863  1990335816  7482527690  2479848215    1091855553  0874962280  1631242048  7925269530  1979857831
0697:  5650637812  0268808494  0033513158  2827621999  4558585009    6583731300  3717423231  2202784822  1979641502  6755251435
0698:  3321173750  7098627030  3717248498  9002709613  9469461439    0163725075  2310984366  4996660088  9051165364  2960713729
0699:  0869815798  7364784774  5332436450  2617662005  3853029512    0483781425  8681509794  0416761100  7350650632  4283925179
```

```
0700: 35010 64692  12467 54298  56295 37297  07846 35456  55193 37370    13735 85898  03699 67731  04403 23520  79830 10541  15639 52226
0701: 42133 03775  86274 17752  35781 04284  76548 97531  38781 49712    92203 01696  67576 54972  87345 40980  19197 35457  94623 02572
0702: 01796 62587  16132 75329  19726 88405  17028 88349  49009 30697    38604 14329  94101 58581  86168 66790  56017 82663  58346 84428
0703: 66762 68626  90522 10926  79018 20269  91824 35703  98581 98223    52114 32160  77477 36431  67299 38889  45241 60457  13957 86026
0704: 76616 54895  54699 00446  44666 77902  75548 29931  80249 04890    97034 16936  40635 96216  46118 89634  33689 24263  13542 91600
0705: 96290 24723  18268 29215  96437 01027  14645 29741  84050 25386    89322 80561  16225 95386  30822 75960  69518 77593  93654 50985
0706: 39814 19195  32040 09204  27692 13046  43995 62594  14281 11061    75702 46512  87000 15327  24449 91216  14821 87613  43435 05948
0707: 31148 70217  20563 86970  09435 67830  43559 09322  63987 42472    66926 12272  53377 60403  17586 73891  43262 75949  60729 23155
0708: 97015 79240  61091 69274  97272 13779  56375 51591  97568 25323    98279 11094  53517 89320  76899 73157  96786 28829  09336 45148
0709: 63023 42949  74320 45769  87280 25604  47416 41928  23264 04104    57862 56342  61297 17049  93216 19566  91169 76491  94022 70064
0710: 69200 40068  67778 49882  41101 72475  66958 10933  53321 24901    26336 80383  84970 95177  27724 01662  58234 51359  54537 99152
0711: 47793 63054  44447 61940  07264 51149  23190 60102  33887 44984    72083 44678  31674 57543  68221 23870  93547 93043  48214 58890
0712: 86396 04102  51061 03372  54453 57057  40385 23777  64581 35764    09371 96040  12766 09265  38246 32422  68130 00953  15191 42163
0713: 45506 62687  98914 98710  58064 23439  35680 54267  85834 75244    82651 89388  22137 12872  11282 64072  39060 78449  91552 53483
0714: 17454 88592  55706 12202  64337 23612  38964 58729  70664 54428    07427 59384  72391 90101  32414 77734  85506 69263  69378 67949
0715: 64940 87909  86851 08761  79154 48404  50269 40539  24773 44014    68777 13881  30980 22200  04937 61359  40305 12700  07355 48232
0716: 55110 96605  96038 38146  68340 09875  56991 05842  61945 20940    26259 56024  75794 18451  24449 44229  53446 34515  41750 37506
0717: 72249 53556  76972 42283  93302 54693  77072 62913  42879 82427    56337 71623  41859 66399  50446 53745  22264 18228  66245 68313
0718: 88520 74949  48204 44514  37638 51906  33464 52402  50842 81984    69801 92688  27125 27370  19331 19897  20739 29691  49725 02495
0719: 98046 40533  81956 03911  35715 04460  23811 27043  06505 87425    71726 99918  22173 79879  36033 78642  56156 93725  37399 51915
0720: 89942 12590  17806 45683  80438 42622  82383 83891  34484 27733    59125 24330  02059 59793  20019 26660  26911 81263  28853 08013
0721: 76128 39962  75377 76009  78138 87589  48468 40160  92927 79584    49794 12928  83105 58841  41811 19719  43298 01418  39314 61437
0722: 15696 53556  52023 69413  35565 39316  02491 69876  36114 01251    57196 70920  81953 07275  89720 39783  52041 40490  09674 42967
0723: 30338 62912  71646 58067  62363 83546  98946 81776  27107 94953    72114 63539  88017 85506  55075 88439  87460 09989  61052 52727
0724: 01698 45614  55787 77025  70141 68826  93360 33019  09996 23934    15479 30844  45888 36661  43532 32898  90695 21087  46109 79841
0725: 22591 84232  76738 78268  44221 66008  71339 26455  62593 52934    51995 96421  04036 32034  30932 32722  77784 16471  68202 69051
0726: 26589 10395  13978 09424  31135 14472  42900 31475  06419 85373    51214 07668  08765 23499  96936 55697  16401 74441  17632 73743
0727: 33863 89560  31348 73375  83138 80940  12849 63118  86071 21115    37831 86095  82733 52617  94234 58363  45496 08639  99690 36119
0728: 26303 18427  99708 57771  84899 63017  77841 54525  84987 59300    38997 48587  66436 73246  76273 76303  53576 39668  20271 55841
0729: 11991 03451  16003 51644  70889 09838  78987 25160  09306 33548    29832 54614  62595 05993  61798 64384  34482 32664  72666 57807
0730: 18998 25822  50506 01013  23534 26110  77143 46398  25816 12344    98733 73507  63759 77435  81423 12333  00425 60725  72810 04214
0731: 66683 06709  52776 08577  77647 66625  28488 92598  23355 44182    15220 17897  03632 57090  83357 18854  70514 70386  45795 29941
0732: 41548 98174  46458 80913  55464 89871  09737 00910  80498 04967    03096 10120  16577 45758  38366 40210  74289 22303  80967 17975
0733: 99582 77929  18097 09242  16841 09464  92617 58221  94071 22028    36245 54071  12599 68795  96637 62586  07027 18122  74510 37532
0734: 95457 07061  31397 42543  61338 88076  94152 16493  56412 09835    16913 51988  23447 21763  43713 23042  66823 83994  29117 29287
0735: 64497 54311  50956 97369  32729 55620  05822 12988  68656 80493    44713 73727  63161 39911  94189 19104  03474 38834  49501 20266
0736: 91168 22219  78223 48323  80827 41422  77387 54685  70038 53304    89364 76054  60942 60902  38615 81408  20865 38044  04991 39166
0737: 98122 70801  24873 79649  90526 71046  72265 88691  49820 72135    15093 76703  68395 83290  64548 01639  00124 18512  81157 74966
0738: 93017 48767  41918 96716  18477 32365  80568 42229  74678 98805    68547 90820  52612 02740  20381 10119  12177 09149  46113 80805
0739: 67068 81673  77929 89255  36875 97627  14537 53163  88397 36717    08532 54850  74370 41113  59286 67217  29548 72728  64187 84125
0740: 83807 22093  18752 82882  92574 18576  18417 09600  25468 83374    87840 29569  03230 94702  53366 28772  14214 28449  20318 07586
0741: 53437 23757  98169 00852  36243 19316  74940 06227  80974 25142    44756 23199  97515 09947  12354 43479  65413 81503  98333 81816
0742: 00020 52449  02750 55385  71430 10978  19341 69249  14001 66916    32037 86034  84786 90805  16528 14783  61172 52748  39540 54574
0743: 39888 42381  51706 99385  17512 34086  27080 33872  66423 45333    13033 92851  39067 03582  86110 04557  61350 34831  99828 24288
0744: 99264 71551  55873 60119  19803 58372  27146 85570  06087 42497    52888 69006  84692 00124  78231 42457  56386 24312  23526 63690
0745: 26929 86893  08998 99889  84326 55659  95880 27315  66094 31043    79374 96966  62896 13927  99579 20999  99363 00890  79296 46410
0746: 50111 62312  29726 37246  11520 68800  28532 45648  76403 31971    39072 17769  70961 42919  64066 30480  81371 54347  03951 53009
0747: 65320 54459  17174 25834  29157 44151  21097 65943  62185 24981    60320 72377  96470 15835  35399 40829  02972 56562  09327 02582
0748: 82089 73572  99349 37787  11960 04940  47264 64446  54166 77985    54802 67016  53948 82617  43744 18840  66151 32183  21917 38781
0749: 74885 24973  18106 02615  82419 16729  00830 21894  22103 77059    54338 46757  99421 76641  28904 21160  69051 31082  02453 99853
```

```
0750: 39081 27490  31944 79667  32437 17478  42856 51463  80365 79133    60889 94000  38721 75568  93825 02501  27038 38869  78891 32309
0751: 89813 25656  97715 17228  01201 18775  55530 60132  12526 33108    04906 11630  10808 44031  23777 88988  20377 77511  60522 50362
0752: 67970 55911  52344 97635  26318 11907  87639 02408  23441 39051    76901 14796  60677 22505  74244 50117  13854 53289  46370 66789
0753: 87974 65584  63553 83332  61260 80904  37162 22664  61840 63833    94114 12246  17876 33326  37715 03007  81622 69390  86800 92602
0754: 38129 50301  90821 00981  46885 15616  01510 49072  67932 52557    34526 90558  75280 74136  23405 82901  07983 57774  34166 84560
0755: 24516 97659  88731 50597  68359 63309  35721 16555  78182 92917    38846 22307  52497 24334  80412 55737  81462 00261  72823 59645
0756: 76338 75994  51550 69237  27235 35911  90090 69161  04868 04452    85986 17743  00926 87239  29376 63331  61846 76545  21363 90758
0757: 22619 60456  95567 30049  87823 19335  86331 39828  22994 89102    63625 46414  47763 42897  62893 92003  54090 93226  92278 59853
0758: 34716 72210  87061 58194  04093 93792  56708 64855  43329 86887    37549 98210  89992 41207  65602 51165  92802 27929  52024 11911
0759: 36772 12058  17636 11134  87976 60993  98735 29202  67655 82952    19797 71128  40414 11727  22443 95397  20285 90141  74368 11049
0760: 47182 23185  75910 41103  28724 81640  34228 90783  85456 32928    00850 37106  94088 05753  24244 65686  26065 32447  50728 97469
0761: 40243 96566  22937 67918  11434 16559  35470 37837  26892 03045    60630 37992  74770 13065  50721 86190  32497 54586  64582 93237
0762: 39884 85030  60961 93059  36669 46535  55956 46428  13444 74512    74157 51690  52537 17268  29933 86329  78573 76851  65411 12297
0763: 58082 76943  71744 09645  43582 01809  94312 51582  90338 95256    60111 29991  99903 40698  94195 07638  79496 87212  77168 70842
0764: 89745 77734  86613 89310  97822 08307  37794 56797  99129 47159    92999 56888  89405 87831  96541 82576  74738 79740  14695 12571
0765: 86019 19632  59287 32954  41237 99636  37922 84248  79807 93459    51070 94650  02327 20920  50458 22551  29826 08688  97923 24861
0766: 59705 37069  23049 79971  81335 56037  17744 29242  35851 63859    30262 72720  37621 24170  47062 82466  70594 39310  69657 00361
0767: 63836 72339  24474 14631  93708 41452  03086 41701  85420 46750    16178 08723  57787 74265  76015 76626  29331 27326  66962 56086
0768: 18226 46008  17390 44170  01218 23165  35435 42073  08853 08028    15255 71640  18872 02437  48581 29789  09356 48717  60132 54851
0769: 64954 21984  59891 47059  49378 01750  78240 33312  90949 27208    36942 14114  93768 34330  38656 98109  19846 49994  64975 66214
0770: 09350 01841  62803 06308  42510 26891  92833 39837  00146 48586    46176 41798  44690 31509  74674 72496  21697 20916  17913 18215
0771: 42171 21899  95074 01022  13809 35609  78701 35615  74522 84587    54857 61984  42762 06967  16384 48593  05967 86055  00987 47046
0772: 58058 32215  00628 21281  17250 09228  78063 68858  49973 77792    90322 62245  57100 80042  11538 45118  46683 19975  36969 29838
0773: 67361 93530  98680 42626  80639 69278  75240 95523  41421 87788    69035 12665  80753 58211  28338 19682  96765 24127  54645 90356
0774: 46138 94239  62348 72370  75955 37151  14545 87719  49911 65804    30238 19327  65137 68322  90539 51028  23507 77011  46276 49691
0775: 51420 09364  15424 92423  63671 82204  25109 74013  22097 28568    67248 79031  07603 21253  17736 32595  25389 95992  32004 92066
0776: 90116 41695  94595 40080  97171 58675  23791 38841  27190 88512    39915 84326  88272 36563  70296 65851  74349 06342  76235 18518
0777: 58978 75196  20071 51928  11628 30558  91786 55090  11291 88897    68517 73551  11787 73256  62345 91599  01425 69970  69366 75097
0778: 02533 60776  56909 19989  80873 83201  78941 11925  82847 07142    68231 59482  31628 92457  67074 43697  51371 05487  42280 53260
0779: 13709 64504  53774 69379  89569 41196  25979 90519  50374 37196    30568 33702  59444 29657  37458 49283  27044 08307  00083 12421
0780: 69119 69303  55375 85001  22351 61116  13572 45269  77133 92389    45297 56868  45322 27634  74860 92103  40865 59568  87284 68482
0781: 23196 55981  29895 64601  70087 45392  14296 40395  47261 13443    61554 49919  44112 90501  50164 68763  31336 57662  56384 30644
0782: 72031 85880  48147 71557  03187 83059  97096 76334  72972 22955    06079 67951  50894 58635  14141 69325  80877 36704  76906 29342
0783: 58485 76303  77411 49851  61439 38926  41230 57770  16527 61112    76345 24895  83157 79384  86551 62246  06043 10584  95352 54121
0784: 94628 26307  87030 61534  63128 39767  05799 02203  90416 20607    51247 90538  92384 79508  11354 19885  45108 92771  96526 03656
0785: 07537 94598  95807 46129  52335 03962  42941 53712  25927 42541    12716 99902  19657 19948  84265 74335  15804 38763  29949 55141
0786: 82238 51690  70828 06423  42787 35589  34029 20000  56844 24273    61070 48269  55924 33262  05145 88005  91717 05365  01054 96168
0787: 18087 41959  47675 13471  90823 61858  65611 77650  12270 47336    12916 33079  20638 45434  37082 19261  43034 21083  71726 64582
0788: 12855 71200  45342 13885  32299 33475  25592 19780  72059 56109    31108 96524  16686 05740  60396 55236  85358 70882  74573 11178
0789: 55661 96982  87780 53104  57894 53820  58113 48102  06093 95929    58793 87929  76185 77302  56096 29002  65687 57591  44621 21056
0790: 02777 68435  05959 15881  96141 88067  46512 53370  83792 06571    14316 63324  68849 89937  72050 73886  91923 96846  28401 12783
0791: 23538 23904  30594 12887  70263 91789  86543 06035  91136 66137    25368 85281  72940 44407  25411 73367  19753 12446  43448 67223
0792: 62550 19508  81274 82844  42569 59435  31392 54601  61352 31571    66447 25807  09728 64518  04187 95044  01015 55851  02212 77876
0793: 23727 68758  96642 24590  96055 26186  17767 44658  89811 13316    47993 69815  57336 41544  30368 77780  24093 29126  03325 72513
0794: 06618 38219  15141 35106  30681 18178  04394 52693  25808 13362    39880 63519  26579 47507  06291 69476  21435 45575  76425 40156
0795: 39167 06465  12989 59143  57962 41727  68247 98460  56187 14194    84318 59879  28603 46398  55481 75728  37527 21268  33256 49245
0796: 35999 76493  24945 14278  03138 90803  87028 86364  24982 04884    90184 20345  97869 42213  11363 38315  04568 19439  70669 12384
0797: 41904 82427  24658 02103  27210 05615  16639 98668  08135 86741    21894 67924  15416 10340  44359 93197  64826 38053  23429 19871
0798: 22370 88025  32484 98665  38230 48476  86619 33674  80839 83664    62816 49558  48262 21618  62209 66145  30386 50529  30625 73273
0799: 50507 98439  46891 87612  47181 59276  11379 84325  55417 32231    78386 51304  00215 51537  56991 42739  02893 02237  74337 58775
```

```
0800: 2415495117  78440 19353  85451 67922  29571 60364  81232 77542   57765 36279  61221 14442  98989 22405  73473 58471  56338 36227
0801: 17578 20625  95525 11543  87316 90434  02503 43052  44690 21287   99894 95813  81860 52321  19166 85001  54553 97771  02232 24042
0802: 26591 96112  49051 12676  46394 20064  25212 25322  04637 06538   21522 25024  20992 68734  98612 79131  82236 43837  72267 17513
0803: 70721 48066  93026 21282  60297 33755  76820 48683  95802 58171   84005 49982  87463 43924  32130 54212  67808 97497  86002 43295
0804: 62741 29625  45812 75463  86267 86657  53189 61354  12801 84805   18380 58774  38800 33067  52533 97678  61781 22479  28806 77244
0805: 43417 38630  69321 86834  64710 31375  23970 93057  48943 06888   08135 57494  13002 87219  68726 31798  37092 26140  49861 06903
0806: 83166 33249  39487 63647  26410 50689  81525 40656  44030 18993   56605 47329  06635 33503  15680 55073  36815 22945  81486 22393
0807: 51174 38384  73862 14795  81302 04938  54010 66004  26535 07733   36844 19418  67794 10810  19696 90871  41856 84413  84819 85743
0808: 10018 96458  52139 12868  92101 88779  23607 87624  51097 49421   98137 73413  07515 07360  78333 40151  50043 19938  17943 34991
0809: 78865 10700  37054 26674  73323 83546  76974 41565  60075 31692   82378 58904  78456 71474  58230 61499  66005 66938  19559 09252
0810: 09453 02557  82969 29394  66085 56201  42921 63879  45685 05931   07995 64551  73881 66028  37582 90598  80108 66028  60950 21614
0811: 86285 38074  26814 45876  52679 88598  31383 61809  58187 01286   60058 90502  88866 48988  15068 79362  81816 28984  86305 07034
0812: 38376 32955  55861 26437  09906 34319  05446 97963  99222 17456   97922 46973  49448 91486  18882 80587  68178 32565  59678 39674
0813: 81531 66059  19419 62669  40087 36133  00664 72055  39792 00065   97582 47429  30154 44382  43884 61324  28019 84394  98358 87945
0814: 40500 56334  08491 53086  63463 25157  64870 93377  51869 66959   15341 00075  48959 66954  31678 38204  76994 62232  39850 98206
0815: 52752 06049  87658 23936  40903 48104  53784 80021  78983 84622   77820 86883  42697 51530  05705 05035  66536 53753  95945 21569
0816: 63646 06039  33791 76694  11225 17055  51372 08257  62200 05302   04102 86774  19850 16881  58213 18079  86709 19516  87286 05788
0817: 01120 62162  10498 46781  27292 21574  49710 35606  64965 43272   83964 23853  51600 44435  30886 75378  11773 35732  78709 02288
0818: 95976 75574  41624 44361  26412 05279  73001 07834  07249 10450   80643 47442  83291 59908  61110 56336  54308 39274  90390 50810
0819: 10368 13554  83124 01412  51939 54665  60723 93074  42295 46191   67826 06910  67440 14430  66213 82725  25323 55148  61907 57442
0820: 34689 02066  17467 38355  35532 53761  34660 65314  53924 83088   20528 71098  87461 38603  62518 60914  51841 83001  21991 07349
0821: 03631 71524  51645 92267  10367 03226  87393 29967  66744 87658   23983 92441  13710 27094  67801 84380  93677 60865  37016 32133
0822: 43381 91165  95355 49153  72389 00916  16866 62360  46998 52538   60298 36125  32015 17714  71898 33631  77940 02979  15429 83771
0823: 70910 36972  60031 30118  48934 89544  81291 50310  01098 44399   43608 45469  47298 42770  28275 99676  41603 63832  49593 86907
0824: 52384 65532  23889 54714  48594 03566  56030 42848  64693 05745   43014 69380  90901 75633  60018 66440  21095 98595  37819 54029
0825: 02199 98189  81517 82575  94835 24844  25332 70733  70372 44748   07526 55352  21939 51770  69438 65402  87752 22209  88924 53762
0826: 53090 87260  41403 34382  25739 99957  49135 43357  66872 79260   88316 39424  15830 19482  68746 78684  97435 27441  71986 82982
0827: 97946 70494  99084 12758  77878 00485  13589 70763  46321 29543   72275 23093  24892 57548  66904 53842  34734 92899  60569 33995
0828: 58081 95553  34809 75181  24942 96307  73934 75452  50788 52065   14384 47703  75897 49913  61209 54594  75152 20329  60480 69399
0829: 32371 58862  44518 06423  65992 52177  54987 65822  70799 11653   47644 48348  48777 83991  16934 12358  85845 31844  89405 96397
0830: 58511 12565  76194 22475  64162 21173  93129 91615  33174 26023   53779 26308  15174 20414  80126 15643  73780 81400  69263 77134
0831: 91303 71300  46280 18371  39720 00670  86440 92161  58424 40737   44534 53566  30110 77765  08048 65956  20276 96943  86903 42741
0832: 79128 84844  87279 79804  39117 57855  37038 15136  39323 73437   23711 18199  96105 06924  80242 81999  27134 61602  18669 33171
0833: 42165 55433  48563 34891  88373 77575  11947 14020  45613 19689   78172 93644  10935 04094  99443 40563  40989 92616  67611 41267
0834: 84146 28086  21511 54546  56293 80430  41784 68976  11225 41041   30537 84944  59298 84079  81845 12913  50095 40262  02394 98059
0835: 50121 06309  12188 81637  14059 58390  51982 40559  57215 86781   11477 81204  10045 49256  35776 82520  40812 72013  48362 73996
0836: 29080 61408  12102 31997  60537 46706  24010 02941  40718 88581   70218 74901  57386 26299  92020 72141  93608 85835  28446 45654
0837: 15244 69463  61149 22442  73333 92867  95204 08293  34525 42771   25705 56960  10395 84081  13272 79242  22354 25545  60108 51090
0838: 02050 57988  49552 59038  47890 90150  57065 00861  38492 23951   95424 42998  20233 70896  51419 06413  04456 45049  71766 17951
0839: 77519 31767  56633 13384  96377 94554  21700 39059  32834 67255   53120 34220  55310 69114  46863 01992  37302 49681  77407 39580
0840: 05701 86361  99455 71479  15789 26134  68056 70928  68581 89559   40567 31461  89625 08253  80371 40914  41110 26786  44936 33529
0841: 59588 63658  25316 26129  32118 63217  16766 75962  61315 75976   59747 13364  31208 88662  89677 00146  79938 22178  30072 04877
0842: 21009 82133  73799 17434  24537 76792  14998 68624  04389 15758   85972 66283  96643 03275  87863 97417  46500 74383  65915 11808
0843: 47780 39653  28928 77629  23940 15828  82111 03620  41602 55270   30948 64677  17477 17007  14618 33406  50243 08119  21789 82237
0844: 88750 67447  84479 45920  71225 90326  65129 58995  32479 26147   41578 63058  60983 61000  43232 12382  29443 96028  54281 44932
0845: 24580 91656  62631 06955  83018 37500  13539 21169  10106 49280   35091 82382  76655 88114  24978 94643  56685 62668  01879 90743
0846: 45487 92355  59833 47570  01962 42754  15450 00656  18777 79751   35270 51270  33409 69452  87144 73618  69299 60151  96157 56168
0847: 39156 96572  20438 88879  87829 31351  52047 80715  57486 20361   88914 98414  64925 62310  43265 02367  25166 60185  12365 65162
0848: 85312 60404  33550 45961  31302 45720  65088 72962  88080 85286   13835 20314  02925 90068  25279 95892  10260 57343  45399 06261
0849: 35091 84452  54875 59557  26778 90892  94131 82353  25300 36919   50397 68845  39618 39905  62121 98503  60339 71867  08541 26292
```

```
0850: 86524 46823  66191 75054  76090 73574  59439 94144  94552 73091    28548 24376  78381 30003  74242 68571  53839 14436  62674 23665
0851: 59052 78320  75154 41315  08231 70537  24411 66689  72143 21305    70983 44086  20468 18212  42387 77989  51917 44408  92270 34327
0852: 36311 91436  36097 65144  08210 94476  99053 88365  19911 17958    33689 13700  32574 56238  71794 79467  54615 21588  47152 94213
0853: 82133 89065  55847 88129  78013 32362  69856 45497  63813 86265    99760 31447  10963 74911  84592 92660  89021 19276  00681 46851
0854: 83199 69192  35059 11639  78526 77154  33216 49585  31957 80321    31872 59864  60069 13079  48779 16565  92032 24092  89264 26873
0855: 88372 98278  71311 57164  33564 36492  93732 99593  07118 30707    50418 94631  70282 89207  74063 22732  54058 77062  44204 58058
0856: 13495 47310  80558 99650  73078 89588  32305 37683  37897 12439    00370 53146  72037 94451  18085 28381  23001 56115  04458 20980
0857: 09613 23829  55224 39319  46206 36808  79324 89155  00652 92130    76630 50090  98602 54220  31261 92049  43553 27660  86888 38230
0858: 87932 77408  90769 02911  32473 76081  04645 24475  57016 49360    91150 99600  84092 60293  24578 91388  40350 43925  07613 51358
0859: 70278 45112  59737 34031  81595 67233  55621 20400  00368 84918    64667 41395  15543 35278  32476 23754  58123 39454  99302 14731
0860: 83917 03703  92948 40845  42433 47254  09161 93386  96751 63425    28580 66615  61670 68020  71562 04211  26062 41252  32096 89689
0861: 90770 04159  15862 60625  62639 28656  22315 58957  24678 58341    30505 48903  84898 69806  04419 90094  04077 68956  84500 60710
0862: 38168 46686  18353 42359  23947 11506  07446 15474  93317 83771    21522 94100  06067 29291  93673 65976  47213 50422  68599 27821
0863: 35407 13224  58194 05525  00709 73704  80843 97953  21521 13204    12336 16266  15479 00332  73368 81236  43942 98484  75696 37609
0864: 83803 46662  82088 07412  09369 41349  77342 08794  22753 07302    34457 48844  08409 53246  23561 27755  85999 49562  38356 80664
0865: 27461 16373  59913 41726  29029 34105  22053 32226  23189 50867    05580 04847  32091 79582  73142 63543  24575 25675  40072 79771
0866: 15096 67832  48985 56727  18009 09829  17227 78227  64982 02479    90005 58620  68456 01829  96548 93479  02418 80840  56218 51786
0867: 53675 24671  06850 02188  99953 92199  42326 94968  00769 03647    99840 98495  84118 31552  67671 57800  88989 06468  91182 80309
0868: 25495 26721  17775 17634  79034 67578  27707 40157  45567 07643    00241 42638  70863 72657  56871 53379  44121 21756  14917 56245
0869: 49029 13673  35768 21624  04688 07062  44704 95073  21783 04842    08732 62103  53565 05678  61008 56021  03958 31047  29326 29906
0870: 42651 10581  31478 97204  27240 96071  07537 82579  25375 68516    71502 04397  81069 23535  47827 31008  44372 15020  02151 13555
0871: 09820 40849  12350 08826  01176 10501  12793 76255  66508 63257    88786 93399  45402 31211  12028 38157  44482 73162  86126 51441
0872: 71247 38817  31009 93184  78451 05488  03475 32728  30119 69473    61561 96793  24560 51145  81765 45937  28762 29331  09840 12907
0873: 40135 83944  60147 95999  16286 83240  14571 04557  03095 38869    39530 20445  08709 67065  76234 34128  37046 32855  90480 22655
0874: 70752 19240  30432 59561  67642 07979  29941 01352  17681 06398    00214 53837  69229 57523  18878 71125  62980 64396  23906 76009
0875: 67816 72187  69049 73612  36325 80951  13545 59004  53864 04134    03528 93173  68382 85379  56016 22315  47700 15721  82526 64326
0876: 93886 69357  80364 73440  59232 44038  87299 12222  36826 78629    71972 25652  44836 75746  93698 76762  14244 70488  53186 67626
0877: 04810 28073  10264 80278  02845 21626  96499 97614  09967 63731    55635 18597  53325 88371  92496 18299  75327 92467  81878 04924
0878: 29765 13204  03713 49933  01533 00230  32246 48499  61029 80436    16598 81002  00624 51213  49323 83917  41698 35056  80878 91096
0879: 54253 72634  08036 39682  85869 26105  40441 58911  41947 12008    34759 71260  74960 45928  21458 48315  74998 89017  73093 55792
0880: 84595 35887  65963 72669  99967 00329  37564 24962  27167 94715    53957 72218  21329 57800  47519 84865  51492 20491  39462 20828
0881: 85348 37211  07488 45291  90011 02937  84137 70882  14504 72344    19718 29779  75865 71324  64287 17300  51846 87814  01754 47993
0882: 61803 13777  06348 70114  64498 17188  32990 47393  76838 48171    83572 31741  07309 23353  38183 11101  64558 53364  56135 05607
0883: 79387 35394  29792 43721  80465 50678  89028 57423  64589 39529    06589 46551  72433 32447  15771 03905  51124 11962  07455 24671
0884: 05427 92523  11336 19144  24509 00890  96151 45122  48402 15391    61763 57322  83838 02188  84959 64801  41208 38300  46509 50942
0885: 71569 09564  72455 18357  93111 69703  47381 38640  75716 96615    60970 03525  78590 72434  52654 63065  71910 48439  13025 17035
0886: 49748 70560  06144 00741  47683 50699  97539 23412  06772 64049    24276 76976  06344 13849  11418 35681  31632 61047  94676 99811
0887: 24546 96785  39896 34394  79769 89262  41055 27489  74860 96506    63537 41379  45172 63559  83862 95471  68171 99545  48750 55552
0888: 77153 00095  62551 86372  95145 45182  86552 74764  35305 60624    79084 11663  54466 99455  93072 87617  24458 83167  27513 97981
0889: 57718 46218  23599 29791  76568 75663  20685 17676  53123 11958    02780 20158  63741 41111  12361 25937  89760 37856  34671 65315
0890: 87976 22853  59264 61932  07221 86207  16863 29450  38683 34881    31525 71210  83847 93145  60758 21468  05709 15802  58710 52504
0891: 94256 95196  83502 31648  95724 73939  36683 56639  53098 83983    63962 41440  95796 08645  74273 57127  46060 45897  05017 34749
0892: 68995 61120  54257 02590  98187 16681  00348 08544  09765 54086    42786 37663  99648 19373  50779 68647  80893 18257  89005 27172
0893: 68661 05270  28499 08051  81992 37957  96025 53053  77039 72562    02180 39117  44346 05747  96024 10022  62768 22812  99790 02004
0894: 89032 62885  45358 34895  01763 05431  90277 71593  76841 84519    67542 36769  25577 32125  25292 10052  60806 07022  39937 08806
0895: 11539 68698  63918 46787  30000 52247  50810 34479  61507 30208    53140 78906  05594 81817  75801 40182  93465 85371  75234 68939
0896: 09471 86418  38066 38273  28951 45008  34964 14360  87878 50979    89699 81492  90653 70213  77895 66068  45886 35044  15024 66613
0897: 40724 87961  99352 16172  81375 82293  33209 96291  10844 89015    24855 74881  49263 05366  52053 49172  66845 61476  73428 86504
0898: 95042 79844  87941 50819  36626 97856  44905 31116  90953 22158    35831 67852  09576 74289  49779 13309  06559 87947  03472 11694
0899: 87058 76087  92106 14079  91686 24899  51882 97340  22158 11147    93699 85670  07746 22000  22404 40196  85253 62293  33296 53100
```

```
0900: 10134 87426   24527 55644   66934 35656   88164 71073   88237 40697     17958 27070   00928 44972   48971 74919   20676 50296   41955 86648
0901: 10662 30763   31312 50503   84756 77914   72790 75536   10203 21772     89095 04036   79479 68252   79067 12714   44046 55586   32280 61195
0902: 09900 99314   27846 10296   73349 84867   50558 37850   14261 98138     44415 79205   54476 08114   13949 39363   25878 12843   28713 72828
0903: 80099 73761   52561 13524   99234 20118   79810 12739   39753 01773     46482 77069   95304 22909   31989 32811   94008 03252   66506 82956
0904: 92734 53958   50551 79386   19867 25963   31968 18991   47105 40653     99680 70840   63498 45703   20647 76199   49760 39091   45627 72920
0905: 81561 19271   09717 93525   48088 34670   86073 45954   36869 61544     67001 38366   15901 09555   80622 87339   48362 84722   00514 32572
0906: 61365 78315   55332 09618   55582 37201   41834 93364   32158 62476     83119 82588   14733 77183   78674 49873   95043 51550   38210 43970
0907: 94216 20515   51675 73020   51232 27191   40123 18310   79759 40064     36881 09628   35386 99115   69912 97867   56307 44383   14824 10419
0908: 63470 70467   34750 11017   51113 99818   21871 95722   00765 73133     95758 20902   24181 32678   02215 29422   70404 78858   52853 15192
0909: 51899 94102   83743 46996   63925 37760   67422 26820   30028 40009     25647 93466   98250 72268   74294 26580   44202 15459   83655 23923
0910: 20339 24622   34536 71111   23599 19973   59054 44249   32308 13215     12103 50071   28672 47424   08965 95606   38589 15670   69856 32096
0911: 25816 46630   59836 12636   56008 99610   67921 91196   83406 39901     36395 88485   27741 84874   19528 60439   82343 70758   24520 91235
0912: 56902 49978   41340 01339   87710 84864   61477 91995   13431 13428     86118 26801   12254 30254   66646 18223   54874 54821   39258 18772
0913: 36159 42023   82370 05042   53131 83769   90185 74169   63029 18382     39616 55569   06952 83216   58015 22883   31840 72105   90998 64080
0914: 67999 40398   21192 46734   26667 48915   74960 12672   86584 49171     18486 25974   81251 17176   84510 27639   03263 79451   89617 10519
0915: 13818 20668   79128 30606   67556 97843   23812 33557   28147 72234     73726 74223   79866 99263   39412 20139   18124 33617   70037 83217
0916: 44962 83528   03007 20217   09081 96316   76399 94588   75096 41387     43333 37772   29323 94412   60263 47048   63486 93257   26063 08246
0917: 24161 54354   95611 88812   63991 10305   09617 21430   53465 19243     79839 97969   34518 45045   92157 75664   48242 85922   11610 41773
0918: 10201 81846   55640 66985   66840 67051   77723 72490   42952 95940     98856 81314   62016 88222   97453 91328   99784 72314   60951 27565
0919: 42910 42156   54904 53663   72835 63372   36754 01874   97647 71093     32143 00349   64571 33459   66035 69837   46343 16460   56317 73389
0920: 47574 21925   23204 24717   91475 66078   45776 65693   98132 63746     66049 51982   90689 69193   55481 67954   99212 66285   95626 67524
0921: 27817 33903   18072 72823   93382 82242   81715 03831   90956 08828     75378 88215   02611 82164   28972 48518   38267 56677   26150 65069
0922: 63758 50608   20162 24156   43824 05809   49632 86363   66387 07008     53766 11555   76879 24585   83689 13485   08899 95348   65204 20586
0923: 88603 75699   99812 98988   34312 82008   73212 76232   46842 29118     64416 74949   59199 89308   62784 49153   28023 85214   03918 82265
0924: 31166 32447   76552 96447   89123 74844   41127 06372   35164 34328     68298 06229   79610 59086   45151 43965   83942 64769   74618 10854
0925: 23496 14770   07815 26778   69393 78052   23109 90608   77779 71520     84189 82433   51551 34291   23168 28837   73820 36551   18784 48107
0926: 68334 59566   33913 96759   56613 46507   28137 00241   11645 02256     01100 84943   58897 51147   54585 68789   48548 11241   60688 93830
0927: 95542 42710   59251 44141   67211 86523   51462 62189   83675 46419     08842 19284   43405 76088   65821 72153   43216 38344   81172 74878
0928: 22430 59597   85631 87021   31849 03866   61108 42571   29884 74190     57905 25084   51550 38605   50456 04775   82244 84480   63752 23792
0929: 26731 61739   89327 94905   01426 41484   05457 51650   58792 77179     92798 30752   41350 18879   58898 04027   47618 45603   98396 03151
0930: 62698 74899   65351 31215   92989 86799   98378 79386   57102 51566     33123 84938   16246 31064   08654 03112   21731 44797   01203 38246
0931: 07553 58566   36275 56256   64984 41384   88288 51926   15368 34513     27347 69609   63648 34172   71159 50640   27048 20335   58062 88942
0932: 22711 83871   57192 01559   40359 24007   83874 26429   82878 59402     86829 82283   13546 36969   80574 00098   11626 23308   35799 52841
0933: 53294 82543   39369 41886   20695 99923   38867 47229   12727 07553     27219 96976   72617 00051   89908 88237   96010 23028   58250 86040
0934: 30587 49225   19218 40234   81727 66078   57600 49551   73769 36632     19074 55366   57134 62213   71824 29293   47022 84733   20149 13629
0935: 08108 88677   80768 82820   96437 36546   05097 49505   07062 16891     95807 83225   05373 61163   37600 74958   54427 04346   56652 81585
0936: 11702 96339   35366 58061   09774 54115   81227 04903   85942 23071     87879 02882   61576 55658   89915 68599   23073 59456   42240 62284
0937: 89939 01098   27192 15338   06057 50460   57996 48728   24668 65523     35677 50787   05415 30701   90947 61295   28789 45989   83765 29712
0938: 00044 50225   28643 76452   32161 17849   40728 81062   28402 21112     13311 72482   74587 18594   15778 46420   62334 41791   57137 48488
0939: 58872 06742   86422 40399   78422 15710   51743 45969   88393 15023     55474 93768   15446 04952   59826 22397   83291 43662   10869 49387
0940: 12147 83854   94910 38301   58656 20548   30809 13694   78511 75237     34927 13574   35336 36678   13022 87045   82740 54860   70707 87290
0941: 47188 46904   67494 81329   88553 10466   56451 18882   65240 80159     57128 04036   76163 34070   53630 80932   17244 30043   37863 99985
0942: 62125 42143   83783 95921   97674 12117   84217 06587   83462 50151     39671 65110   97278 34729   81677 97447   61490 72436   71762 78705
0943: 53527 85447   41430 90255   27090 59061   83824 44269   24567 07168     41368 37159   24054 08525   88061 42353   23903 54566   56925 54364
0944: 35626 94900   96832 21007   07763 24612   30912 88014   75496 57799     93287 03428   09365 66015   81652 82590   74464 18393   58060 11466
0945: 85996 16446   23127 82349   00850 35721   37217 87553   97968 19438     47120 52070   44540 91596   21295 02019   50802 92052   94899 71995
0946: 93243 51717   06154 95992   99509 49890   49975 50737   05619 43604     69353 22520   86089 78308   92964 29993   61205 10415   09907 13621
0947: 02199 45122   07567 05446   01050 36607   63097 41725   87251 44247     01709 31466   53943 40379   28932 43640   21279 99232   19462 88266
0948: 94976 76378   08983 80065   50556 74113   87460 84197   52345 05450     29311 45522   84206 23786   72889 10560   93200 82745   88577 36634
0949: 05233 45009   91077 95485   96153 37215   37300 90098   03334 37016     12930 25864   73384 23481   81174 28770   97432 87546   98521 05007
```

```
0950: 81862 99996   26388 15370   47931 59454   84748 57706   69677 96452     52866 54075   27909 56322   13047 71292   09537 37778   88904 85379
0951: 53886 24034   51889 40432   04515 15309   82055 26017   93973 15525     76522 30254   75664 83674   15085 08409   16059 19571   72236 49475
0952: 68295 53800   77872 72259   38942 74331   58734 63876   55821 23907     65207 14598   66757 70110   73470 09277   37765 53521   90303 18270
0953: 44767 53873   35385 37930   98228 31110   01709 89128   97017 76475     69158 93187   23999 54679   29476 56071   22813 33490   70102 04537
0954: 58553 26081   33624 19373   81328 52231   68744 50214   34597 56869     85570 94090   42767 29304   06475 52293   84497 82926   24677 82487
0955: 89154 99714   29442 70595   20246 59308   99011 00957   77164 17807     65746 92643   34210 16306   41844 19353   64164 61380   89643 62610
0956: 93826 92660   22861 84631   16053 23731   42096 46075   27110 10130     73868 33134   59793 60484   74000 51056   35811 38501   56659 83775
0957: 85554 83109   45271 63409   76831 59039   02153 62935   98682 17011     43625 43120   25653 19582   74877 09974   48113 86101   41954 12693
0958: 74201 13845   73392 87222   24944 82323   20426 69399   36041 88168     82481 25444   71208 90064   78114 86218   05582 60468   25267 66607
0959: 91544 36390   74601 98055   84119 04066   46520 62777   56838 28343     90070 76067   45478 60331   44261 25528   91157 43272   75550 42554
0960: 90643 61717   60410 83547   51513 20868   26619 81106   10415 24413     58655 67617   37902 98824   66669 19839   09396 67325   50418 56950
0961: 34540 79459   36105 41798   90081 87727   16070 33587   31690 40654     92357 55454   70382 72993   20892 54277   70940 05277   36804 58176
0962: 48975 45302   82940 78179   70553 25100   61984 68525   23814 08717     15916 73297   49976 73049   62146 34790   92565 44082   01729 20794
0963: 11634 10814   03905 54865   37368 95205   42177 89533   01179 50076     51265 60866   22939 60253   92416 77307   00795 76442   91339 12856
0964: 35408 09088   23723 85244   09450 73190   96588 96339   08207 85339     34546 06565   87055 12399   60697 37862   79420 81136   06916 68488
0965: 18413 63917   81193 49203   63040 52549   43189 49889   88176 70119     62411 22780   09499 69834   60641 88011   77091 27817   36181 78727
0966: 49360 72898   68173 41900   52965 93055   88679 66245   31408 05391     30441 62392   12164 04514   83992 50021   63405 91633   80366 26882
0967: 14138 43276   13112 85185   28384 70760   14311 67614   92549 94613     04305 25266   55342 61223   67735 92264   13703 45375   60515 66528
0968: 09745 72432   74713 61557   29521 92203   84750 90019   02847 19009     98627 15374   79897 65117   33013 28021   69847 31223   06842 52560
0969: 86052 34311   83490 88957   46556 73253   76774 05367   58742 29438     47418 26399   56869 87726   69928 62160   86539 66306   55016 51897
0970: 79172 99076   63215 91185   71855 33428   48264 85838   38209 55067     43688 32785   88120 95340   99164 91119   95528 32795   72947 02299
0971: 17763 67488   57521 14366   10487 11667   94119 75835   94050 08146     10423 27274   36897 14872   04062 54868   64269 77694   36079 29441
0972: 12376 87704   90549 59995   80605 76888   83666 18918   23557 67611     60588 89820   72581 17410   08448 98145   13958 28997   10453 95543
0973: 70602 28522   47602 70586   64597 68775   07437 71904   11851 67726     96561 04067   05214 91574   18827 29698   88744 68068   99765 82236
0974: 33803 62934   85768 83680   21863 40215   19703 52892   72201 54245     28239 92303   16773 72892   84851 10550   01563 23619   56962 37230
0975: 54685 15577   82894 35968   15230 89515   33911 91380   81906 52940     87756 33269   63164 48306   99428 49720   71367 73472   64816 40172
0976: 11892 07277   20584 34094   53547 75780   27399 79481   08708 26154     67913 31972   95410 05540   33795 38734   07095 87223   88192 89441
0977: 38592 70087   79193 91102   70241 50323   89237 29698   12829 38988     50327 96925   95352 03357   48188 61980   28699 48613   72091 28185
0978: 52171 61418   91140 24859   81941 60314   57685 42014   58580 31884     63893 68018   59588 85246   58951 80314   99625 11306   81369 95198
0979: 78187 06299   12887 56994   78775 79146   51062 51459   75657 43658     97314 02119   54341 31837   90634 19787   83456 98929   62536 84727
0980: 74384 12210   58978 26923   32156 31982   91554 60572   57888 83646     50447 55326   55718 64312   00194 82652   84684 75359   95605 62170
0981: 73441 03943   28681 45874   32379 92060   57234 60190   14231 87067     37356 12636   80915 47149   13442 04341   72467 07334   86088 16989
0982: 97704 47643   91538 31146   78441 25562   95357 53721   92636 42576     24186 51573   83860 73612   76824 25947   32155 12446   81640 24853
0983: 09439 06909   23728 32992   84045 98126   96427 94637   22022 41812     93321 88032   79211 64182   42984 81739   31173 06431   08288 64722
0984: 29091 60768   24791 94164   45518 70815   55317 01545   67247 15539     86057 61494   99326 36517   25927 98583   81844 99194   20082 08048
0985: 11269 39564   17956 36299   97374 69269   49571 67410   67987 68137     23088 28244   81861 46688   41743 13935   48582 47569   68704 35215
0986: 02532 14101   71425 22027   11312 67137   62968 50866   30602 81830     75842 01519   02723 99139   37580 35156   72984 17000   91170 40454
0987: 32999 66463   81987 05798   80674 58999   42363 36152   61874 09488     08728 99737   02512 36309   08942 54119   39446 09608   47273 15168
0988: 33457 80096   66241 94720   90731 13533   75730 39743   33140 33067     72983 09111   62832 78310   89881 43433   58907 41132   71954 45636
0989: 63099 47596   73455 54422   77242 95308   85691 02838   62279 31793     05059 78097   14848 63136   42406 65925   40717 38143   61339 29872
0990: 80630 00775   57760 20272   82523 86841   84567 41276   28221 72640     23708 77184   39633 86274   78575 18519   30525 51437   08909 30449
0991: 35706 58506   67897 25247   69513 71065   20404 53849   65851 72096     67850 71516   22991 60793   52478 61443   22128 81824   80129 49416
0992: 92405 52005   85888 76627   15226 94303   57839 78522   32830 79276     02257 79931   62872 17926   22742 05754   73433 23534   28037 94378
0993: 74941 06711   23043 19529   68276 27902   35833 87646   63452 39633     41285 74770   63143 39466   10576 89589   74716 03231   51771 22080
0994: 30413 37791   03014 08810   90610 24220   33485 97808   20534 43016     75191 57869   94228 47618   10534 51686   63972 63225   03520 37826
0995: 97902 79151   74697 93198   68644 27086   38001 03538   45686 48266     08331 47937   52189 28703   32082 94359   47659 61139   14850 87827
0996: 49994 87683   53817 20780   37959 54932   45829 26004   78857 00706     09393 48723   18705 45218   78049 92411   96658 81022   60112 32649
0997: 38799 99513   88326 16304   79720 08458   86617 50630   66518 93025     59965 10848   13207 71803   71725 60413   65019 05028   22487 69267
0998: 94146 03695   32512 20424   10026 93714   07120 71185   01868 72418     23546 82385   08027 34220   33687 26923   34733 93482   03703 50035
0999: 71167 38923   06345 52994   51588 86372   34490 06723   17532 25134     43177 54975   76126 42069   67625 41196   50620 79161   09123 09566
```

```
1000: 17813 93757   99859 17518   63312 72051   16505 50113   01832 34359      72805 88409   37325 78314   72003 35693   10675 95379   36756 50492
1001: 84936 28704   93620 53158   43885 74724   67368 00999   77819 54632      25098 75982   88936 20271   16081 70837   14678 37916   34403 23415
1002: 22515 49956   70160 58481   19932 37929   98510 62836   78462 69396      61077 48309   49539 58320   44867 04404   78380 82790   05291 51492
1003: 79919 21408   24865 41709   08763 32244   70914 03701   31839 56010      45779 58875   05101 25645   11251 93763   19616 97436   55053 35212
1004: 24710 02845   67569 10329   01413 29407   36529 67679   60267 11016      20248 61376   70940 19565   75380 95525   10484 31937   31827 46336
1005: 26372 68460   95110 66129   42996 86975   73289 39193   92829 50442      64071 34371   24026 54686   43098 28283   38157 79604   96134 45012
1006: 54939 47013   12423 51110   41299 99217   63139 68545   58984 76943      03793 66853   67327 33618   83525 25864   06197 77925   57141 92808
1007: 15755 07684   55625 41697   30306 27087   70199 39676   35771 55696      20324 22790   60134 57371   53110 57248   89006 69286   89137 00228
1008: 75421 85841   28965 77538   54109 15382   76691 54413   34327 37385      76337 20157   53231 02161   74964 87067   11054 59021   74086 82184
1009: 49472 61292   16656 50988   29921 33918   62138 43450   96144 65622      91216 40399   00730 28369   05895 01681   58911 43196   92102 61586
1010: 70334 83683   15054 53924   10777 15261   57250 85651   31061 12414      07870 75445   93203 41662   28992 77291   03516 49375   57117 56143
1011: 84647 52968   32631 48605   52731 33294   66802 86846   30355 48495      02374 79759   04433 71112   49990 14561   35544 03519   62275 84555
1012: 99269 33532   21609 68164   27276 57854   33841 55530   32133 05613      94724 22956   33968 13014   19297 30959   55018 66229   87411 53981
1013: 74458 43373   69260 09223   88019 60957   11055 16826   58837 02480      66268 89016   19252 66456   26928 73907   17703 58609   74213 62218
1014: 02892 47072   82549 02624   29623 10105   94110 58577   42822 09389      50460 65082   75855 86372   49004 05067   71009 53655   84323 54851
1015: 18470 84344   39978 10583   89574 96072   44028 25565   72658 52427      10223 49178   99626 35508   15444 77707   73432 10940   26883 03419
1016: 53473 43739   75884 40681   99282 73398   26954 64593   13028 52287      87641 54256   16503 68265   31150 18417   82421 91559   06758 00212
1017: 37687 94956   09004 54615   14103 23297   75579 06916   33577 54479      82619 17813   48783 31146   09374 13687   17310 75091   72867 83283
1018: 83430 00753   20327 05982   28383 56823   73716 46641   56130 69286      03272 90355   93708 59966   21725 51662   50648 17840   88020 98624
1019: 65251 25762   84134 12358   26787 97503   64076 96502   34693 49984      67783 82041   86692 01336   46363 21546   27252 47412   48663 26357
1020: 47391 22501   61411 47180   96517 37524   15996 13202   39667 80242      42740 47956   97149 16567   60945 87290   36808 17471   71300 94181
1021: 03234 72803   69748 56511   72911 59392   85071 23271   52376 01247      04037 10047   87999 49523   57425 67492   73218 73999   15172 93449
1022: 27135 70978   76706 55113   72052 82085   48133 91901   94344 17038      32593 28769   11047 89561   21223 75955   08772 33903   99964 68652
1023: 55911 13902   11264 09455   71101 68618   89547 74813   03621 57786      03764 74459   14707 26714   58258 28997   84218 21704   08273 84671
1024: 09527 92097   37697 35307   34839 68693   51425 42373   48873 09010      30312 02269   08924 33233   30265 48387   00182 25557   26297 36365
1025: 18544 64936   78057 90229   78192 48437   46052 87650   42401 72651      74388 21041   44583 16205   97042 27188   69502 59454   14968 17428
1026: 15816 53288   45506 11575   65448 57590   54151 57902   67657 06704      69200 19990   30809 94477   78479 22164   02761 27526   51601 68787
1027: 43280 02620   83946 02213   46342 35271   62305 33942   69197 59649      54318 43566   47190 94954   27990 84813   59538 46540   70063 23912
1028: 07919 97095   61127 99148   58103 45447   29441 89764   01276 71023      66140 74200   19684 28481   34440 14926   20668 80950   57726 79154
1029: 45592 80993   79420 84250   12899 90982   49702 94668   46994 66862      99210 48233   82514 28927   33752 33726   68760 83608   57136 28316
1030: 04963 10518   76148 83634   72566 49557   43736 14214   09661 04676      87298 20057   32121 64215   83791 78013   66546 37588   97542 76210
1031: 72328 93227   64859 66188   67181 97227   36632 90538   12600 52389      25177 83524   92948 80492   79941 59696   67563 28164   43297 50792
1032: 55718 89752   30562 16718   09323 76108   50301 25797   98963 37401      57511 43215   34037 98320   66779 78547   98744 61678   22632 43076
1033: 67003 32797   32555 93461   93062 17690   63852 50553   60992 51639      42705 63471   24852 51776   25726 89655   25579 21278   92462 61258
1034: 42086 16581   41604 27354   67038 10405   58634 40659   97069 31223      33303 06104   24204 52410   23532 14803   32583 63958   28732 04499
1035: 77611 91396   59277 50945   41003 17394   16669 33752   11831 08206      50833 85942   34761 79893   18706 33629   07441 93427   18637 59053
1036: 64093 57248   31473 33095   12810 55792   72345 93853   53247 76693      66381 00642   13193 59372   24201 72033   85882 36309   55435 59445
1037: 25832 55501   42324 71659   16875 84526   59872 58220   91594 14667      78548 63201   89372 52681   61673 45773   20595 16526   91372 53120
1038: 57073 59353   42177 50570   46495 33308   15499 96497   38755 21205      37349 28074   28013 68084   03261 46602   25785 62971   56179 66154
1039: 97297 12015   42012 16866   23810 97995   83010 07167   15924 91397      32411 19271   50878 38753   76543 36863   52174 29425   49917 14965
1040: 56890 31965   82094 42634   55238 00959   39888 28598   82640 05945      31347 59543   45937 47862   14284 38120   49459 24752   50781 05420
1041: 39886 11459   69753 16260   92959 22293   85263 10580   33471 30394      75910 11046   29547 26929   40374 73486   81547 69393   83073 05018
1042: 15360 77972   49512 91020   61154 31648   96910 15288   29539 63763      86482 55467   36053 04876   83959 06071   98742 99786   46225 05797
1043: 41500 02755   75878 36704   62415 22087   99442 98684   62828 04823      15637 27542   10573 32812   57167 62766   36624 51236   50866 06021
1044: 83133 55404   12202 59405   21416 96684   53212 08914   82680 76064      88135 27108   70052 61816   63347 66539   33999 37041   10274 02219
1045: 44439 55029   56962 33412   44393 81963   02401 52677   57073 37371      26097 67600   18961 43087   19174 80501   25957 22409   43899 77612
1046: 41647 46764   38473 69120   18889 18075   74752 69217   72010 99989      77086 38117   73204 78287   13832 28487   92253 73852   71349 85903
1047: 07750 26957   48470 26321   20203 08909   28852 38987   43091 65417      68911 65265   60070 07404   57872 66624   17400 96451   77138 17499
1048: 43823 26990   43481 50041   61455 60317   26159 75024   27254 77564      68939 29589   06928 89085   67754 17898   70574 93914   26959 87507
1049: 18378 92508   00776 30444   72180 56515   46992 90454   57284 83597      39560 59067   20323 59377   35925 62357   43478 01886   30088 39565
```

```
1050: 12264 13386   77605 34630   33672 81840   92606 37839   96217 49079      24188 47746   46487 19482   14146 44715   17177 56509   33926 30794
1051: 68541 21771   95380 80601   70224 45131   82978 71458   48358 46692      52810 94578   70901 38030   05718 64245   12466 28915   60319 61711
1052: 19557 82455   05388 33944   39922 36929   45402 11540   45182 33787      65844 07549   22225 73684   45779 59857   11530 75113   21350 88957
1053: 92478 79578   22388 51613   30168 90453   52691 83119   78500 08610      30686 33728   22495 11451   48757 10974   02965 16099   27484 30778
1054: 24664 83571   43261 04318   44760 65409   22305 23452   16040 14191      22050 59803   64035 54112   15065 46461   44070 11393   68269 47546
1055: 58219 77164   76395 85060   66695 09662   11245 08855   05507 93093      74154 04739   29150 18546   40711 22013   32132 22093   01382 91626
1056: 60630 45301   81497 94631   46470 05763   55295 75324   26636 63354      51053 46560   33176 24947   04759 54290   23247 01724   52104 53254
1057: 50547 97863   10199 01972   19743 98728   60828 80445   13177 82155      38936 09942   79396 43463   69170 42524   90556 06888   88886 65657
1058: 45522 79547   93573 94473   17163 24674   41379 70026   70072 77570      95262 92679   00380 75235   75430 06750   57404 40807   57211 83550
1059: 98516 27549   81460 44112   65402 03538   83686 51801   22344 10886      68415 10138   66813 85624   62278 29921   59755 08121   92290 62735
1060: 20850 48812   54542 92027   65698 09646   31451 28664   94502 46260      13760 60452   14140 81751   36267 56443   38697 81144   56091 26583
1061: 89611 89487   24309 66426   21317 41287   45684 25931   29914 23967      40212 89089   47888 65976   28100 68870   00687 70882   68888 29621
1062: 52254 85462   18322 16409   52413 42026   47575 76223   18438 69373      43847 79139   98246 01205   24932 24147   81975 25814   50027 55582
1063: 00574 66737   37087 59725   03803 70243   36904 79015   98090 06358      38476 22426   84224 50413   33888 77921   95123 65550   89755 60630
1064: 41470 98993   69730 16090   86482 65630   39683 93724   73714 57692      07921 74599   02934 40814   09814 19239   23242 41770   69737 27939
1065: 10545 42557   52214 21715   39386 09878   01869 69685   01205 10077      75204 24145   17817 04679   29606 20098   69157 24950   94328 01634
1066: 05676 97401   29833 20651   17680 55199   70720 33091   56630 24558      46379 49499   49535 32980   18497 87958   67951 73407   95591 51051
1067: 61916 09149   87640 49894   20021 05648   01401 41159   03605 84673      73273 61934   67903 61289   89865 76065   42570 56059   82541 38399
1068: 45385 56243   85904 77392   48511 14869   98338 56154   54619 10141      95573 24146   00271 19474   46901 98842   17891 17102   48949 21689
1069: 50462 60876   94141 45142   41089 33268   43915 77724   70073 64279      98616 67928   76453 28596   36572 92415   73887 56141   86754 66904
1070: 78488 86358   61382 73926   79340 24079   69634 92955   07461 62190      39806 02457   22534 68701   43563 42458   86270 24106   52957 43417
1071: 86894 43626   10126 84574   12756 25178   48926 26521   90189 05886      88341 03832   25856 39197   49498 96943   63679 10746   88386 12952
1072: 79008 66305   37883 81837   44609 83171   64425 46476   37674 42552      32526 85355   99543 13934   81917 02897   28276 81616   72900 08257
1073: 60195 46152   37648 72333   64829 07638   02090 65882   68097 38879      44098 47176   07558 96283   90101 00397   62669 42240   92150 37398
1074: 40960 67905   85049 95961   14146 46758   93691 70600   04575 86921      66894 96366   27111 07333   76570 31312   70293 45928   20156 62050
1075: 43567 14330   70769 17340   60875 42001   05940 90731   96193 53991      22868 18955   50375 24041   94693 24825   86074 06756   08568 11694
1076: 27325 19118   59589 67697   64761 79521   54634 19060   30131 88159      48030 79084   55912 37503   36865 20400   42183 35368   03547 81168
1077: 72645 03482   87163 31930   75401 00344   17067 66577   10162 75093      01668 02474   16845 67104   58650 31213   80010 99263   13281 86822
1078: 82501 00886   71884 76262   92607 12265   25845 47396   07272 31831      53016 05141   38639 57787   54264 41637   49850 26201   09329 47435
1079: 36946 08826   90935 59985   01991 23214   16691 45198   22001 20693      28923 13309   67310 23592   78146 63487   18137 82614   53419 33135
1080: 16671 92470   67269 53356   95155 34512   30114 86080   88977 05492      19506 83192   03662 11874   61834 66358   06377 54140   86317 73934
1081: 97770 01570   34086 06412   04093 18428   65910 48356   39666 33672      12911 88762   43613 82429   84618 25014   27267 61323   79751 47019
1082: 87140 21893   98499 11700   95264 17136   36292 89813   04543 74689      91889 03134   27851 46841   37487 80993   22963 07149   48175 78033
1083: 40241 92527   02125 74732   15373 61964   90146 26600   88631 02256      87281 95279   95509 56680   21663 96403   10422 31245   16696 11081
1084: 81890 11950   81728 80506   27069 75929   16758 04411   42822 40762      37146 23114   03467 66755   30419 17412   54569 04219   87300 37527
1085: 81862 68342   94188 82505   84980 70284   03316 12742   91210 96438      07018 60217   41668 42749   22479 79288   98525 56716   72834 47855
1086: 33655 03301   13296 80092   45257 90741   36960 98741   15281 17616      08121 51299   18386 85002   36096 33807   29790 89210   37480 32316
1087: 09736 41655   38407 02043   51600 38246   51807 76013   32891 94155      55724 35019   04542 83760   00532 15864   99888 56383   24387 92064
1088: 63957 19367   28803 20847   80007 10261   94926 78297   44661 36843      34653 46813   20764 51624   06700 91651   60778 99710   50852 57428
1089: 90660 84203   39607 28855   22429 57392   20970 94062   24896 07637      19974 01518   47928 91345   82644 85099   90697 55339   18031 67811
1090: 32355 30412   59856 08815   74967 23666   34338 77131   50310 14490      98345 16364   35075 94451   23212 95493   02500 72844   57831 83408
1091: 21480 21468   85165 48030   93187 29273   57429 93291   49830 92215      53208 82439   65627 59068   27640 84986   51560 74548   56711 17093
1092: 48070 59600   38594 34441   21650 88514   39604 70607   31550 16386      37720 07436   35367 47957   70117 62654   43501 59581   88292 81844
1093: 49539 35245   38202 52293   73421 04781   11164 97917   78660 40954      92205 69834   92451 70431   24663 48314   09500 92159   94817 09521
1094: 39858 98511   72555 00420   40970 22442   40422 67169   20190 93841      02381 92823   89275 54747   76034 29391   44537 01884   80330 62105
1095: 44938 48469   99619 68433   71067 33019   39847 56891   79457 04710      66280 95429   25965 77022   41599 92497   21666 16158   02220 73364
1096: 59180 23619   68405 14717   73726 34326   34376 13646   34326 83611      78486 43921   12657 70508   93654 90950   70358 08943   06605 74059
1097: 42141 91693   42928 88526   46671 95653   25311 44456   42669 88636      47238 78676   59003 24149   06078 05412   34946 84973   22437 02591
1098: 16727 20324   78774 24968   64171 13991   46080 86329   16024 69754      42147 39530   50761 16094   77594 15434   75910 25483   19014 35870
1099: 65265 21525   01045 09532   43302 98360   95775 22224   51982 90125      29581 58305   47170 35026   70145 77776   78188 01207   24424 52752
```

```
1100:  88355 82778  55039 52155  70281 09283  11261 93641  09257 05435    42940 93934  95607 95239  71255 18339  05297 66765  28847 31436
1101:  71996 17676  83472 28013  94123 62965  29653 18315  66979 37867    76705 67193  83566 79136  75356 39524  55344 91129  83311 13237
1102:  34530 67339  15549 29309  72775 23143  37066 63060  60582 99206    34426 84743  08364 77083  81449 68627  01502 15120  41412 18151
1103:  39605 83331  81301 68813  96908 66835  77540 02793  22728 08053    59540 95535  96034 02876  89287 68082  50489 30592  75233 05096
1104:  30844 08456  44953 65142  55421 57882  65124 94138  17503 58003    75123 99801  61105 59256  59721 20207  28973 92772  54973 29220
1105:  73639 44510  90669 78821  66958 86148  26763 80950  33180 43822    84343 24226  37620 14109  25103 69467  20027 91727  63430 21672
1106:  36922 21038  35594 98656  76804 21896  00352 01437  87666 24627    08495 15301  54995 85866  79645 35144  60609 63607  40506 78424
1107:  10073 96196  37121 23640  00979 50781  53871 24887  41769 93318    09426 90507  76101 29684  10532 27240  89721 05406  53657 34705
1108:  47376 46484  98655 40406  85274 74341  38541 79955  43564 92713    15307 69018  16445 76965  93805 39483  64477 42821  26922 92755
1109:  70416 71152  32826 20437  27081 02474  77184 05979  35552 94238    18235 22503  96255 08520  98194 52670  78705 12809  50151 87897
1110:  16150 81915  81622 22225  96930 60324  31580 81684  65478 92761    13864 42775  60090 07166  94902 29308  69470 76276  69468 43179
1111:  14373 27667  76195 05594  60586 41163  32649 63624  97987 56493    77354 94132  15717 47212  84250 33986  66576 69860  24561 07113
1112:  98821 21971  90630 23531  36056 25402  36919 17218  99994 81699    92143 47718  46013 14430  62993 10558  45936 63139  89639 19502
1113:  67683 01059  87759 34565  38396 52106  19702 23568  30931 74794    70577 77781  52314 88335  11320 28141  01528 60984  17952 98443
1114:  70874 95468  06531 89157  64421 58173  80866 73416  82198 33130    75127 14677  05344 41579  81704 64888  46426 29368  63141 18175
1115:  34536 17681  89587 57731  06727 60985  90106 29277  74357 74878    33241 29604  28099 26350  66163 63494  50828 47106  74891 63423
1116:  92080 71293  12570 25758  11625 17077  27790 87392  08211 22009    39485 41904  36912 15965  73458 88566  53662 17826  21326 63788
1117:  04312 10095  89315 50066  36896 34809  88536 30250  28162 30734    98394 63633  06439 35315  25703 90903  55984 48085  22685 20260
1118:  63725 30912  99001 92807  69216 92568  98811 86124  90663 86148    95393 37603  55575 45655  33856 43578  99047 58557  58031 06731
1119:  74605 82326  00534 97029  67576 58596  52893 16597  38028 38365    00832 53268  75313 01352  98415 70062  26080 46342  76686 27666
1120:  58330 40314  86534 45146  37295 66855  52701 90329  01367 11033    80135 12436  75572 87152  23448 53956  17123 48119  09640 99418
1121:  72537 87304  51461 62462  75923 50367  22575 85110  32185 72082    91533 85474  16764 51916  88545 18734  26949 19971  93690 46829
1122:  46696 63785  06891 91197  44134 57147  41329 40701  83162 81708    13085 09652  20174 74763  78174 64702  24585 35810  46228 22394
1123:  88519 42389  86323 61641  84982 99025  96547 91728  52019 87976    29230 24026  53009 63536  35707 57116  78555 41442  75245 52906
1124:  62737 63015  37841 99010  70161 00047  69650 55250  64493 67366    08402 40370  23131 03928  61020 47035  81576 69542  24619 17268
1125:  74466 17204  80613 63618  95583 09866  51107 42992  02060 82053    07512 62295  03651 45729  72409 61343  57626 13169  36292 27247
1126:  37948 43182  49491 72209  59599 76526  36557 34700  21822 10644    15265 92146  57396 61866  68282 52015  58775 42882  96662 05697
1127:  35238 15450  70729 69870  37278 82845  16943 05925  82991 99924    11978 11680  95321 37774  70231 99262  35873 48711  31689 57014
1128:  60009 04997  27396 67831  89060 04978  34569 95734  47261 79037    69096 78873  29236 26041  81407 35953  60519 67643  02496 64641
1129:  33302 89301  51121 34823  73365 65504  91994 24543  29008 19945    73913 41463  72051 78599  56656 88011  66189 90610  07237 90463
1130:  82148 12943  89071 06597  37317 24713  87011 69225  94703 74828    22071 02344  13714 84013  46499 19975  37435 66066  36891 98193
1131:  55650 84829  38464 89085  87259 31552  43089 39554  42401 23322    12011 98205  50917 61730  79951 69606  93105 07548  46446 54355
1132:  16423 43040  05897 27422  23153 32591  66728 12134  14391 81863    80851 38658  77278 72824  77825 69283  66456 97341  60478 99234
1133:  31894 86652  03091 34212  37318 92596  40961 09798  49927 63668    40330 59694  05180 50486  40395 39555  81858 72924  20673 39908
1134:  35619 29900  05992 66312  32878 33238  96841 32465  45966 41161    48227 90005  63042 61672  26545 67098  27094 79943  33368 43918
1135:  48205 85819  27020 47607  30288 10662  74736 10453  87234 51177    36637 27135  83699 95825  84844 46567  01257 84612  95151 64066
1136:  55882 59597  11748 28428  56149 24516  64154 64809  58581 66870    20253 84847  63056 74118  45974 28422  02055 93029  16045 52888
1137:  39032 12448  64839 75344  73126 28470  11545 17637  71162 17506    38492 72441  39938 38717  26829 70784  89414 17450  62714 74784
1138:  89598 26429  85465 97448  12722 71353  79818 49349  39349 64094    53836 59610  76875 50459  57581 84046  97330 18285  90923 68931
1139:  88025 62193  39772 71123  87809 85670  95292 43807  46601 09909    22515 14838  14768 07883  60453 54246  45779 92553  96578 57349
1140:  07953 32897  36689 19269  64582 77468  26961 02074  51930 88564    11640 50795  09985 86955  97841 85098  31076 57349  10854 61645
1141:  16669 08446  81766 65720  67818 74987  57439 54371  27981 70435    29231 97956  17539 52194  76973 87835  68988 04908  47302 38983
1142:  45554 19848  87292 42519  74773 44446  12840 47762  56806 45052    18813 61643  21066 82686  54501 73150  94833 08507  32976 32714
1143:  04342 64898  73391 31595  11323 52201  79988 55873  26755 74615    47231 85868  91830 48488  08463 52370  58900 01742  11905 03632
1144:  70554 75075  00365 63253  27120 17289  63014 79660  80032 30783    32267 02317  10234 95435  91500 76768  07495 26343  62611 40038
1145:  10141 16190  53154 58589  72388 92385  40109 49672  69855 72118    36199 29636  13434 15706  11061 64161  11032 91697  34838 22583
1146:  41290 97386  74870 44084  87068 90209  68060 87647  21765 64796    15085 62663  53674 01895  62237 32965  38145 24383  11740 02129
1147:  72667 84664  56912 97129  42840 65272  00215 34661  04984 32892    81101 82912  70554 85095  80932 58419  55664 03301  60531 13319
1148:  05676 43889  53038 98387  57669 12362  38467 78962  44153 31493    67953 86775  16210 99047  83470 41827  35509 78310  11376 51007
1149:  27171 06145  02347 14620  85900 82588  37766 87578  70232 72497    83745 33490  81429 83567  85118 10315  33133 92139  25626 46427
```

```
1150: 03256 68436   56996 52139   99639 65555   96017 50210   62506 59083      44683 27403   77509 94099   18065 11131   53667 74177   26166 76214
1151: 21959 41217   99156 76460   87303 43744   22695 69510   69049 39097      59518 19659   47298 90617   14311 16590   16989 26278   18320 80488
1152: 83175 01758   44028 42699   05826 34436   33044 11351   94888 54964      06613 03598   00096 50290   94625 53733   84451 20725   76187 56567
1153: 99576 25290   74167 99314   71365 79832   74128 27018   84739 68704      56322 98488   23239 97720   32194 24021   84806 15622   82932 49216
1154: 55260 02328   62986 79894   66135 96647   12669 06624   10598 81829      82699 57257   44441 95678   24334 34523   64980 58629   03120 85753
1155: 64418 36206   40372 57483   24967 43726   08614 87593   35403 34018      03184 33598   41808 90507   12888 08371   60053 98476   98473 78006
1156: 92579 82534   80414 11669   88476 35830   74970 08427   84829 37449      23596 47051   18966 00071   02402 09146   23487 71724   07871 70415
1157: 80369 79636   13235 42360   66683 41944   19973 17338   49349 20774      25269 45102   48693 46444   65745 44490   85226 64458   35820 87414
1158: 58619 51288   06800 39529   59712 80471   87853 18998   85666 39483      63176 69799   89491 56280   05267 54527   48373 69090   07210 66658
1159: 42035 23406   22093 19788   64474 83272   55182 03690   61468 80917      32654 11888   10065 50054   21135 57306   03005 92701   64202 73704
1160: 80057 23933   35855 52819   92959 99206   66300 64952   86812 21318      14358 38847   95577 60255   48609 92623   18805 22748   30358 98659
1161: 65449 95523   30731 83746   76496 62975   57457 72343   96656 29743      29647 14261   89331 19417   98834 59212   28189 26063   83734 08543
1162: 78921 94287   25639 40641   18643 78504   26154 65736   80036 65973      04381 17201   85224 04054   35919 78793   89115 08122   24873 78923
1163: 25498 10804   04764 91516   83438 66577   44582 55620   19367 77636      52522 46745   48469 83480   87101 13955   54343 12291   52253 19754
1164: 91052 83600   80611 01997   47621 81907   46456 30027   82035 80005      95195 40101   19109 38246   41467 57455   40259 50586   55027 93968
1165: 81917 90676   65462 96998   19319 34003   02357 78548   24290 50080      68007 01033   42821 08770   83400 77176   70443 24365   41514 66290
1166: 35329 56886   35067 71518   74761 22766   08825 87745   10069 01297      99504 84095   28701 33216   17947 07810   94085 23441   68048 85844
1167: 85271 41121   71085 77217   86409 70757   67859 92029   49265 90295      99258 72689   91981 76368   38337 90985   03421 41771   79671 23743
1168: 93941 34204   89021 12845   24080 92304   61975 57966   11360 27426      35646 67033   91184 59380   84873 68287   83013 41994   75469 23411
1169: 18856 26831   64299 82670   70750 76329   98256 00246   52459 99897      35382 16371   91842 95131   58038 84364   28631 32518   68353 46906
1170: 77460 10394   05009 23326   89714 27050   72680 51495   84540 76484      61536 93610   09481 68395   79607 12423   28061 71472   53738 72585
1171: 17364 39686   56815 85779   40997 73228   81802 86352   70061 04216      42668 05495   51772 97877   32880 36922   03888 37212   27160 08885
1172: 86105 82758   41673 55682   16841 88479   25459 58750   05127 96982      38787 52720   69580 97234   01467 88224   92581 78718   51446 25910
1173: 86824 95283   66358 62983   64291 77463   33248 80673   27684 73100      18196 27421   02216 32567   35932 33820   96595 09953   30575 07627
1174: 70548 09387   08256 23096   39358 63384   37884 53838   36043 87312      13558 71294   36663 43854   55021 68189   69166 93321   89313 58992
1175: 12304 46169   17859 00302   33355 65628   14206 59869   58449 87574      45192 11359   41673 24923   91327 70662   90648 89844   93083 65362
1176: 30721 96931   53208 33142   01437 51973   54267 73845   82076 73541      27804 35274   38044 14822   67585 40526   81494 89242   98149 75977
1177: 94438 29751   49525 55772   38495 21841   41823 57859   02770 84143      91751 87735   00550 10723   80187 49338   39581 76658   26562 67501
1178: 10102 90763   31439 33690   40569 11282   94930 85319   60295 32363      83262 44391   93374 76344   15046 35520   36894 92136   30142 02705
1179: 68289 97541   36682 10961   57257 82718   80033 34849   67113 46022      19141 85448   43163 48127   21372 02854   95626 00484   15571 58680
1180: 69483 54969   47561 29151   77131 39847   43741 29187   06314 67489      11092 71210   11866 32454   80185 58575   41910 15731   58216 83679
1181: 35057 76606   28665 30583   37659 61236   59900 09993   43580 00803      93614 26337   28557 99700   12075 96518   15863 47698   26913 25768
1182: 26294 64703   39092 76117   07702 52945   20714 57976   54292 44804      53849 15053   24276 35099   84674 26755   31388 15665   60641 76641
1183: 89494 03037   68377 59546   10894 69169   67317 13129   00820 29185      04635 80821   86103 62199   26731 92755   89251 32149   31356 91990
1184: 43939 66314   70009 71421   42046 66881   71797 59944   77006 28635      17003 18232   31662 14384   16058 57765   25823 33810   51445 11834
1185: 01372 87778   31280 47686   17237 98061   69820 54813   18243 43351      33532 53592   82320 66612   03437 59341   99780 83484   06390 84826
1186: 41265 92608   60021 44432   58809 36892   70452 81697   31222 49532      50512 76746   36815 34711   66747 78923   51447 67197   82006 46188
1187: 01669 88974   06277 45442   56499 84692   89856 41023   05492 53646      38322 42172   85051 16246   66427 94592   65961 66335   24056 44480
1188: 42532 07528   52942 59434   65884 54653   98367 66090   12663 66624      83485 77301   36529 48717   83428 24510   64158 75397   08788 46386
1189: 94369 69805   18306 14688   73188 42812   20140 65377   37964 98670      95409 62064   45358 66648   88864 26190   09842 07128   11054 48060
1190: 66813 49569   75730 08686   02381 51601   81359 10480   56105 46183      04154 52487   06807 71210   15349 50477   01033 60405   90150 60817
1191: 45309 99149   56015 47953   04951 12176   82132 99878   18830 44811      22820 47490   57969 13401   79494 05526   31847 14044   93792 91546
1192: 29359 21186   77602 33544   61669 98720   33972 27638   21242 11946      67019 89625   23485 47396   76219 37963   11015 67156   93109 54030
1193: 19480 61833   87855 50001   91096 90666   98448 57773   69401 57590      43650 31545   81659 40876   06195 44943   21431 02343   52081 38938
1194: 12671 23914   85784 54892   42469 69344   70721 10255   61922 74675      26410 11237   33924 69516   03552 08387   61381 77471   00774 42464
1195: 11306 90363   58034 82238   18846 54206   80317 95021   41987 75204      49174 14650   21093 35350   59857 13371   61385 41674   07843 35729
1196: 73505 81203   01795 45974   43698 22644   98887 81986   81209 04166      41734 97287   41427 04040   99164 86672   16325 71605   95930 39733
1197: 31140 04642   38366 81159   36749 26504   26693 29836   04807 77142      16709 86009   79040 44552   14825 67103   36621 81413   63140 39671
1198: 47080 90888   20806 74641   17266 51288   88043 12899   14912 75676      61893 29820   51914 21005   77357 58986   63777 51300   33817 73109
1199: 03676 12714   30111 22086   46078 35360   93402 56905   54440 79227      09030 23253   95148 08240   78589 81270   84281 43684   13256 31349
```

```
1200:  14888 46623   10207 77644   10576 30394   77280 03095   23693 41207      30162 06716   13029 70023   27451 16658   68704 93974   79488 22604
1201:  07051 34870   36455 35764   65722 98627   31467 14293   60300 97270      31151 65536   56323 35649   19814 65855   73040 24785   91349 53923
1202:  12631 70973   46265 98007   47976 69401   31161 24290   65837 22211      27959 17558   96108 18888   46471 43872   89904 34385   28019 76935
1203:  27553 56501   98941 47697   49055 82442   15868 03787   65665 65212      12122 95068   41065 44004   91984 33671   68449 15376   75529 84076
1204:  06857 17814   40098 60479   86582 07228   33214 59177   64830 03577      33140 79443   74812 00097   85101 24392   64786 45481   55817 26279
1205:  01654 45900   98267 66617   83580 47511   43192 47222   21236 64627      03309 08465   57136 07361   34169 07501   86427 19290   91812 15718
1206:  33697 42523   97926 18952   18213 19158   79644 20930   95789 02450      78130 80405   58662 75807   72219 78515   66447 02753   22622 55019
1207:  53314 10057   35225 52033   95931 46231   54566 39370   72339 48103      78559 54490   90094 49390   79890 00482   89853 70493   99237 31299
1208:  33290 42171   13587 05248   85273 58117   36737 77515   13952 46086      97499 32473   32969 83238   30681 02981   62510 51975   74901 99923
1209:  26737 73484   54387 60501   78619 53781   54265 30548   02522 07172      74989 23831   46264 33318   39693 81074   90742 61295   22885 25441
1210:  96059 21646   10970 59810   42167 20731   88387 91537   58429 64876      08126 92769   64544 06416   41216 51523   73938 94735   01982 01308
1211:  47769 29632   60630 80248   82232 14038   72027 50907   81434 36091      53355 59060   92315 21263   20117 66987   04000 35662   65907 64418
1212:  34462 38359   20450 86942   48203 38173   71510 24469   59075 32600      67480 00501   81301 09872   11447 08028   83223 11733   63784 21725
1213:  54600 45373   52374 59780   51856 68214   82782 66456   54464 72117      49853 10920   97373 24370   93815 43091   33096 20340   25030 42711
1214:  84896 12359   10974 36722   51081 21116   97134 47291   00773 86523      25241 11829   14259 06239   01259 35561   28987 67554   73843 05187
1215:  65319 59048   46247 71959   24846 17420   12442 85026   43836 47884      71754 98337   88346 09417   09746 28926   74275 03765   34091 99463
1216:  43969 85167   62204 49745   44584 06475   91702 96308   32744 39891      71710 52762   55684 98283   61269 58927   96654 60423   06396 89711
1217:  07006 10569   81402 65516   58742 26601   54670 12860   59061 76636      18095 25864   41565 20298   48695 47700   09259 51954   35622 90379
1218:  42198 05343   25092 86772   29683 81672   52398 45896   69898 21623      82944 38439   66949 41268   63890 60871   00295 08464   91973 32702
1219:  31975 65653   40976 72533   43299 25897   83270 03261   88677 31239      20830 26255   50828 74530   37738 71954   03131 66820   64777 83807
1220:  54273 74983   69769 68821   37447 68753   48402 82099   63623 44517      69825 73512   33713 76133   58324 70833   83026 88030   69664 09363
1221:  07246 51619   50636 81267   26895 78125   43527 65015   08751 43151      77184 15148   24804 99982   25115 28082   43627 99955   69460 85340
1222:  74229 95465   06761 15585   46081 93394   99063 19552   70496 66662      34556 67989   81174 09809   25098 76588   20961 59448   13187 20481
1223:  74727 36889   83977 12311   20233 39828   28274 95030   05452 20309      24022 97386   36686 60781   38801 92771   54577 20438   11730 55457
1224:  02367 64832   47972 81465   25763 07911   25186 78913   39436 03839      31684 67628   24382 33087   41309 49681   59899 83126   59732 14432
1225:  15289 96588   19462 76995   63190 14091   42016 63906   57150 54279      56778 27943   76273 94368   25810 04431   15897 91241   30112 48921
1226:  47316 59830   92426 85971   85904 89227   35772 09874   87724 80800      02381 06189   86841 04637   75886 19566   98767 68128   78260 06613
1227:  39780 46876   98403 88132   47823 92280   40336 02411   01207 16051      52327 00463   65069 45345   48335 19864   47963 96614   24638 77093
1228:  27966 05637   38968 60463   43713 58970   51486 62962   72962 88721      23936 24580   42235 03016   40508 60714   04590 06371   43121 44345
1229:  84011 10980   31199 96381   97381 12491   87172 64394   07671 60895      12731 91812   43681 66613   99055 03141   79397 26096   36387 11431
1230:  06854 77042   00340 35510   27941 95989   39806 37944   78856 83596      25565 34173   42632 46845   99267 41390   78671 76081   85920 29747
1231:  32611 27941   45497 83208   22997 21298   45999 14735   08866 79882      80239 01709   91506 52230   93708 84766   37516 37276   61114 46368
1232:  32481 56115   55152 97288   35530 45124   94838 70804   24411 75792      18027 12929   15120 78089   09860 59626   74131 79407   90092 33022
1233:  30110 85755   44649 36619   73483 74223   57643 71632   64276 51825      71105 70125   70347 27459   58954 17857   30952 46659   20915 90134
1234:  47600 58956   54561 24131   96060 43103   95840 05221   72154 60331      21986 30636   63941 16133   73994 13105   58233 24512   49550 87657
1235:  74287 17062   30074 64947   53750 47783   03339 09168   52992 78094      53234 81524   15853 65427   34061 64986   64180 71812   03564 49908
1236:  55660 42223   56589 33575   09278 13242   97865 77638   59763 52457      36271 44429   80824 61393   33455 68402   76975 24562   48011 38880
1237:  90938 45894   03872 92614   95952 76530   63519 56163   19154 75253      72997 13131   38646 42149   61867 54438   02528 24467   67131 94318
1238:  41532 49444   48709 53901   08563 14298   91447 32168   73329 10632      94315 53383   04706 06964   23206 87675   29616 89015   01145 30838
1239:  47465 07586   39049 74058   68665 35403   64097 69773   84404 50370      48675 41386   41604 79058   09015 02001   16860 02774   74345 05491
1240:  65521 23713   78398 64814   25373 05743   09877 57173   40885 43545      76364 92147   67952 66416   07358 15007   86904 40476   35230 07356
1241:  04642 72923   24337 93630   93814 91831   26302 71051   00533 89072      79783 35703   90781 85016   36093 96372   57016 15771   83599 12132
1242:  80998 95197   04300 94220   28636 98710   38866 66883   25315 52626      40443 91629   43096 18980   07793 60454   21403 40992   92537 36520
1243:  86201 67989   30496 62293   00080 85537   01343 18448   72447 61984      20925 76668   38788 30948   31716 65109   34056 98502   19116 91275
1244:  15797 52537   09933 77713   88755 84022   99805 95392   93500 91869      64416 63047   23807 27784   75732 43096   45087 16827   30378 14116
1245:  65142 22907   44939 80048   31851 76621   64624 02928   90739 58238      45529 94255   42950 82952   13101 84999   97210 89395   31335 92102
1246:  55682 11440   59872 07307   33342 05086   91401 51602   86755 21445      61997 62279   51929 66018   71389 25862   11836 69284   59015 94403
1247:  10696 36148   89179 91338   47565 57796   10450 85288   51090 45797      00140 52530   97509 12861   76217 83470   86725 47898   84526 70006
1248:  06826 79541   40616 34825   36766 31115   64343 32696   81433 06530      66727 58972   40427 38018   61304 08441   44919 87302   38619 94946
1249:  75266 07741   33955 55307   65820 85804   35799 07742   72169 73229      41431 19831   82035 27405   21027 51182   86446 90587   09374 41264
```

```
1250: 72173 48409  24505 59049  06456 23618  76496 07954  50218 45176    54021 21459  23082 15536  68181 15083  98363 59080  69815 52804
1251: 52065 89494  20072 57152  91770 85727  45420 87638  41232 98471    89320 89848  00067 20889  23009 84527  60830 02670  92893 87793
1252: 99020 98304  76130 24498  77416 11314  55195 86515  11649 70706    81122 32678  38503 78157  26516 21375  62302 86211  84740 83614
1253: 23570 75322  24218 31713  01601 70386  14794 71412  23987 61079    08297 62888  76254 81706  41663 57281  22528 96521  36594 89946
1254: 07299 11894  83612 78223  42302 76982  75025 58332  58786 56801    15091 02863  20279 49798  38165 09399  52382 45581  38184 98113
1255: 72934 11912  22961 01842  97876 91498  87607 84463  21320 15352    81789 01206  30712 65820  91824 72173  72302 78759  22840 85013
1256: 26836 13600  84690 03596  70745 81617  22093 30109  51969 25598    10323 54078  85512 65153  15051 97575  10087 53423  57258 21585
1257: 21532 81234  63579 30833  96871 18375  56881 09419  86831 45286    25931 12134  99992 35427  96007 29391  18272 43999  38236 57819
1258: 78751 08557  81947 27356  44477 79640  32299 61270  50885 37676    09725 44956  44548 98900  83650 61787  56736 94675  72113 86020
1259: 84762 63589  11756 36626  90642 66073  64394 02835  96811 20058    00447 04899  49160 01771  44508 94443  27398 26992  15971 44309
1260: 30846 16922  53708 56350  35209 35775  24459 28810  40030 37522    29906 06203  87552 60607  14245 15439  32946 26323  86331 58083
1261: 17558 48058  02237 20683  52358 07149  87148 45623  47665 42993    76071 79105  16220 71393  88935 53533  53199 58604  41539 21921
1262: 19924 47891  25967 93741  57468 97202  78725 00464  49471 07823    46929 05705  83286 79648  11538 68062  84999 96688  48687 45999
1263: 71005 42479  10793 39634  89330 37018  99146 40332  67122 77885    20605 63341  18756 20320  98623 13780  64856 60659  29368 78055
1264: 60297 13544  18415 86133  18357 49146  13212 71888  42952 13772    05364 25218  22566 65203  32196 23083  93435 56246  11497 57250
1265: 50041 93352  05045 84958  32980 38062  46484 30315  56247 37778    23799 00744  67575 09472  25179 33334  54934 96806  00614 80406
1266: 89019 13351  13737 14461  07948 67147  43998 58505  56822 95067    67328 66380  36094 15369  26602 35955  92504 27152  64368 25689
1267: 76052 90532  31460 91694  55993 52533  38544 50901  39246 12846    15824 53178  23081 35180  79052 73191  34205 25070  97111 93063
1268: 80274 10379  87379 01338  83068 66850  27689 81519  46587 49581    90005 65348  76423 45725  31118 02343  48217 60766  05981 79715
1269: 99778 95827  55024 50131  50464 07989  74858 70888  64998 93944    25373 45618  12804 03882  97389 53234  71555 89335  05853 76665
1270: 21250 36853  27507 87857  14630 65714  33942 02589  63094 85209    94680 15387  01164 48648  52265 08125  44312 50582  33102 99774
1271: 16340 18707  21280 33950  53192 45626  30424 47901  86599 15070    23822 07963  43218 48253  61686 49711  84744 18018  84715 20290
1272: 75130 96554  12797 09888  82364 78624  75482 41681  24701 91846    57914 43092  59000 71918  17961 61191  62910 62772  97126 39892
1273: 48172 02188  77728 89443  20276 87136  53115 04666  69760 85427    50345 22455  79852 37238  63595 67896  91047 86761  39364 49250
1274: 26427 64022  57889 83156  90702 89938  35160 96116  28015 83293    25452 50237  72405 66434  36607 49354  69258 52674  14223 34132
1275: 11742 57358  80752 23540  44343 34582  60370 46455  85324 48123    16676 66966  05340 94009  65180 48591  55514 74849  37248 15091
1276: 81561 46517  12924 19551  77938 02682  24225 96784  08988 70114    21497 95950  13471 75046  62460 69868  44019 50890  25703 05835
1277: 25457 09417  43623 62534  58243 88978  27975 93551  50278 64748    92466 00423  88132 81139  85492 81417  31209 62199  04696 57139
1278: 04222 03380  84544 72995  68612 71258  62711 28302  42044 55380    71838 03706  16798 86731  85210 18306  68961 08996  78383 34886
1279: 01448 81765  45392 01882  25789 31872  72079 33907  89552 22194    17995 94600  26983 84719  74637 68151  00517 38806  06723 21071
1280: 72981 63019  59937 72458  82443 36146  10853 67149  62703 58345    71994 03873  18539 64687  31767 26166  46418 81111  02935 24484
1281: 75588 77801  31751 30462  75726 18657  16909 88102  14727 62636    78847 75206  30303 11907  93144 49086  82912 23310  22804 52232
1282: 82490 83467  25923 76097  04177 01205  05930 50166  95960 26908    89863 64692  26135 72187  84445 32972  99964 35430  57865 55974
1283: 01546 66817  19825 24357  56019 88941  12488 51869  45041 87450    93539 34705  23701 64407  71365 54992  55189 76484  41388 00831
1284: 87038 89851  81955 79195  18733 72385  38345 51862  61875 91665    10524 82411  44744 14814  79191 59383  95254 87588  03502 90230
1285: 60693 78756  84047 02116  71194 56413  27069 13025  04506 93196    62618 74142  34103 84610  73800 33296  26957 58417  26183 68477
1286: 05332 57311  91286 39100  63663 84763  63907 77891  00589 85074    59527 03359  08288 06653  17004 16993  01539 07283  02051 38036
1287: 78509 10443  02189 38750  59304 96395  07349 83267  45260 88871    49923 13873  45845 85968  34413 38371  61119 59905  74404 83595
1288: 26521 80175  45999 25118  38850 31026  61203 57683  51920 70809    18168 96438  82957 21417  37174 79804  96074 11723  76016 92615
1289: 82189 15171  09233 96718  71112 53270  31775 83131  91919 78739    81345 65142  66277 38718  30783 26103  80661 67345  39452 96640
1290: 62724 37925  31014 23556  20011 83807  44080 23864  93789 46451    87228 48617  46327 64148  98299 97497  74229 34940  50274 73294
1291: 12272 25841  52665 24449  38413 75588  43691 40954  52596 02961    72316 82378  20484 37192  17057 64011  63786 50961  23744 90054
1292: 43171 51418  21058 82691  04034 93303  21386 74163  23640 46927    76417 53020  38387 70376  77813 95203  35996 31796  28635 93595
1293: 57417 69202  86236 34776  02879 75053  92073 97784  93850 19713    11671 45322  48770 83416  51321 16750  43986 22475  28439 76850
1294: 34441 65503  22099 01784  30228 00282  14012 32419  38770 87093    94714 26842  28155 79339  48939 69182  59386 66321  94784 32082
1295: 51651 42157  97745 50899  67622 18049  62542 15712  26463 06467    93822 80139  47051 10086  00287 38111  18804 61483  67435 10553
1296: 59875 28204  61947 27389  77154 36905  82273 80454  61050 81403    04662 80164  79617 38479  34724 06033  40535 16072  68025 16254
1297: 22796 53544  05163 32848  13389 93736  26849 08559  87615 06328    59621 55265  56264 34873  59461 22662  79857 19316  99387 59378
1298: 21040 08571  99252 11424  59589 86892  24220 02411  19640 51757    07900 96944  57518 57160  68749 15297  92499 85727  22564 76095
1299: 83668 22557  70025 36230  74071 59121  24888 68904  90348 73911    82758 94112  62047 76813  41564 82229  00486 24388  66170 82362
```

```
1300: 27009 85346   12518 42985   44824 57091   79268 08918   03084 42934     49441 22917   67991 93545   83998 67935   69246 60191   65197 79894
1301: 64273 90279   95656 29258   55736 63102   33063 95258   65939 35452     61884 61556   22843 11570   54285 39205   66791 36985   40920 56892
1302: 82748 47189   67172 76724   21045 60837   03442 89713   24226 06579     97455 74103   58503 44098   00104 83532   85099 27811   60238 12419
1303: 17009 16387   24537 66847   50715 83784   73748 54744   47178 58714     64242 26442   99349 36762   56571 02239   98448 32006   86124 06614
1304: 84948 58837   87023 88919   57377 24699   02555 69138   35221 22250     89445 37703   88484 96841   00762 34569   23147 64354   43111 20733
1305: 35522 95299   67133 36674   77667 51503   47594 35516   92023 62833     51389 06861   86294 32604   43829 29220   05913 96561   62415 38803
1306: 79749 07241   26423 47712   50190 25446   05921 29460   51999 93790     17419 76090   03102 64361   09573 41806   04749 60928   00477 75525
1307: 08301 82053   75864 72584   53357 06165   01431 43215   17094 17756     39477 44084   11525 59912   73054 60455   68005 96220   40888 05050
1308: 63662 41491   80550 86920   75494 91773   79338 53458   45271 43259     56957 23158   38418 49913   93811 18135   71358 86051   32064 76967
1309: 43831 49196   37083 32605   01105 86213   28185 89041   01021 25738     31091 49633   24032 22617   91013 63372   24857 43936   58352 40961
1310: 64056 38594   70391 52282   88034 70742   64558 99624   77581 54069     16349 46093   24877 00658   42234 62521   16279 25598   06077 56636
1311: 59799 98127   18842 46967   45282 00706   73382 62645   69012 71159     51646 76196   09776 99838   46126 76365   52977 49532   25173 52195
1312: 96422 50257   20406 91336   85480 77719   48593 80864   30017 45396     39684 40273   30311 50357   00058 98410   28055 62916   99124 19941
1313: 79297 53455   53286 04974   00453 43477   79289 75421   51721 44996     08274 73893   79274 44045   84616 28459   74061 34397   93962 81088
1314: 97706 44860   55878 69440   52094 55211   69786 61362   24437 98159     67735 27433   19267 33587   55012 96383   45393 02326   04999 98360
1315: 85357 70081   02162 67587   61702 61920   91000 88972   83771 46645     11672 09496   14306 18324   82695 61260   38763 64830   77400 03865
1316: 35405 96318   99085 38368   37811 67784   26124 74920   37510 35466     00184 40562   73298 39874   10370 61337   29795 91259   56880 56365
1317: 98525 52765   88379 77670   34417 13886   94023 62253   91771 62682     46454 63110   15773 94282   64966 79890   23238 13330   47294 30840
1318: 32043 55851   12406 62892   57993 71759   99986 35401   66823 15132     49987 71981   55386 83489   83142 24228   36670 63785   58256 18436
1319: 69827 37353   78412 51148   82869 44501   55057 85250   92580 72412     80065 98034   83957 51423   85069 07726   59611 64647   29993 25470
1320: 36669 19363   05359 54943   23049 16938   23592 45503   90799 78206     58529 87868   72400 86779   35902 57375   23812 63230   50363 39812
1321: 85775 76730   78455 36303   47653 43364   03056 87686   37194 55809     78168 95443   42292 42071   28515 59469   25845 51605   61784 52716
1322: 78303 04529   76160 34636   23947 70562   80141 47167   90209 95091     04344 47038   60674 58350   49705 93441   97406 88962   82633 70576
1323: 89272 30319   56195 34796   77109 29909   37280 04013   41639 43321     92203 46958   03967 52173   82819 52096   41415 93812   09538 75011
1324: 80157 77577   70894 22798   91176 15010   77399 23098   34863 68096     63736 89127   67119 25569   97112 48563   62932 15413   17580 63506
1325: 66792 06903   31758 72971   58790 39801   76566 00853   66561 14076     80588 73246   19417 28480   24619 47629   42859 58154   27955 86053
1326: 44564 58959   01689 78182   23058 41402   88789 94884   63571 66222     84818 70273   19558 61123   42627 52837   87221 72442   95056 61797
1327: 49620 00058   82576 61980   90721 44204   69469 67655   55696 66128     61528 98826   31222 63634   02450 22943   77582 74670   47519 10762
1328: 99561 60387   07792 63907   99631 58741   30084 67693   14446 08224     56151 85290   69417 27504   18197 64133   26857 45823   25350 04129
1329: 82659 43069   13140 78902   24448 95418   08321 49233   89435 83780     13275 22867   29149 42070   94745 86194   19031 79835   60779 77158
1330: 31869 49050   80850 95367   94767 71869   75890 75641   44763 94024     67198 31283   70671 25737   49080 88441   38362 12596   47362 78469
1331: 24490 63343   28965 66594   25265 34878   88435 54777   52502 31184     66155 15155   54196 28396   63138 10264   93621 23334   48757 26772
1332: 58757 28518   65952 49057   55393 87256   76118 51686   14372 65556     67718 35513   71886 61189   11847 56048   53036 22640   91119 62508
1333: 12812 02651   31615 31798   85120 90150   09147 08786   34959 63733     48380 25158   41900 48541   45729 88470   00919 95991   17614 52327
1334: 14259 81882   37158 62826   32100 97865   74690 00742   25819 81561     45358 36918   58599 47005   21654 67740   40890 79357   63678 19689
1335: 72164 89654   52508 34152   20826 50009   46829 00600   48212 96929     70977 88927   55015 08238   74205 81079   40910 95330   04842 54556
1336: 87886 97683   86138 96092   32968 85337   66455 05402   16182 58653     90283 42698   70445 81511   69784 69454   36137 81854   22814 30938
1337: 81813 70763   21916 92927   27727 87824   99852 84126   21340 86652     34057 42874   18463 71923   24208 16721   01778 09704   69618 55621
1338: 06815 66129   48887 18264   80321 13604   44552 35435   76039 17742     32592 12347   91466 10552   61612 08241   13857 53287   48301 70132
1339: 13093 42653   00058 70485   72260 22821   38665 32013   39332 53426     78944 67319   72303 78260   76375 71795   14048 87779   63252 57215
1340: 56675 20692   86420 93664   80142 99053   99652 91934   02708 24053     42708 15491   57202 50607   26225 88914   16568 04617   42595 95048
1341: 77863 22787   93105 33050   65020 02819   99243 90269   92866 22526     97960 10856   27540 40232   91954 44626   70674 79959   51824 75222
1342: 66636 65892   85958 52499   65460 02599   36077 24301   41724 57614     44523 36501   14476 48968   58882 54830   74415 11134   79892 85361
1343: 64688 18264   88830 40836   51241 35218   64927 73869   75028 70486     50750 36277   34134 28866   68953 25709   36242 09626   41755 66632
1344: 88903 28819   59964 66112   26575 96919   74192 10037   23429 26731     89185 84519   13863 52674   36069 67658   17733 72543   22512 50025
1345: 86512 88356   85075 09325   51698 46422   04607 82581   98980 20977     28808 65631   79627 33455   71435 49963   40669 01718   16726 07528
1346: 79120 09779   41122 66388   44560 76302   29686 39313   57915 35768     44170 58982   31517 22141   99589 97006   94621 16033   42549 79002
1347: 40932 32176   82189 13714   93060 58869   14784 17207   15779 30406     42989 91184   27835 17166   06359 26946   11605 07275   92088 88877
1348: 05937 31957   66157 66646   55427 18047   83051 56260   94251 56059     86690 01767   54638 59241   94340 42513   95585 32191   47861 64812
1349: 90889 92408   62315 09596   05566 14887   89973 77552   26551 43811     17265 37177   25069 06211   38979 22892   37940 32426   57714 96789
```

```
1350: 70906 64535  92062 31167  44768 29772  76960 21201  16392 73913    37175 62816  33599 33139  21837 81706  51134 05140  93729 46168
1351: 40866 36142  98142 30217  74382 59472  98893 08168  83896 20666    49238 09837  31164 93039  94635 78838  29708 51368  63176 88956
1352: 41812 37012  01581 23878  95449 32890  64446 81476  06136 27980    02420 30352  63242 84207  50111 71123  34121 49583  93668 29903
1353: 04949 64881  88305 11191  72532 54549  83239 17196  89966 67053    28238 61538  01336 69087  13663 96598  34875 92635  24512 24617
1354: 50171 84825  58672 71427  67706 24328  09902 87666  10785 83662    85507 03266  78974 07579  99066 78588  89375 98464  07968 16371
1355: 86573 72497  30958 44937  43843 00505  88244 06993  65677 78834    76585 56594  24572 55985  69520 70071  80735 51686  03607 15685
1356: 22818 90808  06131 55712  05314 23811  15476 43561  47248 32947    34400 31979  63154 09020  82548 27463  57590 78972  76696 10572
1357: 46586 16552  50483 31892  76861 92264  38281 18088  44878 26070    84658 17190  57523 43475  77390 29883  50396 27627  32286 77181
1358: 84774 77230  47561 68369  59716 82430  76358 62469  05062 03290    75358 77499  78201 33435  67580 23902  60143 42522  35748 00573
1359: 59575 55881  74829 76262  16909 86653  69917 73539  27403 01641    34785 80870  42936 44801  69810 15770  32885 56776  16269 91947
1360: 54564 35095  13112 85765  64562 92510  48479 38170  19848 57930    17342 65227  72875 35411  33071 77744  57398 15685  83435 45303
1361: 54139 10076  53940 93519  86821 13598  29658 37428  15242 89768    87020 27770  78164 26999  63489 80917  75066 64062  42942 01886
1362: 00321 72712  53112 05376  18160 48942  59289 18724  17836 63655    07167 16341  44625 08541  69427 45338  59972 11386  29574 53689
1363: 64556 05219  57795 17048  94491 05315  27811 94094  04744 80638    30302 25607  45583 89189  28641 14106  88903 49977  98825 39313
1364: 99311 78537  33181 85002  93538 10744  70360 10423  04400 56672    92269 04469  74286 00650  07545 13999  81094 32638  41187 65754
1365: 87373 25831  02310 16871  42551 36160  85457 95949  83950 81610    53366 14459  05981 05426  07779 95712  66215 20602  11126 05334
1366: 17927 73690  68859 67974  47917 42568  37601 71847  33089 09567    71209 10105  31656 00608  83577 59699  38462 66883  22783 60183
1367: 59324 69932  23725 56360  41727 76629  54841 00518  50448 46555    66684 69302  34017 43426  53759 20796  95759 82845  25628 68687
1368: 87940 30777  68747 86525  27453 16028  79727 36239  50645 69837    29309 58259  59895 86329  47682 41087  72900 09639  44131 85521
1369: 48477 71634  79585 68749  69041 81507  40211 08182  08860 45722    60166 22423  35997 91062  55280 80362  31231 94074  60944 31844
1370: 18065 41249  96266 96854  28350 66002  25018 10032  68469 20095    55139 03090  62717 32155  83277 56003  54158 32541  40932 04971
1371: 53995 77768  63392 87567  05135 44174  32356 64812  83384 25303    02539 99328  57583 51668  18600 24758  98188 07089  98339 13047
1372: 22838 39207  91068 43184  16583 38038  68285 40422  69119 22113    77373 84943  78275 40842  73422 77131  46554 39754  58344 13338
1373: 72594 04407  33007 91108  11617 00354  38692 34519  18602 84873    10914 37370  27215 02333  11499 35184  25773 57639  03595 85177
1374: 93132 22197  64930 12987  37959 26319  71192 76807  25208 45125    68355 81194  05664 90753  90127 13242  51079 03134  54775 79187
1375: 90172 70937  30795 57436  90605 51390  65675 11900  49762 93879    11914 81532  77798 73356  98931 82304  35196 35511  90074 14140
1376: 40391 99378  38002 73056  70446 65202  04738 44866  82209 48516    26297 64998  56841 13460  09281 45962  62843 20403  47041 13991
1377: 56691 33345  77461 18669  26686 53677  80324 76226  16129 17775    35138 06455  77305 32334  51850 36542  31693 62544  75882 81867
1378: 69668 82387  10942 38772  90589 94372  79212 67696  80281 25020    65035 56013  71529 21925  41554 99954  03323 58414  12074 04131
1379: 45465 28823  70198 57533  49968 74743  52309 03405  25497 55262    04393 86094  34841 24776  56373 16525  68833 16110  97643 21338
1380: 90666 06679  39384 99135  22019 74529  86841 18260  52988 11380    30083 98420  93412 86711  42158 14121  88194 66846  64847 00652
1381: 04539 93397  58584 16322  91714 35766  43739 86248  95697 18031    79377 95230  09568 02644  15135 36623  38288 03665  53808 55878
1382: 78711 01781  05926 76132  56089 43674  21274 78735  65887 29269    73013 01303  25721 30354  90118 78290  68808 48336  67525 99875
1383: 13783 52525  74513 64822  99787 32615  87353 64634  16545 97366    51782 29100  86763 09690  77108 89031  66860 69029  67062 20362
1384: 25944 36091  40591 88814  49499 28856  97833 26880  22214 45381    37933 00562  66437 08677  50767 87354  47581 31404  96551 85183
1385: 80521 31274  49322 94400  13125 21390  25280 71529  41771 52719    12028 44042  43255 10066  99833 52536  41341 71182  62466 93797
1386: 23045 18124  70335 97891  69848 17535  81151 21944  65841 60127    48391 46347  00269 89881  39064 20839  43049 02969  94634 35270
1387: 38100 99578  75324 73841  77757 77243  88657 43673  49681 63292    00158 79133  31870 05289  49999 80281  56771 19356  00608 80367
1388: 28502 10380  15282 44503  25404 87981  31558 52953  81044 47269    05510 29126  53395 34146  78639 26534  39234 70825  57332 07108
1389: 68377 27055  60711 13612  11758 22034  18744 80096  66500 89333    86299 71126  10726 21051  40612 56045  37185 13886  53771 43884
1390: 12779 15675  01828 03841  36289 52205  15983 59534  24920 23401    51031 23901  26536 08294  92784 30334  40195 86282  17562 66440
1391: 45024 00382  42033 62506  59813 80967  73853 10059  48968 39566    26703 74384  68857 76625  63645 75959  35673 99155  68889 67803
1392: 37754 42425  85091 11560  53324 19758  81986 42661  53383 23297    64695 52041  40915 17911  60878 47093  27171 79034  68986 08981
1393: 20302 03402  22323 97246  59236 15677  61050 10971  87503 42779    51341 24060  07364 06100  07421 75869  95406 37408  97919 46140
1394: 50069 19913  19814 58980  63457 48921  91216 91808  53937 89794    74946 95953  00798 76368  62343 81551  36481 16493  08293 47872
1395: 66289 46163  66001 45850  40495 82448  32619 28495  78531 06431    26735 83226  73999 47361  97575 24540  18693 14865  46726 66274
1396: 82209 49228  47229 90670  35728 12084  03371 78734  64512 86395    00372 23168  51251 92919  64543 87942  48021 14712  66499 76344
1397: 73609 18496  76504 93109  88678 71313  56688 11853  18337 12504    40542 63809  18519 26794  01088 24591  31287 36542  76262 40388
1398: 35434 90896  40035 57350  46341 54962  09396 96297  80227 12097    21064 72439  86169 24517  52370 47959  35901 65623  69531 76018
1399: 36882 91032  81908 43128  78422 39906  18914 50392  01009 52355    38992 07396  88658 18915  02125 01754  08503 24048  88064 27113
```

```
1400: 90126 69302   95403 82085   35614 70521   59914 43054   44930 77950      49930 69154   11302 96491   58110 59807   87153 87655   86072 99070
1401: 40824 99687   33440 06387   05371 95839   50258 24781   03234 35542      77798 73322   80205 88913   84875 74862   60506 60344   72833 17596
1402: 13986 76386   68450 58512   09450 97974   87107 22607   59469 12130      29473 83709   59949 31848   54911 35608   77397 13020   75840 33929
1403: 48384 57997   69543 73270   62212 47757   36611 41844   06397 34266      96655 49125   51776 22042   44009 61569   40661 24317   89429 83292
1404: 04378 30603   69578 98128   95596 51210   82892 14365   44074 43415      00235 66453   08366 42256   95875 68315   31970 24810   09053 26231
1405: 18564 80244   40644 11424   20117 53591   14059 50115   05427 31206      38694 44380   44406 05067   89209 09723   84689 30892   81545 43736
1406: 28521 38735   72496 31319   20523 49050   03058 51995   27844 81565      71313 84900   49217 57892   90523 29575   64809 87958   85372 68063
1407: 22218 15773   90474 56673   90606 40233   39905 22289   52408 55147      96104 30491   76773 06343   99772 09449   82431 13433   65500 16635
1408: 61723 45626   26251 19802   64695 33947   87437 89486   86194 80367      85044 95785   77558 88216   93804 33713   97929 63103   40118 37199
1409: 81075 90177   65389 76629   59994 32903   18410 64274   33190 28471      30796 66911   14783 54687   87409 96794   86519 11382   37398 88581
1410: 64533 72913   43126 49548   90533 77648   26221 13500   32452 53833      48044 29312   43409 79542   35838 88666   37997 13544   21958 55219
1411: 81842 70480   26530 16850   44093 22977   33441 27011   35295 23884      31786 72313   65734 98516   23025 13945   10618 69886   27071 51237
1412: 51798 55606   22738 26498   25437 25976   07238 69989   96804 12905      09930 37432   16418 58284   53120 39666   42632 64306   01389 90703
1413: 84930 31556   88302 90398   39219 91118   01452 26778   22350 46446      69051 84778   73482 63564   34067 94088   27967 75317   98604 03025
1414: 60681 28329   09869 27955   48253 84024   03727 46519   85153 33122      16291 21849   54184 17418   55436 72552   86733 10349   20889 54655
1415: 54938 86676   25456 16720   34518 01921   69138 59741   59677 99322      77852 29365   29061 77065   48176 57329   18864 88961   19982 08802
1416: 45072 09190   80746 87390   47523 48909   14765 35579   84893 31874      15168 00123   71988 14216   25324 65923   01535 28608   47360 77025
1417: 65770 31706   55202 91857   71899 34630   65704 21168   42471 47006      23492 12674   13440 92096   93101 38483   91266 07360   47891 91533
1418: 02396 16294   08202 59576   81961 91654   65100 56790   53340 78358      63689 27927   44290 98610   51647 55632   18599 75103   93323 61569
1419: 96912 64628   55376 75139   13313 86473   41614 62047   91174 01460      21999 27532   72986 98606   43768 94031   77081 45382   14817 91195
1420: 59503 34272   29997 29455   07274 76355   34413 16583   40514 81390      89161 92597   90005 06669   23244 55643   57167 98486   97631 34611
1421: 04011 53639   15134 29319   95489 04663   60600 28429   80760 29067      59958 83118   96324 16036   57931 98685   74439 31523   44331 58672
1422: 53922 62228   48577 20251   16074 04217   03932 76647   69011 00815      36033 67399   89183 13101   81165 23163   02435 15666   78048 25216
1423: 69899 95064   68287 32434   23573 83246   64407 51616   40241 64377      90240 17191   24336 10612   86313 22574   16230 67979   72805 69915
1424: 13269 63891   99171 28039   79948 79694   37001 01678   38643 99237      39256 08082   22895 51360   79910 25572   04950 75035   74549 60361
1425: 69642 31259   33302 81191   73109 56927   14425 87971   59251 94823      05152 20344   90223 90445   30151 17237   69327 21051   23480 21839
1426: 81393 62859   16345 10311   47688 84352   13288 04022   10182 48918      48812 14580   63942 07963   29912 17743   74667 43436   56214 02407
1427: 04936 44344   80345 49110   04765 81033   71539 10881   60110 55923      95568 18762   88168 99446   43668 22267   57604 75703   36880 45285
1428: 44523 95314   91910 10517   99235 70141   46077 41085   58813 67748      88179 81509   87529 97504   58804 88466   59727 19120   87553 76282
1429: 33438 39270   45103 04506   52300 51547   30330 35923   56258 43643      26448 11772   92491 69069   10274 10730   02700 49368   04245 41954
1430: 96922 81500   47452 80388   59045 50778   67236 56481   89980 94229      33269 34840   77283 96881   18547 61227   33173 02133   49533 09579
1431: 55985 65356   27320 86275   04999 92699   25276 51780   50043 70955      48838 24986   15804 97035   91620 01918   01365 88033   64638 84594
1432: 69388 94472   45036 46461   04705 28254   54563 49101   15846 66674      45959 05061   45563 18843   05697 71105   17216 55798   77229 02559
1433: 78631 55471   03285 69632   71660 78297   67625 56145   36318 49199      60260 15347   37050 82169   75747 95167   47585 29980   83169 12906
1434: 25366 58873   51933 30982   10262 99497   99479 18099   78505 13932      47582 50963   38792 27356   22615 54522   78397 09617   18739 36079
1435: 34583 68189   98615 77461   51301 65177   35112 13756   29655 07420      76803 06119   45735 68905   55831 71094   30130 75973   77324 38189
1436: 77539 86322   57967 66809   61047 59844   52247 04240   92867 30258      19988 15907   93543 63996   17344 26184   96438 30438   27125 93797
1437: 78476 44775   98973 18099   70581 45900   54221 02813   31255 43254      06605 62487   26155 33295   31075 03133   98121 37947   16088 39166
1438: 37113 01792   93495 35702   97754 77991   24520 03922   80458 73528      38278 85805   57910 63393   92494 40176   31110 47221   48191 92856
1439: 29066 94202   63300 60480   03683 70381   32588 60949   67696 72317      31529 35754   13397 21624   84732 15955   07816 10329   95709 10584
1440: 56683 28662   53476 40003   87115 67564   26338 64168   90608 58600      65252 35754   99440 81649   88117 29756   56841 18728   45504 47465
1441: 42330 79086   17372 31793   24810 26977   15611 16636   42108 52386      02976 47364   36547 26760   42982 73939   20918 38899   84310 60851
1442: 05249 14920   82719 26172   47743 54339   28640 13401   18504 76845      19576 67770   12753 81554   82785 97014   47553 02394   44577 09749
1443: 28584 99724   35243 17356   68804 27788   23252 78210   73039 00784      91507 51586   60558 18116   58920 17553   27045 65852   92998 13882
1444: 41922 67723   41663 12448   76110 35353   62832 17386   52903 56747      91243 62079   12895 31527   03018 98779   47422 05919   28497 75608
1445: 38126 63969   41438 48375   82957 87004   82710 63802   48670 11461      36333 81687   57994 55373   70529 27855   46948 54913   29945 51617
1446: 95344 80101   21882 45881   96078 45944   91393 01557   44927 34897      13520 74327   56659 23133   15060 43999   19890 45093   29254 88561
1447: 72614 76918   48853 54114   70651 21161   12940 65595   82317 00470      38534 21846   30436 36865   86693 46224   11985 44490   45645 68951
1448: 97901 43523   28784 66702   94763 89294   61827 00475   35482 68393      45555 74682   04021 43371   11438 09985   37712 95017   43326 17094
1449: 57313 22541   19836 24692   43129 48665   88064 21822   03241 85088      85710 13586   99798 81272   53064 86325   78609 04163   98688 88826
```

```
1450:  65723 20347   47816 97472   62091 90711   65127 88107   14217 94674     45027 32398   36870 17986   87445 50571   50069 09211   86187 00794
1451:  95200 33335   97585 48105   28444 60532   04424 83631   14684 68953     30857 15247   26139 78679   95576 92340   66386 31351   01754 32709
1452:  50844 67454   62803 86516   51585 35778   25544 38538   52354 49115     41238 54305   50714 31979   07911 71775   74916 08063   72351 46253
1453:  90858 22948   51280 65744   10096 10930   53708 33177   98478 19576     62439 40100   99608 29617   56794 03538   95648 55729   34685 77336
1454:  46699 93566   71266 90973   03746 09886   23819 72527   46327 73922     91023 39970   82696 24754   08440 36493   66541 72781   99432 66797
1455:  35438 72483   49919 49208   37806 02223   28352 25933   26223 88510     80374 10383   47450 77361   81668 76430   13043 96660   95942 48084
1456:  81131 38246   87497 38985   76349 40082   75522 61474   88472 98007     06168 97145   81872 30095   66277 33847   22662 56973   47006 65508
1457:  21158 59411   87569 55576   43991 74135   73509 56057   88923 16226     74191 62278   34247 50179   17488 88252   82691 27356   54088 25807
1458:  67564 18509   88157 70865   83366 95942   16608 92439   60593 08888     54916 42879   49235 35060   65046 06125   55372 28131   94509 41340
1459:  18954 22010   33133 17717   73966 80921   16225 32278   45776 84693     96347 88120   18081 45412   67225 35166   79722 85154   72054 32518
1460:  69800 31004   32997 92322   68168 55883   03052 66690   54231 92929     94280 07273   32799 27743   88188 04582   40542 21346   64916 99019
1461:  38103 50148   92164 69290   88001 12396   30530 52935   84152 12738     16515 64157   71696 21601   86999 48128   68755 76725   48545 89646
1462:  68025 76972   48443 59213   15913 69949   62875 11218   57738 38073     16284 91151   40717 66814   49179 43866   69735 46148   85524 88340
1463:  47046 76665   03092 07683   51467 70824   89155 99875   74893 27303     76361 19192   55903 13661   00809 08296   53424 38689   41107 49437
1464:  58406 69906   74814 13393   48813 69502   33406 66808   69677 07226     96055 73174   47409 65235   44124 59195   06815 78872   04903 10608
1465:  94996 90293   10894 61045   73701 93548   13246 45547   29284 98503     00144 78359   15659 49285   73943 44568   32142 07098   55683 27618
1466:  23719 22650   77564 29409   14325 42870   74476 87574   20865 39843     36652 70704   58603 33488   39076 47172   21229 45908   42630 51057
1467:  78699 73943   04172 64562   04033 29638   70164 88032   37342 62641     57876 87834   40961 46200   75999 52150   05618 16005   94908 41464
1468:  93259 54108   28019 64883   27940 18235   95044 37937   82847 59857     44860 19197   67111 53965   28899 66851   58174 23332   23583 28427
1469:  12478 59930   96027 55345   57443 93374   56024 25450   96343 82954     83629 80330   65508 16001   61234 29563   39000 21545   45089 49452
1470:  42209 50711   80900 07896   75236 33575   79198 62285   46481 77045     25503 05504   54543 81525   26731 19656   45633 57256   93019 32470
1471:  00268 91303   48410 08630   16385 61852   70837 13154   78225 33981     59140 36721   87354 97243   07214 30504   36935 22040   13234 87665
1472:  12573 64235   43433 33039   54146 15643   42738 50930   39525 79967     48497 23446   15083 20704   27634 42951   44733 70610   54357 55221
1473:  06165 37962   67742 17563   31744 91006   77571 63719   21461 08035     75280 80144   34209 10572   29788 52442   54545 46445   53777 21451
1474:  06165 46605   90942 51459   57569 16950   52883 83716   93757 52042     58414 72310   42841 83359   43644 51523   35309 59890   09936 56798
1475:  93021 59475   23049 92836   80707 33619   35406 94652   08678 05306     62321 86225   37200 93954   58475 55091   79170 10692   63793 32250
1476:  01264 97420   09127 71942   66238 24844   80683 15602   20888 15386     64633 45069   75860 14591   23873 65256   41698 89461   60952 76520
1477:  50976 13785   30977 50469   04851 12974   70252 22751   42731 90257     38318 60377   47172 06785   31564 04403   46278 82814   09020 30859
1478:  95500 66223   39473 25587   29309 30645   43520 14021   86813 37526     88227 85691   03889 57885   76991 88430   17546 33343   44173 09159
1479:  54922 52266   75833 92327   57551 96014   21630 50051   07650 54685     57402 40131   13842 91319   89080 02656   61859 96744   43308 73983
1480:  05342 22648   75112 02432   57843 97435   23849 07560   98171 04369     41677 99797   15922 98160   55002 89087   85895 18903   23827 06004
1481:  57947 12405   83908 51730   40411 64066   30846 95083   24516 84173     37447 84539   66970 89860   23829 90913   97580 57408   42587 73245
1482:  92367 32013   10110 54140   36524 79716   30464 44173   72070 62431     68378 90275   30623 92991   66010 82535   76362 16494   93701 13914
1483:  08187 48621   90528 77098   05014 60221   90638 82025   88711 19267     69275 08039   98182 86925   05819 94239   01204 42400   42735 57848
1484:  41754 27364   07321 40138   23638 25673   20058 42062   03952 69053     51152 85661   52565 82465   39919 98706   50255 82406   80000 94733
1485:  78625 08543   66520 05463   65197 74534   86000 31760   90873 56385     90490 73652   85940 61943   30074 45232   32901 99178   98650 30030
1486:  77908 84182   05425 22150   30693 67899   37881 89289   41619 77289     43905 99365   50572 57738   89417 82438   92037 84795   27412 83771
1487:  69491 28893   70921 26247   57996 62798   18332 61049   94805 45383     48874 30763   54879 12507   44855 29073   16718 44397   07815 37424
1488:  56188 49177   54454 13568   05566 54300   66597 15042   54275 28927     25946 31752   89757 48402   47924 51789   24691 53828   45095 86696
1489:  48072 43543   09668 67490   34115 57357   87164 58163   26355 09309     11969 30277   77749 23090   50049 82609   82596 68656   68765 44197
1490:  48219 30861   35897 92296   07870 80512   89074 51866   13091 78851     42840 85910   55463 96167   17640 19119   61951 25664   36473 07165
1491:  16721 96714   72444 01438   29286 75893   14844 37536   23453 31870     66650 33624   79944 48005   96448 44123   61182 98290   77777 78808
1492:  14710 89366   81951 62669   56211 00902   02117 80020   44793 06562     47254 52240   81152 50742   87988 64235   70407 68284   49999 69713
1493:  80446 74779   37812 97154   15912 43999   46706 18391   57218 20636     64210 16847   36867 05161   28311 16419   81143 53901   46642 69667
1494:  57597 01476   76881 87323   54024 15346   39811 74231   39179 55067     16832 01470   64008 99525   66824 41268   58320 34832   44764 44316
1495:  63293 80382   14268 55388   59364 89328   72393 29585   82743 20242     27313 36216   01505 74779   06099 10496   39813 75801   59024 82882
1496:  57095 35973   37770 39243   07164 98528   66008 61838   01129 25673     92549 19156   41421 44083   69955 15281   52362 29305   80974 51595
1497:  11246 41471   12020 88058   31152 98686   69438 83063   19848 26576     42010 74396   96568 86393   30321 73714   24615 10669   21535 04687
1498:  28877 50638   24897 33219   26343 47996   83879 47486   28549 98574     94613 11374   26498 37090   16936 52046   74348 64899   18085 48011
1499:  10275 25223   97711 80963   36134 33813   62189 98811   47996 45426     45579 30983   35997 71675   23000 72336   53915 36569   86553 21128
```

```
1500: 22935 95671  74983 54510  89247 26222  64805 56155  58476 94116    37411 47347  00531 97382  69962 90294  16055 85958  51827 73547
1501: 90081 12982  44245 69153  81039 81895  13489 54388  28808 04044    79478 50900  72290 62092  49983 05204  33621 05969  29512 41725
1502: 05312 58043  58267 69371  94539 35476  20843 11215  81984 06321    37875 46792  14418 33544  04974 14882  93605 13400  49829 91250
1503: 75382 68534  69051 79775  99766 58580  97670 18316  68905 93321    45710 34375  75571 18268  65335 74165  00367 41870  03880 35653
1504: 45632 67397  58236 52509  59545 47612  76057 84625  80459 22340    95954 24671  91501 64895  47773 02237  94646 48549  42606 76262
1505: 48913 53147  99052 89353  62596 57957  44513 94730  21596 56134    25860 69555  66300 26848  02355 17184  43898 59571  13000 53250
1506: 50788 88206  32710 26703  75033 32209  44430 62014  32221 57300    00985 61043  72869 47505  18382 78968  99716 89011  93069 85598
1507: 70921 73198  62319 44416  29812 89479  15029 25184  86340 20674    21779 93070  76426 37053  54682 52148  43191 24716  70086 14449
1508: 23485 83033  75291 35146  81102 88701  65299 67434  07912 94632    87350 56668  29274 76271  59448 45220  70647 21622  87428 88274
1509: 69885 43050  28521 40005  87917 67458  03465 15539  97476 18461    76939 42053  10233 37359  35356 17189  91325 38146  32133 70875
1510: 07344 63750  70500 50179  56639 68860  98972 39239  68800 73157    65032 51049  52145 41078  05448 04685  97123 05953  12938 96482
1511: 27088 77918  93272 44095  60674 75793  74621 08913  60149 42057    20375 32077  86325 35610  06521 72185  99310 26609  13812 96631
1512: 46620 81472  45965 34996  60182 40289  15816 90474  63751 06263    53881 45489  25655 27267  02112 45840  08303 06898  08147 77685
1513: 88189 40084  23951 45182  84442 18385  63417 38811  35220 71942    36384 10425  78790 39417  57694 74776  54704 68943  28164 28427
1514: 61593 21275  59673 88261  30508 19561  70353 21635  22788 75751    43910 52442  09842 42479  95652 57162  69395 89545  49087 61111
1515: 97154 83107  03725 99964  46604 87445  45568 10759  67846 27579    74333 08138  25940 37142  71017 89868  98886 24225  57101 99051
1516: 15792 73305  26175 49497  57877 10096  36574 67683  91912 09192    85259 57737  20054 46556  58383 74900  61074 26977  21109 19866
1517: 79838 07182  25537 65532  65442 47733  19729 65836  54308 90056    09787 51595  20040 45122  63190 94612  17722 90072  00877 77261
1518: 58888 79683  07282 47547  52608 19037  77295 85415  66117 94021    97763 44624  24581 80006  74806 58357  98392 89058  35233 82317
1519: 75419 86038  45000 34754  82463 11890  69675 41017  47350 35164    68456 24216  89979 05364  38332 42909  48070 50513  80240 03728
1520: 30532 01040  74861 68418  55985 08344  05922 33324  14856 46483    95497 20323  40278 79220  07437 76989  94755 73696  75189 18923
1521: 40530 31888  80367 75106  93511 62790  04613 64690  32007 47569    50744 94827  51569 38674  55260 64383  09352 42898  99413 63509
1522: 11795 84466  91935 27835  91171 81334  66220 34791  76644 33426    26674 23491  46158 19908  01759 68725  88838 70796  62038 68650
1523: 75900 72917  40802 07179  13018 86060  91747 19011  20883 16773    52285 82358  82832 40029  78583 47374  08333 97030  25285 80585
1524: 23838 06187  08282 94006  39130 94772  26650 30390  93496 04699    14205 25207  42739 16541  62348 74047  45774 31364  34464 82292
1525: 59018 08444  83354 39789  75522 53509  34014 70007  58897 95109    10596 26117  56516 04355  50378 57630  37384 88946  84187 81801
1526: 43990 26510  01993 56319  94956 11977  31501 10865  78571 67513    71538 94222  06865 15239  44035 21068  72227 35506  74093 16136
1527: 29395 85418  71672 36984  65626 01468  04225 61481  95161 17178    19649 13300  91957 82308  38863 03867  80443 54700  06833 04436
1528: 46814 63328  35153 88738  97630 82041  83934 83167  97793 24399    36117 26876  66324 76037  47408 28462  62488 11644  19990 19314
1529: 08250 10030  10642 70732  72354 49093  20338 17536  73113 86945    65048 14511  90729 01869  19788 80491  80198 96628  21702 97867
1530: 76657 08543  18824 99439  50518 38542  34905 58037  02853 48293    92229 65638  59978 33002  23253 29145  73624 02116  35596 54137
1531: 68750 64079  20528 15174  85498 17140  42310 36597  00911 38746    72815 10170  96815 73386  16370 79692  07423 33287  97159 11286
1532: 86290 28711  97838 79621  28737 70367  84511 82100  76146 54717    79410 04105  55711 00906  99901 94650  05796 13476  30433 69728
1533: 65187 79527  76417 70914  55827 91712  53582 38102  60625 84959    74662 77352  79250 69604  41676 25295  06529 62101  61134 62871
1534: 15069 23053  37393 16451  45000 80039  17528 27435  96873 59809    60989 60389  33279 29042  87962 32739  18365 12606  69246 83969
1535: 28624 46030  65922 99183  05783 06326  79141 81716  80849 97615    74631 07028  72409 26796  55048 82042  18013 98513  20222 62419
1536: 18890 81983  38986 96286  27098 32781  55087 45839  87365 18796    74685 76846  70543 66217  20805 00941  14561 20217  22705 68750
1537: 79950 26830  83890 08221  72521 66401  06436 89462  11200 78123    02132 80992  49613 61797  24561 94162  05139 77268  37448 06394
1538: 92360 29845  39554 53373  80746 33598  73130 09971  00059 49937    26574 80716  90298 95115  40639 21395  78307 76853  12111 52864
1539: 16927 49177  29701 98347  77472 81824  22552 42339  08988 06156    26887 97159  13024 34569  40058 12360  25423 28959  47282 24970
1540: 23701 22815  81772 28701  40131 62560  03520 40681  00906 91572    40963 73452  23546 14692  98862 12610  81957 54089  07334 43554
1541: 08194 65670  35238 11264  21545 49942  77640 81662  64580 06405    34468 22822  54622 12668  29645 30481  31919 59732  62137 21463
1542: 00916 82458  61999 30795  77559 55265  93842 59554  57628 85328    58936 75884  40677 32550  35904 67913  91335 84225  54809 24441
1543: 91996 99846  87986 78210  87975 06496  88404 85325  90183 87998    38698 42643  98186 65945  30762 70366  55537 03688  07714 67876
1544: 06833 95164  25798 23351  16317 57346  90874 14128  46431 47743    99786 69750  98723 90333  16308 13110  05799 06917  49138 76598
1545: 62523 59807  71876 06104  59713 50624  91993 24252  22694 72298    31034 26085  21363 35217  96360 54546  84521 43729  78619 16429
1546: 80335 30448  15295 53737  48657 25527  36964 09628  47238 00932    85581 33473  67794 61639  83765 14933  31285 96511  07181 63762
1547: 39599 48839  26512 70529  56828 21395  18986 29021  28115 15885    54155 25294  26886 76448  93659 58331  64450 29688  52575 47808
1548: 05344 24696  03175 81682  68794 14138  30188 31810  96082 70535    25851 78801  91803 90132  03785 38305  44485 71168  82056 66221
1549: 38718 09285  23820 14971  31662 27159  45351 21949  71853 74303    46486 81528  19038 85111  86458 36226  68560 14737  32725 78826
```

```
1550: 7820433996 1370560927 4417564096 5967735701 0727051625   6006823212 3865258206 8425630667 2099186959 9901504964
1551: 9867886186 7082635618 0329959433 6437056649 3952012677   0201507221 9129318724 7605414333 8138367505 7940603450
1552: 3867330473 9323369449 5783690279 7195561284 3002330218   5566458174 0134407973 5391063346 8895414964 1023241578
1553: 0202794452 0665543242 9821435782 7579724147 6734105025   5450022818 3414999639 2035162025 2073969053 6251015780
1554: 6046864786 6818025038 7284939143 8445094458 9803447448   4905158788 4054748050 0427113100 8734038770 9131439157
1555: 2161573586 2703929984 9280232248 1510276223 9168781854   2464330259 3576712778 1859076398 1077873214 1761738749
1556: 6720507180 6638733529 2248744537 9758485986 2510589007   4034033881 7016821244 2698087952 4096459144 8942606671
1557: 6622485032 5140673879 7506157843 3412004479 4176886952   2167139350 6329887415 4722154530 5897665781 4487642654
1558: 6677892583 3225663812 7177331636 3977566668 3855248224   7661991791 5088733379 2983964671 1672459285 4775415444
1559: 1146134041 6687946360 8062314730 0845116259 3172494862   4280381276 7135313124 4789822217 5184503843 2100865437
1560: 9315231160 9873584833 7744377589 1000100350 1521361904   9626465633 2863319670 0386750177 0314691044 9806259125
1561: 4505046585 3611558073 8315888777 5553170226 4128226712   4671841833 5743309312 8973792153 1161891142 1469358719
1562: 0047052436 4163873682 6049628619 1035085492 0881343721   9523540317 9594542180 0661202397 4854308370 9625382197
1563: 9050343284 0119608788 3709990677 0646720235 6870831324   0111610162 7949265984 5676867235 5089039996 7175386068
1564: 0864355852 8788592262 2381014346 7230822810 9134838569   4329278704 2091714205 5165414534 7393407640 4852575183
1565: 4246657929 8010131031 5854797800 7630218352 0824411091   8293037826 4288108766 6104501682 6594927371 5103753439
1566: 1442098372 6984764662 5068719922 9148395622 2468075984   5939856740 6660833257 2643198912 3385660574 6287129830
1567: 8701584173 0732374473 8364375422 2652558620 0218933641   5296596869 6837310347 7467630019 9779820520 9710193766
1568: 4633073731 0251008561 8699701085 4464583063 6092137600   3379210699 6523014575 5828623180 4357763202 7485135430
1569: 4787543147 1870798133 6365767127 2052907734 2283376291   6366696926 3101131640 2939218172 7143822961 4395987446
1570: 2932612215 2877473733 4101749465 9259556455 2029809547   5170610057 4444807773 4318866850 5002198774 5150743555
1571: 1880829900 5846975419 4681646182 8809195528 2186110338   8049870908 3318736031 2754713091 2678050586 1770917822
1572: 3214472435 6763241920 6694317791 9534548047 4931932079   0237742625 6816120471 3542179449 4286849298 4887462567
1573: 5865012511 3577655654 9196078740 9625176698 2963325749   3537246471 3184473249 8784950949 4315916341 1633490570
1574: 5962255410 8076384068 8727523259 3495176396 0879326141   8512579167 1236308271 0455708077 5468962796 2214541540
1575: 8251173614 8433331068 2153894020 5351821175 1422791034   6365020847 1436519215 5938601739 1618461136 1180770544
1576: 4491774953 9240318270 5657365214 4426910128 1209923086   4803272349 7457549599 5866460881 9468233858 4147333104
1577: 6487654011 7839482713 3870560294 6568670149 0080805426   0149925254 7530622626 6413708193 3164335016 5886049438
1578: 6708589482 6477798951 3373770366 3818704465 5229656215   8858520394 5484775002 8614859407 7781354772 1490237203
1579: 9805786912 4378870768 0523849490 8120704489 4356396113   9961960985 9237974202 1758327764 9611817455 7719519657
1580: 3738786819 2058661351 6864881149 6643711760 1985237752   4764485752 7126084339 8245482939 3510652864 4010353114
1581: 9832844401 7712227774 7263943123 6267391207 1794174048   3756587391 8588919131 2233932362 3631163404 3882751145
1582: 8342927157 1531978167 7144118046 1131347122 5077236723   5419832598 2070839245 9384890063 5259352426 4239560457
1583: 3674400752 5664734581 8051115291 2022315326 7928521041   4311165792 8202180481 4862444348 3435185490 1837418038
1584: 8556044238 6883738113 3133984713 0789819695 2626518607   8809084530 5622616996 5316309747 4061832662 9802178403
1585: 9266786444 2990789936 3131521014 5635101538 0647779527   4959522675 7651693033 1123604776 4082577210 8044037987
1586: 1397991160 5237134654 0649141428 7364438693 2417102630   1361649238 6651681925 1866238251 5239977492 1130689179
1587: 5251374298 4626427767 4317657648 9419499022 3402919709   5632838844 1079351347 1612197779 7837987849 9967324414
1588: 0369913405 6219712491 1051439398 8634463705 6562066786   2867463234 5034385684 3780464127 4851208338 5595192890
1589: 0850792930 9990691907 7565613479 8461413205 4147537170   6934354747 2769192085 1617441069 0700740003 3888563745
1590: 5995426159 9618361682 7375652450 4467220604 3683211868   1586864474 2348997359 6945608457 1095883237 7494664418
1591: 6625106283 0067601977 3294775007 2523486260 0027851719   6808471899 4827362747 8687529674 2814074460 1125964863
1592: 9676633753 9021547828 8634378943 7085011801 9590636465   3747127432 6230743571 1104830303 1089755315 9842230558
1593: 6024784111 7462282747 6125325088 0636122346 1706358569   9257956061 7199074991 2093508161 8403199590 8367344552
1594: 7711575252 1077192411 4458081707 0266391503 8210411258   1648486513 8737377723 5923922843 4267779945 8268848692
1595: 1274700296 8651241873 6083543363 4091830857 6311453852   3806438524 7700641716 9437292479 7365670351 8678073274
1596: 4344351501 2188681275 7940154579 0178843553 5636485926   7511618960 7355983355 2989516735 6382377616 0620270371
1597: 9047017990 3858987084 2574866643 1628952001 9354506609   1760021484 7507873589 3931656867 7142608399 0364443512
1598: 2384132061 8708712879 7221192925 0141667741 7992771440   8065811231 3911986665 8350956768 8414100058 4980426775
1599: 9853121305 1314870096 4403429589 2894365691 8126011999   5715277018 8125621855 0525656870 7538347303 4064939497
```

```
1600: 5725701761  8112258748  0409902344  6445402269  0265050516    3720685701  6911663770  4598600453  4860789892  9350279258
1601: 1474421665  1096215100  0818065827  1734955847  7191636630    0726351668  5442695299  5029707781  0684904716  0275693703
1602: 2844647140  1638906588  2846074297  1327050187  1698578268    9551051677  6392421118  3328708087  5174708068  6500105424
1603: 5374836969  5986574432  0002205391  8411817638  3937646609    4456040934  1213126489  4314971709  7582181464  2609812204
1604: 4663314135  4547260552  9995442722  7318194568  2793309314    6256843107  0478973855  4480231121  7395071311  7493666054
1605: 0306143706  2099976039  6401077046  2275459238  3826945752    0050675112  5419023486  9967154524  8981367065  2547502946
1606: 2933291380  8159047398  6588498831  2047182562  4228118038    1181733176  9369684748  2750934794  0513400221  9457784831
1607: 5688974900  5711303630  4222732865  8865115103  2374352579    2254127408  5593654443  9369436300  4545767081  7099619903
1608: 2985902973  4553970429  2948278245  4232234760  9071598644    3791376141  8198499797  1489207039  9153928425  5993411348
1609: 8163701599  3801777449  6560327777  3697392448  2976751874    5737240531  8711496173  1803831371  1923259406  0911589278
1610: 7379576764  8190935241  8369954695  4763810321  8750273653    1156726525  9322842285  5160286286  4947100873  8796218020
1611: 8938926135  8542209918  7954766260  0241458689  0249814145    1449660581  4504975789  3132372223  4128119372  2275807490
1612: 9612386024  1918508577  0539850632  1254914433  2264941991    5987636283  3019105293  8306054482  8530003589  4576888011
1613: 9330864116  0695683862  2739136367  6925555153  1377045778    3864714814  1979593914  7644127857  1098628547  8556639077
1614: 6785459803  6380852170  6760214630  5000707070  0662195592    4586473520  0712516478  9941265003  3504785189  6725567606
1615: 8108576202  0958751670  9832012214  7599401775  5376381052    8889773761  9589450251  9707722690  7946474822  2708594150
1616: 1556102690  5405927403  2234610314  1146064420  5215004181    5819386953  3795895036  6745412465  0468731059  6248406191
1617: 0923449339  4156762234  2330426004  6039316746  5051533057    0228068490  5706436947  2265900313  8984313401  1178294756
1618: 8076985785  1699653714  8757814212  1307552103  6349081657    8752690007  3202173008  7885098628  4045300532  6696815151
1619: 5362958909  3143490966  0875902177  7816492623  7114906174    8035772290  6291696119  9882678691  3578441290  3391438132
1620: 4443652526  9632003859  8672904870  1138247852  6172239635    5619041193  2592076698  5626557888  6609944544  8894741501
1621: 8831160325  0572160205  3872492165  3284054543  1030350295    3843628399  0143751145  8915209319  1807444165  8711550735
1622: 3696277409  7177223887  2615222974  1256448838  8408268618    5402289142  9187226574  3109830970  6230102333  9458967193
1623: 6947183403  2229584264  9243242060  2154889115  2135258689    0738675524  7144428074  7796565004  7200337622  7497006283
1624: 2757570159  3796045434  4342934571  3841629162  5543777808    0198365380  7957504327  4753659703  0737899318  7505282180
1625: 2211215722  2818439558  3979681579  0844161838  6730247727    6396347147  6059695497  9807165355  0060263524  7236502758
1626: 0045603946  7059166417  7804537843  3903564769  2184521386    6349133588  7285785448  7091446072  6349099977  3313741986
1627: 5791827374  9124318700  2742329046  5056540056  9850035885    8521166558  9440951138  4621016636  5685818202  3517865665
1628: 4123662668  3673446700  9330441604  5669531189  3973344465    5967232657  7091605199  5047027287  0080641138  1032642182
1629: 8384100433  9182036440  2800281202  8992770197  7455457207    4614512480  6061263749  4589885732  9520994305  8905404742
1630: 3229271239  3223828624  3060906054  5242999000  8179194692    2830119852  0202376749  5139091300  5563755829  9292506803
1631: 6243090680  9701847833  3611941851  4541497950  9395392283    7534999256  4693321261  3015627395  1939762928  2327863718
1632: 3886088370  4460010606  4979905089  5411356526  3533611149    4541702864  7001431535  1223323305  4878283970  5547981053
1633: 8030328064  3535010705  6660135970  9002773607  8745074976    2395695940  0622756819  3465348185  5806106596  4820795112
1634: 0463601953  0138584491  6169650861  5457641833  2593969559    1893156271  2551046551  2780883567  7464905862  7225702173
1635: 3123619142  1569805335  8401639783  3559187560  5077227117    9588379445  7590369962  2341092044  2749524840  5041272447
1636: 0808205387  6898717004  9955854439  2964169006  4438146341    9692999775  9441926240  6766776227  0474396120  8903992249
1637: 5802067519  4312160484  9318551196  5247412592  8895408650    8759141447  9525398952  1413000863  5182827355  4025396092
1638: 3197241373  5381222212  3592535897  0298081217  0141680667    8788149723  9343323558  5051073320  2824300504  2592491068
1639: 8926756885  0022155087  2852609651  9886342316  6643407634    5704736396  8317020754  7753864972  7956663230  7340839824
1640: 4607739703  4133197769  6655933425  8187980900  0112727535    6474644059  9041112387  8626104921  2083612611  8956276878
1641: 1767229497  8757873200  8519087458  0198327723  4005440127    5400824368  0634440784  6354320416  0034193308  8911091003
1642: 7171115106  0738188144  1705363097  4134668730  4526972555    7267527774  6044252244  0998609294  3531305327  1350506489
1643: 5657975531  7235206042  8912320219  9007989575  5514693520    4735370548  2386994994  6399093486  4747326151  3287188112
1644: 4561292444  1671974593  9800276571  1370060803  7339788819    1559395530  2876427523  0278948691  3282061125  4867098974
1645: 4433984889  2218260287  0446700135  6686780607  8348903695    4191738906  7137574244  9806807920  7520031433  1460561167
1646: 0392596825  8984865601  7476851471  3360349581  2279260802    1107695075  1944380572  1159235015  5607139544  5538078784
1647: 1900912433  5642345776  1495829002  3080773570  7479807976    1080794519  3475113161  0307567197  5731273703  9853650296
1648: 7768403955  2061487472  5225887061  9712549500  9005775472    3476977657  5473486184  9918899077  7583216132  4706369462
1649: 7225987572  0053186187  9874907496  4080421065  4898230590    8547064479  1456158447  7568567930  8777246544  6749899204
```

```
1650: 48650 27318  46315 22590  66967 14932  01542 23186  82614 30267    68898 18424  85655 07910  94853 08770  60031 42509  77140 88593
1651: 78315 19264  81329 68224  91393 17319  05977 67000  34521 95347    26887 18868  06021 71704  74585 72677  33332 28994  54453 49588
1652: 49562 14992  57966 23921  80385 44331  46492 95856  98781 53196    82491 09554  23529 08435  61584 88678  80212 63163  12818 82740
1653: 92064 71793  80158 04838  04649 45934  96974 76303  06698 50711    27904 33677  96855 54612  09391 65660  58815 63629  61750 41485
1654: 00408 85157  20992 16300  36285 50427  63587 99994  72688 26610    09275 09652  71126 73639  19888 90148  65032 94467  55362 43362
1655: 96751 60887  38793 65364  09011 95100  48532 60256  53239 99382    06871 42228  84982 22849  42165 97608  17540 00178  99357 12163
1656: 43715 97688  09331 17512  57779 73752  66834 58495  57899 95241    54054 80895  50835 99117  47065 29236  50520 47679  04686 80987
1657: 79957 21588  13565 95481  63736 96563  24600 57619  31669 55720    60772 62336  39475 42131  98727 40543  12459 94140  85152 23046
1658: 73469 48083  40558 43661  16633 43771  16742 99382  32707 04897    21482 16236  99313 01845  75760 68865  70631 48255  39457 10848
1659: 23701 63704  61889 57559  25196 83103  61957 20588  02400 47553    03568 35241  53413 01940  19855 88873  68065 75126  89625 47756
1660: 83155 08452  22819 85938  80883 79435  21325 60564  13598 82438    46778 03685  83056 66592  04403 42958  88549 00268  60542 06363
1661: 44738 14496  55817 44677  20388 43573  80289 83905  74657 02272    09397 84819  34268 62249  79978 36674  10846 12951  03466 28141
1662: 89578 04493  44457 80232  81894 34876  30110 23558  87508 97263    84185 73051  75677 63257  83940 63638  27290 23526  29456 82782
1663: 68495 56740  26358 50723  82132 54401  00777 83962  84304 24582    40180 63301  33579 63652  01899 14977  64540 46672  67186 17250
1664: 67032 24763  19330 03566  29568 05546  85738 46465  00275 89764    74186 71584  10981 06577  89647 71980  17392 56457  42279 29365
1665: 00114 70831  30781 79483  18636 30045  08689 52669  23762 87766    28066 83028  43144 74056  89334 14833  46996 08406  39018 60995
1666: 54464 43021  43473 56534  79732 43564  13237 53331  05492 71725    61446 71454  49068 57249  72214 77638  70838 21627  16323 02272
1667: 19316 34191  34657 39774  32976 77021  34294 45386  99390 15672    89630 73257  94775 79732  22613 08203  44397 43929  53160 95460
1668: 16006 93794  10879 15056  32591 92875  86391 73543  39103 45673    33780 54632  09141 64539  61295 74833  60275 58325  69883 35167
1669: 98349 98658  28651 33405  76016 52398  22366 76617  91424 48961    96791 84806  14876 27257  02032 64501  68801 67584  17274 07866
1670: 44631 70639  07204 24490  28453 06996  80675 33730  30816 61211    29010 73883  23419 96973  43731 12800  48910 80001  59523 64441
1671: 46479 02512  52449 76924  14505 62426  21917 67709  03044 63076    78134 60277  53577 50640  57894 05100  34557 25412  51408 09353
1672: 12091 46508  86948 63806  66748 48386  16213 22094  35510 95522    71111 30088  11668 74080  46984 28777  58365 07702  12154 00603
1673: 77452 75700  62182 87198  12709 37480  84573 03108  93415 11905    56732 27417  05200 79474  36961 59239  97176 07919  01273 08282
1674: 90787 90983  92548 92347  13556 32304  06173 92074  59254 17992    58964 56966  80339 48932  41902 07301  15073 19947  21529 28626
1675: 14626 21890  10206 43586  17491 76723  35546 26166  90784 31268    38792 39218  71698 89563  39413 26353  81848 33657  03117 36852
1676: 31775 69005  78998 44674  51415 31792  89953 39596  53616 70842    90030 10758  29553 63482  94183 84694  78463 68443  54760 06223
1677: 80151 18008  10557 10228  99333 06255  59647 77723  98334 78535    52100 92481  10239 10216  75809 37098  69439 89340  45929 90067
1678: 71311 93275  75038 07657  16699 46441  07612 87447  02745 01142    63760 21955  25167 42068  57974 36004  73903 20141  68273 34260
1679: 70883 80397  89084 82304  03815 88101  09341 08433  16064 87667    05757 52410  61307 59383  33608 49369  69405 31487  46025 95389
1680: 00405 69753  97863 20858  60518 29804  72454 31462  46992 67095    43394 49419  02892 24703  08570 75897  00772 64502  01703 87291
1681: 92664 99024  80443 59449  66902 75460  11778 71687  26143 83461    45352 71769  88373 31255  83182 49020  38112 87549  36981 04460
1682: 17707 06287  68612 43538  19062 78418  20636 58086  79269 06374    93770 75712  14451 07922  63233 41049  05485 15715  03470 24757
1683: 07534 28481  08692 41075  15963 26775  00280 35881  99995 63232    71633 29825  88940 57129  28443 80439  13950 54779  29987 77270
1684: 22003 67611  63079 67757  54580 89164  11933 42530  47884 16143    77901 97762  71413 82958  30199 39392  02149 40804  85875 39388
1685: 43819 19726  04875 26935  82282 44091  90102 14681  91407 80108    62748 47532  32560 33408  48450 25262  93748 46176  18157 47100
1686: 83604 60128  89303 76443  83086 93257  33257 80527  02841 96949    37012 15489  96641 15293  54015 02003  93531 16893  00492 15646
1687: 34268 39453  52418 82663  99893 11259  91927 78854  71270 60906    07136 15668  38089 88196  04241 27721  41431 01189  98905 07236
1688: 76490 00880  85491 40873  95718 06957  25068 02861  00690 18391    46620 38341  80331 35184  65638 26507  80732 31602  70887 27334
1689: 40201 74340  14287 87151  63229 60056  28309 57616  49000 65219    50039 46156  82879 00201  05285 51593  88567 85192  73309 13321
1690: 31698 00003  60190 20311  07077 90553  88257 26309  91885 18872    71156 34914  05709 18640  91629 16097  29955 73635  97066 57584
1691: 38886 29903  21740 15435  93807 31970  82440 17700  44392 45541    83286 81682  98706 01036  40853 43791  62687 98343  25025 68744
1692: 07111 48744  19283 31162  83388 51227  93890 53764  14556 81381    95850 87043  68347 56105  32583 15109  38522 57982  77001 39195
1693: 78789 58881  70591 80997  33389 22672  18137 72419  23486 26989    55914 27936  41187 50984  51432 16606  35996 76612  65199 54002
1694: 79528 52660  57142 16677  70422 28203  01045 18531  27940 96797    44732 63649  62311 38872  25599 65373  04495 41381  27568 16005
1695: 85190 21361  94027 85282  64553 84835  00805 04354  85553 59467    97604 51765  21299 67797  72997 98137  82069 95244  91402 37094
1696: 23823 10681  51059 80427  26461 08533  59738 38453  32956 38255    28755 64353  80285 44458  44258 57063  41787 46345  60217 55073
1697: 94879 21812  42384 11353  72194 47068  96596 17746  39434 76834    05052 49367  83454 29372  70738 72141  98921 49938  01457 37879
1698: 92107 64922  16802 68666  16414 80256  34763 21514  46893 18428    53753 81895  60410 28238  08015 92542  20355 19390  39210 47648
1699: 69316 83812  03048 27058  17459 66560  00574 93709  09225 63937    03233 00017  43046 40139  07745 79526  48127 37123  24659 45463
```

```
1700: 2563176985  88976 27466  67340 92251  28510 42678  80397 87771    84614 26123  55297 15494  68315 74061  94948 13299  43564 57019
1701: 1039381451  17479 83531  70286 97576  44424 59163  72759 44635    23511 99006  33537 82963  93068 20722  25414 87369  31274 95938
1702: 64979 16465  83778 48046  55462 89143  55898 92292  05772 32391    91419 53221  15939 40763  04991 21421  24184 90697  06660 42033
1703: 50360 77798  07280 36099  57017 06298  75723 09148  24078 34508    56773 49064  72057 04442  32690 00910  14718 64119  47999 06253
1704: 76648 15391  51340 92251  67994 74602  12807 55880  07762 85935    31975 58701  06019 13653  55763 39879  80422 27739  57155 12468
1705: 67807 62846  46204 66145  82710 61211  20072 62243  76257 24411    23029 18439  63030 98586  93344 60076  00187 37182  40263 80391
1706: 63402 38361  59203 01700  67195 56180  60851 84689  06765 93741    21307 98259  22208 55256  25470 03051  45433 73689  41856 33522
1707: 40737 24838  62724 83197  35173 14188  23286 76163  59633 98075    38434 24963  71471 26920  46095 97813  32880 50732  09830 04469
1708: 34207 35914  15888 97203  27068 01468  31205 34882  16192 90616    02319 93411  76614 21206  26868 40264  62406 50805  69138 60763
1709: 37236 81843  04502 89985  87136 34021  95695 68217  92004 89843    33633 36982  19614 66747  88626 76735  67986 15826  43218 57550
1710: 25833 65354  51738 36965  66226 74532  68844 17968  02216 68949    29282 67682  65475 23181  63369 26156  47911 42773  57081 60961
1711: 42156 21877  11519 71738  82464 56212  25208 10199  44288 62663    32928 88172  07037 79442  65872 31585  73499 26672  82532 63749
1712: 92312 73899  66242 43359  72126 36270  21037 79179  52182 37037    04858 86857  63553 63069  14794 33910  17527 56534  88862 70792
1713: 83565 10799  09251 53897  98061 03695  70740 19941  24010 90078    24863 19552  60670 48529  58347 39902  05975 07761  21370 76638
1714: 23939 92704  86984 58402  77925 32000  36225 54112  89657 41334    77620 77126  33965 42133  73247 85962  45886 03947  63104 64288
1715: 82839 83275  09607 38791  77986 17604  12239 34954  12712 28765    45728 99476  55467 63695  05632 14969  67190 10423  90450 50042
1716: 35433 32425  20494 30300  20449 57184  22302 98671  98685 19486    66835 56168  40066 86470  96186 13991  66203 03931  49998 30953
1717: 86713 48777  34254 49265  92489 65221  28408 69076  51083 32202    53211 75804  45579 82434  43076 68065  10976 38192  26875 30405
1718: 13900 61352  45822 26786  52472 49147  69472 15247  35976 70537    83346 45445  75418 96932  40942 52569  58588 08988  39782 64689
1719: 27729 12704  13656 42465  64318 01377  74412 49009  42317 94905    35119 24254  06508 26582  75233 29354  03457 54244  61681 98017
1720: 73406 01111  78616 73644  35640 82902  50839 54328  94172 78047    15416 17312  08842 90312  87563 31815  43487 71279  15556 68504
1721: 02353 63321  55568 21165  55277 32740  44175 83762  40541 01499    09656 41421  13774 22012  76058 00163  05393 15923  02047 15412
1722: 84529 25738  66919 54915  61220 78487  82203 38772  60691 66346    64530 87390  33653 93614  84402 62183  52389 52214  57572 75470
1723: 17432 62482  28715 94534  89608 46233  44495 85060  76590 29533    64380 25964  75792 58683  35816 53161  78726 62602  60601 69540
1724: 52519 82269  67045 87141  02164 60638  58678 80930  39138 17763    77958 77129  23415 16450  45069 61278  95916 42202  53135 19334
1725: 23085 95947  87470 93177  13366 44862  63152 69561  65920 41021    71937 78853  17612 60310  88898 64739  64977 04361  76343 76054
1726: 91912 31586  68254 28041  13507 71850  04360 28437  51325 83377    31680 66098  42239 22571  64131 57148  01094 72819  55137 55743
1727: 82532 37637  70696 34572  84311 22478  20074 65504  60476 94270    09215 63754  07660 76481  03727 81758  74753 99229  01794 95955
1728: 15587 98054  59270 76465  46731 56943  17299 23444  34133 10796    76636 59599  04224 82446  29993 14031  75465 24849  72528 16949
1729: 86076 35799  87219 75148  31630 88301  29687 31534  43208 90050    78885 88378  09855 01096  42879 81016  56461 57800  42270 01409
1730: 08876 74155  43765 95300  84933 51316  73852 82483  26747 99895    84840 51757  56413 09399  16801 92006  41636 63109  74426 32224
1731: 50398 48466  60938 15801  94684 16698  96497 29141  15148 14885    78660 75821  19675 85546  35413 86226  87281 28736  15594 33679
1732: 74423 12398  19873 59254  06926 24356  28264 55774  50340 23794    86753 36065  36982 25245  28679 67605  33893 20872  27914 85789
1733: 41728 91479  58891 88663  01399 20464  64246 52780  92040 34500    73383 95056  26260 87307  14002 34166  06282 44853  78809 85242
1734: 58050 50204  06931 25034  31457 99179  09818 78455  50207 12121    21883 45241  38513 16284  17135 91695  00499 53765  57204 33101
1735: 60078 31000  06589 07404  68694 82391  99785 18392  47308 88717    05398 87031  44210 11261  85491 39301  35357 02627  03512 58976
1736: 27334 02723  83115 56683  02157 46487  93582 10603  03465 97617    52933 26449  20300 89338  87608 46118  02617 48528  88657 43343
1737: 51720 23938  61590 33676  33436 51110  94768 98323  45090 19586    73812 26682  67868 15859  00820 36978  64433 51633  36657 52752
1738: 14842 93874  32027 9567  09243 41101  75339 54203  55481 27848    74146 51776  24682 64227  49006 81860  00949 22558  22383 70340
1739: 93751 56496  91670 35476  85121 24769  03801 78409  18974 28452    47736 67584  62077 97350  87758 92360  87988 76879  53408 28083
1740: 07123 97884  32447 71235  35697 44414  15275 46971  74415 99943    51716 89199  95076 42251  62071 67410  64330 91711  35114 01697
1741: 02133 63895  39986 32983  18733 99362  36072 48222  63040 73874    89281 63305  12657 93435  44133 53723  89926 39612  29299 60056
1742: 21317 29403  19783 31633  81732 04094  17813 62334  89926 16558    79617 42877  41790 22404  52454 39057  34306 81538  27034 34223
1743: 39737 62663  98300 60266  82010 28100  78486 50659  99333 69424    83771 80367  40493 64270  39186 87255  12636 26886  75529 89307
1744: 53507 86316  85110 69920  97927 49607  10763 77448  83535 02978    07458 04001  07665 71521  97173 11233  31475 62980  98619 17459
1745: 35176 97918  74839 75684  17570 00575  17820 62058  09086 50269    65125 08069  77464 44760  06886 84940  97765 18837  98127 97910
1746: 29730 84564  20219 47126  26411 15546  64609 83006  65995 96834    40947 96951  11899 12427  01269 26543  26678 00826  22234 93291
1747: 04285 53885  63149 62973  65018 73336  84020 60572  02483 39206    10939 63675  55795 82343  19271 15268  26171 47260  98867 14545
1748: 09376 27454  00236 69585  40244 29363  41352 99456  52022 25632    87623 49832  84486 21892  67776 00642  53722 62845  34001 99871
1749: 22898 25242  12776 10049  15604 77091  58978 36978  18475 48870    94048 25898  68548 47878  13337 07620  69422 59832  49307 98325
```

1750:	00815 21459	69045 04962	68175 47163	56810 54282	80936 17849	88745 06082	45786 66931	08419 17034	47633 86035	19652 66078
1751:	13015 63384	47786 50077	83650 68015	85643 19243	75605 25847	03536 30549	33408 15010	09858 95830	38820 32182	84195 08749
1752:	82164 92595	81404 96551	02538 71393	47176 37121	63811 42527	01906 62008	98431 61783	56594 29062	78974 61221	31871 24831
1753:	11563 72078	64805 03448	54168 65753	50591 90567	07718 94269	21133 52559	12445 61361	35732 50001	38655 79152	69235 22294
1754:	83290 30719	22378 42314	13627 53193	45428 98619	68610 75836	78171 82030	44925 72688	15889 53513	87525 66116	02325 53493
1755:	38700 89424	84012 20285	48885 55442	99500 18456	21100 49996	19211 35550	34878 60216	21984 03233	83059 40126	66447 12909
1756:	25611 44996	41558 31045	30961 91606	45103 24964	30537 14844	28014 27903	78627 11260	40622 97179	74706 08419	85591 85337
1757:	96890 27571	37007 00967	31359 80517	19832 27115	53158 53242	59120 49200	75023 03746	75378 79409	51397 79536	73932 33720
1758:	69691 13048	95077 53364	61209 04847	40316 85953	21888 99357	25757 07108	36933 76999	50995 83512	40700 90018	03631 61381
1759:	39869 13257	71778 39459	69785 99429	14669 53724	05880 47067	60277 05605	62718 24621	97050 68597	52109 17670	93036 85036
1760:	49926 20698	03337 79955	50105 87838	04421 97748	53677 12061	60070 20470	42825 56836	70038 86072	74959 31472	22435 36287
1761:	30385 73512	20572 49126	04663 86720	58937 81703	01048 29158	18056 49078	33175 85844	43227 90088	17369 19427	30904 32810
1762:	51879 79106	04195 04781	90494 07470	44617 69564	50855 02849	24451 59897	03831 35728	08034 04609	15779 70904	87154 94638
1763:	24560 80450	00201 77929	34750 25862	33852 91243	57276 74815	93479 60590	36801 11452	92125 00334	57171 75427	25999 49467
1764:	06283 53167	29545 88482	66042 06983	60627 76716	68775 30924	19895 74157	62225 30370	13707 13796	65524 41500	29130 44025
1765:	66662 27211	46032 47707	46476 51511	52725 70434	51623 24553	70590 52085	18463 66234	92274 92847	11637 62744	41066 60907
1766:	91770 32901	54149 78361	77325 63595	03879 07164	21286 09953	87345 87629	89599 77480	81550 44099	41018 47341	26517 49981
1767:	57271 08510	94727 54785	94431 97734	92665 36066	67923 54959	51348 26428	65501 62515	42137 88040	72412 29478	87462 07251
1768:	65510 77943	91570 74273	24419 50012	70747 97011	31019 25586	66938 14350	76911 34665	58024 90038	43117 41059	14564 73351
1769:	10469 38643	13643 30890	49675 30417	00664 45552	68193 53311	54160 24880	83425 12697	70290 02397	42784 31107	48648 12199
1770:	13365 86750	39693 07236	02337 24491	67639 51629	71992 38699	76812 07011	17185 06998	10318 63611	93850 09185	68663 89835
1771:	92394 43920	50464 36964	98328 73812	76430 12051	32211 77051	05341 92257	34613 87253	48834 76970	20300 79792	67987 11078
1772:	70865 95353	75105 27450	64144 77325	98391 29069	35300 15875	70548 30831	37397 04255	40207 23229	19162 11407	59871 78529
1773:	36624 45390	80388 99093	04450 18038	72437 24014	37641 73630	19139 60144	34725 33343	50924 30871	42327 76005	71634 70052
1774:	64888 44394	71499 77367	82821 92638	60012 02678	65792 83355	17112 66452	38078 52012	67028 84482	81836 76870	77318 36700
1775:	53377 38934	27219 28725	26821 70097	42734 80534	66577 97224	36556 65634	50221 69536	95160 93908	57847 30683	18638 45296
1776:	89984 74924	51099 61875	73646 73909	22111 58721	56870 14291	32323 58575	19126 49395	79330 42051	41673 62505	65584 08799
1777:	12923 09641	19369 00942	41512 57334	53359 41128	55902 51939	42245 03109	46231 39700	40621 43399	52376 80493	00752 70560
1778:	59057 26479	87075 36492	60540 30545	95861 67791	73935 58835	49574 59677	22997 27139	14742 01318	32422 22871	95936 81795
1779:	35465 35092	78976 80600	99536 41304	49876 51376	91082 07882	75270 46151	97129 28451	36999 90546	01558 14234	58986 42077
1780:	15142 93858	72283 33321	25039 67223	30479 75203	10866 82553	40180 05975	64226 76869	24278 44352	82145 63478	84349 30041
1781:	88219 40122	55925 46837	95373 71020	63311 21237	42487 18822	48440 01324	83597 82406	60254 92855	22500 52141	09735 24212
1782:	14243 58481	03276 49762	27471 21317	69789 12679	97815 64563	77371 09644	14997 08135	95522 41518	78553 26355	59332 88545
1783:	84556 52122	05066 77259	09448 88405	74715 99516	87511 01416	33326 99996	65473 45057	20817 81780	42523 48965	98324 87432
1784:	02651 67668	69900 95061	83332 68488	70507 00100	95111 68096	45442 87213	88561 09633	81995 42233	21624 91072	04939 45456
1785:	61969 62490	95978 85987	33473 07995	97973 67697	99103 24337	33064 25093	28817 61761	49726 68186	76457 91135	97260 20233
1786:	97582 44916	24106 75261	84303 20095	47760 18051	90000 12753	71756 37957	95586 71608	03523 07352	35882 35779	29418 54851
1787:	94799 49027	11754 27011	67717 21123	95421 61390	11196 07263	68348 54818	68975 79591	98549 85264	74749 30765	73588 71150
1788:	91621 36707	72653 91269	24517 36405	68751 67808	40888 04713	56547 84610	20366 97867	58571 03045	48746 35993	43471 85210
1789:	04041 48886	58006 70305	39258 32719	62211 69813	90706 43353	54449 56600	31865 87227	31930 93443	33922 15635	34675 62400
1790:	64463 58989	05723 91095	13634 46940	60713 78238	08864 44407	84893 91701	67000 96156	50355 24867	10339 37749	98952 61467
1791:	05314 05856	89178 84622	39292 86641	89606 05118	52079 07410	09067 95198	26551 17997	02308 42349	36768 30762	33598 88890
1792:	15712 24929	89761 53852	82414 32463	17929 60811	90678 57466	71774 39288	66988 00167	83077 49814	65778 53628	44183 03011
1793:	24676 56183	04506 43099	04543 15168	45593 80346	13989 97133	18409 85862	11405 21418	32297 74188	51369 05646	79397 30619
1794:	32454 16480	85219 31661	94188 40130	18888 22665	09465 02164	72908 71482	66795 24490	65420 08870	17334 72368	51230 47781
1795:	82893 12675	74426 75193	36816 85660	69397 11939	44999 29968	73845 59400	47667 74908	23307 60021	69137 28394	16759 33600
1796:	18006 63114	01461 13800	38637 82051	29029 52011	44518 27877	58729 29438	79964 49510	93111 33604	65486 63092	63146 57218
1797:	31145 23812	03376 64044	26168 10188	99516 58680	58338 07497	39316 86953	48130 11904	37025 96451	82599 66916	15787 78682
1798:	82674 55479	56293 05126	65735 81192	92071 13324	62534 56142	93042 18027	69777 48730	80823 05267	86649 73782	20799 18267
1799:	00931 55917	32488 03466	22773 05334	67426 80773	33625 42205	69231 80486	09119 18655	19255 63233	73319 20638	51978 30847

```
1800: 73308 94973  50272 96331  62166 78150  64057 27119  71216 04626    83057 77959  27200 49950  88049 02674  61608 85682  66479 87134
1801: 37822 69053  11022 52540  75150 77564  48595 42847  61235 67695    33924 28085  12563 02053  72341 60093  80139 28847  19692 36456
1802: 17274 63220  16449 07207  16300 29317  03465 98571  71893 94450    93839 58165  45519 17842  99189 60822  25165 67116  75305 51580
1803: 07111 81437  22922 42040  29367 94190  15484 67694  01823 89722    66961 27803  17332 73176  40241 80966  62914 59336  49961 09969
1804: 49260 77697  87491 17407  34393 14140  30795 61541  04775 82541    97508 99546  38732 20689  97751 64548  95408 50242  79979 88002
1805: 48888 18014  85209 98857  82737 12438  39391 94264  61478 93555    39956 51068  01741 11607  22108 49190  30740 01039  57516 86069
1806: 42963 26728  30388 99886  40830 10863  20024 37292  07889 08680    88490 04760  54379 24014  78946 73310  41372 02178  00188 11488
1807: 56547 68534  86932 03499  38111 04263  54371 93504  37373 90705    04651 17127  87874 53554  94134 24073  85339 61808  03764 49995
1808: 39118 11832  69499 12962  37932 61392  56897 44842  66579 92106    35060 64060  48342 01534  42628 15515  37847 00282  81488 46692
1809: 58388 53207  89739 19849  80426 44967  26019 68333  70396 90260    85034 13520  54925 44806  37280 20220  77485 37332  16461 52123
1810: 85087 99479  71655 25337  04402 23386  85515 87777  87439 56343    50543 56686  24119 72293  48333 53769  05303 82968  01054 93206
1811: 36951 73730  55130 73392  52299 77619  28379 25997  06822 72979    67851 04452  82611 54573  96773 34014  67808 07159  83682 00408
1812: 23827 90832  22871 65596  83590 03957  91216 86032  22862 20071    97199 52606  97642 37739  73225 95543  53731 88716  62822 92875
1813: 39244 91737  12381 66162  17203 48944  68214 89612  43802 80881    90800 64638  79971 28596  54507 87592  40181 34658  90631 35182
1814: 32583 72113  81060 97624  24100 67845  10115 60114  58959 30102    81903 95360  90232 77017  86818 22301  04616 91471  12547 89158
1815: 32821 08675  44937 23158  63609 99510  94027 24462  06196 93649    67127 71654  89710 05074  25849 59411  05868 01968  76935 55194
1816: 75729 31061  77662 85236  46000 82387  22445 16749  29554 99304    19268 46482  23106 48207  29256 45185  96301 20758  34158 31831
1817: 19617 76913  84235 05287  66018 85788  48554 26723  26605 33736    34884 37307  92069 59315  11730 25780  96981 19119  67864 91582
1818: 70696 71663  65658 51527  99992 65077  69383 91161  11548 28762    30431 77604  06024 67021  39040 10213  33756 78910  90919 07755
1819: 41303 74460  35266 45719  52745 55555  79329 72322  35739 07678    99201 44525  11264 49265  87386 14873  50841 13312  99019 62322
1820: 54503 71091  88833 79573  77728 88167  73578 93872  76160 29577    93255 29135  17860 70403  09764 70012  31838 30533  04254 98760
1821: 27061 63944  81230 98915  10817 76817  06511 93361  02907 16426    50288 93996  74455 40178  64205 80019  69344 99458  20722 04196
1822: 42322 97389  78312 58060  17939 91586  52903 41806  48223 51781    92350 23613  42985 03470  26645 14744  56824 88411  73893 53906
1823: 91233 87723  36243 26649  84594 65479  97244 75101  82919 68372    85265 99966  46869 11921  60210 74734  89441 21476  30626 86183
1824: 14225 45342  06948 46353  61590 17105  94437 76138  32709 09168    26747 55888  94687 67058  28863 07093  73660 28221  31816 52447
1825: 64456 51304  19655 28814  31184 15229  02953 99644  85545 86731    03335 99289  40324 31833  25810 57741  33035 46782  81841 79052
1826: 06276 72273  72918 30049  20167 24115  83274 54440  43319 68032    12968 79020  57116 36993  95012 69408  12512 21629  33933 43970
1827: 70272 63939  74091 77317  61636 00895  14744 43916  01589 03354    72772 13701  50326 69663  11055 64441  55138 51278  21771 37790
1828: 31755 59593  14461 80624  77217 03428  86473 10949  37824 58398    13836 84744  89403 12890  01321 59474  13781 41415  18645 74702
1829: 70838 01307  99959 89155  27014 93926  28872 86560  18424 71638    32620 74839  46157 21846  75690 68511  57311 36395  56441 01285
1830: 30158 43510  57272 98841  96946 97751  68904 58114  90131 20599    10598 53786  48507 77964  69638 21204  56880 41516  16704 60750
1831: 40452 93231  68804 71950  93631 83071  56025 77660  33437 54257    25831 74450  34095 83624  37540 01034  19651 43860  59528 94613
1832: 50377 93147  20539 46025  36629 23303  25172 06628  57304 59982    98707 27258  19096 00927  29732 09383  44160 71795  41133 43705
1833: 30877 86414  37765 42454  46236 05832  15804 84251  11563 10651    58592 70395  92542 57086  32682 45363  11398 31303  53713 26966
1834: 24904 99956  69598 45688  67901 07306  48677 98318  29838 68957    68764 26090  25628 49268  60640 68522  83393 84194  20568 20429
1835: 15357 59814  40037 08168  66559 82995  41729 76267  89753 56833    47791 62364  99586 65885  31662 36874  03234 78649  27655 89255
1836: 11940 54405  78044 81889  88418 26486  94707 56528  77866 21105    54333 37723  57100 03075  37312 41653  87874 14395  71805 88716
1837: 51295 21680  17864 53879  65545 77493  43068 09339  12888 05259    31807 83499  27454 07936  44171 64771  10750 92578  35927 50692
1838: 15235 25282  62476 52271  22568 75289  72234 90868  83139 59198    89935 52287  81755 96683  95947 91405  76188 54604  34656 06508
1839: 91884 86959  24033 22461  90554 04287  63548 76589  88828 37047    83979 58104  60438 88208  04482 48755  67719 98606  24741 53429
1840: 38997 88825  41175 62267  63181 96220  78577 68898  29840 94914    41714 84789  39230 11517  51494 18828  71101 49520  48756 31530
1841: 16396 16092  55561 83146  05373 54125  86507 65490  66638 30435    08835 59661  24145 42334  21752 05935  59755 66277  37587 17227
1842: 69889 63495  40432 37354  60928 69368  06765 09480  36867 26070    52122 97661  35692 97078  25777 79626  08083 05058  99762 55312
1843: 69703 64095  85114 97685  90220 75696  58529 18267  13859 12598    71516 69563  33320 13859  07680 08951  43630 14671  85299 63956
1844: 44119 79384  92206 27730  67318 23616  02274 68112  63795 59370    95062 55003  65720 89166  18714 98106  56806 61408  83615 08808
1845: 59659 58157  97863 70228  62253 44145  59669 23779  69211 29904    39715 81964  47400 29175  01493 51660  68362 44824  08770 08243
1846: 50645 45052  07234 28769  00286 29998  05058 31219  73494 24796    81580 19238  76764 68794  51310 45491  46112 83052  23794 42551
1847: 34174 78759  85188 96641  44493 47518  94125 61967  36453 09529    04450 76862  98496 33374  71594 03295  25714 12091  12436 97104
1848: 91452 43036  01142 41646  12538 37624  04341 19815  32533 44013    69567 23724  18708 43539  75820 17394  80605 02950  38082 89673
1849: 00371 95538  11892 94418  78263 36969  45784 39695  00590 16876
```

```
1850: 09288 97613   62822 63858   58590 78797   39293 68672   77448 32860     77878 43658   63497 80912   29379 06233   72306 70940   62875 24990
1851: 47689 87949   49608 85185   26991 86820   11257 24094   79109 56128     22327 40189   72736 91340   79954 46306   67474 52397   84186 26273
1852: 96312 28555   95336 75285   65921 72372   56917 11204   84464 81848     51718 19831   52643 52588   80599 17379   95707 12557   10768 51341
1853: 40910 12712   39429 77073   91103 99643   24817 37566   45190 02232     73168 49071   04344 16625   80380 67897   45354 00365   49466 52294
1854: 91157 74844   26409 05415   80827 62349   96925 34158   54077 84303     34605 23089   88297 42234   69363 20821   05138 55096   16275 43818
1855: 41630 44681   74459 79880   87489 14193   31162 52951   62546 98205     09169 58128   02792 47806   24471 29137   32926 61265   20962 30608
1856: 63203 38812   77520 58455   65992 84585   29361 52150   79255 36910     12144 35713   72491 42229   84848 95367   49781 49794   71104 33432
1857: 93110 75426   71675 31680   80396 21425   85457 30779   43285 42666     60013 42741   67097 36438   74881 36964   68137 33842   57556 58415
1858: 49423 81946   33432 57494   97875 40884   09912 25231   77736 64373     74405 29166   99332 43959   30305 66135   84915 38736   65389 24287
1859: 58458 87157   84998 37001   46416 84670   66481 80209   89629 63795     80340 82828   14358 96520   00153 80534   29778 83690   12289 80991
1860: 49850 98135   42418 79115   21523 11641   55388 30168   23431 74735     39722 57212   37832 92706   07131 55937   43036 71769   18447 03911
1861: 67383 08644   84153 42554   12748 86018   94178 21284   57069 69836     84334 53810   86141 12047   93258 93828   92335 26657   88259 58609
1862: 90184 56652   29933 04646   09762 33324   01246 00963   09739 90853     01415 83436   30292 72771   07693 17499   50278 95837   04080 37161
1863: 73966 99857   33138 20788   50948 26472   22741 68824   50609 44629     24835 46812   55603 65175   86681 58977   34717 80613   56611 37542
1864: 65827 63631   49363 26637   09924 93791   67122 45950   35871 74262     18933 82210   05008 00787   36162 58852   53207 32905   21057 02737
1865: 54716 73448   87944 54751   31946 05702   90446 10878   47438 17667     68625 19290   15470 69752   84522 01503   54079 31846   27546 11346
1866: 68600 26403   00366 34001   73061 51344   15849 37187   43012 59339     80350 08349   58955 42327   89159 83278   08449 12547   83627 24275
1867: 31146 60492   50028 02642   28717 26470   70285 90631   43157 62843     23734 15438   42340 16541   29830 86308   08004 50648   65108 09455
1868: 25250 51888   95798 04149   68577 77554   32863 77964   58248 14008     94369 75109   73907 82077   78018 09245   33296 76397   34445 39090
1869: 20014 01956   28578 00648   21248 98770   03218 44694   82344 69743     47819 52233   34821 32328   78983 05107   96859 02810   51252 72111
1870: 09063 40903   51320 02110   84102 47138   03971 16747   44857 46060     72844 40133   63916 33201   55317 27443   35525 96165   87747 70502
1871: 94929 58802   69784 57027   67466 20078   66071 87364   77543 21784     54389 39666   37112 06215   07175 37255   31075 50393   11446 10027
1872: 76975 27779   97371 28232   34680 70388   71132 45627   01896 06305     42268 44762   00966 27379   46955 53386   91738 30683   07937 48437
1873: 56089 47426   35551 50328   36732 83344   66349 71445   22720 37056     19240 76753   71700 51941   54127 03229   13059 47987   47643 91629
1874: 44218 48974   20275 60614   67698 72962   25810 41533   05761 01186     01998 36266   41113 21330   65114 44624   86713 82980   68209 43863
1875: 36944 01004   98122 52595   01861 94539   31188 97210   94589 10708     97988 00588   75869 81128   94041 66253   45655 35268   37170 88642
1876: 08880 15739   34920 70404   15618 71672   38703 86102   20979 00599     49007 77891   15760 49733   90779 53083   34059 06184   33576 36504
1877: 47099 07661   09585 53068   43289 53667   71329 87219   51022 39931     12575 95082   95797 55658   09359 74139   53425 39285   52556 10987
1878: 12184 16745   75995 11887   21070 30008   79467 86379   46367 04395     66416 47287   39932 95375   91346 26568   24442 44249   38556 93469
1879: 67253 84422   78950 41480   02879 84485   23056 30286   62363 34876     69720 48152   80733 86577   84193 34656   76875 80744   46980 89418
1880: 87927 08776   44785 77889   02077 29979   72905 78462   42716 08368     53027 86726   67278 62022   22274 31254   92473 75277   50224 06822
1881: 37657 56297   07014 75466   31592 10209   89607 31674   93531 93855     66187 73751   44813 36894   31348 29611   55778 87740   81509 93922
1882: 99320 68349   60318 12992   18395 13141   47216 59717   57483 59589     92637 52131   77061 03490   84368 40066   05867 98644   23304 52360
1883: 51375 52565   04279 60618   67601 53822   92142 71955   92555 77952     64221 69326   32850 95719   92065 42652   68298 16041   89652 02904
1884: 12421 18401   91301 57634   31556 74461   92519 03220   94675 45977     68688 53001   65843 32960   27504 08395   99794 89456   47267 47584
1885: 73080 31493   83995 62141   61879 90638   18974 26801   71880 21261     42026 89087   72822 47649   17936 03241   96287 77008   69532 74188
1886: 87955 64001   59217 17433   68611 98913   13667 62460   32424 98077     63988 38091   91626 12622   58205 26075   72456 42953   61020 27312
1887: 04225 89590   07031 26842   99724 39071   07708 36713   26297 00951     32835 80077   60316 90445   70070 72482   03405 73947   44083 93159
1888: 14826 93504   65241 71935   60360 64312   40551 90326   47781 70277     71189 47876   02973 17076   97652 98204   73932 67103   19677 49521
1889: 60268 04017   08875 12504   01390 93193   59445 42650   97351 47779     53063 27931   74309 38594   83091 71040   25196 97030   75400 10198
1890: 72849 00521   96709 65902   99453 81720   88689 30608   86129 83395     83991 80158   89200 37158   57582 15887   71538 48969   64199 09364
1891: 51739 56640   64011 11476   37115 35410   86568 44802   30762 94076     93877 58910   47431 17255   63155 65326   20655 61511   66174 93308
1892: 57068 98320   09611 92705   40131 73569   51807 65023   71747 71701     89515 79131   45066 85438   89712 27122   71873 40890   58998 46334
1893: 60393 61987   38161 25565   14307 49863   84238 81161   49641 31215     25902 05997   80460 93648   64383 33214   48430 36626   72232 61953
1894: 80200 79108   07171 98523   82066 54523   27278 03668   97666 97454     65049 48933   96859 76213   33785 72328   40285 03308   99620 50602
1895: 39373 66554   32001 71074   34929 68642   30932 59446   98429 65584     55929 71647   05907 93439   96772 78987   66834 73147   60213 08048
1896: 63907 90967   92855 44573   25553 25905   51160 25485   05079 71692     54297 24974   39307 63359   25783 01952   33847 10306   88257 66362
1897: 64462 39795   01215 83631   28873 88996   08900 55134   33788 84024     69557 33983   66259 93488   14732 29946   70576 41301   47016 50648
1898: 67248 74661   61387 78317   96558 81281   34560 96260   80128 47437     35460 03578   35743 92728   40553 31905   91667 20292   07117 05548
1899: 36872 58616   71742 39453   22645 30587   44574 85081   66431 75611     64516 14231   21729 53031   65050 23831   87549 88003   61947 63590
```

Year										
1900:	24862 88596	25401 56198	41146 98010	10651 47187	62020 05089	32167 54010	04531 76527	33508 23426	36302 34140	05849 85857
1901:	94880 39437	65629 05203	46608 05677	49906 62477	36009 90526	70836 38990	52186 06540	38106 72970	53470 78952	84195 92948
1902:	23736 02989	06385 09450	29157 99996	19722 70995	96101 15191	72737 08701	21462 19608	16056 94825	50767 67170	88642 13043
1903:	98140 33272	92148 57264	67856 82577	91038 09497	84496 52636	94511 74664	85689 45069	02900 20997	68185 31338	75580 10633
1904:	82307 29251	18990 01280	06989 90464	76024 22893	78964 08258	56696 41262	83950 25204	14697 07367	58347 84657	07212 31766
1905:	88608 90643	92782 19375	17918 92985	04957 36057	38718 38976	00677 69842	24591 00006	19459 99173	87103 89644	23744 03454
1906:	96828 46986	72455 87277	09171 91469	64733 73721	87120 71146	66519 20209	65540 74386	40502 04498	38131 40960	25564 61829
1907:	75962 17951	43125 43360	39518 79938	31262 33363	30573 19850	86448 07000	47340 89418	94143 60801	50448 98584	65264 90458
1908:	78942 84367	22626 19553	31534 93170	63930 65255	10463 28777	27767 35763	89029 51612	57788 43985	76093 70532	22497 48161
1909:	34899 28593	16160 67248	05196 05768	17086 12757	74140 84856	65357 99595	13228 70405	70202 34380	30478 36624	20724 40781
1910:	14280 73480	35740 12416	99546 26012	70527 25546	83691 55228	70845 93195	55814 23110	66486 43394	50217 12270	67703 99624
1911:	91955 76820	12220 87216	22853 35566	67475 53113	70298 89073	90835 26380	80692 09553	20502 88028	83514 08396	67567 84600
1912:	72034 55784	58121 89382	00466 41394	31032 42318	27379 05082	64457 84086	37453 41751	31051 30003	78128 07956	87123 50953
1913:	31923 01860	40933 81359	97216 55038	34720 83623	14109 40387	18504 13641	56577 55010	18863 11785	76512 71667	72133 98247
1914:	17823 57860	96147 97336	63172 45983	18845 04267	34982 64379	35992 75389	19676 36443	66427 52836	27647 01538	30909 80328
1915:	32204 08879	59196 45210	61113 79534	01535 57239	88064 91511	36846 64348	58151 31627	40620 31386	93903 41540	97654 47367
1916:	70799 92853	92163 41114	48321 22398	49280 81946	32232 75225	37514 39944	12394 84655	35525 69816	90912 09264	08774 27594
1917:	40008 81576	08289 70504	51496 78243	61267 36088	30269 55323	66915 61988	77507 32087	02812 92930	70410 65170	31101 65791
1918:	54338 90409	04054 28794	83603 96500	46742 29503	66786 93324	80440 45299	40117 13669	31283 60864	03328 35810	02511 90985
1919:	22891 53077	23783 13571	17652 12379	61475 23004	27442 38388	56587 24471	50503 01341	14708 82519	09269 37196	21863 03240
1920:	25970 01702	49527 44537	93792 01184	75726 91862	61076 49711	75385 66997	48368 04032	15762 23606	94287 03716	88722 31737
1921:	15747 45911	92559 40805	18676 30574	13747 66922	00743 27227	99973 20206	78963 57390	65186 79333	45775 80663	65149 04707
1922:	50315 68527	57013 40187	21650 32394	36158 85775	48965 04875	75885 62863	91215 46623	59921 14041	75393 11966	33883 90594
1923:	53732 94519	13692 90271	05668 33453	32901 10780	00049 50136	91042 82698	63610 81940	86379 02373	69498 17495	72493 14429
1924:	40046 32010	96616 91941	05019 25308	58538 54827	63659 47067	75711 47260	93001 51308	09436 17902	50063 26191	34573 96929
1925:	18613 12157	32444 58648	34231 63239	93545 65413	27934 80010	28612 95781	70350 10870	07226 83108	16613 90625	86703 41239
1926:	56163 95337	26210 12305	35245 30454	68378 72565	15319 92771	73270 94258	62281 79476	00075 00291	88805 73438	26848 74000
1927:	05821 02403	47326 78793	88617 10307	81697 81794	26825 01436	26952 19508	44015 84600	13402 50794	12969 11109	32385 14663
1928:	64977 79678	13171 02699	19621 58732	20051 16880	34917 27241	07230 57682	51837 98529	32070 97874	26943 19525	57165 25152
1929:	92956 56135	30976 82891	00077 32543	79211 83909	86983 28745	29725 81395	71426 23466	99161 12593	51231 26739	48313 79949
1930:	63848 22080	94932 77460	10988 48415	03053 02724	43101 60554	37719 88103	31279 58670	86902 60645	69173 52320	96635 01529
1931:	29440 45258	78066 10308	64975 43708	01373 29365	32157 98663	29663 34832	31094 91531	02888 08054	80782 52028	05213 10714
1932:	70652 26777	67088 18853	67886 16723	97921 80548	34368 74463	91263 23001	53796 49778	86887 81149	52899 38156	70823 23351
1933:	61537 75696	99000 31615	23717 52722	39715 32340	95553 35861	02766 35648	59287 97656	72304 78655	72674 34383	00835 48901
1934:	90621 01705	17135 47029	48262 53579	31393 04208	82602 52461	74649 32575	42187 35306	43577 03886	41381 67094	96660 88121
1935:	15203 58838	65778 18609	99226 74334	90304 05741	73403 99992	14962 06160	44153 77026	74704 44889	41730 14550	26006 87441
1936:	12142 22624	16584 12313	09978 36120	37312 46084	06224 88525	28534 31933	92529 21096	15644 61862	80524 81993	36416 28706
1937:	01499 82916	66701 01619	08768 63634	79904 41117	70019 51854	15817 05313	64146 85531	41971 12409	82318 39570	78549 08377
1938:	76682 57194	06168 18772	69880 08965	69480 08773	29009 05892	93184 82397	07870 02114	45085 13927	99749 22063	84182 34246
1939:	72728 31073	95348 45022	27746 21874	84254 90389	94456 59486	65836 27440	50071 85460	88497 11530	16825 52236	94908 86239
1940:	00973 69204	69213 04010	78225 10627	50668 45403	92638 14697	71203 51437	03118 39647	61823 98749	66457 39323	73073 07824
1941:	27559 27180	07810 91338	30578 13860	48854 79811	20636 26687	31878 62521	16071 18665	14042 92496	40741 09891	74627 05569
1942:	23381 12027	31626 48460	94307 54024	07159 53358	12628 79802	66330 08257	46135 97426	52638 48391	74295 29805	60945 27171
1943:	01799 32814	55605 16888	43673 10571	99359 53973	81074 51912	48449 96139	60979 98839	80249 63303	03583 05909	56003 18537
1944:	18201 23890	85593 16564	29911 55817	81196 83992	72032 46264	92413 08072	32869 55134	58192 59807	14066 52829	30639 01412
1945:	57302 60113	42595 55324	71364 64668	78448 33471	65028 58057	35817 19713	54033 05624	85037 67641	23448 91991	84525 07416
1946:	27952 28670	88336 05426	54581 26962	83047 89712	37372 83432	95996 71778	56091 25677	77808 26593	95396 88608	04140 15953
1947:	75131 52433	70394 13119	88686 48550	03937 23800	55745 88529	91597 59883	48920 52921	98041 59257	57699 44293	02826 75350
1948:	08081 70080	36879 91795	81196 12662	29558 25128	15166 05943	42467 73707	74401 76041	38718 49891	79286 23503	14621 62975
1949:	81420 56167	93311 70762	18352 01943	95039 91617	13149 63612	54992 32406	26150 03554	07505 36837	07296 84191	88182 70505

1950:	84699 52260	0005413582	9281464268	1042817631	1618270208	3440753030	7122233905	0139946431	5588889747	2333791821
1951:	2902886725	9289946123	9226648617	0324676192	8036231373	9797739649	9636494217	9283449977	3547778646	2455107098
1952:	0868040828	7225780055	3935848299	6414827962	1457672862	6229321710	2227239641	7670292457	2385098032	8846389131
1953:	7715136213	2954334898	0212705152	4495144402	1087279257	2591448821	7857597477	2469925247	8712773109	3138538155
1954:	4326493995	0315526203	7909424645	9092187202	7409095988	3076540953	7276504176	3781632269	5918735425	2724445472
1955:	4394039138	4498975571	1987461029	8854493501	6558970132	2468387783	4879612060	2848969837	0696402668	6091949424
1956:	2670683728	1095848920	7335001544	8794147979	0134908178	2021198957	5988291894	6102054605	7516069351	1503976907
1957:	4150901277	7230216367	3149540543	8618447416	3161563629	6612692533	4751662341	6169355409	2671485967	0136300858
1958:	9528575022	0963450944	6318633383	3777198029	8230454130	2203815257	9074324771	5070846504	3546827908	1625832076
1959:	1834915240	6267159037	7766508252	7567916893	7584106535	7636386853	5564128467	3977886681	7674850316	8839934304
1960:	9777978536	3243568791	9839532866	8514690683	4212604646	2648449447	2178698348	0306678991	4191137365	6532819575
1961:	7865793862	7721420387	5133221079	4521400899	2925246813	9898834588	7064210726	4052166486	9337754618	8374550034
1962:	3457123356	0058427791	4311002968	7460558854	6288357546	6311920015	1066500928	0137660284	3058697422	7277516327
1963:	5913313809	4864047157	9786809319	4287669896	8835230295	9504526404	1880368929	4563488111	1429849181	8253529301
1964:	4702290051	0237716739	0786790383	4309730103	4276342177	3692702758	3404712693	5395264669	8656179702	7030343918
1965:	2860434752	0242462715	1161820924	3200480484	2853061002	8046829179	0390994179	0335290124	8051554982	9889868883
1966:	1681394647	9853067033	6293699765	0536293131	2867167426	4404129626	8091300043	8100238466	8896422925	9979273998
1967:	7039948125	6702879953	7654582587	6241372615	9949997023	2296509306	3938544688	8457831024	0137179458	0933871284
1968:	3030465152	0190526875	4867392315	9004246119	9271717918	5509048223	2093474151	7684562749	0029146915	2336522737
1969:	5764378101	4900495728	6994943355	4413015538	6230285216	3132503758	8747690179	7081135887	9658190564	5408209981
1970:	7130317502	4967651732	4140141460	4780677844	3800411496	4141608869	5199555970	9655135985	5073018215	1098373306
1971:	3114172025	5347987680	2799357689	9655317620	7080150905	5980672106	5167393539	2613167442	0085538837	5609964053
1972:	4530530297	5204042952	4988166505	4376737652	7592096881	6807829814	2787900780	3073843662	6861397388	4685248245
1973:	9347646979	1584485910	9642077404	4905506793	8479723223	5002406738	9054468801	1377423283	7799355482	5214627172
1974:	1050807736	1554161727	5719520024	4796507881	3098358710	4443591531	6208174085	7855728844	4645218150	2102217234
1975:	2926910881	1835472316	2855230065	6805151707	6724764027	5452989292	1785530041	7969372681	0627341726	4402479557
1976:	3198676861	6609295619	2628101387	1826230720	2817482411	6019003459	5872583871	5378584319	5383388880	1669008458
1977:	4991142949	1542048767	2662275558	4226118140	2868406733	8982520070	3856594863	6536486858	4989353135	1912691103
1978:	1787188348	0859142914	5992759237	0453837985	1909917031	7645021661	6151809702	2830402251	5691764096	6155632240
1979:	0837724253	4667153968	2218558430	7136855103	4931498622	3936865094	6430637697	8012322456	1599814811	3517749456
1980:	2801998871	1437700730	4902307095	5639712161	1122662951	0871121410	4011069066	6231889833	6019927162	8571324092
1981:	5990384635	2356304921	3136656504	7063288942	6230533447	3681548581	3359923023	9304202631	0360437766	3387094466
1982:	3057366067	8018639709	0297806842	6692534765	0651375330	9934632292	2704775100	5388886881	1217652709	5626174239
1983:	0214810565	9360915028	2086254691	5203672075	0476994539	3990660062	9054921128	0398963244	8904410181	8224923931
1984:	5447986620	6197548189	5376385619	7577922016	5099144004	1802385069	6514798477	6714673128	4947269105	6427878490
1985:	1355296237	0924734701	4766370075	4688074902	1520712297	0230164645	5993134267	6521337502	8231759254	8048312997
1986:	8910527988	4018869479	2516936831	7668388135	3536383743	4931305689	7626820754	2719465587	4880157356	4712437350
1987:	4272928255	8890518629	0732898189	6933353875	0190299583	3692137610	9673334082	9445082944	6819933794	4248242885
1988:	4471726536	8270114097	3571538342	3742285871	2466546900	2735149993	6359570172	7513993277	4354946493	3252506889
1989:	8165491679	5088800988	4222818491	8596098514	7601494642	4030499862	3960471172	0620529070	4606639353	1584512618
1990:	4319348805	4281577006	8410347955	1889703894	7524852574	0563888141	7168685217	8338739642	9891266095	4639889692
1991:	7605456597	6162117500	0592098657	7371670597	9803404542	5941757320	4279368124	4338784118	7862304608	5303515516
1992:	1941977853	3166376652	6395687036	0289964400	8692954203	5802925982	3995315409	5148907953	3092820806	6488031124
1993:	9907077438	5875184445	0020220846	5803503388	1953456235	7004326126	4200147579	2431719374	1604784525	1124001735
1994:	3806696099	5501523038	0031172350	3150850838	6152133400	4691528311	0655342298	1421123325	5279208077	6227018043
1995:	3795871054	0548323139	9873689991	0251030375	3428700666	5125329883	4701253485	8410935790	0265156209	4613198660
1996:	3451329485	1264232577	3960842445	6658872722	7564977102	7371513920	5761808405	7258135873	7891847821	9304473577
1997:	2010897231	9449651356	8549696821	5741967888	2699471680	3733979843	6468905716	6384318622	1699820733	2619547289
1998:	8660869670	5615491022	9683056553	3446665965	1015861937	0351713408	8267482266	2744273049	9557539452	2976257401
1999:	5892341570	8585840678	3823586417	8583567313	6962099478	5359986809	3065511221	9000388395	3623636549	0656666768

```
2000: 64518 97921   02135 53161   70252 73075   29107 67191   52420 12255     43440 30747   78726 29296   55797 63491   23441 23904   62266 45631
2001: 40520 42780   36294 93161   78957 98152   37675 83553   96725 85751     88587 96765   36634 31817   81949 28797   77619 44459   21560 12912
2002: 37339 41235   89268 36378   56746 81360   35260 10480   32044 49659     50281 25039   25924 60392   56287 86709   04328 30321   58310 42891
2003: 49625 00069   61535 26452   15936 30602   53663 05909   60858 47788     50157 35972   04370 36717   59051 79276   31794 58918   38985 06632
2004: 32710 64959   30608 28029   53736 07042   01077 31978   95365 58357     97319 44649   21637 30582   61171 53535   74934 63063   52663 07542
2005: 55357 95184   44254 04006   01262 39565   31587 55311   47991 29645     71173 61921   37515 69902   06097 79119   31804 98770   42382 48500
2006: 81940 84695   86955 11335   50882 09716   55707 71117   88532 23524     23328 95381   33615 72446   46746 69208   51370 95418   76711 15177
2007: 87793 15934   97293 22582   35846 34369   61151 76060   49303 91098     30581 36803   83517 16635   47041 16689   38063 74881   51706 91032
2008: 48607 37904   02085 11839   52752 61809   31233 43203   04313 11830     79291 68507   61395 51267   91923 14690   13853 28277   18043 34443
2009: 64410 21907   96952 64116   61215 67526   80685 27815   80964 34637     32689 52251   57317 43493   17882 42363   79158 83519   05819 19795
2010: 45993 12156   33492 00208   34911 42735   73202 19969   18619 92188     03974 33947   89193 41176   06930 18824   58074 31547   56577 37902
2011: 75108 09369   12221 64676   46000 67216   34782 53431   25369 08057     34463 48241   97984 92169   41267 28543   03794 69069   43229 12816
2012: 38249 67220   11744 82909   25963 14646   90177 38204   58361 68755     55096 69623   95067 57192   61676 39285   84543 04030   52774 00445
2013: 13966 77274   28858 61335   95390 52606   86566 19127   65104 08602     31334 21983   79928 47482   02236 28019   46358 73668   28412 77405
2014: 57077 81265   26274 18192   83008 22157   04349 42199   03052 25372     95601 54858   29386 77061   57972 81461   16143 88976   63195 46197
2015: 84647 19641   20176 33555   16276 84876   35755 29275   03055 55015     72164 09683   70360 29100   10403 20444   52214 38573   57881 21942
2016: 82752 71599   66703 67784   48835 86126   01574 12592   72892 89755     62797 54321   46259 65390   96460 31005   57061 51926   19693 83414
2017: 75952 08842   04231 33242   14147 10977   43628 98083   94758 62579     06708 53798   56816 62400   78151 72233   04941 76405   70973 34322
2021: 49250 74217   93626 46408   56153 93990   92794 86686   99475 96142     04927 80591   96041 20495   55115 10403   05054 87102   37903 44284
2019: 02783 24106   54290 95775   58265 01579   49919 49609   85174 28053     27815 59558   24124 21077   62102 25259   56171 84884   23319 45205
2020: 17081 13153   23647 53241   00349 49002   92792 63595   60833 94312     19028 82493   49455 95995   98917 48398   10190 93814   62516 59566
2021: 30236 95709   67353 69680   41136 74034   21305 35946   61087 05352     01589 27883   06893 99406   74174 17907   25322 18004   73504 70277
2022: 07104 56136   87114 82990   02082 80167   27954 43533   01019 34126     27850 62204   40454 53081   73081 91192   51527 97394   56555 96324
2023: 22085 04272   35119 92032   75182 24423   76571 01389   00190 88204     48992 76837   10845 96259   73347 44072   74416 70989   73585 29049
2024: 18844 60838   31061 00654   52379 48939   35060 24281   38920 58280     76753 37491   58683 76134   70315 84290   64346 52855   77327 92179
2025: 31596 75021   89276 38483   16986 25994   00969 37716   22487 84766     00683 97068   68160 12844   45878 29416   94438 06218   04417 10737
2026: 65266 37749   38016 66581   00445 23139   58001 61987   85556 05023     03301 41708   74951 31813   74444 00633   08862 89144   41840 89972
2027: 90993 13238   97785 42040   40783 32747   89536 43625   26521 33440     53423 90016   04387 64360   81340 68083   36610 29472   50274 50755
2028: 36268 94131   45958 50620   81759 58029   51327 89184   62893 46433     92106 40194   98862 73801   55347 30087   98616 71540   28234 36956
2029: 97397 73885   19263 48563   42368 57317   44610 53137   50617 94903     87570 19416   45277 04068   50731 61687   47482 87416   64924 57437
2030: 49798 07783   85036 38959   01846 73740   57236 09867   96178 75017     00721 02577   08751 70318   59869 12814   74123 84378   57719 46780
2031: 60815 56347   74486 44311   38053 69661   39565 17584   55139 18525     99162 34662   38364 71911   87686 95718   71297 94389   21014 72877
2032: 87283 15606   16705 98450   35314 14408   13491 15689   56693 24904     47089 01838   20833 71749   59910 01844   37159 37094   63905 13554
2033: 27012 87775   81831 38653   83790 47607   78468 56634   68503 57142     11437 75292   98218 03448   47723 33617   24915 52961   75849 80781
2034: 85336 69202   19071 95914   93652 81227   54822 60045   75024 13210     28497 37078   43621 36073   09790 49285   38302 21862   23916 83448
2035: 73991 37685   10842 68974   72229 60487   76964 77626   81230 08872     14156 17012   50851 50267   30210 98590   21389 80703   74730 67680
2036: 12348 38394   40815 91295   80492 57677   51151 56367   50987 93513     69716 51859   28352 18381   58916 03522   19796 09702   84426 11924
2037: 33081 10208   69468 33107   65417 65313   15050 59360   00580 85303     38077 24547   05959 44644   31800 28766   75226 04165   27232 31164
2038: 53107 46884   87345 82670   17010 58679   49110 66486   31316 47441     73962 14431   12346 03383   55307 12838   09096 03907   89091 79978
2039: 15933 52486   05678 88286   64486 39473   12672 63800   29674 85849     58230 20115   89352 20510   12952 91127   52909 55504   10513 08261
2040: 64375 87425   38289 00181   27630 48665   19270 85802   09256 05222     60009 53787   60645 26188   41299 70748   07752 67372   84452 83809
2041: 02330 64556   61635 42545   45054 70291   55595 03142   86509 92774     13105 93732   13607 41687   88101 80893   69812 88638   85070 01226
2042: 99290 12836   26870 10824   96493 55089   32863 20008   55206 75166     84705 80334   56754 04999   68551 53338   71099 95610   70842 47080
2043: 61182 04662   75653 88790   11515 03961   67131 32467   59036 03150     93061 76271   11045 96252   47862 41367   40726 79440   39022 84820
2044: 98148 11775   06474 08349   44981 70259   90766 64828   57377 68190     26778 19177   29900 47831   35528 37509   39025 78857   01656 66631
2045: 52050 89503   82038 79898   66446 05012   97339 82799   88426 58610     91604 95681   90116 52011   17207 51069   87577 26683   79487 81776
2046: 59155 38019   13492 14610   46688 28928   89415 10960   60221 74305     18637 37835   10343 85821   28792 75938   65349 52274   57838 21141
2047: 63383 96881   32983 18177   00270 43379   52114 74968   16590 55372     64647 27034   97863 96211   97131 92009   65349 22274   80076 99414
2048: 75983 48861   89346 64317   84048 02241   01661 61222   76795 23270     87966 41777   81353 76081   25307 18973   65197 01786   71152 13182
2049: 65345 41000   95674 73852   39815 18529   68327 71016   96541 70319     42972 35715   33046 43087   14001 96989   25796 73651   06328 85023
```

```
2050: 36533 59234  55357 22592  47278 34179  87190 94453  84231 52574    97026 20064  13148 14694  54337 48764  73099 80040  12457 99385
2051: 83276 67826  46116 85439  22152 52754  15394 56903  99025 39073    21821 70606  13332 81865  14189 24333  77955 64982  14334 03883
2052: 19758 21047  75988 64438  73936 45314  79551 68639  22662 20886    13304 32206  61793 31786  93809 92717  64108 09540  63689 73040
2053: 29056 79302  14473 62291  65842 09821  99501 71738  37717 01017    17877 43798  57227 18906  92089 07717  73912 60876  24825 20790
2054: 30067 44677  22536 58813  73045 22255  28754 66375  16122 48663    69208 35070  24738 73175  79832 95722  21995 12722  22850 22546
2055: 82362 99692  48698 69774  22040 68456  05931 65818  19944 51941    64654 57670  20400 64397  75137 74485  09179 82709  00785 87600
2056: 94854 84315  05721 24177  75534 76487  12481 43717  27586 55894    83132 35811  08731 80897  03714 66545  11351 22812  67844 30749
2057: 12856 89325  18970 85345  14740 96558  62303 33577  04434 11001    05983 37823  44601 55805  57124 42935  77190 61561  56164 18737
2058: 51651 18454  84853 11435  61125 04913  05633 70777  04283 64180    18159 61059  52430 11096  95483 56966  28027 16790  65875 93154
2059: 62224 32390  87805 41170  05536 88227  08627 37790  55032 34775    33428 77880  78668 14771  84784 68364  41092 34452  24747 21459
2060: 45779 87861  68886 27567  36019 67713  31237 97724  53923 58428    59795 82011  38924 93314  17394 79666  31844 79631  80748 57928
2061: 34188 39820  93330 81967  50998 95620  36068 66705  70561 48739    06892 60141  41466 54851  69881 26718  90153 38848  37301 61396
2062: 78003 66836  63807 86783  58074 03931  31516 83203  94834 52571    95376 27838  92175 23349  36417 75983  60355 59928  13955 75291
2063: 69201 79519  97302 81464  55650 75251  05346 60187  58731 76523    87265 76568  89935 52193  05709 59371  27179 55835  52397 84128
2064: 41033 12774  31456 08216  01323 41498  95727 64811  47634 54016    94731 35567  47592 22282  32569 99632  74124 01957  41033 60212
2065: 65999 19346  47338 73478  31651 66799  90311 15379  44899 72766    49494 50809  65902 14332  68524 84867  55829 69414  34861 21615
2066: 90436 98482  02995 15882  07160 57992  15691 04010  07313 84217    28605 23426  28396 90052  74205 47108  72188 73827  42216 59117
2067: 39043 29551  13239 41723  00551 94815  63993 78383  29715 29567    72555 64389  72541 75236  37930 27390  58688 12982  14705 05318
2068: 15924 60847  10782 49833  65964 08260  65435 62638  51640 62878    99856 52865  52653 50766  19656 59428  82316 88260  46833 38872
2069: 68053 20254  87803 52356  83237 87529  86832 98631  08698 05489    55773 22691  60395 22425  62683 04293  00177 39634  42834 46563
2070: 09423 30771  45350 83021  47833 13286  05475 77841  95183 30164    26602 05738  79079 37735  94959 83822  43515 75015  99157 57657
2071: 46007 89255  73579 56848  27635 33447  43771 62048  63860 02474    76623 18066  81816 23457  88776 74287  76268 45284  12613 02892
2072: 48680 67290  54513 23032  78995 80909  04478 66842  18433 35218    12491 84681  54780 06229  11663 28425  69546 35358  72929 31443
2073: 62976 75461  09907 18421  55086 45716  75254 74228  65152 52923    70164 44459  57138 47498  17542 58290  64189 63688  04342 85077
2074: 39776 92850  56825 13685  28753 90704  31590 69335  43039 21431    13159 51205  83569 08124  78722 49690  61117 09514  17288 29359
2075: 73916 43982  42217 11324  58938 58013  39156 83229  26624 59620    09192 81437  35050 01484  49959 98826  00576 45167  35144 03578
2076: 21457 49674  04219 70466  05839 84976  38145 53978  16117 35008    81143 15139  66850 29906  13158 21839  51203 02069  61892 81090
2077: 48233 50749  71784 79100  91652 22573  33570 61537  48513 97704    96953 92956  46925 95110  39036 29622  60972 53442  44562 09989
2078: 16227 19516  72227 01638  94605 59497  22534 75770  76005 53047    88937 58516  04507 33225  89192 72369  60945 81856  16235 56352
2079: 36113 67679  27055 20997  97760 69037  84800 85693  10462 79521    04171 54849  20804 76304  97382 50912  37270 18896  54266 81858
2080: 16997 93116  87772 63817  03270 03745  38149 66996  80283 84140    49553 53609  57912 89799  77818 47037  07824 44639  10035 91598
2081: 31866 17370  19949 63638  23505 60561  90898 24977  09870 42036    23346 02004  44821 87293  65742 42511  65456 61054  70633 50495
2082: 31840 94542  35431 26896  99568 86724  42210 84858  17508 47356    07433 31075  94466 12003  98787 50683  33500 27848  73449 51310
2083: 13459 77699  57101 87170  38211 92192  88497 40207  80254 33188    64489 86405  04362 44555  03908 77343  17592 87643  84318 08341
2084: 36674 09853  55981 87954  18372 91145  29928 70905  62921 18264    03265 68657  24928 54870  46692 12689  05743 09390  02355 76427
2085: 60653 02679  78108 73414  16225 02126  62102 92238  03909 78241    94501 62511  74170 60743  39227 36916  43293 75086  81944 34633
2086: 43580 35731  66072 87953  59701 91459  88000 57337  06887 39881    14242 65223  85590 54406  22545 40883  67790 91079  21897 42982
2087: 98394 67801  83322 03600  15232 67647  44385 83950  66936 65471    22740 20033  65656 79734  01069 65418  41648 50514  60278 81419
2088: 88451 31994  43735 83831  59379 25038  82503 32922  48799 84493    21333 35523  61614 31171  61056 65552  78671 78443  41540 25726
2089: 27919 14864  34413 66774  08249 26588  18460 54993  80801 09182    72070 74049  20792 55300  03769 31839  88313 35143  11498 19058
2090: 71999 21211  25330 94160  95906 95415  76822 42566  28519 73840    20308 99429  62510 20859  28939 59679  92215 01747  21076 46248
2091: 14124 99952  76511 21435  98209 16982  56391 87439  16275 08689    14501 36772  76012 11907  81085 18397  48422 94228  03129 52412
2092: 79810 23727  29003 01798  83681 30625  95717 41247  03113 19742    92761 58765  81402 76464  50099 59810  24538 61679  94313 96703
2093: 58051 41373  89733 74921  91523 49293  14092 65046  15815 76188    19047 71424  36415 94316  12748 44204  82386 84938  15337 65677
2094: 42524 06761  44197 10284  71505 89779  68198 31193  29874 81121    30747 09587  13976 07464  38204 89076  74900 70846  63009 77217
2095: 05219 76712  93189 02306  64929 92513  64019 08495  35510 00124    51373 87761  92588 17324  43440 17829  35518 49402  68811 45351
2096: 94036 38646  69831 16689  99356 87404  68275 63324  96737 38759    44093 88919  25681 48052  77532 80733  27091 01537  12817 19435
2097: 03277 28994  70692 17090  92984 85036  97145 08962  41456 10432    60624 73563  96043 57500  71109 74206  43420 28860  39130 94702
2098: 33122 06402  64116 63352  11737 04877  50910 77283  61489 19969    55455 75158  83687 60353  39254 80023  89400 77359  42391 69894
2099: 87633 84652  44486 67847  13326 50159  67926 20260  24211 95397    43791 45449  39036 05603  08298 51994  32318 88821  14725 19053
```

```
2100:  23830 83719   45472 43636   56120 78433   23641 38393   55062 98703     41987 11803   36738 72338   15107 28160   27550 38701   39248 33256
2101:  87764 34949   75910 56318   79017 94136   10802 27679   73346 67644     97832 83738   54647 41606   85031 63224   58125 65805   68670 19889
2102:  78331 31349   61520 12989   14047 18117   00796 44050   61851 18180     74110 00772   62290 41880   65403 52204   68177 80782   63521 31579
2103:  67231 49852   18519 76046   77741 67975   73666 60548   54515 87574     47416 25647   21200 35822   94580 36166   82317 01522   05005 18172
2104:  01659 56820   29225 09653   05123 00717   48223 15174   75716 10014     69331 47966   87659 14266   97631 39639   01494 09770   60552 49772
2105:  60553 82118   09927 61157   34770 85803   40556 74005   90923 04824     81416 39476   23080 37335   48811 79738   76824 65779   60309 67562
2106:  13229 00385   44690 37591   97140 29346   95013 35226   60861 53903     42626 92263   03451 52220   50264 59055   63182 26315   42242 57226
2107:  79950 61378   67479 50109   90757 89729   16806 38855   05649 19683     04381 38734   22935 40958   76669 84809   15430 93622   76553 03900
2108:  46957 83769   48746 72898   28074 18864   24886 57846   62310 24405     14268 05844   49858 03133   55111 17958   18757 82000   15623 49012
2109:  83480 17974   67963 39630   04960 57179   48074 32194   55139 36040     92097 16911   64348 94098   44572 96378   69169 95038   05219 81477
2110:  71715 35999   98505 20714   68899 23357   02326 56544   77508 03177     10846 31812   07601 47163   00074 03767   24780 80160   29461 84334
2111:  50724 97276   30981 14149   17194 61800   54946 67211   29647 86151     36616 17832   60072 81923   75998 26045   54071 85737   30300 41488
2112:  94947 26912   78302 19396   92230 07570   27367 79018   68937 66019     59727 47204   93497 22389   12539 49666   95201 37442   55756 41430
2113:  53338 77178   50427 74741   39289 08443   91852 08748   46086 69010     60090 03539   20716 53901   18866 25029   01660 06213   69567 96867
2114:  75427 36006   99821 29942   35673 87434   88535 90363   76520 22943     01397 95735   10255 79734   88136 51412   00994 63299   09829 20871
2115:  87147 93604   66817 28898   62959 54867   70513 96085   43731 62691     09273 46293   58696 94024   74876 18165   94178 61437   25185 30153
2116:  14344 22353   98341 59265   38912 46201   97534 85828   24793 46854     23402 31874   54730 98539   03303 33550   60031 64242   59340 95708
2117:  74557 08724   65563 13466   24645 04461   17743 23722   97105 83956     26150 76035   11642 51074   73318 86417   49569 82040   08293 72409
2118:  46859 22791   89121 24218   79505 71154   45934 89699   75402 25145     23845 36535   38280 99531   23707 12176   49308 44139   08306 79456
2119:  92460 38358   18631 36094   59280 86163   75517 26622   14369 35011     41301 76131   10189 10461   05677 78899   83264 25039   61360 30384
2120:  73380 91275   33787 02191   11046 37071   18865 01162   92078 67916     77569 19711   31761 88344   38610 38634   21148 46617   22836 11653
2121:  79817 41085   59569 95687   94964 29918   71823 67867   14549 83590     05167 00892   09806 34698   04264 78031   05006 94554   28385 99250
2122:  97829 57026   75393 12987   51047 40995   87416 21572   84847 04574     75389 66393   49074 57089   75724 90785   44677 61330   98136 81713
2123:  38382 39693   88320 93336   35857 77329   51620 83052   21666 39245     49704 73554   91882 44250   70974 91627   88730 57561   52529 44863
2124:  94497 43465   06067 68509   30404 58783   06051 03820   69978 26553     74523 66957   09838 86945   05538 76598   23129 51356   96756 93530
2125:  20074 31971   04748 70497   65614 84634   36746 76807   99987 00553     32404 10426   42394 95360   33066 00736   62109 31780   75437 44590
2126:  56123 10516   78029 69344   80474 06050   33584 52416   61981 46805     46223 22489   87789 05229   88182 69669   83333 08238   67184 45759
2127:  39174 92436   17385 40266   92950 58208   94945 89576   44475 01178     29839 53385   69099 54567   36603 69434   06235 80211   18032 88566
2128:  56358 01164   30913 13755   25860 77668   96091 74400   16793 95652     48777 57785   99663 56401   42988 52861   92955 73159   70809 12855
2129:  67527 26233   17554 90404   58276 74266   77525 17367   78889 45865     20098 33920   19548 17919   79704 69997   59265 84577   07038 13435
2130:  06461 28086   63382 23395   12121 06566   42218 69697   70989 00499     20980 13619   08995 00899   26546 82149   35430 76523   66873 65309
2131:  68891 14828   06762 67485   35147 07239   66002 04695   31236 52867     92812 88950   60892 13625   17593 60430   92684 50993   35558 66680
2132:  90600 21206   29318 84966   52026 21462   35982 94446   18745 23940     04467 00803   45977 32095   97251 33069   01096 18268   49381 34661
2133:  56006 14006   35782 40119   90584 99259   52084 29344   13233 67864     61185 79707   70511 47950   19591 14638   54154 93778   98352 67275
2134:  04846 69026   95440 00358   41160 53960   52344 84661   47215 12144     97851 27193   93304 14668   51545 66255   63076 10968   28510 80232
2135:  30526 29667   01970 45931   13750 85916   85428 80778   43189 62588     01935 98647   69815 84910   00305 19576   41828 74521   08739 89241
2136:  24412 68914   83248 27415   83000 02017   90311 89503   07108 38434     22406 00000   70580 31667   20012 09659   97477 16760   63797 63093
2137:  80069 19186   73394 09650   01539 53124   09119 99314   75030 72950     55122 43128   93680 34082   24042 33545   49572 15685   63438 81522
2138:  85059 80264   63026 80608   82802 09013   70089 36913   22129 33372     99326 95202   79569 17416   78954 50373   80054 72878   44962 76781
2139:  93089 36625   94954 70276   48296 85233   28721 39289   64842 84206     63532 06546   50706 68815   54328 97857   62334 74929   71281 45764
2140:  46800 20817   73915 73722   95122 47182   79889 07904   20923 44289     47926 60277   85543 43957   59239 11044   21425 00580   42407 89159
2141:  99406 72149   26945 06149   03240 53040   72067 60387   36032 68200     33095 26497   23470 53411   96113 54365   88639 31274   99951 34805
2142:  02520 41334   29366 09893   35205 58590   69132 60855   34093 38933     90994 18770   00788 62324   50231 53110   53496 44008   67977 92785
2143:  19769 49288   06601 33214   23073 32695   41183 70607   61644 61886     12031 01439   89220 50548   82030 12815   75391 64575   98214 80534
2144:  26629 44622   69277 44109   83177 29535   73368 57360   62395 60311     07535 41092   93413 64924   61464 72278   06631 56061   94695 53499
2145:  95183 53925   39445 58883   07597 85126   21831 71424   72163 61633     94637 64828   10512 49506   35153 55812   99376 19879   21225 85285
2146:  40398 77744   58126 55509   12135 94684   64987 89568   74483 69704     92242 45528   91908 95191   96925 44641   71956 48975   98924 22006
2147:  91821 59206   26297 16073   78419 69404   35058 49182   93328 39686     31607 47967   52655 84579   84256 33909   18542 10017   29877 35147
2148:  25869 78908   21281 41956   32887 51794   15251 30681   53538 93959     07399 47466   61785 09563   71116 71439   26481 42457   61720 32023
2149:  04863 75577   47698 49100   05059 61688   15820 27240   77368 03790     30498 47882   61866 94327   95855 42902   44658 28778   55023 11567
```

```
2150: 82883 59084  60186 06096  56287 34322  95198 02427  83690 69320    50036 89936  63625 52865  77605 07714  11217 62563  90427 49756
2151: 65691 44738  64069 40134  85787 77055  41387 26468  58957 03199    21483 72613  46560 75998  79809 52576  52154 48421  15500 30309
2152: 08004 42280  05196 09507  10163 15471  48260 71936  46146 16447    95716 42610  15234 52071  05544 66094  31509 26833  41307 85966
2153: 60448 40301  65816 39492  89263 84141  75737 14923  34187 89738    73182 82886  37730 28270  03198 37077  21523 65512  39678 78916
2154: 20057 45411  29518 74316  67149 19451  01508 80030  83345 05206    98049 50563  94528 81231  04148 87232  80051 24668  09392 74625
2155: 72931 64287  55953 11519  34851 97139  87568 23876  77140 81788    85801 97094  77498 14528  91106 61999  46208 67009  47767 28637
2156: 84476 18810  22156 14222  00156 68465  71102 19924  44742 58204    31501 77477  21991 13975  25350 89875  92846 66176  12595 46676
2157: 36698 84069  96514 12576  87217 30995  85775 49345  74881 76276    94873 53362  83985 20418  95951 58084  95496 34916  54527 29748
2158: 27156 64710  93534 70623  86031 90659  74968 17511  43292 43997    43901 11131  44920 32252  75311 77909  78173 74006  03043 63569
2159: 42526 28456  30929 77218  65568 76359  08515 03996  72940 18120    27247 14602  43258 97099  06652 75147  58217 19334  24368 36209
2160: 57931 36022  29433 69460  58556 64978  01874 66688  13391 86286    58131 44086  20626 69640  78181 69935  37574 41830  35498 96327
2161: 38238 45820  08486 81174  13398 18061  92699 14658  34940 55957    38554 53955  86014 12936  67788 45423  74671 57786  60611 29642
2162: 22482 61646  85344 96705  75999 19678  02490 25707  63852 78390    66499 76746  07135 05872  11280 86577  17529 63232  89365 36630
2163: 97286 55913  59056 47969  88815 79556  22749 50089  21676 24613    14807 07770  02774 85595  16978 24652  64204 83185  77944 75841
2164: 27686 01008  73595 40707  97709 78546  99335 29788  80310 13136    01202 97780  15254 56135  69045 68914  35552 09079  13226 38023
2165: 37994 69899  23450 76139  35773 44412  10027 43646  11546 21255    14592 89382  30838 92398  77915 40479  33951 88399  46228 82936
2166: 48278 90134  83926 51946  32288 53780  00201 84991  72500 09933    01444 30291  82975 35194  64200 48117  86043 81853  11198 19970
2167: 85145 87921  24463 64177  15097 58501  94082 13550  73948 30679    93109 80542  03877 72527  07566 77597  84789 54271  91079 48507
2168: 71654 06812  68445 82993  41613 60792  38220 48898  42208 49311    60386 33765  80951 55682  58276 84467  82492 50661  62332 19508
2169: 28624 56126  48908 08037  96929 72317  31631 43609  54423 54640    67115 59006  58303 99138  83576 96262  18235 07112  54316 56672
2170: 92668 08316  47347 35112  33057 88968  97506 05839  99565 47662    69781 56013  63643 38417  32332 05331  09686 55027  06532 67610
2171: 51231 17102  32313 50883  71870 96920  77797 66421  80893 44584    49081 19703  01112 74853  03305 96350  85547 88675  93813 17879
2172: 28769 57237  91919 26341  24091 66124  82685 12264  51308 04562    70589 25752  33243 22518  44633 30413  50790 26075  98717 74162
2173: 00437 82708  99876 46896  04506 05199  30151 03580  74483 76072    86302 62889  25997 75893  32552 10398  34714 63418  54482 08554
2174: 97173 93050  43224 64665  94373 15661  16051 94876  58763 72556    67330 90774  14510 12121  19789 34541  24630 12330  42277 59358
2175: 34927 83503  61414 42678  77199 29596  46235 21174  27427 67507    00722 32965  68945 55298  00751 80836  33672 28481  63518 10387
2176: 23868 13832  14893 14027  75458 94307  90505 06522  11960 16485    66359 43453  14897 73416  55654 30434  39824 90261  38066 80917
2177: 59646 26144  82893 59303  79993 54474  20590 19200  04556 39654    15626 81431  95986 06409  48844 63820  26382 61126  45761 15661
2178: 53632 60855  30717 66287  56921 78549  01283 61759  00620 83403    55325 76210  76899 43173  93407 15770  48489 09201  56478 62773
2179: 18116 00170  19366 88781  97928 85687  48374 59250  57807 56519    43958 93535  88927 28453  05252 35930  95339 31967  13341 12504
2180: 32231 40283  21226 23653  44003 09144  94823 58696  32128 87515    24389 79691  49890 40210  95846 08761  13914 39931  56546 44075
2181: 48908 62708  19247 75845  96675 06337  10177 25660  42506 51052    30946 67935  20758 45491  54110 25371  87422 79441  38689 20045
2182: 93071 29654  47469 03043  16668 13827  25972 81496  41816 87802    66007 49912  06623 47853  04968 19774  71253 30861  49249 86462
2183: 66007 49912  06623 47853  04968 19774  71253 30861  49249 86462    26249 60432  29333 44217  14008 53479  57682 37750  40774 69434
2184: 41410 23811  24437 38397  70470 91123  74451 07075  71469 80529    94820 58806  36954 61310  25241 79774  91179 90879  23422 16007
2185: 30397 10202  63380 66644  90243 47524  44750 47981  52347 87723    14883 91945  64194 34504  37499 06755  02696 75695  06260 89229
2186: 46633 24180  08908 61258  75856 89261  57352 08722  56601 15354    44039 49494  22300 09233  89490 11906  33618 57755  09953 74748
2187: 79702 80514  97594 90151  74163 84510  85607 30408  19447 85900    47779 75100  00233 58975  57795 52192  29026 39259  45800 72117
2188: 37980 30130  43614 28925  01012 26432  77058 75759  13533 84863    66404 56497  10517 90067  07835 14889  09265 63970  46975 82273
2189: 29029 02092  17103 77128  53189 19878  44829 21388  96526 31599    10891 12163  80554 85392  20001 13719  99985 75248  42603 10857
2190: 43496 37956  77733 84481  29990 06885  34915 45856  68306 92840    12173 03186  61737 59571  89689 73100  84985 07434  69673 94521
2191: 82348 87975  09348 65304  55307 27077  43154 93539  55866 19481    52790 76847  32263 69188  15783 57995  89665 55022  88143 21137
2192: 50147 04874  14419 67518  95697 53553  28092 11790  96774 12158    74963 60432  37975 67009  68450 97380  05290 83793  85810 30999
2193: 43997 87115  44767 02356  97810 78278  73464 30865  56608 16833    42243 84410  01128 95761  87329 14368  43822 78079  23040 49577
2194: 65412 80603  28293 57890  22672 53455  06046 05044  80066 99875    63535 24732  50946 67053  28847 88279  23405 89687  22329 77052
2195: 98625 51803  18096 84990  99003 88715  43213 41521  92231 73902    67418 00764  07149 06008  86695 02243  01404 17881  40611 75609
2196: 20580 58232  99256 27434  88060 15668  75521 60407  92926 63564    63069 45655  58527 12865  27518 83020  82879 84701  33114 56884
2197: 04232 82758  78134 15006  51165 32493  07336 60570  23861 42704    79267 63560  21034 72827  16163 31849  80399 53250  35480 29108
2198: 12438 60037  55720 96961  80023 62507  77904 71359  44190 77204    25749 93598  60763 77448  84842 75378  95310 25899  37917 02224
2199: 93370 12623  53620 93115  26548 08724  17794 49587  43990 54303    25749 93598  60763 77448  84842 75378  95310 25899  37917 02224
```

```
2200: 79857 63504  66399 38183  44215 30335  67845 19791  68122 76923    87543 90702  44087 40684  74150 24991  53520 68329  12586 51958
2201: 66169 80158  59280 90890  32435 15894  61800 79637  83419 53627    34866 31563  04106 38746  14528 40081  63736 86201  54801 47609
2202: 43757 86928  72880 69131  76527 44764  85037 53497  88441 67839    10693 78470  52117 96065  48077 20445  22588 77593  78881 48156
2203: 77224 44203  79169 38128  11183 37110  92957 21128  39291 90276    13610 45629  25736 93444  24777 87617  32317 30791  27571 30310
2204: 93796 49654  82932 04163  07532 41453  70818 34540  84248 89231    33691 31404  24275 03832  24833 91176  15680 09352  79892 26716
2205: 48679 21923  34339 11692  86011 11325  50434 39419  78997 65612    84299 55723  95803 94170  38846 41775  99195 17915  42843 23468
2206: 01034 65080  41803 94832  24011 07995  29594 41525  72269 48544    87418 20584  15715 36516  84079 52621  63621 30379  92068 43814
2207: 52551 75788  52676 17564  19051 06926  31124 11182  14237 47335    58535 01444  08027 29529  94658 15376  91421 67892  71902 65092
2208: 34585 54168  93854 43832  15585 55296  29419 90284  10749 28440    32527 32745  57994 65243  97043 15354  17674 22633  87818 90612
2209: 20844 62324  97924 69492  78183 31456  93947 45231  89842 55317    83276 21672  00113 87099  49477 61734  35807 22383  17576 73445
2210: 07021 47899  92357 99603  19370 42086  64718 30809  50212 97490    76396 82740  45719 89406  13711 86317  96408 65369  17447 98476
2211: 43939 00579  71426 29895  23870 43794  70918 31731  88984 29593    43765 97457  78560 24112  70932 66404  80989 05220  08134 77443
2212: 93657 59154  45357 40538  80659 94830  03618 67704  23479 47094    19969 20583  37698 17198  84098 82189  51540 14714  53374 98348
2213: 84679 58223  69876 65303  93720 60125  43578 89409  17081 39757    42965 20578  14735 30692  60660 12546  17327 14983  98398 94828
2214: 35285 96884  83077 63073  39232 78858  37526 08272  23270 66415    15171 97479  92781 35904  74536 73027  68116 61087  99977 12973
2215: 58047 23773  42007 24251  32068 50888  38975 66097  81982 17064    88320 61651  14669 47166  67505 48543  58082 98742  46919 06122
2216: 18143 42662  95558 58114  26032 10694  79922 48744  25436 65357    52507 57802  87915 59976  79128 05842  54565 02290  84137 75569
2217: 36184 74513  41548 52610  99570 74791  45501 12434  25189 55176    88216 13770  55048 54496  67646 78233  16921 20688  46113 94715
2218: 86342 52466  75543 22463  08129 28467  23641 90207  53875 70621    81338 18387  94383 40105  78607 70666  83379 16401  40594 31654
2219: 84074 41404  70355 90332  33129 32067  77221 85480  65054 29788    86697 32294  24241 75166  82979 99736  28836 97720  30257 97392
2220: 68320 60112  47884 03921  14567 99173  76117 83133  02500 27028    10594 79276  45422 08766  36347 16790  21615 30400  57883 69599
2221: 37577 34920  98969 63976  02543 04398  73058 92645  78170 42238    15742 53822  20012 15972  16041 42024  20560 33549  01341 88134
2222: 68945 22551  95848 95997  05266 40899  71408 62032  29096 51627    43235 08500  65578 85106  97944 49341  35009 75153  43082 64127
2223: 11481 45764  26003 48464  49020 88829  07971 43464  46490 36718    46001 97016  52677 44468  73014 84669  80730 44446  50083 46552
2224: 47707 59048  31305 96341  98963 59808  57423 85027  81723 46376    25948 79141  87876 82343  06157 43550  70451 00816  44101 37263
2225: 46478 33486  61281 72833  64653 56728  88924 39052  31877 64052    94459 32038  56390 37736  69950 09414  29782 33443  83956 33600
2226: 89426 47425  73458 85209  18127 47564  57230 60025  03285 70338    59415 71117  49071 18489  57450 70584  11178 78447  83490 95178
2227: 45033 97801  18775 02860  15262 72799  17070 38051  13156 73411    71025 95820  19659 47953  07703 87294  20398 85556  67079 63660
2228: 64259 63234  11289 58946  03790 10548  15902 00260  65631 02677    78199 32366  55164 37709  46407 51893  85420 02078  27221 28467
2229: 94173 86739  83951 71364  55545 27972  96016 49492  77547 94215    17148 54917  45619 31551  58658 14678  40324 87511  54558 31361
2230: 98253 17266  94280 56238  08059 90030  31536 34182  20824 16329    45645 32760  71447 67735  64039 86160  01769 13594  97153 51067
2231: 78418 48791  29608 71855  94152 36517  15403 04097  09536 56175    78242 02575  21975 99446  03967 06159  49674 00366  57081 75520
2232: 40044 14313  90116 33207  35703 26863  94834 60798  94115 13461    45434 40789  55413 53679  18811 07384  38753 18836  24094 82803
2233: 06271 90901  73514 01833  61172 03857  29695 48210  12496 91843    25692 07663  63992 17787  74005 67790  27845 39526  47103 69886
2234: 71134 59297  63782 32893  15303 98278  63709 41111  24714 12379    47476 64835  87188 11799  57169 55290  01519 47390  89042 61484
2235: 50466 65836  55411 51552  45088 36659  11340 47323  82875 95515    10291 01902  51711 54909  75795 40148  66755 13314  67566 22907
2236: 17549 47859  46672 01508  24609 11663  51894 43603  28198 25662    47627 80286  57082 17146  33736 55824  11649 36380  65339 17445
2237: 97729 20566  61454 47165  08966 65402  42427 34869  04526 99860    55708 62753  21150 71548  00430 24888  15580 01226  56182 14168
2238: 41476 88461  92315 04946  11742 54070  76858 51198  87512 66974    53527 43853  16602 53112  86952 32532  21646 06228  64296 37758
2239: 65278 25333  31997 84154  13559 73222  48053 12732  40031 54061    32367 70777  58575 56748  79099 05191  26291 79168  07879 98703
2240: 66616 97868  50913 72795  90725 39993  38687 81109  03864 45058    64382 96305  70559 52148  68028 02530  74850 36957  84912 46408
2241: 95588 37756  58245 43844  16490 96825  10637 85166  37412 83818    72389 80105  48612 36856  49424 93674  14517 83463  43069 10357
2242: 38436 18052  13661 95126  79550 57184  24593 26388  29987 61653    93600 15728  75175 08660  50774 25081  70188 36753  83320 04482
2243: 72262 69024  95249 05468  03298 72976  59435 24550  46135 86148    83865 00612  98257 87488  72739 95209  02650 41762  96738 90336
2244: 95938 37685  57691 77482  96472 34587  75006 02387  31250 52632    46291 82263  53649 64591  50727 51632  69575 20502  06187 04658
2245: 64910 64431  99482 12694  60505 78805  29272 09376  68297 51703    05667 08019  29074 53025  64677 71255  08585 77396  84256 89068
2246: 22010 78458  83981 59521  58783 76397  52477 09346  90942 32847    40345 41223  83229 12926  97180 74693  90825 29679  98468 48578
2247: 65942 67316  28050 22169  71524 58098  51001 63117  78661 39962    85995 42079  43090 51316  33491 20638  46274 88030  51525 42175
2248: 45997 49992  79847 43032  12970 77296  59368 67752  49452 26487    97621 26554  94615 69891  00340 09632  25638 73863  93206 64918
2249: 86361 66993  34246 89885  68882 32639  97341 48355  31841 36145    12916 21776  00029 14736  47426 14253  44026 01587  43556 83087
```

```
2250: 0560235009 8932324562 6819215997 5524719313 2969441193   5245006954 9894081817 7132983987 9800851429 3851392094
2251: 4587220032 4811800126 0014278042 6502015301 2027500024   3149792289 5290586440 8516838351 1012109089 9761769144
2252: 1835669051 4013073485 7923371813 2397701936 7532166040   9525259022 1371063374 5798322154 9781394328 6489556337
2253: 0986261062 4734640600 3782347901 3119327204 3608695589   6676742398 8041234205 1650693246 5439501591 8506707092
2254: 8010509963 7394078555 6871350164 6238403571 7750109927   7252108716 6603069544 1767327387 7848333707 8319271670
2255: 1129458728 6376148221 1673772875 5591560123 5610935940   8357351770 7960845701 7072136843 5900603018 7758387782
2256: 8378326571 2004272780 6515368165 2000590650 9935877412   5333179610 1693107374 0930699870 8276210767 9817182481
2257: 2008433726 3667341995 2644521298 3074253750 6002106699   6099491860 2464514667 0698780597 4352808964 8780141567
2258: 2988268707 1405801660 1848216303 1216020785 8567872559   2421024815 5143774519 9835493009 7165799603 1640663672
2259: 6861490012 6751181898 5958673258 1427902562 9956305642   8565940064 2762922513 3418945823 8059716157 2777864389
2260: 5360868592 0316211271 6854322374 5936962377 5459912947   2435598895 5762402095 8552311629 8564666720 0442502320
2261: 6651739518 2056074020 7577686136 3737894415 1994312005   4397526717 6729169890 5155554205 9101010636 8312135504
2262: 8347183261 1294299676 0731289276 4151199173 8925087453   1453698552 4105371060 3020278512 3458399205 6714061366
2263: 6332342046 3454971406 1323864265 9030700956 5212972740   0418745040 6931716430 6137917377 0870934348 8184595549
2264: 1049322222 1877681594 3129014007 5364925834 5702329375   5494257678 1481525955 6476738987 2336208329 8159343066
2265: 8316116526 2495643226 9222300657 2032184840 1766804427   2242407878 6239425524 2449245005 7418713187 3302038208
2266: 5268495327 6497534062 0654102392 1196883998 9403031193   1179202653 8740711408 2726524406 2174350381 6571287180
2267: 4163198568 9109692907 8885557144 0216499240 9621328895   5328488872 7584521579 0609803788 9538522987 5921814974
2268: 5770828287 5505944595 7907730196 5551035386 6691528375   0827113422 2380469861 2055898680 2114209299 1623691115
2269: 8790198201 8335943291 2486708843 9418477264 7578924805   7132116184 2479513025 6446976046 3178103639 1920909565
2270: 9006871520 6522629581 9350407804 8890879401 8887760197   0913613663 4947720616 7384593071 5639359829 0562845886
2271: 9478457905 3140674721 2722998313 4253344567 0384491773   7617759063 2833057318 8107359875 7484043805 9693792062
2272: 2136827467 8576304389 0649672914 2202111696 2437054599   9046226873 8065512950 7022508414 3441842600 7746458312
2273: 6150054502 3133058098 3052996535 0567149298 4581239400   3407522940 9625470826 0053590771 7167945627 1055524773
2274: 8905097957 6716879559 2780558037 2014588087 2858083887   0183687067 0813157477 5617404244 5113755950 5208088119
2275: 9126051880 9605744798 5962098409 4570464660 2658682008   8197794572 6095305790 9826576560 6018457881 9371978309
2276: 5874042369 7527482624 7481156917 9318561306 1977944220   5638189315 5490305804 4599599197 1702980365 4049341670
2277: 5461007427 8839121832 5568811216 1650733588 1731130407   9380235419 3704822023 5663752271 5502237856 5280566855
2278: 8891455902 6446766720 3166360707 4249958451 6651591625   9754152436 2830614692 8775356836 3893116923 7443170951
2279: 2862980683 2771259061 7948393635 3850320787 6036865992   3889553645 4192463279 4223639435 0688095798 4717071823
2280: 7350472627 5061983063 6916208928 3908316711 8701629549   1125992057 8983719563 6202376365 4309793040 4788848026
2281: 4659709206 0389234180 0613588908 8453034892 9581649654   8507219555 3488095808 3596458973 5016407751 2556332260
2282: 8155122581 9480003925 3450135176 8422196305 7126252323   7519974795 0545892469 4286859180 5540432938 7765178897
2283: 5506565674 1250847951 5081816949 0861561003 8625097995   8316722444 4121823311 8342082831 4815229201 5254462895
2284: 2952644122 6688768376 6569013482 6959318253 7193450313   6016760435 6244608129 6172099424 8829349597 6194247807
2285: 9979501179 4981959734 6362499059 2237687269 2205043023   8138873210 8428035857 3096590311 5398460522 3511884172
2286: 9066502316 5510849370 8192010104 1585797840 8259581658   7493576760 2759132244 7052037032 1428642475 6927736791
2287: 6702971763 9049681714 6783494243 6393726494 6890015996   1101173755 2536339273 2972666788 3158598431 2798470407
2288: 0642750117 8830914657 1218512832 2413297024 0381985742   4655229282 9735621670 0868786163 1783436802 1300643301
2289: 5727806162 1993433803 7422412408 0805814688 0872104487   1059227206 9507858766 9872132215 5980678838 9365632986
2290: 8066025881 4243229802 0531160743 3395408655 3322016095   2795070202 0890050277 2567256662 8646668437 9946701773
2291: 5091657582 0372445906 3311131096 2726887133 9376576426   2964338537 9928302541 5410950503 8143950783 5992192126
2292: 2161134596 2842248889 5791942187 4886748398 3902187048   3921692116 1548323708 9210996775 6927359931 5933530256
2293: 5487263206 7781475962 5703872626 1157653203 7467710206   8068480324 7726570189 0104896363 0843656868 9325446005
2294: 4451785975 0554197464 9995232701 9417099921 6176783419   7174116363 6101029440 2479066194 2106073865 2366758426
2295: 3787704361 5141390706 1154994222 9994815024 0409975713   0972852573 6177935959 1785460451 2482955569 9906372403
2296: 2872041813 8835653564 0964649170 5831387736 1399534430   7946689161 2601784389 2550209383 1235951127 8978175026
2297: 1069063758 0234037813 1002970122 5394697408 9198202800   3529192756 7943376164 7851974922 4417551185 3220324564
2298: 5298052708 0832025022 7479855946 1862822447 9355278267   0470913284 9782897309 7921961200 0623378231 3420241902
2299: 4133381190 4294109423 5195477571 7649297056 6284637570   8853548655 1378419015 0593319539 7762177886 5064992778
```

```
2300:  53015 94369  81589 40491  98418 57106  40391 14787  98433 78791    75412 48271  66217 25040  85276 73955  53513 77148  91910 34960
2301:  05132 40234  71346 42670  29378 87275  89235 76442  13743 94390    15874 19893  36897 13832  81806 05645  44912 80447  97251 87539
2302:  25710 95436  71471 77979  67760 35927  48518 35550  72318 59926    27959 19717  46001 40354  87213 04949  36098 03904  07001 51730
2303:  20972 48181  82163 09073  92802 92021  25572 97200  48835 20499    66926 56224  30508 48067  32272 08722  09087 57050  58372 39045
2304:  72721 27565  97134 03856  62816 71677  79437 59277  17260 58713    18641 84868  24429 34758  30810 96330  48366 18442  54943 76078
2305:  50915 85814  74126 74264  22050 69467  72831 77687  99120 91351    05557 81221  42642 39111  88947 93380  97958 22287  48044 24541
2306:  34918 47360  47389 42476  45428 79456  96523 89818  02757 79547    56803 68727  70951 21316  17967 50774  06890 05821  70850 59303
2307:  42285 46842  15895 75844  79619 73523  59794 91930  45146 79926    11960 36288  54397 87198  11654 03870  40646 72723  76946 58090
2308:  49272 81926  43305 33684  73431 88554  85885 67517  59154 05779    20602 11523  32629 94444  20165 14171  99948 01410  71388 39276
2309:  30420 71688  80254 87688  96745 73657  02270 37406  78115 84482    00316 66631  28025 36487  11538 57309  24173 26566  87614 36941
2310:  51216 67147  89313 62114  40362 81688  04964 26952  57304 50994    91692 70255  60295 56070  04633 58651  71646 24250  28202 36283
2311:  43152 01480  91996 60385  12802 24017  87886 19314  97889 31296    15790 50234  28210 97021  77419 96829  93358 28797  16581 36925
2312:  98317 58886  25718 76413  60806 82012  19487 23842  63132 61097    71080 68749  49901 93011  43313 30669  28800 89359  79478 42868
2313:  68047 06127  91137 98897  53487 67148  98624 44777  58970 40889    42986 26327  09227 22048  22569 12576  93662 49778  38483 36070
2314:  36326 88973  85398 25665  03227 04208  18785 99587  96986 07224    02519 45081  75729 89741  60298 07368  58540 68096  28611 70884
2315:  11524 15031  84622 29320  67809 80488  32991 09022  93168 85199    04478 01220  84821 79339  71933 16840  35937 36678  25109 40657
2316:  38765 80120  18732 47448  44143 71828  04035 26859  10291 77972    57902 79139  09574 37472  20457 90345  89716 34179  35103 72397
2317:  55776 17293  52385 60568  89320 63736  34214 70138  73352 57187    61669 60221  58519 91350  81535 39140  53100 39267  69920 42628
2318:  97093 78532  44577 17572  96999 87592  40286 11101  07738 76373    97212 64389  64441 41571  83918 00072  82453 99284  89906 43507
2319:  44301 76059  67946 24406  74095 56888  00969 69778  47288 39455    15416 68537  51142 33234  62165 46763  83900 26815  33664 29938
2320:  90799 54755  96761 54237  52099 86470  78463 80382  55474 70828    41626 91929  03977 74941  71456 05902  40182 16703  61400 44969
2321:  90273 94528  96663 56331  83912 01581  95476 97945  84741 03921    96172 58226  24626 94286  26222 21548  93462 15066  71055 08863
2322:  58379 46419  50232 07052  53200 11350  03341 51855  54300 12070    43556 98558  46213 40859  61511 62319  82613 70693  59822 11886
2323:  85412 20598  18589 98951  21151 33605  77967 00939  13996 40307    97100 46379  00517 05077  63145 48458  52129 76956  83942 49765
2324:  64628 77117  16458 44817  83407 11467  58121 97511  29812 45096    01792 35604  89352 28167  15230 74848  86908 13723  31692 89968
2325:  18153 52910  59457 30272  09528 72840  91188 07555  00126 47015    85628 29668  38181 44018  91512 45412  04987 98127  53263 12150
2326:  47281 38174  00459 62409  30069 01731  35038 65467  43214 15452    31155 72992  12481 16512  68627 71728  11328 58111  86561 26507
2327:  25640 54273  93914 96942  93587 70628  29415 10989  23180 31235    22663 67824  31325 96520  26848 73229  08691 01439  32156 42442
2328:  00835 03182  29255 03205  85981 15130  97651 77208  45925 57682    07753 05977  89031 69843  70402 90169  17269 26650  18282 63438
2329:  93166 59650  51916 18880  33707 98622  39208 19934  99476 25502    01783 88369  56791 65550  68120 18418  35521 51253  54087 07762
2330:  29239 62886  03412 54156  70136 97689  69430 68093  88498 18537    90939 53144  89872 68335  15926 85677  56696 74957  72361 44678
2331:  33492 52245  64157 81682  17093 05566  00160 74467  64108 32820    17840 92897  01670 03391  05321 41373  48554 75163  42040 59953
2332:  05391 76342  55654 04076  37106 73945  11289 75510  06533 78167    16291 97701  13854 15372  31341 81504  90966 64902  68608 23694
2333:  41398 32153  22509 52895  04199 12858  58006 91218  53365 47728    58213 50790  72170 12568  24085 90778  73044 59740  93587 92763
2334:  87262 77464  93369 42852  52382 56475  48378 82255  96319 63966    62658 00289  50753 15248  62639 50063  51848 26740  36956 58058
2335:  56347 10654  44211 94078  44021 88169  14618 76257  08194 87685    33827 88930  90745 59416  74308 82619  35140 17674  07266 40637
2336:  32867 81060  88817 08262  25630 36024  56243 75987  11756 81804    38295 20346  86115 32139  19694 12821  54388 39163  27768 90821
2337:  14103 61240  53029 10499  15419 43486  81288 16089  26464 43595    24747 95249  91492 93432  84607 10087  64259 09557  68301 48372
2338:  71758 09655  09847 31274  99406 80299  40752 95321  09053 01371    51329 35815  78391 37815  78436 30271  27927 03349  91174 51282
2339:  17038 16633  96670 32376  75596 99295  05776 43964  61299 05142    00478 36428  42604 37356  07876 67238  86179 53280  20340 90828
2340:  41606 58237  30361 97393  08160 72634  44298 35881  68938 25113    43370 91029  32428 94734  32064 71243  73399 72726  26456 59341
2341:  12470 07317  90070 12463  87209 58875  88536 57140  56972 17061    20565 35172  76381 14104  49820 70014  83649 18642  79684 63084
2342:  34825 02758  79518 31229  65784 88910  97909 40224  69385 15993    50582 38629  52613 21451  41093 72095  24366 64955  19667 88204
2343:  93448 69327  16718 08730  68994 36314  70451 98341  17941 54460    38550 69939  82875 95589  94322 21531  02433 96189  64957 88078
2344:  33388 56842  97608 70120  81074 44091  64982 35800  97117 60695    88112 66792  99546 24576  58143 53139  58527 01664  61658 45280
2345:  60221 04968  19313 11044  19700 41708  61218 23176  53963 85446    50136 58237  33441 94391  80577 46021  28429 20519  26897 64470
2346:  01128 87338  99921 46800  58074 01045  80712 68232  44974 10601    57842 76271  75538 74967  56081 68447  63506 50367  07603 35899
2347:  39966 49864  93730 79460  28992 35820  69046 42587  87072 10302    33534 28244  36721 52072  38370 83380  62787 28905  23928 62111
2348:  27540 06174  00099 18502  71402 47732  27513 96025  34976 26782    75246 75620  58951 92080  91921 02970  07448 70893  43503 04990
2349:  62183 33876  88841 34675  39410 18901  47672 71092  88358 56509    46665 12139  19369 42277  60594 12886  84938 34142  88419 04102
```

```
2350: 24155 12224  0058159249  84249 90312  99463 47993  59993 44597    5996284705  1384483384  6248593015  8646353976  8296601103
2351: 66961 35581  94939 32378  3448493020  97990 24124  85821 30436    83750 53668  3268187178  58593 28675  60090 29582  19386 03826
2352: 65750 31068  48924 50618  00870 34544  25947 97686  90369 23793    9791174230  12949 64805  90232 40221  44185 55777  7365488998
2353: 49887 83485  0862573849  3219208408  69258 01495  39971 55797    23300 84742  76598 23060  61128 27751  60407 52402  61186 78440
2354: 13749 49110  8006698995  72333 29662  41038 38375  3801309444    8391001826  24944 26993  3675374304  30462 67052  19170 32404
2355: 08491 32742  38311 40260  9183404688  1612198290  5701675215    4835591904  9991064940  42725 07579  99435 42966  10233 34787
2356: 58001 05814  3953690317  80838 20055  83558 59185  1450694573    52227 30439  21822 46212  30031 05435  2769678510  2777874981
2357: 61353 60226  13105 18142  7936467935  65858 27384  1203797259    16227 83997  46142 23874  78201 28587  45696 21950  0685097439
2358: 61258 25713  8724775322  0147577165  7730429019  4361453850    0246083960  0454881830  3703296910  33122 42493  27389 97004
2359: 73276 58384  7483389368  4982265975  3906883545  0186412829    7777393354  8740098010  6166198528  4318382572  5130675051
2360: 57118 46833  8353210217  7935208298  38406 45061  48998 20294    38880 39679  12384 33653  54304 37014  7985081672  18730 05387
2361: 97879 89579  11418 66123  60617 14228  6015213250  0502207558    65148 51221  8242570653  3265182713  34475 43129  05509 13514
2362: 46468 78181  88597 24164  24513 24437  9688973859  4054374058    35775 24050  33436 87000  0733107127  9074928777  10695 60467
2363: 33049 27771  2999592376  60487 83723  1694332286  6539413624    7700451658  8971280232  62851 50878  36369 65015  56287 49216
2364: 69591 39963  9237230939  29289 29519  12167 45027  72899 55499    2802097685  2804341286  50809 59000  3430761954  20818 53146
2365: 91899 34651  89445 66368  85161 51200  11399 55889  3961780496    2141863215  2733575443  8767433332  3673503114  2396973754
2366: 65920 76872  17179 75747  0282167129  1534491577  5411255862    7252083718  0015136825  9949424755  0646462759  2212016258
2367: 51655 93578  15770 15511  71389 46980  28573 19416  1183178764    87636 46293  46679 95831  6178437626  17608 80304  15286 34546
2368: 40220 14066  71678 44468  75059 81248  10287 56824  41285 57665    69150 20032  8545315174  6574572605  38408 31696  7448708954
2369: 67836 82931  14839 21924  58506 32457  78146 12818  4486481970    39737 49132  8484294014  3775081835  78061 65921  39567 53029
2370: 97809 07709  17611 28073  69286 61384  90487 36926  62196 43552    8913675931  0106170199  7228992674  32421 34307  32626 17084
2371: 96354 80198  70809 19942  21027 51448  0148604937  4770571134    5837389692  2534276077  5121175999  56545 58707  7717974883
2372: 89131 40626  40012 52476  0214656663  8616002699  0369462543    2998016496  7894043138  3680018328  55937 50470  0646818567
2373: 64725 50157  79728 58765  69236 82275  58038 92742  4567172326    48284 43796  9138877579  35376 01815  0117490408  47293 44208
2374: 66729 86426  6342579749  4259297910  76470 44948  3574038790    36917 65276  65268 68586  5147243699  7744465902  01541 12773
2375: 42627 84952  0305302127  8368431594  72710 62036  1458879413    20425 84538  0557038003  3891400284  59743 51096  33268 26628
2376: 07373 43210  1618495930  8785813627  55688 01261  4892465712    42799 30705  5764811851  4470145865  21097 22611  7836765812
2377: 69024 84367  38247 51317  4443674877  05098 68650  68929 70479    85585 60831  9200582114  2102114333  21528 11189  7900415117
2378: 40000 94252  39789 43465  59580 01816  59743 27473  5598075798    63496 16058  4153215277  4799411614  38408 57900  87075 39022
2379: 23016 84039  04087 49781  6333477998  87945 35291  67726 33017    0093023212  7498462090  81757 21830  88441 48299  27149 32141
2380: 59612 28915  58639 03556  50737 62804  0438579187  3175393719    5167071414  50439 68390  3696472963  1963466631  0441113648
2381: 61865 01456  86677 79002  3717193691  15029 17839  76240 13847    46514 84089  61133 46648  16427 57554  93171 30530  56494 29187
2382: 04937 06705  45479 12407  62724 83402  1357885542  53669 69328    03107 44945  66129 07113  51741 68286  6398696045  93398 11223
2383: 15255 17742  9193275141  17478 48539  4563014248  66910 38012    63746 17708  0792024371  82844 38203  50493 26415  62519 34845
2384: 56977 49301  40553 07530  6711887823  9209652340  1336063426    45252 66330  3141996764  9177107792  2399251091  32734 07218
2385: 1058273721  10799 04084  06267 64920  7817491890  11888 46003    82315 83040  39107 28568  13429 67910  4388319287  06249 61210
2386: 01509 14165  08359 52302  4369124977  57924 43385  2415971533    0841672507  73801 59475  53812 19673  9052759423  6924079654
2387: 38694 52860  9189566360  40000 01722  71627 35343  4153736978    4704200693  0226015315  8248093448  91853 30027  00427 59672
2388: 22183 88529  04733 30889  95709 00096  55115 48181  95231 15713    5251282844  07587 06099  9295163994  1494565419  17815 55222
2389: 88129 78071  00929 70874  3042314849  74521 05997  96880 81897    09948 58695  97999 03912  74999 64659  61755 50086  42633 62219
2390: 26296 78910  8798674122  74374 75434  21376 76232  6277897697    89268 68692  0387635170  59853 64839  97582 11611  22301 44134
2391: 93533 07704  05081 51918  3875050291  05070 29543  76602 56682    3111986255  36353 20288  5722461593  00089 53001  68340 20565
2392: 03079 95356  25071 85873  72014 84738  69071 01653  65140 86156    1398898861  94834 95851  7159101017  51258 42201  9786267135
2393: 54894 84680  04787 43169  79275 33683  15839 88515  85194 86074    24137 91795  99705 08434  31989 34383  17103 05429  2119365838
2394: 93306 94392  25248 76873  27277 47134  72118 83736  05751 58928    42584 30859  0362357706  4823293422  9560696464  95179 62429
2395: 34212 21303  08631 22167  86543 11649  78104 35880  30533 42496    35658 10370  75299 25972  86722 52906  0631498829  98338 68171
2396: 45894 10472  53108 90526  82598 38831  4315270536  81299 73584    10119 03830  47073 46572  53303 93029  58645 06729  52075 97778
2397: 82679 51754  80239 19192  34976 40332  04126 56269  05969 00516    55239 27668  7617306446  53094 36986  13659 18385  75123 85976
2398: 98886 09399  51907 82436  60952 86831  69507 67789  94780 11862    45930 15999  9604044617  08020 34581  61059 68463  86088 98691
2399: 67739 65079  8596393846  1248611752  93561 43050  1371988874    6296166347  16569 21368  28080 40352  72050 06947  27721 16286
```

```
2400: 49991 12486  93163 91064  91472 53990  10764 04929  21496 44908    50069 12495  02521 67109  99563 67021  22296 55502  15357 32718
2401: 66623 96206  80065 20103  59162 78350  88764 15291  26351 17299    25287 86268  03005 75688  20667 74488  64435 96353  40363 61097
2402: 78762 12703  63187 04611  93148 27481  23463 49406  50612 52048    08946 45235  51866 41826  72490 69986  20593 48432  02998 81636
2403: 15660 40204  36809 70653  33911 85343  92571 77030  47665 39286    37653 37253  29384 43484  58065 10615  41932 69323  46492 75852
2404: 93726 98663  94706 34751  68446 52742  42661 77038  62396 74001    60625 19156  98324 31814  08829 45111  68000 39525  63989 40544
2405: 75221 72820  09127 06519  84652 48392  19175 84190  13254 37686    28358 67483  13244 00205  87645 74828  67434 73854  67648 05691
2406: 40980 45927  59098 25223  40916 14337  34070 11207  73019 59880    23622 98668  98874 45810  45807 26927  50918 56238  34131 38562
2407: 58489 23781  92520 21131  88779 73649  90211 52269  13140 88219    47053 48135  51051 71438  99414 62779  00588 30678  45684 69178
2408: 47547 52154  64973 78335  56931 66599  48855 89411  42709 70465    27368 76837  09741 64384  45183 38010  73295 02563  15254 73793
2409: 96092 04754  27290 36237  90915 90763  54534 57634  82525 57720    05850 22847  24563 17174  17054 53263  30714 40275  11643 72554
2410: 59732 82759  00335 45996  86349 08986  07651 39534  41089 59976    72743 05405  94131 93374  61338 08094  13130 67685  15039 76638
2411: 53195 79702  37482 83934  49813 15961  36661 11558  17634 36992    95740 64528  69563 24573  69304 21502  64204 84560  74648 22148
2412: 92288 91922  41492 85870  97475 11523  11882 23748  87573 76862    25777 72752  94139 00048  46898 51066  43630 13988  07672 13688
2413: 95899 69384  37475 94677  30224 33826  73223 77737  37596 98930    82303 85150  40452 28723  26872 50852  95330 08869  14281 72932
2414: 29952 40096  33070 46047  66111 00544  51258 03987  81303 82138    53380 91414  14958 08700  43697 76618  74320 72655  86350 72127
2415: 84809 43524  07657 53536  63085 89897  38879 01890  70233 49797    35108 01721  02863 77688  47008 25186  81840 01171  04206 01140
2416: 68377 14216  54695 20205  61778 35095  76016 95235  36560 00799    96807 81287  55292 85537  38063 16024  18001 19344  16066 55825
2417: 01502 29065  50942 69445  83122 41274  09802 33478  69737 34232    91001 21750  21807 89549  19539 87931  41007 14119  42879 34493
2418: 29259 59322  90085 25708  92856 52759  68674 24452  85180 48580    01664 40709  59519 94273  42260 54111  94750 07158  38475 55739
2419: 02697 75387  74352 35477  89323 70485  27413 95677  16532 01609    20850 57459  07646 67876  30006 28838  48187 92376  59402 88954
2420: 37988 10913  07612 55793  68192 28066  47623 56077  53978 88017    49885 65555  91758 38303  40649 01623  31366 91932  30519 26979
2421: 28148 51585  33294 89909  10470 23855  98746 37286  47025 44722    54604 39166  15455 53327  51587 62079  55582 28842  44158 15743
2422: 78532 38884  61175 18284  36701 62162  08155 22605  87503 16770    05255 42304  44631 82471  21112 32983  52657 90516  51796 53975
2423: 75532 05506  39746 61909  30402 16823  45779 78447  79159 47112    97864 81944  82734 65084  77660 32814  47126 69327  17119 90278
2424: 71025 11709  32945 77733  39643 92419  96705 42606  60628 71315    37146 60572  11042 35738  67195 28329  71302 38671  44171 41349
2425: 82725 42115  55003 66890  55193 74725  44789 97442  20575 83641    34727 02712  93979 67035  89231 51399  19457 87476  20122 18756
2426: 71450 81516  92418 11065  40955 95710  64747 15346  54364 28104    71950 09573  87076 71191  39106 63799  07713 76433  26046 29926
2427: 84968 06668  00804 56801  20431 20289  49004 21876  91199 76348    41225 61431  98287 89752  03195 97005  63047 47415  10516 71201
2428: 23492 25136  75503 17927  03142 22354  43029 57938  61387 37871    88380 23998  07526 53519  17227 41140  48782 57429  30856 26256
2429: 59261 59545  27214 74695  59304 53906  67453 51445  25242 33942    48159 53102  80122 67067  63617 05454  72000 47096  19926 54849
2430: 20576 96171  36046 75412  91469 39560  00654 42869  83022 58164    85264 18415  97221 82557  22951 68481  68911 98767  92681 46680
2431: 17643 11658  54268 65187  56090 70522  94253 72429  53902 34498    16463 22975  09544 97615  55201 05637  79754 47403  60191 40704
2432: 03435 24407  66946 47678  00515 80780  67284 26498  85877 26389    72384 36340  16396 55380  54018 84511  42497 80296  12282 64698
2433: 15917 75604  75149 30056  91095 75891  58553 65547  89425 42918    11286 45364  57157 87224  92962 34053  91372 41562  32218 55206
2434: 32897 09313  00239 72477  73719 36084  78645 36399  52724 89988    89913 51696  97557 05398  35241 30479  49469 04690  67233 09514
2435: 50092 39494  47726 72536  48975 49188  77397 08347  05272 07918    34355 61559  44813 56632  85298 73512  11901 10701  89387 15283
2436: 67953 70500  18722 26201  73838 54075  76254 80237  93023 27824    18739 87602  50940 90152  56535 57937  54326 87772  59902 11670
2437: 90795 44858  37220 81496  23512 16071  45049 43969  12175 30312    92064 81106  12501 60220  47359 77250  45895 41850  04680 58613
2438: 58158 04304  88656 57395  18744 40238  05395 24572  21786 91698    82948 73629  17539 26579  85515 47888  31723 56485  01083 48881
2439: 13926 98289  31759 78643  92508 65164  66672 26239  81970 76548    91588 01893  28764 93629  08863 83404  25655 11997  05346 75263
2440: 53529 15308  40109 64731  78984 33732  10773 26845  17890 57509    07021 31779  22623 14604  63287 70599  48855 87412  92587 86564
2441: 91819 76997  66536 00801  31541 21776  27723 53973  32525 45859    52714 04260  24534 09393  15798 20144  86346 10381  36107 24584
2442: 21083 72723  83467 85541  99780 87498  44942 81751  43879 64131    51538 65849  78347 49156  08604 25396  68024 90118  81101 94688
2443: 38281 52395  91029 89285  55530 05300  62440 32482  67604 38383    64051 81047  67125 59640  54413 83981  01644 49801  36285 62412
2444: 97972 37212  79835 37039  18943 72260  76120 54535  31066 75626    88182 10286  07119 89592  45060 69164  37309 83777  19382 14654
2445: 45428 87776  42120 12144  60844 81351  47565 79798  11762 01044    78097 01696  43303 96046  54378 76977  70152 98435  24146 68226
2446: 57947 04643  33935 30820  51778 18465  71739 00515  96760 07325    71948 97242  55936 11588  81576 28265  98454 55941  83693 47838
2447: 36024 52204  23645 82436  61683 12711  60294 28118  12509 68782    13565 77628  63923 45426  02881 16949  91379 06962  08107 37374
2448: 53435 78258  38386 17683  32994 37269  46076 12820  40259 07340    57753 68823  38704 01463  78796 10294  83145 20007  08869 91625
2449: 38982 47375  16975 58409  47423 02335  95245 80212  89222 00542    14912 03075  45613 46023  57846 39800  54230 14990  38928 54157
```

```
2450: 20066 21294  61017 07845  09915 02440  02429 51973  75952 81096   53338 87481  63640 52680  27251 79225  34124 65496  88612 04514
2451: 15630 16845  51504 11797  36213 66320  97349 59571  01175 27172   39764 54641  79806 68157  86255 91307  91115 52880  39878 86008
2452: 07808 43049  77407 08375  29835 28586  07770 46480  40973 34272   41833 71907  42771 81983  05867 26972  12412 15149  45618 91165
2453: 22015 36740  47604 80763  65903 74901  67996 15223  46233 57860   70610 46520  82268 48743  48973 68200  42724 35749  63597 70900
2454: 82172 52308  32259 23108  38447 52550  75053 89288  01573 03210   96000 86560  43527 29270  57789 74251  05417 47035  17574 10601
2455: 18118 44850  48774 66004  55820 96012  42574 81549  37392 29936   19453 95031  39133 40713  09423 91917  86306 92496  81380 33128
2456: 59159 60468  85750 74462  27610 07007  62271 55582  34383 66076   03005 19764  10134 15584  72681 49273  32480 09145  91741 06830
2457: 34714 48562  22730 40992  46222 92175  51176 75741  87020 29542   34200 01117  30794 50002  34780 31533  92193 38734  80806 07350
2458: 89592 41645  44745 74176  55293 22233  40906 26086  13005 83615   70643 43511  57821 74307  14849 13793  08882 72707  32589 62947
2459: 60053 53130  26671 44010  81011 56274  77634 03265  56970 52713   62149 96490  62044 61320  45560 38935  88641 95629  57770 94730
2460: 27046 74255  41607 91525  95014 22074  94131 92990  42774 98150   40134 98545  81527 27519  15418 29526  98478 92564  78287 21701
2461: 72906 37944  16853 27325  27020 07013  88361 22550  22499 33073   59588 43830  76229 30040  91044 86866  30537 36683  93292 59958
2462: 29821 95628  44608 39524  44287 31854  99964 51826  78456 17774   99500 28712  09321 23450  72828 96654  93446 38329  68689 72710
2463: 63468 53303  06842 47093  94493 37341  12846 80149  43131 01659   83793 43320  51757 95808  49448 48322  58041 16376  52393 92594
2464: 68240 05593  15099 78618  48430 31024  52064 78458  74867 96940   55817 96907  24524 11571  77826 04378  05164 28374  60301 65853
2465: 71784 71396  18183 30122  86686 90181  17913 58280  17529 49284   45398 90071  08398 25543  81077 85467  80583 91669  49984 20061
2466: 68068 52956  25756 18644  91087 40455  20058 11275  22553 59381   36743 30378  70804 49484  75827 66701  15603 87695  99037 66688
2467: 83661 72785  94723 03251  33494 55943  96007 77566  28637 88507   51601 23495  60446 38475  76640 26678  41904 46881  42689 33861
2468: 35993 64501  27677 91610  83420 69360  09046 67808  78684 48479   15542 84920  37075 09454  34474 65718  21779 28656  12266 77758
2469: 07096 90492  95279 46565  26393 92962  38026 43332  67304 33757   27950 58738  76282 84549  73774 23877  35734 14960  25548 59348
2470: 80769 25057  04410 22133  59529 68103  10507 92458  29215 24983   16856 82462  66857 55898  73189 09507  81613 45431  39346 82391
2471: 61789 84277  61753 81339  99646 35740  17051 12283  00440 50623   69743 20829  16432 97484  82999 39596  15900 62871  67961 54683
2472: 41342 73292  40333 87733  82256 59191  53168 43800  01423 44878   46772 53432  32540 53167  50792 93650  87678 64529  89901 77281
2473: 38278 62394  87019 90077  95165 94961  84253 67736  51465 07406   80325 64947  25298 30116  71897 60532  86650 76763  09359 65236
2474: 13894 57572  66633 23411  51617 15053  12474 09066  60850 26331   68660 84630  26931 34137  38266 64498  55070 75492  46952 42704
2475: 29830 13039  63722 52236  05935 79451  61055 00653  58052 68940   17179 50126  89030 60834  00718 41915  11876 37053  49595 09059
2476: 91418 75871  94602 99660  26175 95254  93970 77535  15166 50747   43127 39498  97215 48046  96065 01012  29943 73820  80080 45851
2477: 01597 94165  79260 02861  29188 74875  38893 37186  55358 74891   71008 91816  80270 56352  81879 20254  17544 51868  60057 11711
2478: 76408 86488  64215 60840  90881 72504  61407 59193  19239 35668   55128 22291  01983 86069  47970 34638  18634 46145  78344 32316
2479: 45013 08251  47500 90550  78666 90485  95669 02841  14542 32700   81430 97071  18054 07169  29210 12540  23187 53650  36074 03096
2480: 14848 35577  10806 94082  61845 51993  56171 07609  77677 33740   71756 19518  99366 08095  14756 77868  84717 51351  30770 46402
2481: 88249 46294  66098 03296  77412 92704  23886 09103  65226 32164   28378 25915  04203 65244  89769 97305  28799 03029  96748 69545
2482: 76446 52737  02131 90664  88242 02592  45884 40583  17972 31841   30989 33064  38681 75339  02765 34412  52295 22665  19012 81679
2483: 08690 35126  97034 52837  26305 07941  74293 33374  35227 01745   35316 81199  78811 86939  43919 52137  25151 37708  97990 62692
2484: 13189 70081  35008 16527  46513 51127  71959 97559  04304 31639   52210 22245  97222 89378  08039 85789  70445 56639  28963 16857
2485: 97287 83373  47033 15075  78268 15288  45924 07266  89851 48022   56029 57115  05326 31851  06121 18812  08250 78158  09167 66664
2486: 26584 15528  89092 59359  45414 70406  19699 77715  79817 67249   69169 80397  20267 31141  63869 01945  15112 52369  42867 04056
2487: 96933 00036  95791 52171  26208 90638  22195 27290  60975 90329   53796 08668  87129 62765  12616 41164  88154 51535  66504 21564
2488: 80916 07487  36299 07705  49387 30250  02051 59250  67245 19648   93992 59208  69981 73006  47337 22203  65026 89522  69885 88540
2489: 12599 46925  97992 91273  63292 04787  02038 25210  61659 85583   22028 89054  41577 38655  63169 99550  72740 43995  13981 75605
2490: 65435 80033  22282 87752  86884 15183  15835 70092  09206 07427   66446 11993  93203 26938  72428 77881  50096 66996  35569 34310
2491: 43120 59001  81888 93145  98821 22462  83831 14339  53109 94962   83413 29230  72015 35521  39758 48721  31846 75448  65491 81884
2492: 77953 18138  98579 63278  37777 32296  81899 84970  52357 41021   41084 63981  79319 22657  63170 78474  34875 99932  61482 47026
2493: 37906 89714  53149 84121  26227 00576  96404 01831  17421 77171   62720 94563  66371 28280  23986 06588  99335 88989  76001 74334
2494: 05159 76975  18504 26584  09758 10864  88343 09913  29339 59949   65287 87987  43930 41598  23713 87399  71951 70081  94840 32412
2495: 31544 10118  51776 93211  54652 62350  84139 91249  92802 08794   05546 68472  59668 43073  06903 25432  92117 66828  56870 83866
2496: 39525 87290  90379 50604  85936 97690  74486 17427  58294 23540   45477 77618  45836 47859  18294 48045  71021 12971  92953 32370
2497: 30065 54478  16543 11826  61437 41047  14294 99328  64412 03008   68058 08414  29846 11599  49879 16472  75861 62531  15163 98388
2498: 85747 07018  15215 62523  20473 45578  28847 98466  63960 91391   81298 72087  51690 51874  79029 15625  51708 23023  04553 46388
2499: 86754 29325  31092 39788  10744 73830  01097 46679  13179 37773   81057 84957  62600 01122  59737 39969  99100 72138  63981 42315
```

```
2500: 71078 06572  13341 44519  64482 17250  16074 51270  94048 67031    22001 49730  39878 87410  64750 98345  11264 41258  65520 52179
2501: 57940 57439  65812 46481  39072 13742  09342 70265  97604 07223    20842 98767  32721 31376  25375 89543  21555 49193  11344 96519
2502: 70999 12845  12643 65431  06787 90839  78182 01166  92592 04364    97780 00673  28436 19895  40062 60420  84097 83111  50008 79682
2503: 18457 24713  84992 47798  66159 09335  98315 46716  18114 58164    73541 12945  45224 08613  36227 07046  18938 28561  32062 91442
2504: 04627 33268  72330 84094  79245 00910  06359 95090  28742 36790    08377 34177  63109 02843  73118 57949  13810 90301  08251 61913
2505: 25434 14117  32414 16418  50947 98996  47334 78361  25054 97206    48266 07215  06316 11291  01148 35613  99132 79434  07672 66173
2506: 17983 92683  96199 78058  84108 59000  88699 20730  85314 97212    98854 07811  45073 82452  93580 99284  50777 98715  07083 44993
2507: 28087 62894  61236 32830  58023 51590  27917 99750  77876 38942    56356 07622  65576 33134  54258 52958  77592 58291  39254 21077
2508: 00468 82990  19991 08673  16830 15767  55732 00825  55631 85176    35342 09709  84974 88171  70877 24519  58470 41083  08210 24726
2509: 62240 15977  34654 90403  64906 91879  65621 21544  43799 37466    85780 71334  40758 20012  30518 03496  27168 47898  32971 28620
2510: 93715 39153  91561 50182  51756 07139  23654 27297  43016 78668    11445 16707  76574 78279  30303 71161  23431 39661  33048 24649
2511: 87297 19205  43230 04621  14750 09278  63398 63186  68518 82083    64123 23212  31426 09676  28642 69610  46122 94641  76028 70599
2512: 77642 43552  03192 60046  14856 60978  45052 66801  37554 23920    34276 88288  23849 99945  57182 47375  16909 21448  22196 73248
2513: 17927 85052  34153 90123  11212 65130  57018 16703  90737 45665    58715 33537  96094 40687  04272 58870  17311 79718  08294 99153
2514: 96738 48354  05103 71254  96074 59406  12206 30401  90358 86656    29249 49831  43929 96479  82684 81804  48168 16612  04933 73921
2515: 28860 68508  11953 78511  51830 24134  29904 93798  92059 67162    11226 78328  27483 78328  19165 86721  37676 01287  09833 94604
2516: 76238 27928  96650 17113  30959 97028  12066 53075  68432 50009    24709 62157  73121 86906  67678 03684  89391 15834  14744 04521
2517: 60861 85875  61599 14645  73659 24539  54860 34277  07279 54796    32924 31123  69375 35442  75796 35954  52440 21232  82815 25588
2518: 82006 97212  57713 14058  77128 57947  75549 69298  90143 71478    38123 01104  86077 59502  47554 59749  42021 55829  06387 17360
2519: 07510 35074  57580 46299  37018 93101  57068 04720  28502 89583    92196 99731  88895 74743  87460 91005  11327 82897  97208 95184
2520: 79619 43672  83821 48225  23325 48220  12029 07637  14116 66723    29713 96103  75782 65762  31301 82341  03401 72701  11347 03162
2521: 97753 11973  91299 95402  13628 90792  76850 19535  19396 36588    51210 59502  01765 19041  77195 98457  84555 06307  82276 05894
2522: 88401 76822  07970 89207  33301 12763  51973 30396  05848 94728    04375 57914  07989 13471  01003 23584  00030 27711  39982 29986
2523: 41685 45196  27306 77358  62721 88267  74693 24598  76903 47043    95629 59680  40810 63321  28849 75747  48269 11570  51017 67117
2524: 91683 47832  84854 42887  89475 03776  62829 20167  67660 89483    78068 91783  26500 11705  92189 08650  55365 88412  37934 50088
2525: 02407 23458  31922 52400  52615 05013  36338 48612  16809 12266    73696 18027  24417 21865  59533 43161  64304 35919  93025 78147
2526: 69034 90001  99387 45180  25090 13490  71477 80031  93954 17336    12932 84053  46320 78370  94493 04905  74582 20634  59495 76601
2527: 29613 91053  41104 72468  36411 10289  69895 05005  90906 03219    79097 10251  87441 61227  24824 10419  85543 85334  21102 93423
2528: 00960 08494  28628 51815  17198 28662  51250 02772  55402 22612    91826 46467  17241 63404  09864 34782  94164 67820  89424 04758
2529: 80575 45304  29458 77417  71508 67736  49896 60636  46111 00305    64499 20218  37509 18547  48922 86734  33431 42353  11750 59580
2530: 75079 98698  73741 00180  12241 36633  31160 38471  01722 36354    14792 96892  05110 85287  69434 50363  28152 34034  11281 45942
2531: 32421 22000  01005 55430  96118 09743  77118 12400  42269 12813    10166 32823  91739 08440  23224 52278  64031 69788  81089 86477
2532: 78749 88350  32884 24391  59442 60714  72408 03452  19067 48799    10228 66771  45124 03273  69928 71232  61400 23576  68903 25747
2533: 57143 93984  84306 46808  67252 28906  38145 29265  07817 79323    24513 64113  19336 11595  29469 44959  86908 24049  06449 90063
2534: 33386 40760  66261 45404  16052 44045  84113 93188  34609 11575    33254 47245  72317 39933  72739 01228  10061 85843  95900 80048
2535: 09470 44053  11090 55163  04825 41927  03279 12090  45389 23561    06257 07367  78695 61617  37223 18124  74404 03009  40380 41250
2536: 89933 88411  77224 16672  65250 22473  86071 44677  11752 57385    10992 38917  16270 36078  89312 15098  18750 47277  28631 16185
2537: 63446 67802  33103 58153  31152 90768  41842 72251  07544 80523    23524 67521  39362 03400  46286 22903  24864 58721  93662 58402
2538: 39730 32404  33139 71736  33744 89646  39352 48981  73560 28901    80071 36371  47424 08139  48841 43243  13743 93200  21890 81854
2539: 83591 61597  04397 85156  90686 44232  99576 31602  36660 68021    82035 29434  51407 42232  07397 43843  87186 05148  80103 25512
2540: 79213 86793  55169 97113  90534 92949  92991 77245  22621 60320    16544 44797  01271 23498  54479 36657  00180 53235  60650 08600
2541: 53002 07099  75680 65916  46931 22428  19457 50573  10103 67453    80750 48428  81540 65134  25486 61619  72730 29360  68675 40829
2542: 67886 49060  71996 58535  69009 47337  59733 55744  71595 92400    68644 88884  55977 08306  19241 04546  53617 83834  25132 38911
2543: 09215 04429  32271 36459  25768 14159  29123 97485  00044 09017    66334 44377  28441 60060  78860 28890  81708 01081  66467 80638
2544: 37091 69146  34040 26006  17516 36835  75823 97465  97581 59128    07892 27406  32839 89454  62161 01711  07169 46818  80814 65211
2545: 97758 72695  83853 71679  47716 31715  46249 77151  73327 81232    74151 14735  20067 23807  73327 00959  38833 95531  06257 94110
2546: 88181 79512  96734 75811  97985 99912  22682 02461  70965 26774    60246 67432  98627 18224  97843 10971  35637 35969  88900 82949
2547: 04663 38449  06235 68807  61786 84867  46866 91822  14723 09925    57215 37173  97301 66440  08496 14842  91827 01016  26173 33944
2548: 17897 15628  67201 92827  31446 10317  81871 81562  06632 45692    41445 09616  77434 24803  47309 87674  89821 14967  07732 40497
2549: 47690 10292  31986 75217  17478 79073  88567 06532  33197 59488    89800 63499  86870 89333  43168 49069  47640 44745  45239 01997
```

```
2550: 0584899501  0872899247  8208398833  1147833893  8433547111    6149149717  7548480963  5313769183  8199251275  9923784775
2551: 5109636054  3497044288  5387962754  2924512791  8837051754    9154391987  1489102059  0690755249  6694697172  2073718082
2552: 6671172764  0492451201  4357223757  6275538507  9964690611    0542993683  5167536655  2898492480  4859069925  5799392070
2553: 5945013770  1778684775  0523937011  8410849184  0125685750    4940944584  0973988183  5736688192  0430923813  8619014370
2554: 2085040916  7063231631  2835465807  5220450788  7378959692    9383217374  7282497497  9914333181  5730277255  3007663560
2555: 5216557177  5055179330  3749303729  0373059889  0943209182    4053577767  4322709493  3369066969  4369212517  0405912524
2556: 5682588924  2576052362  2583511351  9616607974  3230302157    0761954949  8530223336  6365954773  5362760817  0395987436
2557: 1042610177  3146974634  2714711380  3116733507  7768855120    8501448733  4232000664  0842420851  2517404695  0163339506
2558: 3889170553  2343054158  1770910324  9971304285  6260699761    4058238164  7221260685  2610605503  2913105384  1363122915
2559: 7596382656  2669201675  4871095987  6666060296  0262094253    3566125528  1199329631  0904535315  2735091307  1052665779
2560: 0312844863  6839429067  2145455968  4800860531  6438322548    9867674981  7476222679  7613434663  4094815437  8682433170
2561: 6638986878  2949898997  0264854041  3047778191  6043007773    9538025764  9379519443  4015178770  1257999696  6170969838
2562: 3047956013  2981222987  2979661863  4449295369  3705903238    3003327617  7365002432  0170146819  6523720323  4058768964
2563: 4879854399  0934814941  7080389089  2439145286  4962758397    9532877359  5086852220  4863151040  7866499812  6435435070
2564: 7922480346  5297712584  5871580939  3164924836  7479879149    9864788586  9432895161  9097375529  1329221200  1348253685
2565: 4612631213  1420781376  1793262462  7784537910  7274034376    1582751880  6401753355  4364238860  0938770797  8410977320
2566: 2336662793  2938390757  6543157174  5010642530  6646460153    6317441711  3555262402  7021092613  9292889811  0345635430
2567: 5171492055  6337208948  7550390466  2857509061  9364450526    2960251048  5924504375  1519469771  0106938919  6582397760
2568: 4016586741  2595966558  4981405602  5549107101  1966379323    2673666716  0202240350  1120305933  7137981487  9005689403
2569: 1575198143  6150068790  3022475892  6555591606  7218527879    7046094552  4207189657  0433944648  6531488889  9842467875
2570: 0397293648  9293424324  3595783441  6364517346  1616112705    5344703927  2070990857  6461109065  8079661066  9972184551
2571: 8240505902  3178889902  4860454848  2452853996  0062963570    3262555302  9152075387  1332907799  8426231079  3542832150
2572: 6874456798  1567985995  5066436404  5063564119  3131023193    8566058784  7105727011  1230773822  7712794429  5352473942
2573: 5338293883  2235081802  6909762463  7764627891  0901182498    3943245719  1206835173  0286920042  8990538326  1931381974
2574: 7721759816  7508500646  4787071129  6158288489  7693322504    1764815272  3090701559  8239545559  3215688125  7925745144
2575: 7393111784  6955062567  4407032690  6678656842  2807242804    1415289367  2991508811  8989838201  2377899770  7315321466
2576: 5122182661  3973189272  2617613686  6987636095  1813804363    8214382058  2455160633  4066552807  2197187994  1903940812
2577: 1278776300  1099458579  3454071741  8929194330  3799044926    6819808317  4913199580  2516424169  9154400779  3738037833
2578: 6609055853  4489758112  9474391984  0274021698  0020787634    3575369357  2752310683  4194734826  4608316140  3655527147
2579: 9533660300  3320758599  0256061237  5553950879  6915172839    9498521555  6811670324  1737952085  1437827490  9062600548
2580: 3039692449  5991111716  7122974200  0714672595  8940631682    7460104556  3101314576  4765160746  8623356784  9263655907
2581: 9128712058  0479268463  8897641389  1359165313  1606129952    8515436740  3803132195  3341417923  9954343527  3609884221
2582: 9104422659  8101880122  7776327628  7886153786  4854766763    0381239480  2688996117  5938723343  7934981221  8236024397
2583: 9777116266  1592725525  6719770214  5020983341  4606020338    2267388386  2694576865  7500777645  0588828273  0307259295
2584: 3374521927  1591890263  4641481174  2491712321  0233973488    5344939985  8480339461  3345841243  1546411559  3542271270
2585: 1962832975  4301939842  2288329253  5342761121  2189732972    1671297832  3394301550  6221278135  8331321166  4736620418
2586: 1103463602  5800008675  8284124354  2183419520  9956111255    0025457437  8092962266  7104392281  5641924485  2880575090
2587: 0531381735  3829718347  4069561911  0767544274  8221910309    3706519644  2044716419  4246106595  1732191626  1544940162
2588: 1547451693  6061117483  7966489957  8296791391  0078777458    3908463123  1873664423  4729222495  3068517199  0622152770
2589: 6607878934  5055771571  6878214948  4113208421  0588025874    1940383773  0435826179  9530098725  2240556477  6722639792
2590: 1325929681  5308945025  4370618720  3605120715  2119958876    7578205226  6109308003  0592566341  0836368516  7672623419
2591: 9028155673  6462697677  0034152829  5769566620  0133226663    6693769980  3163623939  4580503977  2065769019  7632533208
2592: 3308457920  5602437554  7463813943  4519365365  4007150424    3975309733  2352813750  6899963511  9536360064  6690085209
2593: 0729306625  8480623263  3711215587  2027864797  4874102681    7918984398  6561811682  7809261523  4252343598  2283871662
2594: 0467942082  6590070812  1601583627  3429196261  1223612497    5798807054  6161337015  4365522462  2708088530  3643872264
2595: 1233709294  3558463399  0585464818  6347744117  3363083529    8074214696  7997005741  7994729994  8589975689  1310340911
2596: 6109920628  6984430527  0828007622  4379134982  0871869369    9157990978  5974782660  6441085785  9832769043  2012672697
2597: 3062076840  3582255589  1714719221  7173552597  2617005690    6152934233  8030745177  5870382634  0474095663  4544582257
2598: 1409882521  9193106051  7303169052  4859041286  0997694373    9223157781  8477383949  8062767328  7323323301  0941439305
2599: 4138366923  0633706071  7755779226  8768620598  8461161261    2868455625  1942481918  8542097914  5269414557  3907885389
```

```
2600:  77797 33824   84930 90691   91217 41671   90534 28161   59717 95788      68160 53942   06949 80479   68779 92333   90819 63009   85088 94421
2601:  97036 92860   39184 79173   68281 19362   16228 96952   90151 96189      59966 33390   77598 83574   66517 74968   51892 84068   71658 75092
2602:  38940 95402   18162 68760   86461 47114   59542 34918   20305 66607      58117 97189   15606 23780   92638 89771   73487 44060   16241 16254
2603:  60826 69088   77313 68903   65362 76547   01388 78424   79603 08058      63953 26395   32720 92658   84947 87785   22388 94208   23826 97247
2604:  91852 59771   07545 35123   03065 56013   49685 92922   79779 59952      72844 39186   91606 88787   80172 14109   93092 56001   25577 03031
2605:  55051 56123   13934 77344   64451 73422   55576 23526   89448 36476      11777 32356   74596 30874   71461 89870   14515 64044   69315 34175
2606:  58580 15152   23985 05732   51834 06025   53080 01601   42776 27666      55989 95561   72674 98673   50429 09663   62702 82745   90224 72338
2607:  96545 62400   26428 27168   04443 21524   21725 56262   04790 52294      24717 32338   57117 23229   96326 08286   33106 47439   90434 74989
2608:  16294 83706   58453 22837   37509 31097   69212 22610   28083 80525      55439 52148   08774 26112   41286 73623   90109 32911   09090 09500
2609:  25914 40149   17335 18168   59262 25110   23595 58144   27024 21044      27749 36642   39949 81891   65821 39339   01987 37862   73958 82041
2610:  97721 88261   53984 16171   65858 78470   34369 93941   22068 76054      84824 57327   88551 99886   37622 02791   61733 71648   97682 98049
2611:  32178 03572   33951 36707   00827 51707   34375 67873   40001 42972      35698 46626   59448 63434   48965 18190   18254 84140   68482 16357
2612:  18271 84187   21266 82581   44480 32408   54844 19777   55846 19849      80186 29177   53906 15877   69006 12321   94844 44473   33242 00981
2613:  74097 69361   09594 16747   79165 30234   67456 00329   62814 76889      16243 48657   22917 71166   66074 68708   69357 58514   42411 43588
2614:  62020 47260   01916 20223   62588 75899   63450 07972   33225 11349      22800 61118   00714 86577   13268 15881   63742 46730   47243 36206
2615:  65930 14867   68691 97486   36259 48826   95358 77104   16937 28632      90044 27968   86288 53308   97407 8067   64005 79078   85818 55789
2616:  27609 69250   03395 85903   82763 08513   59873 80974   96412 54212      79978 93582   23236 93463   85848 14018   00606 37910   70665 84556
2617:  62906 76519   36698 06823   59398 46861   09645 03807   51569 53016      36651 10055   49569 76095   61586 47339   41871 79463   82615 35711
2618:  36582 19142   41211 12483   95997 58615   58094 23856   01500 28739      35465 95953   52453 27189   64274 68924   74656 14526   05688 13544
2619:  44847 63683   78650 06262   64528 45544   77609 39454   23687 47469      85827 31474   67414 02473   77807 99456   91282 33005   76909 7325
2620:  34172 62477   50650 32955   67188 90958   52342 45828   29914 91510      16849 51566   59953 68417   96636 63575   32077 32766   50518 92645
2621:  99139 16533   85001 00537   20042 04770   34523 02701   80357 68299      82740 71685   09494 37859   20697 19562   06483 92126   51337 02548
2622:  35250 15976   55646 46627   43196 31651   47490 51028   24495 99247      60705 66753   07521 75677   07602 14156   99605 48593   80838 58906
2623:  91772 96315   49341 43944   36204 29084   87656 25923   92181 35739      45878 62347   14362 17118   85410 97840   64494 53285   11289 59769
2624:  06011 18488   01303 94300   81357 88776   68046 33262   64769 93596      38324 65471   01530 94558   36680 67624   61666 48141   47313 05908
2625:  10635 40068   30570 42162   17882 18839   20642 26287   74179 88123      71304 31473   24916 61880   76304 18374   60114 00375   71620 08065
2626:  22455 36092   52752 72693   23326 58721   93499 87850   15578 35478      17733 90121   09019 32414   62389 02453   81113 49004   28653 81704
2627:  48069 69823   50943 19717   27115 19954   03004 39310   08467 99698      41286 62196   15517 48706   77321 25977   34770 89197   15837 32818
2628:  52397 78492   70052 59008   30371 40749   57072 51589   17038 28495      46028 90992   02609 45762   06738 89908   00803 41452   54387 70074
2629:  96259 95759   55943 51080   66442 71153   32338 09962   30165 44861      47052 41716   92781 03533   78729 69108   20998 38335   04865 74854
2630:  92317 88527   45609 11117   82310 19557   83296 49449   74546 82980      44661 87032   89547 60938   89303 24471   71250 74294   10228 24413
2631:  96772 70757   60706 38063   85759 87775   72895 69192   83585 87722      80866 39139   11014 64575   61179 76328   46978 37794   91218 60651
2632:  56027 80543   64420 17684   99862 23380   02494 28554   86685 38157      36593 01083   84148 00780   32288 81667   53889 34867   45093 88580
2633:  95216 12404   68013 72939   95976 75094   62412 11331   75851 25851      48445 72858   68211 62212   52053 70093   84267 86354   04043 74355
2634:  55045 86420   98138 69360   05314 53381   38209 01569   73125 75797      42264 74686   69358 97791   44617 51660   51040 73292   71757 52878
2635:  06059 55378   26984 20152   43922 95195   61122 80804   23348 23108      29230 19622   39195 27979   57981 97649   61442 87232   16960 23943
2636:  06431 21055   00973 71700   72920 80077   15949 52497   67890 27379      39996 50546   72936 05503   62714 51624   06714 20601   93343 84698
2637:  97871 50819   20447 65951   23528 23974   03300 92842   99857 17273      35939 47322   96468 50206   55244 80927   26628 27541   08605 69509
2638:  83188 35337   55026 23274   92618 64192   75377 26436   66207 00566      80937 09648   80614 06388   78094 52426   15389 38036   94558 45674
2639:  00436 30150   28045 99477   22003 61341   59676 40096   54814 96236      72719 91916   08458 19780   69723 37273   49779 83733   03873 06751
2640:  55241 27540   71084 09488   71468 69490   19429 35603   56249 60388      54902 34429   16782 78283   64352 04366   10261 88533   74026 08401
2641:  57279 90452   21988 97136   99281 20191   14796 98126   50220 74035      13448 07564   41034 57531   22853 80911   95006 48681   11285 04519
2642:  49444 99804   49868 15797   20262 84357   17594 45329   96893 92844      81156 64814   98903 49386   93393 94503   86915 63630   82119 26873
2643:  22171 90529   16231 74327   31225 95256   07694 50381   26020 73829      48419 62173   71564 09521   59694 97837   85146 59478   60334 11112
2644:  31641 14056   61485 33723   17260 26312   59157 22085   51318 04515      76437 45900   03260 42984   44804 29859   87320 56710   03563 47988
2645:  28393 67014   07799 73077   22053 41648   84531 83812   47027 43891      84233 18370   77259 88515   71727 23529   18022 51256   91777 47548
2646:  38502 52134   46178 75216   69401 45082   58774 57213   34758 89290      29416 48551   17790 30268   63746 90515   94281 84808   61010 28762
2647:  46797 88701   59969 24611   00343 61372   24271 61837   39905 18379      61419 87103   48550 17188   37160 48640   98854 29597   90165 10256
2648:  66063 20251   36912 30842   69439 77190   84133 75290   72096 99242      21809 62864   67329 53347   24282 17845   79037 02806   34376 20549
2649:  32807 58698   09114 45926   16079 36176   06177 25829   47176 09945      81799 11822   59955 07350   46334 92901   14190 79938   00851 09994
```

2650: 24211 73829 36698 61937 60055 63612 63358 17072 45249 21544   84675 37508 29409 62025 11555 41773 57206 12245 57069 90283
2651: 24247 44229 07557 91588 22052 19591 21975 53228 32561 07476   92092 62051 08410 01072 70215 42522 13785 87110 55912 76244
2652: 45348 09583 53228 68742 28514 81290 13986 55627 02486 24779   67742 05915 11271 53724 82098 93086 10518 52606 89024 49771
2653: 12337 86625 92207 31994 84206 28676 97644 53416 41617 94378   37523 44168 40880 06935 93307 94558 12314 05840 65180 52365
2654: 44221 11599 09576 87665 46722 10479 43870 66877 55016 99712   67063 44796 37048 29280 40916 28339 94796 07361 51959 65990
2655: 34580 24227 45822 35211 27740 33735 59609 43585 29732 25558   11552 27235 58977 26607 81128 63570 86970 45204 82589 37780
2656: 59561 50123 69883 53578 85703 98475 21406 03692 12794 37760   84315 76810 05919 47351 04897 22555 12775 88364 20824 32911
2657: 86422 30189 89633 95040 77237 64809 45978 49078 88805 00290   53443 67434 11896 63377 79032 48460 25432 61916 39183 26043
2658: 86195 39324 67979 35737 65558 62719 89845 99762 86440 56150   34427 73752 92049 20820 37405 85786 32612 01408 23728 76721
2659: 77311 96959 61608 49454 40920 51419 26547 03831 96351 85230   58443 63994 24530 37283 01071 47528 28869 08137 70917 92335
2660: 63465 20964 92191 30252 00925 42066 42416 41801 75576 60630   08270 97835 21867 23680 21438 25601 60812 13060 68949 04360
2661: 26112 18695 54228 66756 64716 48486 78738 69350 85030 32509   63456 29788 75269 66115 64790 87315 92610 76789 92066 74852
2662: 88816 31985 49321 96231 27058 95172 44229 92187 75882 00504   63050 98695 51021 53806 11047 52801 19824 11079 15395 95680
2663: 80485 04273 00364 62200 29972 60398 84189 79703 96851 25432   94552 79704 83097 70327 08396 86632 38022 46716 61961 45996
2664: 52729 38716 28629 77536 20635 95286 62314 94605 94976 65735   37280 65699 85361 57117 67390 19434 39370 90054 35376 08276
2665: 68810 36036 64687 05767 67782 10390 54161 00357 45219 15295   73088 71596 92758 75025 08253 23119 14452 21698 49951 78883
2666: 39636 41109 06056 81513 36509 03123 13882 94195 28746 03095   27871 93553 56463 23081 64834 09471 20984 87530 60033 82531
2667: 87417 61403 06055 23381 30761 88667 44689 20479 01969 58685   77743 82864 85057 25763 24022 96523 00531 38606 98166 16483
2668: 07880 02613 07071 66514 36690 65587 95055 65173 64708 07394   61074 24789 08082 33849 09098 75322 58074 86723 11556 76739
2669: 61798 03141 68911 23606 87596 27356 30498 96699 86491 50864   63226 28489 50015 72105 70079 54460 56788 42077 49204 48690
2670: 76015 02012 91635 86575 02201 33648 59845 77126 44116 72527   94156 16462 51196 36553 51958 81826 35458 80669 05987 69759
2671: 38092 17056 82979 26068 77861 52932 01861 06644 62688 95133   78792 47853 29127 39323 49263 55544 61662 53846 36630 60820
2672: 07274 17056 58326 62830 61054 24093 24168 54228 73932 12310   44690 93703 44898 25670 61162 84865 60183 59449 00348 59105
2673: 99788 09578 98863 52404 99971 82734 51358 90475 24172 76533   43031 76253 59913 92359 03444 37346 95385 87536 00841 39724
2674: 31519 53009 14190 03865 43328 93777 06178 94057 03197 21617   50785 39216 30379 54617 38188 39908 59922 90915 96250 31966
2675: 84003 59226 12965 96234 16485 15091 74029 09136 90085 80593   03257 08413 11787 30305 02894 49267 01940 39219 29483 80627
2676: 59006 48377 28137 89134 06926 60147 86890 92498 21810 93940   90933 18845 70046 09872 90491 74463 91024 87620 49574 93145
2677: 39118 01219 54117 61720 19419 22068 79434 26938 08522 21086   17004 71615 96544 61512 98429 38535 52615 28020 18492 44653
2678: 30036 68035 85681 24800 90879 41443 10601 71459 58576 72250   80131 11803 82336 52343 95988 62369 87217 75351 75908 64285
2679: 70816 00991 04224 10783 01480 71570 74588 90287 13781 89077   05014 96434 83173 43987 04102 87472 76101 19691 49673 93631
2680: 92614 95530 05163 17597 09438 91602 80794 79804 18899 40168   28579 74619 44841 80516 51697 14514 30051 08791 20256 02759
2681: 25262 82450 79171 82547 49768 10036 66439 55096 56724 43800   61021 72044 26243 89615 65622 31894 36774 34398 33065 64249
2682: 30444 92572 53782 13818 35153 71397 57036 19943 02914 79453   85127 00143 89175 32443 18537 87728 04366 98124 34748 92025
2683: 42438 82311 85571 43396 96934 32434 38566 05750 20220 19557   50065 49560 97093 58194 77843 68048 28822 24366 35415 52150
2684: 02199 96371 42320 37741 02213 02580 03564 09507 55877 11055   85334 88951 37045 23628 19105 89469 54267 44526 25611 34784
2685: 18571 46393 45708 25364 90330 32005 63100 53386 42174 69039   72415 92324 30124 62685 42090 66010 05288 95455 49301 84530
2686: 78603 04027 54890 98426 46471 84583 40201 87803 70442 59602   64690 75735 65568 41889 37669 30403 67749 98792 25900 15062
2687: 11926 31678 25599 63288 34188 70191 37855 20196 14621 17813   58973 98721 59724 05966 98185 83643 05996 25912 46803 31777
2688: 90995 37214 40771 69046 70658 91216 40921 01467 18718 23751   61584 41797 79981 27633 17544 71950 84018 31811 54725 72555
2689: 69213 30605 70366 92497 32473 80238 78887 47409 89390 97593   87385 48847 32107 45195 76983 66191 05504 39506 47547 89663
2690: 46763 32858 05748 87281 74221 42127 45582 49938 41255 52289   81854 80474 02825 17205 13087 54834 42069 98720 54484 71361
2691: 14704 52437 81395 29886 38576 69733 83345 15754 26648 92420   40863 32778 91563 51442 70957 50100 56305 81431 08571 88668
2692: 49750 59488 27168 82072 62996 06700 62866 98812 79414 77421   12366 19434 91658 88227 29001 31192 25996 43049 92933 82211
2693: 39441 21335 44865 12349 32936 63952 58596 62320 56579 92437   09320 39696 92825 77775 64171 96310 85491 18316 88612 48328
2694: 10298 73711 63845 41294 84204 76738 38890 95554 79496 08976   92073 18174 92647 46240 30144 15218 61706 33882 63728 70378
2695: 06876 15288 59248 04637 26634 25680 26185 75629 81478 69253   84256 48452 39398 19250 26380 10015 36117 25680 05564 35103
2696: 84948 46366 73280 18347 39795 08996 79732 88240 70360 71364   57328 81088 23985 18124 48858 60470 43835 85974 92385 80683
2697: 25429 27781 04101 58341 87039 43228 27914 56259 52303 59604   94535 75411 59854 87593 55168 05113 26769 99816 98957 20344
2698: 85054 25635 36503 14881 96186 01656 51478 71070 67607 81005   02020 02817 58442 26432 98142 19253 16269 07766 16158 70773
2699: 07239 92614 68659 62269 56177 09241 86109 96915 77637 70393   81174 62970 07464 28728 51653 83836 77037 81934 61219 02446

```
2700:  16448 73167   52390 89783   98080 52595   95164 55704   58712 50638      66070 59178   69582 17881   49754 02807   30288 38151   20699 65290
2701:  46749 24063   96402 37960   11896 49281   58904 56354   24135 85369      97181 45988   04103 17310   96458 78835   73193 98651   95220 22515
2702:  41315 21073   07274 19306   17963 25167   92195 80393   71412 40575      14965 85637   90884 66061   55340 16241   38432 11367   77586 54359
2703:  04625 99397   12058 09909   72814 73442   35411 35603   95466 24222      40683 79125   83772 36048   13145 10646   22223 11989   87494 43799
2704:  65819 13505   45159 65577   73898 09009   30893 85158   70803 25044      86932 48001   53188 46872   49557 36741   60387 70453   93211 57204
2705:  08917 72664   44936 70036   25218 02493   65532 99241   37427 63251      99835 22330   34772 95336   05676 11548   64362 49421   80896 46535
2706:  79272 57471   95338 93541   35736 00573   96154 13669   65846 04773      77701 04444   38046 79311   52412 78130   02264 63477   73960 14870
2707:  85402 71162   68594 88198   36931 06947   99351 23962   79323 63945      36586 54185   25201 17482   59640 52511   64144 43072   90750 97834
2708:  53931 48117   76114 01964   98812 76866   57022 28017   57420 62324      98644 86508   36841 93495   75837 76115   15871 20717   71076 95734
2709:  36722 56695   76371 44419   44310 10777   05445 14626   80360 67390      86483 10269   32841 14095   74708 14718   43590 73839   91981 17127
2710:  72213 68372   02590 50391   45604 63830   74717 39088   68416 67647      29976 42570   47493 04180   34316 67542   56943 97628   74839 02058
2711:  61248 82984   60420 93961   85360 46431   00337 67752   71287 70387      07825 85170   03737 70629   22479 80706   76226 85369   04726 89291
2712:  62255 04301   45965 70556   20765 09115   09774 52677   68044 10728      17931 05647   99394 91183   60765 76829   28570 30089   25543 72975
2713:  77061 55537   09979 40990   81585 39078   94706 32254   31981 32635      35595 50190   27993 36951   89666 19350   29401 72567   70737 46999
2714:  88364 74791   96498 54999   69018 41591   74765 18350   84076 35813      41297 87876   81321 91806   30580 94710   50966 85464   35676 85011
2715:  52611 52928   16876 73188   93826 89507   16064 74326   34323 82575      66155 75250   14363 24191   40468 72247   66642 77414   84321 90222
2716:  57522 95356   40933 13325   36487 10894   05855 58811   07572 55673      99273 27203   24129 52663   00783 01672   32026 58714   79932 02038
2717:  85318 06282   60649 71756   11005 03957   57037 97740   41826 52709      15620 85960   69438 63625   52534 10168   85824 31395   58691 51343
2718:  09442 11519   42663 55154   49893 87346   97254 95607   07780 41858      78570 04615   07876 42942   95833 30680   47691 22268   30954 34219
2719:  29947 75486   69162 19135   06953 60587   44352 98897   86264 47534      59399 25192   13829 67840   23865 04830   13794 05335   35643 01520
2720:  49308 48124   96561 96804   22835 12538   02857 58736   63008 39528      30480 37496   53389 32514   44211 08803   15104 13386   26693 12101
2721:  04728 26110   40683 55369   76090 48036   41486 83190   31552 98874      62925 58288   30933 24610   06381 50882   12634 37962   73067 72514
2722:  13460 04094   61039 12783   91452 98935   18038 24988   19379 44057      93771 99425   72827 91301   33919 92614   02483 26489   43659 90486
2723:  58262 10690   42016 37700   21222 81957   92781 79810   15062 91703      69448 21025   53667 13305   08762 80008   91747 71090   28387 97767
2724:  11622 24249   78103 84256   32568 14078   92785 62408   80344 36465      40373 35596   83792 47124   46245 35155   58555 03526   82049 30699
2725:  98257 33276   21963 08739   60573 31532   98533 31494   75838 33476      79347 21908   51470 86064   68753 31929   14648 33371   38261 93803
2726:  55704 78418   70659 62382   65244 20809   56118 24383   01020 24486      46336 57988   12850 69320   49436 73431   40462 53655   52485 27001
2727:  70825 32418   25284 50587   44132 84455   59741 57379   89901 48538      48684 01434   13149 27943   11070 54671   91413 59836   68678 08177
2728:  78512 76756   25763 45608   35916 62832   94091 42206   12332 98619      97471 30001   76916 39924   36711 32837   76838 50638   26045 69549
2729:  85221 17973   26898 52316   01818 49179   19780 48325   58094 26469      05267 47358   25388 17696   60591 20539   45223 62190   49668 57305
2730:  67942 81024   76256 62687   07781 74930   71657 54242   75770 92739      42399 64194   82623 15979   21395 13266   05793 57816   56387 84774
2731:  02828 04347   49317 59352   22678 49018   31952 07393   67212 11988      66580 70432   99419 68887   88862 18024   71394 19860   54147 65831
2732:  26309 71098   66535 03521   04672 25087   43193 06926   19322 70475      30712 14743   40614 34035   25223 59790   88120 09941   24232 02743
2733:  07690 34366   92843 97634   59667 07683   01588 78606   18426 49372      37891 25781   44130 81787   36576 25473   97000 16879   46111 63007
2734:  19024 83975   82334 60278   25491 73141   56329 38957   33182 66243      27911 43551   06529 53363   30133 72942   01687 63380   28193 23058
2735:  55637 72049   47253 20877   02090 36470   78736 77933   48802 32711      98717 88904   67915 32436   03780 61479   30960 96650   47426 99880
2736:  97175 72402   09030 97579   48759 47949   02105 21547   46017 62254      52750 53806   47630 05359   43348 39819   04518 68127   09333 86602
2737:  66949 99976   25498 64712   55198 89799   44128 82387   20281 70523      01579 00628   48254 00250   24565 54631   58997 12371   43184 75038
2738:  07601 50471   27707 23251   47810 13881   20424 21472   92273 80553      06618 88382   74855 24440   29940 06017   37393 55853   39826 56611
2739:  94671 93577   93947 38267   85014 83553   79663 76487   97376 50193      98355 13917   68482 85217   95513 94555   65274 52697   35374 98992
2740:  00504 75044   80775 51740   25265 18666   33886 19255   73598 39136      71827 11537   42885 84315   09412 76391   79729 78021   40032 29045
2741:  44977 09858   40246 02862   12071 16717   90786 85552   16751 74251      78298 76393   50475 60203   08098 91938   64283 30212   89583 02982
2742:  87006 82184   47721 66771   50510 16999   32774 88490   71852 88909      95300 02157   20750 44675   18755 39892   42474 18404   57696 13492
2743:  47866 02149   78855 87001   67454 58290   76876 18443   63750 53334      34018 74806   44847 01634   05166 02238   52946 08993   49774 23576
2744:  70922 31479   46747 14816   53817 21606   36231 67142   94998 28286      35657 48520   17823 19176   63034 59640   21380 94579   00793 35510
2745:  52791 90295   17451 55502   09330 48097   84718 49423   56282 30056      73745 94504   48959 25697   39295 43807   20551 46049   70558 18656
2746:  35576 72943   86088 33371   65464 22400   89773 84703   69082 84526      09448 08387   24435 12049   20975 32385   77079 26440   91469 73387
2747:  49713 50421   56882 09295   40862 22191   87148 47351   05392 35467      57086 62930   66459 73743   76637 78134   49869 30432   96865 89896
2748:  86865 44338   41448 87143   43282 66217   93975 37319   31189 14222      69734 25893   06168 72912   74951 04817   44517 76357   08097 04751
2749:  32379 81029   87579 58294   18974 68441   02960 85664   93303 18371      19257 81535   92338 15545   14591 49272   33641 77541   02118 82494
```

```
2750:  83031 41135   05464 94424   06431 58533   80812 46363   62398 35242      96233 49880   10332 60663   78325 19021   96324 40092   91560 24206
2751:  71546 22583   46825 28228   74385 78335   21401 78116   44240 07548      61555 52729   42119 30474   96980 11291   99214 38513   34221 63123
2752:  26856 99455   75036 92159   63567 91855   80950 67434   46997 37087      28515 49741   33123 15211   10446 63121   48291 79592   25521 56910
2753:  82478 60235   96755 64438   17489 87956   80701 73497   71607 26116      14787 65426   37935 45864   73863 61626   22210 13784   52404 20739
2754:  24025 72794   57219 90956   39319 53368   62991 15890   89389 66523      07368 72574   64039 45798   95326 89194   96565 04921   45166 36117
2755:  87632 57836   24070 02039   27650 55222   07909 82475   71870 71226      20616 78893   47806 46609   18286 16323   61380 27060   88421 61720
2756:  86478 15718   22877 81002   47211 21089   98669 14313   24534 97887      94874 54423   37592 13337   95526 69608   03003 87187   48736 14308
2757:  53603 45282   53192 80773   80874 96460   04988 56796   84958 17594      15721 35314   19870 41116   70995 19261   50332 82332   12353 00924
2758:  62564 95616   16961 73175   75363 15949   07107 79640   65090 69679      99406 03660   88456 66014   24438 31018   72551 05331   36154 13141
2759:  77722 58649   66538 57293   32534 42471   43471 83621   18265 31146      03842 95809   37842 69606   91469 20062   23979 29594   04164 08833
2760:  67172 09063   53927 11813   50575 16829   66432 02175   54122 74917      00000 63081   45643 47832   68245 11462   25556 91632   54522 61998
2761:  87536 24091   53191 73600   14651 19318   57316 50610   16659 08842      51261 42085   12778 74385   32745 14662   91781 15344   37239 88767
2762:  91827 85613   76752 69450   88782 28048   26763 55159   59534 07979      06933 50878   96914 70115   76627 82310   74233 12451   25822 75070
2763:  52465 39162   42369 32287   35581 68867   19746 10664   71967 30169      34641 89189   20412 54689   20655 59245   83458 29085   15954 09576
2764:  87601 64499   14440 42248   27792 37955   27302 87192   47384 44200      87272 36391   65233 82720   89901 86006   92844 67663   78729 39083
2765:  45070 33802   99129 96446   19641 67711   79638 55903   87290 36058      29510 00515   46657 85097   27437 11758   59264 82221   48197 53165
2766:  50205 50422   75398 94382   59969 27056   06030 64335   99205 06281      38586 52694   48932 64170   94275 92057   34481 42040   12689 53429
2767:  37986 93645   29027 55984   97803 07554   33265 83207   14427 01630      47806 42635   89987 83467   64685 16200   62346 20088   27717 12579
2768:  88226 97156   39628 06605   20792 73700   24358 74514   95040 40139      33430 67975   38274 13383   54600 75122   95130 34591   69932 39937
2769:  63782 61225   55097 82483   03246 43180   16567 42595   59178 56771      24634 29172   99175 76441   03921 44103   09766 83418   56362 13601
2770:  08246 17124   26484 56262   61604 38886   06002 51480   96767 30095      77841 98099   38007 06182   03424 34431   06305 40056   87848 23882
2771:  93948 28244   62827 25400   27518 21998   59634 87724   89020 98064      88936 98995   66063 61655   87893 95180   94028 33813   64396 05053
2772:  33325 34783   61479 60476   25737 21464   30853 57981   50424 01722      76001 28270   42394 26119   82539 93116   37149 30795   36636 77546
2773:  22965 94809   45166 26945   90451 63480   24367 88370   17622 98171      78604 46760   23313 26172   09485 74401   07726 46135   06372 79301
2774:  75620 68474   64079 92940   67398 64778   42916 50103   14258 12121      43688 91610   83129 75569   59877 74844   39685 77660   23062 76791
2775:  03462 36093   01347 82617   72459 56743   85680 86410   37895 99505      93014 58632   91159 34774   13960 87733   55863 58383   26059 13726
2776:  44556 70453   62209 53738   76128 44486   18471 23646   79180 53701      02999 08506   46309 62688   07762 65134   79790 28960   04386 78575
2777:  73662 01663   69259 93468   27662 81269   77002 19711   69322 97310      04729 19082   06547 79782   03044 02371   18696 33589   65023 73125
2778:  43088 87664   74638 47713   35912 46103   45411 80126   08733 91292      42674 06432   19699 24900   27607 16573   42687 69211   16793 74448
2779:  50295 08059   93045 79796   56729 60730   79934 45138   42024 22899      28437 06275   96256 39875   77209 31561   01404 08392   90761 93845
2780:  92916 99604   68146 01445   13797 83600   47418 12197   01957 03872      83417 88348   84123 44665   56259 41488   37533 92913   10585 06492
2781:  22246 64928   37176 98153   96739 53339   75472 36437   79527 64996      10640 62444   60322 96240   72905 52208   86877 63973   38765 02763
2782:  49601 92687   38777 33600   41681 83173   04766 77252   16246 90187      10576 24088   54574 67713   57721 84103   47991 41122   66508 51103
2783:  91025 61247   81807 45287   12685 40781   14872 71955   14562 76856      67099 65390   50529 46864   47205 09664   65041 28324   68735 38877
2784:  88760 20653   48583 39569   49229 39549   70076 16450   95833 54607      90285 60026   19244 32994   62531 83977   73099 33587   53743 57403
2785:  62706 56864   97665 73747   81760 94579   96282 01768   64954 01608      25135 24316   30945 34010   99329 78313   45485 66346   48065 76371
2786:  67425 04323   98332 52594   18983 48849   37077 07199   47846 92791      20292 45289   87364 30518   72447 79752   14244 49584   61898 02714
2787:  12439 05362   40810 24269   71412 17329   79709 40480   93524 40193      73753 98955   16499 59758   33206 79031   66201 44858   17630 61662
2788:  31083 89930   03673 47249   49154 69713   82674 42075   98741 90656      57076 94228   32039 39059   51984 10124   88926 25383   46296 94284
2789:  48151 87854   07580 65483   26391 80089   93230 40450   13321 40086      39318 22695   96848 07951   85496 86537   51908 34683   92027 12915
2790:  14855 03729   57278 93871   17930 05634   04518 73231   69906 48944      01706 05255   27244 77604   22083 17955   65940 97748   61634 28020
2791:  03600 63934   20832 92675   87991 83512   74424 96645   50982 51075      96911 23065   93782 45571   08978 30517   50851 61605   47254 61101
2792:  74990 39363   98273 66322   17019 79252   42176 74562   47280 44461      17565 42879   15530 09053   38349 77328   88611 24245   19082 65537
2793:  67543 40492   76534 70450   72094 82421   29699 96008   51995 27895      59340 59174   12765 79679   90496 39248   79391 37835   70063 57317
2794:  50530 87457   54804 30971   69675 73362   68965 83013   91065 16253      07011 05504   67607 04874   79059 32198   60160 13975   23679 50905
2795:  96230 30342   77889 22279   91281 17317   00575 81969   72754 63187      46275 32895   82054 25426   49377 24505   53608 30895   66255 58382
2796:  39836 60475   22820 39603   95124 63191   97310 49147   95785 11518      33297 11638   85322 15964   86629 62715   86855 79966   40733 21649
2797:  64744 32259   11800 24003   21578 33822   03433 19626   92771 80141      11740 06563   40666 53808   75307 59777   33449 20037   52611 52081
2798:  26966 46036   66729 87794   42180 81691   58841 75641   49312 19943      83248 43922   46762 79045   57882 91767   55735 14428   49753 16680
2799:  08797 04179   16422 19091   63115 31098   99217 80295   88717 41284      52337 70624   64075 95652   44556 54723   72765 20756   82494 96297
```

```
2800: 76430 31749   93854 18432   99564 67633   59065 11127   44201 27953      64825 30331   79017 50673   55993 93796   85153 36517   28099 66660
2801: 53663 53603   57249 68845   67618 01746   29763 74074   24328 42428      41250 76999   57448 48199   73837 58873   10509 23662   89207 15183
2802: 01241 57243   83678 18126   00387 29683   25801 54912   28805 76657      77848 80161   77415 54917   71970 67618   37123 59782   64650 87871
2803: 79226 56743   63760 47791   63308 47198   03241 03782   75741 83976      91614 91244   99734 09823   13254 85613   60256 02504   42105 84268
2804: 22880 55354   26268 46620   59986 86634   09450 29175   74967 26986      95818 37602   22138 58723   35265 62827   40851 52726   06252 13129
2805: 20097 73496   84210 47675   59162 86467   02149 49474   90023 27707      72816 83265   07643 90945   91835 86313   81257 36409   41952 07779
2806: 14627 61658   05274 54791   04892 53609   11174 97001   45102 79486      15289 72632   11037 14105   23668 75708   48459 26290   20511 02532
2807: 62938 08026   22287 58372   76003 02236   19579 58962   53719 64145      44521 60754   02553 85932   89911 83958   51914 26532   57062 34405
2808: 04003 78903   10174 91717   73183 80842   39851 15695   92576 15660      05363 18187   76960 87692   45437 61396   68618 16275   69903 50051
2809: 65373 99048   42198 56620   37089 83940   98253 21120   92362 04184      97850 14235   70300 71464   62023 97034   59000 71101   19254 33767
2810: 52256 74113   35955 04841   95585 50565   80954 82659   94629 60186      66642 43844   77783 32815   36452 60655   71779 08067   84560 27400
2811: 34883 08223   80969 34645   28538 41966   86681 22892   70888 14796      57713 26414   25788 23516   57931 37346   43120 54125   65541 26903
2812: 74290 74017   65379 45499   31861 95868   59153 98091   18582 34095      22122 26175   64152 81899   56639 92150   78138 21614   63875 81095
2813: 43481 78649   51153 05084   04504 53546   96659 68276   81370 31771      10046 38153   42993 63393   58791 62649   38543 41544   84709 53720
2814: 59625 64928   41001 88138   41696 47046   05242 72460   63903 14621      38878 48124   20719 25424   44090 57556   62972 63686   89642 01545
2815: 42089 72750   43745 15869   49783 61759   47618 18986   98598 19517      92190 96136   10998 40858   28535 78383   58936 08689   40919 86674
2816: 44118 81185   45274 88302   78766 85145   92030 07619   57174 81568      81504 69588   63568 63808   39221 92242   65707 51755   84768 18127
2817: 58228 59342   37184 73552   01468 10648   77987 37390   47123 53307      41982 35896   39299 55085   42222 47032   29852 63754   87352 44252
2818: 87754 16459   92351 03089   98645 14582   28328 21479   45214 39152      09155 70931   41537 14389   05624 32069   67852 17296   94101 75576
2819: 99901 98708   99103 39975   64693 40985   36894 25343   75935 83513      60457 41908   09060 16088   91854 77916   59727 83128   20364 40250
2820: 21651 82049   18782 04535   70265 61754   44693 77289   15111 16457      08053 12524   66015 59672   29690 08851   11273 88563   14261 31734
2821: 76824 66553   55488 66274   57787 72086   70419 37460   67077 18535      48592 42843   36797 73079   76539 94680   22277 57132   90908 42651
2822: 08156 37502   38991 33161   85945 19104   36075 26665   73139 41846      51015 48421   88838 03034   47545 81128   90244 38065   07234 82199
2823: 36932 37323   93142 54546   60565 51994   88034 18440   55181 24813      77451 35032   71536 88721   35598 95172   06638 29641   73977 53546
2824: 81661 39776   53503 35251   30019 54429   26110 81982   26476 39864      17992 80703   94950 81533   51040 92463   09329 97639   09732 15832
2825: 59474 75271   45670 50330   60037 00948   42254 63354   89088 20379      77478 61757   48899 73628   80046 67638   44106 54724   04588 40204
2826: 69743 37349   07918 61198   52537 03913   02204 33452   32219 48111      61677 32386   53470 98291   21696 38742   67648 81838   02794 01861
2827: 22581 86512   10802 07885   02077 45806   91363 56375   42314 65674      47114 06508   72966 81119   74301 19026   92259 52653   89299 61213
2828: 35370 33395   69800 83012   55507 97758   61702 87450   29903 60511      57350 11620   66018 47895   99037 20665   92351 63322   73071 45600
2829: 98016 38689   22981 08290   34839 63010   92687 34076   01404 23556      07000 42411   31524 23428   09435 50885   36060 14548   20586 67134
2830: 98651 86760   62677 60301   09596 68954   91150 07869   15245 60262      43472 19507   30501 80726   77758 01526   57006 37836   75055 97012
2831: 77530 30139   90800 23910   22647 16926   02121 99371   33305 20751      49169 97098   70074 51631   30544 52731   45838 21526   54454 68412
2832: 27378 75757   02748 94570   77492 50721   21327 25124   70809 53572      66342 80310   55262 83371   04158 84034   90922 91740   52160 91957
2833: 20325 36976   73216 45129   26595 77347   01275 39724   53126 93966      11349 53053   83385 05356   01984 43073   65766 55767   71511 51752
2834: 87970 53312   31829 57249   26436 31966   15788 09312   76168 61243      90732 89931   74471 24961   35892 64797   36738 69276   53744 11363
2835: 65039 58672   23783 18947   13115 43815   71505 24390   44371 24225      83683 96144   88141 72067   18699 05448   71589 78073   38219 83379
2836: 22463 50850   46300 28691   40198 91799   54595 05186   22540 37273      21442 56147   29666 37978   45223 09836   93617 39915   27037 52807
2837: 81461 44907   26231 38743   09791 36847   99168 11475   56729 49011      38691 95872   06953 59434   63200 57992   57921 66872   91074 49014
2838: 84841 60187   89688 81767   99190 62713   31682 42976   99747 86385      59285 19007   52378 95595   38892 55481   34603 94271   42234 10334
2839: 57001 36121   16058 71738   95813 50176   80671 57416   94768 63528      66432 11320   42887 74057   89334 92135   83053 72881   89594 42554
2840: 15949 61721   04717 65452   33807 98673   58299 97711   68488 98517      43719 71226   84362 16404   45274 31903   23291 28097   17866 55789
2841: 48087 50896   03658 01724   04510 39249   57275 83336   33119 31947      59148 62682   13973 01323   93237 21395   32222 35697   79587 94958
2842: 84918 98971   51957 10060   36321 27058   47750 83724   55688 77361      64998 73980   35847 61027   41990 82261   04541 23896   00671 76419
2843: 85068 37665   54937 45021   53416 46783   18159 25066   06000 53566      41084 38772   99376 94886   60686 99322   52325 82262   59428 41795
2844: 56654 59806   90268 65703   92454 79915   68895 25492   50280 21499      82947 82317   48597 65907   87219 89568   05486 61981   77245 15550
2845: 24002 21180   05114 61857   62597 97178   37396 81701   98394 47757      77113 88999   14156 84197   73821 82406   57294 71593   17486 29116
2846: 90664 73056   37951 18805   49275 85893   47331 90722   74349 51772      29113 87805   96630 13087   94581 41388   38232 29188   67204 17280
2847: 09159 62200   19681 58701   25965 93346   46586 46500   36089 28582      95801 31023   97664 51216   66525 71550   84790 72912   32414 54469
2848: 21712 22761   78368 65263   45429 15149   18221 71157   18255 08019      10223 26138   50501 41973   74817 33019   05862 19771   72907 11019
2849: 69169 53691   10121 41404   26000 52554   95188 95420   43072 05379      40635 06583   29885 77967   38059 51549   69154 17268   75740 57104
```

```
2850: 64095 33825  39307 83449  09861 05264  16551 48126  09905 08086    80740 83717  14370 45825  04438 43466  18445 09046  94294 15978
2851: 82863 14410  78108 30457  17180 31501  40879 27221  36031 92520    95403 14887  28955 86441  48788 66461  55391 83649  76193 98979
2852: 11290 24605  30977 87217  68992 74695  49956 79748  61357 42628    95732 50984  67243 59159  52934 90160  07617 12946  98618 35070
2853: 81563 04469  91624 80529  79877 26974  00172 27254  58537 61305    18609 27534  47097 98767  80014 60355  64238 71572  85814 79767
2854: 76932 82759  49889 70097  07931 46951  37903 38299  10543 91210    65993 73159  20002 26307  03178 34130  72395 79618  74538 76300
2855: 40688 15654  17574 21359  14333 71436  06839 66501  08210 69823    86338 29449  98656 55583  78520 08579  78352 37365  66426 01181
2856: 60478 88386  30099 27654  70063 80534  29104 82850  72582 86277    41544 22236  47463 08609  43382 37514  90331 47705  35496 93712
2857: 22112 31658  46954 93065  47741 85059  90800 16239  57446 86340    05523 16577  65242 64839  99159 42340  54982 20921  76316 20102
2858: 81799 17300  51932 72259  93074 98422  81478 62298  31187 50189    14900 42319  02830 36567  63778 73505  63816 90484  15117 24341
2859: 40712 33836  97062 65732  84015 41576  10962 36034  05428 15348    72998 02918  47872 73994  10576 63635  47997 58577  79548 16777
2860: 08665 04917  22144 49366  59144 30602  45097 21201  67741 56517    03236 34771  12630 80080  06547 85054  64950 51608  61107 69834
2861: 85392 31658  64120 70435  83212 15357  25423 81559  93473 21201    32778 66193  00799 05085  94555 24842  97040 07049  33397 17742
2862: 40705 74559  69910 99351  37593 40663  32679 58548  29318 02071    40211 67455  06468 26662  04082 77278  29529 07108  31046 44752
2863: 75905 01335  31765 64810  63624 11819  50892 29517  37855 37790    18404 67092  21722 43920  76961 69662  37045 87986  13070 94005
2864: 65894 85304  68612 03404  21011 63484  87700 31832  56769 20514    56579 88467  89286 24428  63348 50195  47216 06706  33928 74175
2865: 84660 66153  63113 06040  57264 23231  82433 28820  90466 49677    73169 47320  98711 78516  25905 50602  16931 93550  83121 58066
2866: 26627 07366  91000 08620  08424 10376  06984 26562  21249 45138    17135 04117  28852 95197  87338 53923  68521 93442  81784 26581
2867: 96057 16033  77978 94206  68292 11995  17716 83227  99741 43102    11345 84533  65239 47984  90146 76493  76553 06859  72145 63226
2868: 61150 09916  96281 91066  41365 67885  58946 42592  39558 26108    62999 62249  61546 69940  15784 21810  77141 30665  52631 96466
2869: 49145 22563  12883 56340  19117 24980  41047 49456  23924 68019    38641 34381  92783 10039  93968 01343  54619 32975  04683 66649
2870: 53407 08837  56955 64263  33328 23336  37237 20035  22361 59005    33340 79132  35239 29032  73226 27971  64617 62173  48468 89431
2871: 28416 13604  20500 37305  37326 62090  89895 39740  69506 38803    64436 35362  79054 07088  87536 68834  05714 42417  69142 77078
2872: 58457 65620  55759 17853  57196 33944  91450 77665  70964 33581    92914 02544  98319 16644  02232 45663  54006 13674  90524 29955
2873: 74733 93087  99105 11303  16955 92290  69503 32957  53242 56344    22665 29958  55008 49274  33188 48323  15667 60777  91692 76234
2874: 40673 43898  64329 52618  33656 65144  73721 69028  16347 67087    19699 46589  88127 59442  33013 68049  29946 45945  41501 24629
2875: 19408 33112  63506 42042  05337 26046  68440 14926  90073 98222    25634 64786  40187 00315  08942 53643  14950 58632  67445 84879
2876: 65793 36637  32114 73590  83350 97509  34637 60787  15309 69521    20463 84620  36488 13508  10432 69145  92428 70788  85779 38721
2877: 39450 18451  14606 64591  89810 79501  99341 52174  29821 21430    60780 60843  10233 59698  48952 86320  34412 16907  14659 81275
2878: 27135 83747  03190 99689  63493 00208  02773 12340  07858 66132    16864 40454  60805 93438  20480 91993  56837 90637  91199 26888
2879: 68952 94818  64022 23858  05786 98511  43695 23135  51869 20875    30502 02178  61985 66307  42530 09345  91397 91190  46438 83949
2880: 14361 90597  48019 46711  90395 51361  01388 26818  36999 56984    42257 79607  20751 05513  40840 43716  09321 69578  33672 74964
2881: 71490 67007  39979 98101  88020 00417  74217 69675  57709 65438    90163 49267  20648 08782  42967 24487  97755 23996  71535 48773
2882: 58582 43755  43024 64461  38323 94418  44060 16111  31778 46896    11342 95155  95355 90781  74873 81838  04942 79384  86673 12025
2883: 02701 06779  43186 74179  57849 73741  43940 94201  06509 47807    68075 71383  77715 21535  23015 22495  92551 29364  58882 43078
2884: 91318 70009  53633 95496  87606 55770  88083 98972  09810 02971    63932 31588  67979 45097  53173 92443  12658 87493  68188 61151
2885: 36378 88532  30323 81150  03040 40841  53335 80563  74922 41429    17994 92246  82333 32281  13683 24735  41741 17034  79443 18586
2886: 82096 43539  52355 83990  91065 11618  19281 56988  05770 31654    98428 19565  09291 28380  48323 82629  84402 20633  36909 79704
2887: 45631 30101  62571 53176  73399 40938  05482 18956  95443 19692    68073 01593  36675 56647  80018 92215  15114 90181  24647 69475
2888: 59826 82876  90174 04219  96649 60642  32126 24658  47165 16830    23587 05478  92516 27291  77791 92904  88121 79469  35136 93985
2889: 13989 34910  90162 10898  14880 73228  20804 51872  71633 85712    43414 96389  38184 81395  87679 55151  09263 65442  00943 90479
2890: 57745 45433  99767 84612  24728 03138  64820 67144  59143 31197    39153 33055  62076 94565  85707 09598  71966 35785  77479 80859
2891: 13630 17413  43998 99475  71745 64659  72273 11651  99034 77641    42862 27054  35153 86680  76152 32981  74027 51341  86096 13241
2892: 91291 61912  30950 66721  36493 42072  81209 53256  33024 73261    62905 79288  93446 79964  40616 37197  39235 04729  16503 31969
2893: 03538 11267  74928 85078  41981 74472  31914 89223  38534 73563    13893 23094  45472 01076  83277 02558  56077 13032  44922 36606
2894: 86235 95905  26922 92307  02435 29774  73527 75910  86881 22423    45908 51575  55913 67340  97854 13133  22048 34844  05545 92435
2895: 39188 32134  42538 10105  06361 59120  67776 01116  27453 16128    85744 15455  44757 46816  75853 45290  31586 90089  51112 74329
2896: 51482 95916  02611 86208  95584 69701  57706 42877  92377 24097    74031 76398  02744 60371  16697 95495  78844 18915  70547 54678
2897: 95270 94365  00346 89810  93556 19524  18999 06870  98218 06398    59993 99786  87051 86131  94697 39299  23206 45153  43389 51479
2898: 13570 87104  48427 73737  41667 99693  03523 53156  06633 48187    92011 10174  59574 87792  94797 57602  78906 52005  78149 45983
2899: 59135 35307  45545 77528  68769 85318  27813 86175  58999 04970    89121 98341  77900 04108  78401 27897  16880 88848  26659 85870
```

```
2900:  51209 20428  60372 69876  12657 70643  93616 14123  30472 89718    26679 90761  43282 01448  95180 06500  65237 75967  05778 04391
2901:  54605 87033  94414 70181  10860 21816  90579 11809  65060 75815    88518 85061  18116 19106  43583 72588  78409 86276  16754 29396
2902:  28900 88261  35179 66293  15968 76644  81035 37332  02194 15399    97574 80652  55915 05087  89231 95537  78074 08878  18780 13793
2903:  51056 78948  96530 33368  96186 56699  06212 83032  68209 28091    26226 45965  46695 65668  71445 01332  66996 55128  69294 42454
2904:  53597 67091  21077 40037  40658 45666  12725 16253  00793 13570    97239 98103  24660 99476  30605 38465  10066 98941  07413 40792
2905:  09987 61378  57473 93583  35889 31556  95292 33602  15176 90277    68401 10203  15734 27266  08049 83215  39832 49030  71118 01161
2906:  37882 66710  85695 54619  78976 62300  43801 61916  65770 13599    37799 91765  15959 54122  61402 98049  34873 40531  80470 78521
2907:  30853 26329  92026 66588  52960 30295  16161 56106  55422 95651    19846 38773  19218 20049  99719 80458  19331 91849  43982 29901
2908:  88608 57723  86358 30752  52022 48551  49384 75385  20702 57946    22015 50219  90476 79701  97627 89600  89331 41788  35942 77014
2909:  51613 69594  33629 43347  45220 61816  19169 88103  21466 80901    99329 47348  13468 51313  19165 42416  91837 80937  00935 30060
2910:  61753 69663  11288 20779  79868 63778  86691 02162  49170 54907    44827 36819  88454 68718  55271 03505  52907 38404  69291 40460
2911:  66075 14055  97827 11005  68357 69058  99596 17049  42839 63432    70719 59609  48261 97512  31174 37374  72249 83719  23906 37804
2912:  73994 96474  64137 10685  46193 56379  48347 32538  30521 89658    51331 84426  65274 66005  71375 01883  84804 69731  17933 82330
2913:  23239 93899  85124 13653  09612 99835  25508 14789  33631 41379    66603 47731  48799 61916  56445 33763  88306 39330  86750 46955
2914:  07810 46979  01284 37565  97965 57981  34474 93949  40434 46888    87748 45379  99810 20363  84212 51743  03345 35405  13391 32561
2915:  25197 22723  25215 59910  95137 48084  30493 35801  20241 35051    09636 52233  12576 24434  16236 54146  53825 13592  37037 84664
2916:  47536 13179  98872 25106  08665 57977  26709 72737  73467 84178    74918 52763  51264 52472  99742 07956  75954 14954  91781 49042
2917:  85364 98777  29976 36736  26840 70979  61208 69629  36642 08463    35257 60892  27640 49323  92905 29180  39466 69431  38165 42360
2918:  15089 66363  97041 83087  80946 45165  30421 10908  46845 40396    09060 87035  18075 98631  64221 10264  21114 13382  88600 12127
2919:  83334 11902  43713 54567  51015 18296  56077 03413  58463 26445    21044 58580  20692 29566  51746 68973  93963 09309  41510 83144
2920:  37772 32983  96680 77164  56603 36320  62152 89019  24222 18545    10155 40546  39951 52564  41264 92639  35331 16911  00952 42119
2921:  47631 39507  25819 14683  96652 12901  51018 00192  58645 21672    96119 06045  93658 10730  67529 57618  77515 54510  62557 78604
2922:  84470 07816  19792 39356  19747 26622  99711 74231  68227 05400    34864 99377  34916 64770  65498 00728  32886 05429  23985 21629
2923:  17418 82107  56298 26123  43125 09784  65576 13125  77160 54757    57770 30743  10294 69123  86601 15508  97837 09070  75393 97611
2924:  40224 51510  59048 20314  60066 33878  89114 75155  80242 78706    87215 93603  34326 90724  74251 37968  37833 98663  35517 48824
2925:  34741 55546  69545 69157  59413 11926  67365 56262  73647 43691    42756 91884  15330 56878  43322 49026  76001 80159  89839 39326
2926:  01493 90900  44532 46182  92833 86807  89926 87325  48435 07182    24997 43188  81529 35893  61125 04315  32409 51113  90231 57473
2927:  59912 82860  56065 44175  61277 76452  21967 79382  89386 77115    36675 80406  39399 90868  70189 10824  08214 89056  44160 16846
2928:  23781 41797  65881 34301  74189 03591  50205 62693  95027 74890    30171 85322  34783 22955  90625 89446  21082 31745  56738 60557
2929:  77022 88029  43259 54170  23990 39935  67880 59771  77025 29407    68547 44356  60134 23083  53424 36810  27327 60914  36682 42411
2930:  77295 75952  25850 47471  34319 47057  54215 39527  09930 91634    14345 08441  38430 79640  22599 45814  99954 19197  32679 50781
2931:  20374 72653  31432 60163  80033 67880  75901 83671  49615 19477    81130 61092  03023 57357  22753 01351  05569 58369  06221 88311
2932:  37618 28631  06673 79006  74907 88622  15925 83649  39399 50279    07576 73059  31817 19273  58923 74497  31812 71239  85157 74577
2933:  71734 89771  70205 71335  40225 55004  08551 98190  56312 66351    09186 52001  05652 96095  78068 51478  04914 93800  56953 34630
2934:  39847 98263  49607 16402  68750 89080  86392 02459  60160 39696    14867 37519  91956 74563  81331 29991  41689 64616  64704 31453
2935:  30045 19928  11963 07235  52481 76906  39562 12029  66081 37095    74472 19605  66518 19243  32297 40089  18951 80888  82209 53575
2936:  44659 41075  02973 89344  00581 71735  87243 67237  66279 34943    70398 48820  17921 14163  49453 66510  99420 34783  41299 01422
2937:  31628 91849  32425 79881  70478 14145  35740 66213  35923 48452    58833 66011  12217 72672  61331 57895  00446 91778  82387 73139
2938:  51780 42521  45579 46211  88537 17975  56075 42369  74218 12849    75846 64072  19507 85249  91925 09560  84862 97344  42919 61844
2939:  88175 82463  23883 89174  66118 93991  03637 46572  91583 94911    13045 36123  15398 47303  40691 70264  33335 22945  41029 36179
2940:  52440 69706  34847 08948  47727 97397  79481 95658  64635 66834    35914 81659  42567 98161  14065 03562  33481 47625  63254 31388
2941:  13126 90921  16273 64517  89327 65492  98357 94884  53713 07456    85474 90032  33215 95523  81759 20009  03236 78366  44346 21500
2942:  09598 01621  14629 23187  55697 80595  36642 96183  55878 91403    21724 95582  04264 62140  42520 23318  34851 45544  59896 84732
2943:  08820 60138  99875 58786  58486 06333  43837 33537  13104 64728    65764 09175  60877 84807  68802 32716  13346 86345  06073 05417
2944:  84625 46596  85205 80783  79309 83876  62175 32975  31512 03466    52114 66517  33679 07987  24221 24459  97698 95178  83877 36552
2945:  29301 93549  99430 17419  26785 48916  60499 32735  64364 43794    70319 86982  16179 59721  90472 68428  62588 39564  68956 11625
2946:  36059 99660  18737 36056  45699 17999  66482 85298  82850 83590    81026 22583  99532 47375  42401 54656  18167 97490  91606 26404
2947:  75453 07124  61330 46955  42561 34876  83660 91792  68099 55762    72396 61611  16506 32299  52970 00508  17117 96137  15406 16726
2948:  53511 65123  60125 73254  83828 74899  01946 45359  37966 62820    90704 52077  65844 26348  27613 71258  22927 14234  23403 98215
2949:  75673 13644  77884 32074  30395 78550  40307 90646  12889 73903    92988 81734  04360 92558  17317 66536  02780 89008  93349 29577
```

```
2950:  0180362170  0010482612  7417664532  2314837556  4384222318    7784118882  7911752603  1667477753  0465914051  0603716598
2951:  6386179019  6037967720  7424647094  8066475188  7770108541    2624123221  0902585782  2975512701  5951643113  6595916703
2952:  4427714020  5069302352  2766065075  8133489926  8211358082    5622616976  2376108624  6872409914  5078882679  6044039105
2953:  7660830450  5548572211  2766161001  6757395044  1696121517    4294196095  6211868699  8358278215  5391881012  1898620118
2954:  5531083585  0353151268  0401400364  9571341187  9578108966    8320270558  7596919185  4823085682  6416867580  6477869734
2955:  8250736469  1697692860  5346580488  5793720644  8456679149    0561807594  4158043014  9518587889  7718466828  1123612152
2956:  5984298152  6281201718  4605094378  9793601952  9661919847    8492254719  0166519478  7140959245  4996326254  6768761220
2957:  7282899509  8465475295  2569579112  2667233708  2944942740    8993031727  4881764918  0245607269  6908438283  8498758989
2958:  4299888561  1040320291  5087125964  9124493201  9359406305    2858785047  5661724869  9993314877  7973218127  7721318092
2959:  7803638763  9885649881  7218074089  2349956697  9075270104    8091654385  8434086249  7008557237  2979649806  9473233753
2960:  2393996799  8774381626  4033112323  8347719694  2493747298    8143266248  7638314387  0917138180  6321174074  5926109851
2961:  9680080311  6940330909  4498873975  7561281413  8227165538    6080465952  0150848266  8984907515  5406274953  2644314808
2962:  1353123543  8230450008  4863889905  5939942929  7516153161    7886541140  2839692153  4299380827  7199912726  1433823004
2963:  4288724194  5692542875  6030598839  9289365002  7234066683    8796984600  7198024567  9587772209  8030054477  7066373803
2964:  0939535446  1709334861  6030812754  9077249922  7817929590    9471736472  4792615323  8355431618  3070704865  3203803935
2965:  3360924736  8640646494  2879155832  4789749964  5601274498    6436334425  2864323013  2522633400  6873095214  0971151255
2966:  4303528188  9629769414  9658903362  3637276818  9949983169    1764969806  6789174299  3096086552  1447840310  4374650828
2967:  4107637267  3793704325  2791507858  9771829423  8544722276    4526976474  2024144265  2666110367  8983583599  3719369648
2968:  8056800458  0121998041  0454101885  0325651885  4252084413    2272709960  4337261920  9046606857  2167230317  7583108465
2969:  3239860888  6839612784  5451833487  2613332244  3739336959    5032508044  2768253920  3565673379  9079956678  3152124303
2970:  8438278985  5262493233  3143415881  0243596091  1639267336    5533329826  3901096962  3816997768  0947837789  1904150515
2971:  7093606124  4222113346  7379444854  5947031559  4860092533    9389417347  8107084517  7628251492  7930853733  1814716448
2972:  7772485195  5367991228  2030128684  3208625339  4893112344    6478039559  9157162025  5877955490  9180405945  8806089373
2973:  6071376271  3834733032  8370023903  1624443190  1808492371    5709731734  4561802273  4189803636  5244842041  9162342164
2974:  6642194250  4134141420  9907837220  8831658629  2391303412    9326715138  9563785594  8956081804  8354710792  6387305578
2975:  1871304311  7908200191  1381043085  0783642327  4922109876    3039665592  4661115296  8856482418  9240814278  9597965350
2976:  8132111974  4773809739  7465174394  4456196487  2071147238    2834807416  6104072744  6996316308  6046836414  7731866218
2977:  9936232485  6712320353  5485526255  1163254328  7623949942    5492667723  6270912684  6583730399  6507683120  6427752320
2978:  7085283043  1785783229  7904624162  1376849924  0393627527    1136077824  1130206399  7364596444  1971664388  4145534838
2979:  0200724625  6292702283  5197321808  9289983171  9897303286    8208448748  9047060026  1924209922  6067335294  8176794128
2980:  9809457241  2218538159  0726359440  1420875158  3518257833    6676621588  6282871698  0809780953  1891611644  7594993570
2981:  2242546586  2818888486  7121681475  4183381255  0902343070    9923731162  6737110251  1002262658  2505280965  7453941055
2982:  9304389695  0250517858  0321825931  0338863704  8324985127    5126550039  2703980789  5651542012  3791412211  8619805115
2983:  7663365808  6394414529  9358062789  9307538934  3702727240    8674534525  2447945856  4961081478  6253053372  5717629960
2984:  6324362633  5832697673  4581182296  2772073216  2508669834    7496798913  0946618230  0281517804  0624043333  0075915167
2985:  3795234474  4946129108  6596589033  2279883371  7993521669    1043638512  8519412688  8392440714  5252045053  7567503467
2986:  7799239916  8394731911  0847983509  7003413028  1305695850    8347449931  0070737498  3295457595  9709690712  1052680926
2987:  7611826288  6844875462  0276947159  2542674572  1644139624    4607641914  1102786283  0948256087  1876308298  8071554828
2988:  3022229743  8836735537  7810435002  2096870320  8450456258    5412551221  4044417785  6247471923  9107968286  0683595974
2989:  6386644202  8767160514  4853574454  1028941716  9531534219    6926034548  5337385625  4873994637  5179552205  5694085267
2990:  5310253181  3605016403  3132292192  6423332217  1602096776    3123343538  0687153811  8301857181  5855525895  6129058422
2991:  1221780843  4450649369  3132292192  6423332217  1602096776    3137294269  3188831620  9455694591  0168375963  7451992925
2992:  5839032866  8688923354  5908739826  9416819266  4489898624    8230870198  0864625643  8471506131  8567692038  2738275012
2993:  6249847770  4172936211  9069588655  9186729565  4989601217    2561669629  0435893562  3822075931  0953911178  1897564656
2994:  1237697731  3633455230  1453241057  6976844174  2008560872    6865977429  9504364601  1924450555  8690403118  2602110344
2995:  0869429871  4300194754  0245410580  9895579079  4520487766    5930591507  8168923325  4481039390  0108880419  1132718322
2996:  2766238942  4363673352  1368686425  9630135600  7066577309    5482912948  4883761895  0583759463  8628529345  1374479248
2997:  0022668765  1731609624  0188739401  3518687524  8334116279    5135773171  4013861475  0080861976  6022355436  9214764540
2998:  1502667590  2890377913  3786714954  0308451204  5988580475    8128323607  8227554991  5456944802  1633498818  5997420737
2999:  2538999629  5356212044  4919258354  8063915870  8066230978    3752378986  2805139503  0795274201  4067164687  7500517421
```

```
3000:  22406 74063   57155 58113   86810 57868   63812 48050   93942 83937     67618 42822   71150 29154   50851 09983   95322 69528   08491 17533
3001:  80193 94359   88350 99419   29318 79148   20108 90615   05696 22507     07174 82005   23117 88906   11617 01035   06890 70278   86625 34880
3002:  31020 48870   79997 08081   67655 26158   15189 87695   22006 61055     40771 73416   57737 43627   76011 18205   53561 74495   86965 12352
3003:  67787 18421   45905 00649   85298 78354   53315 10323   04059 29750     56772 44213   28130 68772   96673 94364   22434 25246   46681 07114
3004:  80674 94680   31657 64059   99598 36981   18972 69011   04180 59395     92535 03737   61498 72271   67561 04535   93453 32945   40096 73930
3005:  88204 39843   65386 50411   64318 59236   49528 07174   03959 03220     57155 74444   78757 80141   95864 84580   92791 61959   75832 80321
3006:  47338 13752   61196 76928   33807 28456   41913 77933   29152 53001     48576 43649   72613 45084   01031 65525   49053 84427   21568 11848
3007:  10730 10263   89798 64783   17105 80725   21816 26752   88852 46847     60400 75443   52300 78725   31568 84224   81727 41441   79406 05997
3008:  53472 50881   21548 34177   34904 30066   61643 82327   32629 39248     30395 80764   02147 91614   14336 80753   84235 05883   42194 79785
3009:  74341 25624   30683 52224   07179 67143   40094 85374   80852 90717     46637 76949   25590 21630   23916 80213   66133 73005   48125 17205
3010:  53387 44810   65239 44014   11545 25777   98273 94614   48787 79217     54140 98447   24499 32173   75744 94411   06606 52749   23572 79387
3011:  80307 36609   67600 39514   14142 76514   61410 35855   95703 97529     78788 11412   54563 08925   57399 02830   93503 66375   69184 83563
3012:  09615 61596   23370 66178   00454 88461   17817 12113   02361 65757     27519 18296   15170 08829   97753 93879   07119 33242   94524 89113
3013:  19078 21829   76028 14358   50437 85611   17923 52221   01900 96287     17363 90405   06456 08287   25769 17670   94467 67928   92601 58373
3014:  75747 93463   09887 78191   14419 66493   87559 74801   73478 54513     30296 41513   15578 12835   19395 72869   99898 70837   56315 81165
3015:  81548 62995   28438 24097   43021 80140   51615 82904   92236 80162     24142 82008   40820 03757   62803 97357   89167 28923   27261 32250
3016:  18874 48837   30249 75900   50915 46728   32024 76208   61588 55653     20129 09044   14862 80962   64018 90286   88790 86194   27171 91896
3017:  90456 45054   41330 20997   69114 06729   90831 85938   08398 22843     30334 24984   47865 80224   86883 37330   77679 83474   92580 00335
3018:  26456 56159   98612 98550   86225 02094   53057 16126   95237 13190     32204 00551   39234 53427   54150 43856   76578 69052   12085 28674
3019:  74968 42016   38928 96582   42025 41808   89169 74234   65260 47168     66743 91459   53006 02290   46366 03887   15752 10956   51746 34873
3020:  98033 21596   95006 99987   06420 68420   40494 32670   15766 48942     95872 97441   02856 70328   05707 19202   59468 63115   89407 57591
3021:  07708 48345   39236 33764   63260 53883   98947 04234   97828 19058     38301 08541   13043 92007   25392 15390   93874 96393   30221 08678
3022:  89695 68869   48761 68103   59344 75325   49133 01210   29370 91760     52122 94876   98548 01017   87830 57710   42364 90003   66821 90674
3023:  51660 06046   25776 38666   35718 51921   29714 01913   93253 76829     71843 26983   58804 35904   89062 38567   40840 96813   19930 36826
3024:  37999 39863   83190 08122   11892 73559   81541 22218   04503 07641     40357 71729   06563 45664   62529 27327   87151 92473   45274 97066
3025:  70898 44897   53381 54937   33014 65155   98062 83903   62137 46975     00349 97468   90564 12756   22498 89711   78587 38829   64443 92222
3026:  61846 26173   13659 22367   40923 38911   71154 62143   00621 77603     49429 32316   59681 33687   04904 45648   62255 50512   62267 31896
3027:  87688 19861   73887 93512   41251 82853   21567 98541   22643 81531     51010 58122   52508 93701   11874 40030   18956 00893   16886 72092
3028:  26283 75603   85996 63658   72267 67734   62221 96320   94549 31751     55071 26807   70298 66079   05391 03620   33576 73564   25017 88876
3029:  94406 78907   95922 75718   40434 00060   33745 12226   45151 31780     77742 87108   42744 01814   79365 89598   70528 36383   15981 53805
3030:  33626 18641   26783 30875   29705 62006   19431 64873   30248 48617     99248 73340   16926 80989   08323 82487   62840 13586   31507 63163
3031:  86489 07294   14192 98431   47675 15985   11890 91386   82527 40041     50036 18234   71602 55543   73466 87213   36779 49328   75869 61066
3032:  77948 09373   17957 84130   44943 40430   93710 35481   59388 28216     74360 36640   33929 91830   64099 44691   46440 41256   06929 90341
3033:  72399 25039   30308 92812   59172 19798   07794 40361   43484 66252     82205 30012   90100 69571   36411 17616   84492 36351   75898 83629
3034:  89005 65962   80296 79990   37379 51090   12858 10092   98532 14527     13519 00876   95119 04006   44022 40346   99913 58411   59033 36081
3035:  68871 44924   28341 84125   39162 22695   62490 65404   61785 42633     22547 66632   45789 05716   24933 37860   50201 62522   82091 56032
3036:  88847 62436   86189 00977   05056 07676   24073 55527   92673 53026     91605 96780   27035 51904   90282 45564   65635 27357   88050 24157
3037:  47491 14052   79961 66540   16107 20991   09575 58989   44952 06177     79181 15777   37687 58895   04805 11062   92888 98531   54292 77866
3038:  71751 98120   85709 34724   40175 98134   05283 20968   45804 06260     67407 50102   89555 05056   48726 08670   45580 14199   41558 83318
3039:  57104 25920   02382 27681   93821 50104   22207 24856   57889 51066     10846 76619   10886 19592   25278 35892   09441 35098   04917 42830
3040:  93742 52197   10183 13965   01774 20365   60654 63550   98151 41229     81881 21349   10713 85324   95542 95244   42523 22698   77430 66035
3041:  17129 40026   95009 22609   17680 96124   94650 07583   40135 87138     86739 51392   91403 14844   96688 23695   36572 95741   57517 91405
3042:  55441 45161   57997 42214   51899 14097   75462 44993   36785 40294     05070 65489   27874 77917   90601 80726   82992 48641   29418 82228
3043:  52097 00948   14796 44147   96593 41983   41892 21802   23726 45800     19819 18582   74840 89818   46731 41365   63453 73742   78386 95832
3044:  67967 02554   81803 63926   65675 60739   44256 14960   31226 76963     68637 13015   17823 87316   28356 99841   85143 91946   34334 04135
3045:  11846 58501   71369 01112   41211 44106   62063 93199   23426 30312     96413 19775   78136 76824   83215 44116   47761 74420   14167 18417
3046:  07527 39143   10195 83663   42560 22234   09074 55312   11594 05073     20668 20210   22030 01600   27404 14009   77606 69654   59491 41994
3047:  58140 40275   44721 74628   02383 87497   10145 56891   29959 55747     05009 10648   15273 03581   76894 88519   27537 24415   61994 44276
3048:  46943 28680   79931 47864   64020 85637   26428 89139   23727 70269     52773 95390   76635 91497   62647 15243   39460 84636   70881 27174
3049:  39026 30754   36015 86694   89913 21101   91277 01419   52426 17332     85768 65533   16618 36749   17676 94186   07381 84740   60489 90478
```

```
3050: 52315 27509  34792 34233  73523 71594  81413 00652  67499 96489   72468 98685  85082 61977  77903 68561  33845 06360  98945 43256
3051: 12840 89440  02914 88064  09284 08082  84589 18439  94400 92051   76427 09724  90092 73486  26762 79097  36060 31675  59503 65090
3052: 53188 34795  02659 39853  44954 87919  56875 23876  90841 88347   69040 40808  93566 92939  77303 16039  60154 64997  27212 33783
3053: 95482 13398  72890 01395  84774 65939  20122 36408  96513 19220   47126 92602  04207 22108  79551 94198  15864 03123  54331 96682
3054: 18789 76683  63859 38265  86559 81253  66835 79345  62454 22016   47566 87356  94878 18518  20138 46871  74843 15052  59713 65702
3055: 84727 87364  27370 11254  22950 71883  26219 64771  50465 39247   97990 37983  27223 61285  68385 67969  10623 05288  49099 45988
3056: 67005 19836  29331 57376  64957 09228  85313 56874  40865 85232   30786 78182  46577 44224  52686 05245  23735 91705  91899 99540
3057: 65598 17369  46458 52528  59248 80278  56121 72763  78661 99636   71617 89303  30728 50183  80565 87418  24836 16747  84666 73996
3058: 40720 53486  95562 04347  45406 42767  40446 12872  81669 20342   48486 61498  89295 69431  49389 76132  13059 21508  89279 30060
3059: 65599 64766  71088 23902  10404 94707  64202 80664  30754 35574   56088 80666  54383 18632  97052 18170  01416 25511  27941 61751
3060: 02480 17172  76138 62286  04876 80377  94386 90393  66946 37710   95551 02733  01154 97341  82729 45700  87151 50619  75552 45547
3061: 42421 11721  42706 73027  96776 21015  14430 62237  66237 16089   47416 19560  08817 44932  13890 62789  02288 18855  66461 99753
3062: 23208 97948  40621 63934  15803 52848  97578 38844  84715 14875   46051 01793  55792 13794  04651 98283  20819 57022  94384 96229
3063: 57606 92656  23950 05710  21880 56776  83670 91227  69336 52305   74499 22931  74974 28871  85204 01669  39343 21964  29059 74562
3064: 41651 80681  98423 66540  73333 71317  28799 37689  77660 68158   69808 67570  78248 16927  90894 58859  97722 80150  11638 84227
3065: 96025 83202  09879 48263  79821 97574  41861 08565  35134 23467   33478 61004  76875 45143  53811 62569  94020 95519  32556 40314
3066: 12075 74358  09961 97811  78455 61388  29366 80807  05978 12936   43276 06939  29827 18906  81641 74896  79993 91359  13537 46549
3067: 55071 47123  56281 64062  80423 52174  80875 41897  92134 01385   28287 72230  15621 14134  49636 92128  98045 82539  84622 36150
3068: 48123 54151  05489 43785  23975 06565  46714 17201  93447 78050   61098 48208  80220 39187  32987 22376  04853 94770  95119 30901
3069: 29089 88973  77624 21274  82304 16956  66230 29645  92132 71512   67626 11986  23214 82703  84715 93365  18659 53436  67856 81773
3070: 89155 07528  63388 29808  65657 14317  50272 91241  67702 03471   42368 56447  25992 80382  60903 46465  95672 15139  22530 05264
3071: 11812 35426  43140 91017  87890 76491  83270 32327  38366 26010   54030 85161  84368 36316  57414 69845  40754 25704  55191 86600
3072: 38332 16122  27868 07670  48305 41889  63026 51489  40870 08918   59641 82444  45703 79713  98870 76769  01549 81622  34719 36302
3073: 05978 57553  21088 14595  65508 13022  13095 88638  94861 98329   32356 44425  82776 81877  84542 15420  03802 73487  11265 68205
3074: 19925 19761  98095 10419  69450 05479  34627 44268  74414 79534   46543 90707  47857 96589  69943 79715  19478 95684  18494 85983
3075: 59459 65416  17594 57742  45323 86310  24530 03630  89924 28691   52650 53042  07978 88343  83729 69088  67778 87206  58462 22790
3076: 82797 57532  37009 58515  83206 72481  50774 47780  53457 62914   98987 65695  20204 22623  18365 73730  21786 03501  68312 76510
3077: 54044 38302  41581 03330  89772 75188  39974 90677  11219 35750   63371 06314  94741 11573  95977 68548  48739 15529  27658 26045
3078: 51813 64826  96928 00914  89430 91135  30992 78689  47317 56105   37340 88991  96096 02249  80916 13948  48764 51268  79337 93386
3079: 64759 83230  45669 33839  44187 25436  13505 23177  31365 59250   14353 36247  37893 30072  88290 96415  36699 68925  74816 03339
3080: 89447 88780  55293 07655  55953 37643  51850 14947  73712 63413   14156 65287  34752 88606  76976 34188  71595 55998  68540 40837
3081: 40773 00691  10695 19995  89417 51268  43602 71148  97874 20173   19547 80183  37105 90357  13770 09030  12133 10651  02488 12985
3082: 74119 65330  26183 81717  04496 95310  48264 69080  64282 36648   40145 26521  32321 94849  20296 89179  04078 59730  67076 90957
3083: 13297 47732  56766 25562  37077 70953  55747 85162  41987 17946   22105 45246  80047 13245  89433 80098  47071 86175  16721 49249
3084: 00474 05647  07939 70544  86983 57522  16463 28435  99402 37993   98535 79986  16318 71897  74075 21728  52590 15303  43410 70166
3085: 45163 77562  15718 78887  43660 48157  63581 59975  46833 75337   60591 87637  69257 65689  89076 47124  30945 95138  58064 68323
3086: 06724 89247  54890 32695  31332 16650  58835 58324  76921 64928   77124 24541  27248 90748  36354 55305  77316 59242  74915 92635
3087: 19528 90170  97364 99219  18594 80179  56769 73361  10796 95439   95766 84370  99033 88402  63194 16865  98787 99428  84054 49794
3088: 59183 66619  84507 19177  07318 81103  11728 67458  83172 73986   68484 88407  33982 45385  91159 97407  57610 00948  76052 76932
3089: 41273 44788  47304 68532  12056 03208  71424 24103  69254 55681   17617 87755  59514 17091  18033 29858  00059 22195  23820 92983
3090: 84917 71727  51054 60489  80135 37521  86697 63657  98294 91813   45588 90345  26428 89686  86440 12540  99966 38239  57735 85388
3091: 85914 84530  09736 56010  11035 41792  42962 64870  23772 11735   75874 26449  73449 59126  37052 95131  39484 22019  05591 46743
3092: 41514 51998  40457 37060  05674 99572  41650 53082  66942 74592   74082 98602  16939 18563  13528 65918  06572 91817  35110 86801
3093: 52748 88176  92181 93589  46205 73436  18900 41103  72258 07622   23332 25718  64094 65141  08032 59094  87291 90485  79898 16989
3094: 13419 55762  04368 36647  56890 66639  67616 44082  64593 06140   08157 76282  09741 59902  58436 42447  96662 47542  40436 68213
3095: 04006 19504  58616 57490  67972 88317  23811 17479  42248 47297   01552 15624  26836 30269  56032 90625  14259 64932  51511 25011
3096: 24835 93827  31270 79765  24643 01366  04562 29714  54814 85820   84408 70989  08609 39832  21658 90601  28278 13365  69938 92118
3097: 96022 62064  54626 63352  10945 63756  17187 57033  12613 78312   85316 17203  26137 71705  39306 28053  80905 20300  02696 91647
3098: 62080 54504  75898 88402  59925 12375  02290 80699  07935 00198   94223 87707  93939 84106  74165 70434  80256 26089  31843 59089
3099: 69389 90769  78436 72118  17654 50100  03908 59928  04047 94965   04957 53631  16739 20111  53548 10871  75771 73836  01297 14684
```

```
3100:  72691 81812   02805 26237   92917 27267   06689 21896   80825 68418     65146 11694   58487 19320   80933 42734   09921 29231   55101 09301
3101:  93167 42349   28706 77585   91919 07173   77985 56645   98802 69837     22979 28342   48586 36870   12594 71514   32227 37957   57115 51891
3102:  92790 74840   22694 23010   35833 14537   67273 15910   45825 05311     63306 77950   50673 14643   53713 21984   78206 66684   76337 01048
3103:  05827 64950   85354 35474   80477 01513   91456 12826   91971 89091     02566 01854   98139 13955   27149 59236   20074 99278   97209 77142
3104:  85289 17084   50130 76298   22336 30975   86331 41992   41974 54636     23928 52888   66622 05704   77746 01614   60338 58615   87144 82622
3105:  15130 48314   27604 55612   60010 43300   58914 22557   88399 69299     59219 46484   26573 67210   26167 82040   14048 49225   03266 46247
3106:  63307 80375   92068 78999   60918 04827   93890 13154   57132 98948     41181 14497   84093 75513   14136 43130   26698 87943   50430 62582
3107:  98966 22983   26010 64218   62205 31102   11099 83713   20726 30755     50549 83890   52991 51464   63120 27945   19118 04045   53333 07611
3108:  29225 52884   89441 78114   44771 11150   39248 03737   15717 27209     43334 72680   00529 04906   84865 20969   44198 35260   01551 67884
3109:  42045 57492   41185 36669   84149 11619   46349 99602   96711 45827     85482 87199   35201 65719   99747 42922   10882 63600   63298 31530
3110:  28046 92643   07446 92285   78013 62667   89490 95099   45384 21540     31255 33511   21980 67576   78397 99486   00016 48216   84261 83960
3111:  98867 28756   27602 88682   85991 82052   02009 62410   67662 74337     91700 98989   08748 70548   09590 85758   45730 18818   29449 15009
3112:  22358 90629   58744 91270   93831 88062   75101 91061   69355 95987     73449 91239   13653 87098   41831 92163   81845 75741   87007 58887
3113:  16809 46693   57527 47969   92895 69975   81496 67050   44541 50256     83632 72157   23954 71775   88782 23110   80773 52713   99777 11160
3114:  07189 46249   62665 52634   56526 48370   43231 53187   34286 69700     81605 65869   07194 91709   56605 14887   59669 26096   32102 27528
3115:  00696 23536   81835 32251   85262 05854   52331 24472   72220 73436     73181 04674   92385 04069   57346 76929   01761 95675   21641 34523
3116:  19226 06642   50493 96895   61979 56765   22877 18068   08842 33326     08957 45454   34333 63609   96578 32138   19990 63877   43832 44825
3117:  51532 69443   06324 29934   86764 54295   33117 23835   16330 35532     91859 12831   72985 22309   59852 46469   91246 31568   73799 92442
3118:  35187 02843   47949 71813   13577 92834   90939 42567   13488 06665     62391 58146   82013 42571   74012 76216   80834 16022   79201 56440
3119:  12738 00940   47604 49305   68849 62445   93838 85932   46601 76073     30538 24391   22184 43483   27966 95523   19663 31611   84368 88543
3120:  59554 56802   53896 48442   78611 15387   53116 92346   32098 87289     00613 22602   01727 66245   76413 26593   30375 47582   37831 47819
3121:  69913 82656   19963 54920   63694 80869   07745 66630   97594 42107     33527 37598   01079 88602   71280 21454   87656 29705   59275 58017
3122:  38923 91772   34587 66680   68603 77432   65053 42808   32261 80407     39394 84160   77621 09403   25588 16263   75404 72522   48735 82139
3123:  96996 79270   38760 42772   57792 05127   99089 80613   67177 92455     97873 45389   46958 45563   51515 90636   56112 29565   28783 73282
3124:  26952 30332   03315 81538   85216 91202   10502 77345   95862 17066     46950 92267   78644 11438   83201 25126   38044 04993   14203 09298
3125:  14985 30881   69521 03813   26228 42385   27317 76145   94675 46143     82517 18740   92917 48824   11774 58352   32440 42099   45601 45755
3126:  77658 07773   36498 43509   63808 79835   37007 42488   58770 24396     35144 39031   97648 72755   08609 99727   07298 01665   70724 66984
3127:  78907 39687   99282 78989   58419 57083   24473 15278   00673 77556     10850 66227   24070 31654   16265 63389   48666 69218   62671 87802
3128:  10040 93385   20991 60286   86391 67138   05134 83434   18381 06150     47075 25884   38717 96905   98704 75368   09887 62414   00051 54655
3129:  85954 06608   41008 42377   65396 18223   83322 78430   17522 65500     68902 95953   58790 07920   18430 33149   06580 77183   41170 38872
3130:  83832 14115   12278 69351   23627 83166   48051 63073   89847 92195     89748 50792   03492 78562   55476 00100   62705 33294   65647 22290
3131:  34068 76261   26491 35168   38306 63333   60879 41107   81541 64557     10328 74374   53068 65950   03544 92961   80105 98828   50032 17553
3132:  84409 00725   69399 31190   15733 40158   73033 87556   53244 97218     44945 77454   28466 64717   95293 22943   86155 57903   89477 36447
3133:  04879 22237   24218 95313   13530 94588   90433 26708   77677 58214     85880 49412   99710 50162   18533 55646   63443 89355   66013 74441
3134:  28278 98912   96076 58390   06189 19965   86518 06510   48506 50691     63319 75890   22148 94673   71533 25700   07617 82967   24863 70254
3135:  33597 06624   62181 51178   97991 83266   14674 99879   05872 41997     14672 44716   06660 72458   56565 37901   85906 35566   02588 64858
3136:  95401 69933   71514 94385   37863 81693   26372 99837   24508 85281     89494 04362   32085 05983   19990 44459   34572 09266   39130 49597
3137:  89658 93643   07406 15255   96319 78644   82604 12963   83414 48486     83158 48342   49791 85389   16092 38308   18372 57448   99510 41221
3138:  94604 26161   63553 47249   97430 87110   46012 44907   83313 32202     19686 43927   06903 35801   65480 76688   43471 09135   57694 27405
3139:  86587 37675   95634 37064   98181 18022   84922 29126   08677 76322     75200 86805   53619 33838   05704 50413   54868 12869   78595 67894
3140:  37483 67186   24276 08766   06606 46321   79983 00399   66078 64840     30500 89614   97765 51473   10474 74124   51645 79343   04224 04814
3141:  13342 30366   02229 15637   85021 76511   99052 61305   63501 87081     76474 79321   97803 32558   33023 03517   03075 21052   67001 86764
3142:  00058 41845   38766 91104   29910 38447   85651 66434   75070 49953     72516 10700   23362 69409   95887 07947   80790 30325   53599 81885
3143:  31674 25745   07135 09183   34541 88579   51131 47556   01449 94547     53460 56180   74734 25480   75229 22588   61545 67061   63877 17265
3144:  52677 10260   86130 02084   23023 89719   82734 62619   53017 88056     49207 15636   82862 75109   58202 28165   00483 09668   19675 88941
3145:  10702 85042   42830 15053   91577 00015   40661 49234   36370 62195     26300 43118   16634 74513   87103 33583   06402 42504   75016 24570
3146:  36384 37229   67788 14348   72758 91045   91671 04699   15555 90671     73151 10065   48398 63247   91364 09458   79151 35727   10253 08222
3147:  25616 10250   82057 64472   30433 69918   04802 29171   48380 88145     40085 77699   69703 79858   70350 22617   51939 75470   35644 85264
3148:  09817 97495   88552 79451   10118 61251   76502 11124   64087 10765     85618 27605   50959 97274   32844 75320   04230 71548   39237 58169
3149:  68757 27903   25635 79433   52769 90616   58684 31507   59460 59488     61089 69417   84319 87177   31220 34056   51422 19450   43436 59755
```

```
3150: 27128 71464  36601 51484  09077 13806  89317 69101  22163 68024    17352 65170  14045 24098  91540 43867  56206 85223  09157 12898
3151: 05111 17338  90230 09325  49339 66411  45894 06887  88679 06226    66115 37390  00021 86194  78190 59178  79211 34017  94340 82765
3152: 42980 27533  28352 83022  28756 97175  83836 02276  04567 14706    64584 64268  27469 44102  20315 96167  93301 66765  83017 11284
3153: 00871 47027  35035 62385  73565 54384  96903 70606  26468 84582    22488 39311  19311 67217  68828 45945  16346 05597  49217 86492
3154: 74412 40733  68736 44685  62854 20986  65651 46261  37436 05476    62494 52013  11556 10031  45501 81794  52475 32312  45928 88074
3155: 59783 48035  28478 43840  35404 67729  69960 44578  73948 24389    90287 48873  08611 09075  80755 96504  65280 31411  63976 82001
3156: 11760 94177  31652 15682  24079 76868  47432 57943  54736 23807    22536 15022  47070 95512  20417 57833  56539 67613  40637 07436
3157: 93061 72383  98596 22636  58686 60067  26823 82648  50098 54555    85242 06111  36989 47473  87424 64734  95392 77089  26739 53665
3158: 26372 84721  07254 36194  78419 67162  82319 27217  11751 72983    37452 55711  78126 00967  98109 82808  83425 21898  50548 59771
3159: 96146 37647  00737 60290  17187 84849  50951 93876  83147 24075    40225 87339  62657 86228  41876 26316  93156 44673  10304 53324
3160: 96747 95116  53192 71447  75169 42789  71933 97334  99916 61877    55159 61936  17072 17353  67672 29200  82616 99666  48300 56969
3161: 28670 70003  71186 38994  92758 44786  57949 26469  02339 37837    20313 91238  32161 55008  83401 15459  63079 59875  78557 47333
3162: 58836 11542  01181 20712  68884 10495  54588 51642  77896 88193    46336 82061  43562 63527  78773 22406  30548 19315  91367 50212
3163: 39633 33377  39439 69829  11580 05770  35359 80287  45203 99412    94749 14963  11847 28956  96176 24651  43303 77083  55563 72839
3164: 91431 53186  42460 97213  93920 42647  88583 71826  76643 08229    31991 58893  12681 38149  76761 99063  45594 87131  25369 60653
3165: 92635 54823  70250 14445  75209 01460  10697 88923  71772 05582    81447 71407  92568 91184  98323 01517  51808 43329  85086 71131
3166: 04310 65622  45312 15661  05450 16219  99463 12800  41965 21728    88281 25294  02774 15979  79762 23604  14924 59550  57710 12687
3167: 83217 44763  86420 29515  36307 41389  92735 38277  97055 85963    81381 82541  70077 93784  09649 89230  46588 75267  53362 13388
3168: 54172 75749  01221 86094  33137 88894  18573 84834  05751 16019    54631 35576  84627 26842  84527 17618  97477 67431  73250 90268
3169: 36580 38491  69079 42599  05862 04456  03067 53050  03226 81495    06170 30581  99132 16204  82762 86359  78278 58148  28056 68693
3170: 48553 27423  83675 60699  31461 83848  06738 83119  59475 42007    59095 56237  39961 72005  76804 24332  96294 19524  63523 64457
3171: 63536 75459  68177 89096  68577 45526  55280 95323  44522 84313    14791 50219  36192 40443  45298 54810  08787 49346  37070 56343
3172: 80311 55439  00707 85491  62177 54242  86359 11172  99908 13220    29393 01840  18314 79689  23696 89763  03794 88463  58035 04235
3173: 97156 11152  95701 24259  87757 30069  98749 89812  26750 07335    32228 82369  94680 97707  91746 75976  22419 14068  64071 33793
3174: 26914 64527  67169 94206  31317 00600  12266 06513  60462 23761    15770 79957  25530 22762  32685 60382  07738 40657  50176 36378
3175: 74522 56552  82419 57560  42328 87532  27063 46959  38680 05117    79333 00542  47015 21499  01497 02182  59619 63815  20134 28979
3176: 25844 03544  35972 33431  96096 51939  74044 46807  03679 46861    86024 80751  86513 90565  43113 70334  00053 46056  31979 23536
3177: 98676 71453  08776 23780  64164 69266  17615 38743  97268 94318    92899 17809  37351 71523  85465 75127  11019 20915  50899 75471
3178: 01358 05324  32740 38166  05663 43666  11360 01351  23432 34649    43307 59195  34726 36387  85827 33538  29136 00490  70130 52914
3179: 18825 87662  98814 43767  54692 20691  13882 58246  99659 86511    37444 85625  01466 98075  88443 76027  53734 31980  58159 72082
3180: 16375 06011  53883 03394  83002 51483  85449 81812  62944 31038    39792 46911  65071 87612  28742 87178  28584 25722  47555 53705
3181: 14784 84146  57542 53835  05590 98924  51807 74064  94296 27747    56771 79969  23168 92044  84917 66626  74613 23953  08884 79310
3182: 61879 46720  77536 71580  77167 72430  03672 93249  16187 20013    58619 85292  02395 14482  27718 21871  08375 75524  45380 91424
3183: 52335 60265  72483 03256  90277 82536  60869 08598  75921 23138    42532 50836  89973 56761  20697 87058  13462 81962  37833 68175
3184: 83546 53620  15999 03992  45034 92232  54264 72976  69274 14657    46732 50403  73034 54833  60589 47419  82625 22979  82940 73929
3185: 07017 99883  67145 44171  01757 29907  92861 73086  35026 57640    67607 75474  51483 58505  49151 13331  93630 87781  81495 02634
3186: 61600 47679  17977 73871  45260 84544  84470 82314  87029 62049    29380 46082  49872 96642  38292 83887  26312 05635  34945 33442
3187: 16192 57463  12287 77702  55197 05391  35987 32932  24151 98517    29072 60904  30598 82258  47806 89017  28438 27399  27188 65878
3188: 59175 81433  98300 70281  34313 93800  82905 57749  08401 55213    89763 36303  43849 03172  40029 02477  36421 69119  18040 59384
3189: 22073 22236  61738 01322  41582 04992  70232 02580  34284 14476    61849 55359  61635 05321  13488 42960  36623 43860  48782 11306
3190: 42645 82910  17278 00542  01566 56227  16515 44500  79018 65497    25206 49297  37993 89487  92575 94081  44038 32055  02881 88598
3191: 40579 96305  32187 67816  17240 05203  10242 93528  74570 93779    71184 12639  52320 24522  62690 13101  50439 37906  25639 47935
3192: 05352 12021  16537 64230  05963 88467  04747 75025  39079 75401    58481 30892  87564 27439  42821 13181  59614 99145  77194 36813
3193: 02180 96885  21442 09316  37252 81416  91775 49932  16966 79167    35886 01219  48060 75274  33637 09374  30517 70170  77420 94537
3194: 45321 38557  46782 54869  87593 08784  89707 14200  70676 95907    85462 25636  86204 68789  97487 52170  03710 13733  93011 72979
3195: 79872 94329  20348 38441  21583 24247  83146 56005  03775 63396    32173 66465  81758 78148  38808 88362  48436 80638  86760 63156
3196: 95723 70829  41304 57049  73504 28001  47016 35515  96409 33044    70296 60103  96807 72002  94992 07334  74895 35327  70611 03844
3197: 74860 80775  36192 06343  83218 05185  38918 72590  81352 63785    23555 17132  50922 01516  08016 66821  15395 45389  76147 18145
3198: 83108 71411  81532 42465  62555 75824  60160 08396  50018 28206    50329 34826  60411 84055  00760 34234  25140 41962  25631 45466
3199: 54532 67825  08223 55445  52643 06897  78880 09823  51603 18503    28207 15586  31013 39253  91853 13551  03757 55111  26617 04233
```

```
3200:  38596 52834   44912 98935   82561 84515   27244 12149   14331 43326      88543 52755   57239 31553   95777 77229   41122 62892   43097 14089
3201:  30027 79280   89295 68433   52350 61894   38157 37825   96476 35149      90699 72862   64187 78808   08148 34369   02419 00237   21852 17825
3202:  90954 79516   66228 15974   62581 45114   38936 85455   54199 41551      09945 60757   02300 63332   80788 35184   94956 39040   25333 30716
3203:  80194 89224   04995 72755   45685 70251   94839 67887   36931 60763      57532 89293   88436 99721   27657 04114   06624 70833   99649 43048
3204:  62890 82133   70708 01087   05156 17085   90229 15007   32418 87243      81705 59783   26056 07685   91984 06159   60217 94484   26038 61123
3205:  90200 49722   46248 33502   37022 57222   66697 29748   33318 08186      15569 43928   79372 70183   23506 51644   58119 46119   70260 82007
3206:  58117 26985   97661 24172   71364 37253   06540 37480   88211 58409      48301 78501   98547 66244   19082 83560   48243 52883   78783 34489
3207:  81839 95394   72592 52212   33529 31102   29512 85579   04352 99335      55045 17599   44890 02448   63989 54511   08467 72057   78140 22610
3208:  99894 41286   20910 82760   74655 88081   76878 70387   85853 77214      06242 55860   33361 29708   36451 36549   03943 94186   15468 92608
3209:  47673 97862   66076 29567   25439 25000   53183 91637   07798 68111      78159 73525   04907 44219   96186 74146   64131 71871   51443 47565
3210:  44745 44225   34050 79584   45580 97302   87971 76620   76780 88095      58398 88579   19717 73960   03817 79892   67831 14787   63391 69154
3211:  19179 61762   29683 96342   84440 10720   04237 74413   63475 83440      19226 25023   35215 95933   08564 34918   37851 44517   12313 77422
3212:  03856 84214   63147 73474   99434 51917   38273 25254   01243 74335      24963 90365   56532 06574   27588 66660   78911 67832   28583 27939
3213:  32985 50589   95110 55226   35567 87840   52011 73766   41244 66156      35468 00919   85326 91830   67264 52633   29978 70197   69121 10626
3214:  31680 50770   02419 41828   21258 72588   76178 49659   33169 47564      45437 94434   14464 21037   20376 39872   34044 75959   74162 89217
3215:  96505 83257   31690 08411   50526 39479   02524 91781   08530 37573      50656 45441   10180 59221   37356 04404   23798 43090   92122 91920
3216:  88826 00380   81620 14195   08197 39660   32290 55481   36264 74916      38752 28942   98132 74348   58529 53435   00158 32101   38912 17830
3217:  69454 63598   60228 77502   85709 35318   26888 51744   60294 96756      86077 35045   82208 82878   05875 42460   83850 16510   73618 46549
3218:  00599 05505   60555 43446   86895 07757   31919 17776   45128 74696      38839 01564   85931 06752   10344 64017   46556 79978   77212 78012
3219:  37779 43077   54499 84681   54033 99865   32977 30957   17690 94401      57192 72832   78782 51527   27963 37860   15887 96187   82675 78021
3220:  44143 23524   41442 82078   78977 72937   98822 93306   09163 75542      33041 63870   12824 91474   54986 77806   41386 96644   73935 49630
3221:  18740 23300   80316 91906   77162 69062   79133 16265   73706 37217      11171 37501   52144 73719   65355 01345   08196 32807   17580 75085
3222:  70337 96247   10800 84528   80913 79715   80249 31183   08373 28693      25795 54962   86842 75303   21482 61663   15964 69781   89836 73774
3223:  05726 59442   48998 72661   65007 26929   79511 03358   68622 85879      19281 04178   42042 67149   65884 31722   45706 94302   63987 79878
3224:  26380 99512   01654 01739   28719 29993   59722 51769   86238 85550      83272 18922   20565 75413   93872 34416   00752 04633   85796 71517
3225:  88230 24525   64963 96301   15822 34800   37661 00252   61233 62279      58124 45713   04564 72196   62872 14669   75007 86775   06109 07183
3226:  01786 23265   00932 02841   03225 99074   26101 12216   72605 55403      93885 68457   09358 54871   72999 71847   31906 71153   71906 43351
3227:  85642 09158   90801 71741   18206 86809   85891 06520   48861 68941      43318 26070   04217 29258   85002 53004   90581 17806   31485 95379
3228:  02936 47605   80476 57117   43993 13970   99095 77905   35712 54661      07809 07954   95264 70495   28340 55058   60954 24773   90044 64986
3229:  39976 47565   13897 59230   55133 81033   64410 58847   81226 87617      58293 43946   26718 97164   15928 40196   19991 79179   13756 33096
3230:  21731 80707   06736 69340   75364 90371   22652 18815   74656 09524      94341 86534   14751 56279   40341 07959   30146 65847   26156 70425
3231:  02878 68758   82707 62792   34860 34627   10688 49664   89694 35652      47471 60996   92813 31988   91215 91009   32776 95212   33318 67495
3232:  19583 04781   17450 65316   21886 55017   89528 58270   11748 92376      71969 22360   50787 03935   33419 07278   40669 86068   33152 21055
3233:  04260 95699   62953 24026   97606 51738   75269 37801   77971 47621      14377 18807   84450 87915   97024 51132   74155 17324   06835 26917
3234:  82860 43559   57029 26790   44905 09816   73658 15254   47398 86401      10419 80382   40184 42536   09420 35158   76749 52997   66813 53315
3235:  18354 73653   06968 21201   97329 37089   69995 20142   41886 88912      33274 37807   66529 72019   31970 53096   35338 99179   89328 56271
3236:  17514 99352   11219 58211   94406 79805   16734 78043   80330 38252      07659 17023   01066 23168   82857 87675   46660 56277   21824 01383
3237:  81869 84679   83165 20068   76140 34231   90892 11314   13083 73113      75741 82505   19728 01460   85876 55859   69847 46115   40796 99618
3238:  17855 64982   61247 97068   78259 21890   20093 94160   86546 03440      39667 42254   38772 68805   91065 92240   22205 38615   18537 89078
3239:  40920 46326   10667 86832   71424 88356   83124 78236   70459 68784      24172 72913   64993 01775   39204 22025   76132 87765   64073 60451
3240:  76582 13088   41086 55999   24507 67505   37045 83846   24820 62222      35038 35653   49707 59144   33784 42564   63352 60528   60322 58431
3241:  56246 99235   19431 88653   27644 64095   70433 51031   41823 33764      82699 50622   18271 32474   22367 77099   75007 06553   67960 34296
3242:  77591 31448   11149 77444   38896 21611   80826 14676   32645 26852      30044 46899   23350 96937   47982 97594   81251 12504   65618 35407
3243:  84516 46924   68332 58375   88464 75707   82324 47513   33516 42959      10163 05795   34006 17589   90119 09901   61406 05465   01185 16942
3244:  15776 74662   79931 60111   96197 02125   61934 14724   37475 06228      93444 18119   00664 64386   81324 00710   51405 94368   35084 71566
3245:  29113 75406   75319 82436   56998 25745   04925 07115   03302 07922      19533 82933   02608 09014   04945 04403   50378 26843   67570 49554
3246:  93299 15561   62469 51615   34235 78125   05963 61973   81621 97742      88817 54021   79358 65079   05916 44889   74593 64988   50167 02091
3247:  60671 05209   98533 51691   41841 10947   01438 01050   12339 56604      94072 68095   30977 06119   02927 73709   45007 18097   73369 65570
3248:  15175 36732   49694 61495   12648 30599   22522 38758   55405 40685      10383 59833   12070 20212   23952 04600   36404 84065   30860 49212
3249:  26051 20027   04301 40443   53483 54938   25682 90302   74774 22887      70288 22304   84835 06867   98401 20616   10200 36896   66909 31222
```

```
3250: 38458 21559  19860 09884  92527 46487  85089 50189  01559 39840    03758 80171  02216 92205  86450 94146  26015 32204  68034 34377
3251: 01291 24005  27761 92650  65588 30837  36089 74819  60693 15568    35751 37413  60863 64019  82042 91493  01994 44930  76752 03497
3252: 04780 57793  63432 45760  62262 18092  30133 65143  08742 69475    24713 70826  34192 09202  98224 87043  05201 07088  98190 44803
3253: 99591 20239  74295 47369  19526 97821  64490 82995  49927 44143    71431 17623  88217 62185  72417 52890  95892 58515  24133 11253
3254: 26145 62189  51069 60470  57156 00795  52927 62453  55267 77732    19908 95146  98491 60109  32162 24252  75065 79545  97223 19468
3255: 54139 49689  37260 01034  04513 60350  35165 09661  81814 80597    30028 33504  72523 02899  47696 86512  34601 75013  20797 44728
3256: 94113 77344  84667 36375  09397 59960  48295 49127  15552 69117    72307 62429  84216 12881  04248 85227  84719 85334  32746 73516
3257: 94941 00882  58724 80247  62116 60512  50348 65439  59398 82540    45521 88659  99221 79941  07832 86410  43270 58904  72581 83996
3258: 09794 17320  74735 93085  00906 77582  23706 47942  06931 69800    52275 57251  38221 80393  26757 61500  13925 90834  08698 83474
3259: 68544 31363  42905 22347  98797 83409  81831 47133  37784 42046    92959 57098  87046 54627  15982 85867  21720 14305  19761 94303
3260: 07033 51050  66110 50526  25723 65118  60560 52548  79982 31267    64812 05316  83295 00604  49728 14442  15049 25026  84718 42912
3261: 51236 91739  39034 36698  73765 09615  53566 37185  83561 14606    61537 50768  49692 12378  65562 36170  07770 84257  65379 35640
3262: 95798 16999  32656 41009  76081 08687  34031 00657  73350 42140    35120 43830  02348 26993  75698 34567  50249 35120  78538 28513
3263: 15903 63050  53126 36315  83131 41184  41628 23371  51056 62255    09908 65785  40926 69815  13922 46064  20451 88190  12673 70077
3264: 59923 37191  10286 22510  52108 32677  94575 41888  51553 22358    97923 19597  18451 19142  46265 96009  38847 47509  91926 64802
3265: 24859 67166  60911 52455  86262 94389  15249 89118  54095 45386    72107 95808  88209 30890  94077 01412  91386 47394  21445 06263
3266: 57987 69564  60246 51313  80220 96463  65166 42778  97011 45995    66585 45070  62964 16703  02977 78708  70659 99043  07468 29014
3267: 52749 20532  87374 68437  12308 73963  87614 84590  25672 79367    05310 40110  07659 23046  48108 92013  97946 50456  04917 80083
3268: 39876 33720  21772 29158  19961 01637  17244 61430  17273 46172    44747 14029  14117 27691  37182 57604  71785 30261  55333 72020
3269: 21207 66270  38040 47201  31076 18379  51081 02907  27193 36486    83772 62197  40268 90047  41160 25708  50743 55379  20897 04508
3270: 68022 87232  36046 62636  14736 56643  61053 77147  59853 75388    76518 58105  48237 54077  33590 88152  57604 74643  33442 25522
3271: 78234 55212  69740 07543  85275 37837  13121 70967  28682 44159    65704 34596  03302 50275  47701 23106  53040 50770  66667 60936
3272: 22103 99307  72055 53297  54107 27990  04990 64560  27800 95286    22408 29986  76232 69757  38031 30390  24502 62771  80347 37099
3273: 50523 63786  17762 85702  14273 66928  85102 80255  33171 79889    56275 25858  35444 61695  36458 88133  50385 96427  96757 09518
3274: 51300 77840  48018 36040  37227 46699  00365 67910  38415 93641    69148 86688  30462 85849  00102 02584  70956 91170  40328 50461
3275: 64134 85988  94840 17420  67764 11146  94344 90429  59164 91325    63013 75930  71425 07631  05376 80635  41057 24055  34784 83097
3276: 36795 04506  97878 29465  93227 13251  06681 84020  03743 12836    57475 47401  30041 78275  28848 11652  67562 06416  64719 23335
3277: 80620 44112  22885 38216  24790 44426  52232 72443  61152 80900    55197 52009  34328 34462  01859 24758  83521 62247  73146 74694
3278: 56526 26266  00739 38188  34468 01546  01487 02706  33856 94499    59222 70506  49263 59122  31482 49171  86883 05051  00547 41228
3279: 39179 30095  73312 96196  71541 40135  69277 13380  41823 59540    48544 71795  15680 24946  58890 88809  46494 40126  39365 96599
3280: 01805 91933  18162 90391  71743 43980  81957 84881  34562 79399    61108 29030  06332 96410  09801 91034  44620 28413  97207 88790
3281: 75137 29782  58596 49527  43310 16008  15189 90601  14036 31787    21905 64153  99593 26933  46470 13114  34595 03849  00257 95444
3282: 66148 77810  67646 70113  87991 85893  64757 94578  30502 21813    63440 36373  99995 14741  00978 22030  53007 63474  07271 01524
3283: 48212 56633  51625 63697  14285 15048  49502 24633  44002 97447    66039 78554  62773 61455  19972 92433  68800 68496  98863 34312
3284: 69601 63594  83374 86651  79836 37384  55638 70770  24565 36508    89292 62594  67929 56762  85781 65656  03923 85596  78551 18621
3285: 88377 42944  54325 50036  59055 30785  10087 41933  63321 86276    51281 07095  62158 11729  07631 43906  28939 89560  51095 97470
3286: 18074 44521  17129 89884  92269 67575  40253 02779  35877 05615    99550 41920  28002 42413  94517 85725  89131 76601  74829 73742
3287: 04609 36077  28561 77984  95803 28250  66842 44084  29419 28238    58171 36792  32067 79389  11336 59683  96457 44120  37568 26513
3288: 40331 61703  44032 28723  84536 19713  67161 69616  40621 15493    73796 04389  18842 81140  35338 71938  64934 91555  85290 45002
3289: 94354 33947  24118 50847  55279 43961  67193 11041  11366 70973    14808 01258  48621 21188  33068 58210  17077 83977  36217 37663
3290: 18800 75789  50495 20713  11171 94734  19846 12454  91989 14032    69673 54290  31256 33530  32827 78463  19828 30638  34808 68223
3291: 00561 90633  22091 74618  46857 92207  89348 78763  75401 44538    59406 63368  59713 97302  25379 90464  06057 78798  62026 55196
3292: 02061 57421  08376 17971  84308 52953  16344 44898  46511 05340    41851 44873  60637 29553  65829 09215  62183 26758  19323 22650
3293: 36368 07250  11752 67325  06158 71330  57022 61671  40615 49854    42340 95634  57881 53512  90905 23412  53582 02184  59521 15385
3294: 10069 39841  45739 61000  95906 40179  21856 40514  81266 22176    14859 75201  78359 88176  30802 00669  52928 52688  43327 13204
3295: 05775 92862  55784 24110  06626 30886  96398 56581  48975 28743    57708 72690  13176 75315  62640 06173  19548 20212  39499 79972
3296: 48320 96452  88039 31636  30463 34744  72802 67654  85698 44967    34036 39723  56017 93558  80832 53896  16669 69778  88907 95248
3297: 66678 53805  60158 84061  72649 53450  40235 47503  49378 49326    61397 35280  64344 25925  90561 72767  09643 66751  09597 12910
3298: 77873 73267  80313 17482  98825 27246  03609 39763  49446 47799    13647 38602  06926 75640  62045 65303  00436 63024  39503 98678
3299: 82248 29912  53549 69847  42877 77175  63664 08129  88101 88612    28823 67986  85502 68409  47666 15759  40003 51200  36071 63892
```

```
3300:  71998 60302  73437 92465  41034 07076  40779 16766  41693 91152    04424 05971  86755 07590  17956 03599  99671 18566  28161 46514
3301:  27161 69153  92718 57416  25900 38019  55484 82025  07170 01224    55708 57429  39077 42774  86097 40180  85828 43483  86444 89593
3302:  13061 03646  65376 34411  43227 09639  90024 51054  22089 62603    42574 16889  70933 09167  82696 35568  01749 87355  99863 69219
3303:  25488 02534  10657 10977  21763 28513  92870 57069  08321 76623    63831 90132  92906 40138  62826 12743  46765 43453  05677 00016
3304:  31623 03914  13344 60574  98720 64844  62557 31036  56553 59051    04345 65957  56455 54258  93639 94206  98517 84223  54442 82422
3305:  30988 58591  89310 41300  12346 89050  35416 32471  74774 40858    33739 19945  68029 09241  07660 99649  03480 25605  84250 19890
3306:  57034 44632  96207 16456  86918 73152  79586 05870  44900 02486    39729 11525  86310 79891  03718 00975  72221 62430  51914 23254
3307:  43217 81762  49240 96197  74856 55954  22623 52465  78882 81033    38369 06226  38466 70095  59841 88121  25190 16964  51025 20843
3308:  68566 43196  50924 17657  81521 78868  57147 40292  44851 35183    95994 39704  63628 43747  90556 80835  95464 56917  07161 14784
3309:  27629 03855  42217 43702  88023 43838  63913 17595  19825 33059    76541 71133  77296 15162  93476 22197  48090 00876  15687 34192
3310:  68573 49606  16372 49539  84843 06284  60192 81178  92795 92150    32578 59565  72991 73596  43850 19878  77916 09823  67375 12005
3311:  61081 04353  05416 43901  42519 96686  86565 56402  93463 71434    68143 13477  67451 73368  48181 86517  51911 00362  70841 59464
3312:  55687 46848  19133 19976  41662 15684  68875 86299  68312 44340    62130 89421  93362 68775  35126 39138  18429 23457  69675 37228
3313:  34277 37032  88781 60487  80380 04975  35711 66385  23945 52776    48327 22256  07986 25161  57116 06642  45871 03858  73376 31752
3314:  17605 83714  99109 99657  38364 67633  50619 22561  83941 77991    48802 15235  24314 20515  34895 26813  24137 58230  38145 76051
3315:  05628 20263  59499 42907  55150 71666  30724 20171  95103 16236    41561 45105  77974 34247  73468 75387  16561 15883  67441 53504
3316:  36251 76086  60042 75591  75945 69550  01078 95675  02685 99930    95937 13558  34202 55453  84928 82845  91854 01581  88715 14250
3317:  46814 01552  18926 13442  26859 04331  39927 18102  40783 09888    91685 55377  80326 80307  70826 60075  61932 00939  19612 38178
3318:  75590 70342  27431 13622  35479 04787  63755 26469  66537 13797    42173 71951  20577 34133  68691 93717  93380 95148  63673 15649
3319:  47885 33049  59481 16772  27126 26060  41806 73615  53830 20820    66345 00769  03776 72488  19016 79768  74164 84388  94005 71568
3320:  24637 96466  37362 24300  91619 65044  95305 65065  49807 78497    91937 51346  73260 99894  08612 33781  98173 24962  37341 50791
3321:  20061 97617  35385 43734  92243 14426  68173 64530  87747 20916    96946 54472  32444 47929  55866 00344  82117 36732  69695 59487
3322:  89548 09477  41013 42990  44854 59283  94542 36877  24469 94371    02740 71048  98593 12145  75090 09878  08851 98424  56406 12783
3323:  01802 38952  42153 13581  50997 83734  81191 02784  60114 18717    36462 52511  03815 54183  25342 71995  83572 45746  31104 87818
3324:  45432 12098  27447 17476  07578 73524  68263 58844  30456 08027    84078 92182  15679 65239  43575 46424  59007 61694  62734 96080
3325:  57418 64554  26893 71370  08351 25558  35365 19640  03750 71511    60799 37328  69345 58343  26626 44421  86739 66530  12319 37378
3326:  53969 73368  61829 87796  97379 49088  53157 83070  12924 48624    15861 82998  09608 22951  36124 15781  51005 11685  02899 12766
3327:  98562 08008  85934 28642  87618 14044  15480 91107  99607 00426    82259 07600  75091 64313  43830 03676  65667 09152  36904 35253
3328:  43046 60618  08705 33591  36112 58757  95314 40276  94400 32032    53150 54483  37396 22190  61246 99570  63321 61735  99832 62163
3329:  31559 59500  18063 03984  14686 81033  28105 31892  96545 16174    63653 60742  91696 19214  31753 42128  82021 99226  98085 89546
3330:  62542 04744  77466 51635  26877 89541  62280 42568  79334 93857    09533 09808  85420 77080  20394 48883  39099 94568  80479 85520
3331:  50029 71343  52499 07568  78637 16612  53320 65191  42596 90838    44452 13562  16954 98976  78776 16360  50109 66607  89199 08733
3332:  76957 14440  23676 96323  91694 00121  83947 28232  95935 21592    90382 59243  65506 02909  33654 58507  30666 89133  81854 65166
3333:  57842 36416  35183 71922  79417 42050  64914 29996  24196 29042    49980 07507  67650 39063  70037 85668  78656 82578  70002 96477
3334:  19899 22994  97511 50860  68175 38761  06163 14101  32929 87674    28206 47373  97853 08533  75650 30048  56393 97861  96588 03790
3335:  94905 55417  45279 89910  58310 52649  33115 74708  05695 33852    32774 44300  03885 41373  67603 18868  04617 55620  40837 02914
3336:  45944 35810  84339 42969  41023 81258  63396 80065  43462 56586    64637 89088  74244 04880  59253 54826  35153 08479  53278 13633
3337:  16879 08918  71627 18757  15590 99554  67176 17574  17712 74028    96832 10878  00643 61529  58654 63017  77424 09809  76775 89162
3338:  85289 45560  71428 99346  78086 19875  62509 74814  09444 94263    87290 40629  89314 21812  49690 18807  58872 28425  68291 56112
3339:  53018 00873  96362 42022  49551 70235  14051 73539  92532 20542    21236 15940  50046 99125  35567 54415  86482 02678  40530 72594
3340:  06091 16734  32987 15620  46689 84066  94039 63946  35444 60657    63547 66403  61236 69831  14417 60234  19050 56178  29772 27527
3341:  67265 00266  09230 14062  72109 51423  12826 04933  90166 24763    20355 76237  89899 84896  47593 98062  64527 44767  02914 32008
3342:  39717 02408  70533 75390  77487 66747  37400 93531  72021 56228    60448 65503  75464 71066  78631 52007  74026 69453  65004 97725
3343:  70858 23469  06214 30618  34774 33801  07304 28076  61190 08511    06484 79488  17606 95297  03340 45629  15927 57023  26271 68051
3344:  42544 90013  20339 64033  92193 08538  67192 18795  65832 13305    46474 29702  44164 89940  84546 15667  16394 42115  25738 30477
3345:  67541 81392  74012 91268  58654 77399  82464 24026  46726 44867    16319 00000  66755 37043  91863 65601  78873 37570  03553 72250
3346:  62867 69208  83909 77086  86567 70148  60350 89097  06501 48752    49027 42268  61033 32648  69719 29036  51782 75621  01332 15615
3347:  79550 64445  65276 87928  61053 51872  47151 75988  67674 66791    34795 75815  22427 75228  03232 35657  10835 44303  45872 14589
3348:  03216 35963  41406 43281  79309 12258  23021 82440  75635 27887    01484 24515  96653 51321  33156 37449  90728 47864  83280 89130
3349:  75779 13134  41658 00495  61236 84883  12781 94830  37152 93432    07124 04448  25465 47135  62919 83517  61474 40396  91593 21532
```

```
3350: 0452566098  8003814861  1929002965  0706491612  7171070906    5056257442  9088201821  9992114499  1998569556  2587732284
3351: 0865443766  1310322468  8880098150  5897514350  5935728564    6192841488  8374323690  1538528644  0403175062  1162293370
3352: 8087324389  3116122360  4872867818  5147198262  1662470993    1603100046  3163518387  2490230063  3912046848  9326250882
3353: 0690339679  6692846008  3729677390  6642574619  0507150582    7281524583  6267688620  3043184957  8583223177  4329283949
3354: 3946423454  0396537080  2077493836  7601677455  0731898516    0647617846  5594696459  8055221483  8224544782  0404172304
3355: 0938956228  2943568550  3054696465  5871885745  4162601133    2913968982  8918096693  6267769768  4397873454  1073497611
3356: 3905629470  5704196224  5641710635  9915042274  1858711900    7581134685  6864699240  8245384284  8106284985  4751490892
3357: 2662790429  4136759574  4094804086  7456281343  8361666650    6404842845  3186623147  7784194590  4335104819  4431125647
3358: 9106427632  1080533787  4181676017  7340798499  5317506580    0305169492  5520230582  0196091144  3813548400  9861506803
3359: 3448829523  5936306715  8378979120  2495077659  8248361680    2432057456  4232201556  1117420147  6388904503  4945324627
3360: 3593674910  2725191970  2351252031  4886520367  1825421039    1241138706  3719756583  7336193876  0971966947  6100736192
3361: 1380780785  5400823020  7764409671  0372320058  0887187251    9035272227  2392203198  9002995428  7903844593  9945363916
3362: 0546464010  9953488722  7833515218  7233218229  6528666516    2263728275  6217474988  3381964965  9483685401  2989362096
3363: 4442531871  8777727254  0800344578  5944605533  5427825788    6776761063  7318603891  5539785106  0773463665  3316937884
3364: 6887806524  7655577808  1777353520  2897930939  6970480226    1830311592  7565966985  5450609580  1811599301  5411217413
3365: 0482286879  0449542134  3767056131  2179088119  4575565945    6108280728  7357136118  3288916986  2986709808  6679310199
3366: 7119821728  3775692983  1434516558  8524354740  6995813628    8443340735  6089840881  9631912011  6255643043  2173106378
3367: 6093451810  6446088563  4353754025  5842560314  1959481180    5509523901  3563997564  3082649184  4412692490  8106876215
3368: 4466145070  8790597915  2202625264  3427582931  9076021523    4060601281  5230669605  4159988879  2558751610  1402475040
3369: 4444860093  1164278078  8105323081  6659019040  7251131332    3115457721  6884298789  3539975101  7461082293  6909635238
3370: 3699795087  4524688444  3289532467  1333014170  0220219399    6482499644  1651310652  3519868861  2160079239  2615002308
3371: 2879590938  2426967248  6500788442  4409300277  7577320148    9840208877  6750877544  7489842850  0515457002  3677691079
3372: 7844519040  8743185847  7276749876  4320349326  7737387611    0307878604  7693730611  7667228291  8548217959  5624833721
3373: 3213117471  6317034116  1538277271  9135057105  2452853248    5434355587  8583547584  6657271815  9298902450  2735259587
3374: 0683330201  1508135789  3325882760  6146577172  4871658636    6541044902  6577477164  0207964472  9143292786  3120600242
3375: 8225012240  1041658967  5577157500  3756705918  1786403418    3000381066  3880075529  3535842346  9552834690  5849648691
3376: 6522011257  7903750457  8157159265  4774941951  5223351768    6644200623  8706876281  3513708281  7730906037  2814099450
3377: 1166162384  3894553512  9924881509  3193907062  7292110258    7246107955  6077739316  1429039426  4289571132  6541108683
3378: 5546582587  6070275480  2656656821  4204897391  8477891607    3738850241  1944549013  7706521671  2700233857  3820339290
3379: 3260561711  1331584888  6885136075  9753662940  8795813445    0939794780  3086309410  1951183564  1591480467  6428703691
3380: 4799078999  3402448242  9785495555  9187068510  7847854931    8409633516  7274799634  3527670809  4067391889  2348319055
3381: 5046236464  2492051635  1817752352  0631285775  4935096158    5824234800  0210681628  7861070810  9347604856  2961121428
3382: 7265892069  0295076375  0577551119  7573110485  4585912070    5300746264  6250993772  3353210612  7462158934  1055249597
3383: 9761125625  2906585224  9564129264  7113614799  7253303359    6586409134  3540175332  1701208046  2944016426  0439493263
3384: 5867652927  6411189220  3174007034  1645589326  8219463391    5422289426  6266780725  9387819207  0697071562  4279867806
3385: 1249965194  9840639802  5600788299  1406562595  7650732518    5713117284  3237694475  7227716854  9361605315  8015435018
3386: 9040269656  0252963925  7360332360  3203169697  7739137827    4047977239  1106574575  4689617913  1549543028  3816540465
3387: 0918120325  3214499943  2675382492  3735348662  3386209108    3562298918  4830608792  2700173967  8700004177  9946250990
3388: 9570912802  2406934291  2331482594  2817495346  2538347590    6772145648  7926864809  2248062172  2946585542  9845937638
3389: 0969003733  1511780659  5397576485  2266458909  7236058861    2509626092  9069137714  0423506647  7199302117  2157523041
3390: 2509626092  9069137714  0423506647  7199302117  2157523041    6627914771  3082208352  2530930129  0652627092  6600733719
3391: 0629716225  9360288410  1245763842  7618490019  7182767843    8360036277  4847997636  4520184428  3187374761  4665084364
3392: 7999360031  6273174411  7613613076  3635815702  9883728177    7639860533  5124008800  3506303874  8930296410  0192509042
3393: 8624781617  8459593503  0622576533  4061429675  3484375892    9850384500  2979822024  9571351088  0934440206  5693810693
3394: 7379375041  1432917214  5010289738  5078551235  8072669835    2747879488  3824726735  2657366262  1906105591  1268019990
3395: 4117915001  6690100815  7485143645  1218300856  1718979655    1799158503  3819230288  9288004140  4264765508  4822678579
3396: 9439750058  7148185852  4548043397  4622317727  2941149598    0179346687  3198289392  7995403075  9854276284  7890763628
3397: 4764945439  7335081926  4530297630  0696768238  9517561150    1397300028  0768376844  7659868886  5722974411  6328505372
3398: 7667292138  7053617184  8685858355  8267333398  3510911741    6402380417  2855555991  1674691013  0593503480  8414740616
3399: 5513036863  8122660105  5615226127  5775729536  5485642889    3834082574  7467509730  5515829519  1513022285  7704747283
```

```
3400:  3721080936  1686566038  8453892336  6266115174  8418248881    7349535665  1056706896  0531529375  7012779684  2660814042
3401:  9956669658  3926534594  1298942036  3874016019  2962658612    1921697459  8298656429  6377170411  7583028258  4431902173
3402:  7942282141  4946823600  1570907500  7821826620  1180941525    8670665838  9779127090  3164784006  8184293199  1554329518
3403:  7278644612  1368422059  1689700551  9519812504  3061870368    5595879640  1422903201  3177023199  6758903758  9011412016
3404:  1246174073  0659829555  3164534607  3920700696  5703780097    2774501070  5873832278  8355070583  3093296600  0835922597
3405:  8442783493  2512738232  2137452458  4005839735  9556899188    3270581196  3352671394  3225634913  0210433868  9444489029
3406:  3934745994  6129063342  2173235009  3118126886  9367090285    9234755114  7555932186  5740762663  7596909634  7490724706
3407:  8471349953  0357541250  0293988333  2820034386  9397515483    8568259024  6076191742  6537788760  7904920704  8127858764
3408:  5229645569  8122754779  4189838371  6078508107  7848671192    3379278401  3538887066  0294782486  8691304457  5525028866
3409:  7411741528  3245020154  4228306554  3099031963  4572571002    5652603139  7964907825  0803279387  5464855226  0125427702
3410:  4314207650  9877549976  9684606243  9501642843  4838841636    6173040709  1687429801  7885586408  0955523565  4416801734
3411:  0282591241  5945490760  4411390975  9394974430  8294315111    5133364302  1982088525  2845317422  1777732229  5096949460
3412:  5761823937  5523161620  7284173728  7202169645  6382791200    4444699446  6439285256  1220201465  4268593627  8648156755
3413:  5950951782  7888115096  9662688312  0125758082  6312333582    0771566806  5603863336  7341667178  3034260140  9154006687
3414:  7306193771  3064530229  0655766379  1957320432  5138859818    9921252595  7830807296  5886139900  0894501684  6374498565
3415:  4351302988  5746419798  9749740512  4867004412  8694070969    4314384459  1545346886  7606617779  2325586414  6708988320
3416:  1149843075  2677526363  7617930735  7212295313  7089249723    8441614840  0576549110  3594924301  4368394023  4095681913
3417:  0736747963  6291060543  4369115495  5196252395  2628424268    8396158272  5902189909  3803116340  4628754204  3455285076
3418:  0346623449  8386530232  5058558426  5955882183  8223991046    3927888425  9993776124  4906195757  6514525196  9210286547
3419:  2772107368  2445288640  1044983715  7392834797  4489323404    5059732090  0042246532  7006022023  5773381408  4918696804
3420:  6355873952  1380485833  7507919275  1487401748  8151312289    5161665637  3600689142  2327056930  2219059957  4731449170
3421:  0109753919  2255430531  5324850591  4564169419  7575600051    2023597917  7254445269  5301663492  7559303963  1998374511
3422:  6478077373  9818707696  6820873971  1509502974  4524037610    2049695730  4607730839  6461082248  5978118740  2298123598
3423:  5583892834  9800465610  2995437755  9372834543  9101981895    5180512662  3219936496  4657866000  2666735399  3814570091
3424:  1467508529  6637166996  6582791005  6051360439  9590992034    1230101619  5723460667  3959425563  4413790059  4921543190
3425:  3082974732  7260812562  0436731733  7217528108  6090057708    9207385682  7541111801  6236546343  8251935577  2786689369
3426:  3916011945  5355601980  4938474969  9798640124  6492794112    8118982407  4607783260  0828183913  3216518594  4056612303
3427:  4352809830  6032481271  5726908335  7095183411  2260873699    0612025581  1796060070  6372093864  0762224356  4460099696
3428:  0888947590  3263350817  7209768981  6966587057  9990765248    7766784523  5666837295  3807820232  1273111534  2287221471
3429:  3517505447  6221079089  4076762599  8232242041  6650241576    7660316383  2758153570  8033343780  1961868750  7951489608
3430:  4145448408  3100358561  0927603159  4765420884  2530608424    0777982417  7606089576  9091227028  7578208316  1296722693
3431:  9022594151  6601164817  1599959845  5555719027  3004553678    7459064892  1883481857  6901890583  0898198916  0034960346
3432:  6504131138  5900148818  4209488165  5352837833  0814627691    6354119027  8580349366  6332292598  9834126911  0030072971
3433:  6760632395  2067548303  9552132634  1745443914  6571004128    3193017034  2135368886  0396278549  8946981580  7982323487
3434:  4347404278  1698917153  8479411764  2321874125  9938954735    0003252817  2603783296  1650846157  7527315007  8190192637
3435:  8633876803  5374873468  6170858676  6729674225  0526275639    9134432331  0790568206  4480173242  7571156231  6011003912
3436:  4724326402  6999545879  5113001312  1517642915  1778296823    2651383776  0979552216  5908829162  9969493377  2135412916
3437:  4011304266  3581739428  3342811305  0331249415  0304461363    7494846365  7522298947  4425925788  1479529067  5056819244
3438:  6249212571  6153241960  8127169843  4603349587  3833557174    1413345866  9730809229  2525152037  5009179135  2559656336
3439:  0993318318  4063556310  4420043993  9655316009  8796683827    3077852599  2616027207  1461237970  4950969711  2274155032
3440:  2429672198  7757999131  6610951744  1032263036  5112714276    6300747668  4230811991  6139310090  8308740843  5201992403
3441:  8942666266  3381353547  1127910708  8414061601  0719781160    2821502239  5279672403  4097772145  2887120861  4468023193
3442:  9400348140  4786168494  5925509429  0062183028  3377520401    9293071465  8395172062  9205769859  0734820200  4691158460
3443:  2372899200  5127316198  7944297470  5434696995  7137174382    0103114401  8909248205  7754633729  7425919042  5343104220
3444:  3398534862  1973887578  1404018085  8727103200  8262043658    9781638400  3634838696  8548319080  1001642809  7785103526
3445:  2003378664  0619112534  6025774255  2005706436  2622985617    8791046436  7940171516  7392951478  5854369890  7071269705
3446:  7363207878  0076856729  1969305093  1053618702  5564035579    7858277853  7621486413  5805395914  6284946015  3931030301
3447:  2258959879  2435273855  5693441985  2020719019  2236120389    9029377165  8522598279  0288206619  2503916303  5073081076
3448:  1004322495  0403561442  0659646078  1594023829  4934492729    6959982169  2542347321  7509992145  0916429057  3256727776
3449:  2713376291  5312317829  7828981323  8037017923  2493369482    7057826732  7116987492  8119270360  0809282882  4193146307
```

```
3450:  34460 55910  87156 43947  60089 17761  35239 97525  37828 82964    81996 28679  79459 84668  61793 10612  10169 33202  98070 56784
3451:  95669 80555  58429 65505  85049 08725  71913 35451  37998 49617    16370 66544  21491 51116  11124 80131  12354 75597  40996 21625
3452:  99311 63990  79771 86342  59291 87310  24681 25546  10664 27047    60230 04966  08081 72109  98183 78658  97812 37307  28571 10778
3453:  92472 54127  06187 54464  74349 66242  10894 70600  10919 89353    19151 99895  62324 35294  61521 67995  62592 18348  49571 81107
3454:  36843 52379  63524 00203  31882 24698  27718 40099  08521 73900    79811 11402  19328 94046  52912 77376  92327 32162  85278 84486
3455:  58371 56084  58992 66687  17110 76481  97649 58068  75719 81434    29957 56053  78687 53752  66661 04788  84888 32692  22156 82870
3456:  59780 75790  61792 37775  55379 39282  79238 24301  52323 83721    48571 27392  26286 10984  67189 48315  90611 50788  95473 91612
3457:  24671 48963  91596 62640  29066 48858  02511 41078  88350 00672    93103 35768  98618 18824  72366 67903  94519 56271  68151 17773
3458:  46946 95753  19280 11084  95595 34325  54442 44612  47458 49995    12891 20655  18914 62195  43567 85330  05789 89818  21296 77836
3459:  15991 29087  50521 32262  39931 19700  08663 66758  86212 90412    82703 61752  02339 95352  40171 13271  42624 75630  39018 73262
3460:  24726 35003  76915 29641  66089 67011  79056 58495  21876 24675    93927 64616  06037 36646  91660 59663  05490 71113  31756 23983
3461:  61158 24981  74708 47810  62073 14482  35902 35199  00723 93978    78600 57077  32089 45700  53552 09384  20627 94027  42246 75850
3462:  01536 02943  37544 67658  15864 58247  82248 33755  70226 95494    00574 57048  95523 51506  62936 45641  90580 69618  01573 14918
3463:  90903 44157  50808 52530  52996 92379  55157 41758  02797 10386    49172 09504  73309 50385  93456 53645  53718 82161  80075 54474
3464:  96288 30229  74562 92283  70107 77029  41138 59414  56568 76032    83334 86196  15178 25848  29873 58056  30555 57434  19301 17585
3465:  18272 30638  32733 17920  94689 00912  70937 15640  93753 73505    71030 90302  89866 10909  52250 78547  89399 42622  58952 81174
3466:  20854 21589  56500 99621  40995 08733  57930 51214  45878 02555    05796 47102  72263 00725  15345 69772  71778 62832  25449 43483
3467:  78743 62777  72401 14131  03667 30515  90045 36307  07441 00197    75040 75124  36248 97948  15768 25182  72378 29117  71653 87176
3468:  09455 19980  02332 94020  88303 07662  84727 50628  93558 57602    14931 53633  47615 29135  25354 70175  92182 71484  49185 14212
3469:  54312 70472  81966 49098  11528 44374  84669 89884  55655 77032    35464 19772  15131 22079  45382 20256  15409 47549  75855 24473
3470:  79197 43742  91762 12065  96744 04906  11657 79358  78530 99146    44885 09267  21137 31250  41312 42012  02597 38276  91550 23321
3471:  00561 43025  42274 91255  49308 50164  10822 34869  54960 63816    47454 09107  87225 16018  85288 57391  54147 01526  44774 78299
3472:  04198 62278  93467 98237  15276 57220  74929 94720  58988 81342    53943 31738  55569 33270  40972 52528  78540 53883  42051 87575
3473:  29601 13513  33426 50868  79959 46439  46739 34365  13984 59811    83920 07428  30353 94913  13361 61496  20689 71741  01173 76756
3474:  89339 96231  19908 38884  16444 30020  78754 59776  15304 53165    01872 49556  51804 00040  99356 65392  76132 54153  00852 37607
3475:  17327 17203  13170 73941  11934 42038  30797 52990  71726 78628    96016 69722  29177 88628  89965 62649  18971 54919  06354 75199
3476:  65645 80043  73570 20921  74504 93951  93491 62758  80031 63974    85638 68673  51363 48305  99431 67755  47504 16945  45210 05742
3477:  96561 06760  50524 14619  13402 07389  82288 30002  55962 58104    53640 69621  02677 18168  14598 10437  62612 19988  01997 65084
3478:  19012 05503  40562 26756  08751 87256  95778 50061  52603 24663    98579 92000  51375 77820  05302 95251  20921 18552  73799 51403
3479:  67349 78659  96355 38273  31844 44262  99145 67191  13692 99013    29228 27022  85494 34700  88087 10580  65043 18229  75703 36407
3480:  26727 96479  39441 15043  43882 01030  56878 84193  88387 27757    93622 42142  78921 49479  01894 78607  03831 59426  74502 48012
3481:  75930 48255  16755 10420  95507 17672  98388 68083  96530 87204    36934 38858  10424 06111  60278 07533  82535 40997  52574 72867
3482:  35168 70455  90048 78339  02610 43091  43904 11376  79081 14968    19863 94060  87156 99867  85011 05130  93504 18680  67845 88454
3483:  41469 12000  82586 64569  98642 89410  12145 42303  34179 34595    30687 82318  68631 94411  60150 40372  22532 86916  23390 44109
3484:  46931 65623  67647 64225  81739 39081  55181 36458  53382 61313    89659 27669  62590 69819  78291 43054  01483 42007  41438 19677
3485:  72154 88097  60392 69252  23188 81977  09590 37008  40021 80830    07356 33050  25019 60020  37913 78374  39927 65951  47027 68866
3486:  53771 52110  57287 30684  83705 47486  76900 97062  34537 20008    73286 54740  37372 88621  01332 01852  58430 15054  67095 78639
3487:  33120 27136  43023 57062  89011 11870  48939 84747  02790 27975    52296 86635  00611 21118  47001 97033  86976 34455  63850 65663
3488:  58563 48772  37030 27510  36785 76022  02722 54132  45147 85303    42642 12357  58173 99887  01657 80679  73553 69320  40218 02215
3489:  53554 07183  22743 88604  49895 62746  25715 47008  16230 84711    27825 00963  17427 03905  32991 56251  86018 38584  37405 99764
3490:  13373 54889  42351 11196  23221 57609  71199 44431  60484 99387    03290 23353  13132 02026  51913 93212  26232 81678  40350 36835
3491:  72121 64580  14263 43746  49243 95028  93616 42769  26930 03965    77137 30469  13482 95728  74131 17796  21601 29825  62715 67444
3492:  05339 46090  99679 08704  56559 80287  54580 34251  89617 43800    03324 43403  60662 62715  04232 96019  08579 05447  89594 37583
3493:  94903 45752  58061 34223  45524 53733  14841 86784  94621 51392    74852 08662  76178 78291  64150 31093  19029 81530  11541 03218
3494:  82230 02961  17219 77290  62429 33801  96145 32306  44675 00075    66258 93519  19127 07358  39795 44008  64477 69044  98533 79291
3495:  14051 03422  88441 77178  74588 33808  13651 69513  37381 54699    64645 84503  75890 77957  93810 17269  55792 14946  94580 19540
3496:  00327 37433  64978 45813  25395 93914  58108 87062  72539 41693    86684 12300  76193 68434  34170 12316  38793 51530  62897 58536
3497:  09916 24503  43800 04011  17850 79898  94551 41953  66532 20912    24210 87933  98134 14325  67108 66544  45560 48017  70971 15454
3498:  06706 25623  41428 94821  67814 74204  93661 95525  67085 79369    23393 27428  19716 40246  94182 25010  57639 42436  48055 29338
3499:  99319 42845  02477 00381  03295 81414  39046 45953  60615 53056    93602 02618  02598 43604  41225 58191  55511 62740  53170 73815
```

```
3500:  59587 10029   03816 74353   77062 47777   72185 74358   91331 18018      92327 52000   71305 49041   25139 84345   64288 39129   83958 40169
3501:  99038 31714   56576 69868   88523 21959   56481 12358   88143 01047      03051 70347   51733 54753   48893 93785   90064 08351   62857 95634
3502:  96510 52667   32322 06795   78270 25604   59482 72488   96452 84553      13366 97359   09803 89960   61947 54466   59285 56420   62834 87434
3503:  37526 26548   12713 35662   90561 16207   63300 26086   95547 86264      44030 37212   30807 53624   70866 73687   25012 92887   41434 59173
3504:  65343 67295   86023 01944   72321 69568   81083 96151   99245 56395      88503 01873   06157 91849   23030 92959   30192 55416   59034 26301
3505:  13454 33635   82045 87941   32884 74978   94699 73809   83902 37534      94997 47684   38850 30531   44477 61859   28066 13982   87843 58106
3506:  77631 18589   21426 88128   72386 19125   16311 75599   56074 81097      69039 30641   74860 18885   25880 30761   56574 25484   21972 68459
3507:  55630 12869   66587 53974   27845 47576   84010 21966   62622 11546      17200 42793   71360 68914   22941 45690   66809 57063   92624 27713
3508:  98759 36497   92912 90312   20979 77374   77558 11239   31254 08407      23599 91745   63565 44980   44979 42277   00848 65805   77975 85172
3509:  90540 84184   76207 26236   47066 78400   71335 78388   26859 15809      42830 82312   67747 97499   78976 36663   37555 40026   79926 99059
3510:  64108 89971   64621 17684   78411 44563   25513 46883   36470 44002      96453 61649   79567 36183   83955 06709   88629 47728   83880 75407
3511:  39245 98971   69052 58526   02701 51671   82110 39900   46213 18286      59125 07226   04369 98466   65582 11496   96264 13709   26609 26161
3512:  91654 68769   75043 23046   11392 08556   93426 29819   30318 24889      82599 26739   56229 98518   14500 27483   02449 91321   90863 78068
3513:  36825 39083   81285 99093   44597 57860   99273 65834   95620 44838      47200 16176   33954 32606   36430 15627   88501 44527   99668 15350
3514:  17114 47839   26432 68275   80266 70071   14145 78454   52445 96857      31299 99004   40007 99055   02264 12578   47270 39181   47528 94050
3515:  37260 35398   44359 05479   30205 68856   85391 94652   12058 44708      39709 82747   70083 90820   64539 90649   03736 99966   74157 11748
3516:  92989 91494   95210 60307   83385 56842   50899 16491   68190 83015      30934 08023   78743 09343   49662 32870   38241 00352   36149 14171
3517:  09109 58997   80072 44327   37862 87196   78810 49262   18973 99078      49102 78267   46070 38748   33111 70972   97163 21024   03402 53419
3518:  47780 08940   10239 33501   53764 50323   38966 56549   86892 12917      45403 01893   26766 70547   10338 53282   92232 04166   86355 65921
3519:  24132 29230   28114 74536   34875 12538   37016 30474   46962 62701      85145 34872   47262 95410   98728 22043   58908 70796   36924 90926
3520:  38192 39878   57186 47474   24192 31330   35721 41826   11058 44058      99970 49180   42658 15985   18887 47191   70613 13948   46026 01888
3521:  44883 38588   22469 17658   25925 89822   28843 26016   02306 31819      24874 78437   82500 54534   87878 75855   96584 90782   01909 28380
3522:  43074 00296   84344 05437   66772 25228   10169 39456   69499 15360      39104 56948   72774 79190   55012 71312   50162 21424   51418 48869
3523:  00065 98716   31535 54193   79822 57740   76514 11967   89769 64086      96400 51500   18493 88567   07671 09400   76335 21135   75025 10103
3524:  75137 10546   76554 89127   11040 09965   48849 70092   04221 56516      93775 46287   48903 02994   38167 80102   54206 59242   60166 27995
3525:  48956 85143   23562 54638   03510 25599   73643 60156   74433 08491      44312 22616   05416 35284   16365 93088   46361 26729   54028 31982
3526:  20483 45238   48112 62464   63740 36822   19651 68387   45905 53588      28096 72207   28671 20690   66351 33301   87862 70365   50216 31320
3527:  39645 92511   14378 77108   51925 35111   24653 16773   47771 83989      69601 70414   69399 43884   50285 72794   91634 06890   42225 75560
3528:  78473 55102   45924 32478   32954 41940   59290 20208   78213 46498      65721 62960   36660 81911   64631 43741   26520 83003   74696 96840
3529:  73962 81770   01400 20216   67830 62308   43211 08179   76209 08547      33362 54679   77546 67977   12043 32101   43301 49926   49313 75955
3530:  63084 26521   22417 07065   89699 50828   20609 07028   54644 47321      01143 94823   25533 94465   23988 11727   12537 23341   19507 29083
3531:  78560 39452   61256 59426   27692 80850   93102 61451   83654 79866      83982 00872   27082 19511   37044 59992   99652 47009   44076 70913
3532:  74742 16270   91485 25764   28761 45034   96003 82445   50093 11820      10600 32315   64977 92378   32193 94844   33382 20557   34813 45892
3533:  37980 14194   69188 13237   08147 67544   29469 35785   16225 23507      83976 18452   48950 33866   66077 68724   06067 83268   21270 74859
3534:  36709 52358   38070 75637   10303 04682   87677 18531   92855 02329      76438 54445   12873 10516   26769 24005   83230 99225   16709 91253
3535:  32280 02275   11251 05227   55608 72760   70331 71130   27893 49110      96341 14916   13940 02936   13239 76804   64156 69206   30974 51032
3536:  85191 90455   86621 73512   29252 86511   72584 57541   00697 95160      42004 61449   37801 34077   80538 12921   05001 46940   97700 79206
3537:  64744 19428   94925 14862   61336 05647   94006 82636   87434 90051      55489 84829   36755 43515   98349 70536   98181 43784   98553 69256
3538:  04816 30530   19529 95418   11014 01018   20612 78049   42075 71967      64662 56913   62006 59812   62201 24214   68314 07437   36158 90738
3539:  94421 00071   94644 30389   49092 02284   24264 73710   63079 56052      92942 04985   90997 05762   06254 69762   75891 36038   34753 30905
3540:  97016 84078   85228 07924   66014 33251   64343 36075   82017 87620      83537 33065   44779 29295   55065 62090   07278 77798   14259 71409
3541:  70632 69873   39015 08374   74647 23910   15194 80959   82403 84041      54088 84469   20291 63352   60686 17024   91193 26235   40170 98241
3542:  76930 51389   74670 85397   99792 96577   56646 89679   69251 98310      63964 32506   08429 50854   42903 44902   23829 11187   59774 80707
3543:  45642 61341   60418 41766   31216 24664   28437 54334   07663 44079      15618 43409   24869 75512   11270 20127   83958 52985   86647 27841
3544:  10865 85349   19931 70663   08535 38951   86796 59464   06973 25520      30836 36008   63416 69098   25134 19963   35682 03955   60790 58092
3545:  91141 91623   95133 92718   79881 09381   76451 32990   80147 88564      68177 92836   18286 50281   79966 78615   51690 44746   63613 59267
3546:  15679 97406   99388 72090   22783 25281   22325 14777   66766 20979      66218 65982   15458 81507   67149 96015   62029 09830   16319 44690
3547:  76502 18173   94984 61597   31466 74816   60657 71984   89524 18291      58406 88707   06836 84402   41005 50397   85697 44982   84978 86877
3548:  99623 06924   16613 47900   66782 79746   13618 42410   22162 91614      19552 48058   81225 96931   63604 27290   14266 24567   59185 58210
3549:  42763 88106   65944 58437   00011 42226   03181 52768   87685 12950      59801 26111   81719 55409   54345 31132   92435 56794   67745 59955
```

```
3550: 2658178386  2466003440  1052518446  1732155481  2245007304   3961632772  7728906623  7869372231  8003530693  5167821487
3551: 2347915708  4075251190  3346912509  9436153055  0104299869   7569235463  0784366877  0701398890  9443875363  3005652720
3552: 7647788357  9754835582  4246628717  1723681164  1629584663   4308380793  3107632629  6233257796  1183713396  6610493674
3553: 8390939886  3962560559  4978248071  9762435076  0626876382   8426493544  8418206176  7152991030  9396187298  9961878376
3554: 3022665575  9291206550  2617041442  6647502703  9360333529   1188080786  4194213529  0436705809  4525340963  9275040182
3555: 1487488473  1467984683  3138052422  3648319182  0085033674   5043372189  5534365735  9929413566  9625807236  2937119975
3556: 2470574563  2588381884  9325523484  8274216263  5774223473   9967779923  5576433848  8462183473  7822154969  5671885047
3557: 0122431231  3717363584  6005903384  7958222599  0206966660   4730159868  4554614163  3542759847  5427293329  2174193983
3558: 5790147789  5546493118  4911246569  0204524251  8569133153   8145598426  8294423337  6249768369  2082733561  7680902287
3559: 8818309010  6702738488  3584516356  8370743058  3453535226   8185203955  9771257137  3868799804  4370036013  1215411508
3560: 7400239435  0843894844  3367579891  9882188498  7478862035   6543706439  6920850630  7498136564  0658705201  1969212007
3561: 7849566318  5376362434  0022528070  4352746958  8651806446   4706865680  7137002072  8413532918  1734350851  3378676515
3562: 1919261664  3795280230  5918268409  9584765301  3812467842   0542301412  3575312124  1526346030  9026778307  9437139694
3563: 8447455571  7646381079  2119155315  9434725614  1397893226   5743979376  1649740697  4173071806  7815093761  4409548985
3564: 6425330278  2427920142  6167118463  9111672122  9072410961   6738032920  1171240087  7171377429  6633602155  3758380031
3565: 8805661493  3782093071  7143089725  0428776741  3946991109   0662037965  0909737125  3250177213  3723055421  2945006586
3566: 3389635296  5546371977  1854975128  4374988792  1295876757   8565168363  8365276857  2959843271  8694692173  5145165873
3567: 1003436285  4715350377  5213679940  1594004851  3032352030   1300750298  1531226823  5477015283  8395661747  6755188712
3568: 7794429528  0244505621  2063156056  8782997534  2243872349   8962145910  2962483326  7865594138  9034043135  6726345478
3569: 1108710386  1930380176  3621952708  9391324601  4255205352   1373039186  9071310261  3993552466  1960262105  6506398132
3570: 4513446127  4450732124  3237207337  3692973173  2536493589   1353023854  3263813067  5083956541  5822148203  8201103778
3571: 9242782925  0488865129  0457470322  2210991877  1672579491   2531106619  7499320962  9438858917  8936855766  6910558711
3572: 3829917120  6563806191  1344563517  1520787089  6982216055   3958787109  7037218805  4083390876  9624632244  6131882911
3573: 7459717523  2735095546  1173774013  7387863713  0953092152   9866815837  3215878421  1700168318  3266347243  7960447629
3574: 2510379379  7718351172  3422686129  1816148994  0480870721   8354627998  5571848753  9968600799  5024507657  7062138261
3575: 9605769659  3058367039  0839686087  1210772995  5357868327   2155160917  3073191886  7346405069  8328980571  6233561506
3576: 9799976194  4433253462  0801257658  7057193839  6415202493   6920368112  3376546746  8205411725  8571171671  5799929466
3577: 4884774917  3179900075  5753667511  3513690423  9101421208   0359462587  8814644802  3148424513  6014979169  9290856868
3578: 8798272711  4349700422  6292340610  6350986116  8602753370   3055995899  5898979130  3348781575  1217875061  6710569256
3579: 5310725075  3570385060  4532136113  1335921270  0931145499   1356060478  2348583014  4524909991  0993405797  0255524200
3580: 4636959602  8332329142  6490583858  0904301563  8437421745   4802790353  5319877249  1465367175  0063401153  2604137995
3581: 0899973407  6330846093  2461736399  8304028979  9017875877   1740581603  3770733397  6796885687  2057942770  0120880394
3582: 3852610607  2864559818  2210312438  2130011564  0978350822   1107033688  5220140227  1648346245  9281861705  1373949004
3583: 0257150240  5088048717  5019471401  3047385520  0868744018   3340162991  9291076275  4641838390  7854827711  0873988221
3584: 8184304213  0238765633  0202556636  0238038600  6048412494   7910623571  0483688435  6199301750  4862268637  9668916245
3585: 2045375699  5717680834  7516241743  5967880859  3931399341   3215811730  4207042305  6449745032  8916846445  7633653121
3586: 2117682529  0270185864  0632703529  9089017577  9036279289   6971909795  9630555500  7591697222  2037156170  9328421376
3587: 4655274884  5903996802  3253186359  9329790445  2423470811   8900397933  3527010853  0115740310  7036626386  3852338555
3588: 1476187832  1758455810  1983972768  7385004220  2023852160   0460898804  1551254097  0459395604  5705087941  1499216855
3589: 1749092219  2176416091  2722056055  4360415633  6174956076   3480037844  7571654827  6826397379  6016206620  0327081132
3590: 9246225875  9476434441  6218331799  8230854782  1538160438   9897932062  9448304153  8097940670  2029001710  8505010692
3591: 0572268058  6274918208  1590133545  5513109254  9631060007   5102977323  0383324177  6067441613  0495776674  5389079790
3592: 6034719808  0359732577  0059072731  3978088915  3211951573   3619040378  1219742043  9731098879  9713474302  6201022732
3593: 9295810144  0234968644  6943950553  3787312173  3813947903   3928252134  3809449550  7605439458  3811495841  9850796204
3594: 8338405283  9759783805  4569141022  9278561589  6867533815   6458776019  4930465554  0481933236  8883134299  6688849175
3595: 2247980269  0533800618  8330613482  4233952265  7166480674   6808652086  3438931679  8403979391  1267993439  1078240913
3596: 7775465696  0066788326  6047846354  6379895746  2950550556   1726768919  8887067673  2161067674  9856031820  7094789647
3597: 7830118043  4918510091  8388062015  9343259546  8043174141   2956916472  5915047361  0719746534  7210592580  1072400264
3598: 7686413024  6369377100  8659909128  7208308246  0138162226   8663732117  5382013611  5519159989  9637863287  5455829919
3599: 1863093653  1566903316  6300918482  0495684600  9030427310   6973697415  1576038208  2850099689  2817387107  1896118243
```

```
3600:  52742 77577  44092 44607  25448 63804  44833 37115  39057 79415    47892 49555  90211 04809  98956 11758  66611 93323  54230 48763
3601:  77157 47434  22367 57566  72376 95918  62558 68788  56107 65606    15660 33440  76416 08275  47039 47528  80742 47187  68361 18134
3602:  33399 35028  58818 47879  23060 27977  26047 55420  48364 36952    09892 15752  17627 12318  58995 87300  10467 27905  81517 34524
3603:  32530 29943  68028 92625  44880 21828  24296 15097  06711 52599    60234 24271  39540 46973  29548 22365  80252 37829  91892 73287
3604:  88906 26185  54521 54488  96937 86141  63096 03215  80507 33274    21914 26057  97720 73618  34237 86586  25775 73123  33436 06915
3605:  33958 22805  02162 87063  64033 48862  41596 46139  52412 81034    48039 49373  20585 53763  41090 86337  42728 47530  17768 15370
3606:  41969 42104  85150 86029  61916 14870  82289 53593  68002 29999    90829 77245  83616 85480  95443 07111  86180 56291  75486 78374
3607:  03181 40543  02600 37061  00624 85267  41526 74894  02610 25273    50399 07617  73038 70949  34426 05693  17267 53820  48742 44535
3608:  80547 11598  50343 58376  24534 88777  19597 51369  73073 42006    44907 44401  52944 55663  94826 59583  51429 47574  26588 62067
3609:  91600 93347  78590 26924  05220 85230  15090 47689  60926 57031    83836 06680  22550 99951  85161 82737  42342 39969  09675 46265
3610:  08729 48788  74214 76131  81011 67577  10994 66021  31433 01760    83194 68168  22436 10518  69258 33392  41861 15877  18227 21437
3611:  65529 60159  63025 21769  34807 88701  33220 74362  00529 32189    43675 18136  71285 38003  98649 28561  48999 82896  35681 46878
3612:  87024 44282  18777 86579  59364 73351  14227 08124  80364 68303    03190 06213  21583 80135  17124 36933  99165 37862  50012 68219
3613:  30365 40956  09446 72057  28990 53009  84899 28885  88256 83922    91158 26582  22967 76894  69213 84234  38115 25393  53906 32556
3614:  30931 13473  85878 34755  25249 59785  75122 25541  36084 40030    10451 12235  75875 18683  58241 76689  35228 51886  27325 45093
3615:  27778 54368  19404 84885  80510 37709  37641 44664  97764 84596    05222 90413  50737 32272  62588 92583  39980 52343  81441 64325
3616:  18910 64613  53220 96060  54898 16070  53037 79566  54517 84029    02935 66437  78722 95846  02160 43478  83758 48104  39326 93358
3617:  79313 28860  08196 56509  38423 38755  11352 71578  93773 91900    62761 24142  91133 90114  17220 33551  87304 91082  54071 93157
3618:  87058 78120  42814 30788  24640 28768  87753 58616  44376 08854    73028 71382  61624 11226  55814 06204  85070 87322  73396 60512
3619:  22172 20764  99019 05562  69448 33879  64963 12094  72290 50319    06163 17693  38393 76496  60732 20419  70644 37662  39951 70808
3620:  39497 10903  74289 96725  01822 94038  80096 51902  71129 88347    72211 52331  88279 49949  89546 84397  16462 71223  22805 23959
3621:  45046 11566  87349 15450  50710 16593  76908 18998  27619 96097    94752 48725  28337 91783  91109 24787  30348 26502  88552 05429
3622:  96573 63511  59604 14178  85485 43609  88275 14543  62674 57629    52606 05041  11672 27057  41317 14791  52523 75163  32691 67401
3623:  08542 62491  70842 19677  12390 38046  78125 72177  78848 86885    31904 04965  73086 10615  04774 19422  22809 77841  63945 98896
3624:  52362 91074  93578 85277  85975 79788  99607 35790  73406 96090    97714 56082  60906 62061  75045 21049  89309 46528  36489 73275
3625:  62732 63153  45973 87148  38024 52461  13656 92270  28783 00539    15955 44328  20571 87719  29121 22995  43006 65618  36668 78551
3626:  98537 19430  74346 92782  58385 70660  85010 21029  83092 98185    73285 07468  31410 62998  55102 77702  31402 22470  72704 20062
3627:  25915 40396  08560 48125  64139 87909  69777 58970  93028 40291    71862 69269  81218 19322  09565 98753  47717 10667  95762 47482
3628:  98444 44491  14605 84901  57554 30236  87569 89847  93707 69892    94411 70963  16923 99265  70211 97631  99999 71559  22986 03313
3629:  43366 40012  91026 52963  25310 84662  67608 46077  65368 09512    24000 33946  67558 09592  38526 64661  10108 03260  48709 48740
3630:  96687 42762  16248 84495  22795 44450  90283 02402  12985 93957    75001 13616  04030 75880  18581 47775  45442 20221  91459 70972
3631:  81647 91619  30456 06925  19399 46065  80748 42165  59092 14670    20220 98206  00000 69955  26125 42294  03600 87007  60963 87323
3632:  84686 25000  59458 10049  29707 49720  74978 17223  04910 48895    45523 29750  65574 25297  31750 12992  85895 84468  87946 64286
3633:  90745 06152  62569 05218  39883 31051  55167 07307  50352 50286    15486 06116  35685 78584  22841 65061  05957 14543  96300 77008
3634:  28909 62729  74445 74824  43378 72406  96886 08138  75010 25831    78937 90840  59349 59043  85985 89517  29124 73751  80261 90662
3635:  91748 98790  38764 77893  04209 31739  09963 17773  77697 21711    28720 45292  81590 31862  51197 75753  24399 37524  37309 12430
3636:  39712 78209  67376 16802  20000 76450  49152 87923  74096 13352    93439 37493  24053 44411  11510 38476  45599 48572  02750 05642
3637:  46557 50033  42928 18303  50275 53851  10411 97104  00885 38930    19987 12518  88409 62084  90181 10770  47665 94257  00027 03961
3638:  24545 26194  23117 10609  95655 98841  27641 02447  64867 32068    54448 01468  01885 06405  47064 82147  95345 03447  89803 88053
3639:  36811 32827  91635 11588  20087 58654  89541 89879  32021 04266    84760 16759  08052 30901  08891 52360  48174 93544  23945 36922
3640:  60951 01346  98585 74903  58777 11695  70856 96735  21325 71817    17000 01207  62090 91257  14789 88509  44608 27816  75344 17602
3641:  75913 17788  65751 95536  35390 24373  80616 93392  80319 58653    64108 70722  89597 26062  87970 00515  24907 78539  62071 90789
3642:  66838 80550  92192 62842  78737 74376  66339 81839  72084 65583    88820 13334  29828 65312  56440 50283  74028 27896  59736 01903
3643:  36367 33678  42882 22995  84356 51403  65831 59851  86556 23921    59512 88570  75588 93562  99237 80591  65862 61180  13646 69466
3644:  95855 85253  96894 73899  14539 56495  23372 27341  98243 69202    48487 07644  49349 61104  22669 77335  71742 70666  08467 30557
3645:  97815 84172  00608 20911  95673 43029  06309 66400  25283 08875    62899 47642  32682 42225  90851 40248  63876 08849  40076 64724
3646:  03423 38663  55609 29280  42267 30529  50420 98390  71098 20201    20192 57188  17163 05609  89902 78092  15044 13901  42501 16538
3647:  78537 14285  86833 88642  28457 67412  60446 46423  47300 57156    72143 74795  43290 49981  74639 41882  38349 04556  81416 35510
3648:  81742 19070  19100 34971  10516 87982  63265 63754  72760 64786    30173 78896  72296 64893  68722 14186  48617 21466  67482 47512
3649:  22269 71588  76776 72107  85001 63001  57342 23842  29374 02608    68845 44935  72836 03297  67198 91281  41474 58829  02558 87934
```

```
3650:  14936 01064   03290 17301   43070 12069   82969 19105   72001 97183     45521 51449   95536 52417   06091 01517   03389 47968   86182 50672
3651:  93790 87655   30400 47104   45939 75498   47524 18891   14539 47707     84900 72888   19128 13853   56984 99368   49256 83421   82886 31301
3652:  48513 31430   13913 15567   39591 56853   95657 46397   80417 35995     84289 26300   08374 00305   26078 44465   24385 29481   73474 52048
3653:  40746 64356   56763 37859   19569 70540   52968 50975   19698 34590     25403 23534   50336 19594   74447 07741   00558 36540   61870 88358
3654:  52285 58295   60343 23157   31770 10652   98530 27574   77001 88016     58821 73275   39870 70519   11057 69355   22111 30768   81880 77159
3655:  52388 04465   62399 91987   78084 93246   24136 48453   38216 29177     05217 28660   15645 64560   01410 18201   87926 05995   29167 03764
3656:  83274 15298   90176 91518   13628 37810   92191 77444   69652 95664     32119 02325   38276 38692   63168 15345   60132 06668   73089 47606
3657:  22721 42403   21433 11285   77160 84639   33267 04399   67031 88478     39446 27395   28267 08734   03361 91792   85306 71668   77103 91146
3658:  60105 28773   89473 84345   14835 54404   18778 62223   61615 45515     64687 17644   98202 29341   43658 90340   31573 18469   30918 82205
3659:  19396 43289   08651 26819   31009 20953   19966 25145   46217 68183     53834 28540   76587 92390   94883 16767   25424 04766   28177 32200
3660:  97001 30182   60833 19768   39895 55788   16348 27258   03627 27413     25314 74157   04053 15083   51321 31092   34574 23895   72051 95242
3661:  70869 59590   70286 86323   92316 06455   52265 92698   86267 35313     91086 50073   85410 88704   16781 01083   40179 90707   20028 81127
3662:  60066 42297   27198 78254   43278 00769   74770 63304   89337 28390     70105 94857   75993 25775   09593 57255   59037 49256   66559 12136
3663:  81322 78921   69487 54417   66555 96315   31027 00548   41707 25270     14747 10118   82245 84127   10569 67111   64262 16957   16624 02408
3664:  46339 16880   34269 49342   77513 91688   43624 87555   42790 04855     26730 40561   63509 09032   60138 16698   44323 91693   72546 25841
3665:  64895 32307   44979 56517   93632 33933   71333 11452   74904 55750     28369 62436   41209 59786   89071 21602   75675 19766   58670 38342
3666:  16210 86365   84864 73996   79443 61287   04553 22349   23289 65483     43023 92634   93255 83994   03645 44368   26189 05641   24835 42591
3667:  85382 50633   55551 80831   18579 42661   93591 99469   12835 85622     72145 63142   77193 02360   76021 25126   84446 64832   09438 88597
3668:  17710 46522   76987 44782   66132 27736   39519 81123   50436 43631     65520 90246   53061 78729   07367 32745   49209 50543   41613 45902
3669:  01471 01432   53035 85133   66717 32865   64063 44772   29569 62541     55551 20843   11192 16933   76671 58401   99355 32522   30968 50812
3670:  50947 17565   14847 49347   84949 80914   55074 85485   09910 34314     34645 65486   58858 28102   48333 70373   50293 37104   61362 47312
3671:  54038 40651   42360 97458   60353 91223   61976 88934   58708 02052     97784 11587   38292 45529   03077 92586   97481 54941   25931 52492
3672:  81201 52220   61574 80150   93099 10569   89803 74624   48330 77250     74047 65648   62233 33303   66893 50901   18790 24225   16203 42540
3673:  82118 11229   79203 70485   38136 09200   91299 18219   51606 57681     89419 11827   19472 25764   50555 02011   33207 54313   76354 57262
3674:  78466 70644   58320 31837   77582 56419   32502 56342   57275 36441     13541 07957   48884 82513   79626 08934   03945 25796   40662 86882
3675:  58661 36378   93503 70599   32219 26322   74572 49271   73040 93621     84465 38482   49536 14746   50711 91394   86726 89088   62853 41997
3676:  22419 21585   38963 11333   44519 24967   91290 02106   23616 60306     66282 09729   95002 92846   82664 99219   21700 88614   91790 69929
3677:  46419 86413   02060 40856   32299 98803   75268 09111   73509 45250     82350 32321   94347 59527   45781 91059   88707 22275   63745 56493
3678:  04416 13548   07622 03082   60754 79281   63886 67511   57249 89570     05085 42071   65190 30399   50357 67482   95201 83185   30426 72377
3679:  58078 92393   90044 13975   35841 97474   43241 58432   50819 95200     27121 72960   54394 19258   01468 56220   86839 89877   62353 33012
3680:  72014 91190   42255 08006   88217 53072   06016 32407   10729 10410     31684 90197   47635 29196   09068 17629   46247 21131   25850 29354
3681:  06402 00693   32490 45377   00771 45948   26234 92024   78370 94446     19037 49056   44197 32519   94499 16191   22779 07931   91373 79901
3682:  34629 30224   33556 55849   18826 17544   85713 86712   84782 36868     67071 68051   64890 76171   65853 94747   80283 21503   98382 58843
3683:  93260 65476   34339 58836   78836 69757   55960 93289   42514 55148     57318 70837   73056 00040   28746 25296   90032 05111   08282 47341
3684:  12528 70315   40427 68890   87198 62738   96188 22288   21531 69666     89435 79979   03259 36043   62195 59743   98948 10959   72991 90429
3685:  53529 39584   05841 86384   85888 63284   92082 36115   17334 34066     61947 03991   82490 55094   14002 48833   56261 97360   76442 20334
3686:  60508 19785   07663 74158   71936 29039   77453 65250   52880 84130     44864 96176   42402 04740   16690 57773   77164 64422   00585 96734
3687:  71968 99641   32725 69475   37782 91556   32083 57288   05872 93663     65191 32160   74380 69652   62872 97637   86894 40829   94625 50811
3688:  13577 35901   55275 64210   94563 97941   88765 84959   26700 30299     29379 15983   33831 96224   18741 14471   94123 28137   51665 64591
3689:  64806 16861   10792 53116   89261 06919   80826 89425   51901 86254     28312 09033   04399 85090   77636 96979   57995 44909   72021 52755
3690:  58682 40529   98979 00450   10157 27623   13924 20503   85850 75056     49614 76635   46991 47797   57959 24466   98008 52941   90182 76103
3691:  35143 66119   43862 33003   25434 43623   83532 22947   73787 41658     55377 64135   51258 80524   85755 34694   58847 99668   53662 28553
3692:  83675 50010   73034 83356   16328 25767   42262 41548   17216 78902     74579 89362   24096 25946   60364 03154   46842 45682   38337 86226
3693:  36945 45495   65312 34115   70293 24360   30590 41798   70498 32825     54890 59487   78005 65421   68826 82515   81591 82153   94126 83021
3694:  57720 31857   79712 97432   03757 63289   09659 65000   29645 06674     96875 20246   59794 60385   13932 34976   22454 79343   93734 81980
3695:  58080 27664   28096 16085   08570 31396   48352 62065   49522 29178     81048 17311   44289 51362   31962 92367   09655 68469   06546 96605
3696:  33909 33519   76510 81083   88494 10696   05758 39252   16042 28762     08683 66852   93592 24266   23029 91390   21561 74719   63261 34440
3697:  85536 13696   78128 15437   61994 74167   77906 87547   34454 35243     46638 59187   74296 92461   99311 63653   08554 54317   78818 09975
3698:  51647 04974   04917 82806   94090 35752   83221 87476   07989 37382     34091 95016   78725 01857   69218 72179   26513 59998   90013 53999
3699:  54331 95684   23045 87809   27974 60931   68035 25192   98410 64460     49170 71348   05429 94147   54996 61516   42449 94740   98590 29983
```

```
3700: 83566 32258  77509 76505  23972 65837  66720 45820  59496 58216    08515 40536  43142 83133  99844 06221  76015 53598  39310 70055
3701: 52730 96080  22070 11694  96478 30747  61843 69624  83375 72900    58613 59756  95534 56603  38474 34106  24833 23597  37401 30507
3702: 23375 56315  48547 09287  40843 28310  71588 53854  89701 07000    50083 37311  29373 23978  16154 32798  35153 54244  54387 50302
3703: 66791 54114  80493 06164  06959 95623  33995 35701  07044 24410    41829 81112  03669 51534  71155 14643  07519 86102  51494 82763
3704: 47787 85881  22634 50803  01872 05770  47063 57483  43490 40779    08524 70167  38776 60738  35087 62729  24702 71098  00727 54285
3705: 94126 89406  49548 24978  21436 67292  92932 81882  23920 14432    68789 82221  58426 07378  99897 82004  64283 78666  76600 89087
3706: 28530 15830  95311 52184  90127 15535  08275 17927  41673 94333    26052 91165  21603 53155  11035 83059  69863 79997  07314 18932
3707: 91010 76275  26459 99562  52879 49270  98410 16115  96634 13073    56131 86142  62782 47014  05843 76848  39984 90064  78388 49198
3708: 90695 24015  13450 00205  90164 42358  77958 13274  47772 46414    70872 63486  13500 22468  66874 63136  66058 46064  72758 58029
3709: 24408 89509  61276 24017  48833 63801  31695 60152  14466 94436    90860 24877  15999 04866  33291 16543  60667 52368  20231 62784
3710: 50680 82190  09337 52337  81211 22452  76194 21458  98624 91412    12473 47862  79054 18062  99346 97563  82723 05645  35138 16612
3711: 54919 13348  32399 61813  77536 33944  16121 43956  42379 32847    18164 38919  18871 13141  14755 72044  52554 66782  77640 54839
3712: 95176 40490  53688 27966  97130 16514  82070 69107  73268 00195    56015 49155  98682 95303  26184 75255  86990 93224  18527 97074
3713: 56041 84345  66975 07547  02628 41863  45164 19940  90963 68649    56219 83607  09720 33694  26757 77917  02950 27663  62027 47306
3714: 94733 77982  84773 51011  81210 29435  99889 07002  80082 66599    80679 99437  55712 78894  18080 79094  38486 49879  42505 65838
3715: 30781 48903  05065 55200  36359 20377  84372 38042  02557 41537    52404 24711  60521 73782  99566 55886  03648 25137  49134 46439
3716: 90519 56231  46599 77352  11027 16189  83033 02587  99532 74536    57257 91742  78150 91258  47608 87429  76707 16134  12069 77612
3717: 03239 38360  93171 57966  52507 74539  76969 57698  34032 31427    68907 30458  85333 79164  21319 71820  46568 53748  98388 12465
3718: 09238 18611  91236 99035  66704 58833  38247 75280  24392 06789    38526 34681  49968 31611  01805 68438  68094 08419  83107 10165
3719: 49650 79912  13608 73504  51665 44642  91211 00985  01855 02526    01450 62665  91866 73485  40803 87299  81920 29154  76096 58169
3720: 72896 48402  66319 86334  40957 40833  00266 09034  25887 90795    06987 88988  77151 53143  93376 31697  56206 37444  79557 13110
3721: 35400 53086  97073 85731  59750 43243  13301 34717  06698 15441    19655 21896  17249 94781  85672 87652  28082 45808  86256 68134
3722: 72445 51143  62700 26270  63065 10417  63981 64205  71377 61538    09746 89263  14498 09604  53901 99406  80468 59302  36463 99457
3723: 68647 77031  00254 50424  94399 04492  29938 21831  70379 85292    22777 37499  39699 05280  93849 57351  41297 15109  13170 62920
3724: 87552 06654  23065 98341  16846 46795  39550 47945  88970 26374    71815 63425  18906 72838  91578 60435  05659 67709  70863 37819
3725: 94324 73106  12706 45267  01232 85891  37271 21064  12242 46500    66568 97745  92470 58878  68103 50118  71675 61563  74938 67383
3726: 41936 05851  01684 50189  28046 81317  49303 37313  05486 58280    30983 82887  28029 61607  79539 86825  18940 14666  73727 57938
3727: 38953 40885  78092 92108  31211 63961  38838 09512  64636 24337    15403 61744  26488 89409  60891 61693  87795 32969  35082 68669
3728: 19049 53040  11010 29752  43882 31979  60346 15437  01970 04349    57137 58079  01758 26836  77722 07985  90427 45727  58676 89958
3729: 94198 54029  07653 70582  32410 97913  91123 03562  24355 96679    24311 82108  62667 95778  93807 23023  43632 81933  35471 66118
3730: 66939 03567  95362 96031  45568 81115  40871 09182  19669 43381    05155 82477  06709 74849  27090 87364  59721 88514  41547 97601
3731: 61963 02331  43699 66565  52956 88201  58770 25263  89757 07249    81814 75821  15097 46049  91167 50335  80639 16877  42480 43690
3732: 89275 37931  23117 23533  85333 37207  58779 05099  93273 91117    10889 27941  13180 64598  75664 83700  96621 71523  80917 76527
3733: 01760 49602  43645 55287  08850 85726  22516 52867  09526 16565    88236 88632  12175 82995  69105 00068  38050 32064  08949 95900
3734: 63237 75645  93026 19018  10595 57010  04333 07205  11652 93224    21462 19135  63801 93188  50051 52155  91135 96633  80746 94077
3735: 70076 76431  52307 46754  66125 02869  64581 72116  70286 80767    71885 22349  09648 77978  67858 38728  78458 81693  96462 24967
3736: 67810 71920  52462 69791  81991 63332  27239 21314  14767 98198    81651 28108  40100 13537  11727 57834  19924 59071  46469 43386
3737: 77616 41684  86324 21250  66139 88630  54394 32379  37523 99044    92701 18160  15992 80552  12900 37508  61150 94490  97884 48657
3738: 63483 73013  99463 27146  25412 31801  58188 80141  67979 49022    19128 33782  11266 86127  03786 90797  87811 26949  35862 52423
3739: 79324 21392  70017 97965  89863 99737  78799 44903  80774 27758    65801 87381  54300 67883  20471 64584  94009 08026  28412 98449
3740: 76427 62943  16931 91704  08186 52524  87923 40332  93836 95243    63756 19072  57929 48421  36971 76073  78249 11650  94208 56187
3741: 94459 10480  33190 33002  58497 25432  52920 01642  25478 82290    73499 13715  88329 65402  59466 40681  14039 24135  16160 79944
3742: 52004 09919  79216 68629  14838 39166  71723 93440  69483 08676    89986 44416  29260 25868  41112 75536  62577 31378  57700 84556
3743: 22671 90625  34586 50318  38513 29843  23715 51590  80988 46840    97870 87166  65466 95288  96366 68905  71526 65873  28053 86125
3744: 99346 43236  41806 21205  76439 19856  36631 82100  72671 24562    02324 83887  88156 78433  88396 17251  17326 67390  16928 95252
3745: 11017 68130  26392 68723  72089 69060  04299 51353  26242 34427    52011 46529  62680 25580  34674 32513  94993 35716  19716 98253
3746: 40425 91345  63928 39685  45945 25073  15062 37800  67266 66702    24968 47196  63072 32010  26974 12542  14893 02150  05129 35463
3747: 62158 66594  86504 35260  56899 99042  89275 67520  34707 08812    05823 57456  68403 76628  72021 03062  77294 10811  24742 59195
3748: 70712 50181  55953 47211  15590 04405  68407 66422  24162 32258    80123 11274  00491 62319  66338 79952  72581 00783  55712 69647
3749: 99285 31205  01427 03576  33075 62228  74388 15858  78807 13478    56330 04793  65224 19999  95989 24783  46539 70756  07452 78121
```

```
3750: 9281392673  5107909775  4965673206  2611842227  3238155407    7929181612  8089070590  9952890094  7752991459  5381517322
3751: 1384483347  7809990011  6945618641  8316040433  8372813100    1429086288  1274213235  3553445081  7779813362  8784416215
3752: 7305184462  8390764876  1064048894  0020490277  0722235732    7824024605  8003263228  1884167639  1221607262  5829681832
3753: 1120835144  6974919234  8930159047  1912814511  9267641305    5762574295  8604016397  7978978414  6987809163  4533789826
3754: 7749680032  9091832947  8173771142  6332227177  0363544808    4320933365  1007635378  7595840688  8906590460  3425530303
3755: 6887882856  3448151444  4729306666  9354414876  0477086542    1658844514  7452352507  7883046140  0752314151  0308075790
3756: 8481514292  1712999966  0397287929  9472896702  8278209156    4385853481  0533044251  5052914579  5228686532  9758531355
3757: 6654734920  6336700233  5135571131  3201554955  9088571049    0407912108  9773152485  4741958788  5537110351  0492806550
3758: 1299706454  1252005493  8335680072  7273868726  0595665227    6261567634  2568412946  0290227614  9979747385  3412497107
3759: 4138929750  3035921254  1694212674  3198504073  8076886459    1721864132  1251118926  9405819753  9697151261  5820007818
3760: 6018568256  6929758875  3283035071  3975653210  4492398605    0492925131  2658134098  9298829573  7055945997  2918165930
3761: 9618160092  0617461062  4786478951  2807649546  6695275934    6123951705  0850497819  1323013111  1068330834  6630577780
3762: 1743571058  0877249603  6108257514  0136644697  8124483143    8066173922  1296124124  4550497256  9497921635  5620028318
3763: 8770198185  0057095561  0142593308  8651721800  9278280100    0987415570  2009342163  2267889845  6244602971  1889295891
3764: 9413422530  7798707828  1347897265  6951109265  0466534587    7628624596  8013150381  5006217452  9838142001  4948874767
3765: 4948598901  8407977716  7922814025  6717026692  4687774006    2891965670  3612881361  7179089213  8565840531  9265094300
3766: 8272059329  4510407466  0758530954  9559324478  8754601596    4313105772  0837025052  3285151173  6999252388  5851287871
3767: 8355245890  2092130478  7618847924  6014268913  9119168020    0147771928  5337722905  2565853290  1337594941  4850815120
3768: 3630746552  6108427410  3870090187  6750864453  4188301380    2695625686  8382489689  2401716075  6825824134  1860976664
3769: 0920817743  8980581836  3663344422  8178057775  9477765712    9797387371  2791071963  8898109176  3670327118  8635355211
3770: 7570054099  8273028658  4364722406  8448697398  1622374412    9729375160  3506220102  0674132343  7531774206  7590873830
3771: 6976824514  2387587413  3545024186  8961766402  3037295149    9510711829  6328898385  9543651083  0514464519  5619066590
3772: 0336841358  5456602571  0248203326  5911725115  9769668386    8623821250  7577615262  6042475645  9119535504  9810666171
3773: 2761935503  2379509502  0987592484  7841881165  9871824214    1175106463  4635202211  1654520937  7219410170  4353920249
3774: 7729110700  0683538998  9265222986  4793309121  4599456031    7332119784  5531259012  5184840248  3662705033  2916678378
3775: 7521726689  2742991234  4502768568  1766758075  6500044387    6333028300  0327438360  7020579116  8648857062  1841759616
3776: 5675138194  8754087874  1333714710  3958123073  8521072584    9507416354  7704027966  9111354419  1954027650  0884795695
3777: 0604142518  1736654242  0952759430  3634242646  9707925221    1538695280  5364620957  5858434970  6321811904  5211202949
3778: 2922990170  9282285407  5654520049  5665603343  2544130638    7588415038  0830462528  2138027081  9112147491  8921912438
3779: 9388756046  7233490037  4118366495  2906772654  5672399858    8255471749  1036540880  3663477681  6555460493  2536947147
3780: 6920653081  5385294512  2861298399  4184995770  1787642563    1756705861  4491251234  6317315081  6538889572  9805511590
3781: 0635040090  9688248235  2730895586  3078411696  6795386529    6224526601  3303417586  4542490985  2892381638  0700038400
3782: 0025556754  7294167301  1869638922  1849618015  4014669467    4742649291  9617673897  0421950856  2416142219  7526244563
3783: 8897051541  5459214503  0878718165  7774022453  7722866004    1023484406  8061110563  2727554033  1083003583  8214230552
3784: 7991543181  6700617821  3246620211  5408061707  9257137587    9562907751  7071496184  7443163867  1214878040  6014892765
3785: 0904311724  0100978447  1299901779  0328533399  8437663599    5745778814  3619627435  4673007284  2940399692  4552449365
3786: 4301255787  3371025349  4594317551  0597083482  4731386849    7126333404  9361218186  7864518533  1560612913  6893112923
3787: 7451840630  9302789350  8367820810  9584696344  6203640044    5830132733  4833397907  7942697774  3175429403  5672233820
3788: 8654806218  4956029828  2045528560  5194111670  3126093629    9602817206  0095159730  6398706793  3315528162  8856677722
3789: 9695487595  3924727857  1792754146  1663622910  5993518877    3136297520  4462746603  3252446633  1715092491  2015313774
3790: 0828788815  1258001230  7692657127  3154257029  8329705864    9387656893  0739972353  1643727362  9248792558  0896396322
3791: 3018664950  5006048328  9253675217  6521268734  5279138242    0067175029  9659775545  7670281659  3043215134  7946129801
3792: 0398388106  2505902978  7170214794  5638320210  3522209157    3032190137  3161349153  8094280111  3373742168  1361940527
3793: 9417320667  9056704741  5370939308  9538836075  5946660695    4954300408  7872715010  9701749724  6733770612  7963408638
3794: 7655445901  4313458096  0340091287  1290573363  3953717247    9188885568  2108672792  6079481991  9726643945  8591957608
3795: 8973985022  8297130446  0605023359  0297014372  4016473566    4330795626  7949037930  0976241846  5757929619  8110232043
3796: 8112562631  3358649002  2675737133  9850151962  8568087131    5546965808  5297060264  6763843627  2996850019  9690596926
3797: 1840800246  4067839294  6562635547  7784330411  8893721077    9468948843  2491389776  3544136764  7121774657  6988960575
3798: 4736095494  5355521922  3119611304  1669798401  4643793946    5082377218  1868877115  7335817569  0029086409  6661125275
3799: 2754419567  9967979294  2061884888  0696201041  2003933516    9723754448  6877259305  1174534537  2038496553  0143330199
```

```
3800:  1250673243  4536776086  5685487067  4847759128  8706220955    4990368346  6740723829  2494071501  7744292714  2066604040
3801:  8651644845  0405483574  3961068042  4829034191  2234559601    7296013731  6511846062  7647056297  5780737876  2653415363
3802:  1655100402  3669577311  6473679634  9719815433  8181447078    2021749913  4752967371  1719375823  9467750466  8688706122
3803:  1269573404  7324062103  1296289886  1498109179  2210805772    4669563384  4173616670  2424814593  4608299983  8608327409
3804:  9581115417  9175861440  6188230164  8261431944  6223226968    5703320275  2584770472  1966622311  8323324094  7917864419
3805:  8092852351  9748764498  2128908700  7147216968  4592678315    8222716338  2989613171  1047651857  9513737320  0080147312
3806:  0708863993  4364436576  6923120528  9931904804  3307936096    0410111308  4643439011  0373317387  4689777950  8294623466
3807:  8658998053  1165517455  0254876420  7052913187  9973612195    6392497330  4624513626  0003383238  6360588051  5895884765
3808:  3884467271  9147164033  2222312109  6641112995  2350757495    8545963277  5383872521  3923519426  9335160716  6115642223
3809:  3709380312  9516725474  5514984820  5056494516  1745181951    3693067641  1823037890  5665817215  2528408434  9769375400
3810:  4390039412  9382308824  2103213486  7084169374  0295796574    8081883004  5997557433  2935666794  9882405829  4292895632
3811:  8589694975  8644795350  9898351790  0117887190  3029171469    9541216397  7061768608  7606696640  9473702494  1949286972
3812:  4022359135  2477202033  6110752206  5617640170  1792092922    9933186941  7850103548  6436714155  0808578530  0098390636
3813:  4294204551  4065765678  2525408491  4333807145  1925574221    2069694324  6952671821  7570145841  2681029191  7970946029
3814:  4150107624  7745020240  7185878873  3940074077  5497635736    2553313736  8829850106  5276710591  4085556836  2804569282
3815:  5526449127  3659897450  3684645293  9885158244  6451954538    3829234044  2283958215  2809850900  6074510148  3034289668
3816:  9840995316  0360045154  6746129960  9888890040  4476575801    2615217180  6445368442  4069783752  2254669197  6297048010
3817:  3871202041  0102766959  7094708286  5633127071  1527879201    4950650489  0716305395  7891401989  4412501270  2493605599
3818:  7570243999  1127859794  8120214999  7477705717  8688588980    6741221059  7791968663  9153174696  2320266057  3593862060
3819:  7917814643  5402874002  5258878405  8034632131  4366351862    1896730304  9187068972  4555896523  9549048039  4799590188
3820:  8107485475  6079862472  8110551343  4201632250  0865290484    1387682130  4268102996  9326537028  9924413187  8643928779
3821:  0070214734  4020511700  8777333478  5608378371  5573986801    4603707673  5038478507  5087570026  9667422664  2395337171
3822:  5803276936  4065967034  5939918156  7841557180  2218333305    5560995743  7748412501  3821698754  9144644968  1927167872
3823:  0583340679  4062763123  8090475032  4851991662  4901952742    4758784096  6065229412  5373948823  2827280367  7834898109
3824:  3123460792  5663026729  2810086597  7485019926  7809769653    3389166538  5307099858  4543578314  2034614630  9823863811
3825:  1081873668  2632253621  1414407379  4705585932  7378784183    1347948141  2288438419  1684423033  3023015448  4717622607
3826:  4238039095  7770504873  7211260565  7968944552  3091160241    5927960746  5248612401  1611825048  8025176533  6799005556
3827:  6440295621  2899758972  6118333222  3863742572  6190618298    9587226296  7364598980  6625704812  6290076675  8744803229
3828:  4374570128  5218859861  0449221073  3307339303  4692845773    1298315075  2169155403  1137910313  0818436736  0477133906
3829:  4463041700  1085833810  7116652774  4680569208  7492584952    5452224620  1768751523  1902487842  6586118190  2141860436
3830:  1880610051  6869906638  3636132362  6239722651  5998722897    8678939447  3489034105  9295307133  6373818902  7181709909
3831:  0652682673  0108311370  2929520338  8519124556  6309257050    8817754963  5097223401  3209720219  1588464453  8585725610
3832:  4576431916  3563036823  9263916122  6121347110  1254126257    1282011411  1196226444  2532466994  0839020067  0749069876
3833:  3568946365  9010999889  8452331174  0844879007  8632129156    6581798483  6446505611  5390099446  3060532251  3096760611
3834:  7617413781  1864701988  3160197644  3476613947  5058769442    7236623545  1591878854  1952762742  8392904835  0939661754
3835:  8078375380  4217232073  9282871968  0735598838  1152832502    1536361700  4180136275  1938202528  7064402191  0064952634
3836:  5103912150  4983617713  4637572767  9923181813  1621713911    7721513084  3688698393  4000934071  2021430291  8996309269
3837:  6473293672  1767822128  6767408778  9577602762  9925045941    7186277197  5644160275  3514640350  6486034399  4494484830
3838:  0473299930  9711377570  9648422296  7528739230  5879227163    5615608592  7343569802  6377993585  3729692787  1082286402
3839:  2475113384  0096528965  4424273098  7228856745  8361994768    2262992546  0557750336  1424053684  9421799006  6541709073
3840:  9128395031  8711136668  8697217371  9836419513  5865130130    0942890130  9317806229  5975458852  9835165148  5226896696
3841:  9172493488  2333424368  1848375841  3337186281  9049245949    8296064574  8152632952  2050284639  1960375037  2763310313
3842:  9636873707  3819441987  7072948111  9490733508  3467606784    9422662581  6000184221  8834170372  5929446958  4856834736
3843:  2041066675  2078442789  1113476440  8412352521  9160540803    0872012216  1113128964  7111023959  9382940597  6706445198
3844:  2481314256  9479497677  0124484335  7006277043  3652690633    0777806944  3456932712  4656084339  4242668238  8129121554
3845:  3867691393  5210176031  0987817822  2306140066  4608522202    3794551593  7154000019  4428527696  1615403683  3399715029
3846:  6405712327  0036851729  0884669844  2726513192  3180219877    8599042107  0059692921  1217172886  5162388797  2294864313
3847:  5358745726  8228963998  5944404206  9722910903  8101020391    0564252697  6759770461  7351676202  8657362936  8967778086
3848:  3580990077  0917386802  7257176140  6385550767  8825330909    0929277848  2521667214  6255641445  4825571242  7600163239
3849:  8240914269  9257016013  1509444115  9848504869  5316696558    6439907966  4319674965  7208835047  3329508689  0669372726
```

```
3850: 69943 44118  92730 35838  17036 05759  18585 86777  39516 24220    58540 06180  09567 13340  78060 47014  61459 33435  39633 30608
3851: 61491 03625  62266 57515  04491 71994  39850 64016  47212 07646    21931 60699  92745 21965  11090 69572  04975 72085  83363 19522
3852: 82674 26183  14973 98635  93078 71626  80188 86624  09616 47661    08469 18615  56953 17428  89977 91064  00742 18789  67816 65650
3853: 67950 88311  18704 62425  75334 08463  46615 00286  58400 98338    56095 14278  29002 52153  56363 73704  36935 03423  60844 35309
3854: 41544 29661  96500 40429  55984 96334  62281 27353  47402 16189    73202 05854  19183 46295  11910 27283  86418 83261  39229 58633
3855: 22804 47387  81502 46152  94649 96760  44207 51771  72048 61437    78174 69276  80321 21631  70824 34751  12847 54893  31668 75632
3856: 91831 36250  10134 54637  50457 19558  19562 33361  68163 00316    96945 08204  15336 89821  14740 57057  64037 85105  31460 24111
3857: 53264 17317  82121 29926  80233 29268  84970 97086  11942 43803    83492 62452  46165 88863  45802 93059  54445 92515  48003 27928
3858: 59401 17940  53786 67027  79065 02983  65425 57869  98430 68043    96001 36329  16472 88448  76005 05663  58996 41943  87778 98686
3859: 82932 87624  02992 70636  76685 30893  02073 05418  51636 25117    85465 16013  93110 56669  83664 64704  41661 23560  90386 06701
3860: 56857 53435  05440 62517  01118 59502  27547 41383  87030 51877    18343 79111  90756 71246  54894 00363  55076 32940  38482 60453
3861: 24704 15272  03250 69156  46809 52987  89727 87609  57238 94835    75740 80269  76905 69150  88217 00992  50199 89978  50930 35169
3862: 90171 02582  56629 36381  93670 89543  03164 32701  99255 51633    15212 78044  01527 72038  45480 28397  28056 50269  35921 78065
3863: 40933 71479  15114 22538  73465 65581  77585 38464  05051 55238    66398 68629  24451 10361  31025 52146  04064 55680  17815 32644
3864: 21195 14407  58477 37237  02765 18517  15394 95416  88180 19478    25194 98292  87891 38098  36207 61567  63223 85732  64525 61435
3865: 05591 81542  25479 64894  05235 83885  07125 85521  85658 27000    03641 52691  13474 18855  65318 89637  94405 34738  05626 39695
3866: 26768 08767  15315 33759  31464 37039  01290 68890  39317 30897    05411 18886  21555 85403  35499 62782  62949 54131  07045 18327
3867: 60299 26000  14764 05650  47121 37401  58888 05872  09170 64749    71881 67982  39122 83240  33200 00685  04690 43939  90115 32767
3868: 76908 48234  55705 77479  81530 56837  87296 89058  29743 45348    53708 94350  84449 44207  34028 63899  04702 27532  04711 82069
3869: 39608 56377  02036 56646  32367 77213  34233 90547  43331 38621    99966 85722  43237 96375  12189 00397  85911 57090  75596 67707
3870: 20864 88291  47349 95724  17961 66557  65055 44352  84789 95262    33342 51424  33396 74158  17261 87981  88140 09512  99287 51184
3871: 06374 79350  26469 03588  79096 06086  35095 36898  83663 13304    05451 25168  76947 24533  90809 74094  26645 37630  92943 67893
3872: 51418 63048  96817 86285  98707 43639  57291 10854  22061 46267    51307 37104  30286 12411  10715 65240  50823 75971  05957 21501
3873: 89674 95055  72620 27571  23804 58422  85547 86415  43493 20976    70150 02295  87753 20735  58034 36063  84159 28346  91484 12253
3874: 23119 97674  16488 30706  99863 01859  49975 63662  09258 72648    90764 13188  28461 67943  74660 18771  58463 85128  64864 98725
3875: 17686 06870  94392 10740  97803 30748  74248 32611  22753 24563    05469 94788  30105 25636  13970 08738  54410 80256  50001 70564
3876: 01171 77118  21261 61753  98302 19255  48292 71817  22392 80065    39388 25756  82990 84158  55733 60820  85726 51495  03693 40285
3877: 87952 30900  59584 04302  87964 71120  69136 96712  87919 47354    46829 27378  13326 42173  46779 29132  86678 96886  88700 88895
3878: 56810 65854  06793 20047  25976 58724  84440 90588  53867 55664    46453 73589  40534 51184  86123 47346  62091 54890  37114 43889
3879: 23269 68838  10737 80153  53547 99520  22533 68029  17074 45029    56740 76982  82805 98545  26106 42434  25365 65935  26354 97293
3880: 78101 29674  76687 60412  41440 64955  75037 10613  85436 66240    29936 36155  06602 79027  23099 95664  19044 90975  14370 30018
3881: 22780 50056  42615 67976  25103 91340  14607 60350  98328 19584    29265 86175  28812 10214  38829 81332  72569 17774  19976 78393
3882: 33830 74507  06219 22142  75364 47902  42435 98339  91287 30779    68142 66157  68067 45406  90251 90063  56939 47307  25834 52688
3883: 21839 89363  53126 04197  53301 52042  88688 45077  27557 89282    43124 93978  89879 34735  53340 82708  53873 94937  80070 46750
3884: 01728 39985  24268 31642  19731 97619  20192 32349  38126 38026    83011 89205  02625 27193  59985 63142  11613 37073  25509 31588
3885: 81895 90615  56555 02420  90180 14287  65559 77935  77069 17707    06365 44801  69249 87408  78288 27403  35437 05122  23745 06726
3886: 61658 01491  30705 31544  26739 93111  87998 67160  48434 53173    83025 95699  76896 29829  47173 75285  95758 64313  27097 32746
3887: 37411 84974  99680 15746  30798 90239  74505 96482  25818 39603    82684 96601  25221 25936  97521 98408  83187 44056  50647 76164
3888: 70619 40067  05952 72964  97087 83754  98637 02025  45881 22832    79195 69394  01982 16130  44510 22736  03278 81625  98537 91080
3889: 36024 16279  61170 82423  39759 05928  31679 07294  76918 68134    16213 58377  86596 42173  82261 22394  33866 01712  95956 12821
3890: 92739 82870  51853 30573  98605 40834  30177 62273  69785 32934    00756 29527  95209 29450  99892 89974  57944 02647  80712 97043
3891: 35845 55919  47291 89804  87243 16837  30038 00030  71656 68754    92369 61607  41604 01403  15751 07963  38037 89015  01869 52553
3892: 97589 04218  95810 21086  90126 17493  12631 53296  57786 54564    61963 77405  04040 19205  18471 15695  49833 34460  37054 47970
3893: 13390 42611  55844 74777  20570 99715  85491 74304  12946 36198    08510 05744  43416 57163  90557 73102  44266 51454  27055 01458
3894: 97898 07394  06040 35756  88239 21809  37252 89819  92642 51689    96810 38585  67741 07390  28790 19203  92383 22903  19598 56499
3895: 31174 43752  30628 65635  69104 62711  71738 89118  04211 55013    14455 86374  09294 03292  72741 24000  11813 96145  32036 63812
3896: 51722 42045  26510 20396  30122 07435  80562 43012  54346 70226    39270 28758  39612 70839  90345 05225  40757 60868  39969 35469
3897: 39800 96423  35202 67026  97054 47010  59882 51748  29517 56849    45538 69535  42127 73680  78927 20949  86461 34141  85597 32945
3898: 63921 00235  14132 45272  17424 48816  35204 79711  07052 27878    57127 42701  00267 15004  46371 45359  86280 63413  11945 52622
3899: 77839 17961  11274 18402  60084 09173  85100 20314  12302 32785    68816 94724  65587 47576  50270 83317  34585 82248  49345 95217
```

3900:	87596 24941	99192 03620	19877 36756	76454 03049	79119 89679	48377 61532	66763 74028	67156 07085	09324 47587	83676 12445
3901:	77552 79879	60496 47526	03098 10411	49611 18636	17422 10879	71137 75890	38503 58086	81093 80505	55438 55074	72944 95979
3902:	57206 40784	74917 81963	51544 17738	37930 15892	89804 60316	99595 21245	73891 33122	98839 21397	09260 08057	82548 65522
3903:	57940 17667	53290 25731	55578 85637	67117 59010	35809 38807	92290 53926	64041 30192	20324 43406	87249 55032	60192 48656
3904:	30095 52181	97238 12905	82220 41972	64824 13951	32178 34382	10465 76372	20915 41638	14555 40858	63555 43627	91426 43923
3905:	99525 83156	40813 76099	58716 99402	81151 16282	44103 02721	17778 40891	52669 29797	57234 53780	87873 18286	31875 13257
3906:	28432 65516	78469 72309	65440 77150	84909 91315	76291 47516	74140 45806	42149 93892	80475 64097	16427 64175	27842 09133
3907:	89586 77670	72616 04285	04530 31918	28696 92400	40587 17021	19428 99352	05356 13546	62179 41463	90904 99728	87289 09511
3908:	23215 55313	16007 84421	60881 65717	92710 51201	80858 76718	46878 72019	00703 26380	34075 53705	41019 21860	31784 84761
3909:	25338 11110	26755 55058	86075 88680	83462 85117	44126 47473	59904 06356	97015 90547	23591 86121	63676 28365	19947 42222
3910:	58782 78377	93438 58697	68035 51204	58479 34173	48905 22706	09408 21933	62921 02336	94754 06284	71482 06343	68974 89232
3911:	63059 48046	05579 66451	10915 15194	12154 02873	86645 53104	45749 82715	03736 55386	53958 85752	72344 60626	68178 32393
3912:	34979 35514	62689 01568	86219 55551	12306 09497	98512 47563	22234 36818	90814 17883	78147 94979	22060 93552	92842 91707
3913:	89356 62348	46333 51650	41896 64928	76410 03678	21082 56337	35532 01736	18655 10330	20353 74357	01064 16886	53195 00876
3914:	59047 50702	34075 83265	84187 53383	48805 78707	27714 89628	36248 01560	41019 27200	47057 29625	21285 47727	59057 33440
3915:	83646 15674	10201 48099	17896 15503	88465 40333	48349 61375	91078 85497	21643 27632	28224 66097	96546 00350	20949 91176
3916:	54351 02336	20772 38972	94301 29337	64810 17909	18505 32023	99726 56751	54476 41717	64059 20288	66718 77827	94810 77125
3917:	21588 36533	42565 23490	13535 99297	30699 36136	03093 37619	89710 98312	03386 21763	93800 65726	28873 78128	12149 87387
3918:	78660 49516	47295 75098	08985 44580	64066 27841	87651 21004	91309 56319	53916 51337	23145 03310	53081 99941	26200 33263
3919:	58833 75070	77262 14119	10472 38541	81389 14552	35242 40230	44869 37651	94826 16676	41007 01410	92604 12818	34927 96966
3920:	20838 42678	08502 75899	27932 76324	65294 26485	69602 03227	96126 86464	20146 40615	60333 81549	34204 37686	96648 02744
3921:	41468 58556	17773 09325	09788 52565	95142 19842	56784 22128	19874 92865	37781 45956	22524 09415	12566 49067	63131 99506
3922:	74996 16141	70961 31748	14158 49636	06488 75296	24498 48670	55716 84801	09650 44463	31696 46200	42947 68898	59710 93414
3923:	29474 24962	64741 45490	11134 31394	68337 33929	13019 57759	18773 61650	73470 11921	97626 60891	83120 45994	56291 74896
3924:	23588 96937	49108 54854	61124 84226	49886 89207	63586 36514	26754 15357	94484 59376	05688 30917	61164 61312	95072 46420
3925:	22190 38371	45193 77062	54399 69799	89887 05067	67535 43781	11129 09831	77371 44587	95986 34401	69025 15403	64118 64837
3926:	79347 99749	10050 22123	86958 09556	87699 17759	16344 65124	94678 71036	78805 38037	34895 53183	27063 77504	94465 64306
3927:	42227 61710	58980 06853	23733 62137	76370 41410	08480 94007	75097 28732	64461 21842	68546 30929	66729 24377	37946 47256
3928:	92865 29874	93033 77994	22955 87955	49762 32810	37173 10426	87817 80743	99112 38641	02409 03976	02592 58309	70719 03731
3929:	97817 38986	13926 68439	60650 08976	45482 41205	00656 52489	78937 18288	19305 41576	05114 19374	11103 25533	75867 08158
3930:	80635 64212	13958 86832	94426 73470	19326 32683	32594 03769	70678 65758	48716 67451	99053 43991	62935 97347	91748 36588
3931:	52703 62036	71361 52240	51550 85781	40274 59049	92716 56119	36673 18418	47958 14989	08821 97628	72021 47308	99032 92071
3932:	71232 77348	42303 55037	22165 88043	20714 46735	45366 78858	24827 76729	69884 98814	58247 77160	16851 13840	90962 13744
3933:	14411 91617	53628 02777	09541 17161	91346 04196	71706 04845	75030 85178	20227 67634	83115 71343	20636 87257	34601 90507
3934:	73168 44590	63472 10840	06270 81932	17566 98668	62097 45257	28728 08165	32410 68311	59416 29101	30141 13908	37514 64409
3935:	14990 01488	23869 27374	95641 90426	55192 45198	44480 43615	66488 34035	08710 78436	57179 36435	01596 77820	39226 76411
3936:	96982 02495	45744 12241	13405 79236	15451 12874	20469 91285	42028 95108	41248 92619	60527 44154	60165 91339	22021 44332
3937:	35200 20622	42916 15965	04659 60558	27141 10849	28339 75439	02293 66752	76574 29662	31630 52209	90120 00373	29446 72721
3938:	22381 23080	56921 72881	90366 78211	84859 67974	47071 17050	31802 84225	39151 10082	46497 29979	07563 71399	39199 22491
3939:	05417 10737	79153 59289	74810 21038	32135 50927	35771 39893	81667 06421	32693 00181	60082 22164	15776 59537	36042 26018
3940:	08035 13638	19409 15525	02326 52054	88562 09698	49096 97245	40874 38970	88105 06223	47039 26369	02822 45543	49062 86067
3941:	53507 80919	47576 02464	89920 59874	17326 92674	76934 12489	00457 04382	20833 79576	48134 99424	72068 60473	49773 60548
3942:	82206 56454	29617 30414	17366 27318	70271 01711	60193 73241	34025 18462	99409 16406	06333 77429	98306 65735	39554 26875
3943:	85418 29489	01267 36392	28817 77892	20749 82292	36798 49994	81792 74253	61273 02557	10515 12606	28513 36020	12240 55138
3944:	65797 12249	41552 46950	68564 24020	31265 98057	02716 37326	15165 20277	41058 10468	60371 02666	84736 45598	02122 44961
3945:	69593 66336	46894 68961	95289 76405	01332 69239	13573 73693	96818 98507	31185 16706	25675 78499	07068 84878	40935 76419
3946:	91981 63013	97111 55365	32841 45146	22845 27037	34242 38209	07056 28635	49333 46056	89089 44897	02705 07527	94429 63758
3947:	86498 55838	87231 02865	54064 37095	79601 79063	61426 89697	96828 39398	64843 48961	67438 09992	60654 06610	92524 75601
3948:	03454 16733	63904 65694	71924 40428	50807 70548	99130 81532	59657 46651	87181 30915	36016 41422	36673 59990	23053 30867
3949:	21260 37240	18487 96102	80465 24728	98128 49808	64103 28695	20106 03397	11746 84512	96059 53361	17618 92984	15584 45897

```
3950:  6098377020  7459817053  7207924192  8831021822  6671805712    5625307721  9675607365  9508489743  4136100512  6674742385
3951:  0989760981  5916295282  7926577983  6853903908  9514407309    8050881505  4215850774  5427571105  6490708688  9412335820
3952:  9970511015  7580655968  0622161370  4626407120  8483410312    1332119702  1942297217  2060002278  0741064531  8867291322
3953:  7515774547  5402613685  9795475170  9426108811  9475043919    8293801244  5741532590  9792005014  6574737046  7592354799
3954:  7064301494  6086032028  4274998588  5867098689  4144096848    2446725920  0268469807  4098462904  9329368266  0566425973
3955:  4983273332  5633418203  1681512155  4122283113  5436851908    6236805832  2240957260  8266020734  4680718950  1360047060
3956:  3429911418  3583286956  2858036174  6964544938  5479329019    1204429304  0953718735  4180270270  6681006655  9357130383
3957:  4934650762  8787965058  8494663270  9117130571  4251234147    9011671605  6251364660  0650953731  2440680957  0297223907
3958:  9688662870  5831354094  4998563269  2965078047  0164289041    6875442261  8154535301  5388182627  1660695636  2353837473
3959:  5927886474  4526738475  6277919255  3099825362  9394417732    6884896970  0770869292  5518071055  7234923648  6018578186
3960:  5299992543  6738438363  3136981443  6663297073  1884223683    1855499871  6081138339  2421208291  6698627718  8069299285
3961:  1939128631  8928223873  7521313737  9479104428  7633191746    6247150366  9972614000  5426767600  2948851905  1698073635
3962:  3671683468  7669346012  1537413604  2905364792  8556709248    5210762715  7428127115  5970976885  5045538186  6569741368
3963:  9675640770  6016811837  0362934882  0696263590  7856788006    0019980398  0782413255  1281494886  8926620044  8373738691
3964:  9083359712  9004817166  6388197409  8527527403  8688943934    2729269297  8836634215  2470901753  7757674043  1395068812
3965:  4830523484  1714490986  2057846633  0674148800  3339972332    1752668989  5792443162  8986114998  2561452497  2914463146
3966:  0443774041  9663541873  5573927428  4667399919  8725328809    7459519810  8751699489  4508622070  3678922209  7356299075
3967:  5937121689  4410421244  8059852021  0231329202  1677047416    7612104758  3929564734  6833721162  5348438001  0499042843
3968:  0528351888  6977514114  2323385791  4783852028  4143566128    9935085763  7564551101  7787618343  5687428986  6110753361
3969:  7468285785  6166615568  5598698298  3625356479  3145455668    9065567272  5849387072  9056220879  4140582623  2302463959
3970:  1040865353  9955544907  8773263764  5752112869  4114906556    2758031142  5337369865  2912924471  8324316720  4470053025
3971:  2499412751  9924956404  1373399189  8360480573  6296969030    9565689115  1435858986  6722372105  4298903366  3454040751
3972:  5832304697  6968337837  3448708923  3286726745  3174212918    3280880192  0178147740  3017173802  0733770825  4604505409
3973:  6652996154  1380514143  4479710876  2811178336  9945688821    2256557965  6656136838  3694365148  3989879138  2464044104
3974:  1373892755  5567004272  3254879238  4295901901  8043584027    6626778732  7603619210  3452291451  5046865810  1521600933
3975:  3216810557  3239167913  6332238865  4409503931  1381331839    1083775652  4558074423  2786775891  7910016135  7863398841
3976:  7985834391  7259265239  3004308165  1782928113  9354626044    7550278473  7354282252  0495760439  8799703922  8407327303
3977:  0553061000  1992450119  8724157516  6787021884  3352974578    5154693098  7828982646  1724146634  2591968177  1246665590
3978:  5680506254  4529779273  8157675072  5889408570  3015121655    0688542444  1212298818  9325953151  7820659476  2446605373
3979:  5279792215  6848184098  9873335503  3869005945  0178804595    4755026884  0976668899  8905397121  3727665573  8579354293
3980:  1454268349  2009432229  7906447413  4674400884  1114655634    4722741782  5970470326  6317362445  3449435636  8514999311
3981:  3577922595  9744305550  1160416816  2987895218  1146482107    9515110392  6203963212  8181522028  3390096707  6781911267
3982:  5892472566  2292039146  0699069560  6691674875  6982233487    0366823392  7962572532  8942702598  0031718068  9130472224
3983:  2599600790  3142575062  1402117496  7439138929  9580352780    0781300691  7788394493  4841366044  8405510084  3526151273
3984:  3863846424  4754676273  8570534415  8792219887  0595051814    1890102948  2336734272  0799839227  4293668886  6498931401
3985:  1691281020  4004349108  5449614198  7851092967  8866510522    0321607654  0076420537  6210803512  4713351070  2609175467
3986:  9793088223  7791185265  9538539012  5443422929  7899885145    2865236317  6284955092  2399521768  0693363536  5853927644
3987:  0751501552  2412311058  3555763481  7031459197  0014620671    4255593579  4774439711  5177467357  8484460073  2207761166
3988:  2570331605  1902513498  5828154820  7834868966  3417837328    8429674715  3540551331  5076842398  3339949426  3337901733
3989:  5045012229  1838402788  9235672557  2339228080  5151864340    5338519405  7816288096  6251410853  5254196370  5128643307
3990:  9075823630  5769057084  5932093260  6124415984  3831304388    5629876057  6318783932  8782415129  2000540311  7442085473
3991:  0895520375  3054003697  5025340238  6917697438  7633604325    4118182623  6857567037  2873636630  4299837943  0819044866
3992:  5222433060  6044065545  6145043183  1854973240  0295132804    7673973183  9333424083  1339952956  9207036722  8994024112
3993:  5578647093  9560022085  1868367159  5352300723  2752073589    7131624936  9415716261  8774830879  8940509603  7957206690
3994:  0926639270  1073348635  1779516585  6344369331  6808942399    4921058866  7403053466  2969370620  4264278583  2614548026
3995:  3326384561  3704145734  9714753060  5935552716  8013317529    3108176938  6499905715  9261831463  2234540199  8556546475
3996:  7144092467  6208604835  9524943384  4937707259  4457954000    2852994963  4992420020  4049246113  6616542169  5828322256
3997:  7765288224  3602329560  4325732399  8135856564  9669240411    8441277850  1144748242  0129162348  6106292100  8143706719
3998:  7366850258  1742608664  1094458943  6744614723  9592441493    1741000915  9518269208  2035111030  3202136211  1119117040
3999:  4248148066  0961932158  9267315072  6195103210  2706857576
```

```
4000:  87446 73770   83176 69457   79162 11107   85378 61289   34653 84157      23019 11047   13564 33684   45049 57161   90145 10999   11735 99207
4001:  20757 80303   44651 64364   80619 68336   63654 56700   28178 42387      68876 43024   93117 46825   90132 17720   82728 40628   41080 05716
4002:  81443 43254   51351 38681   43934 98479   55058 12691   32726 58634      01490 35840   49293 73941   56537 59203   83053 45620   20089 55946
4003:  50308 64721   45552 06630   15137 67398   25126 90525   24696 66859      58655 79363   25221 04435   13349 39643   95467 91142   82584 40559
4004:  76699 97768   06318 23290   14686 55667   25355 14829   62081 90617      16899 65876   22908 56997   25145 03631   39941 48248   71597 35894
4005:  90746 93531   10885 52560   47754 51134   81178 51981   00589 51863      84875 27851   25595 53473   83475 91485   61062 86833   16950 28401
4006:  26855 47872   38298 61312   20310 01797   30636 08815   42248 78724      08149 62322   21233 71052   42856 31458   01451 38967   45791 96360
4007:  64307 83712   08272 91620   33062 09535   76318 10381   48260 68490      25324 85243   23895 20658   31189 73950   00158 50928   23923 88841
4008:  06146 05467   54115 98148   91802 62681   73132 96059   53362 58246      77518 00012   90625 65157   18612 91114   41704 09745   81067 67733
4009:  54083 02167   65441 97598   24735 40148   10592 82458   23828 05961      94396 79954   29117 21549   90238 97154   59594 14000   83622 85776
4010:  63769 49195   88433 33886   06000 57534   00715 46173   86964 50542      54269 60063   89001 18957   46446 87291   49089 83679   95305 83105
4011:  81649 77080   95095 64264   17473 47953   39962 05225   26170 68188      73394 40135   46952 44400   37232 21291   66365 61081   58743 37754
4012:  01192 05360   60046 32613   44437 95739   29416 20850   24992 98946      75997 54740   38373 07883   21350 96691   55679 69793   66918 84919
4013:  87391 35190   94528 83499   75737 58203   45284 73005   04968 84624      24950 57044   40280 56087   43534 27139   68852 65814   09386 12061
4014:  86404 84796   56508 05884   82137 42316   07976 18669   64872 78180      08247 94424   57115 08593   38731 31034   90760 19161   63231 28118
4015:  64660 03165   51207 29553   07810 94775   20375 98737   47383 82147      31224 88930   64419 70508   19966 12985   00801 12513   53395 32567
4016:  41920 17380   16737 05622   79912 76494   26153 75403   19322 85948      69559 94682   30634 05169   98604 57217   02090 08372   40586 51824
4017:  83299 43668   59355 02125   13431 79416   12661 97499   39542 35191      57649 06007   00796 79808   78755 62761   37045 84713   32247 69956
4018:  52824 66639   17420 26162   77275 01636   96909 75436   27078 90845      57061 62576   11581 02352   37351 07570   00327 92713   40216 72365
4019:  94698 35800   15489 28982   71075 82639   34277 09645   57425 83959      84692 28208   16880 36662   99944 52898   83764 69117   55508 37102
4020:  73017 08244   67648 95916   29433 88999   57918 36693   63982 02374      31292 74551   60862 05359   61138 03114   45363 70732   41878 25134
4021:  74463 14901   50720 21600   71705 56828   13203 51415   08852 29789      60207 99556   51367 44027   48955 14214   07323 27690   71291 82644
4022:  79438 96534   05106 14830   24455 43353   67979 54576   26550 56987      63048 72937   99090 49593   59596 44863   91455 72270   62029 35090
4023:  20555 24151   16347 42579   08563 26331   15693 14110   57676 10945      62109 22302   44473 19636   55433 75745   21065 27645   41611 58420
4024:  87159 44970   76836 12437   32267 06045   73311 89249   89175 73312      45528 30599   82868 85161   85141 46873   12660 48995   19398 22418
4025:  43430 85085   79023 20829   41995 10204   40677 83160   75233 59639      20462 92223   97692 90867   98383 18232   83597 76368   43207 78063
4026:  00540 08588   49215 36645   34920 74532   21481 10494   55956 85704      64243 38628   40069 45038   54734 67405   29393 45837   34115 65810
4027:  40524 51080   85376 04037   78730 03400   26906 20094   61740 12465      08362 72478   53734 69772   30992 58427   77635 96689   96069 32749
4028:  43302 70620   29490 24490   87134 77392   80889 17256   50390 90468      19999 39554   12755 32871   20586 48319   91053 48746   82672 19972
4029:  21256 04666   67700 79329   83711 29610   33366 89991   83085 59299      88397 89798   01342 09372   76522 25830   67180 73390   02563 74433
4030:  51021 64367   14260 61495   88688 46079   53462 64996   42613 27480      66133 66470   95315 89882   85201 84126   71748 48676   76974 68134
4031:  02061 15636   93621 39914   68927 97145   31450 80199   82938 92381      59605 88593   16185 19045   48437 92545   80252 47607   93304 05896
4032:  61483 49768   62095 54556   28931 56400   34941 42639   89978 58909      40225 53357   00522 92871   08492 65927   56759 81328   76437 70635
4033:  14281 09307   11669 30155   18591 45926   54677 82257   20491 09487      85412 03627   09685 75407   32353 25536   44935 14279   47633 19404
4034:  86874 58875   31523 19117   98724 62468   82354 76222   02311 13096      43807 92588   21098 89551   14597 52096   77187 23513   82601 96547
4035:  21040 27022   96711 92792   79039 32659   37050 28675   39030 25601      21855 60718   85576 53408   23768 08398   07356 02254   48061 87790
4036:  74836 29126   07468 61130   53885 46650   82906 26082   91563 39107      26802 37906   81653 74678   51135 83107   58719 81992   02589 25147
4037:  74545 71316   41003 10754   29675 64790   87248 62533   41360 27546      18327 58056   69094 15259   00082 77279   57421 40461   44743 69977
4038:  36765 47618   38765 34996   92010 85245   89166 33884   53715 93062      16483 88962   35884 71695   37190 17300   76674 35256   00058 38319
4039:  96751 49633   85603 22917   98556 51419   81911 70169   96425 02531      54372 13381   77152 56831   40626 17102   79661 54318   64328 29453
4040:  25799 73697   67399 96099   31456 73700   63882 85052   24938 61021      36540 63336   24846 01386   95284 87373   75943 33593   96131 52651
4041:  60643 59979   89149 20549   70111 67958   83629 67120   62670 95962      59408 74275   70957 81961   15890 46009   78606 71439   97321 31696
4042:  46954 33820   13965 04802   67877 04950   50490 87720   22945 14622      18623 04849   74860 17233   44244 22870   02133 83695   87364 77806
4043:  22451 62921   12418 61051   73823 90332   39891 17590   43573 31391      42834 98922   19451 01249   87335 86338   92499 29845   91442 65663
4044:  83137 50601   26840 16873   53141 67023   57732 62582   98719 74068      93067 07543   80290 48483   34661 94110   99098 52990   95620 84092
4045:  09304 89156   58872 66692   22557 39399   12069 79664   12619 84469      89728 23244   33193 58880   72662 90828   32323 58929   94348 36610
4046:  04432 64834   00896 24663   37673 90781   98112 73795   93031 19336      41847 32421   10441 34203   21510 31198   77473 94702   92448 01031
4047:  64746 96602   59807 19120   94576 41781   68129 55598   61088 72124      50370 27943   86192 93558   40778 41929   40484 03946   08954 00084
4048:  38050 88990   49980 53865   89945 02119   67849 23330   69822 59810      90456 60737   32512 12431   23109 88474   62999 67021   99484 16859
4049:  02841 24321   38528 12763   13463 46951   84379 71533   87617 73755      13093 05342   87800 18907   54974 11088   11810 17554   68938 73471
```

```
4050:  41524 43174  80481 44357  73457 97669  53499 76181  01815 19952    29485 28250  04695 87419  92433 86347  14392 94124  84475 38016
4051:  46766 58521  15649 35110  89943 19697  02868 90531  09250 09435    81514 94219  49215 78816  83201 98267  66497 91411  45882 88288
4052:  75682 23980  93103 08595  09764 85705  72108 44741  10820 69350    30371 02983  13721 06099  45705 10761  83830 55404  75757 88512
4053:  20000 09942  27513 40588  25072 87972  23249 29215  44770 64925    73319 73396  96369 50320  14460 51497  05012 16584  38953 49297
4054:  35178 89489  74996 39257  95690 93078  08860 67546  51177 79514    53165 87201  33147 88627  45601 59427  72383 06563  81872 83850
4055:  85529 58644  20993 46418  76416 00326  78278 54255  66656 74578    20166 61082  47779 72198  26775 81884  92540 17230  88744 67551
4056:  89257 11978  01927 84254  62299 51509  70251 56125  41760 50367    73634 27783  10142 76445  05890 94303  67400 22226  33268 47301
4057:  99269 32396  39906 29967  17336 45757  54843 28195  38033 78403    51943 79639  25003 16573  08315 97180  42980 09088  74225 68394
4058:  53477 46196  05178 62759  63007 45087  54758 95781  08631 06195    29958 61193  65432 05229  24026 56047  87816 86614  59912 79304
4059:  02189 51639  15936 56715  20338 04958  83599 68748  92101 71907    26061 64016  64023 86931  34606 49858  95742 92151  18595 45407
4060:  83612 32622  34557 42965  69773 93785  61851 91453  08973 83352    69562 30870  56603 75368  34025 90509  07111 29773  17027 57136
4061:  89082 98596  29426 01911  06723 69274  13697 48442  41174 92090    31207 31318  45878 91295  45800 40624  43792 56709  86584 30864
4062:  68157 22822  03424 16910  30454 97190  90110 66873  43548 22778    17279 47018  95321 63877  77716 54686  52210 95234  16974 47203
4063:  44773 62920  06342 63881  67663 82986  80572 17766  22706 45740    01805 25470  32406 87343  38506 50574  06674 47173  54318 30325
4064:  47726 51990  91083 97165  03747 80009  72647 46938  35574 44790    62158 96789  68939 80943  81344 99489  11509 13080  73163 18336
4065:  23445 35860  19284 21676  31746 08206  11581 61715  20166 83238    55171 63726  29181 71194  90054 58534  40311 74940  38984 89892
4066:  12516 69823  65618 69081  82501 98759  82439 33340  53776 21963    47262 99673  61026 06893  83668 21509  61101 79978  63155 50076
4067:  62576 38255  69605 74317  27235 63845  86178 86971  61056 69935    46263 83017  00271 32463  65179 27823  51743 78320  94283 13294
4068:  29032 80891  82088 35423  80801 09049  51967 75858  55259 11269    82637 33513  15522 94845  79093 64446  33485 37779  18397 71639
4069:  04582 33686  20419 33684  78421 80673  41363 10065  90989 77721    30225 80824  42500 71575  80776 70797  33424 72898  49255 03204
4070:  71741 46642  89236 56090  84982 81714  13033 35736  43551 69294    00327 48289  64772 94783  13392 06200  16482 34063  32580 11939
4071:  54557 27225  32266 89891  40120 20223  62945 80248  42519 33438    28490 69674  31244 54959  85702 11923  83428 02656  13183 54703
4072:  85429 24787  60717 91176  37070 87970  32520 16771  85119 55940    36283 53912  29737 60959  08877 56457  77836 81851  67609 60334
4073:  22796 76399  44165 51183  01843 41230  78015 76696  93471 38511    16925 10304  33624 55678  71835 43221  00827 33104  10626 22083
4074:  00940 44326  77923 02685  39546 68865  37910 42723  32850 91774    26015 36327  92754 10162  04179 15728  63487 81118  78484 89014
4075:  13250 82901  98623 47512  84900 69865  74624 78176  49542 11292    20145 26685  25371 38454  78042 84503  42836 08175  69829 19622
4076:  12850 62920  27804 96266  47606 74392  36820 68958  34118 21479    22560 91786  60848 57397  31625 43136  99331 40228  41265 19413
4077:  22807 73934  46187 65437  24096 18798  93745 18684  41272 18652    13641 84496  49916 11503  84444 48936  45853 07812  33073 66848
4078:  25081 89821  60382 22122  59181 45526  41264 26566  63444 83701    92389 18269  29929 44599  84025 35475  90144 20691  37914 11498
4079:  34482 42913  57747 72600  13720 69357  17640 36979  52603 22286    92389 18269  29929 44599  84025 35475  90144 20691  37914 11498
4080:  56550 06676  45574 40492  42671 55357  84050 29030  64079 60033    59042 50698  12028 14388  69449 22696  86761 94765  71326 72823
4081:  73631 08415  13593 02510  92011 84461  97787 09847  67327 90980    47577 78901  05907 28687  66860 27985  54510 21124  93001 78891
4082:  70888 83691  34611 34627  57645 59588  62220 11275  50744 50499    91789 45834  03999 45431  34382 96399  85777 38309  86964 79867
4083:  57903 90554  36816 06800  21622 55404  00381 65503  71537 84155    86474 53216  91096 96216  07320 27718  52746 20177  85598 85340
4084:  76240 89588  57838 68637  08225 90346  30997 02163  05737 75247    81899 85932  95980 48964  27551 54782  08962 10569  47281 33921
4085:  42124 46363  91151 40906  65898 54158  08459 00831  93858 89887    10303 04349  53240 74650  01032 74503  32734 12215  54285 49133
4086:  02327 51110  85131 32461  29042 89442  87594 26944  10860 88757    86343 08003  41822 66017  30835 36467  87932 30007  80085 45044
4087:  82532 63416  11752 47072  27063 37277  70104 63094  33811 11510    09534 62042  36815 68922  47431 27589  82595 76766  94247 68037
4088:  85627 71807  78926 64375  19460 56364  42365 58653  81922 53504    41285 56157  76618 14468  63075 37189  76809 85656  52861 59278
4089:  73302 15339  32440 93930  40636 94554  67486 16100  21234 58938    40152 96103  24732 40029  11446 17230  48663 36839  77814 59263
4090:  75820 80277  59735 52320  84462 11910  07153 47890  57050 89117    68962 37742  29613 74201  40508 35802  32029 41483  81293 90501
4091:  79391 60770  66271 40405  25435 66259  18542 25198  84508 55674    32277 66242  18791 63527  32858 70448  58621 44128  49264 31391
4092:  30899 86784  50805 84507  33237 93519  91663 88184  26880 45693    92924 66244  45824 11217  65978 62247  87541 59736  96464 42697
4093:  63057 91320  29000 16426  02942 96467  49396 81553  21685 32175    20445 80367  01747 25278  85453 88269  21955 49051  83723 59987
4094:  65606 23246  78019 96217  12212 54513  07044 98049  38656 31108    03613 08959  61066 02267  97907 55290  03848 67151  00103 05763
4095:  92590 26002  89907 99751  29024 27425  30982 38186  56175 11483    78885 31189  37625 31875  29275 91572  54784 24456  27743 91872
4096:  16078 43657  14407 88544  64268 75580  48185 53965  55806 32477    29388 41709  15902 99363  42728 82208  66012 32649  32730 80902
4097:  18642 20034  89389 57130  81016 53206  08656 88064  12124 97660    60502 72721  20354 46363  17217 81652  58106 70483  92159 14412
4098:  06952 95889  86549 03369  26795 50387  08638 21369  06762 17068    94078 99617  96899 74473  98639 63590  79585 13997  97381 52525
4099:  82748 66147  61307 63364  34633 90777  85896 68288  07057 26579    89804 90741  88558 27616  60411 21669  84035 78633  05404 67038
```

```
4100: 8147275256  81297 54588  36305 75697  32046 99173  03579 38342    65501 35040  50588 38488  52832 58898  80803 83625  33877 86637
4101: 98077 94066  38196 89604  52013 73061  24120 87617  99826 36219    74177 59741  57267 47648  44668 69911  15855 43177  61078 75871
4102: 09135 16565  00656 82151  87029 79588  19157 19250  20807 64541    91307 50042  91285 74791  70576 15886  79411 52247  07402 90140
4103: 27778 39059  63996 81142  52941 52640  08563 76008  28907 44357    76032 77805  02021 44182  53675 51562  20331 83758  07502 03384
4104: 83026 92917  73604 05505  93559 59980  20688 08793  58914 35270    82774 13256  94589 16368  30327 63263  02951 27809  45800 88016
4105: 68186 91622  21110 75490  53813 35093  92375 30401  99120 25190    76937 89883  15754 33908  27234 75989  75571 80284  50056 28958
4106: 22571 55433  50549 70811  65901 93458  34542 34597  20038 38870    31469 88773  36511 42464  01310 42378  34161 27086  90703 87177
4107: 85483 70823  78473 03968  34574 23117  83794 46314  84062 93878    13881 53073  86842 06500  72620 98940  43242 83973  76970 19880
4108: 91158 71672  64770 66788  39992 21865  38807 89585  32002 33445    52803 89192  93548 13687  96436 82181  37038 02387  30148 68950
4109: 80900 35001  30353 28418  21973 47984  43372 32925  27668 29700    58033 21092  47530 40954  75035 29159  73784 60416  91379 59706
4110: 35324 90062  55069 09965  61088 64511  35676 75189  11609 69882    94086 27409  92087 36741  55199 76213  81824 43467  00789 21581
4111: 55038 05866  96819 05913  22815 33735  73597 92870  96883 49019    14285 08444  49726 83479  35088 13593  12429 78312  92066 77132
4112: 79928 50992  83451 96283  05442 34273  95911 84131  09617 05965    98716 40620  63386 35601  05678 91967  89909 37203  93306 56126
4113: 53836 79288  00297 57623  45969 88541  86314 20840  35810 22571    98925 81178  92668 93338  93075 77113  24816 84682  18458 08098
4114: 42129 64863  59992 19077  04185 75255  79215 22326  22040 72329    93377 83742  68156 83450  89279 12510  68379 42049  08549 62278
4115: 72315 55908  45011 40899  86872 93882  86093 49408  05363 84434    26579 09841  75891 06223  59076 08610  79995 44240  97741 95919
4116: 73899 12185  10902 95827  98227 44645  15027 99747  73296 10121    46255 83898  52059 02392  38281 82422  68754 99933  48759 14473
4117: 57036 15424  56568 36651  11324 17409  45085 84987  18445 49329    24594 40377  23967 28182  62485 40969  23469 21435  51389 63380
4118: 93259 69930  02773 71487  65994 82898  66318 08565  46203 20099    38880 15606  67272 59038  48670 38489  70199 67564  82906 03181
4119: 18004 14459  30111 73284  60946 61449  21592 96369  66056 92684    30529 44763  40887 90996  31966 10644  69996 35128  27750 08851
4120: 15302 99255  21306 11031  83023 88043  58306 44173  70154 17163    24422 23072  63758 58188  17387 20779  55044 97148  99195 54064
4121: 53292 68802  62537 79715  34515 90680  09435 14066  34855 51981    30207 84600  82907 49810  29465 66279  27547 52612  00579 81005
4122: 20035 90598  28381 33556  97716 62201  39983 90244  13342 40140    94515 58072  31798 27846  85343 97429  85524 56961  16179 37629
4123: 94223 40989  70894 17861  79259 45546  42215 68551  21751 54380    08409 06033  52225 63845  08298 68917  10469 98704  88442 10627
4124: 89144 91193  12494 43669  83920 15981  42464 54882  17636 81659    09734 92342  90013 70260  29443 18086  52794 91780  47260 20900
4125: 49385 26486  15181 52063  93481 70375  62320 93460  32076 40698    89263 75805  64594 72281  08841 11783  45467 56623  89124 36060
4126: 91998 92683  99173 15475  54378 51369  24873 62426  33189 52780    23508 29453  59242 11173  59729 99898  26821 30903  60824 74225
4127: 74741 69427  91553 13862  12431 25067  39617 20982  31596 34519    94125 49613  31937 93753  88219 21223  51935 37548  73947 28416
4128: 14658 53272  16661 82531  01299 35931  97438 50948  97659 60166    37176 04292  38910 85432  21867 67847  72196 37805  29341 03499
4129: 41478 41391  20134 03260  21579 41061  94212 83096  81419 40052    34768 47140  84665 34242  88652 94709  91042 60179  22354 40443
4130: 00203 11312  80068 32505  12697 06861  58453 04885  95736 82692    38302 35002  84417 99102  44098 63764  37088 06886  85924 20004
4131: 41926 81493  47781 38271  22915 56113  80155 41308  88308 85572    87047 23379  95250 10813  70357 63588  44331 27413  31422 34873
4132: 64834 50842  04722 55688  98214 99844  50272 71303  53642 54417    72734 22369  91214 68711  81297 62327  44040 54319  78683 62844
4133: 59680 08496  98354 32994  63652 05957  38973 87101  10765 00198    87500 56594  83440 08991  34877 06199  53410 64550  60667 18631
4134: 37512 39730  96633 21156  08373 70488  77968 46518  27495 30430    60523 46300  46073 86306  73564 26396  27250 49612  80936 85240
4135: 24953 26012  60677 89134  26687 15926  30863 66058  54130 07986    90150 01882  04427 36319  05018 87242  54034 08676  38044 81464
4136: 52707 52906  52414 33308  30833 21860  19705 93985  97319 56440    94298 53535  36055 00698  43855 89639  40712 82487  91189 65312
4137: 64365 60046  92782 85328  64177 11162  93540 71406  39990 19823    88860 49123  32123 50785  20989 65173  68524 23246  02878 21490
4138: 25566 71128  44928 11519  20929 89054  20768 23134  17615 70774    74463 16316  41279 13651  30772 36093  83441 07986  38180 18540
4139: 47818 46657  49949 88580  46155 39201  48142 57288  02921 39640    06654 86918  89118 54146  21020 02188  71511 51056  65125 68253
4140: 08210 81354  03499 83808  18950 53228  38331 89958  78000 94361    42800 95306  78567 14624  05064 99426  52361 07615  05688 75314
4141: 20343 55599  76392 90287  53329 92043  82813 70354  89130 56495    26352 34056  20166 03315  27743 19487  14315 54197  01236 84310
4142: 02267 19375  29274 35271  34182 88165  82796 69661  33335 98921    58437 27680  73422 89081  63555 41056  76432 41867  19855 47748
4143: 84861 76616  24905 45857  63530 80385  85062 95507  13047 09573    86410 97015  74259 66488  05125 65748  75687 61097  05502 69543
4144: 89555 28981  55042 08088  36369 69791  50842 03951  05448 30928    50360 83906  83745 71922  81051 84221  41423 51486  27250 03136
4145: 14346 02845  18818 64023  46711 08239  18799 63140  14352 49527    24020 67950  91089 98416  01267 71247  50511 71955  17333 40335
4146: 52236 46609  54060 70169  44585 76354  32772 16137  58178 62436    19933 96174  72706 87651  12919 31681  38390 25913  13686 59532
4147: 78722 21283  26372 12237  98835 87440  54477 21769  64794 41537    74357 72717  98239 71900  78501 17844  50838 92004  90197 39623
4148: 36588 99765  67597 72512  30145 26144  59175 60465  81605 83116    17622 97218  53044 18259  60572 95280  08088 14753  05201 84227
4149: 90433 96835  75208 66070  12560 02391  13354 09701  83438 29999    37425 95820  00554 99178  64122 61494  22792 34221  92302 60271
```

```
4150:  17687 26260  49727 88737  53032 55542  24688 05578  37946 69212    47429 47395  64912 90068  18337 86879  89195 66708  14172 82925
4151:  81678 57994  81599 74964  32516 70925  14811 68080  52300 43438    01333 52685  96496 08659  13337 38786  50152 52578  72690 19898
4152:  29095 90124  02895 66098  25502 13905  81272 32323  39354 67662    41095 27754  70647 95178  85545 55040  80941 09068  18356 92696
4153:  61758 88510  87064 19010  32952 16247  33991 85709  80370 14055    59368 63003  03075 10594  42032 99907  58230 25482  87227 76340
4154:  26914 28756  97351 59571  04977 47894  47694 64674  02775 88450    62429 38981  45249 18795  46971 64337  02206 53640  88953 49590
4155:  58289 37551  40526 09702  50351 38883  19058 67630  79543 80592    74281 76179  31849 84355  18701 03571  57652 10956  55939 02635
4156:  51424 17211  19492 10715  34643 50433  18111 12492  83530 03779    60239 22133  53450 85143  78024 31385  29896 24994  30111 93997
4157:  15919 12745  26994 53086  09704 21849  35248 49564  42042 12278    96351 02939  37226 63523  81405 51911  25336 79454  85335 48182
4158:  91926 04297  30065 43002  45842 18004  30442 63037  85124 00632    41086 55220  50647 78422  30238 24833  36608 61294  34468 88216
4159:  01629 34709  55922 78623  71330 60951  42619 37842  12377 64700    54618 73891  85329 25189  46554 58554  50567 30124  33358 77769
4160:  42827 73448  01812 09123  46983 66814  79316 53250  68211 74308    70456 57121  58117 59010  10102 26457  39344 55849  96635 22854
4161:  60522 05084  10020 82847  75706 63778  38950 40805  56692 56171    44428 30942  96275 43089  09667 86584  42367 29077  41486 22876
4162:  71421 32106  45302 78661  91500 40792  65171 26374  67294 24208    23709 52174  99170 69919  38978 49799  01254 32685  29280 12509
4163:  97206 04740  70228 09435  97738 07433  58680 46191  36637 42224    84946 37309  67093 24076  43985 41715  58010 39061  21765 09516
4164:  93754 67569  97374 23253  39430 62764  26250 64435  39454 04273    55984 17977  32947 57459  78269 63189  56224 79624  94478 32107
4165:  91175 78367  11318 67167  08183 07936  73178 19225  88459 43215    57157 53120  54676 09245  00427 91975  99287 39154  42019 14019
4166:  77440 94696  91066 36159  72877 07444  20963 03884  58518 65401    84962 27560  34133 17497  16452 48098  19149 30785  98046 14745
4167:  99690 28398  96811 63546  41648 12374  26341 96358  73324 52323    52872 02358  82859 19234  66451 44526  39224 68256  75698 80847
4168:  69841 00005  18683 70265  06680 28933  37112 00737  53362 61940    66037 09954  07367 30154  48162 65558  95276 82899  99451 76840
4169:  73515 65702  68036 95738  38912 07350  45652 95053  87942 59287    40720 68701  68156 01838  41033 57193  38021 86328  28591 99881
4170:  16362 36492  32118 35758  36976 28574  15533 42226  34298 61025    80464 83520  88189 91843  05289 20586  75622 50183  60038 41985
4171:  81563 81771  68983 63795  66791 54285  29461 81333  44344 43319    15271 13381  12546 20973  72100 71466  89276 49452  93988 73445
4172:  51162 80170  81901 94680  30928 59383  13079 61095  72952 35102    29529 24678  05981 37096  74676 26661  97001 47926  27524 53686
4173:  63469 95675  15896 87773  59009 24304  11297 92380  71587 42407    80836 18205  38252 96225  25086 24233  55257 60358  31651 90338
4174:  49930 51878  41706 73692  83554 69980  62377 09575  73670 56796    72593 68967  80509 48050  90687 59845  15858 80379  68334 46387
4175:  98538 55560  35915 36183  54616 98710  85949 42775  10916 86000    93528 59789  69447 47916  43416 98529  73588 54847  71777 36232
4176:  93749 75921  82926 00308  91351 97188  73435 15726  39695 24268    29548 14989  28230 37756  67827 49004  68210 34755  45118 22405
4177:  04371 62486  85295 97243  58652 59168  63593 61213  13749 04740    00732 13505  57032 49727  12302 76292  55254 86110  62307 69354
4178:  77995 80509  00758 68831  56088 63603  36216 39996  14662 26056    35639 08139  52822 82660  81870 76876  77889 70756  32913 67608
4179:  58775 78770  71313 98948  91913 79662  80447 51400  10128 00879    53399 05423  96490 37237  87908 91639  20829 36301  42114 94695
4180:  74959 14402  85131 16722  29325 16202  58900 88260  22343 26546    35314 56283  73854 01708  65840 40309  77748 78510  16061 76025
4181:  57625 87239  26902 38031  90994 01961  95584 47504  15673 14659    94685 84996  43667 13453  31158 01832  55095 10293  68115 49393
4182:  21805 11477  06273 79492  79232 09137  49513 33342  08972 43163    34742 50208  99452 18890  90900 89178  55018 74110  18017 80498
4183:  05145 64343  77471 21433  83267 42117  44228 42221  58049 52073    49323 14552  99889 71067  96893 32954  80731 68787  55385 96370
4184:  51836 59454  08025 39807  55527 85217  20045 74870  61060 25445    18933 59158  43278 67044  23849 50257  19174 10529  88861 72662
4185:  45615 51612  24397 83961  62360 35552  35358 16110  72209 09206    78031 15722  43940 51607  25351 05079  40794 62500  24223 60527
4186:  41199 98865  59502 96994  31083 27486  16983 37574  24409 85778    79992 32803  23111 47339  31710 81227  01219 83815  07331 88740
4187:  48840 41325  50409 16479  23200 82188  81420 24189  92665 84088    23904 59876  30041 74209  73715 13118  47526 03988  66588 29844
4188:  48783 43247  89985 91260  58347 79156  45785 01849  93763 82071    59958 42411  79392 86561  36456 65186  52542 33065  68533 69786
4189:  75226 78368  57921 47588  26163 70321  12639 72040  65231 39060    43229 27758  33833 56345  22793 08981  19169 85380  03987 94727
4190:  29801 12017  38873 01103  53115 26126  22376 00791  88617 10733    84159 76723  68610 03464  77979 74196  70379 14745  73494 10015
4191:  83039 52690  72321 55686  32567 67063  90025 98735  31848 33589    85893 77886  11902 34609  01453 30224  79884 34771  01051 92833
4192:  05330 49579  66361 39953  46739 17660  56302 42072  98082 56089    60288 78738  33269 62907  65196 66848  47042 72254  34912 67215
4193:  61750 46784  19827 63260  61899 57166  04540 44156  17422 38418    24437 67632  04534 17899  31382 48961  71291 26828  66389 14103
4194:  62106 23075  18380 97642  09954 63804  33685 52972  88309 18645    22102 93456  47276 55414  16460 03588  97092 26438  07078 45555
4195:  27899 86833  10868 20118  51419 56318  16076 37800  15951 69337    94590 17434  77361 62668  84789 56877  71221 92708  00226 91154
4196:  00786 60597  45312 89310  17783 20505  31601 34505  35354 39286    69002 40849  75277 54525  72235 33513  16161 13864  24157 77173
4197:  85182 89000  90939 53750  77672 01862  42932 88968  50340 78857    25903 36913  58966 81356  04381 40696  12402 53590  13502 68733
4198:  27739 35936  16149 65592  41892 37449  99988 99094  94879 63227    26354 64726  27589 56235  05146 34215  54092 98358  95633 99614
4199:  53650 86552  95746 77266  69700 93586  21088 15047  97006 88221    92918 56938  80161 32139  54784 17425  97780 27200  91261 04062
```

```
4200: 2441389930  1367134593  2296974420  2307425616  8985561454   5584759458  4782526946  8298974292  6823381400  6023600795
4201: 1356713993  8445834544  0681895279  1430774127  1770551433   9823188689  0214727844  4457590352  7829061469  2115013987
4202: 8820835104  4716830176  7121032995  1337897042  7217718496   8774173940  8199432201  1277680554  5718535329  1226646067
4203: 3318857548  9059806627  9684492731  9733539953  4621909773   5288012093  9181231722  2536691211  5539198907  0112080611
4204: 6846780441  8099774846  0920444199  1548992641  2080059990   0644974983  2040174104  4793348541  0856923662  6408716875
4205: 0118974534  8856504750  9953237467  2365735685  6767701622   5085397045  5601797170  7951587486  2724627728  0504265500
4206: 2136964345  1487739938  0176791868  8621056413  4272676051   9723853870  8931433693  6024820097  8028275658  1502216100
4207: 9282462093  4336337735  5545570408  4688658832  6286468261   7096960155  8528789070  3321332653  1253622873  7265059911
4208: 7544115532  7640261332  7775814258  5425698004  0393615813   0970121672  9933077348  2423405576  3209754010  9596269199
4209: 3410975976  8365962012  9434639657  3047633851  0218619280   3258339407  9493567947  1741366566  1778066839  2841102512
4210: 9888744780  5490121399  5415790764  3981139061  4483000135   2413464739  3267751256  0631348726  2763649533  8014413159
4211: 1863356514  4226874855  5298214248  1047661065  2651193523   1584302242  2767586790  3537523974  7085784924  7557691296
4212: 6968764198  5234904004  3737314082  8408966583  7628002535   5065404819  1636907773  6878857199  7208867017  9151908299
4213: 2206871993  8181529479  5716109364  3375615175  9116061275   9129335760  6741536011  7368308247  5963126478  5217739415
4214: 1532329654  2398368321  4707269292  4968241520  0362083421   7302097552  5276092517  2334982948  9133356060  3560031985
4215: 0765566590  6709191068  1902934859  9129455359  4895058586   2594764024  6713474745  0801791770  3712630260  2158472594
4216: 8006236748  2415894114  9584335562  1445193600  7957784489   4377318285  2874523913  0623567248  3916823398  4493381782
4217: 6154817919  3408694519  8038642651  7125788893  3459371495   5790314559  8946819981  6152716007  6041084649  8566028916
4218: 6763270334  3729234835  9513780805  1795573664  7750673625   2980626409  8199451249  3907918435  0160612754  8475672556
4219: 8610090361  5499224680  2584323027  3730566231  6675926679   4206918293  7863501961  6391361283  7383173835  0755302717
4220: 5057041741  1931572339  3392345764  5102153679  8332138922   0510652458  9828817483  9331122191  1647515345  1933584993
4221: 7404013167  7251758338  2223174849  5890722525  3826559064   4548491582  9825102191  4515447335  0280798481  2097491078
4222: 7913675023  3759840055  2810748608  7901693365  6471047429   3705917149  9700148767  9969623858  8474178677  5128651720
4223: 1620042438  2154218245  7862191520  3098388156  5920284766   1794547948  0183765489  5206209553  9477333864  9907787922
4224: 0931556524  0604112446  9538210935  7002970209  2819980545   9793901339  8341313078  3815853926  7604330798  8783537290
4225: 3530101217  2800706706  7347728838  0183396025  8793323158   8292298830  9379154735  4904804307  2368695210  1854326260
4226: 1820455829  6934381875  2454437626  5354937514  4294588163   0996504388  9606746371  0447382474  8406184272  5456491304
4227: 1218505143  5326699217  7325817809  0565452263  5428981107   1403117263  8664150625  7055282355  9054722763  9570235501
4228: 2881120827  6005346575  1859120500  1555643655  4385924629   9794906140  8244480961  9539925164  3522257944  9665768629
4229: 8084573652  4258418710  6620146748  8334954695  6207897025   6938327425  7833924713  2812700534  1601884321  5928445500
4230: 5956011470  9200428852  9678375164  8182415880  1845205486   8763973382  9420269341  0101967176  7496884992  9561128876
4231: 3599037555  2221656168  7055140377  7193196699  0089616466   6629456707  3826810350  6054436154  2539084185  0287130501
4232: 2341829756  0633871232  9578091048  4519716115  9453740323   7531306031  2021520949  1845248362  7709747313  7695870387
4233: 9718854832  4472588492  4673232419  9229289257  2660055677   7909972054  0621065249  3682728250  7013315237  4015431916
4234: 1783831826  4414560476  7169455535  0968269599  5068979036   5369719894  1214474989  0199894515  9947734106  1922239331
4235: 6797875007  6807821104  0508253866  1735274141  7395527468   9858307742  3568369625  9229682670  3055666003  9168410441
4236: 6693002447  2903217154  2514772758  2634690346  9067451847   6674814067  9676314003  5123216082  3234477840  6936945802
4237: 6828994018  3814517572  1106096009  6902553909  9446501113   5640248633  2481292955  7404869831  1850305870  0155605810
4238: 1309039798  3538702270  2472167013  7507156062  4617396500   6143543333  7453227881  5895176368  4805818452  2455399263
4239: 5256977293  5764983229  1820188492  5044437408  7012759682   0044887487  3977807922  3889842765  2082772365  0474055876
4240: 6526050884  5466862697  3779314888  8369436970  6511643161   1777311247  9622146114  0260625601  7060713635  5038754059
4241: 4542334127  5557582053  0077518773  0757284798  6064585697   1340139835  9314179963  0950931364  4176483444  6714731343
4242: 4670879094  8500033142  8430803139  0813799711  5366579309   3960633649  7878514395  3565388736  7265410369  8418616911
4243: 2690425293  8584012730  6936323401  4107169763  6181577912   2928006912  7503575955  9786944945  5287408424  5818676688
4244: 0344884348  3364652692  8143376311  2874622242  1417574958   8680882265  1970224931  4064030172  3066229127  3352842134
4245: 6340614207  0897407915  5115679363  1222356385  7674160240   3240284286  6708508702  9299379563  6346233238  4808373175
4246: 5317955154  0649565334  9827736154  2763628734  6444735584   7042426461  8389881721  8417672974  2235591746  9067787290
4247: 6246538290  7155020533  6803092482  7372781338  9216538684   0682732578  1258235735  1065610832  3077866414  8536990267
4248: 7552976484  0299883341  0771391209  2237535712  5577477947   0073971205  5307567652  9188670676  3094767850  8627633316
4249: 6753044080  2714279976  9865881335  0995098804  6524560642   6736984653  3087591057  9214180269  6588662105  8757014214
```

```
4250: 2975474691  8250978013  9157988607  2088935550  4907943797    3830720457  5226581823  5760598457  0767616010  8680249659
4251: 4348353356  3210554391  6521179420  5290201063  3822972339    4451377263  9277461880  9152196085  2653860845  8280774513
4252: 9425525583  0441181917  3902846443  9987085103  8037731254    1641203403  5059990995  9367797065  8652389099  7274963262
4253: 4654336043  3302461681  5505366035  7032926045  5196719277    0671156580  8091581017  2596603829  8405776873  4021854660
4254: 2224421430  7718518469  8745807349  4061239993  9202364971    8094840893  2036367695  4972892775  9530592973  2861543351
4255: 7065647729  8044336852  4420764448  4672491029  6033260703    2567148771  2317380524  5538012094  8164652284  9729858451
4256: 1271482115  1693955006  8716049901  6114541011  0590045158    8194165770  5430005980  9010945826  9230472521  6455527196
4257: 4057637964  7714870047  5825038567  4932294604  8136507228    8285188718  8107497539  9752900598  5271824846  9637273135
4258: 0412594096  1465907175  7826019370  9126296938  8292013947    6629200902  6672531715  4490728409  8388084458  1071199782
4259: 8358601720  8130575068  0618282950  7830978771  6694875629    4780628052  1844565973  8290696027  6638906670  4405359484
4260: 5932306523  1291818752  3457277370  5434494918  0569794724    0092798800  4410671888  8643683112  7012923802  6451470430
4261: 3610536963  2736023148  8882084735  8571606198  2565953712    5099597076  4174704284  2919932534  6943213900  3327467551
4262: 9383615525  0455948905  7903302768  0240444688  2114055566    1659654383  7098383726  5622923469  9716091782  0576078850
4263: 2123814967  4086298656  1185655830  3897683408  8283881105    9396209730  8618229889  4056276476  5887661782  4939728190
4264: 7761148415  6998993343  0134077774  8229771232  9847082501    8194513786  5788047523  3329775153  7211189388  3760998646
4265: 3568962748  0238407057  1484975537  5026492402  2082530363    8795523812  5215033996  5404599613  3413020608  0278947646
4266: 7754611362  2881026852  6753634854  2066574371  7712145501    1907941971  5543759040  1948353652  3548561001  2231323009
4267: 6667846767  4310276325  9834838535  2634453997  4609095840    9850701762  7233449141  6952476105  9446183641  6172855258
4268: 8242194715  9926513799  9336300527  0420337790  5052769053    0086874207  0658665092  1801337035  5680654824  1904990435
4269: 2352983297  8574441528  0219285754  3988274991  3256274411    6351719182  2916335955  4273620342  5849812248  4388621686
4270: 3874127699  6714401377  8831272578  7699379501  8977643084    7351471261  7649223498  6612568886  5364472351  5025919699
4271: 8131937449  0574905227  8708938931  0529933543  2963834597    2557869864  3020735916  1484283060  9763710884  1483110099
4272: 9319341374  4906269445  8832421807  9981259084  7166330269    9361718123  2263559829  5890347968  6475463387  4167730821
4273: 5744960541  7063115740  2460384611  2173891214  8843332531    6882784341  1558942092  7536217289  1724500817  9853777280
4274: 4162889449  7138018219  2348099684  9248106173  0439945038    5012163505  0466408647  3806993192  2210551311  8801862563
4275: 5588932536  8825418098  3139846115  9963539369  2134765967    7380171695  4401402179  6312713414  8060140050  7774756083
4276: 7539444406  8267725186  7236525596  0057639931  0926082666    2778404301  6971851798  8443408046  3339440326  0456530096
4277: 6307741496  0821640398  7189243798  4279860679  5475372805    0655262847  0846435326  8457141981  0356322452  1011215524
4278: 3319507008  8783658294  3720637431  9474684655  2271523320    6866046093  5838503700  7973886617  7709219587  3436367988
4279: 5042088752  5408508332  2495948515  4034304933  0002102909    2065944174  8821299409  6208028330  9328601869  8310381371
4280: 0756469398  3210867223  4840106587  8552942308  6201203677    6357843314  4285510812  1237771021  7101401438  7183907281
4281: 2961107291  4387504699  0133052291  0542390345  3869035757    1048982672  4596002057  0303100267  8066649960  9837408954
4282: 7499508311  3296831992  8489176624  6987944289  2401155677    8368872171  7018988454  3845806923  8398401312  3918593469
4283: 3294772627  7430155570  3800427756  4719166148  7038200740    6955830681  5765735516  1169138667  1066758416  7883411742
4284: 3359979088  8248497214  6640513303  6625921317  4982200586    7431631471  5865121013  4556423979  4606513957  8111049092
4285: 1008063118  3906634597  2747542008  4699169242  2304008042    4438675426  9607264593  6458453904  6219041656  0119783963
4286: 9224985134  7693821415  7178944709  0469083162  8592160274    5023696815  1782029230  0567076411  8399197925  8429671242
4287: 3532983749  3556741471  0034963395  2573340457  3455444024    6076150362  0217630705  8395175865  7878314449  5626930835
4288: 5879946263  2257338046  8992478575  2869582621  2726181824    9224456785  0225598365  0648509486  7617632871  9864379879
4289: 4647193093  3480072334  2386064617  5608592994  7936002313    1110925556  1581797293  9669767049  3109470083  8749774741
4290: 0969736372  6576004156  3721239355  8951128367  5764271606    5010739551  5350430509  7828880231  7361625383  7455543852
4291: 4470496739  1591425153  1583829334  9159552677  7868012829    1425578161  7717370939  7824024570  2412332324  3272369184
4292: 4726528671  8924596813  5762850012  2793730862  2735193675    1536339670  9274275286  4725169631  1986058667  4802959166
4293: 1191354638  0003348785  7133291227  9588698015  1619350489    1890573171  3264221539  2946710863  8078752062  4039653591
4294: 0851496391  7755839336  9076689504  0156869917  7722651498    7143653499  4214997093  0878706000  5370084539  7042120194
4295: 5159406694  1559445849  6475901872  0994206409  4324098095    4207513649  7354312740  2722777611  2917963871  7171186999
4296: 4927356347  1900733831  1073583626  2412308543  7362398929    6250171897  5420346564  6119606492  5052763429  8448393918
4297: 5662119068  5566336680  8094423909  9557938012  2668598381    9548283529  1597446476  0404612823  2516390139  0240370445
4298: 6480800667  6763228143  3529833637  3708869078  4353296142    7553977962  6098854106  5769505852  1375593907  8435986493
4299: 5683503571  1144032323  2870325361  3942732304  5284184780    8249636595  6746617093  3910542611  9839334728  4958952915
```

```
4300:  63541 81957  79201 17328  33834 90308  91286 18909  10162 74199    17109 58025  80633 45240  38719 65291  68903 58950  07588 89009
4301:  20220 11764  07583 48176  37327 36170  88231 43601  87522 03425    17549 24935  29841 72376  41601 53077  32236 27332  13846 36066
4302:  71573 11875  49910 90494  60212 24975  49165 39478  58188 55156    22867 54867  26793 14382  96816 52999  41584 25665  19532 75258
4303:  19024 20426  81292 08187  60870 02454  28472 80482  32322 58392    76716 08558  47585 98472  18757 45543  52218 05700  70032 24036
4304:  09859 89958  45521 72043  13632 78237  23221 61506  44439 23633    73432 41098  51366 50369  04557 78360  91855 14824  58943 96322
4305:  23750 24164  20050 93798  14781 09408  27644 51358  88400 56869    35337 29188  92783 42545  61988 87831  32014 98031  84504 32714
4306:  90646 76845  61266 96288  51279 13831  73882 18696  43592 34771    02457 22369  02290 44161  68849 21121  65426 89683  31826 23185
4307:  68020 83855  25600 08767  36934 06579  32640 95244  08103 56888    02444 48515  46242 54480  53389 77656  68480 83313  57846 31925
4308:  83796 98016  92944 77948  14757 52250  97935 34031  20350 94569    73073 31356  00272 19918  44402 31509  03756 41833  15010 76511
4309:  53905 07280  58082 65586  13687 15862  85221 21186  87580 17840    36324 89831  00698 90520  85998 62739  06887 04061  20553 88760
4310:  30079 31062  53241 49510  72881 25781  93454 35416  11509 43138    41175 48378  97295 38418  04275 55009  65822 80691  39505 95398
4311:  83677 93924  01662 26574  81889 67758  88925 52019  53206 82526    76247 51600  58286 11750  69068 10612  35945 72973  30524 62112
4312:  93334 60946  22119 01991  37787 38487  49249 26832  04012 86008    19539 15925  31937 86813  33633 66523  32050 55222  66014 44806
4313:  03658 30167  34416 39414  53296 87234  47160 13487  63981 20357    47602 08811  13625 65416  52953 10685  71438 60779  93084 87174
4314:  00423 67106  92500 71470  34744 53683  32955 54930  73748 31505    86087 07094  62463 12629  84760 35459  14095 11985  54144 86869
4315:  55924 04658  50309 83312  35329 53872  90548 58516  48618 37409    27184 93885  18153 97290  29889 60600  94634 47144  71253 26134
4316:  23625 00824  24334 15457  53595 26975  82960 23753  56515 70574    56010 19135  35652 43125  31632 97478  19513 53920  55244 26460
4317:  63175 58991  51262 26954  44859 66419  48959 21614  26752 84844    57609 07608  52522 31654  94519 36893  15958 83144  89700 19801
4318:  92719 27498  14748 10874  98891 06222  14333 78383  87095 76175    26629 80761  90377 04495  37523 45558  33987 25080  17458 29387
4319:  24078 07100  68749 43133  14738 52741  81458 03771  07240 01460    99671 90248  47217 82874  45739 87086  62247 74714  99350 05781
4320:  50689 16661  29402 50940  90679 65921  54772 25717  20236 17441    26495 67480  34952 75032  68637 95713  86186 79468  46342 04156
4321:  73441 06288  26659 86874  82731 67132  43486 27153  50125 66806    05768 70124  78527 22917  94953 56304  69374 59944  61730 89527
4322:  96545 34582  90060 72673  72461 07671  88686 72577  11487 99966    27512 57730  91916 44812  87487 43088  92108 92405  17066 92088
4323:  16550 67432  90962 31177  18466 19630  36309 37486  28333 67011    56249 12357  03089 34362  21514 13623  67038 02008  30024 26344
4324:  52919 36422  85371 44102  17611 45734  65188 95279  35408 83163    47751 96712  44865 04425  71969 00970  36422 15399  06162 98092
4325:  20381 00501  61703 56439  85358 80533  31239 62844  30989 88784    42278 45591  00032 08693  22620 19907  89289 35314  69955 37819
4326:  94267 64663  66175 89867  37975 98752  58786 46879  26399 99448    12531 65566  58147 94082  59760 63470  45199 69626  70998 13241
4327:  90657 04753  47245 65252  85831 76952  35889 71986  42395 04434    58480 40050  58568 38503  98294 83068  26125 33442  02060 97934
4328:  04311 89378  26789 31134  99393 56492  56870 01279  56495 63003    23869 30545  00418 91969  78214 12150  75900 97403  30909 95330
4329:  79348 65851  60358 32840  58956 79015  38545 21751  99183 19940    40954 24079  08726 81153  95425 95473  93478 97188  15079 50662
4330:  14624 60655  13778 57954  09328 70655  46712 49445  95060 16011    26820 33617  93937 22323  88794 29322  55951 79314  65535 08313
4331:  40080 38954  49527 28654  54437 99067  85823 69581  08748 83876    01525 89322  88901 70605  19085 60990  80607 71960  40887 57921
4332:  11917 05812  57391 60961  50417 73148  29959 12998  05431 06178    31555 19513  27054 23261  08059 90077  11279 36626  19448 48814
4333:  14763 43160  87646 98753  13133 95985  92465 22083  76107 42393    77228 65803  47705 37474  81630 98808  10506 16032  99558 21833
4334:  07329 83330  55480 96052  93401 86397  75025 59705  74275 17786    96876 01933  28926 08782  60144 51081  00322 71573  19018 39609
4335:  79134 46985  87390 63095  95271 34969  18294 39438  79882 62693    38252 96769  95529 35994  93952 77657  13497 40941  34171 19002
4336:  15356 52918  53868 97059  38583 91807  34341 12207  05769 69205    43943 04774  61721 82501  43582 24029  77100 54671  07528 39844
4337:  19457 78013  34441 72992  74605 42340  55343 05165  07523 05172    94127 23070  04602 14394  13322 79060  21907 09980  03931 80076
4338:  49634 29738  32696 31847  98228 05524  31170 23763  22599 94139    27244 77854  36447 33899  84927 51711  01774 74621  91529 50799
4339:  24167 64685  88952 18142  38750 84260  29838 97030  43359 52175    62171 80422  10157 70861  69178 99224  62522 64140  94098 06600
4340:  40432 30755  64266 27168  97823 29754  64472 81859  33374 32309    89081 62073  88086 12873  51111 48700  70999 19636  19570 67308
4341:  16083 31573  72436 34654  19557 34050  27789 15042  09178 12779    28036 31186  87527 06814  35761 44920  77400 08031  14178 07847
4342:  34221 48875  94691 59841  84269 99507  90760 62163  02290 39186    07627 69338  66234 76282  85165 57443  99703 08120  59205 39817
4343:  06535 86888  49179 87528  21122 41742  65511 26870  50856 42369    68637 33487  53133 48722  63609 21806  54057 88577  31731 59859
4344:  94156 98154  06440 62675  25201 06202  63250 88001  74323 47032    71854 98115  51679 58976  39823 05496  13606 02792  56963 26411
4345:  22313 46532  59576 73034  61450 75960  15207 33874  56247 21745    85629 62194  74544 67518  41477 66801  98293 27285  56627 74639
4346:  76475 25927  58433 39022  86596 33349  98691 33818  78169 99813    48922 59580  97766 82784  10594 72921  64432 87249  97239 67715
4347:  00956 41060  73195 68554  33457 67339  11231 27007  33909 75136    27535 16797  05838 37426  81219 84857  60529 98491  56367 08556
4348:  25201 51371  33411 18269  87562 24980  45489 66551  63541 86004    11266 90048  00232 20454  12529 84547  54178 49900  57596 85230
4349:  84430 82802  27586 36774  68281 62865  16073 02022  25120 21811    80603 79267  29199 28645  81158 46362  06400 01020  66348 15663
```

```
4350:  07718 60275  64548 41951  19317 36330  70195 65673  36777 52438    52186 37714  59675 58597  93622 99021  11249 37364  72564 04298
4351:  11137 86730  20944 35912  99041 53577  89600 26463  37586 61429    67273 56015  17615 36886  05532 77754  31161 26184  89306 47068
4352:  82219 42486  44501 01892  50170 09485  39767 64903  98499 19649    90383 16429  10828 75765  47978 81457  59698 89400  99739 00640
4353:  63520 07848  99329 77375  26459 26789  30595 82162  72889 58306    08264 65087  34244 33582  60528 55914  35407 88879  18018 15588
4354:  15511 48625  15054 28832  76406 13854  96839 06181  67439 54815    37076 28615  73473 92486  86495 98521  70629 91066  83333 07737
4355:  68003 36625  92638 15640  46228 96333  79675 29300  98603 53802    34205 34018  40462 30814  25026 44772  09547 06663  26140 15934
4356:  96356 77995  56614 86527  99247 58617  60379 50990  60571 04978    38458 31462  78273 23400  99304 33165  69927 27215  53231 80060
4357:  54766 33830  56263 64521  45298 79067  79863 92132  38321 44150    10992 80794  37613 64906  92788 00514  22991 00739  91781 44195
4358:  77907 41325  64058 26732  04503 08522  53228 28673  47145 11853    02968 97368  08406 24591  23221 44658  01691 81438  82329 44219
4359:  98200 38901  86015 23838  61631 04611  02584 91393  58109 73214    66708 44384  19349 58694  75140 76291  81866 41459  04078 79437
4360:  00064 06878  65570 16500  37259 92787  24270 42986  07713 14900    10185 25708  09290 05329  04787 86946  11300 35496  17544 66875
4361:  58988 01690  36710 93308  78957 03854  46692 71122  79520 57065    23471 61986  96890 19716  31734 10241  46044 63198  44203 63609
4362:  72268 54897  87902 76349  93116 13501  25408 87644  61292 86688    18006 22116  66230 30730  70721 36291  07002 08169  36368 19459
4363:  43279 22397  48577 33256  30485 86461  70010 98023  28106 64133    64049 63392  78421 41095  89560 99253  05509 42811  88602 57496
4364:  51202 05977  46486 64339  01225 45747  93736 96384  48659 48480    95634 96343  74588 75179  50938 63338  60150 21322  16253 23601
4365:  20242 91427  84090 41315  38281 02410  34918 77557  05545 61501    52615 85991  55172 53742  97943 85813  13282 84879  68111 11763
4366:  61237 14481  76143 25545  92907 47197  62798 82907  73348 32868    36686 25879  41637 42785  28972 03675  91802 11203  16316 62552
4367:  49621 58834  79433 30082  66688 54318  72868 90694  59858 89533    44887 11085  08752 91040  55342 04954  96094 26658  23314 99164
4368:  64060 33680  04550 77455  21704 14107  34636 92687  99874 78621    55096 93886  89659 17735  05533 59859  76865 86561  32464 82060
4369:  28540 00875  52391 95633  70802 87200  03586 56331  13973 09240    06015 42680  80389 34965  15393 81476  62133 70823  54457 01895
4370:  15015 77507  97829 97619  13897 53097  16453 58725  69547 52891    27249 58906  13347 09984  62043 59416  96415 07109  51893 79513
4371:  73265 81411  61500 52401  23503 65422  38716 34489  06660 11347    60285 34129  81170 83414  35889 96266  10523 48469  25234 87564
4372:  85269 88702  33928 92719  11261 50510  09837 78272  98617 41955    82208 41127  52403 34947  57174 15822  85150 81232  45690 78345
4373:  42659 38286  31163 48784  92828 26127  31112 68086  67130 30118    51304 73774  55133 44396  42607 85514  49604 77265  93259 26817
4374:  57536 46656  99075 74002  96869 04727  72342 06776  32626 77480    73738 99259  49306 13482  00563 17226  08337 61916  52102 86568
4375:  74295 31903  10511 75401  74495 69467  79400 00949  15359 28301    73156 20510  67289 85037  91126 36795  89859 81723  74416 46761
4376:  64701 37773  67082 00886  55283 47166  20401 72839  43832 30566    74370 32694  86825 34088  91993 75437  70838 70454  73575 55545
4377:  98437 77350  98394 27310  16905 23680  05371 09933  28561 14171    01046 05505  75342 50693  91781 84866  04427 67134  26645 11365
4378:  83441 33250  83069 92377  81480 90047  19711 23668  85064 65835    74596 09686  12861 99025  70745 07581  16279 49638  51132 61608
4379:  66522 65521  96988 51073  47604 40286  02065 96187  53068 12460    52829 13592  35926 50197  06899 27402  12004 25732  72128 11061
4380:  83148 28418  42005 03642  21267 25768  08790 42084  60962 55936    60413 32104  22613 80502  11483 89032  81072 50550  22680 83759
4381:  95368 32547  10066 22742  65307 61763  22957 41477  26459 29606    34080 27818  28126 69678  19175 20724  51977 75250  81634 07143
4382:  66037 73379  18378 58676  70551 76697  46967 01099  58678 76743    29387 66740  69185 11697  65252 52767  42060 54548  57839 61602
4383:  04109 14342  47069 34248  38027 39954  44552 04795  84350 04700    08078 11970  52483 73523  49228 42987  38500 82827  00281 83710
4384:  20709 20766  90776 09653  85600 93311  95937 94569  28401 20445    55746 40495  44966 68551  73916 34478  26854 70577  50366 87830
4385:  17471 67242  79595 79543  28257 43372  26551 40184  84337 04111    66817 84287  70720 46982  10542 82865  37975 96079  62867 48704
4386:  62017 19281  77078 38955  24193 62168  84760 34549  93910 85454    97811 04157  02746 99177  89712 32710  28539 87685  45398 64022
4387:  84809 85489  40101 28963  41325 89499  44804 66434  11340 98425    36227 71301  64847 15049  07833 52077  11742 59753  56205 27989
4388:  83263 30732  19322 57471  41933 40629  13673 78488  05409 74677    00537 13163  31267 42823  22810 73244  88060 77071  28424 97718
4389:  12271 60547  38070 39388  27869 06253  51822 00870  93574 94101    67473 89167  51520 49985  97211 59772  57844 53274  55841 37851
4390:  18333 86734  13809 67630  15742 07696  96595 22893  22641 05077    85381 38589  13207 61380  09326 58318  92629 01620  19029 22698
4391:  33637 09804  17245 75331  94482 39358  71842 32580  54492 62006    64609 90419  75013 77826  69590 72321  12906 65764  70347 78164
4392:  09928 44311  02846 75679  39979 20923  54814 67716  46506 50587    88494 03064  89627 59633  44356 13005  79120 85138  39844 79149
4393:  96098 16028  90047 10537  21811 61999  74447 16396  09746 12643    74116 60702  28580 57390  66586 08522  04094 06279  28259 17840
4394:  93804 14196  20367 42791  01935 59046  90233 45749  76114 35745    05753 25021  86164 43113  14542 76428  18919 48362  09270 02221
4395:  42729 98308  02685 65855  21195 66522  80090 34487  51319 62362    32468 69844  16939 92692  75347 69721  61818 08645  27293 16111
4396:  04870 19945  94499 45348  77563 45490  65664 68585  58134 71902    49996 15985  74443 07379  19229 08998  23950 05988  32852 87628
4397:  33482 47233  86720 28806  90098 10210  64029 15114  46142 72452    09971 42803  85170 57030  44407 30526  47960 82634  43414 74210
4398:  98844 80527  12865 45875  42371 25511  30379 28238  20946 38665    65908 17607  71887 83664  72019 53587  83244 85576  75272 48467
4399:  70234 41035  83265 10116  90999 58521  01945 51835  92302 64217    57842 95421  00374 22839  99085 85914  06574 80239  62564 28239
```

4400:	55697 74195	95488 81850	13307 25578	93772 76824	77882 83136	37894 26809	95889 91858	11832 09766	71790 62439	46785 39575
4401:	51929 69346	90590 75363	74041 51001	77416 29946	93080 18112	38915 21130	24730 35400	54879 01003	59084 35201	98410 43297
4402:	19317 85276	43954 57011	53421 61494	90363 51725	54218 61247	94012 17294	34783 11997	52025 27403	74500 82182	04197 88082
4403:	93944 62564	04942 58559	78994 37664	63597 14316	64566 81007	96565 75494	00774 47438	34824 89114	29080 60756	36927 01253
4404:	06467 67489	88294 54204	46673 84671	13237 69166	68701 06995	41589 72420	55990 71917	00998 16815	11102 80153	27174 13466
4405:	46429 53033	52793 24919	09386 59638	78100 63958	35674 18736	27923 25152	75453 31813	53301 98360	74866 87751	04380 93684
4406:	93079 09888	47852 73393	38330 27331	35176 94464	15154 81159	37240 34209	30623 35948	80961 79860	20305 52729	18119 13516
4407:	69248 02978	07547 69360	28322 73588	21173 35432	62493 38050	76224 66721	32798 16389	28263 52985	98000 12607	17049 83009
4408:	00368 81360	51753 01190	78763 58842	03085 80348	92346 35619	70222 53276	18081 40243	46898 79032	27128 13483	39574 56579
4409:	13613 57648	39802 12702	24601 92703	63070 65321	94218 46377	72847 28585	87310 02440	58715 23512	78918 00231	46245 86672
4410:	26686 05351	02091 99899	59514 53549	97267 97846	92171 38400	94783 43650	81900 40359	19695 78287	67891 43168	25400 48882
4411:	35979 75487	46150 53809	12208 35330	22031 86433	80455 13040	96667 64589	17333 36769	93321 65951	46341 19846	13582 22418
4412:	99881 20269	59090 45076	77375 86105	05467 78916	62452 62274	35206 25491	21342 87337	53059 90469	45302 85321	00294 44420
4413:	04496 39467	27858 13420	88224 38046	62755 10727	66743 19440	71009 34172	72542 74864	66874 35853	51073 86489	95961 18379
4414:	74758 42028	24118 17260	56911 24964	39071 70503	83783 09574	04733 24722	43729 69359	39610 37583	22102 50709	52613 88362
4415:	48084 80591	62560 00706	13322 56410	02606 95705	74948 03014	86466 05113	46747 18380	01570 04190	47599 60179	12281 39158
4416:	57058 48108	71783 17445	11464 38087	22454 59905	09726 69730	21210 59204	37744 98758	08971 90097	89444 04107	93229 58633
4417:	14487 14859	86005 17599	20326 11603	97923 28503	20991 73053	27443 84285	69016 23199	71744 61114	46228 35476	46209 40786
4418:	92668 37844	56184 48039	80761 91035	21447 54814	47992 43010	66414 03990	64789 06676	45429 58515	06660 80682	11324 57022
4419:	81122 98950	48640 68227	52583 46231	44499 59505	52708 28656	83032 13489	65736 48363	94550 18318	29914 38946	17528 43500
4420:	04073 35181	48968 51637	72767 29197	13550 83315	16919 40399	94437 09481	90937 53468	16928 17257	80417 58666	40011 22854
4421:	36885 00856	82775 71288	47245 12714	36652 20394	37116 08025	28208 72044	96024 62484	13203 96633	95214 98266	89411 83881
4422:	66976 49466	94038 43785	64906 09286	88143 23806	00022 11752	94516 12073	06190 91085	66943 85693	72658 52678	83551 73154
4423:	38009 95410	50299 90176	89792 17723	47817 74923	98596 40466	86337 48242	82952 01834	82130 47046	06165 70960	17369 20615
4424:	29624 07589	13542 38866	15952 54158	32755 08582	58597 04804	95173 31594	03204 12414	42684 99846	16346 23435	39793 81500
4425:	43235 06977	14651 75966	20214 18444	30273 12862	61101 42027	93249 33138	85026 68687	78564 47158	76935 51415	83684 26291
4426:	06124 23920	54212 31614	04418 78938	07802 28934	21053 90627	30490 46760	25626 13118	68619 92633	05075 00865	89669 82368
4427:	12883 51194	19081 72962	38956 06704	65316 79206	28985 73647	08480 73560	96888 89394	01053 16186	60232 02601	56330 03070
4428:	34980 96824	98402 88549	96302 44166	54417 18385	88835 61470	34047 91136	93151 92794	16801 70921	77591 55781	31574 82816
4429:	35804 41344	42208 84248	54084 83752	47182 83209	71761 65916	85947 74292	05932 09661	93289 98136	36745 65375	33822 95000
4430:	26502 95865	04127 76366	29586 85606	40081 40252	70336 64247	74833 33203	03562 92496	22327 79109	68712 89107	25397 70355
4431:	84439 91165	88710 04829	26101 97902	64358 59844	28723 45536	85467 77321	76886 59325	46849 44747	70260 88890	62882 77207
4432:	00284 90430	21273 67390	00820 97567	62892 57779	38624 09504	97245 77015	13306 78660	68356 46060	14237 75883	71736 63248
4433:	66321 32511	65203 01857	70779 60572	98406 47212	91829 89943	79515 51277	58775 21654	72108 14815	77643 99507	12489 62306
4434:	63588 24086	53443 75301	68640 82176	05301 11988	28071 85656	19814 80086	36290 11881	62836 21044	97076 14232	13330 33415
4435:	88269 80631	50689 14714	74524 16004	57187 77515	19586 06403	58360 31258	61722 99736	25476 40050	26640 02965	88229 09141
4436:	08941 91664	85759 94388	97765 87726	11179 53410	33197 80797	58725 12927	92100 55145	86780 40166	92925 80371	74628 45292
4437:	24570 73467	23163 98031	20009 93648	69083 62338	90394 71324	05058 98369	34747 39705	77602 25276	28274 12427	99557 43114
4438:	66759 36037	24530 88585	97062 14228	59984 53825	21798 24952	15266 89375	45339 90492	98179 16476	10027 80189	43741 57791
4439:	36242 38591	24508 91814	24441 24529	32563 60184	00872 74025	79023 00377	79608 00917	46406 14940	44801 60910	62928 38223
4440:	38330 74583	20519 99724	66308 98699	72186 04276	24988 98799	45407 69053	04237 30595	63618 81647	76624 96303	86455 87231
4441:	36090 39516	35667 16287	62818 21819	94709 70688	45757 69690	75314 64477	83918 08815	92377 96001	47022 97911	57799 65051
4442:	11819 24347	65623 06651	27116 75912	53251 84862	74019 87302	17448 70835	09203 09449	40193 52009	14060 01918	17074 37740
4443:	77456 54715	87949 62107	81095 04364	99132 99107	69508 71903	62364 09410	44261 77404	09193 40345	48343 47382	91763 04176
4444:	10260 53077	81180 76151	69351 09473	29027 45832	20926 90748	97786 57001	89944 31797	09569 99931	47314 13857	77238 12044
4445:	82757 76210	88322 93106	46201 00161	13142 91853	03017 30809	52465 72635	67953 70555	50122 46447	35663 24072	06380 99586
4446:	45323 08162	43972 82864	20620 02971	89032 77337	52351 75512	93252 67231	49144 10761	31140 65711	46965 02663	40275 05477
4447:	70707 03141	16283 89553	42485 67766	29866 50943	26484 70467	23666 49154	71944 11539	03536 84422	90354 37085	45004 73659
4448:	37913 29713	31716 23429	55787 18278	02640 6220	92952 13382	06292 56002	22537 3111	15465 20819	74082 71141	20224 95635
4449:	83848 05570	96980 04524	35550 31937	38065 10442	04469 77429	67505 48067	62346 81258	93210 96363	53862 35871	02106 51806

```
4450:  1850693071  9473073686  5136622143  2213388667  8093464168    2016596991  9687931202  6498681200  2929601886  8102407486
4451:  0882564625  8133986499  3258873136  0354551666  5995699492    4154850835  9160765554  9485211564  0188114106  8514561037
4452:  8497530860  9068802198  4840123941  1940840713  9334199422    9275086711  0142496538  1345797707  2612866252  2029470311
4453:  8539264829  0212887595  5860507892  3256276407  0395683290    4333859165  7003259301  0117839138  0936361021  9263937312
4454:  2900199525  0204554137  4619923620  3680757377  8551383215    8697170272  5704579555  8118069310  4901463291  4709019862
4455:  9001162498  3054207544  2152268787  7246479491  1931941853    8731878012  8894653133  4698508562  4225745767  4385733949
4456:  3480080168  7908683365  9575989238  5681524613  5331585962    1943293805  3034006123  6892160827  8899147598  9671175896
4457:  6464031609  3101858174  8735314053  2006675942  3734094851    5440894356  0324075151  3552477388  6631458497  6462761108
4458:  7221341107  4985912780  9336726119  5809967091  1590682819    4539618239  7801129380  8420806843  3737180737  8970011923
4459:  3268057387  4394328108  7510423445  2380613553  8354669314    5449006822  7367906698  5109313177  1750513532  9979499566
4460:  2630783841  9512949149  9760330456  0791073508  7206223469    8254526947  2681740652  2399219270  1738916923  5951129177
4461:  5899275983  9661840623  8610736731  1531735369  3246676025    1496004042  8091136463  2264902228  7494310814  1657048926
4462:  2058508370  3339476096  4886842673  8227293382  3401448857    4737073869  7482517819  8965519362  2409490441  7081149442
4463:  3108878401  0505963370  6422511000  3669622572  0496839513    7291030832  9833115636  0168927570  2070444551  1314815353
4464:  2420587655  3380495241  0921816827  7544961233  1486093338    7114198435  7053701981  2064800192  8988676822  2270653027
4465:  5782049230  6967577100  7477978251  0828744576  6436390656    5976461905  5926213421  0941955303  2591911094  0911546666
4466:  0464553909  6454909051  6926989415  3168833879  4373204931    6552506521  9360956694  2693168152  9856471566  5025899165
4467:  6118633931  0618051891  0622781926  6663824655  4998785803    2893658713  8009582412  2258164252  6468662634  6211673943
4468:  5979048525  4032549975  2567730796  4740968438  8743599446    1492800493  3434222807  8932262903  0960287581  3349646210
4469:  8361716519  7510274041  5697601732  9626013203  6644056208    4763732553  2580822186  3393224272  7993014737  3082829295
4470:  7527433240  2850462077  7122210591  2089783416  0015788597    8737630051  4690661119  8177917898  2400617615  3564670797
4471:  4310790200  0080580834  5987815680  8242770627  0953457447    8381687458  2895593371  3472647268  4492995405  1853041113
4472:  9733227252  0911669736  7304136809  3754520546  9257679961    1179773958  0250248625  5747933489  4059998347  4269108498
4473:  9259697198  7492107629  4211527315  3493945584  0029326310    4825724747  7804429350  5009788099  1860107187  8581577912
4474:  3675175348  7262797915  0655574488  2465496372  4979142818    4741779836  1859460225  7612300714  1791458493  6874620923
4475:  4998090766  7598337713  8317270972  7028919243  3626716516    3721489899  9945513791  3358504251  5339762007  7530980856
4476:  6375887221  7523570093  7809520603  3175203941  9900426211    9011950661  7433738822  3830884735  9602406473  1779420773
4477:  7629344681  8402527688  1086076206  6963214339  0698094306    8386204370  8164547271  0587896065  2310304587  2341687752
4478:  0521908862  8151152632  0139012544  6711467215  8285044697    7134727869  0031714908  4227128187  8299069117  5267038288
4479:  0540806607  5637063271  8884769117  5143509310  9915613925    3780194403  2407192966  9628923728  4355334104  9421029512
4480:  7596807916  7043823171  2701646249  4859134379  6208684182    2813940323  8135791831  0715601531  5960019024  2816485792
4481:  9406565489  8019938119  1438449432  2040044466  4778660876    9097344696  4125686287  0220719185  8809272841  2105826037
4482:  4965226279  5408348273  7182621225  8892646322  3822051248    2899880272  8733629078  8767195633  2829303899  8541012261
4483:  4742694234  6049901385  7692073104  0769668503  6087545688    7680204699  0656143041  6618551589  9329162344  6739763601
4484:  5092625629  2783262242  4234623594  4810488707  1820139035    1309817774  0974263740  1883394566  9323239409  7213151369
4485:  5096990536  7861869365  1356665334  2764898520  9687967003    3281249850  3460248016  1785571483  9087911295  8799738657
4486:  7056511814  6042364519  4028934872  3106545403  8868161958    5367106518  6536106913  1873689772  0883246303  0169541081
4487:  3480731756  2810705732  7463503921  8373004152  9021832735    1482546431  9242102939  3490800023  2650453755  7097038981
4488:  5231756209  4863067134  9221951687  5814557053  8241392660    5267370259  7468967587  2558174133  2309789604  9489019656
4489:  0998411631  7848689475  0594241991  0604603405  0384308817    3215954126  6901464995  4744363028  4084533801  6011181227
4490:  5111250560  5403105721  5927810345  3965780943  9280148952    9514128506  2782324888  0690430985  1620136541  4289093666
4491:  9453965506  8673465828  2465475320  5115879880  0628826729    5721347703  0780114088  8337978642  2463254816  3451949263
4492:  3975024364  6150188979  6046173779  2053035475  0932059825    9848239724  0320439255  2706267585  6545446213  6854413499
4493:  9187851921  8903808319  8884584536  3800238011  4106523940    4544396219  9009370925  5119607189  8239017251  5636651193
4494:  3820745974  4398185593  2219307465  3482105599  6901653720    2644135840  7948601177  6218242272  5038172912  6541966429
4495:  0030023258  7735537133  6036804265  1118446758  3951512600    8653277849  8644944636  1527289237  8874915677  0349506077
4496:  7414843218  3794637741  1201749003  6980732119  4416768851    4009229192  0051049150  4171019893  9646670286  9504356606
4497:  2132098572  3678413518  4244168483  1054726166  6669361909    7245258762  7215031065  5285362898  4475157276  0147394134
4498:  3186259654  5171449108  2077637112  5382500291  0389256640    3683598591  0963845085  6728144456  2652513369  8976091421
4499:  3413501357  3876026723  9814338827  1161722379  3162299019    3388953104  8858572372  2763795864  8810118190  6622206502
```

```
4500: 9793481131  2546213616  4712786059  0329861658  2981860508    1676048709  4753269389  2081951725  9885951727  7651253385
4501: 5388069759  6791512530  6598929981  8933074188  3024139646    1265712774  1304239068  5429554632  8789933354  7018283982
4502: 1517871141  1239705702  9098090301  0470053991  6950316928    8438640467  6600490448  9769552087  6319027327  2252770699
4503: 7599673361  1420026089  7759626354  1735739353  2849478463    0873515466  2886258469  0754010845  8809322022  6050808440
4504: 9228609429  8473700167  9631773332  1048057491  5494887272    5367482639  2929893391  9168615208  8147813355  8932479710
4505: 9165125230  0839461100  6426431212  1510863955  4584206179    2120293671  3368634328  7852894297  8628889523  9675977544
4506: 5259935678  4259522264  3786264024  6171558495  7548911458    0119918532  2923214441  6728747640  1457748024  6477813697
4507: 0770131540  9649195238  9578311190  8314370599  1104000395    7221840549  0881319254  2648479460  1154035747  8451798105
4508: 3363628495  4683464922  2895402773  0898049989  3598840986    9605474952  6969190629  6475285329  9071471517  2096811385
4509: 4460704693  2121707687  9080291217  9994958174  2205442229    0708314922  9207222694  5227112738  3799832124  4181655569
4510: 1424409051  3080410808  0312195797  3722978551  1836329460    7921066253  5782354294  1714149202  5197948464  5933477280
4511: 2474529183  4811435733  4700555176  9740310598  0440312832    1232818331  5781654073  2814651812  9748314782  1171485938
4512: 8336546117  1944845151  7658078651  0642399805  3484022190    0528537001  0347631046  2079375684  1410475427  5547150255
4513: 6581761911  3001045734  6968093535  2652658022  9658372897    3835248598  6373104187  3442009453  4694686687  3412952632
4514: 6388702289  5586677773  1342120243  4659103543  2064259586    2338632899  5045159493  8084534501  8273776269  5477286676
4515: 4764924029  9252635184  7812199509  8692383489  8242603853    0431298811  2843892375  9616633388  5192148137  1321933245
4516: 2783839010  7467679743  5447990618  4244761448  9949424816    3572556675  8318197329  4163186249  9670937540  7725079908
4517: 4343122414  8004530204  2872375584  0950428246  0710931843    8450851862  9125151196  6662739429  0190218606  1062154930
4518: 3534607976  7160953327  1526355728  6184033502  2416664585    9113767495  2480066940  2834796547  0817549242  0159270392
4519: 9587830385  5746635666  8916699201  9575789135  1700291200    4464577333  3322643705  8258044330  9899928733  3037313147
4520: 2185090930  6562336210  0376278498  9461053113  7079140927    7881873207  0435268647  1302945218  9841250499  0185935460
4521: 2285251338  7393449626  0173247894  7913552953  6768261972    6537797264  3492994469  2957578557  4799342346  8877554705
4522: 3611146513  2128160220  3630490688  6033380947  8585459087    7203703985  3337203667  4971373116  9978768718  6012775349
4523: 1892344663  9217943412  6496573951  6382479964  3781017939    9885533185  4392231350  6912864481  2817881748  7526933761
4524: 8035520483  0939200911  4391994695  2363771513  6064369351    6040367176  7608378251  5728439463  4318990720  8203009229
4525: 8506677793  9409882848  6933322611  4754868586  7456188674    5007048248  1581368657  8166763970  3527730497  6918498351
4526: 7946613356  7355853219  3770471511  7047386614  9117452757    6266128270  4635960380  8743038303  1471453542  1390199651
4527: 3041815938  5326512099  8477622630  6054225203  3347750526    1802661334  2764680696  6587923333  2900843459  3111059498
4528: 7475407654  8150751753  5558862373  6252130013  2613220062    5490512920  9685040377  4781450002  7058180207  5607866270
4529: 9686453973  1172572421  2491509097  2823923379  6981313746    3161961507  0824631080  6200911979  0568642671  0701324524
4530: 5607673495  1178950381  1332220304  2549437910  9809091370    6516835084  5946615805  9112049492  1595913612  8462656190
4531: 9773154773  6795101549  1190815952  5435452460  5325710800    9319492647  9824416537  1483722202  7414503959  8199573275
4532: 9251582525  5590341299  8867546640  6214593066  5434168422    4455310167  8554083912  0992275345  5875851155  0132936208
4533: 2579447680  5155415471  4666094046  4604624006  1379055965    2790219569  5530324509  2077950662  3024957810  5516238902
4534: 2486554549  6374776837  5823834068  7258895633  3349700838    7325942802  2450177985  6473516616  6768725546  9255309500
4535: 0738166195  1589683698  5347149953  0834954131  7368243932    0400453576  0176959587  9820671969  3141273689  6885157485
4536: 3260045868  8441103324  3774925995  9634569467  5309578491    5371525066  0195265398  7693359375  3588001525  4562271530
4537: 3924338000  6965851310  8800575738  3515303186  5823896622    3366730856  0719668803  0046697735  6478231003  7695453941
4538: 2656538806  8675324002  5021213669  7396644202  7307563565    0467794193  1857631543  3912868272  1277188359  1065972276
4539: 0513960173  8847587079  4921561138  9446025428  6966672782    9143754173  3686833417  8693901817  3385196262  8088536308
4540: 1732140312  9204612707  3089920374  9903544591  7365141052    0528398767  7086680716  7571972782  7438144858  8182377995
4541: 8314589647  2597864885  6807510737  1048667070  0038748401    4423134367  1339594771  9995620728  5264639505  2581966520
4542: 9595695381  2237041000  5410934754  9791061112  8306078288    0164409846  1056297758  7190834578  9491524575  9739248702
4543: 8226718881  4440929891  0322538767  6097912584  0038487774    3665333175  2626044901  5827850111  9861206821  1232028014
4544: 7494704764  8574202759  9885197581  0461534640  1422248965    4737137838  5702954403  7247666190  4707916535  9252449150
4545: 8661523059  6320463853  7246300440  3422955951  4136049528    6963220432  8246487096  9890727164  7438557053  0400411374
4546: 9169936656  1502191760  7711406989  1477902048  9770136192    4988674700  4612125459  9955446383  9942264592  0340389312
4547: 8287233701  9470085817  9012455914  9794668877  9763773325    6929957203  3289299699  1445582733  8475016520  6294246013
4548: 8557693539  4473186457  6331697230  5556637981  1848276629    0486813283  1705324682  2389877516  4383224573  0998914378
4549: 9023047371  3486343757  8522889225  6338536758  6813221391    8199013958  6391795095  2755023684  4737736096  1908419260
```

```
4550:  49481 56113  86206 41738  50579 44609  36853 16590  78522 18715    98136 12243  55364 42978  28651 63668  42683 46219  66220 67053
4551:  74227 22413  63560 00660  62721 78878  53741 13792  24726 80599    82287 35013  47126 19009  83957 28005  33751 52735  69263 12244
4552:  57844 00720  84052 95848  06088 52537  96778 16232  96955 81130    29020 90328  71813 08503  86600 95102  05872 35611  45341 04480
4553:  44121 49406  99250 62851  38618 96652  38003 43248  38734 58186    83390 41166  49485 11911  53108 67068  44457 23579  76116 71596
4554:  05197 51517  33957 63504  06840 70973  57614 93878  21154 91695    57058 23009  64377 34652  74680 69966  81468 25720  25335 52783
4555:  61506 38207  63889 74693  23299 88496  77174 33188  87949 95893    66868 11848  07465 94446  16923 39037  99949 02650  05078 88903
4556:  90083 31265  99833 12336  89643 45613  12972 34043  61297 03647    47351 63124  72492 61447  44093 58097  50184 88433  34404 14892
4557:  20192 29328  33210 14365  91923 38287  87045 70205  98638 51179    48546 74497  87526 32347  84366 23728  28609 98806  42803 83413
4558:  67440 69838  40139 16729  51218 19238  57993 23116  56203 27910    83505 20985  96616 36716  49234 16977  46999 31279  97937 22678
4559:  87364 44814  97532 64669  12497 97717  18374 63049  57983 29673    66200 36608  44487 91458  23334 37438  34457 23304  34832 92513
4560:  79374 64442  96894 17535  98616 66133  55735 68664  46352 92708    88759 70003  13963 97790  37509 56298  93853 86242  26304 34391
4561:  42742 37461  21738 21526  28132 47158  56656 67054  88450 06436    87773 98516  22314 36622  97939 42884  22531 57021  29171 41430
4562:  96086 92609  83822 91900  04667 28532  81584 98531  63111 69571    53674 57617  83187 48446  13044 32774  37593 89914  86612 30688
4563:  30474 53769  75485 67650  33437 69467  98395 56858  44398 24973    01363 03784  17715 31611  03787 95965  12136 36485  61192 46725
4564:  04586 17780  83216 28010  98248 62656  64616 52812  97787 92470    97322 22987  71599 84085  87185 62254  64874 82231  37194 18567
4565:  93456 76737  96194 30714  69101 25491  65415 87615  59925 40558    80532 19485  40639 68671  24297 03222  61324 66247  23996 21241
4566:  28634 01530  43525 60705  15718 72023  84688 27836  27104 53658    75978 81087  02512 51528  94743 12997  46587 74003  02900 93989
4567:  46042 30863  77285 18871  05918 62008  63403 95240  28536 30764    83180 79808  15156 85124  34078 26448  41964 68915  76238 27311
4568:  95282 56037  60097 30761  32892 01746  12376 91045  43562 24047    46404 95612  98656 41027  06382 75745  66075 27422  56040 23344
4569:  70212 81983  21634 73991  00547 79687  22568 70394  27817 22135    88440 27714  09889 55463  93853 68173  93773 43615  18027 64298
4570:  19046 14057  93400 83424  35914 15207  53198 39158  83097 47617    61074 75362  54152 98096  77705 31283  96089 15814  21319 44838
4571:  50346 54024  50791 04714  96579 25172  90165 96878  50467 99975    05354 69436  13381 81190  82288 13724  27259 30772  61195 47837
4572:  76738 70597  58361 64521  54477 41659  12458 92495  10171 40137    62616 80251  04289 94132  30030 92922  23453 34970  02812 31029
4573:  36219 82911  35588 34719  87423 77910  13738 69122  96885 02968    73339 46627  30161 34514  03736 27928  15358 91958  36159 43468
4574:  03849 50802  27006 18931  67472 31401  49521 78690  37834 16526    62456 19571  98056 11987  78816 90544  33430 00088  20937 70474
4575:  59803 75635  91320 81184  22982 71607  96122 63545  58866 78802    41812 29369  96768 31316  11117 20021  64671 54860  29955 19457
4576:  94443 10704  65069 74653  18289 74556  22543 15210  00973 48049    01540 28259  26108 70385  63679 75432  10558 47225  93915 77574
4577:  42051 19905  48495 62986  08948 67793  79794 77083  46342 19465    85936 50709  75817 05161  79937 84752  95425 94325  54650 73686
4578:  37355 25276  83590 73659  31395 82907  30237 63393  17053 76297    61410 35704  79966 76039  50453 53545  83794 96731  00636 49700
4579:  35046 40652  88747 14230  62508 41284  29550 29444  54873 17534    40285 65514  79648 25627  64592 07966  70566 26497  19270 92331
4580:  56933 51922  82989 21825  33916 43459  56060 10315  47244 40492    15509 50451  58359 16882  65050 66977  47655 50006  32547 14717
4581:  90326 47312  56370 86943  47545 10203  81397 82887  51897 67680    47879 82335  40078 76004  15244 47208  33538 70889  15832 84370
4582:  26776 23816  93446 13888  23705 78040  52899 69450  11411 63261    96607 50664  22743 64655  44442 17933  32845 76798  99310 70109
4583:  26811 94362  02908 32687  39568 15954  95405 04725  56036 93036    92723 94197  82756 53551  88523 97468  89063 39369  40531 48783
4584:  07105 49668  43237 64247  83481 57628  64869 27722  10611 19616    42944 80660  24084 05535  02205 11083  05794 79050  00404 92565
4585:  51290 88114  91839 74457  68783 98540  05838 64563  41786 51920    53356 23707  37398 15647  45385 15989  16922 15473  75643 30614
4586:  35701 91566  03315 28926  24571 70428  89814 33696  32449 29100    68399 42522  42103 52764  86613 23969  74945 84501  83699 44218
4587:  68173 76123  47439 35662  08167 65526  42467 55514  36473 19086    94140 99366  18862 60373  49221 16801  59801 78288  74550 27914
4588:  14546 15896  19041 28956  55027 11279  56438 06765  69273 10260    86757 82299  16787 09688  06891 81606  07197 60253  74224 61226
4589:  09570 34379  84988 24510  14902 93709  43875 28091  56299 11706    39091 80932  41137 91617  82395 33167  58616 59015  20731 07747
4590:  32732 91260  51706 99012  26279 58771  34753 49951  21244 39358    25767 88979  02887 69529  24188 93760  13001 29234  25757 89680
4591:  90111 68954  92818 51998  41577 27630  53025 56219  93390 37116    65892 64846  25486 74173  50963 00672  40316 99233  30982 01179
4592:  15211 62884  88817 37505  61619 95619  98232 48399  29038 33573    47917 45939  29898 96130  76531 45920  96989 14997  74487 94426
4593:  37896 47108  22701 35373  26657 53386  06422 48363  72365 80571    67424 29042  56963 89914  97163 07490  14973 83648  22770 33255
4594:  32398 30052  62594 16440  44448 52333  50027 74541  70154 09850    28629 25531  84652 25996  00355 36173  67659 75757  80985 03683
4595:  85260 09796  15283 06610  35484 60864  01955 25857  25378 88329    44165 42519  29925 97144  17636 46008  21933 89435  95471 97387
4596:  45194 56032  16666 15452  89472 31651  16663 21132  73241 83614    09963 63980  80961 25743  25747 20128  82006 23307  71694 69375
4597:  45516 89401  41587 03920  54406 04841  65498 89096  41163 80684    71767 99175  50885 54929  46440 76428  97566 45641  69000 18330
4598:  12421 08060  24702 61489  49437 25732  11512 92538  66192 32503    69758 30317  10692 17645  90128 86628  07756 96962  94457 51376
4599:  93069 61708  21174 86673  60863 99219  96707 74233  46821 50524    24597 01555  31823 41863  14999 27092  30241 51674  79823 20675
```

```
4600: 7255653502  3849819113  6016207604  3166384625  7705743878    4736921458  7129293169  6132082933  8200164825  1734777762
4601: 3244489489  3242899134  4313143020  1146036766  0851289821    1570761342  3062690373  5653484620  9784820069  5980785442
4602: 5706439921  2707650533  6143934464  2766729074  1393936609    6478547020  5945203837  7853099824  9290135143  1255623125
4603: 8346806969  7888664434  1438514770  1914785759  1885897527    3049112207  1260634794  4613959282  1432923053  5355482840
4604: 2734782305  7752487799  8679705613  2007806597  6701427020    1150812919  7166589401  3254385715  9226724213  8488278832
4605: 4851696628  3334848424  5607831489  2005757722  7680655793    1935796389  8129155037  7659645472  6554208482  2249046157
4606: 3118080272  5739806599  4176233051  9325451354  6742373098    8307100742  2965068684  3110826533  2863779961  8581045580
4607: 2408550587  3536387720  2736171935  0785799604  1067390724    8107677722  0823218139  9453482252  8317998338  3655627040
4608: 6843048378  8964667480  1793575778  9482021096  4600562012    5804227887  2358472145  4188646235  4235488401  2194085495
4609: 2316668831  5846325339  7934910834  3599923273  0511944786    4210922895  5570040551  6335774852  3079159729  8124824838
4610: 8603529143  8833515037  5084151974  5889621829  6175614320    2931902134  1586764388  1921137219  9373763722  8805288619
4611: 5610933408  8585658606  1104884844  5153683435  2773076134    8872982145  1907891438  7865894803  4739993885  9010620621
4612: 8876663230  8817763084  1205052817  7580550030  1560676703    0187781443  8371169416  9825016330  8254427529  3208624554
4613: 5348725370  5116864521  5771963272  2669303152  4418648253    8054553741  4062644664  3248418870  8497988559  5319398689
4614: 1448370413  9105087664  2555057170  8380377999  4356804519    6747773451  6996449679  6693986693  5458790637  3117795184
4615: 8842263857  7311904593  4258651208  4851929484  3774899818    3197427646  4045589481  0232737407  5088103556  0109633662
4616: 1547194152  3087423879  6230808933  0770224213  9198347537    0723240370  8864620335  4195269977  1678254367  6401785335
4617: 9225816778  6091079138  3772303344  3825431161  1212915826    8369027791  9820678811  2685658353  0921891028  2971179508
4618: 3416893782  4797421993  7210253187  6548118835  9464422461    8226692400  7821719540  3902566935  0843290516  0699362734
4619: 4604845914  6760125472  8700138594  1822431335  2763854039    2322129674  7281990855  4685267603  2707394294  2526906040
4620: 3193850079  2936069552  5060694419  3465302797  5677189294    7601382311  9441389184  6600694972  0018274842  1926246424
4621: 3046985292  3919838994  6988673253  0460045583  7384010202    4549266355  4959109905  2200937730  0542444276  1940859628
4622: 3378602400  0292178737  8360559575  7053733310  0589835361    8111054120  0922643166  4652700142  3967042206  0852789418
4623: 7103960586  1102613836  8873183669  9262279033  2373632839    9457214341  7158858292  5632022807  1516385046  8922778389
4624: 4862929181  3186401589  5660040043  6962829051  9525192034    7606526221  9248662930  4230963068  9072679478  6404112929
4625: 3850610445  6763765747  0799807757  8173114310  3107303666    9251216218  0960454109  7554962538  3423078778  6317506686
4626: 3585621518  1097474319  5095603893  2242880499  7946485081    4849580039  1254964923  8532255779  4158865331  2175333312
4627: 9776580458  0710697396  7502378059  6855508221  0352009211    2387080911  6409404184  2602265675  5170758949  0234450137
4628: 0656106742  6159079897  7551510303  1292019012  9473826367    2733039862  1446633035  3417436878  3941409675  6039365834
4629: 1149070998  3485502727  0646909115  2594300029  9292646420    6026472080  6486637446  6382351318  4878196603  6627371040
4630: 5656574541  9300204153  7124366476  6444206469  6782035631    2420626515  2005276613  4611286511  8868349606  1508317018
4631: 9754150877  2585007424  8250800195  6015621126  2084602107    9046346504  0096946302  6102554377  8929498379  2549203682
4632: 0355120560  6600678764  2130940635  4743064132  5620105286    5786382708  0181247323  6158223316  4010361448  0166153284
4633: 2973707893  9537370668  4917773815  5776735428  2632912727    2353195172  3215676348  7383304617  2981685905  6703105868
4634: 8679071048  6681460015  1134392274  2149328546  5038016354    8381389627  2065997122  5561081134  6371347135  7503216185
4635: 5319219995  8388364095  4482407257  7609650635  0097873583    9238462298  9014141761  9364114627  9852344635  6304951357
4636: 1917740682  5111623374  7411532438  2304363885  7880036611    4487519234  1834904043  1550591322  3880628440  5448106639
4637: 7535538595  7176023813  5628189787  9528001483  9697229856    6563174432  2677300266  7807160437  6466464369  0166215214
4638: 9579767231  5991061390  7762895599  5864431615  1198556731    0342051102  8401253635  5137522059  2710073299  2373179214
4639: 9011425357  7116535056  6022209705  9185058012  2798720093    2717763041  5055224269  1114744813  3303432585  4699159954
4640: 2150528822  6119835299  9358124836  4651347813  8442043341    4062854433  6151678940  7898341464  5781825001  2607536177
4641: 8779756833  6275428766  6446912970  9951408280  3384198169    7320390331  8963802208  6759357713  1808897411  1327252419
4642: 5162123944  3553846268  5451049812  8113496711  3684810707    4147938142  2131891311  1594929295  3690951341  2154328661
4643: 2998730216  5328373753  3256634574  8708791660  1629541983    7034385349  0580852175  8898869568  2410899189  9765904150
4644: 3019443348  3807850640  2487936931  4924188461  0921510501    2710422611  7949460648  9183226349  1185753328  6290405251
4645: 9147825078  1680068071  9061961636  5554969819  9108932967    3123655199  1188011846  3144881927  4683788421  9551652894
4646: 6446178450  3481059095  5723098784  7015645145  7105430880    9006035078  8264278094  4921891564  0504193216  5197302746
4647: 1981534596  7440841749  6040911862  4182921129  5119820216    1544112274  8487189311  0735783248  1035696719  6159420219
4648: 5449250833  4588732615  6933242215  5536458139  7037644666    8921800840  9788555000  7759087654  1675508419  0624831654
4649: 4364676110  5327450988  0127264723  8751285871  5554342887    0438011563  4963767960  7211296909  7402268448  3789864632
```

```
4650:  2721208918  2478104918  9306828192  7927891909  4131945470   7819135492  7681464134  8341070777  9470264772  5528546698
4651:  8190599675  3293112508  2820196764  0452493506  8196037758   5016505881  4910118040  9434566927  8530624619  7818389573
4652:  8496954116  5697244152  2044849060  7005198060  1958473110   1854688358  7986907948  7038468755  6998681023  1101650256
4653:  6396965838  9507806335  2647637648  0807383839  6194036027   8427152658  7674587518  3167727150  8683118045  6471859050
4654:  3357945409  4185424092  8660498558  7349648151  7057149229   0517332399  0612923918  9567705528  7854621412  8241322722
4655:  7624649490  9621965143  0412276697  2978991804  0885923267   3358847356  9703297808  2075080290  3559341463  4209193525
4656:  9700976411  8125066347  4851082157  2531645597  2452901882   4214834186  1324788463  5446682268  0332701710  5655405913
4657:  9034744737  9486586297  0839402556  1652417702  8097292576   9168845865  3970559790  7050704994  0146545031  9553492089
4658:  1783755992  9802432690  1589073419  0746624886  6735644198   5854198610  5114914938  5374057777  6984294838  4262808552
4659:  4648450489  2368238543  0142885859  5510099417  1264266207   0776229101  8399630566  4038754790  7234348420  3253836785
4660:  8723727617  4421843484  8063283159  7683561869  5471497373   8741869804  2110711485  7927110823  5292402656  3183039786
4661:  7101859663  3284217016  2149118301  4950245008  0911298985   1145314560  3101494732  7195116454  5797479305  2634897968
4662:  3698898235  6937648301  8901524167  8943612787  1669752330   7996688080  4958167585  9164284139  0855264559  2655610626
4663:  9548197591  0964858347  4047110377  1005981493  1012379992   8391373589  4100454444  6266803713  7263017975  4848642154
4664:  3289410223  8360976951  1745151631  9449430921  9936553879   0117628906  3195764793  6246363691  6590127391  8945570286
4665:  1774038223  8032720688  5284188235  0305091996  6009179609   5228545135  1899618208  8377920204  1267937653  2870738650
4666:  8668355170  2915272686  2036712886  6187846031  2657531154   6828701122  0748448615  0942832654  0592682603  9991261025
4667:  2852257993  2150158429  7184353786  1459209121  3151330115   6091638627  0018384458  6264066927  0242157269  6567051141
4668:  8627474806  5416772436  4136362822  6997517710  3545501453   5212671286  4471417343  9332672486  0814074045  2947799605
4669:  9078253743  1041395588  6744202867  5679256960  3655412021   7262111411  4612914763  7678706503  6272974675  2269144074
4670:  4770077364  5975821231  7908672998  2340572264  0691644693   7652171027  2205274342  6829263645  3009546371  9629914289
4671:  6063595332  3423909459  1617980947  2545354223  3955241190   1798761588  8917923540  0606543428  6137977441  9802983110
4672:  6715536462  1634477350  9751557058  6887059574  1494003178   3234720480  4717695198  7095256469  2675167318  4524108810
4673:  0722551931  6434500611  7778946238  2905162275  2239594243   3360459450  5079362872  0238399824  7473822948  1000503651
4674:  4292323325  2347176029  8985365582  0142310520  8375845084   5367468196  7215099250  0157730553  4459716716  5121817895
4675:  9884329440  3944570899  0981439408  0786944984  6885523555   0375531720  2950617990  4198057647  8901105743  0194163379
4676:  2862377862  7596163986  8109321384  9019757354  5874956103   8546621221  0299845255  1679932781  8192680377  4638627044
4677:  1349602019  7355994780  5416216192  2125439741  5673788800   4226505652  9834116119  1801578033  8605886875  5946762022
4678:  5618711795  9397522862  2939185407  0612023855  4015291794   7878727741  6829751880  6137976459  8760038298  4975402175
4679:  8687923073  5791041880  8675942469  1678155570  2259112698   6577245385  8702300853  8332251217  1480998643  2114179870
4680:  3344444417  3534539848  1201581458  6236866217  0868581796   5644993797  5205815111  9869901932  0963417050  8462555292
4681:  3088320220  1576111711  8093116520  5187619306  8746030458   8494359094  3431980957  6801492971  3110058821  3124631965
4682:  1026685674  7196918381  8408011826  0763945667  3399851440   6200399633  9962925799  4072022497  8236956071  4987779135
4683:  3693796483  8594487029  9751218152  5699139633  2437254897   3971168827  4304729223  7093103066  6430966060  9049793022
4684:  6282390055  3779934413  2399052001  9215864840  7382803489   2817939428  3009675072  1479553538  4497328288  8325150065
4685:  6531032478  7020709442  0677086734  5924378072  4979342830   5155173197  7525947139  5386951775  1287001821  5015097765
4686:  9608998587  5231389214  6233251052  2823375629  5195954753   1512765311  4179771510  5538530596  6126821357  1215770669
4687:  4012845914  0529910158  4829574946  1326937410  3000084321   1231047738  6481131556  5875692854  5288393443  6269504254
4688:  3295616190  4505689916  2072128604  9910656481  9700792697   0370098962  7240477508  0398299389  5983744816  5236594135
4689:  0871613744  9057466091  0553598183  3234479747  9709581936   5164000515  6740687507  0757318660  2528861648  5238566004
4690:  7408958415  6250250005  2831594617  0699008789  4292788315   2876511338  5837726819  2613206211  0881884448  9681650496
4691:  2227666331  1305161384  2891530990  9541685157  7204476029   7540398568  1483842971  1201530388  7453913433  3889609490
4692:  7471501979  9073144616  9235748265  2393516683  7975906480   7112807102  5779184238  9389669997  6756022304  5372825292
4693:  0642105883  7614718353  2525250693  0218419785  5036663546   0384762390  6085277138  0880931772  2522455729  1908318274
4694:  0071139932  0338261698  9391847191  0053010590  9257634394   3194112508  1911606584  3557519242  6981505540  8086698110
4695:  4408338515  8979460733  4452066208  7653345560  1367293920   4117116925  9942571088  4837941652  6840364181  2517028100
4696:  5438833882  5463256357  5287202934  5207476793  7277283492   6987284794  3803143502  3577642979  9768461299  0645838273
4697:  9512888584  5007297478  1277890618  7326639128  5271583975   3758794672  3827678620  9647893423  8872506267  2224237383
4698:  7806030558  0385222473  6446096827  2106009465  3932270519   3829215227  0579129286  0401376003  1017952155  9313428098
4699:  7902277989  2487622973  6986262258  9868637064  9414087777   7033918019  7565937782  7778239600  3787395954  3302980238
```

```
4700:  78787 06212   15627 44942   97342 03070   42205 69995   26572 40650     49725 96430   58090 46570   08831 48593   14812 09542   46756 59778
4701:  52252 32425   16671 85185   38286 89332   39691 41713   93224 25746     56095 92694   90560 88960   63704 78516   21616 63529   57814 37688
4702:  27912 71011   18171 80884   51830 98958   27824 35581   83603 94761     13154 86867   98700 32443   69022 42411   28019 37702   38902 47155
4703:  11526 54621   08885 05728   65484 86556   51296 17678   02511 11827     35878 27908   57505 85381   98420 33420   56605 00917   79022 48751
4704:  29538 72661   78391 18133   09464 84213   27247 59416   23097 47479     06308 25744   98816 78361   03231 49891   17646 11824   52448 64291
4705:  96500 66028   72190 83502   86095 75202   71680 51982   73953 33634     48651 98987   11899 52021   13183 13246   96610 50133   60278 96253
4706:  29213 03669   57406 49372   15271 69833   55607 24341   82088 32162     74376 85844   30079 74541   88601 14828   24470 36656   18105 79423
4707:  71621 71225   80089 02953   38644 41093   50793 51417   62556 54899     31974 50643   13774 31731   24458 21659   84598 66442   11650 38095
4708:  79492 33064   10049 04697   12045 34834   06237 22347   32219 17693     58633 72248   39346 08077   77630 16057   69666 97081   05053 59063
4709:  54494 74577   02264 00736   27045 74955   77794 12801   62635 82588     46163 63294   03571 52268   69309 36464   51882 79044   78582 99081
4710:  99811 00922   42314 59653   63242 00311   77857 55399   88793 35357     10413 14643   63710 42764   42787 77188   49163 99089   07686 72894
4711:  76419 09482   25425 15963   64352 20924   43227 79237   54494 13556     66467 28292   51851 56643   90747 67983   02293 21134   24625 80553
4712:  03880 63570   67012 95054   17813 18141   60799 28606   33583 54897     29985 71803   26355 50809   19494 42929   96463 70326   71815 10330
4713:  85763 39705   76817 78734   44244 93434   15165 78908   04907 54394     47410 26597   54923 50747   65406 59570   78424 24046   68079 04556
4714:  37273 19939   82095 87156   33347 83323   44983 04410   76474 86812     83768 25333   73162 41177   73213 25716   07039 28332   54739 67306
4715:  63281 34940   59114 07475   90413 68250   20475 91686   61971 64955     42771 89316   02900 49234   88029 93042   42994 47915   31107 34149
4716:  63798 81226   19287 12579   37584 00303   44216 06856   30357 47792     47308 87419   15714 23152   00349 98658   69594 99834   55132 63835
4717:  65189 60813   36431 61602   87935 41578   59010 42840   18599 35752     77619 45098   23577 51957   97523 90825   48066 56579   53743 47066
4718:  71594 82163   95557 47107   00032 19977   40826 19957   00318 18708     84805 45966   73659 10277   83335 76924   58652 68236   89485 55975
4719:  48704 90643   34167 08809   02085 56897   54898 63614   68230 28627     41250 66930   54468 43431   19758 98018   62806 60015   08179 61563
4720:  45260 75953   24675 97941   07049 73625   05116 90882   09333 02806     23262 14125   83502 22623   68486 81020   50072 23939   08615 58270
4721:  87500 74014   49050 66941   33306 05314   50005 31673   67783 42121     78580 36911   17575 42466   60448 97674   58739 73801   87812 03518
4722:  84105 44663   11113 45709   66062 57614   82662 17109   60854 87119     18351 95019   36523 15631   56158 23639   84056 48164   71370 44349
4723:  55612 77720   94317 56231   24031 98574   13970 08737   44139 09592     50577 17847   26009 64961   34872 71281   49198 21089   84562 35018
4724:  98207 45343   60933 79484   88115 15488   29523 32551   49741 25325     21416 85848   80979 68893   53123 20583   01757 49408   68590 98138
4725:  45722 64599   07167 81974   31614 14094   07492 25553   16117 29596     55153 42550   32042 82817   64151 05659   89248 83764   71994 17234
4726:  46739 84349   82425 58992   32010 52407   43314 33036   50629 54376     77154 03126   62793 08250   34027 32385   77758 70011   48672 10654
4727:  45223 48899   06004 59689   22539 47749   71661 55999   08259 99630     56684 05565   34054 66685   79698 04267   96290 20167   53469 96816
4728:  77118 30044   48486 74520   07783 57194   69979 01119   79807 74815     17986 71406   19324 80032   68211 53240   56238 18245   80721 68367
4729:  39665 68948   23515 94448   30903 54517   40333 12311   05550 95509     39724 78359   29829 08249   98598 65351   99344 44101   44810 70003
4730:  25478 02536   17823 72001   66601 10677   18300 74088   85267 85637     97177 47956   20942 50167   96485 66450   78429 39529   34541 72790
4731:  71561 53267   01339 86700   65435 62513   23086 01101   78810 20426     90813 24842   46902 61592   90362 67758   16643 16858   66850 02329
4732:  72776 63502   87896 95147   84579 25968   25221 28915   49994 13958     97839 22808   45590 69431   32867 40161   66660 72840   69445 87497
4733:  84464 06679   30171 78565   60295 56202   12166 06547   01159 05600     89260 86510   96945 23976   09403 16567   75201 28752   91591 37718
4734:  73759 62727   93302 14797   15846 48812   52646 41673   42260 00349     99823 67612   64958 77823   77188 67911   05991 78584   38622 49305
4735:  14539 40319   62209 13734   06370 86272   07290 22095   25708 21812     62672 79200   91980 90447   97229 21483   82015 11651   38065 09454
4736:  51247 89294   48977 53041   60651 56345   74760 98739   30358 57448     42653 37858   35712 74104   25307 80524   98557 70301   39588 55991
4737:  09516 85779   85058 28901   35143 73429   32979 34523   88832 22719     70654 51373   62282 41089   70185 18406   59002 72216   78015 10750
4738:  83619 85272   61067 00449   95249 61786   70723 79941   86857 22785     10387 78420   63596 02038   31109 72951   71397 69724   12819 43214
4739:  98414 17186   73978 92230   86330 25541   70161 44022   75048 11220     09808 94627   56085 52852   16290 90936   41625 34218   78340 63648
4740:  64753 92215   53522 99315   55248 11764   48583 88381   26366 79973     82386 14044   97814 65858   00418 41374   23941 43132   20027 65411
4741:  14454 00691   19412 02622   13934 44031   18181 79649   45754 06873     76199 21809   79204 30782   97405 52952   20449 59460   98038 14081
4742:  54063 11052   64851 04019   38179 90819   50596 78023   39328 88615     75376 24590   91325 28878   40801 25770   09286 58697   57441 17440
4743:  03487 31663   18697 05151   32710 86802   70617 04082   30972 77927     36756 32654   12731 93817   87978 28212   05099 40388   17082 10730
4744:  03020 18008   31341 87414   91604 35677   75083 92847   47003 39963     56658 56467   63230 79270   04499 54955   72650 79055   25683 58406
4745:  21731 96451   26431 04874   81911 88211   80541 71633   97330 05031     33647 47379   60463 62140   13175 05951   00051 10502   70747 77548
4746:  60658 67594   77391 44478   31214 83681   38866 10198   03571 09342     40625 68549   26079 87979   11957 82509   24579 45569   27011 90374
4747:  89120 87683   36037 37617   36085 55171   46431 09725   78297 50209     26405 19712   65772 24913   43695 11631   75587 58972   50478 29500
4748:  70105 94777   81429 79553   74265 22278   13537 69006   95513 96555     08055 35762   54357 24415   16525 02377   24789 96008   86386 02201
4749:  16429 51574   93423 76138   34579 51169   21522 09362   42863 64513     21521 94575   24042 18380   74193 92225   90697 94002   16410 46813
```

```
4750:  17696 07688   43840 01930   55528 47500   63059 48155   17827 41558     64828 69776   74637 14185   94411 55526   07052 30647   96750 56845
4751:  35908 52720   11440 61220   29843 13572   44980 07154   71885 30707     84295 88761   51440 09468   50000 67958   86046 13232   26223 45296
4752:  44643 70310   69901 99103   69841 05705   15441 43060   84730 66276     80018 06164   22893 68332   42170 72758   91321 29440   98588 22899
4753:  67190 43554   03507 53146   96110 33734   52804 55707   37793 67315     58521 71629   76139 09400   18564 52374   92433 71898   37997 16695
4754:  72807 99564   27173 45185   76336 14903   46642 11635   42806 56851     96037 45966   30440 60669   03829 73294   86842 97166   48187 12724
4755:  39443 01853   77928 60637   31285 66138   34023 96699   87728 10261     67971 73273   30464 04592   45214 54308   59964 61590   08696 70388
4756:  49869 85248   19719 03576   47066 66117   79376 04638   63538 55909     23343 61895   98931 62234   65414 75033   88244 36593   24340 64670
4757:  13137 56930   79907 66752   71781 08573   43709 61171   28847 79081     08177 15991   67531 59961   33271 27614   10209 18775   97197 94730
4758:  64389 64533   09586 34055   05268 49521   17947 98323   21142 75161     22258 39933   19073 15428   58342 21512   58182 77683   37886 19359
4759:  79726 75260   21971 38483   42224 61467   02571 96009   68646 85017     02656 00245   19782 54318   42113 72246   90846 63560   55754 08741
4760:  03739 43952   07156 64049   71430 55758   12961 40170   42074 93201     16611 16426   83372 41051   70962 89084   61417 07126   15058 96540
4761:  97666 36051   46538 05454   18928 93678   82450 50008   86115 85181     14677 91179   78012 77002   82409 33132   20676 44307   51008 99672
4762:  89331 47524   70546 08974   40378 67221   75480 88742   58869 98755     00496 88731   34259 95801   13308 12613   07109 74361   29569 45873
4763:  87231 02449   42099 34454   54957 88189   22129 27914   76339 95114     29438 95105   36520 64392   39867 96016   95846 29083   74728 03348
4764:  41307 66465   37899 35341   93534 71611   04519 87475   35150 35515     64262 21568   25071 25094   37187 67890   60812 31495   67507 13206
4765:  77430 30110   84976 10491   01869 39923   21767 93608   89399 65550     68735 90959   39631 60590   14391 13413   49681 34918   63300 53747
4766:  36467 84294   83917 91596   41253 05129   10727 42853   70185 31043     90688 56127   43930 34167   81617 00153   63875 96816   37200 67565
4767:  95144 86360   91061 64392   39423 01507   55118 74586   24025 76000     15715 12828   88726 77769   34533 05629   02544 61387   54099 30101
4768:  33420 48405   61339 15783   69412 46821   26647 69463   13752 14615     14243 75946   35546 00523   31212 79536   24572 41866   41194 88795
4769:  32683 69053   46314 17638   05678 16124   24344 81038   59468 49011     69127 05386   34939 84923   71356 96699   89089 20298   30274 87620
4770:  14821 98633   78592 09867   12469 69437   50750 49294   70749 78101     19846 89991   69547 55096   27253 78697   81838 11417   84461 68687
4771:  36995 61835   20802 98314   01522 08753   91590 60035   90768 17552     16471 94283   59950 73350   66584 24062   98824 73279   97736 89662
4772:  01783 85767   68499 32146   58650 52576   98769 42496   52158 90673     09509 35214   26988 98785   86614 14537   19994 40185   23200 78320
4773:  55309 89196   51103 29019   93510 33028   76461 64099   40557 75736     50471 30032   14544 09354   33390 74214   59256 55120   68506 31600
4774:  75155 36995   94229 11938   37736 87491   15130 65833   19846 21942     08703 67357   71403 52660   12303 07846   14709 53252   69629 30528
4775:  60178 17019   35487 96910   40005 54700   64102 85931   09822 19730     24434 84317   70497 64683   73036 13293   43841 66052   31319 36696
4776:  69548 86449   24277 17427   64013 68340   49220 30140   29618 39463     23401 90380   44417 19103   32803 33858   51740 61572   77567 75619
4777:  55014 41915   31487 03402   34620 25249   48708 22701   96766 83078     11066 26267   79420 34314   88055 51551   68897 43782   39359 10354
4778:  84381 17008   92890 59254   32250 26993   46030 03280   36433 75737     73272 78164   03213 16386   24535 91510   88107 24619   53177 29362
4779:  18660 35488   75075 19530   21144 53992   63965 57560   13564 11754     89859 34031   87281 86457   45903 58548   18626 20844   41451 65665
4780:  83126 34337   40589 85273   60710 57748   78530 78003   77629 11166     49783 52937   78073 51698   92403 63257   13901 39004   82381 76178
4781:  96097 64228   06156 44186   26065 66876   04650 31762   86002 60849     51683 22364   28138 15011   59858 40379   84491 63061   04926 75054
4782:  79013 66668   91097 25498   92652 76047   72076 18800   34297 80245     38776 34641   11550 65129   54610 10180   65371 15721   94729 24689
4783:  12950 15031   04048 83621   50312 76372   53824 90032   89994 00252     66782 43560   15498 14216   27258 27120   72456 38731   72449 40920
4784:  80864 25718   48137 43440   87679 50281   78596 68359   62754 76040     87723 56148   55043 52585   20205 21925   45420 52895   11984 82889
4785:  60701 97345   73443 75885   43968 03650   02741 12158   01783 04523     68784 15789   44593 24214   01800 26379   19264 79091   81966 08577
4786:  86050 96247   41311 08803   13406 41070   82528 34341   48269 82017     31849 86246   92642 98915   39478 93138   52886 33147   18565 86211
4787:  65279 01941   83388 86978   02622 88359   92447 52529   27135 23002     64828 13966   27930 30325   42849 49109   11516 05697   63115 28129
4788:  77507 89305   95619 28307   52818 76173   77369 19557   60561 38475     46096 89977   32017 16557   51449 22603   91039 32559   61466 50599
4789:  08637 89077   29654 68957   68917 19724   64369 28612   04547 45744     62314 04576   60580 98635   21502 87937   61807 65191   74365 24871
4790:  19853 15380   15409 63156   58209 00226   30276 62241   94287 36882     03986 41260   50175 73218   56984 02945   24207 17724   14214 25301
4791:  21534 88657   94135 68936   50086 38121   97728 05421   61984 01914     71171 10300   63735 40123   00576 17108   76505 57792   88652 91807
4792:  33737 12037   84654 83939   45250 65690   19504 28379   77503 05780     47181 16543   92947 29475   15223 44883   87340 89205   73124 46916
4793:  51678 36194   44997 81771   42764 78575   94306 44755   76471 17874     94823 59786   97175 30858   39975 43522   94888 54374   90708 19314
4794:  46000 02977   05520 96750   16228 61597   50912 47294   29151 60387     46808 42729   08098 28088   98236 11073   50110 79176   67606 74261
4795:  69830 06783   45910 17101   12542 96754   88793 46362   44162 36274     98871 25132   74235 41793   63352 17897   58351 49397   35265 84726
4796:  82436 62329   16619 25160   58537 98111   08678 95044   20713 65989     45554 41247   45909 49868   82771 26163   93292 92411   97053 22874
4797:  65067 31220   73064 24034   93531 88615   73821 99317   80231 95804     11485 25546   66083 46377   46092 87506   19171 71485   77957 19601
4798:  78050 15793   99832 75410   36445 58618   25269 71366   48983 49992     02681 30319   18396 37761   45846 35145   11042 24491   36887 82791
4799:  42240 09666   43476 06728   33234 15489   14149 79623   98399 70070     38832 44112   04194 91521   32179 28997   96497 23581   06836 71448
```

```
4800:  6865043121  9493187402  7680043214  2815066784  3508785689    1159921246  5711050333  3413096283  7072218740  9050440990
4801:  5304080208  8488029432  9050018701  8757822470  8922667976    7459246735  8923372369  7905876740  8146268511  4947496390
4802:  4812082822  0691584054  3602154658  3139856368  1834770280    4717687940  9079704070  0583351958  2672136706  7456616860
4803:  5527930829  5699490290  3629356381  3653544419  6294575463    5118847047  7228365814  3829290160  0814625963  2039265029
4804:  2491922262  0521046502  3401277381  5276549617  9244354972    0964558951  9383057517  9361972734  1156834816  6052862010
4805:  8581520734  8598602462  8851095732  6046304241  1604075227    3282853791  0284631130  9302307392  1413126394  5780534111
4806:  2509024487  2594478495  7765695730  2327042936  3076286169    4816478661  5323645121  3404766390  3575090744  6277168316
4807:  5754021868  1089578165  2097942340  0089735838  4280541576    6550575192  9711782692  0695344859  5546720501  5773761798
4808:  8567497486  9352747101  6924573259  8574817492  5176547215    1931590263  5356660297  7786674452  0328462677  6086793182
4809:  0169164450  6523446628  9598972154  2522232494  9079539041    5992725016  6987892431  1085582758  8738973826  0291380583
4810:  0569884439  8597951803  6021475600  9936011687  5219696935    3074890979  3723775780  7745398059  4498030107  5538634319
4811:  9606460795  2628522522  5303368351  5781652801  0660186670    3425444792  2990068461  3921856451  8753115422  1601309802
4812:  4983307028  7004978744  0792276796  3572923151  7189137116    1162427743  4945916329  2462420068  0985483487  1460803984
4813:  8870654885  4087989598  0098288402  1601012947  1793421050    8863967351  4384472730  8270612866  9276959841  9102815456
4814:  5768610944  8633054966  4964640144  1007597793  6955745164    2118611754  5707907752  6790251438  9003807254  6570193492
4815:  1243854373  3647906179  5787488765  8652025969  3989406957    9983928171  0023264479  5215020751  9288838812  5719357129
4816:  3983723572  0863589257  0822064306  0329749573  8456156552    3512906055  8218465482  9027418151  3340291639  0825213069
4817:  8864383454  2752824123  4818382101  3529999916  2643035819    4988742343  1897213695  2262779760  3622380998  9559705363
4818:  3174957526  0226366271  9539427907  6660341074  4571328495    5371018467  6826858414  7123068344  2384537441  4403281367
4819:  0292930678  8386361431  2615557237  3638367956  2734306891    2598578625  9207752241  6107721695  3986096044  5579403912
4820:  5557232461  7951026683  7502491020  3178843564  6565853278    3845776994  4976160569  8869253519  4629966447  7832072750
4821:  3083488142  3810622502  3517768462  4957221256  4708625192    4463332714  4917049647  5694251807  8298283837  5995092670
4822:  3811815367  9901103111  1162370500  9016535745  9821072917    7302245969  6198573208  5313391635  0339311502  3306615692
4823:  6220853842  4820950862  5312479472  0975473148  5563991841    3331500969  9283278059  0750126726  6827722577  3587809199
4824:  0656184808  9992810272  5431544571  3240450770  2980910398    8112951558  8240118852  4307458485  5395971928  9836721480
4825:  2158865602  1978315395  0842933980  5522135205  3243613692    4505059449  8727042854  8476031360  2437396534  4441633634
4826:  2968377130  2579150901  2983242161  1910005378  6042781588    2560869801  2791179584  3569688063  8184586718  9031806300
4827:  2699332112  5415902311  1857786226  0056430693  6184789419    4096434086  9076259628  2872017788  9755088292  4510676572
4828:  3104676563  7731828355  0570055532  4100648442  8152385954    3596776368  5786701505  8306887945  3927790501  1206993779
4829:  6335900407  4227718078  4573526586  4715279259  2222446581    1380644029  8443801924  5331836180  8671020619  0415826923
4830:  2253615603  6876029685  9209558103  1057476834  7974217600    5759597589  8381590519  0028132330  6955965453  3529228410
4831:  1133350027  6000712369  7166616421  0144021849  2088309430    9572969016  0225072046  8614822121  0084971975  2279742638
4832:  4012981542  7991135110  9360193450  4448998338  6347025003    0013502566  4810286566  6617226996  5543583968  5159854015
4833:  6634498680  8068816251  6255931916  6909248236  6692312836    8213607890  5274314344  8521068527  9646020086  2952145039
4834:  4880273395  8308482312  4796419977  8367652162  3345432089    1785259870  7631637182  4346817754  9105442686  9536540975
4835:  5812892082  0560130164  5639329349  0251935356  4609093640    9330876575  8326487788  4077805236  1509297892  3226818722
4836:  2271374134  6934874100  3528535751  0105670994  6067514626    4543491923  9625132688  4620730533  8963602676  7865410757
4837:  4397373903  6519790668  8316967223  2958489564  3315770899    8994433853  1012424840  5600984062  5713948542  8836480939
4838:  0589293342  4955049089  9374034675  4327311879  8065110279    0317019858  8900179493  6090880588  0231690930  7784479395
4839:  4477953579  2578860908  2940176898  6519826403  2978968527    1456337577  1300620295  9967691547  5123366965  8807147531
4840:  6377248242  0089301475  0574529802  0572581025  9556141401    5915179212  2746288508  3099544937  9695927377  8833133585
4841:  8165170667  5610428794  3867290221  0917531930  8601302131    3534039351  9629543489  8422360940  7804931818  8704646999
4842:  4770327966  3750151802  9901306552  6559784666  6655853470    0353601292  9520533215  9445329941  9374193616  2924327557
4843:  2814898706  2695553832  7366651672  0111982468  0315110977    0749924583  3970822522  0955547007  6783024379  4208473629
4844:  1092458295  5608810563  4431782979  0772521709  3779556266    3519766478  2712821419  3907487075  5400759653  5426932052
4845:  3021830362  5321820426  7548106270  4755197200  5555800385    2780033867  8572489523  9333671013  5711321343  4824279269
4846:  7606616065  8180083733  6149547782  9037026611  4699231268    1724704258  6297508758  9775282007  1469539065  6052862847
4847:  0153403615  7353710111  8000307055  0574046752  5823121687    8118560516  5989567126  6522477779  1718384381  5050219366
4848:  0305071208  6075384583  5369640327  1887207896  8573593638    8323878874  8067073243  1286071762  2061675661  7273208863
4849:  0406042637  0485547169  1666976182  3931115540  8093595876    9806551902  6550900344  3724832662  4718591219  5182197082
```

```
4850:  72830 97334  60148 10482  02612 44376  57270 24472  94519 39051    55261 13671  16372 11069  42137 85469  61644 11727  30876 27632
4851:  66195 57064  01129 97404  17091 54934  31828 38325  90595 48924    82970 67921  05916 12124  20941 44093  64808 42785  80040 30070
4852:  37377 51535  85614 31020  97749 13844  76415 48546  29612 64328    98897 26573  51378 28438  49096 22192  50404 58155  38132 43911
4853:  94717 90813  28740 99366  15491 41475  31965 95892  88182 89137    68118 77524  11419 87053  87046 31806  24072 85804  57208 18534
4854:  42118 30129  61138 70594  63539 09386  52317 21598  42060 50004    23549 26960  57793 46974  36970 51088  40213 66988  61443 32706
4855:  15873 27656  68639 62707  03002 81132  29225 62242  74935 29474    01349 51616  56556 50899  13680 52855  43999 95973  39920 57026
4856:  31267 40657  61663 74472  49562 24452  46246 01840  66464 18431    94640 77126  93635 49989  11219 96348  98933 91110  31072 36899
4857:  99085 82889  64586 05491  67981 23749  92508 67940  67296 51070    94865 00272  34656 47456  51138 39070  45761 92905  86211 83204
4858:  40882 56620  48317 30164  11587 62248  89743 01474  49299 70208    25223 65496  85091 37732  88638 91515  91591 42786  61489 80134
4859:  97067 02337  82900 94034  91718 19065  47286 61733  81160 15436    76278 31541  85339 57457  65358 99802  19524 72199  62648 45603
4860:  23599 56139  63387 01941  93505 54479  35968 17100  85511 27796    34628 80426  35170 10877  37460 16513  69623 66070  18032 37016
4861:  62289 50935  80167 50098  22689 76617  53765 05577  24140 04558    58228 30232  46609 48581  32826 67447  91571 57412  66233 09586
4862:  75909 26706  72705 50309  84229 28546  80406 50350  60028 89889    42879 14835  38309 81297  30525 11195  81693 84988  33426 07430
4863:  61277 14972  32078 76255  84732 00902  25805 36980  61656 57872    05995 42547  40915 37039  25558 20599  89659 30191  59189 48247
4864:  18437 06225  02264 28064  13260 42133  07332 33266  15748 63922    47207 79207  13095 92551  01472 62695  40550 65532  87837 14180
4865:  67003 15846  28019 20719  11979 94276  53875 71134  81583 31633    65684 63400  82519 69811  71494 66864  41351 27842  25305 71343
4866:  20203 07127  92849 60234  13930 24925  74940 72061  92399 46402    13288 13517  86657 72577  62805 01340  48808 19316  81619 54246
4867:  09236 14319  25981 93637  71075 43707  29859 05911  83225 95161    87239 18541  58852 39924  22694 03921  27375 59466  51089 87486
4868:  65464 07001  88988 98972  31355 25637  51046 13144  07535 99344    76930 20145  94023 51730  79986 43133  93740 75351  62704 18042
4869:  06099 54767  25059 04741  08703 23189  31437 80676  09032 64671    82158 10145  98120 28791  59069 01618  65951 19382  49840 36314
4870:  50563 82159  99902 16975  87757 02406  37219 21698  38594 04119    80209 42071  06008 92190  89554 59564  23850 13930  72381 55096
4871:  76980 79051  42143 26336  89012 92169  24272 71614  30755 84866    53654 00976  04986 74528  20510 07320  37809 50553  93744 81054
4872:  98976 38276  05520 76119  45797 47202  27656 73226  03020 47233    38062 79790  16085 93265  25955 55871  43764 10191  34042 64753
4873:  43906 28989  20114 70590  35663 38902  71478 12114  15149 11899    26906 36549  97669 37770  27892 57218  55313 58556  70569 21274
4874:  94790 49958  28598 39610  61990 12636  62997 68574  97303 71020    61107 39994  43215 68663  41142 06258  18679 05837  70128 19864
4875:  44139 76311  26374 70982  84983 57113  93350 12998  16384 24287    60627 22433  41827 09585  87756 37412  37710 20425  68127 19060
4876:  55248 12204  69832 12871  84885 38017  33134 47165  67971 13240    49218 00254  23595 01978  10005 83525  01652 21094  93779 87854
4877:  15811 93782  30326 77277  65144 96024  53419 62672  18425 26197    00426 72484  84933 78783  17048 60862  69574 40649  41005 43672
4878:  97552 13041  41806 39208  46324 77332  60618 92030  55389 52871    30204 21620  14245 95431  25195 57961  95605 69967  65407 71182
4879:  23763 33073  00314 66691  32751 74376  79597 01636  91128 43652    44063 35440  21962 29467  39070 85289  86968 01603  36237 13506
4880:  19438 40176  41700 31625  88644 41875  24131 76179  09282 40209    65254 79541  95464 84798  03064 47140  34741 48194  84611 63723
4881:  91463 31447  35113 34977  40882 25881  06475 03346  27911 19334    97408 70994  02148 71565  64924 73409  23184 31159  25045 59513
4882:  48263 28647  16045 33110  62379 46428  46786 77592  48432 00283    07180 18420  05151 74221  70228 87064  73625 76518  98758 49165
4883:  64508 00922  26312 03976  46670 88619  46094 89546  83271 39379    76785 66262  45962 25627  50988 17585  20994 54344  39827 97048
4884:  97256 73201  62852 27936  73881 14929  81924 06203  02151 87029    24737 86923  29175 22162  18554 26794  87928 39148  61781 25849
4885:  69813 85589  54522 34740  93401 18521  67312 03683  57491 31712    13616 50653  59922 36301  80174 79527  47383 82610  67486 51839
4886:  83619 32473  37443 25215  43554 84815  23357 71296  28382 21342    78643 08296  52959 84022  51908 06969  98632 37863  51802 48187
4887:  18628 25337  07570 03586  60583 69954  60073 01053  88021 74217    19834 99731  20974 51756  53021 91375  63890 65052  49605 30529
4888:  22863 07955  01971 64216  34901 50314  31015 51737  99616 89182    46230 10822  17789 03180  25301 33390  94696 97695  02279 78462
4889:  61604 73100  34352 90862  30464 02758  28638 43111  75287 61124    83481 64820  89684 50396  67177 18681  22374 09886  96608 06529
4890:  72044 04107  21067 82299  09295 08316  36144 43077  29408 30298    61898 40471  23528 76294  71004 10963  50866 73836  41383 51259
4891:  46131 50504  61474 52097  85866 00684  73774 55674  13652 01893    63423 60435  09178 03660  50691 02934  04149 62948  83822 16156
4892:  83602 50935  39421 56723  42346 73547  13082 46881  36899 78433    65682 60682  70768 13436  75231 60437  68144 62968  35700 53982
4893:  20094 72735  05404 50862  02171 86287  44053 04378  84922 53335    86140 81213  91572 37933  84091 73656  34834 98521  97393 50771
4894:  54716 53951  26111 52218  69682 05422  08054 29158  08164 41448    33517 49923  52678 24671  26776 00883  89108 07451  27560 63820
4895:  51164 58791  40094 51807  24533 32712  16527 24015  29097 53897    79654 58117  81703 31521  35927 16233  41886 63346  19333 70070
4896:  29734 24597  02050 60323  39061 98003  60292 64665  60262 73934    99240 59286  16817 16523  63088 42102  67184 57718  89331 35511
4897:  06725 13929  50555 69745  03629 64822  19925 34178  26048 89150    27919 06240  05467 65316  58967 25621  00070 38566  39377 34693
4898:  84552 60610  42745 98331  90474 56108  73585 62474  23351 65432    25762 73234  10605 52775  99106 64879  81967 41650  61742 00047
4899:  58236 37884  27333 23714  02584 08696  97636 49884  70219 90692    21313 85802  46343 15809  79796 55197  28470 34696  48316 30203
```

```
4900: 02119 65949  80198 93474  14860 25776  91929 87451  73250 24068    23182 07826  37095 80484  84740 09081  48950 03821  14530 84415
4901: 80120 94731  28625 08355  58810 20533  91268 95926  87145 95366    71830 88104  81308 22919  05604 31971  36470 46310  55350 50888
4902: 14910 44914  47975 45780  18109 45167  42023 20248  73392 02744    51096 93644  95658 18654  89428 43854  80803 05508  57771 11137
4903: 76475 87373  87338 89609  43458 05681  77846 72937  90896 97783    28517 72088  60448 26638  52277 37972  96347 26104  88154 30166
4904: 04467 68938  76290 51191  29837 30828  47570 55767  21360 99506    97896 22612  12272 26184  00140 56618  38253 74627  89816 98250
4905: 83178 08677  37135 02834  32275 46995  85734 38842  78836 78920    29495 05458  09781 23044  74221 93839  58088 98936  31856 86921
4906: 25410 47433  69962 16474  77351 15351  85220 37319  48314 91159    04800 36378  36820 44509  14384 79899  20964 19718  38828 92112
4907: 43703 52305  53374 73482  63894 46211  13495 57599  17757 18285    27201 78337  51081 10812  83016 41983  79444 49540  50575 74421
4908: 29808 24686  25442 24143  10160 06067  04507 99649  49584 77061    06065 62059  87918 54639  73868 22154  03822 96904  41926 99624
4909: 51221 94179  53452 54430  69888 60524  75106 95429  46225 45490    73437 16563  43248 84835  73682 75141  53799 17559  57934 16908
4910: 60760 46764  11692 18973  88982 40304  03924 42815  53190 40772    88857 66114  26340 16522  61424 80457  69373 00808  80110 10520
4911: 72147 11295  83437 74540  94188 78043  34391 34256  10111 38236    59789 00817  63640 97041  68672 82386  03254 90218  28144 60881
4912: 05921 33186  56825 48989  33879 87281  67251 83670  38388 00065    33586 91518  94327 81900  87718 28159  18272 96058  05215 19252
4913: 44135 38890  94608 08497  32641 60456  16457 59360  09429 31935    10911 76908  34054 55628  44752 13275  66406 85483  37433 14292
4914: 86340 04347  93736 40900  56410 12193  12522 50358  78124 78624    20523 40018  21027 00091  10164 91835  32773 39458  47692 37697
4915: 86888 19376  39028 14828  47393 95968  96330 33364  45544 45688    80921 15216  10250 37521  20972 65999  96003 88381  69088 13499
4916: 79560 59363  36880 68718  54103 75635  57705 96423  35120 99378    62839 29184  69589 52752  47067 69000  75680 35114  45106 32205
4917: 34853 17355  47991 31726  15566 90533  91072 72081  13950 34693    83066 47169  00422 12470  86763 42929  51225 79497  22893 18219
4918: 30655 90266  25080 34265  78514 12168  77743 63872  01649 58647    20334 41874  93534 57212  64075 45422  50495 41476  06331 76037
4919: 94074 91655  78183 99105  12348 38250  51869 93242  26116 39857    52258 63981  56613 66291  94281 01607  66561 81014  58991 57610
4920: 62060 75969  11825 40179  22399 12149  68983 82492  03169 74759    66080 34758  60595 60034  90702 05728  05036 71768  80076 32160
4921: 44327 04893  70771 51846  34527 38646  66201 32048  30659 61350    98899 32935  51592 18152  37311 43629  56232 66256  75768 17482
4922: 38385 10734  81684 38133  27811 19303  11857 31669  62301 42003    22443 41089  38299 60376  79043 86412  55047 73045  73094 07873
4923: 49391 67656  51483 58358  28534 78590  32012 90500  22732 14774    94316 67202  37992 58078  70517 10273  78896 98423  69893 06489
4924: 85655 28974  92465 99468  21576 58760  27529 88408  67501 10315    99612 70721  39785 39901  66833 97567  15929 70258  12107 28353
4925: 96001 26931  44668 00992  87399 81185  73733 70526  09206 87225    92315 72557  23976 46034  92879 63983  26556 70514  37128 08869
4926: 43199 61019  96547 85512  31561 08403  48002 44073  45027 84379    44811 90444  82993 40524  03256 84611  38748 40417  87315 33469
4927: 69705 97752  16519 54454  32959 69109  49015 38451  44499 87444    60729 99876  83305 71938  84776 69686  68563 33991  62232 95936
4928: 43051 31062  39130 65714  02033 49150  43702 58381  68424 46141    63271 09730  27184 87594  50700 78227  16496 84711  86952 09749
4929: 54099 75723  50701 85514  67924 77307  27096 13868  20507 99236    72973 11356  22861 89582  16380 40789  04200 67886  01387 42671
4930: 82681 97088  72304 44857  40437 28113  94568 22338  90583 43916    68954 94243  07375 35719  76075 55435  19129 66935  24470 41177
4931: 64276 45695  38928 42301  36034 33817  60177 00277  64406 00957    70351 95831  86745 41548  79114 71854  61079 72123  10023 58138
4932: 81696 40161  45774 12691  15177 35471  79896 55669  98211 34594    43979 79433  63624 13457  43681 10968  27463 97182  74158 55977
4933: 32523 80994  31881 15713  38404 35251  84174 98789  69721 00241    92526 92801  79838 97178  77421 26673  40857 18850  94966 54270
4934: 75683 97432  87267 29548  77131 09405  83067 14190  08434 18623    94864 32080  44168 15008  56323 71529  47348 23087  93203 86255
4935: 28988 45099  81922 79753  38888 36473  17227 87238  47624 13706    93917 92997  06999 91572  10314 59800  70146 13547  89535 58996
4936: 05567 54020  78257 76320  30686 61836  67600 74048  55346 64147    22205 27273  23898 81194  41244 80336  25828 65318  35146 86484
4937: 94676 10497  42135 33001  30935 27441  68619 12097  35719 36811    46367 76233  34887 45113  38809 42704  86951 80040  12229 56096
4938: 25619 55999  10318 60576  88556 34206  10831 31347  46351 69290    71516 70915  90358 55962  09354 87334  64104 29127  30025 42216
4939: 40888 36667  54383 00393  92753 05066  74018 17457  49539 06025    50311 06407  03339 72204  97202 67163  27204 39535  22172 52664
4940: 66904 93934  94008 92606  62463 55322  82817 20896  96726 89070    60784 27730  47560 04781  16355 16124  46808 04815  02280 75877
4941: 90224 07634  41040 62752  61168 81041  73429 03127  79763 67267    78994 90821  95357 49427  16171 62101  30445 54569  06390 41419
4942: 49501 12823  63003 44735  00457 08228  84447 74182  53555 59384    70221 25426  56635 59847  67155 41181  13243 32004  86236 81106
4943: 37833 51390  41019 16729  94049 37043  62930 94616  68512 86087    44174 11577  43599 03044  52878 72078  20910 38128  82704 52615
4944: 76515 62155  84331 01985  90633 37571  13915 79248  06849 73298    31993 71080  44954 41955  13383 98921  33432 87455  71140 54529
4945: 50859 47652  07375 34639  64150 70816  02484 11323  54651 61112    64280 08370  58526 33632  01018 28965  87156 79626  24606 53439
4946: 66742 78469  28939 82896  10174 71026  07792 75831  31581 89033    12954 70254  52940 36392  48879 94616  11016 17240  00895 99941
4947: 33523 10572  95013 91347  02235 36222  68249 92850  22984 71198    87782 30859  14233 25361  16692 19462  10249 00834  79981 20512
4948: 88832 88113  11478 54419  71645 89645  39273 30024  44132 85540    82304 69395  24803 68431  12248 12412  79621 33280  37451 44613
4949: 39220 50352  20867 64430  97857 49381  07773 03617  54142 89576    64475 61052  45087 57639  00247 45577  60883 11954  77445 89763
```

```
4950: 72847 11975  41503 29925  71097 27054  10821 88447  20316 32710    78736 70317  09366 08825  20678 02995  34473 57552  82403 17747
4951: 14977 80257  32301 58258  84763 97194  55152 33370  25619 27159    55097 62026  29984 09318  44787 58968  70324 66447  94289 93374
4952: 32616 44081  62969 71944  80219 67189  88869 24512  62969 89372    34104 90188  06698 56690  21389 73882  34489 61990  15365 11737
4953: 35007 22161  98475 93757  07627 30027  74133 03958  49048 61727    96110 28215  44749 96184  12343 95301  71338 95149  63251 47597
4954: 18028 29754  89645 28868  26946 89858  66945 50070  39441 60346    61535 67348  07191 06616  14161 76412  52649 29978  92299 86955
4955: 64279 83769  97941 11946  28750 93125  88696 80664  82954 16307    57239 11212  75078 44556  33200 14998  39673 37232  80309 67587
4956: 30732 84837  41760 81142  53648 03907  19972 35768  48473 27134    42449 14107  65050 21951  91988 51386  11514 98196  74601 50619
4957: 46883 55238  57608 65878  07441 17944  08633 80739  21261 55833    49343 35417  44377 29976  54411 49483  74030 38246  02530 36193
4958: 27251 54639  62803 04930  54410 93306  84498 63277  21153 89957    33426 83308  65058 99102  87059 88317  31448 80980  02488 19031
4959: 46830 86581  35594 59853  68913 47029  95498 06870  20805 45083    88176 08329  91633 94399  54759 95836  90007 36103  24140 71708
4960: 08817 40688  15765 63129  45670 70313  92969 37886  72712 88619    43650 43646  89777 96736  96773 98392  60644 54677  41070 14720
4961: 26646 35792  18253 40332  48136 90641  37227 00387  35548 13469    70497 01500  80518 20626  51518 21017  66530 16808  86563 84948
4962: 52089 66005  45592 04652  67030 81418  36053 32190  41787 58533    51635 37049  66014 31652  27941 77162  67230 22767  73391 72567
4963: 31423 21543  47024 27198  29300 12991  08452 70448  59271 99461    73769 40703  58119 94115  72711 12403  90166 29926  64323 17189
4964: 09949 24084  31797 61510  82418 53289  37921 29583  26130 57459    19116 74707  90873 51643  08525 03412  16574 40266  00205 82868
4965: 84063 94661  61602 59013  98378 53874  19633 08585  23280 60285    15498 46927  56612 94048  37770 45399  81536 03398  71376 60096
4966: 58353 06138  73859 66811  74820 21038  81925 22812  10361 65924    30461 60318  00760 25272  64318 56891  04881 81697  90088 64608
4967: 27498 68697  19557 15199  13840 28468  14090 12138  70457 01394    92262 46281  07703 75513  58913 78069  21954 51867  48079 35210
4968: 41070 65447  77394 08244  56128 90824  81194 84373  02624 64845    14547 77693  20738 90754  90548 43819  34051 16622  57211 65783
4969: 82136 66881  19905 03958  97048 80432  48701 77034  44551 85623    08998 30600  51218 89996  17002 57249  39489 11910  88466 00896
4970: 60132 54412  23449 80946  91865 13617  98190 36144  24558 69157    50560 20273  39513 31880  95166 81121  38263 73628  45087 40598
4971: 04858 36754  69453 34379  40279 74977  72588 17920  47705 72351    35731 24648  75767 69051  56374 30496  17097 07176  13225 35395
4972: 17789 50985  70161 91342  42096 21440  70805 52222  92567 02412    46978 47345  29199 69904  60536 99628  30538 22553  82109 48478
4973: 96600 51384  97528 68455  67967 53766  62399 99738  07276 35730    88029 81365  91603 96904  73101 19398  88016 53901  93367 01347
4974: 39208 14371  07814 73620  68190 53754  38033 34553  53972 53836    31346 52827  34661 56221  40189 13429  73235 83445  92373 91808
4975: 60565 58699  44055 87650  95390 27309  01355 14749  78819 95814    06096 53498  76849 89266  09008 94102  77650 46193  56470 92683
4976: 69199 32207  43341 11091  55325 10917  71614 53760  90924 16451    63226 32017  56390 09099  75364 83630  08130 45437  54874 07097
4977: 25003 44675  80357 75415  85832 76799  19433 29841  91973 71405    97725 71757  42298 29480  11157 85338  63247 34046  18669 06814
4978: 05293 15138  15315 34725  68610 55445  26652 93879  12492 96647    50868 45426  27072 22857  50521 77536  87658 80291  24060 24277
4979: 00987 13224  81290 76718  34408 81042  32106 91571  66437 71912    92679 77932  12722 78503  53389 53003  76145 94390  33836 89291
4980: 14891 25698  60310 87778  57970 58496  51787 30553  25476 12395    90227 19042  97991 46461  40522 40863  18623 88562  59853 64599
4981: 31608 20594  33331 92943  98496 90774  99680 31037  78426 90910    54372 07695  50873 96807  14495 12255  85789 58362  42670 29878
4982: 38500 86307  33586 36959  68069 97187  62005 21597  35574 67863    65864 26910  26063 44224  28817 34941  97800 12201  30279 14534
4983: 59521 45475  90602 04606  82657 35488  06825 98705  34211 81100    67170 52892  29394 55686  14062 13378  40354 12647  30379 20558
4984: 30532 38948  65973 32285  42246 81731  50683 75330  87233 82833    61410 70523  99003 32578  41721 38929  26753 21628  53363 96350
4985: 77131 90414  42284 05230  08438 23096  50020 79328  36426 02981    08536 65824  20983 72013  25431 85076  59115 76182  39085 04052
4986: 97588 70595  20516 40466  66941 56550  52733 80495  00384 23434    19363 60368  20165 84865  36106 45155  85967 38200  74394 17189
4987: 57741 56522  75411 67100  25376 25063  89934 80183  08113 31756    52919 30422  54164 82232  00874 31719  87522 34571  12440 08867
4988: 99758 10594  96310 88141  42029 34583  47295 28848  74378 81309    98610 61125  20133 65001  78201 88398  69335 66927  90765 38755
4989: 95411 06051  75911 93240  69436 21757  68328 42431  24069 04823    38762 17411  34120 73077  38627 26818  58480 62463  58574 76251
4990: 88226 99375  39188 26135  95860 77988  18109 84643  37976 73346    31558 70505  77779 29419  59217 42190  04344 65263  00285 22907
4991: 90028 27025  23366 12613  80333 92133  29415 61923  46485 00266    72335 93858  43732 11592  17595 48557  18722 66988  44989 72919
4992: 93500 74207  88774 19225  98273 46237  41743 64847  11595 30820    56595 02941  95345 36850  08794 51700  41722 71643  34803 42834
4993: 22425 86271  06912 06888  36619 11905  30824 76815  48263 36481    49910 25412  04189 78044  80930 88306  67863 32530  81657 79524
4994: 12782 85366  16998 59270  27035 73024  63307 87447  91376 60934    59920 90296  82721 57480  15078 20796  16241 69963  90486 93463
4995: 38650 04612  78676 57190  81012 76468  70106 13855  40511 58355    38195 32856  52255 01692  62654 86616  67269 52694  83766 17337
4996: 54118 62752  92081 45406  46838 07726  23336 50703  65429 26787    13345 83623  35782 23742  73384 51126  27737 77650  86218 07141
4997: 31251 71398  51119 47111  82323 25396  23918 21573  23291 21135    08292 69925  45622 24490  64673 68967  98286 60703  77077 40243
4998: 74335 68321  08159 24623  68315 93294  37991 98663  24455 25843    15680 94429  57883 15486  19345 26794  98471 38403  57344 71875
4999: 68216 57104  90515 87433  55981 87063  74084 93365  41947 40978    28369 91157  66142 68860  98168 97917  69340 14598  42228 97743
```

```
5000:  75322 45562   84355 77705   88659 41948   73424 10956   74228 10637      49706 53115   47427 45076   07815 02027   46139 95361   16016 97768
5001:  25330 07918   63285 36631   58173 67730   29950 05065   48178 54502      86684 82741   69650 95999   22862 50381   38959 43802   70049 92319
5002:  98007 43634   08482 32589   55458 60826   35791 83956   50111 65341      38789 11046   24193 36519   84801 40442   51819 66833   86737 61548
5003:  19911 34526   62710 36317   42713 90413   43920 59621   70911 31106      97778 81722   03880 07248   25591 85259   02913 87264   19808 48636
5004:  97944 28467   65007 71004   08194 34666   49719 30345   01141 90437      84227 02372   51237 18392   69005 72152   97174 99042   52768 74133
5005:  72130 77342   37925 74136   92146 88531   95811 13258   07079 72058      71425 95383   51034 17777   15302 60925   47769 95044   20427 50394
5006:  44001 10153   51935 56761   55930 09361   31073 51342   12979 84819      45431 39659   46333 31639   87204 49133   12025 98685   67615 48935
5007:  32948 04517   67785 71150   15888 15015   58848 94214   04435 75808      81723 11535   46925 97923   46681 92129   39307 88246   34959 60358
5008:  27441 09504   83069 82872   98617 18194   16425 50707   08601 77923      79637 79209   21050 06388   87453 90389   48175 95524   54122 95943
5009:  96324 19885   11701 47731   47110 31839   87005 53537   08305 64824      13343 72427   75681 75366   20716 41697   90314 72340   34603 86469
5010:  20436 09703   37354 00000   97761 88696   28549 59725   73340 81539      25497 83704   49226 23330   45479 22299   14793 77572   18577 36638
5011:  83022 49030   18487 72286   45588 89558   90521 66395   82829 62097      28745 37164   79029 68033   96092 59104   87413 56368   11299 75772
5012:  96993 89434   52252 78383   64368 10997   44384 06617   96196 78244      52073 90215   74895 69365   43932 21726   78903 42024   01440 21773
5013:  93755 51085   98517 32062   09293 92478   36448 22632   83789 27677      31344 95818   73993 93704   28113 23745   70867 81490   20923 68233
5014:  76279 85426   10212 61849   18087 87958   23878 88584   17190 22477      15150 55882   00338 02261   81319 74307   49791 05605   08975 98026
5015:  77187 90417   52182 19196   18686 69582   52205 99891   49242 33020      44330 58373   34021 35327   84621 14679   43465 65618   54857 42438
5016:  68738 67994   30776 03751   22314 25075   87848 51051   44124 30781      94675 99588   89486 12259   45915 67686   06245 54297   79457 69260
5017:  14591 69737   67902 47172   41420 15931   69522 73782   68025 63944      38126 57301   56006 22440   10086 57212   21309 69697   65998 23813
5018:  08177 81950   50979 35688   58787 61831   13252 11897   52748 71672      32916 57408   41879 80299   03206 74055   69194 91794   74312 21372
5019:  97231 79868   47375 73113   48666 92559   76810 76481   98168 62422      84878 82160   12224 24964   86429 25686   15233 56620   60923 42787
5020:  45345 96525   74018 66971   41955 89145   94726 76630   16005 37120      45647 86211   16107 10417   72531 91178   84481 58563   11384 91253
5021:  80116 48210   34185 77847   90311 62409   94064 10375   67069 26794      43911 38269   46737 64167   91008 06849   56602 94848   21603 86659
5022:  35223 14979   39487 16236   39751 91562   46479 71415   41871 90038      29588 01225   35335 12069   25265 18424   71911 98655   70699 38507
5023:  78530 53332   88949 08241   49067 33350   23481 19884   17287 91021      89161 07604   92104 34196   59972 93705   97065 49560   46341 30324
5024:  50325 11640   64829 49681   74267 40607   02248 45888   28654 44971      00453 58462   98866 58921   73040 73735   17533 90929   17843 91412
5025:  18507 29273   82907 02181   66032 43756   23417 77202   82543 37695      01525 03424   98877 63970   66822 57478   01360 34384   95174 86350
5026:  20196 60924   30089 77722   52899 05288   23984 48509   89092 54972      38176 79744   06728 63421   22221 67830   04092 68683   57280 44270
5027:  15556 98777   81281 19202   20965 40894   53733 46544   43184 52886      57434 23089   73784 99055   48768 73859   19735 12013   38807 99957
5028:  45135 41010   35343 63170   10861 63952   47424 28502   08395 10274      15521 68088   14517 83315   88394 41044   09651 54223   65361 11422
5029:  72725 55457   35671 11165   56733 38212   09160 93220   39318 29133      75945 70828   67785 01197   25275 46874   78739 60134   29483 47846
5030:  80048 99587   70219 64144   68822 53763   61121 83095   65301 34065      46181 84512   74808 34920   06791 48610   12280 19174   13529 18127
5031:  23278 98745   02868 99629   60378 83882   57216 36750   22040 20956      26166 53256   37853 98505   92595 64000   42729 24285   06209 06576
5032:  34688 00389   84365 19905   59352 24927   51329 53657   08745 50755      25382 23107   42782 56294   20540 01190   60575 76089   48658 97852
5033:  73687 99488   05320 43073   71851 33759   45627 57663   93785 26490      58693 24832   76509 98677   94680 84147   68878 03425   96944 55196
5034:  08072 48813   55538 66051   56131 07297   54828 55818   17993 77723      73815 78628   17385 14334   23436 95682   60019 58964   78873 64181
5035:  94665 57391   65989 42806   12753 77664   43900 52099   32378 59529      94382 57773   84011 58683   50550 98217   97328 57367   08408 31732
5036:  71079 76961   00575 53417   44606 72848   76775 45833   48703 40319      61322 11423   06747 60634   24480 97304   09156 84861   06916 91917
5037:  08737 02889   61987 82678   51737 81640   26790 15541   86356 67068      07913 46557   59944 66402   90631 63601   75717 49412   44304 86594
5038:  33234 98216   07380 31680   23167 70919   33744 04224   06763 98939      15830 96585   21322 62198   98510 13164   46887 88415   56207 07391
5039:  30810 28662   84172 13817   75247 21260   70251 90395   38729 69722      69046 41355   64539 36630   73997 56889   78671 98933   96712 20923
5040:  69998 95140   82289 04934   50094 73297   56870 62801   00503 99240      07834 68644   24593 30094   90644 78728   02063 41789   36760 67334
5041:  17800 13654   95945 38567   91989 00729   07840 47502   09553 32991      03117 91414   42633 18599   57043 11364   25452 37525   62002 37361
5042:  85483 18473   02780 49138   42952 19302   26401 38397   93069 66940      52919 16968   10539 72131   54189 81836   34857 74066   10015 01933
5043:  53807 04553   43003 95613   05632 25374   00975 11592   18349 12751      11694 17859   36152 86656   38072 47939   82244 91498   99155 55379
5044:  59202 58789   04666 25287   10379 89413   73082 81416   18391 55307      20291 61088   48530 11034   66033 42794   63592 59317   11852 00388
5045:  57621 47863   43609 13912   06291 99200   89843 83381   70591 04923      68071 95731   80339 32031   16267 45297   17071 23474   55159 59898
5046:  71655 64306   26841 93109   40011 58951   73587 81261   32831 63868      97454 69661   33552 32355   16093 51499   12007 16160   37988 24080
5047:  36703 27368   00392 56536   44915 68349   42797 21457   88238 94871      94437 05352   79830 73271   79519 21178   34960 26827   60952 66825
5048:  72037 24006   33744 50842   21688 46776   06371 01975   95040 91803      96201 21426   27312 31127   78681 86148   22648 63573   19043 95792
5049:  50577 15319   58415 56497   38002 57467   30649 37945   37474 24429      47560 90035   48377 50865   37702 25097   99443 20865   16928 30769
```

```
5050:  09654 40677  42287 65658  00714 67375  28114 31534  93839 93842    35145 30438  28166 55991  39131 05076  86626 95762  07606 47529
5051:  54576 58286  80110 20080  84655 47260  35436 48524  48180 09309    08611 78266  17954 87388  53820 23030  85138 20389  68976 33834
5052:  48458 05715  06803 86952  70164 62218  42971 31807  11105 49872    04274 90239  25081 86843  24290 62371  51552 18447  07711 77424
5053:  08884 25403  80896 20256  31327 65416  10393 77956  93489 42808    36386 06762  23381 10465  61241 30546  20033 91918  12137 81743
5054:  48152 71178  57535 80196  00451 67680  47486 41072  67079 64898    83440 03728  15648 54121  29741 46436  07327 10579  38769 11033
5055:  39173 56783  75755 87757  21156 72667  29952 52240  56223 25535    76654 63126  65462 81937  70724 44523  03416 61307  89313 86596
5056:  65529 13733  75334 95848  47691 59699  03384 26258  87988 86944    91298 89807  86029 36036  18562 08512  09184 18889  00195 85279
5057:  36649 73044  08383 93278  24258 40070  82848 73663  21201 13564    44325 64713  33188 62092  66041 33716  71509 68413  06438 47978
5058:  93900 74666  14314 19497  52000 22033  87138 00043  27664 30376    04288 43021  77913 21162  37284 00959  07394 63660  89400 80145
5059:  69570 24116  06871 64271  25323 01890  70444 43067  58273 31798    17951 71597  78974 12898  15338 39641  90245 14969  80500 15904
5060:  62223 21418  34496 37930  77952 88720  38527 75156  63941 15526    05321 02157  06042 07403  15523 71786  91939 60842  02166 80156
5061:  59936 67157  65289 65765  38639 27802  07811 47011  57007 63445    06550 80166  88939 17290  97660 18414  63575 67308  99368 24081
5062:  47938 81117  87996 40926  11493 83442  05801 84663  02871 78337    79219 29853  05935 74614  39074 81627  93054 12944  92651 46710
5063:  06038 07114  38379 98200  87458 77016  41969 47264  83839 66299    39452 40359  19271 52917  14407 93174  89191 17976  33832 32856
5064:  70131 33441  76669 38874  56703 10065  96912 29624  16857 73989    64892 08068  85962 24312  80413 17652  62079 60446  59576 28156
5065:  36962 79417  69069 68239  97543 62204  71305 82143  77532 77384    48974 44614  73116 49035  72320 23701  69893 93549  94138 57907
5066:  67959 40462  51560 93716  52996 80180  79631 59007  49838 67926    09001 36631  34291 11745  96113 65280  11567 05119  36868 06919
5067:  21650 89848  40386 18725  82984 34114  35669 79396  06117 33204    41803 59427  39407 15187  04698 05960  79073 12903  05930 71163
5068:  41615 79223  41865 13771  70815 50281  57758 07229  77163 42408    40259 19544  69127 20991  94547 30965  86650 87617  58527 63498
5069:  45706 25923  36687 53875  21261 07448  62566 31301  48570 05367    63913 40829  89723 07908  75428 63101  08193 38655  67530 82505
5070:  70043 78899  47623 57241  84928 91822  87770 64679  15287 86350    72803 22970  88860 57400  36529 75839  10266 42539  22761 62855
5071:  16849 85355  04990 67316  00638 48646  57736 12628  04628 54877    40098 87589  02132 66717  67113 59047  81191 08152  67021 53908
5072:  73409 62861  13758 48566  80127 13604  20569 42685  97568 59150    61942 58389  79608 23965  87559 71233  81508 94166  74802 90762
5073:  18761 33325  48025 13094  73699 54949  06957 06817  43201 93735    48423 47551  96739 29089  12009 41562  07026 82672  34107 36629
5074:  82849 30136  55823 20648  54486 75781  39522 35430  50196 97738    89899 11394  28786 25841  14004 17379  93836 48414  78025 42172
5075:  87257 55288  02372 71491  40932 55637  44982 75302  30306 93591    08705 18332  14168 94412  28677 61793  74743 53976  76988 95727
5076:  49186 39613  57823 12735  00746 68280  01215 06159  28879 85178    14681 95242  63959 89690  45622 10657  97352 36178  46898 68753
5077:  43294 75354  54430 80707  95266 66970  52399 75949  25005 30934    26166 41967  08034 87714  29525 09197  59199 95804  26027 03458
5078:  21221 54325  46430 77518  73364 39452  08470 33546  17555 99062    54885 38258  46527 64465  09955 46855  85976 01005  92834 12119
5079:  01610 33227  06944 71685  71924 83438  21729 39795  84922 94041    28823 99720  52767 47016  45501 49319  43651 58062  42124 52833
5080:  26217 91368  11735 03472  52069 22898  87154 73785  18161 96986    07944 90592  88259 05144  05361 79907  44187 45136  64643 50144
5081:  82558 78439  11211 93488  65956 64899  74274 89373  96550 28860    01724 65966  83853 31694  69929 71737  09141 91221  02360 92569
5082:  85257 22529  00555 02360  33301 01244  73820 00557  77297 90502    24445 07760  76506 52628  68179 62219  86492 38956  18965 40473
5083:  33411 04313  94807 44135  78106 43978  75670 15670  53861 67753    86740 38882  88326 28634  22672 35717  74021 69897  64876 56342
5084:  89011 69742  90754 00710  73615 31248  45829 37028  38152 06688    60318 63449  57863 56612  05751 47631  83651 57593  20029 78568
5085:  66582 76178  09883 22751  95672 74223  70293 15488  97117 97217    28336 32676  81263 43041  96136 08836  97702 09431  55307 70839
5086:  75303 83016  84296 70656  57553 52351  59349 92132  71315 29708    97339 83649  02474 83066  72882 42832  66595 29129  63657 22518
5087:  76413 80169  54969 35978  73940 47149  22789 89456  32906 78905    05289 49559  31397 27300  97725 72134  86130 67715  42919 60005
5088:  47389 19155  79405 13666  69769 89813  30862 96534  37419 01002    37825 50036  08617 46154  72068 26150  94282 97351  89768 02248
5089:  16895 99337  12837 13714  62291 49553  00959 01863  86067 87370    83668 06031  75843 85210  70590 56295  13292 96901  78708 31492
5090:  45901 74545  12992 04669  34964 43891  90601 76553  37373 59271    19762 77253  32096 56107  21802 79764  44745 83122  67190 74537
5091:  42021 45232  29170 23167  53192 01780  06824 58823  12472 05970    97912 33966  43738 02003  40813 78062  86480 77656  06161 26555
5092:  48125 93395  20053 60502  69975 54097  20981 82237  00906 78194    75067 54718  35891 42389  37794 99036  69579 10508  37808 98432
5093:  02917 44518  76760 77544  52127 27678  74458 20026  64416 03296    51071 72285  23710 21985  45667 33742  56328 33202  22586 53593
5094:  69936 77384  11561 50293  98126 41847  05802 69209  48463 34022    31219 60589  20510 94934  95015 17392  59096 59135  98747 48290
5095:  62088 84711  59270 72627  43226 43086  74084 73960  03469 55339    56810 61178  75367 95017  56900 36162  52492 79989  51938 72085
5096:  01361 75749  84855 02637  83759 45462  64070 28580  02371 06409    99030 57339  35544 62916  03298 09975  23348 45863  83345 52748
5097:  32310 75090  85302 25218  62885 83171  46864 49608  43656 76701    34582 70713  95572 33118  19046 22635  49908 84649  75138 04262
5098:  37692 15072  47239 67829  81695 47780  14139 35123  90737 05359    83360 47163  86569 97491  80102 79763  17265 80758  77539 40434
5099:  60912 50027  01105 21246  86608 04451  80777 83836  78690 63947    41620 51698  60437 28480  67479 03821  72425 00990  29874 96917
```

```
5100:  80484 46601   20963 68582   60864 47155   84953 35818   21054 04535     45377 53888   62299 90315   19649 85958   64650 79825   70649 35966
5101:  55997 97769   90427 77138   58039 95532   21612 87823   13507 75967     44499 27019   97390 93695   00599 70572   44469 04462   44746 29974
5102:  39901 32832   31477 62583   22170 53990   37866 75873   62714 88474     24411 22937   25419 55547   13889 93728   40570 18723   87663 79162
5103:  65325 80338   25946 14153   80254 45267   41316 05197   17390 07835     14932 58166   04662 98863   96100 24454   21962 24895   02180 26298
5104:  38175 74047   16112 53544   06940 95397   65355 56581   23803 62318     12836 03343   20020 59544   11650 24174   57900 03954   29997 54177
5105:  21351 09816   05213 11936   78625 15622   48706 72272   96471 90811     90929 35212   38552 30802   30195 90081   22689 92348   00919 55364
5106:  49723 12695   09972 84396   19006 87136   68690 32190   51808 01566     38123 45623   44090 45464   81915 93277   44552 46342   97933 02342
5107:  40397 95243   33226 20556   88493 39921   66149 48763   94782 40108     20304 22591   32353 12315   15199 60187   82638 74533   97632 27625
5108:  86633 94693   86530 27708   57314 43153   15447 71133   27800 92361     66574 97160   65059 54009   31866 30632   66272 63228   26589 72947
5109:  72508 70680   88598 17376   45068 16086   72462 93763   48721 08055     56924 08083   51911 60019   23405 42278   51407 94296   24150 71444
5110:  35124 15145   32208 92788   52551 88660   56334 87872   76129 62795     40138 18618   10913 19207   89769 96368   14095 41857   07980 68979
5111:  95127 09437   07789 66126   18395 99426   42001 20341   09311 47636     64755 02291   70070 77152   72473 10390   07189 83259   14021 14606
5112:  39038 31038   46962 14320   46502 52372   39454 06049   80145 21970     63550 54826   33499 10745   88225 02496   34864 33030   35781 62629
5113:  58165 23981   69672 19780   94585 32511   23699 16938   67419 23357     56168 69479   71083 37703   61649 18624   01450 60212   24267 31929
5114:  12685 52857   93838 36147   31753 33072   72207 39098   54967 02871     57860 34573   55176 85264   49401 75132   41942 70638   49690 20817
5115:  04661 99760   89476 47794   95203 51058   37267 78555   18470 41332     17747 37941   10140 80463   45009 64259   22680 59634   68065 42770
5116:  54274 63073   21045 42562   77095 23228   85994 09678   06066 31526     18749 81445   90667 92517   50610 92249   31062 98145   91677 96099
5117:  33701 29441   17746 92364   59059 35426   54308 35678   92951 27672     25377 85988   06787 89996   73965 34254   35841 35989   53047 11416
5118:  84044 22330   88666 11437   86354 00101   20813 28820   26419 42791     82685 96442   32904 09125   20162 33792   51474 17452   24748 88609
5119:  67456 00600   64568 11719   77810 76377   62907 05126   24213 62217     34360 38374   01115 09705   22610 75818   41239 81770   61942 62875
5120:  27990 87440   92472 41294   48982 11861   71470 97520   99892 62597     50955 44060   34657 51787   35015 55776   84714 35611   95859 28650
5121:  54391 13953   78621 06739   85275 23402   17592 31639   16784 43400     08516 18134   19921 19806   83263 82136   68002 06490   11081 06692
5122:  80045 14816   70864 84578   54869 89028   87730 74638   39196 28020     23683 88046   23355 84047   41221 71342   03804 91312   44963 76145
5123:  53025 91174   54201 97525   37483 21712   76519 76580   61579 81781     06641 34613   29880 69445   29298 55483   37168 68747   83565 39525
5124:  73764 20781   62506 24377   99832 01800   22562 35804   95897 18256     64734 38853   72207 74626   88221 32173   73878 27848   63725 00061
5125:  35378 87401   96836 90131   83763 78400   41127 59295   31369 26252     90781 59545   50710 62923   85810 28999   32967 99167   36778 48201
5126:  94352 85627   40411 40138   03716 49595   92564 80209   45332 25484     85602 90251   74739 40054   53244 21179   14454 49391   65347 82819
5127:  20596 83187   30101 41593   41358 65882   50936 36157   66585 14710     24813 73118   44349 77606   81730 16664   84342 44294   70310 31503
5128:  25315 33342   17232 47243   34960 12706   31329 50091   86015 35430     85453 43733   44445 99936   37715 71073   51786 97797   47927 21318
5129:  27142 79565   36166 81440   06744 11598   79374 95213   68619 77689     55778 62801   60034 91726   10871 90247   68521 90457   40122 73850
5130:  38761 80745   16818 88307   75119 56591   88329 51075   38584 51364     11866 08748   80564 63266   36518 34042   27877 90376   32537 60251
5131:  61941 68471   38737 41553   96382 52227   74401 40828   40455 94836     90972 11828   10322 57253   57571 88307   22232 19831   27550 35490
5132:  41177 88182   75752 57997   89467 53255   63045 85128   54784 98040     63309 24413   05344 37121   36589 63355   49076 39351   13693 47473
5133:  67235 12946   22504 18593   23339 79795   61110 19953   20572 40534     18226 06809   83029 39368   67767 40291   55538 63167   20656 16846
5134:  86952 87977   95410 20525   03925 82072   48926 86649   23917 77537     40906 73204   99325 07262   45130 19681   90395 36092   04035 65466
5135:  89238 34113   29309 18504   38024 32346   05490 81918   10496 66964     55964 20631   67192 52365   37968 51344   30608 76721   20799 75489
5136:  74950 38509   62654 11945   81593 09351   60721 84252   11054 40623     76182 02418   17303 30479   25756 04059   96391 34677   71093 21014
5137:  83736 30783   28446 89172   33882 06184   09416 96241   23865 80379     00264 30565   82220 83027   37923 19874   68262 64761   30179 43061
5138:  77494 38597   80717 79295   90217 26365   60298 75728   39211 03544     97442 64394   59721 78276   52631 33003   43933 83974   31988 23640
5139:  24204 74170   14810 19705   87707 60077   03507 27421   62885 16557     31016 68533   09785 07654   27704 70763   71883 12342   19895 38039
5140:  82267 17333   30634 98505   59886 13551   90081 77597   17348 25684     09448 72855   35776 32264   22005 40647   96126 70939   66115 69390
5141:  61240 81200   92103 68599   28115 20883   70058 60976   71883 81452     32235 59233   78792 33186   26036 42306   30399 01956   90617 76783
5142:  42384 58537   37002 46818   91474 11794   57595 84708   44060 29848     40801 33534   89368 26584   88356 95409   42662 10405   76526 27378
5143:  54320 66117   02184 02035   91583 09251   43072 38814   07270 20825     45215 22110   24380 27677   74004 36830   88527 87246   15418 31588
5144:  98831 62753   10253 75444   68352 90891   59991 61107   20143 09775     90944 48232   96839 85814   77811 70470   39114 01959   32887 21910
5145:  09523 82924   82121 74707   68925 25820   78985 26383   38647 05358     60664 97827   85774 55514   50631 69882   60783 36621   98450 25861
5146:  01458 09896   34119 18856   56894 95711   14960 88790   48557 87914     79863 28762   96057 75655   87197 95255   27388 40384   68430 97665
5147:  21102 64461   76743 89932   96124 69702   40941 53524   36842 43336     23345 50470   99952 18456   59266 40857   46723 32391   19665 57115
5148:  96126 78418   75738 31565   28916 38607   25593 38222   34141 03555     66877 15778   66825 34327   47086 72383   57265 01632   67637 08257
5149:  14436 43214   74758 62211   81949 28314   25083 88238   96974 82936     89764 58211   78855 77396   81466 88343   05538 56155   71077 97884
```

```
5150:  79459 82498  01296 38759  70119 41174  52638 55707  00830 60476    28420 88192  54771 47078  36131 64053  53738 01376  87674 16544
5151:  80387 31938  84681 77210  23811 64040  41179 94750  23782 85588    48184 39556  34335 98672  82292 71763  38788 72276  06557 63471
5152:  88175 04111  34773 35154  94276 53891  71465 51141  95291 14775    75248 27954  15189 31212  27936 65786  26194 36124  52261 50222
5153:  67067 73744  36042 16984  23498 54403  18711 67162  04711 31082    64058 48257  54845 81215  88788 93341  08374 42211  33376 88500
5154:  44992 41028  19898 46224  80316 53929  90826 37339  31745 21819    21755 99071  42292 68463  58361 28060  64934 74780  30592 27594
5155:  97448 55198  86963 29593  91884 41602  15121 60189  72730 81317    05382 31353  51065 05366  60774 47121  75766 61899  12589 59377
5156:  41689 41156  09348 08524  52361 63642  82838 79168  26779 72522    34215 24431  00057 14573  19069 34385  34674 42762  11103 60721
5157:  15354 10655  52487 22476  56386 25411  07108 69623  76413 49048    73419 16618  13303 18720  11338 28733  12037 45893  67550 12638
5158:  06937 01052  70649 42613  45173 65516  61257 42049  51989 89009    75182 94281  47101 10474  26846 04136  30561 29054  07909 24787
5159:  58367 01239  41335 03614  10477 88721  93670 37783  11651 83862    63204 25123  76336 23741  31814 49983  43773 12003  89806 98842
5160:  88780 83467  82391 22001  09469 12482  83448 42025  55178 64060    74281 66655  83462 59684  30950 49738  73665 18383  10928 76800
5161:  77775 79374  77676 41919  07657 40479  60417 30320  47192 06297    85930 90770  84285 68048  18075 34764  54340 42724  33361 41000
5162:  51569 36968  66030 95705  70551 49625  93517 23473  80562 89741    73827 73406  54951 21818  86074 38665  58920 44679  30469 43454
5163:  50012 44739  88237 51477  55515 38963  07934 08199  17685 99212    51860 59274  18852 99587  79350 27966  48590 41950  97863 76003
5164:  45553 87376  30584 21363  75333 51829  54844 27796  99967 00843    59958 72064  98778 63052  61296 95583  15605 21428  61959 39096
5165:  65262 99435  25023 37753  82542 50220  20424 66496  19252 40404    25851 92051  70702 47594  98547 78215  89019 12305  84572 10131
5166:  40813 06519  08470 82376  42210 26738  05400 52161  31847 27916    18816 83259  99905 30724  03099 80843  55825 96954  58463 43966
5167:  87214 66382  13893 40784  92811 11754  88897 87003  65024 70670    88605 53126  63649 82717  03515 09539  17253 98542  99235 86516
5168:  18859 56506  43484 95635  99537 31374  92587 10482  62197 18706    14298 68733  16507 18284  69671 28414  59914 35287  68063 43019
5169:  49421 11762  27131 98190  85748 25450  28975 56713  58975 77149    36159 32484  54053 76373  93014 93067  46191 23080  71150 44606
5170:  81324 69653  02304 04963  52216 44516  34385 31041  63406 53962    98245 06651  56888 93911  39714 27687  31010 39139  01686 02766
5171:  44052 29618  88257 20732  67070 42831  79368 85856  58667 13089    79084 96337  45521 90059  42445 10302  58495 11129  85807 46063
5172:  29539 52180  17691 25570  29479 94341  31607 27340  31083 78047    74784 60756  87021 23502  52562 57267  28754 60917  66876 28794
5173:  90301 88906  16476 31019  06303 05691  99282 22328  24358 73263    65216 40460  19925 62718  91030 44722  92639 06476  39651 40832
5174:  82128 02430  01848 02553  87799 84918  55473 35052  75475 46831    66922 40174  50202 16813  65699 67696  48874 65495  53788 85979
5175:  80361 77020  43395 89997  58648 95473  91853 38167  79006 90266    78720 16795  28642 27709  31022 94042  71494 93922  84727 71333
5176:  66377 00805  67615 50012  17794 38341  62723 73394  29608 47327    86358 18010  59235 10889  62973 96602  31130 10699  27954 45370
5177:  41717 84773  41467 63897  69845 38789  99504 81635  60878 87677    60841 05283  52813 12683  08851 93752  08494 44314  20852 31749
5178:  79777 99504  80666 14598  73437 90888  11545 26436  07571 40303    05329 04426  36400 78943  11626 04933  72219 74540  05745 51739
5179:  10065 73414  45095 68387  23748 26368  16983 33045  73823 32766    24246 20450  03538 54645  46443 19282  72555 99442  05546 49293
5180:  91507 26160  61876 04373  81440 92781  56829 20504  55247 80007    76233 89218  23911 35631  04003 90118  17745 35684  54913 07894
5181:  28582 17597  59587 65071  21082 80757  97663 07887  78875 07920    32333 48039  59605 11351  66224 80132  95408 46840  17846 22641
5182:  93630 18641  03076 22852  48459 71239  76831 56639  58613 50846    60577 17618  11878 69846  61650 08168  27651 80924  56791 01717
5183:  33402 94883  16892 42244  20620 21730  94161 52273  29859 67196    20371 18967  87647 73997  85499 26015  97367 84982  05276 23482
5184:  41410 78154  54656 28306  44450 05870  37080 71569  48982 28706    12980 49269  67010 45356  79042 42929  31191 90246  94645 32652
5185:  79028 98469  23572 96349  51092 30805  58143 73156  22928 68548    94346 25794  23878 52375  29083 12065  08231 07634  28730 28461
5186:  83472 14733  65890 55372  17178 43852  13703 83271  35383 95464    52536 54043  50533 81240  92751 20529  83255 91363  96933 50475
5187:  12305 13125  86621 57055  15439 61183  74723 02781  87731 98144    16973 30938  30029 85716  74091 79201  38411 93916  78166 52674
5188:  84304 96230  94792 29591  28018 26928  71695 15774  14214 87604    36310 18102  44111 60135  41809 35624  94276 60765  34953 21022
5189:  95165 98124  65513 44525  57587 20851  23853 53443  16968 45966    56952 68722  10020 28138  83172 33888  45408 84562  94316 11870
5190:  18311 61063  92931 83750  94362 67312  18531 10520  90030 38526    47413 18366  09347 17695  10394 90003  10460 99800  45323 72644
5191:  62707 95055  29695 98650  38058 38410  43270 30835  71927 77004    38945 91567  40505 26364  72406 66429  87998 89257  72555 72331
5192:  50341 28523  02133 48030  71988 04523  68681 78389  16728 79017    15179 85561  30364 04520  96469 76648  38548 65193  74998 04802
5193:  87537 63193  18967 56343  26111 70948  64050 93161  93861 77221    54270 38017  31150 70194  99095 25467  06968 92807  42748 83561
5194:  00615 25652  20451 46558  93340 23478  24029 20723  09124 51815    16247 35171  89654 44824  15212 44533  15187 12343  38832 06681
5195:  77428 85685  06618 08855  89729 53365  77970 98180  21097 89531    78912 15094  69923 94637  39052 13788  07629 83387  71055 48134
5196:  30306 29568  75196 77596  48482 81607  52794 97638  78137 83429    20518 82393  73694 45780  62308 73965  15663 10576  24934 35888
5197:  32548 56650  26366 91150  68568 19541  83159 84225  04362 20398    06645 03158  10401 70342  41000 46238  51159 54441  39138 60188
5198:  91886 97984  36465 72601  44868 80171  56935 22382  99161 57931    89239 04679  56566 03962  95431 68164  85751 35416  17551 48083
5199:  59899 99695  84500 53745  33460 27409  51323 83040  73355 11557    24112 76776  54105 39354  22940 53870  87477 93579  29961 92172
```

```
5200: 79613 75612  66302 43528  78096 12266  04605 87018  82587 59821    59346 37022  12972 13954  17529 13329  46800 70487  70805 28999
5201: 57920 08382  23261 07430  52398 36607  83026 40242  73883 22928    15640 59730  97319 71516  66042 12181  63478 06153  68360 40789
5202: 80337 69122  74774 72913  40500 39755  17335 87679  70383 99640    45289 43964  09294 27434  53771 62330  48421 47731  46682 31842
5203: 92513 73087  06308 80812  95198 03813  57043 77898  47457 06531    55699 38461  55642 92668  19547 19430  15969 21403  24027 63090
5204: 24447 14673  59998 30588  74931 25349  70630 17246  40269 41821    22752 96366  97425 25119  70684 84993  44416 74926  14782 96745
5205: 38882 55196  69804 27543  93912 97738  42928 38150  23378 07099    86736 08939  43582 14358  59138 15874  94561 48093  45696 68712
5206: 11976 69641  31379 68847  51803 20435  42298 30381  40773 08248    35206 27536  63565 38341  64809 91806  43403 91578  49291 62936
5207: 12542 15592  02389 37648  13050 26111  02884 75063  70064 30669    79584 88967  39496 69587  68165 03034  33542 26869  66984 52284
5208: 03140 37541  01078 52618  28673 88250  33101 83790  81365 51547    06607 26531  77300 99986  12079 45501  22487 19557  98711 85310
5209: 82485 90558  79275 07947  50364 78561  71381 60449  57509 05203    71936 83041  37665 79191  97408 74145  72919 60403  99404 90991
5210: 36024 29474  35273 10162  28808 32684  01439 98821  10850 36433    68392 97675  90214 85305  81122 85253  05580 87922  40580 20618
5211: 85425 93349  32913 28367  51979 03739  60948 45702  37324 31555    67591 43754  28837 15962  17776 58991  94495 16672  59813 34100
5212: 63648 23755  27160 45388  50324 24696  20308 63109  80270 40493    30290 30014  25684 35113  06945 52408  59397 76428  71704 98914
5213: 04242 47092  57262 73017  59618 76742  69321 00745  89801 26076    35238 60442  66146 49165  85668 61146  81539 68836  59265 67577
5214: 26160 37936  39600 79992  48604 09713  65911 99978  14966 32294    30037 14852  10586 83890  69269 41147  02610 14940  27255 83201
5215: 96886 47517  74430 41528  13671 42213  53873 32103  18409 98078    17716 22525  99655 25578  41742 15085  33439 47726  42384 34680
5216: 89739 04501  55898 19396  24145 22300  83088 04205  96707 86649    45542 95547  49015 24998  94259 94906  20154 01887  93535 20533
5217: 28785 86037  28579 69733  34471 96078  06667 95695  80229 02544    13925 79929  03059 82072  34206 34994  83691 47720  06441 22262
5218: 29629 37298  89341 26168  85317 17180  89598 18097  61999 80280    49788 17891  13567 69118  11194 78985  43256 76730  73540 39441
5219: 51108 58017  69786 02507  72085 32770  42596 63099  23469 18539    58038 23742  65593 21591  89186 00495  76035 80908  05242 46925
5220: 70365 65260  83142 61931  01917 65865  77776 17378  10082 64301    05119 53071  16323 35158  89135 28848  77039 47415  57926 57654
5221: 02649 72789  15849 70586  04178 15494  72055 59932  13601 09907    19197 26259  26966 72305  92201 58322  00183 53749  25437 85662
5222: 37963 71477  75780 63351  40247 39276  90932 59581  34642 49917    87036 11031  36375 95275  21763 26301  13800 91240  14156 39303
5223: 35129 52100  76568 63869  03294 13835  26781 52544  09644 86730    45697 50369  16228 21947  66667 25477  75317 11705  76828 67745
5224: 63685 36751  78781 95598  65128 45551  44275 48401  38499 57178    04570 58839  57652 70613  23406 88628  53794 45093  50567 77266
5225: 35164 46912  40711 65978  49167 35392  78958 21720  70869 44447    33683 66569  21773 48757  84976 42264  27786 82028  65066 49089
5226: 34542 97871  25679 84066  44971 74818  73556 26457  46738 15865    99721 42236  50674 20963  30744 98963  85767 72509  73692 24121
5227: 86239 46887  44477 82193  38963 33126  63365 03319  65794 76033    53655 85730  57529 20217  05476 01651  23143 73512  02761 95592
5228: 95727 06502  89412 04675  57103 79282  57945 03686  16761 95279    27407 10298  50288 52005  88519 95883  53050 31342  69889 19148
5229: 66918 72403  00316 51616  88155 13227  79333 05596  21760 82564    79342 28457  70710 57313  69030 50335  31134 91544  32301 22077
5230: 95126 34309  45335 87653  01300 05552  80519 23755  34250 61353    04852 70650  33380 98835  91096 22260  04589 89092  55216 91795
5231: 35245 20993  09879 91080  06322 99215  04087 85667  74394 83811    22430 40920  18420 87138  07526 71565  63396 93300  25849 17656
5232: 30092 71831  65744 99738  58526 89789  49426 90406  56222 92317    69570 77371  14686 16257  42785 90293  39398 84720  31627 99767
5233: 53173 63220  43728 46651  23634 18190  29140 65597  20145 35459    24088 30605  69689 52258  86264 86381  51227 85674  92753 00931
5234: 12921 72213  19647 57181  24620 07066  93878 71928  79136 16640    29438 14728  11656 73498  48747 17766  23361 61124  81769 10955
5235: 49574 13038  51114 47356  75035 69159  87682 05493  87778 10845    17778 25708  50428 47871  73780 64896  78446 34796  82450 77761
5236: 23637 75072  31001 60105  09566 30596  40364 17733  05684 32337    97893 01653  10904 60268  40816 84842  47941 96477  26679 28284
5237: 19407 81390  02058 31232  94963 57019  87736 05336  98122 58646    02960 12509  06836 54932  87019 19255  49593 17449  74470 00369
5238: 03148 47442  61306 91062  00428 65541  06143 40717  15196 47137    57160 81055  75408 63501  91928 91793  53505 23915  72884 76879
5239: 09786 06550  98007 40953  37380 82729  29881 68796  84413 29900    42830 20599  35652 34656  76308 16435  06888 99692  71748 89016
5240: 95232 35769  02829 60098  83257 63354  52280 24651  62418 37045    45355 31576  13080 32058  96310 34395  54042 05006  86152 46371
5241: 56439 50457  62370 20854  00924 71154  67788 42300  91300 62234    27305 13562  13729 43465  56936 81080  88157 29547  64540 73377
5242: 11622 97129  87719 55418  60844 94129  90906 14882  60618 21918    09920 52431  85457 53649  18000 48455  75696 57300  93273 99754
5243: 02797 30886  19202 21798  29511 92565  06829 53042  52360 71240    69741 66068  82003 76072  70864 80899  25063 37421  01553 00606
5244: 00275 90898  14923 58280  48724 37851  38286 77665  15592 69059    79063 49531  37367 64843  09192 57233  22423 27216  34206 86828
5245: 74038 24933  07940 50714  98264 30532  60439 50582  45042 94587    68933 15070  55635 38586  58077 86079  02502 40518  12802 39056
5246: 32524 12306  75987 29012  68888 26594  26128 51875  14932 66229    72357 51895  28269 16114  82786 24146  36531 98915  04648 58995
5247: 86271 04695  39323 28792  43809 17925  40617 56369  45487 11362    11474 25068  42719 30993  04931 23456  95498 15777  25890 24506
5248: 20811 48710  89724 67269  04918 74032  81841 27371  17862 63938    02752 11282  23952 52503  98697 17574  75816 27744  10537 15144
5249: 92837 34569  10546 88942  51711 70104  82305 56307  77022 68594    42470 76104  74700 28486  30307 06573  97071 99464  50090 90700
```

```
5250:  48980 38433  31743 71119  33210 33730  07187 45769  23547 48867    70592 59948  93069 72698  97701 91941  99572 39417  52727 43624
5251:  26418 19645  14533 95134  04108 21880  35676 40711  32637 12438    98511 80227  46661 31185  73541 80899  12377 81462  45322 44552
5252:  54111 04320  32528 52464  12890 42754  71916 83876  77423 94957    97608 05669  93055 24091  72343 90682  81504 97657  86075 86285
5253:  94436 39729  27294 85227  86074 46825  97007 05056  97359 97557    55701 34336  78290 99464  81184 91407  78077 30756  71605 29372
5254:  79817 57051  90521 67644  88810 64978  84515 49918  35317 35733    51851 50788  03566 96680  61653 10062  35784 59358  60494 59095
5255:  73874 77503  58947 33520  30431 35116  55608 69016  26579 57794    45821 98598  66989 27571  06939 38595  46555 95880  70016 41197
5256:  34832 35466  05043 65005  97423 01127  00270 09433  52596 38917    82161 80027  38590 86176  25404 80859  67281 10495  22197 63927
5257:  83909 45737  94958 46846  54762 86965  39319 12736  87397 43691    80834 21032  08536 79995  08107 01207  27173 49488  49287 02367
5258:  87014 48474  73402 58220  25066 74817  62420 28566  72421 27852    98620 14043  72119 06320  73822 98765  64050 68378  33085 63104
5259:  55380 59896  59147 83683  47874 55723  00356 03081  21387 28245    29326 25249  37831 11355  34610 36598  58794 88828  74826 20661
5260:  77898 47793  65130 52252  61161 64977  18626 16023  80779 12370    12077 53947  80408 49654  40643 56268  62086 67773  71952 21703
5261:  96099 63674  13642 26409  21751 15614  21973 81604  07191 50663    07674 44751  68903 29177  78909 45396  98769 18689  55000 09905
5262:  98011 22068  50124 70104  33726 52048  15633 57330  89691 49242    92152 17678  44805 84311  11443 23887  76230 40963  41724 98412
5263:  15664 32768  27768 49828  38652 35204  23549 80947  59505 26453    83629 22557  93116 33891  07281 53761  84893 88180  69751 54588
5264:  07026 99091  94026 08161  64450 69565  62569 37415  93163 73789    31477 25044  89207 79039  03658 67159  12855 06262  33245 36489
5265:  68449 99422  88240 91122  06961 13887  87401 68364  88009 67527    81971 24231  08470 33036  08618 97668  03542 46225  70443 49958
5266:  53990 58858  33455 20995  15713 00530  76483 92898  71417 59416    40145 99358  70581 36491  59487 80202  44716 98773  74544 23447
5267:  84945 96088  26312 09304  63699 56985  10833 11214  13995 65989    67315 08341  34432 94372  81525 82660  76737 34161  86509 42550
5268:  00313 26475  52376 32164  12938 36526  42666 06105  65468 94012    14839 54006  13176 54161  42403 41332  53623 34819  28563 10964
5269:  23690 51094  94119 55482  82739 89006  05242 77664  08768 17756    79106 37382  44858 34083  64374 95486  52172 95037  72573 25646
5270:  02123 79810  30172 54837  66371 56161  04037 96951  89037 56037    32418 63544  90720 35779  10727 24850  49024 15847  92165 43540
5271:  24248 89563  64817 05314  75488 32345  58249 01560  61030 77303    89259 11895  11122 94629  54441 18755  81819 31595  46598 94983
5272:  68052 37329  93041 59028  36384 35829  01263 76769  27354 57568    39382 20674  11712 24075  59162 41846  30693 86981  46656 11829
5273:  76961 75607  81619 23155  90762 36958  51412 12746  43868 58237    33213 70757  04997 99907  85127 04210  98331 96668  68359 75024
5274:  09771 42410  57428 35243  98446 36103  14448 78741  56894 46795    19864 08125  72869 39999  58052 24203  34173 84559  76207 12403
5275:  62640 39366  24296 10431  57808 85515  18797 77120  73258 05302    95087 05264  46084 40050  31454 18278  10676 06903  20156 43505
5276:  61421 23308  72433 13284  70611 53178  70624 58829  85742 92326    09630 12516  74446 90543  67809 92006  76562 54793  63259 40137
5277:  18505 99266  22920 18423  21145 88217  13097 55684  32613 13558    73030 34105  53890 59355  30853 25175  98996 75348  59440 13155
5278:  57096 66493  13511 13267  16406 21418  13097 55684  32613 13558    89660 10430  54936 69877  63606 87929  98915 85604  14637 01069
5279:  98965 38686  56936 74193  06495 60406  88198 20858  04909 73439    52014 63106  30162 25795  57623 83717  50256 35821  23186 58712
5280:  05174 93793  38980 33911  64010 16696  69553 80431  34209 70787    64682 07609  46649 88335  84502 68701  47685 33795  87549 72832
5281:  45605 12655  50156 66985  88032 08475  86331 33587  85683 61987    17974 42139  31268 54722  15386 46686  85803 67173  01472 79577
5282:  00561 33335  51519 92759  51137 68887  93996 63574  76794 52821    48315 21669  08516 87219  71235 57872  63299 05538  37338 10800
5283:  09217 17721  92035 29965  04826 36926  42975 11272  82424 20532    45683 81385  58952 08297  43479 54639  34546 20469  68556 27324
5284:  78668 45850  19426 96801  58749 83128  52267 90810  43654 07195    48544 26376  52465 80524  09643 84288  78725 54796  09600 60830
5285:  66559 75174  16228 37292  13609 84813  34019 50710  31681 40758    49389 12213  71650 84495  80501 08232  71587 62603  51525 62230
5286:  04399 39254  95365 44870  20111 26230  24157 99454  58971 37074    47016 12547  88281 74376  48984 01390  93347 70306  09198 51033
5287:  28207 98674  66119 73739  92821 75827  67158 76394  54001 95118    53183 94698  77525 51945  42754 51453  45683 41834  01889 48463
5288:  94263 04345  72791 07577  59275 13462  21026 45766  74235 26208    43591 07711  60750 82186  33329 19689  38649 21020  07927 47522
5289:  11348 70196  99786 72955  48929 08727  81601 47349  14547 22258    91087 91282  16668 33189  01032 36162  38145 66246  31872 46843
5290:  57739 61414  49113 30319  25667 78132  67613 14311  12968 29141    16249 60285  06068 37913  83350 13180  10839 86702  43632 27607
5291:  89498 38656  59791 91373  78543 38861  74877 59964  42274 79159    24474 47068  17609 26837  84496 22480  75519 07201  05949 95306
5292:  61405 89541  38704 53917  75089 65294  59314 58590  07323 47972    15583 26676  42694 93381  93813 44684  39759 67896  57975 36980
5293:  37631 00809  86507 85001  00809 26412  54540 89229  35744 03243    58121 50949  55598 10798  01729 89255  32373 39067  79887 87706
5294:  56325 42317  55321 82918  46083 61775  87615 20790  48218 44642    84793 70598  64293 86177  21225 67683  63942 37664  48595 67313
5295:  49799 01571  88359 88825  65274 24222  22023 64089  09150 09830    36178 16542  43615 50467  93631 78656  61664 72326  00299 42858
5296:  50067 39467  06057 05328  69580 33022  67956 21153  70260 41637    33855 75969  63013 03462  54249 17288  04192 71733  82685 46597
5297:  98627 86279  87878 85557  30508 40144  13979 27797  97337 83952    49024 18155  83465 93632  55067 31315  63062 58539  96201 51079
5298:  57826 27492  01383 74865  67515 61385  84930 66092  33893 41914    32293 92363  80722 48442  87395 83423  19675 33809  04814 61973
5299:  45373 54060  81119 02296  96071 81409  84568 12241  96227 10339    69283 49339  81521 09302  29336 92737  96067 32742  46731 91557
```

```
5300: 36337 92629  74846 53478  97755 84963  26389 68208  25990 81188    84979 29619  09074 25847  81397 82053  31408 14091  20422 15926
5301: 21657 18646  51384 27343  51477 12557  59613 89994  27904 59252    93044 56462  42562 91480  49796 82254  51215 07058  69999 79704
5302: 68873 70545  14936 63257  84006 75743  85453 84080  10189 05091    08972 72596  39081 38158  95927 92645  56839 64299  30230 32756
5303: 00905 87900  42817 07139  83364 19213  00256 91163  70448 99475    26451 04305  12626 54578  50966 81131  85607 44339  00140 48456
5304: 18497 02981  41772 96168  32716 96037  18109 72819  41482 89371    01038 65457  73968 35589  67467 24329  46735 40126  25960 10797
5305: 86695 89709  25699 04273  13632 74861  62213 89285  38029 16872    81842 84455  81567 53513  27009 34102  71836 21224  25869 88959
5306: 46555 96956  71683 24336  25615 91140  20420 23117  59837 32926    94311 44102  08144 91930  43685 64017  42792 88817  39064 60504
5307: 51551 08546  89788 41574  26486 62362  00664 27412  22702 90749    87420 70945  40494 52211  48385 13822  72102 98933  51069 30466
5308: 76213 72187  97186 31238  75421 90509  62418 84998  47743 26556    04237 12867  09213 55199  14917 73003  56021 76073  60811 81350
5309: 81206 88033  27835 37558  66970 90026  71994 77170  38583 66642    98538 39088  16826 60912  33246 24649  19100 36899  77890 23432
5310: 43452 65335  42207 53750  60719 73017  93597 77776  60179 09217    55170 79913  56365 97566  96417 80477  45550 44628  18516 05162
5311: 91063 95264  42083 90096  81918 15627  54488 59971  92106 75398    02630 14065  06153 29221  89010 70997  70130 40059  58977 96937
5312: 99903 18485  61610 93623  56959 99621  44398 34151  39511 55616    97162 40183  43145 09980  66120 78407  94929 42023  64173 98621
5313: 25530 97510  54031 70549  38443 16250  03360 85280  46944 16038    84703 82599  08575 41099  43632 84569  54896 84897  60598 68364
5314: 59856 67986  49074 97305  04981 48283  01117 19957  23004 19795    61522 95274  84044 63597  34728 36692  51558 08093  85749 49690
5315: 72355 47946  97051 23088  52038 73914  11667 82718  04376 94373    69580 37044  20692 85275  13621 63917  59363 59850  24701 99334
5316: 91738 24127  38027 72541  77337 08574  30065 50970  20010 03414    88615 78379  92848 13118  75154 72317  48011 42569  13968 91687
5317: 44493 85253  53612 08781  36947 40221  77255 83594  83348 75308    38815 28297  87528 27728  56858 05724  51443 85445  81613 68029
5318: 14973 48073  65347 98097  09008 70689  32175 47546  04945 82892    45857 71941  21701 91932  39326 76050  10577 45499  91148 70564
5319: 58391 64026  62836 80232  40461 61406  50155 32795  24734 69453    11701 42995  51252 24194  54438 84873  77594 31436  26401 87327
5320: 76604 32024  98396 02787  90486 14506  55245 45629  74547 67777    86662 35072  74236 09611  92056 64642  08681 93310  65113 53453
5321: 06658 21026  15839 27384  07666 25999  25142 61807  35329 93301    06727 70697  57490 78426  39525 09517  21672 54789  10236 97080
5322: 63648 90400  61428 34315  49086 70484  81385 30560  21986 78130    76383 13968  46480 47836  73600 95466  11040 71672  68911 53820
5323: 93285 91053  29274 10573  68929 00643  01106 23526  43976 68348    95592 25622  09338 48009  32669 85737  53884 56613  03699 77665
5324: 12327 83317  57715 10141  41661 06506  77481 49118  92856 98825    06140 50064  09041 15508  66451 39390  63630 98311  10721 42829
5325: 06075 55477  92487 49880  97520 63786  31151 08443  45225 28248    18383 35723  34125 37609  85863 48072  44176 36212  14340 97845
5326: 39834 78969  39635 17605  71341 42864  22878 01200  63074 28386    25874 29650  58462 78218  59398 88241  43297 64186  76113 05686
5327: 48804 45587  94262 51312  54831 69129  77001 43525  92815 32644    72366 91747  36013 69846  56560 97392  82815 05155  02321 42662
5328: 21377 22047  07038 97866  16069 14666  96563 71576  22748 97792    89469 42428  99940 59868  26634 28081  17464 45633  70945 84280
5329: 20779 26619  07627 12911  83488 08944  61320 38708  86197 02063    87279 19401  80643 88446  66576 67771  62108 11976  82628 51200
5330: 42617 85609  01611 88244  75921 80271  58688 63374  02954 71827    51697 31453  91714 87799  14748 76839  76789 28350  89418 51858
5331: 42955 17063  62708 11826  42156 82075  48962 54570  48005 53721    02901 23118  70612 71145  95067 23173  54035 90842  44156 71700
5332: 40555 52094  36291 55653  76080 99060  50012 51215  53835 74184    44309 52755  73230 63415  43237 40670  33774 48210  54012 78275
5333: 03559 82205  23058 46812  54877 67763  22785 80775  80062 69592    23438 52085  37289 95809  68213 00739  95366 98974  08673 58620
5334: 72432 33866  42230 42358  56373 40183  48894 82266  17879 55223    22509 73868  01836 45728  17888 98272  38762 87867  34003 99729
5335: 91356 20480  28964 84226  86631 66060  22702 68433  13160 35655    08387 98138  68805 62271  26627 18704  70986 97689  04816 55681
5336: 91036 67586  76584 47485  24896 67661  40932 53172  15483 83377    70222 30176  23885 51727  62901 01716  34971 04690  38347 99193
5337: 48582 28657  51480 30437  93441 79838  82295 41524  00396 36399    44120 64641  91259 29955  47968 55624  13940 86924  28028 00646
5338: 01623 01914  60470 48531  96795 16058  38163 79033  76643 20996    67340 96890  06700 63118  15698 55068  83866 55036  97038 61650
5339: 91687 56597  48723 53666  43476 87712  38321 27147  77657 48333    59141 50686  95072 81807  99869 14273  98637 95964  67455 32369
5340: 33972 89108  92938 98029  57767 70300  75150 34812  98630 62890    63774 38345  75666 75474  77783 53327  06948 99660  24052 85472
5341: 39011 39225  65769 45112  58157 99507  48011 51420  66983 54588    02424 09644  97177 93561  25598 46506  54564 30969  30553 21508
5342: 20552 02003  17274 98370  12221 72327  32035 80954  33720 16230    61550 10544  65098 89700  91372 93602  33555 80707  55913 69875
5343: 51997 77064  98776 92044  41235 58037  17814 75038  73594 51318    75191 71117  06674 67647  98231 57936  18295 45952  19260 26606
5344: 89549 74769  50664 66774  77538 17884  94028 68608  56041 81671    75522 61074  50271 59016  11824 66685  89464 63310  02988 73218
5345: 55411 95133  48661 28856  18657 65354  76701 81413  02823 24624    36692 79366  60033 39157  81142 98004  05311 49587  57017 23085
5346: 19520 40131  88498 91616  06099 33521  16129 89468  67445 62465    86099 15696  73028 37338  00720 23738  54627 17893  84549 90857
5347: 08843 17544  78188 22496  96264 79431  97155 05031  54028 45123    77123 62226  49271 35780  95011 89154  82407 74005  54071 71483
5348: 14730 06657  82144 11459  92323 84292  16740 10052  78728 97837    07078 51112  13572 92840  24258 55508  63492 53846  09990 76941
5349: 58852 45704  07013 08722  35525 63809  86165 55656  68107 68977    62306 95320  51199 72553  39910 96417  21808 43083  28671 23563
```

```
5350: 49155 11623   86175 29973   37424 51887   20784 78951   36473 23183      23233 28231   00906 91434   75785 66919   31979 45101   02547 13428
5351: 38686 45505   70048 59762   95418 34670   82521 30596   07237 95395      87781 40652   34129 64140   67486 08694   48864 47397   16076 58710
5352: 51704 00330   99801 06206   49451 47094   62013 59546   02750 28501      62758 82965   77758 46487   94864 50372   70267 41437   60535 63526
5353: 37146 11085   77251 48901   59993 95654   08644 81856   67390 18519      87134 68854   94942 29074   75229 25940   26935 65870   67666 21939
5354: 16876 13331   30989 48230   65042 83103   34629 97341   55153 28857      86659 52338   97677 57923   64218 01309   56168 41825   67680 67996
5355: 89639 94925   76619 49032   17829 96707   56516 50371   42905 80697      23444 41315   72890 70201   30536 22581   12581 41516   20495 50446
5356: 80986 09956   31997 21886   49474 32144   73407 31131   52525 53301      02817 26572   68180 85550   55026 96381   77743 08381   09709 18682
5357: 95587 98178   65213 02373   48578 42935   03839 82450   28093 46534      88197 15735   58389 38869   47461 38278   44281 58927   68566 40564
5358: 14427 10133   04399 34319   04730 06622   87856 02642   63770 60692      06183 61867   40583 37629   18933 77888   94417 66876   39328 29252
5359: 41119 87709   68069 80190   14287 15008   45081 88427   82148 72005      04790 07763   39187 63917   56348 80642   47643 73692   43875 67144
5360: 35994 76796   42088 66893   22259 52360   41998 86288   67616 98554      56848 09132   93647 50138   38541 84589   09419 45278   82766 38292
5361: 41519 43478   45695 98812   92289 77885   72687 54360   10837 40597      78625 69725   28760 03720   99268 60495   37051 44740   74556 41427
5362: 96037 13466   64833 89314   18084 62046   96361 57992   21314 02372      23716 05720   46986 22290   35530 20077   47179 20936   21032 32159
5363: 68263 95337   70258 03506   57966 53208   00636 85645   67750 00201      98224 57988   78655 49935   88652 66017   62118 02129   89707 27166
5364: 85693 28924   85323 25784   39159 61138   79842 67053   50037 81963      49236 40494   39441 94174   66339 47148   99660 95530   20756 25701
5365: 04239 38999   93573 78334   59011 52798   68426 96188   85230 32749      71881 64899   84888 94513   49122 84089   10449 78605   39059 78809
5366: 29116 88454   22695 08454   23284 95943   10730 15915   03875 74983      83297 94516   74146 84716   22694 54790   68113 38415   54166 83549
5367: 19257 56032   53313 62129   66443 77488   26746 05580   72746 03450      78109 12506   92987 49304   08378 81275   03660 33602   54959 28010
5368: 75022 46401   11092 44859   24786 32256   80980 98180   83242 75000      33686 32012   03477 73418   49016 08157   88764 02099   66764 54626
5369: 85206 89543   27106 81121   99609 42040   22238 83296   31040 58205      30789 97870   19882 76867   36424 28022   51493 39556   23149 00355
5370: 66817 10478   20508 41428   63281 79340   39291 21488   51885 24030      67348 97954   49913 33207   04236 10573   53587 07792   56904 04543
5371: 94128 20437   42934 57373   97358 98578   12649 13593   62368 81955      84888 53616   13011 79611   05916 71599   96839 38966   11596 52859
5372: 20417 78306   03497 66025   06927 60318   00447 77366   24899 16496      61423 21518   80224 61355   27592 13351   09644 45481   15954 11830
5373: 74985 72655   95435 76750   81708 53832   27756 04383   62928 77158      37160 62942   28083 00383   67279 70637   41418 01432   63834 80341
5374: 61236 82720   48137 76261   28228 80108   20575 07911   63147 83397      09643 92432   95508 47037   13246 70108   75204 54304   85424 53892
5375: 85619 05629   97177 09921   77144 69457   43780 96209   27283 68599      18419 83315   27099 29907   62362 46413   77870 20488   36036 46015
5376: 43898 69702   46242 54151   14133 17832   80437 74933   12708 07063      76445 69926   14556 33473   24005 05248   40270 80130   35616 99133
5377: 40702 70146   87963 10915   45888 72142   16380 27319   89468 47019      01846 23063   76034 37849   52412 28114   76268 53776   78382 88257
5378: 77130 43792   80772 50122   25644 28605   97854 32504   43012 23059      51651 19682   87018 70016   50702 28947   40949 89570   35435 93497
5379: 38886 21358   15741 24338   85214 69344   13715 76676   37055 15287      70228 22652   62711 73043   47421 48428   70567 56993   36565 70060
5380: 68937 50724   47294 40738   32141 39561   27914 16163   62826 70911      92327 34431   31176 84932   13038 82350   65091 80952   71054 84151
5381: 21213 51521   94590 81376   59028 61706   90741 61245   94368 45130      22361 59755   85603 13261   20485 32604   00412 41917   64207 41590
5382: 80710 24720   87153 08559   55622 18929   77425 83400   90319 82588      68998 39505   94084 04026   66946 32757   21731 86642   80199 55787
5383: 89959 92536   64016 66176   20909 93134   21631 21006   50355 19979      96889 95288   31904 29278   67557 00010   18438 22211   52755 18385
5384: 81865 06291   29529 75678   05961 24281   14178 69802   81497 32341      25910 84560   84198 51114   49170 10320   98981 97523   67450 21452
5385: 90930 98226   25045 38037   92414 69968   55556 65911   25438 35720      31597 35426   21409 92348   08304 86206   96730 36133   81139 99701
5386: 75253 50252   05649 37826   46450 99296   00200 86819   03509 60903      58158 01995   40663 12237   27098 55995   35336 93252   74757 97239
5387: 95088 48723   12020 36535   35497 49279   92132 52742   75659 50792      90607 75690   24966 01142   57925 38502   43124 41052   42924 96600
5388: 69876 37272   85468 90589   81725 91834   39160 51212   00248 68055      23896 05609   35953 11429   31512 44363   51838 00152   29284 98874
5389: 64795 74835   57978 90135   55230 99822   49153 93208   07249 52253      33486 51218   29648 59146   23898 00857   86529 61791   13460 94041
5390: 99411 77108   23526 59338   82544 22596   54249 78832   80353 98243      80249 09034   09864 90726   91344 42551   31099 79821   77870 42163
5391: 00570 16096   47946 85615   34411 69239   30959 37635   14762 61875      70391 37677   24368 64707   72495 59986   00522 90654   22646 95538
5392: 23042 92087   57120 82074   37443 38484   73152 62942   75872 08878      18552 33310   91191 51041   22028 79337   77192 54205   11433 33895
5393: 53585 82795   48589 52635   82668 85981   03214 44217   33817 33894      95781 48827   18581 81893   26074 49988   02498 59777   54233 63817
5394: 43560 85140   22668 86463   88186 91771   48452 82064   83548 03409      06906 42163   76564 20103   40718 88150   18777 26198   82377 12581
5395: 49590 19930   62657 48806   85270 05296   91666 68198   99301 87106      97805 22766   05136 21062   02860 35529   41559 38109   27540 56359
5396: 69318 63034   17176 07748   36634 66917   10481 38624   07984 13221      05754 10176   51827 18390   90179 34905   39103 91566   18275 14743
5397: 83767 29978   23758 47584   12964 99944   24088 47575   56531 38555      08504 00681   37256 36867   08464 84386   07767 05621   10083 80538
5398: 15095 26163   86386 93956   90243 51446   44850 16949   44309 15361      41693 11515   44973 24952   41148 09785   85033 05835   95613 61394
5399: 66670 18869   09193 07196   86413 47097   36623 04605   42708 21220      95565 47420   78714 93878   94380 51415   13605 00798   00724 40977
```

```
5400:  77487 59345  57847 87909  20444 19604  19615 77072  58295 74719    03465 18795  16106 56871  82708 39036  59115 37262  87674 26337
5401:  38492 11523  03277 83548  90952 46703  16981 52687  24026 14658    41688 67159  37993 15121  65761 45269  64494 26097  92869 87547
5402:  28787 68241  97451 14143  16794 07699  45296 33660  49305 62574    64908 37686  94460 45319  68431 62314  09826 53339  82787 32013
5403:  14130 80908  41298 62190  73455 73785  61855 01702  71700 52410    65011 39570  54317 37107  24720 49906  37178 29949  60374 67949
5404:  40192 87002  58176 13106  05128 27874  73638 37171  95120 34368    54089 80022  25337 13416  49361 75523  92363 72477  27158 23715
5405:  14805 71585  16398 44608  02101 74016  19950 16474  37710 93417    02101 15996  07637 62309  82337 10537  81558 89767  23650 64239
5406:  06975 52227  08087 07062  78869 43221  24502 37574  79308 93542    74922 18894  38545 09215  19500 87172  85955 05942  76319 48686
5407:  48599 88325  10330 11946  21835 02645  30234 71137  65835 11800    94853 28731  05936 31998  29460 67369  56537 54271  54836 01095
5408:  91213 14627  13361 08595  68947 16724  48007 03314  07252 57980    50319 75835  34509 44336  47354 19420  83720 46839  27229 79868
5409:  50462 51480  51566 74296  38404 52425  73841 63855  58625 59587    50639 86071  32876 62780  31748 92490  39537 75491  43861 23475
5410:  84572 23262  07543 20137  26942 95512  37343 65768  39562 59780    33715 76064  62187 12155  44467 65631  23177 30681  20528 46299
5411:  39238 45962  98211 63820  25094 47031  16632 02885  06601 03405    86083 91465  86700 39623  49473 83251  00144 16839  76437 24687
5412:  16620 03776  56695 83011  80190 10024  78041 54277  67405 28154    05812 76855  03326 43576  84163 40753  34022 01008  13034 34648
5413:  17206 12205  33150 58464  95334 80634  12762 66074  66048 86155    40513 45167  54592 27732  59235 11538  70664 61749  30533 94449
5414:  14983 24377  12318 88628  40078 32260  52727 05520  90590 02177    50336 09168  94819 96205  71477 77533  13587 65076  16334 34693
5415:  46208 07564  09928 94848  42780 67309  24520 23707  63115 64856    77953 64357  16705 28331  18984 19499  01586 99722  25538 06619
5416:  20055 31637  75648 72908  39633 48449  35133 00283  07631 60260    71694 77943  79243 59879  92137 47933  25881 22371  26207 99619
5417:  03807 73984  63695 70590  40767 62500  60338 07518  30932 79971    87313 51552  64275 51344  69775 41724  15901 59412  77468 51808
5418:  41287 43816  93720 68216  49257 98848  49534 22468  25644 79895    70276 17602  79880 60814  30369 10621  28432 53805  64017 70209
5419:  56942 03857  97276 50217  59259 88148  27085 48578  53812 60376    61233 16956  67044 42221  29931 08125  29591 33202  64839 14539
5420:  96127 31522  87989 35593  03297 83918  19343 10059  91074 88028    71823 88454  52021 49124  95496 06822  26955 81510  87502 94613
5421:  13529 41381  95199 84688  37033 57852  61497 39202  96943 44417    37603 40775  34652 10929  27253 14336  27643 93660  43126 01569
5422:  26223 11691  62752 85760  64835 72184  80170 67200  82655 64728    55625 54506  86299 41452  20473 21074  77947 85823  16217 84335
5423:  04365 22275  22413 38801  83652 62610  88647 56419  12335 67982    50107 44485  76740 41835  93976 82043  27480 84416  69685 26434
5424:  04500 27825  79827 95440  42738 26127  53800 25952  69286 04131    56420 14566  22060 91219  19589 07515  67621 88937  19218 90030
5425:  73209 27803  67225 73869  60807 56015  30006 38166  54821 41103    24802 02571  18135 05327  20931 49069  16844 61087  32434 95548
5426:  61011 49561  01066 43020  18551 27459  17404 93703  77515 20011    28587 77762  48697 74788  21417 98945  79419 00972  40758 80186
5427:  77486 08626  27521 35898  67167 67890  80031 33749  84188 34735    62986 60606  81809 36705  04610 69298  46822 39852  03532 69945
5428:  95821 65406  58226 50103  36165 38202  38984 15101  38725 13874    75401 27445  11207 06057  28494 98534  38637 42553  18591 86725
5429:  65095 46738  21362 63398  20966 27565  38762 59868  28413 76706    35851 66429  42216 57339  74057 42155  87085 92787  00077 25582
5430:  62615 03265  51589 84331  83281 87268  56415 57009  62391 36725    81779 42980  16151 56775  87567 82106  79830 42787  04256 68152
5431:  31747 56457  10806 63523  43737 57339  46641 77778  22649 13736    09560 89751  22710 14979  86883 11571  27745 74584  93097 77979
5432:  31979 48191  08862 04314  11742 53388  27291 70814  25523 06869    95067 17177  37709 41872  75281 88710  02998 65542  63402 98903
5433:  49310 53715  25502 77732  23660 08637  07766 73084  46125 91147    71147 76489  89289 64124  66125 43960  85822 86786  19252 77114
5434:  72001 12773  66296 40034  89475 30229  60025 79371  40483 33580    75589 14892  14745 56729  22107 46451  11938 38661  16941 58188
5435:  79546 51654  22945 07591  43546 33359  07577 85229  70104 55385    14126 78559  47065 03582  30427 71107  41978 59251  82946 75389
5436:  96076 27071  48530 63325  36894 59466  59178 08593  43201 45765    50876 80385  97176 55390  97972 78891  90452 14801  36696 94415
5437:  42402 15483  82023 90497  50580 22406  34393 92366  01262 12894    25549 60265  75292 61066  57375 82873  89371 03717  76540 21035
5438:  34076 42149  97437 21092  08298 18575  81352 38493  55035 22848    11033 76624  37471 45258  58450 86326  38069 76343  17757 56431
5439:  11355 23985  20745 75508  89229 95376  01673 05404  36414 65207    53824 47676  00126 34728  96796 83442  39136 92916  08515 89775
5440:  93571 30369  73254 02012  68679 35925  64588 73020  26516 07104    65307 24128  68452 74048  71871 07318  07120 16094  89351 15119
5441:  66470 05326  04603 59882  80898 49670  43793 31173  19381 80354    71647 73561  24259 70162  39573 33100  11307 53861  30013 07185
5442:  87615 52235  83394 39828  73350 18646  72853 17506  69399 84279    04586 03705  78794 11512  16770 85243  75649 87817  87374 00745
5443:  36678 17547  03493 74367  81879 63109  10288 63840  12113 75370    70532 43712  25286 07933  51874 04187  16448 61046  64826 88768
5444:  16829 14113  44927 50875  03348 01339  25000 05159  55942 57171    30074 36131  74787 70083  59623 66451  41546 06230  00031 80911
5445:  36740 29443  83184 41915  02595 31244  42923 88597  82870 91433    93015 19792  35972 90456  77476 17860  79728 94037  82128 62612
5446:  48723 95373  73203 19923  33028 16781  08003 84930  30421 71544    16359 73911  41128 97460  12233 79546  12078 13304  41131 16020
5447:  44564 05351  31476 41478  29358 34233  67998 39039  28112 60289    31248 03685  84511 57147  63967 52960  27550 11693  95552 38833
5448:  23862 93596  69288 16514  08370 01548  95697 59379  41000 80535    38114 52427  40420 20600  29590 35206  52068 01921  83756 86015
5449:  71261 54037  43827 66082  17132 79175  31298 70527  19917 80780    16300 94104  81585 94632  37734 38101  85607 54645  09167 86795
```

```
5450:  0116903214  8070603347  1414823741  2925933223  3600848291    3318075518  5253697841  7114986768  4029682183  4062888070
5451:  5162290963  3359301604  9204182380  9292758716  0830535175    5116414595  8021079466  6091357883  3402129607  8990680055
5452:  4724664173  8327631338  5974544238  0906279359  5886764109    8472455420  6327783515  5566201001  1358344945  2022181034
5453:  5721082413  8017185708  2161885620  8100149343  8351793894    8010271853  8190019198  9270162413  6575868982  7059748175
5454:  5608096947  7890558507  1388038102  3079288745  9732159647    6331619908  7881220394  3423911551  5996384586  1333791244
5455:  9248599127  4754136120  0644348974  9641259226  8463875197    5414438947  6710211086  6399235404  8914547603  0172601918
5456:  7252403960  8466868414  3964912739  0536334189  2298620372    7397508986  5160757704  7923870991  5471280242  0018602003
5457:  1888921035  1968105375  1368902783  9312980417  8155173544    6958058268  5680608029  0565935741  5961092268  7136789199
5458:  8729445878  6910892639  8791532519  0606860973  8786072679    8897596510  9654937267  2619440697  7621111632  7104584381
5459:  2912937128  9576739385  2427921538  5700472579  8099369497    4267406087  4537891430  0109970364  0398592003  1146806459
5460:  6418534802  7220035392  1656816518  8607044590  2422465984    5099743431  1052581120  6374235559  7098357301  0273312368
5461:  4165245613  6932827260  8658013289  8678777118  5727867981    0431664537  4397568333  0479490611  5697175666  5219918730
5462:  9916637873  8356140701  9212362945  1271691124  9923356802    4446886242  3614547050  0450739468  6621408076  1040503786
5463:  8365059771  2755890615  7513944761  4154042421  7030720950    4467836595  6748338791  8388677237  6940867205  7864847727
5464:  8847492718  0391320155  0229569371  2008672566  6674267595    4137847249  6846118103  6833851203  3319371120  4874423883
5465:  6212360483  0950661082  6725332010  1336550499  7713719834    3920514505  3983737866  1812300840  1122545572  0065327481
5466:  6603547245  5287017764  0368076153  3144702940  3748795987    4010807409  3226062745  8939850175  6524584415  6699115180
5467:  1195381176  8638534275  2535422892  1989271092  4466765953    5453852755  5226342588  2124405706  9054396414  3724033819
5468:  1683036326  1140920686  7835208192  3153316738  6032888496    1096877536  4001528017  4617347819  1389815865  6821638719
5469:  9429632642  6052310784  9313065364  4461038361  4685877272    6606269447  8496612376  2266357177  4023768020  0072766531
5470:  9967846996  8136138782  4859187635  6692531120  5373290365    8092984144  8212639749  3176928073  3960474115  3003904434
5471:  1773069729  7742493119  4057909943  1160341583  6555818640    9855438515  2629496705  6544807875  8461888868  7802639186
5472:  2230177485  8296771329  3168330531  3932012682  2581175341    0770320535  7851304678  0049298225  4439224337  1203575594
5473:  7496896235  2324810622  9854876444  3541828734  1409664051    5643987110  5781309771  5756681711  3022790865  2369529853
5474:  7551655174  4572714355  6122876217  6983262143  3232742248    4426633402  5941141533  2026013663  9377911795  8985171913
5475:  8395942252  3996010546  5377910517  4155299685  8617447818    0419551805  8774608665  3878424950  0881302885  6943671183
5476:  6989565389  1103849908  3615313780  0509756738  9026707481    0590582252  7338429769  3776812294  8627858732  5228165622
5477:  6670300583  6574773487  0527737877  5770451855  1041400273    9330173770  8864271062  1105676714  5473242020  9935184419
5478:  6715839814  8213341608  5465531869  7949091885  5140347958    0706234299  6927851598  0559865626  4015324535  4572232340
5479:  0027886073  9391781313  9441802678  0165383706  4294184125    2665913301  6207392135  4602224420  8905911289  4704643305
5480:  5327261982  2652918139  1198267405  2218023363  3428879907    3679096539  4876948483  7574104659  2261218136  5895761078
5481:  1554438204  7891962406  5059023890  4835992176  8649560937    6814020016  4883798825  7445297151  9455800651  0103331449
5482:  4099985613  3372981100  2830416708  9695522105  6852411606    0902381350  2578148770  7603415745  2916177271  2200078510
5483:  1062604777  9055292440  7862567545  8444901652  8633620256    0857375612  0685978178  8969836677  3092798828  6474422506
5484:  3403964271  2802267718  5890650520  2459219224  5847578145    5378959792  0262530358  0664072996  7231699961  8087330120
5485:  4452446042  5523870280  5636821960  8298839880  4775517529    7720125154  4866857843  0750835562  9463339346  5991736263
5486:  3945130080  8089375209  7737685306  8887501879  8404201871    1916986038  8920699485  8267832415  5296137745  9662669322
5487:  9897543367  8779295254  2883531345  9103839871  0979062059    0512025847  7713645772  5785047257  2423265123  3800604227
5488:  9931786798  4758914076  5826591176  1085409958  8098217657    5126558338  5682253647  4162996414  9393498742  3856446139
5489:  6829782369  3721910254  0342228101  2541957218  5978191191    2538702618  2395891849  0283025058  5877949416  0926056231
5490:  7738062678  6045786671  0668108411  7051980566  7832798759    7852022369  0678773724  7408485766  0235746771  1868480075
5491:  8285624358  9320368511  8389319850  7671223642  3673974096    1330426886  0110400944  2182814567  4511683994  6233564823
5492:  4586373149  5033871455  5282686527  7824853658  8038369793    3973068037  6270298880  6917535901  6697687433  5781949091
5493:  0539754416  3800769192  3281485224  7745391266  3192101094    5964387544  7161902909  3350359583  2119406653  7588002929
5494:  5076879461  4396084590  0015997619  4910124084  1246034521    9284507961  8490132525  3241970225  9707072490  4899390407
5495:  5721565296  9120721275  0537347282  9170414117  8086373018    5003929192  2679903474  1534134686  1963450914  0181649080
5496:  4810897448  8583293649  3953923202  9042102453  9113919199    3682724542  1851790220  6111731671  8077186808  0209379683
5497:  2630979307  2251731238  1447052002  6733761513  4233163900    2546738038  7521901783  7219902081  0402286995  0078359539
5498:  5961651896  4824536111  2533416390  2715573322  6406494445    8538197109  3964698942  3681832480  6577569019  9173944970
5499:  2093209634  8733298289  6108019291  6959589017  3781725360    2020720066  3599354789  3034499601  2347133898  1522384510
```

```
5500: 45867 48209  64085 32977  50962 76308  89046 78131  81705 03192    88852 16361  62807 39265  07111 96159  00071 31609  16566 61670
5501: 82376 52687  48260 30492  96277 30454  10820 16908  08808 43534    11857 09950  87526 89986  42456 25276  24127 65306  36955 61464
5502: 49609 87561  66844 17600  28946 47445  76792 72982  58091 41126    11552 93746  47907 22632  05565 26722  73980 70122  62253 06788
5503: 23578 10086  39648 55583  75249 22340  17300 74511  97888 33249    30450 18276  01379 64536  83519 84384  07054 07590  77551 25780
5504: 36251 34032  34002 88976  54251 94597  74064 85167  84112 17738    19758 86444  96282 34336  29856 02171  68998 17374  67305 37119
5505: 63810 48747  20958 52279  35663 86140  32621 03304  34192 08610    28059 96627  24222 64506  04375 93397  35091 14449  39595 88935
5506: 57342 51890  02155 66519  33351 89682  16571 12763  65207 71933    39488 78192  35409 50828  33536 70600  93849 29504  77333 71796
5507: 37222 91812  90533 98893  19053 21967  67669 71631  94939 51915    24670 39051  32127 05391  92438 29959  14305 69774  66244 21096
5508: 90011 99523  50473 06637  52109 83231  37639 99167  58798 17708    17461 58995  50200 14751  76178 01080  88709 19238  78779 54984
5509: 04216 12334  19649 62478  01139 04244  56867 21464  91369 54302    21238 41100  61991 07375  66372 99144  84091 39051  54970 13211
5510: 60725 78113  44358 93520  66261 27371  54773 30801  46197 36610    19449 98825  36727 09254  15355 25449  20149 46119  60000 55146
5511: 87673 12577  08757 14778  78360 15429  22266 58902  63430 40447    31583 78365  03638 11754  37643 82438  94418 72574  93835 43445
5512: 88972 02731  55391 16623  00336 80923  61244 19641  66734 83813    23076 06028  73500 91448  88962 89991  50284 30382  03591 38081
5513: 90093 70634  03597 74192  56534 35441  74458 61786  09545 26697    56096 78588  70387 19767  71239 19402  00202 19986  56277 32176
5514: 45237 71264  09483 78383  40626 88402  92320 04265  79466 36356    79100 32990  31506 40192  68808 16695  61822 82833  02681 15511
5515: 36964 10154  78332 08803  78018 83036  91173 38834  22774 19122    83844 43034  87173 87629  72106 13402  36888 11994  60648 49761
5516: 06073 68660  63691 82384  05983 26641  71971 33913  64023 82300    27191 48694  72923 74232  08698 35394  59949 04269  63449 80516
5517: 00710 25414  31711 39552  93578 24373  89733 93711  76147 01297    68451 57138  52247 50832  82448 86998  88581 17097  72694 46329
5518: 05994 48747  48808 89274  72302 81045  76063 11414  36823 44725    96563 61365  34048 91203  79873 99902  02059 56459  73943 82868
5519: 11153 39377  26479 01280  92260 10150  88488 18956  81337 18451    95872 45835  24895 83928  20714 07259  08647 44499  33241 88843
5520: 44584 67894  44793 66939  79507 37571  18238 14405  71012 82225    50846 37075  53312 69571  63897 30146  76613 46967  18393 76157
5521: 21223 69883  38913 09368  61057 50456  14227 28011  94083 84912    22999 78901  65360 30009  91729 29911  06757 06949  59819 21803
5522: 99622 85413  25950 61979  28948 19019  28386 70419  27452 17752    01492 91967  35504 09675  00744 16695  79277 91634  75382 44736
5523: 90812 20594  91564 24379  26703 39725  30201 18451  99551 81119    28647 21211  77905 68653  60693 13584  01361 44952  50805 35280
5524: 32808 97737  49091 75058  10228 69576  39795 12751  82849 84174    95747 91890  96477 35222  14291 33546  19646 69018  98395 92353
5525: 22442 60819  40737 31054  15057 99872  52899 89893  73065 31198    37432 27728  01482 81466  83629 00167  22427 10994  27497 62652
5526: 43284 25282  32111 23347  13748 06985  17447 22721  04364 48000    38612 43765  88123 89223  94764 73724  31702 45610  38194 96271
5527: 76485 62260  31113 47861  80477 49509  21893 14027  64198 04977    90134 27491  36827 00767  43791 70867  70269 99547  53703 10734
5528: 81486 20070  66037 53838  67438 55242  40645 51787  97585 98544    27043 69264  40253 52368  07972 98753  20563 85594  50464 94125
5529: 67353 23377  79378 94300  32608 82797  72225 90420  36424 13193    83091 89426  40635 15884  00929 36960  48635 48563  66723 15840
5530: 00119 93520  71136 57462  37627 25831  73378 84852  30645 35034    58744 61919  81085 89534  93692 77580  31382 48390  89527 01162
5531: 13589 03171  32273 30782  58087 59049  16821 61685  72792 34169    19661 46660  87621 05101  68455 25879  16036 91983  84290 81973
5532: 72934 85053  14458 78750  71466 78839  82941 92427  18492 42590    17450 36775  81464 24337  94720 54534  36072 72156  70545 81839
5533: 58603 50019  48151 72963  88265 96967  39310 35967  31306 54260    16226 66029  04675 25016  10197 80277  66162 07665  45033 13498
5534: 00970 40495  94789 59960  17521 78800  20237 58634  97981 29402    27982 82534  71195 44506  35214 92194  84893 00723  12300 62244
5535: 56836 72088  14813 70072  96887 09195  48697 63580  56685 98749    64429 69226  83523 24652  33601 03687  83838 56272  26444 40900
5536: 00019 41456  66025 45011  66401 85355  13463 33977  40084 84131    42124 09764  01792 05789  84309 43338  06028 73332  67073 63829
5537: 01849 76335  67196 10523  64059 85723  71396 70233  97910 83026    60236 53929  26108 90937  04707 58852  29872 07823  17352 38995
5538: 30575 61496  13024 33425  27136 24865  15488 73175  03535 90853    13610 75371  10022 95725  29082 97678  83818 86942  07565 26241
5539: 01488 19098  82428 83993  66294 62222  40626 21924  73407 86221    40680 46926  86586 57376  39833 82878  95494 68613  94617 58139
5540: 17568 84612  91052 42303  31337 21326  94077 97965  88801 07646    40728 92499  95279 39197  40413 71389  03636 27564  75049 20600
5541: 22861 16701  78663 54178  59565 33028  77211 58795  44634 22684    64183 10725  23893 17082  21289 62160  27304 85268  45201 61083
5542: 18733 32297  96293 69445  66640 29834  10658 57042  11858 04841    68427 63969  16386 30431  33094 20659  80459 35542  01478 78032
5543: 70117 47965  46617 91703  62176 51807  83761 25829  20167 28544    78533 22789  75256 59978  50231 35937  82040 47322  36054 00631
5544: 23545 96613  22604 93459  12364 51130  69118 35590  50376 07771    46597 73165  13150 95980  25992 66277  63223 43293  02950 93669
5545: 66522 05296  99192 87425  11152 98019  80334 64631  96034 00908    51711 13623  60075 36331  71321 84678  79809 79626  08323 63105
5546: 93846 95003  20526 20588  98437 08896  56829 32768  35984 99788    24107 70705  16115 78828  65564 18089  89792 49026  74457 71942
5547: 01032 50418  41021 46912  68252 05240  10776 98152  91634 76581    59131 15370  55403 93823  00074 71629  03165 88572  52833 13202
5548: 59280 97138  78832 58818  21273 37989  34780 87887  18443 71396    75038 14808  23459 73258  51321 24300  80738 40905  64160 12588
5549: 38828 89920  24054 30916  92056 13677  47113 53775  43068 96409    95856 70550  39600 63408  64724 78756  61176 36067  35173 83274
```

```
5550: 94789 72358  03270 49941  39694 68510  86082 56366  55085 11050    99144 68583  27978 95370  21244 39662  81509 55906  74063 80650
5551: 23609 92208  25939 20295  60350 07006  78749 72136  01952 92877    91086 88993  88312 90017  94188 84321  41512 89036  04159 48659
5552: 09723 89607  41247 50537  85795 15814  00633 02409  68633 06844    28663 18771  61728 71050  49858 44231  87090 18343  44727 19122
5553: 65932 26390  27749 14379  69928 62687  61194 24469  58212 80265    09914 45808  70384 04541  45633 54975  96413 58127  59518 63402
5554: 82191 05698  94842 68520  02986 66751  07056 39558  71175 59074    47611 03315  33440 18679  72004 38566  11799 10634  76047 21337
5555: 12399 95983  06165 02128  68461 67104  94555 61400  28684 28975    63770 34516  95695 56731  69455 13442  19413 05391  46648 39730
5556: 97545 36366  67027 15361  10109 56265  19830 67731  72692 85952    05987 98702  66684 75790  65299 27621  69429 46909  58660 28029
5557: 05906 96989  81962 13301  32934 02577  55321 80858  36958 40872    27077 26239  74948 62445  84951 05047  88858 77513  07655 43897
5558: 99925 78527  23258 18760  85318 95023  78961 33813  41336 49280    18814 53382  75941 78209  66523 08575  75362 53591  30968 73327
5559: 81467 42602  63248 74859  89849 87440  66695 27966  42814 00555    11407 96650  33335 43418  29278 44640  55063 39786  18405 08842
5560: 86558 98660  48662 86747  53243 10508  73420 06423  19602 20463    67693 22920  74473 96955  40370 36517  06035 21502  90714 93630
5561: 22356 94280  85332 99841  62120 01092  91499 42016  90533 82603    70952 85753  18144 31820  44393 18664  54546 67579  55681 84960
5562: 65532 36018  25446 90749  01455 38738  35391 68145  04954 70439    16520 26509  59376 68672  23546 32628  22562 56292  95046 10714
5563: 86678 21634  91376 60123  44468 11361  54185 72790  75279 85099    69058 86735  31728 16026  32849 06858  90599 38417  85302 02343
5564: 76229 39619  39400 25015  41828 96740  96082 97775  41792 85286    18919 22264  93546 19033  99469 81373  98992 90465  13461 28309
5565: 83667 69088  33923 48780  48050 95230  10561 04780  53299 50107    33317 57875  58784 60602  38971 31124  62849 16460  54628 72585
5566: 16507 88827  20869 68725  04580 65895  74776 67376  88437 89689    19597 73852  13655 52936  80679 90063  25776 78135  49442 03718
5567: 98168 15368  42300 93769  35931 22606  89963 62662  09151 03973    07264 89714  22826 12214  33910 94989  90131 70527  78633 48754
5568: 63177 55326  40053 48426  43845 72619  21057 27966  41550 83422    49900 72034  65925 47051  10309 42564  62463 11007  45840 18146
5569: 05489 63428  16240 76652  96773 08969  50967 08093  01741 52662    70417 02155  42005 38096  80452 86326  39792 23546  93477 51526
5570: 22362 03480  52633 52219  63057 55285  03359 82844  39093 66287    01755 36157  29412 50084  98270 80084  86305 22462  11934 45562
5571: 70233 01179  97797 89015  77055 31993  63367 41913  70650 24625    84034 43392  66165 17336  68649 10532  84770 36629  01967 35247
5572: 43907 03004  39132 82514  27796 51031  43601 41864  04780 55858    27679 51211  00461 35953  90270 15044  78239 26734  88066 21713
5573: 23099 06581  61060 28072  82985 27787  39319 01309  43911 61395    53251 65711  37046 95714  73545 09547  78646 05213  99126 62107
5574: 79674 12343  73939 02248  33326 26289  89148 87138  48211 46349    40539 81997  72706 92943  84314 67419  26742 71183  03504 37445
5575: 73556 44399  51886 81055  92119 12398  62651 58495  05343 58973    81046 97822  83185 46969  78566 65301  39070 09243  36952 69922
5576: 29848 95183  63887 27158  16748 28418  18552 21204  53613 71196    56307 52027  79691 49300  61690 24441  66134 04159  34735 08908
5577: 78463 72854  34247 89491  81626 21049  21171 95101  05203 15663    05929 48087  18296 18901  00803 76189  05062 81120  66879 35823
5578: 05201 85359  84147 91265  74136 98848  02382 64500  73442 86519    20219 02052  73667 33018  41213 37332  88760 38635  89823 57086
5579: 84715 61828  80828 52564  47832 49575  76813 29835  15231 44385    55452 78168  81677 89439  43083 81123  51747 48179  11692 74973
5580: 47996 28730  60997 33315  46911 12098  18173 05901  78432 48510    12209 84662  29124 34298  90080 68372  66466 28941  21597 40682
5581: 39978 97034  23013 90946  12307 67708  32104 37046  96458 92921    51516 62749  05886 83530  65766 48064  44221 80725  10346 70859
5582: 79268 21140  78541 72497  55823 29778  14487 69174  15572 06410    17356 77761  35333 80502  22286 64625  80223 75128  16441 91700
5583: 88109 19335  86460 27114  34311 66712  54682 94732  72406 80270    85208 14289  48834 65017  44112 04735  50879 27645  97352 64352
5584: 29034 87601  64124 71307  71848 52593  11759 94433  55677 90059    47410 63237  63731 26527  78630 98498  56805 27886  09665 85134
5585: 42897 61545  61852 91195  04538 22542  03432 91144  94474 82069    33397 10949  34782 61201  01932 57954  66287 85888  99427 55191
5586: 70862 97029  90520 35297  37077 00035  49034 00180  68443 55772    73729 91183  97562 14836  25067 19440  53328 01103  56176 89850
5587: 80667 21506  37469 41276  21291 34983  63085 72908  76391 32690    35346 67346  06394 38262  26589 11776  19903 92121  39451 01259
5588: 57470 49171  74200 91659  16818 17416  59045 74543  10174 88913    57338 71681  73127 74003  88547 77912  50595 15955  68241 34749
5589: 10351 25249  70562 30704  04598 59376  72135 88704  67426 20010    69477 00397  03840 89600  47511 15639  58281 78812  25595 11100
5590: 78104 76713  98023 97876  29703 50831  55886 10888  86474 10774    97269 03235  24037 99892  71025 34211  25405 67742  47310 53858
5591: 25477 50180  09234 02178  34864 86817  05583 62277  63237 90090    29950 62032  92941 53368  93333 28440  85898 31382  35585 71098
5592: 13737 23661  11921 68675  53193 00268  55077 90723  83014 84791    75758 88088  67147 92709  95942 97127  19594 24491  98606 77037
5593: 88918 03199  23941 51506  11862 68986  73107 45234  26330 21831    27546 42174  23951 37456  13947 48276  06471 30037  95444 86111
5594: 73279 28806  91212 83021  15539 40029  80020 51510  53088 24185    48469 96482  03170 40186  18069 63161  77046 71784  94098 78507
5595: 79072 59869  74417 45918  17409 20477  96196 89298  26539 21360    14666 07468  16490 16189  11971 83569  61111 08203  90824 92635
5596: 15517 82690  82250 38511  29894 36594  52471 30357  34475 36820    03290 10102  00231 34725  32624 36146  00882 34902  21455 17018
5597: 25710 92701  81373 42974  20063 68596  81049 11471  22917 01146    82040 64959  78680 19519  68856 43461  72717 89690  12121 11586
5598: 54684 46316  78743 93910  59056 58849  71537 59388  40511 69820    02086 96299  59482 33326  46892 40916  33966 16469  33731 36983
5599: 28630 40579  68262 52840  47135 51364  21997 71073  55458 68221    79249 40999  04986 08039  70546 97519  05513 92287  30302 70039
```

```
5600: 64059 75110  59218 13980  64228 13939  61555 84172  44261 98269   63160 84463  56875 72880  53707 01109  44651 98919  54745 50149
5601: 81817 27839  73044 06715  75157 64483  69452 53779  27092 92132   15227 55698  75693 53351  92091 67273  92543 00050  27290 88348
5602: 39081 81519  74349 18685  54095 03285  65539 36025  92573 89151   62469 19155  07954 49274  21131 31045  54923 69563  72992 65539
5603: 60257 48186  42838 33349  70587 96951  44724 91558  71057 68276   71092 60523  41465 48823  26639 18791  70481 08545  76835 46321
5604: 33042 82507  80820 22604  44268 43188  82181 77292  40040 12999   38974 18426  97326 63207  91749 72972  70203 43970  27069 55210
5605: 83447 52733  50908 18895  02522 28091  41119 09473  51175 10029   66039 71107  48491 25192  14870 50698  27667 31738  54761 40766
5606: 54038 17191  91148 06214  65873 44904  36976 35469  92821 17325   81774 11806  19927 31087  82676 40652  82343 83126  82418 00156
5607: 14235 80358  94661 46333  46379 80669  70292 42194  69035 39121   56787 12789  83015 43674  56161 38785  04999 76674  60530 24049
5608: 05947 98417  58352 03072  84490 86019  52725 93156  75785 26211   04812 88491  39266 44799  97422 83737  40660 24365  25118 43770
5609: 28878 66946  10026 79712  25295 19207  36575 04646  17286 73701   48025 48199  44286 66162  17840 55978  07044 85033  90097 85845
5610: 45017 09726  95041 60249  44724 64022  42501 70131  92060 38252   00870 42243  07690 41365  00016 93921  60611 16854  41434 91865
5611: 23155 44287  23141 28476  07339 60740  63681 56420  43781 07918   95227 43662  33836 41748  11984 13574  95790 02886  22331 37040
5612: 12817 62785  67003 71813  13911 75197  54818 24495  64706 41471   02473 66129  07254 45610  55158 10190  67460 02025  07381 56161
5613: 34341 01962  16345 95181  48195 72045  79438 18884  81578 34062   37746 49664  42137 41293  47518 27664  61274 46578  07505 34619
5614: 60278 44208  68026 31320  87519 16005  80555 92247  25346 03982   61071 50267  20208 24278  27527 43524  98504 18727  12908 35849
5615: 70220 68779  41525 74526  74954 79386  15205 84602  61650 34241   33351 17718  00481 83293  10831 10567  60635 08188  45496 47139
5616: 37091 26362  20274 99018  73856 15093  86244 71017  64175 40002   69373 70648  89118 07140  51547 77114  44298 87327  29081 20210
5617: 26073 35954  09152 64241  40219 28396  88322 83662  37327 30062   20684 82559  35479 82241  19512 12009  85719 40502  33654 91188
5618: 78602 03006  19483 94114  57146 42742  20596 89587  22297 59712   16353 52044  37949 47675  39695 29789  24556 70543  08238 22624
5619: 18246 06051  46646 23042  17285 64430  71159 61651  70884 32971   35598 34766  26229 18472  37671 60846  06894 15096  41903 71115
5620: 71976 68438  33129 90200  14936 48527  29838 26893  15743 32686   55498 47559  53544 81041  64413 11744  07152 41879  68019 40762
5621: 38234 19986  25781 94565  36041 86232  27951 38913  53611 28803   91664 28862  33952 22770  88666 16733  80353 03147  29474 84676
5622: 03591 75970  52214 40153  60922 63889  71968 01628  34505 42407   02235 50771  22631 16044  48870 80084  17354 64647  32060 48274
5623: 56152 20689  32549 33601  27607 80648  21127 51479  97476 98826   56609 09423  29977 47799  84037 17342  60302 85947  84896 95338
5624: 06011 50116  28322 80454  91557 74521  92350 55044  51397 71024   58158 85447  73356 37292  49613 82802  04612 51951  35584 94750
5625: 30621 36639  13696 75163  58502 39503  95391 58345  37776 89078   39462 34403  44327 17439  24745 91616  62215 68501  70408 81887
5626: 73344 71535  01069 84298  54709 64117  83699 12927  44795 01650   35382 44487  37566 82148  93795 40161  70339 81698  05913 54411
5627: 99469 40238  15500 38871  51503 33854  53736 79085  52476 05623   76286 06534  27685 55209  80061 02852  51631 90362  26391 45255
5628: 77393 29222  39995 83843  84653 63727  36597 65968  13978 07586   13454 36938  33320 51510  43797 74769  70749 82883  21088 50972
5629: 49747 51444  91967 52126  26119 75790  34070 82834  89741 50895   36160 89931  06471 64300  95802 91124  62872 20051  60239 13340
5630: 31293 55358  74340 55034  89432 81410  87912 77935  73015 74044   36710 50066  03738 38566  29035 13044  28368 97457  15195 20235
5631: 83637 57010  97303 99964  73655 68708  71515 69626  81217 31127   06234 51167  80054 17989  14641 14637  60627 48854  59611 03622
5632: 43511 41086  24141 93700  17800 98692  09819 08161  30074 00904   17409 53646  48023 30018  37179 71154  24101 00147  79515 35676
5633: 19206 75625  47266 21699  86426 57067  33345 83801  56865 21490   14802 83961  91231 10224  46644 84262  32602 48873  94067 06070
5634: 64283 60351  53090 55110  64278 80823  05442 55906  53597 07188   35093 16299  67090 15000  87067 85543  64583 44414  52200 93492
5635: 09329 45031  96049 45066  62712 96035  86670 74644  85359 12724   75498 11386  02171 04040  74541 77280  79096 97357  71166 62524
5636: 45003 91363  76721 25735  52787 37271  23572 13597  98212 36837   22806 66832  09995 01921  24010 11532  53310 55844  36026 44183
5637: 85027 63569  35774 91186  39054 12008  84664 83776  91514 86255   97333 43575  79008 05213  56422 07805  45431 31781  94960 59271
5638: 07580 30305  31617 58201  87582 17395  05567 71504  95358 03729   54972 91396  06847 25164  55225 97343  76841 01015  49316 31690
5639: 14015 63127  04334 86156  75766 94572  51166 31551  73951 56834   75837 21005  11328 52819  22324 75164  57833 00032  83081 20432
5640: 48504 82987  14313 92199  26481 89729  09068 29197  23868 62249   83450 96010  51847 43132  26280 54770  38981 17224  83912 95366
5641: 42575 41671  03698 08334  12050 10366  10979 27657  41866 61782   60158 17557  36730 70581  08511 00849  09088 29912  65905 90074
5642: 91832 91373  99337 63538  28253 18512  42075 35303  15660 18447   01459 55601  71861 30388  57746 33276  82321 11704  76983 86656
5643: 98595 72155  29566 71990  82695 92539  95180 72109  13064 92406   00661 79934  80376 11548  82485 86339  52046 66195  72442 97194
5644: 86678 19955  00412 80521  14354 06573  54137 30804  01285 13845   67633 95560  87538 28113  09670 54826  75224 02963  49398 06597
5645: 90104 58553  93898 47033  62970 85807  84926 70491  84614 94578   08140 95563  10170 98269  32467 18691  25267 84582  91440 34155
5646: 19209 58891  09995 84928  42769 43200  32250 18514  20009 83784   05853 61202  96560 30999  67483 59715  40825 11197  62816 34360
5647: 33722 31423  52895 24292  92215 60080  21423 27762  97721 35681   35727 26336  48812 37812  17275 96652  35314 30900  73297 17633
5648: 55464 10508  45953 00001  53714 89796  01773 82818  39355 92063   42417 56429  67456 45430  20667 78835  59786 23106  74936 15849
5649: 33019 65268  29669 82247  65100 04421  69096 37252  60500 44757   89661 42857  56044 22134  48775 23205  00503 72967  26480 80777
```

```
5650: 68903 91803  02262 22049  38863 32544  75173 64402  69256 02479    40409 87018  36894 38190  35995 33159  89908 51199  88909 50488
5651: 13612 43570  24308 33135  25117 92391  61675 41778  49806 31030    94467 00752  35093 66304  05025 04270  20796 49887  25327 55228
5652: 26912 33147  05897 00201  89222 99247  03527 16868  00339 90692    96927 29912  62920 09394  00032 18514  70121 34697  09471 90373
5653: 60444 81285  51847 89067  09481 05051  20340 62083  52243 35803    51583 88484  10908 24702  03180 16567  91498 53820  17317 61765
5654: 21916 25068  62847 77656  68229 76668  62130 08933  55892 13513    85105 92901  72040 65480  24947 50671  15986 88871  11911 41327
5655: 68720 43393  65693 17285  83408 22214  32840 59935  62873 96285    68678 28035  31182 89227  08118 14941  21193 76454  53022 47119
5656: 91532 11515  79673 20664  81723 66965  04894 40463  55146 48191    33519 59115  37486 40367  06839 92105  05061 71289  66737 01162
5657: 68190 55364  06356 06243  74933 64944  18563 18510  44823 62481    79016 73359  37693 00994  92580 50269  05950 19102  83088 22435
5658: 93543 00769  17900 70366  87089 34066  57858 71963  42782 92333    36787 53270  06401 93744  77381 95876  67616 91645  20659 78992
5659: 37689 27002  10965 01122  55740 25004  98952 72085  09739 82448    21438 57843  76849 43308  34245 34766  23463 97170  48659 05312
5660: 95131 45923  63943 34215  69255 58444  81405 08115  12005 45458    58457 28183  27551 03143  82447 44581  33020 13649  60540 12663
5661: 52611 15889  16775 69100  86156 55598  70697 74958  53023 90911    45284 78429  34928 35779  17906 50951  14058 15058  97360 47817
5662: 21989 47611  40909 56416  29555 92021  90147 34726  72711 77543    79664 17246  27275 50677  28746 20054  91895 18300  71181 81279
5663: 78531 54929  15687 24636  29110 55381  37466 93268  25324 90956    24520 35204  27997 00506  96771 45625  90822 62348  29816 47481
5664: 09688 20954  24024 90989  01204 71180  36793 32791  16333 19100    83221 97079  78977 36081  22196 72725  36520 42500  97496 42697
5665: 16899 19089  04920 96635  24022 99458  59439 44340  09822 82575    73525 45862  88653 99052  85594 58206  93960 40769  00319 10217
5666: 57341 82130  74671 76269  98783 12083  62234 12221  21998 65881    44807 45184  97796 46323  67813 57944  39897 39856  27688 74405
5667: 47672 90719  30668 90050  97585 17017  55216 31275  07874 45371    91115 76943  03130 28713  88917 58655  90366 05438  44020 77737
5668: 35677 56296  44165 86268  55567 18315  64621 49162  78804 80165    76857 89978  09840 52473  01620 82870  71046 47183  97211 35315
5669: 58138 41241  28285 09777  31280 32158  82296 10519  83533 33336    96766 62772  91182 21519  74261 87336  66093 17400  67002 93048
5670: 37502 21208  25654 65171  50997 98885  92468 18205  95477 38533    54872 31847  63375 77064  76699 87179  38352 89827  94365 96690
5671: 54525 85066  64071 03805  38956 70544  06937 68391  94338 49008    57782 61968  68413 16170  19007 17051  48122 47101  59563 75504
5672: 68259 85279  81072 60727  51281 07366  10041 77945  66664 21008    71757 30784  93303 77719  19425 65206  82459 75327  16306 73919
5673: 68086 83020  19606 47963  09072 42522  20286 00308  78612 31965    70517 48093  13521 32696  22077 14432  71281 51105  49089 14321
5674: 26037 32601  42680 49915  71140 88999  31025 14131  51169 50284    22644 85962  50279 27835  45693 23746  14820 18144  44937 01409
5675: 41899 55579  34115 74191  74568 35355  15734 01052  23351 38810    92617 68994  10236 94357  05646 02364  78855 88478  19932 66050
5676: 88287 67707  78428 92450  55535 70063  32774 55298  76329 12479    88674 46945  14029 53632  58432 15888  39282 31043  83029 32370
5677: 34301 85687  83959 11509  56322 79171  41282 00108  89547 04840    78402 75997  12139 35777  14966 85887  09596 76329  12656 28657
5678: 19504 10298  16965 63311  92900 95516  81573 10595  83321 73654    21812 89306  55906 45738  89263 52561  63843 95944  23668 75501
5679: 45128 62553  16076 51359  06057 59999  94315 55549  12086 52458    04736 04467  72520 65121  68923 88740  59817 84111  45866 51816
5680: 73407 66876  07842 25816  32417 13075  61778 23036  04875 84946    11547 06272  96427 40982  40651 81700  47341 96705  94708 29792
5681: 29270 37292  11893 46129  85250 98027  83149 62868  56664 90828    79133 26758  41757 61827  03900 93899  51985 64890  99406 70385
5682: 71137 79151  95356 89392  21753 15092  03636 95451  59071 19876    45905 26333  04482 92987  19902 76362  42330 73348  21693 15902
5683: 79206 68416  66891 61004  02408 09999  30629 78306  31612 72156    60777 07131  86635 64665  14160 51967  00263 93351  21238 89728
5684: 25497 65707  47057 99407  04031 85805  57185 17577  68272 62620    28956 31763  95896 87247  50027 72806  02783 48498  31630 40262
5685: 42983 98651  98829 02850  11796 76600  68550 48694  03376 58803    76320 22394  41588 59079  16411 68391  72802 50098  06516 41006
5686: 28784 72249  63024 02021  05110 01481  67734 29321  67884 41224    63332 47631  24051 01220  58879 60827  79772 53803  95244 53913
5687: 99583 56000  85954 32573  59462 03014  28894 17380  80782 71368    47452 31023  59532 25140  41754 47238  12161 43505  07487 40691
5688: 03134 10824  41996 11920  92935 63112  35444 56132  16808 08352    95909 65904  52872 28785  17655 75286  19736 94563  60442 20344
5689: 34416 01105  24846 80691  08218 46899  08374 54203  40438 50645    20453 15743  05703 28267  17941 42890  87026 60890  84398 02414
5690: 78168 55903  43863 55567  86320 47875  97794 89155  46791 65068    74917 93738  55303 94836  48773 62050  62419 78408  08684 15548
5691: 68360 49994  88475 82359  57149 14259  04294 55822  89884 50168    79916 24944  86102 42442  02789 00906  36459 15098  46345 73360
5692: 49067 74007  85214 51227  18093 49195  78654 47713  15882 95286    79288 82797  39367 77483  18805 95877  37614 78576  48685 79762
5693: 65940 59470  33209 34738  42827 00980  43223 84215  71025 54695    02383 42671  49757 32141  04080 35087  37650 01840  78374 91518
5694: 68012 74975  27697 17678  88665 72451  45925 26278  91753 66213    21962 43675  86599 76223  12570 36670  83892 11141  66805 60878
5695: 68653 12223  63246 40223  07589 86044  95281 69843  02653 48533    84846 90709  02795 18293  81728 00206  78100 33905  07278 66962
5696: 57042 73475  97528 44890  22968 71820  95624 51927  41779 66533    66291 19152  65479 90083  95983 87717  91301 02835  33943 19360
5697: 28561 42659  37207 02561  65006 34272  37030 11589  06286 41421    47217 91031  45302 67703  28126 28493  08932 36937  69668 07314
5698: 15390 01343  92377 68103  02161 69160  09476 28629  66942 29235    11481 40532  28932 64125  03963 81642  49757 74627  60414 86728
5699: 38531 74586  73594 02355  29204 93505  72699 30247  25720 39123    35809 45020  15326 95932  33969 45955  20634 09065  72381 18867
```

```
5700: 0496 19587  0545 05422  4401 02795  3454 245605  3962 22616    8946 10209  8354 301707  0494 298928  0127 987254  1207 063943
5701: 8549 195042  6666 827023  6620 865471  9621 176137  8415 904772    0024 483487  2012 678957  8285 435031  3989 532606  9562 044300
5702: 8576 593522  4109 076581  3164 424247  2551 285639  8153 872779    0702 146535  8773 809994  6813 660842  3603 825902  0316 437768
5703: 5126 962205  6669 573702  8915 731222  1219 284972  7053 744935    6389 459926  0775 364818  2578 608294  8347 023587  6084 178344
5704: 4617 165332  0034 565361  0188 589082  1800 438417  9804 759974    2210 354470  3655 167004  4580 050964  8097 683373  5977 284423
5705: 5645 760893  0457 702647  4339 712538  1194 180221  6067 556201    3036 811681  6932 053150  6461 689595  3131 683461  0462 123827
5706: 3003 877803  3516 099014  8714 894948  0390 969817  6975 500286    2560 302907  1968 592532  9456 854310  1265 694261  8385 340105
5707: 6074 871518  4778 029783  7423 868135  9369 362985  8551 620662    2226 067290  6165 953350  8045 769873  2440 955315  0796 806554
5708: 3347 883330  7862 080570  6681 366146  0472 676515  0021 376488    1660 328370  9370 720861  4282 121638  0395 744407  7078 912469
5709: 1155 469606  2164 890929  4958 375926  1378 399216  1518 345831    2876 667005  1204 554966  7084 683031  4471 812692  7075 507746
5710: 5648 328660  6866 316596  8319 289990  2921 417715  1835 235007    7256 014207  9172 924999  1367 377170  1780 229537  5317 709274
5711: 6783 790486  7921 166545  2582 899555  1371 289038  0473 703409    0556 447932  1460 446085  3763 145360  6230 304977  9978 850396
5712: 3943 751058  9738 412265  7203 087015  1585 921030  4237 070631    5658 595896  1642 689985  9514 661325  5539 192348  1067 046380
5713: 8736 096752  1866 124204  7275 552462  0274 844575  6691 357917    9157 831472  3514 675452  4958 333766  8590 339525  2950 863838
5714: 6996 715028  8131 178888  9022 907632  6781 807877  9054 170708    6857 763488  6189 857685  1798 519414  6819 555638  9740 115461
5715: 5337 606757  8184 902044  9908 713081  9049 154840  3845 793652    0899 699636  0169 143894  3190 990724  7909 820659  6445 702038
5716: 2304 314535  7589 480907  4962 878845  9927 659756  7496 545712    4751 410479  2838 084371  5982 029811  5539 958402  6705 391120
5717: 7024 559161  1815 492234  8837 598740  4936 604846  1900 239944    8689 051729  7925 741214  2366 124470  6916 296996  8455 199722
5718: 4207 769306  0718 149296  8162 188203  5578 197845  4797 348560    5214 934086  6049 686969  7408 864576  0109 065997  2130 230760
5719: 7202 275787  2684 070384  0800 912958  4899 682837  8441 900702    5748 882515  4190 986802  2638 610580  7559 662610  8930 973842
5720: 1418 150813  6254 077266  0487 056196  5780 467281  6264 323154    9712 420314  3909 502968  4607 626721  6949 048912  0340 742672
5721: 4836 078934  0634 090409  0505 986280  9760 218454  2623 675684    3468 981998  9140 430841  5814 927846  3406 623381  4173 924504
5722: 8432 601767  2509 747020  8297 319471  5534 311587  9159 072367    5324 899748  4304 710942  1655 463609  9008 761960  8104 857737
5723: 3574 880883  5988 929052  5877 052391  5950 251043  1441 837134    6929 834093  2827 586748  7855 437575  7573 297035  6797 355996
5724: 6818 732067  7623 647610  1189 472787  2827 712190  7705 780766    0105 377896  7649 981037  4345 470261  4694 258405  3531 894196
5725: 6740 877938  2755 434166  3920 826645  7593 865299  6543 658254    7438 419124  5926 131576  8641 120928  1193 269218  0190 825569
5726: 1208 511385  5233 366349  1768 807834  7351 675596  2211 863885    8292 669017  0940 475739  5357 012036  7263 145097  7119 293630
5727: 1668 201152  1151 663550  9251 282312  2242 716068  4211 418565    7400 455325  8669 332702  4068 556774  8795 323851  5085 126904
5728: 4474 107310  2200 802852  3859 099976  1089 852556  4071 631205    4077 675629  9421 764342  3690 310939  3906 091197  8172 872270
5729: 4974 329234  6060 176392  4743 113447  4266 085111  9734 329704    2844 569711  3173 705750  7617 348890  6484 092068  3620 104577
5730: 2846 163294  4505 716100  8028 425358  6278 457430  3461 664503    2160 466823  3399 884957  4228 585411  5252 211973  2011 295410
5731: 6193 436115  5845 452645  3624 376347  4077 828152  4125 553131    6244 716051  5444 422337  3222 589622  8025 192523  1746 450662
5732: 7498 867734  0644 327135  3280 776580  3653 872122  1076 293770    5538 955787  8719 614219  0778 497251  5403 624916  8800 777022
5733: 7828 975408  8305 125612  8381 222782  6850 872023  3046 840329    1999 431179  9027 888724  3257 952718  4620 083519  9666 983051
5734: 8095 555069  2578 350479  3638 859302  7812 949878  4145 063699    3671 862311  9119 688210  2721 575530  0233 403194  3452 029892
5735: 2267 541226  6407 443310  9380 901959  9961 493503  1473 540596    4066 088971  3718 508150  2160 232947  7770 183707  6330 549923
5736: 5405 403769  6214 650735  3600 638376  1587 459639  6520 490838    7107 200928  6679 054748  8848 441572  1913 360087  3277 061795
5737: 4736 354274  3417 662406  0216 134402  8472 533481  3014 843217    1583 052019  8857 520107  2411 511775  8837 541406  7912 659001
5738: 3685 643648  7722 142986  0843 469000  1681 973801  4648 097621    1894 677845  6413 755960  5597 839331  0193 580156  8754 770979
5739: 8108 549385  7126 816588  9571 117102  7076 283780  3554 395637    1377 408576  6569 616788  6216 314590  2553 726434  4369 577182
5740: 3423 448081  5655 289279  3008 451890  6478 595782  5204 118796    7667 829718  3037 915355  1513 056327  0455 390169  2848 304108
5741: 9840 684845  6812 183417  1767 616189  7907 001450  3559 330535    6807 562508  2543 930674  2643 445658  7168 215685  0924 653931
5742: 6605 730248  8682 205556  7332 801799  0845 651334  4831 510644    7924 423003  2432 825746  3771 285523  0440 366933  8036 656692
5743: 4238 241330  9806 168453  1155 923494  6518 334683  1281 554016    7746 551573  5855 264231  4104 514899  3845 677441  7220 306512
5744: 2741 835611  7874 461892  2690 664262  5882 908142  4454 552422    4784 345412  6324 292862  7182 803395  4098 326770  1028 860086
5745: 6259 592439  4325 583935  2122 617654  9459 699343  6492 597921    2977 004587  5666 165022  9098 764871  4044 690972  7996 093861
5746: 3114 692086  3084 884970  9174 985313  2106 059594  6641 856317    1900 507284  2778 808410  8628 380679  0715 579289  0137 343650
5747: 3594 896630  9473 378516  1794 084283  5613 235535  9930 614646    3576 386563  8586 823893  1796 857920  2692 032571  9134 058002
5748: 9838 675579  9581 226230  4837 947935  2474 217760  1474 750830    7459 706711  2894 232057  3563 329996  4598 056621  6060 835171
5749: 6397 376139  1880 556581  0084 567677  6363 623159  2581 762926    4760 677674  7978 847804  4544 078852  1471 249632  8250 564919
```

```
5750: 74903 61939  49511 13346  39193 20223  32308 47241  15266 99671     28704 25676  99485 10225  93219 59102  53369 55723  54323 29899
5751: 40892 94263  83865 82565  99469 32194  73455 09530  98680 68902     93739 18459  42869 75341  35976 10476  57260 80847  14807 51323
5752: 00806 70798  74785 82051  89152 30752  49471 15032  51763 36255     43531 33488  87319 39891  64950 56152  48143 79224  74248 61507
5753: 15336 59338  54735 87231  42035 76170  99667 73341  10985 07167     75646 18383  30990 35620  37949 57421  38100 21260  39748 88672
5754: 79819 76227  89280 15270  03933 66581  42169 08773  27866 42492     94165 70075  23876 47392  27683 36752  33229 52892  21599 99635
5755: 12667 65364  72272 09259  23378 52943  73650 79701  84357 55417     61035 10040  47313 01407  51293 90861  90428 04772  06627 99810
5756: 36978 43629  40647 75125  06396 40908  56785 90745  17014 17721     51501 43140  73965 21593  39577 13067  31030 80796  27470 49315
5757: 02498 89858  19482 15125  92499 04204  01199 15171  63416 29226     02911 07111  79855 94921  02853 51794  11397 38113  84211 49084
5758: 30210 19439  97367 77619  88206 36486  77401 18330  94173 60476     57716 28674  72467 12947  51258 88126  99559 93521  72700 63785
5759: 00225 00456  08253 96713  52791 74776  73001 98500  00143 14949     92670 51298  75313 60218  23830 66235  34776 75871  37211 55841
5760: 99601 82904  60688 97497  40382 78109  63331 34858  94781 16269     87982 75824  79485 84630  69881 46095  02122 43401  43181 04810
5761: 27032 60670  56799 07552  21683 55243  55014 18280  93824 13542     63254 34141  11126 79079  64396 49541  30074 53500  67651 05932
5762: 82401 13525  65305 85355  18683 32827  67905 49003  82673 56411     90310 26065  68517 36795  82728 68363  44916 88464  57291 28472
5763: 12065 23631  30407 73246  90926 08949  08081 58990  68612 94920     57047 09028  35923 95227  55700 11965  80713 91719  76545 23814
5764: 18223 59500  36979 40426  69161 55937  90966 85474  44367 72468     98755 79736  17698 91164  82132 53499  28427 11600  40866 43635
5765: 52857 20453  92592 98804  07788 88104  23614 44594  26691 95194     59748 85117  80840 21344  60845 58021  90012 14585  63753 13424
5766: 56517 87209  35950 06522  19303 52592  48735 84767  44196 28930     71826 96053  51823 40312  56160 50817  16423 86398  71069 69465
5767: 21438 76131  23259 88311  83461 83313  09474 51601  37783 96412     24492 32165  26450 18878  59771 45954  87743 96794  18185 46689
5768: 66359 38494  68050 96035  52247 75819  65875 75691  19044 24398     97871 16129  83140 26524  03697 94235  08929 51056  81689 08343
5769: 00713 68212  81729 53250  38541 87107  25271 49289  43225 40490     49268 03999  17239 63116  27375 35732  11060 68674  33467 62978
5770: 96082 60766  30471 54828  46863 22411  46815 90643  45957 54796     39905 61316  49909 25162  07653 78018  42160 43688  35814 62455
5771: 31414 00637  04504 90788  11934 43724  28813 51168  54405 40882     83261 15810  21041 57810  82315 57676  55111 92720  39742 48358
5772: 61637 74169  06071 35256  69976 28534  45263 54513  29799 25809     94403 18438  90420 76942  37621 70790  38964 95253  78358 37785
5773: 89185 35675  29729 79653  58681 26545  56820 41609  66537 62577     14492 55025  91067 38492  67716 92094  23911 95734  25465 03036
5774: 25854 97322  09290 73803  02352 81969  24520 79679  24732 07801     27903 18432  32131 96309  50031 74946  81908 63351  80844 15825
5775: 82718 79601  45877 99107  24893 53601  95477 55222  10950 04116     97300 80913  58294 55178  06935 91121  20491 92579  99356 32026
5776: 14979 35962  01526 72267  43414 14238  95103 92162  74296 77822     75942 49384  25474 87863  92723 63516  23177 30666  74907 53375
5777: 44771 45562  10905 80778  63607 58792  07235 84963  09149 70668     83735 50358  57715 22696  95534 76612  15764 24182  28364 42577
5778: 08512 94593  20279 12774  22081 89008  93198 11201  20081 02859     99040 21367  53834 33397  16405 96204  20230 09705  12839 32939
5779: 22170 03098  17340 03729  23475 57348  60504 06496  08319 14206     70516 98020  67069 14491  06354 49080  39848 82215  72694 66952
5780: 10444 43155  37203 58380  64096 19742  41113 56505  28823 54598     12564 16508  92302 34052  53115 76279  95938 70948  22618 14592
5781: 25135 46023  94836 72373  56403 79425  51861 53062  38704 72738     40171 90197  07554 72139  56058 55664  52684 83373  52651 37898
5782: 16131 38107  70097 86373  78113 73073  75316 09144  16477 70751     46867 22272  64368 46473  41343 37266  10509 95252  67860 05550
5783: 08452 08041  88429 92983  59508 88885  17284 56938  35994 94982     01018 26674  15001 06829  96342 35923  16764 73136  17380 84765
5784: 20769 91787  56438 40959  43179 17482  00323 80103  86066 60191     32026 36932  61665 75876  43156 78267  44011 43195  69297 13956
5785: 68802 27671  55150 26625  48493 76696  91778 64109  09743 50416     68101 25754  38499 56237  62063 06277  98729 13093  31109 14282
5786: 85614 88409  17441 07092  17837 94612  14235 57216  32599 58541     67860 82583  41251 54378  59832 89748  45583 41611  12439 88869
5787: 73597 29332  60578 71878  04317 88196  26439 21193  65991 56111     11917 53717  01478 75350  41565 93560  64453 75037  62930 63406
5788: 47636 65698  12339 24678  19695 50439  70118 24890  65852 46620     79089 73458  75650 83120  54623 95319  91450 74982  77830 26384
5789: 11054 10705  22523 74748  03030 44325  31006 17819  01425 67852     42601 37010  27915 84574  10815 96723  64240 76528  50351 16194
5790: 61245 20495  61802 79073  41287 15277  72500 56513  38576 30393     23771 72631  43469 02972  91689 74135  56666 88220  42779 70435
5791: 40310 38730  26014 22998  35800 14756  00526 65361  42204 70958     66654 46297  83198 08358  71278 10214  05200 22267  35624 04769
5792: 87643 69888  03393 19984  36863 08195  11622 85942  98978 57082     79055 31753  97031 45067  41275 49768  73417 26720  58996 28646
5793: 54892 58415  91353 29651  55668 41626  07151 55334  59211 84899     87087 43087  98459 39826  09910 21750  91405 81835  98619 97502
5794: 93285 20775  02380 31155  58919 59622  19603 93644  20566 50857     52412 36994  83968 46097  19014 24139  07234 50453  09550 32989
5795: 00787 23620  95306 17744  00342 03009  10626 28276  87585 86843     83337 32674  22409 14076  92896 61597  79588 89058  75835 93864
5796: 04304 10020  67685 94792  03776 54208  44531 23329  89472 38791     46839 40947  10320 24076  69656 85780  24736 77169  27547 18511
5797: 05326 56855  11534 04595  95365 51143  31367 36604  90155 78454     74349 00258  35208 06603  99317 55748  91795 74528  46800 00426
5798: 06599 23674  13483 20302  94423 54110  23532 36066  04613 17953     75188 79269  20781 77631  91685 78469  02859 30407  26679 84091
5799: 34021 23048  14974 13998  26377 91033  19759 61149  30816 35525     40668 11978  76310 18824  62616 57102  64262 06739  79979 77529
```

```
5800:  54622 54962   35374 01546   85623 37858   90123 17294   94763 29095      52592 70184   97416 70748   31481 91362   16343 55018   89592 38911
5801:  68910 99768   45470 19522   00782 50800   44069 39573   48040 86861      87666 31357   86332 34707   89140 39935   14630 24058   25764 97347
5802:  35344 14721   18133 35106   99812 38516   27554 79873   81831 78512      07305 14412   32821 31798   58370 57275   62573 60701   08902 61818
5803:  00138 82243   21352 01411   26300 65679   50487 42027   40123 82337      67285 57729   51208 26520   46110 06673   84619 30609   21761 79477
5804:  04006 79042   79863 58438   02753 57906   88277 40332   15113 23405      26535 09487   52676 65214   28134 74235   91605 09827   16403 83978
5805:  66504 97391   27894 80477   17273 03315   82014 42007   79776 33431      35131 85438   97992 53326   11218 89302   86774 53010   65582 17415
5806:  97851 57201   54911 73044   60453 70864   09053 96674   36956 41757      06174 29631   90479 91165   51448 15372   25455 73707   23303 14473
5807:  28188 43952   59306 81851   34415 76969   14680 82835   30711 82885      53945 30827   92689 73033   24898 38209   39054 89120   93363 71873
5808:  06835 26788   55694 15758   44392 06856   93021 61419   44472 16811      26450 71617   56812 12857   72192 77966   93919 89070   31953 27563
5809:  34362 42971   29719 21023   47242 12090   99324 56629   82307 85864      77130 47049   17294 71861   84720 41185   73791 70577   31888 37925
5810:  22221 06376   61475 86622   43880 51370   94681 79697   96267 89571      47996 08286   60254 46590   78592 57472   21482 92691   19355 01964
5811:  41103 65143   85019 74440   62907 27013   95469 96102   53880 09486      66940 09460   86636 28190   37163 91225   17089 99393   60362 01704
5812:  17498 26601   55439 18867   92175 44863   13882 25977   76156 66267      14269 62113   26134 92052   98985 13175   12187 36594   80863 20004
5813:  98595 89027   42381 96175   19396 76924   25200 96402   85315 45914      16587 51320   03666 33000   70448 96396   22745 46727   89282 37525
5814:  62334 50832   20368 23940   82421 90242   10484 34718   59945 75524      47633 32143   51018 87446   10068 46661   27255 48199   16764 52652
5815:  21605 74331   28925 82230   47328 81503   37460 61531   76557 88549      46109 54119   60068 11070   11498 23385   50603 70142   69368 59389
5816:  33492 96168   08819 89922   86751 22381   58452 98728   59328 40401      34333 95619   05861 24317   75243 48622   21880 95000   45832 50648
5817:  07733 39690   29889 24145   59532 57775   79182 52117   57121 80385      66902 82487   56017 87126   75954 84606   42224 98905   39073 31976
5818:  33264 23466   57914 49586   07127 68320   67384 79177   04561 98082      39962 01285   33628 12665   95830 85679   14093 82896   22611 94085
5819:  93417 72667   03423 82475   59181 25337   80122 70923   52212 24675      83624 94663   04558 47685   08187 64592   20831 70858   74548 83759
5820:  02321 80906   34401 06988   39257 11520   51839 02242   52104 43597      96898 20022   43998 83191   64780 91790   01753 78933   30694 73586
5821:  81618 36487   61914 83024   25102 08053   09323 71362   15602 42272      16832 27850   11385 26770   74081 09738   69502 75882   42331 79438
5822:  46430 56872   45100 50830   03847 26928   39631 98298   32121 14523      05037 69304   60416 81033   64602 56579   44352 97927   32287 37793
5823:  65760 56649   00229 42080   17812 13573   81511 98363   82482 15695      03952 72563   65208 20779   81926 72139   12091 81755   92772 89397
5824:  51307 16037   39626 83769   81452 76638   83558 48041   85683 22364      82111 60813   45368 33091   76623 69466   83041 84360   95721 04630
5825:  07328 36251   35114 19850   39373 23375   88474 35919   74578 28562      05442 68249   04797 87663   58666 60725   96285 78199   81494 61579
5826:  01577 28044   93346 63496   74142 82362   61832 43105   27828 99411      48407 19445   17038 67845   87990 02775   66272 60767   15829 99822
5827:  51978 03981   89181 09146   78780 77130   09380 45852   83699 23338      87146 10775   20178 58128   00160 74752   46949 95631   78392 62116
5828:  43675 36293   17493 27875   97615 56690   39065 25685   33161 84901      02276 26333   10510 08623   15904 83343   47753 74537   91744 75314
5829:  01260 41228   98548 67197   43602 19921   14067 43536   31112 51454      76560 23785   81648 68460   91073 09883   46359 01116   13772 86367
5830:  47092 90805   11854 86971   61153 71977   56494 45070   53009 23140      11412 25298   18893 04274   39158 03252   28350 03482   13011 40165
5831:  40611 78884   00905 53110   70446 77099   05913 49507   48022 48803      67973 49438   65279 95185   60429 67306   95549 92442   82529 74253
5832:  39632 18055   34033 23472   26511 95587   81770 49349   65034 33458      29324 36959   16760 63506   18817 82042   90937 88748   35096 79751
5833:  51151 82854   12549 77859   26833 90315   21742 49877   73467 67853      88456 02903   68554 40769   46523 80166   94792 38342   81985 41340
5834:  93004 78461   68611 12757   88803 02373   28824 60065   84019 54837      43277 79078   82997 96991   55280 23170   53944 33091   52399 16573
5835:  48854 08659   83599 72431   56282 53615   18398 61726   14531 13237      17168 53599   64371 46419   62099 08657   40351 33938   66662 44806
5836:  24848 60829   82583 58589   09811 71743   68983 68510   90089 80554      99088 60770   27041 02123   61908 54444   84643 87132   00142 58920
5837:  83201 17035   39737 71811   18343 92626   62446 96011   89289 68907      22426 99846   75466 46990   64669 74430   33786 98335   92098 76973
5838:  38155 00213   86686 43582   01505 66174   58148 23797   47024 16567      73617 58949   55477 53023   50782 36791   12520 48540   77573 12374
5839:  66138 86474   76844 85293   59154 66438   12761 85428   39412 78941      73530 63311   59218 96435   80870 76428   92397 92029   66402 27577
5840:  96094 23155   68321 00896   05548 16259   67344 14037   88416 59560      42075 60738   73061 33007   42759 83926   84530 00917   67008 87906
5841:  40613 13846   20305 54016   86667 79552   32128 78948   23816 73529      48787 63477   16670 37123   78754 75629   71703 82569   77189 19856
5842:  33980 35555   71715 08273   96205 20582   80062 03634   41068 24403      25958 64249   81747 39452   45146 14040   52291 01508   79982 69939
5843:  07248 77033   41358 72472   39160 26899   34505 49122   36902 32812      38194 90626   44729 87007   60657 28389   41517 67764   25365 80788
5844:  12637 67875   47139 24461   48212 33475   57015 86296   00025 13696      16306 14567   22757 25636   29487 78417   27453 98047   04153 82785
5845:  17202 38426   41712 02528   65035 83317   59506 92652   55043 67855      06719 66462   13018 94027   49348 10562   71526 98595   23933 63461
5846:  30780 54859   47619 10044   93818 04887   90994 13407   05959 75886      03372 87928   49801 69936   07069 65976   36120 97530   06010 95242
5847:  59577 19089   38490 18564   42970 28361   23122 32813   75410 48290      70069 33984   83286 42732   52656 47739   30014 70687   09937 27502
5848:  31177 94932   00657 13392   97458 49703   52642 36855   85234 06444      73171 46338   21972 97176   15924 84291   30261 96316   56193 41375
5849:  82611 51011   61522 84231   01789 91112   88951 52548   18645 60607      64179 02121   03764 16545   03602 72178   64568 29739   47116 27254
```

```
5850: 25932 28293   35933 65774   72215 13456   30006 81408   81324 43517      12298 84116   84314 89171   82878 25780   87774 21884   71298 57557
5851: 29993 15037   30849 41724   24964 01187   11933 09904   31258 14144      79370 33394   06775 69815   95829 95248   49958 47169   65245 73911
5852: 15176 00100   55829 75426   21522 00382   31648 39923   54297 30642      02343 00148   48221 29396   95671 13785   68286 87795   71928 94200
5853: 59902 85656   71162 07587   88198 17950   96231 19979   23168 48828      45411 87209   01410 04520   17552 76459   66238 58949   01316 76471
5854: 46203 33062   94313 03890   33925 32163   21646 03390   16532 14442      88595 15880   44978 22919   02225 22110   50617 74401   26743 17276
5855: 81105 32779   41357 68157   87295 75861   84372 19427   28102 48307      14329 92536   75510 49571   76992 51257   82672 95856   41000 94951
5856: 39515 33875   71625 68536   41686 35621   03504 25230   26201 56599      23110 35988   74625 93708   37519 59647   10405 75687   68387 94259
5857: 63017 72003   38242 69570   02871 03321   35214 82043   98461 01585      59716 37500   33749 46523   65444 31902   88674 08175   48598 23641
5858: 79340 80519   47401 49403   33393 65984   16318 77774   59932 62745      83597 30721   56546 20908   82297 25324   25998 18256   17299 87912
5859: 28998 80344   10579 07191   62586 87954   73174 25319   35146 28312      67799 53243   13889 45541   50549 43138   70861 76301   31755 03615
5860: 48869 46150   46946 72643   46136 76083   60576 67608   53741 01263      95438 75250   56195 03623   16680 95003   93471 72086   97118 17218
5861: 35382 37073   01735 55720   74514 11053   34352 61576   56047 77077      59419 25725   26732 01183   30305 54851   85898 39344   54537 11643
5862: 72890 11087   80031 08469   13898 93276   66770 40346   35636 16518      24859 66200   33367 23935   55292 58564   10640 66485   55047 10556
5863: 25391 81792   49404 65005   08627 60516   64352 20864   64960 36501      09727 12558   91520 73452   58629 76532   94430 34882   20442 02804
5864: 07437 20105   16334 05686   40173 57858   29049 92853   66704 90484      88104 47838   50857 27528   20324 81895   19005 46136   28987 69246
5865: 00908 39722   62306 02591   60747 12861   29599 13938   34681 02501      40201 71471   67494 39094   44707 85192   61265 84097   51468 27286
5866: 20995 13656   19067 66388   55653 90867   90242 55454   94964 45380      31146 57050   61294 34207   47606 72803   77177 10171   19103 29078
5867: 52146 60047   88914 79535   45793 77381   26693 82672   09214 65720      98524 26268   44029 00382   43611 36442   38307 57720   86513 92156
5868: 07013 04665   52139 55784   78942 38393   80616 76719   59157 81770      64502 76338   61960 73066   59487 97138   90975 70396   18128 88284
5869: 65371 03876   39814 71806   34033 22835   71479 41175   01474 04380      77020 59011   55707 38527   02820 13328   63371 67165   43933 08796
5870: 81179 40105   82972 56297   68152 59936   48087 84546   42587 27569      10083 48516   79712 81058   88805 66943   07341 14797   37319 53951
5871: 15231 92922   93211 75084   58899 81302   97142 33121   91167 77272      65453 35610   91315 18238   03737 99645   10865 66281   23847 09075
5872: 14473 71057   12784 63094   97887 09815   65762 45705   80860 40764      53367 13631   58589 52280   71936 09626   24147 77727   41742 30735
5873: 40425 06675   80630 65678   81887 41070   26437 02439   17926 93222      45370 76870   94115 79910   19744 39795   15490 01117   10593 55890
5874: 03553 69490   11528 51246   43480 40126   64628 05644   37145 41230      94989 44964   89936 07687   52715 44443   92634 29522   81817 74018
5875: 98679 56921   78982 16913   21465 67148   04536 27241   81774 96143      58097 76338   69284 73077   73849 23762   53450 38361   74081 90475
5876: 38017 56097   57607 80544   37655 11402   28397 56717   87226 16616      04078 41694   25416 35477   82891 30976   26628 84616   36197 14554
5877: 56816 74137   09302 86895   68612 87716   01216 86615   73292 08310      59028 22707   85058 72897   55619 81205   77910 74172   90736 10411
5878: 79773 54797   62142 96968   66082 88171   41977 56008   48672 57260      42286 25264   84326 93786   30244 13311   16240 35034   22622 26539
5879: 80061 18744   84784 78090   49168 43558   42365 93887   25362 16243      72240 99717   96157 44558   95055 13241   74917 03897   62211 48987
5880: 84073 13165   02657 11267   24178 49483   48401 63913   59655 27496      58870 51627   32299 23072   19504 15355   48342 21436   82184 55427
5881: 45325 91833   50311 34425   00551 20216   15797 63916   31049 42800      02659 90522   25527 48165   69837 01989   37208 51501   45386 12728
5882: 54833 75526   79402 26250   02650 09164   24158 35680   14257 96342      53478 19705   09867 50911   32655 69882   00822 84177   96738 44500
5883: 95752 20887   81906 25263   91410 17550   60891 17193   01051 72967      63214 14096   98539 83097   05007 99671   70924 06058   64437 13173
5884: 89669 35949   33170 71355   44488 17714   55848 10523   90914 78326      86960 07936   09512 91706   63291 54660   83143 95858   16946 69147
5885: 76688 82965   98476 47595   68475 15231   38761 27266   29434 30979      63290 87535   85802 23881   22253 11163   42825 78851   69513 57434
5886: 55006 46944   59104 42421   86369 86046   51399 40108   00041 73535      55010 41523   47460 65580   97309 22492   16980 21054   67722 11182
5887: 55327 38316   28464 45004   73436 77874   58514 20652   83320 13358      52255 88411   62059 96510   49374 61866   03579 05812   31529 14425
5888: 16347 58499   39089 87216   37019 02782   77691 03903   26686 03121      02735 15772   13100 58724   12969 67569   44013 38700   51679 89581
5889: 33564 29851   83600 23359   23315 27739   16503 72972   66862 31465      87671 01170   79784 65677   00955 55678   24701 90447   56551 87781
5890: 27224 32765   58806 57640   37673 59433   34304 68746   44944 38011      27272 92498   53032 53514   95830 64697   32229 75973   35828 44688
5891: 54609 03866   08862 49794   52588 72950   33667 67879   69514 96557      21039 89537   83969 64293   52870 27131   72847 55164   90340 74644
5892: 46842 13713   81756 28523   57732 69332   76050 48717   68842 00863      64177 13417   77787 87679   27830 20817   41130 43595   61850 19987
5893: 75135 49021   81605 39685   20482 69807   24222 09929   30736 31280      25994 24783   96887 83119   37248 04470   58979 68911   12413 87209
5894: 16763 69648   28342 66839   96813 90620   53694 23850   62830 71300      68110 91673   76469 66307   90836 67261   04366 29538   97645 94177
5895: 13209 39202   16576 53005   79193 64627   06963 81765   84422 62375      62461 07662   15924 99903   53070 62587   11791 94423   99231 85007
5896: 96296 45539   44940 45958   17362 61234   32357 78981   06809 68510      99515 41556   93805 07485   00501 18735   32594 44825   07163 22699
5897: 39894 76951   38038 18358   37360 86131   70823 85660   55773 28044      38274 97962   09012 80417   22815 42868   86846 50134   58396 77257
5898: 84234 16450   71783 64131   02608 29963   86611 78235   37336 60836      99348 44919   78833 24983   87056 09241   55078 78094   46234 55629
5899: 93197 24672   95499 06283   92358 85800   57149 34654   29322 06669      56710 52597   61775 86617   46518 97218   72158 17885   26224 77232
```

```
5900: 3480752356  3414242216  9291117041  5343253593  5969079852    3201475642  3408323940  2775576511  2067246254  4289700821
5901: 4280890816  6964156978  0801765778  2558709083  4316776400    5375214000  4832960841  1784079652  5093792748  1310501563
5902: 3243159624  0520591501  2821438957  3178225441  7388838231    8851371888  0222673618  5880943538  3505493927  5642153673
5903: 1816818920  0900939071  0828184658  4799663650  5833541806    3107923938  4692497773  1976756602  7189860048  2303202575
5904: 9398665393  7553045227  3953143998  1684415838  0836876983    2165787893  2887213913  9174689651  6220566414  1173873274
5905: 6722141575  5347522529  6531671710  2508775501  7354190695    3964319882  5378069755  4957652641  9988552274  6746327928
5906: 3894596215  8994191376  5707166874  0122672502  2475751169    2084743170  1839138393  3112488306  0223025335  7345745549
5907: 4365746997  1657985962  9439967216  8524316008  1624225213    2972684502  9622332132  3622516871  7462387268  4152110026
5908: 0477471215  6604069761  4258489351  5751863832  7013039976    3394009162  4153926533  6659296433  5845507075  7476982022
5909: 3799574588  7543387728  9555623645  0238178086  9797525387    7447616167  7748344757  2514286104  6788315634  1227118072
5910: 1345687643  3002944587  6782561599  2703068301  1729834845    7541217658  2959381834  6515947917  5396372921  6323650582
5911: 5934466741  1930567282  6220880863  5067868149  3360992884    7393064638  8840226734  2484164599  5793463186  9250962759
5912: 2655177407  1436387763  3271848552  4880026350  0225029125    4753422258  5390638298  3232389411  0021388231  2156700454
5913: 9513217819  6285397815  0145900238  4193070126  8762953588    8169054865  5503535595  6930685754  3874673264  9959844734
5914: 0908193975  0822164721  0753039156  8633633851  4528333636    0905076499  7072694405  3267092570  8283974481  1104354005
5915: 1328407016  8762788734  4091075100  3153173838  3136925504    4777681777  5111153371  7332105561  0580304212  0121324921
5916: 6903327085  6349537865  3295150427  7620457557  3403466042    5176550177  4906276831  6759130641  1721216674  7962960453
5917: 8275406736  4041759813  8614725772  4451085751  5855505706    7730253492  8942404275  9822125609  6773372317  5158910760
5918: 6879223040  9616344494  8551621563  3324179151  1527467571    2734520373  8150089712  7031401511  3874655511  8994984973
5919: 2515419509  9624400888  8090411344  2532679237  3226014427    9240306761  7670885574  2586133169  2037244075  8954255215
5920: 4989006386  9239582944  6988002631  6995640801  9904136429    0114889272  7039144696  9887118055  9480405664  1911157018
5921: 4154508829  1763988530  6549731216  2574787934  2914208368    0943682258  6560877609  4482051797  4147088398  8190240564
5922: 8425970556  7565714756  5607611072  6838871376  6749129431    0196386677  5741871104  4367850566  9538683968  6378147644
5923: 7165961214  0190884715  7699739958  9953156323  5958584661    2137010630  1556819930  1594792739  5606056047  9946303822
5924: 6175710260  8999225890  6134253300  8946682278  9645643728    9453654036  6219750383  5374369298  2854191904  8376482874
5925: 2623531719  2151323125  2861963926  4902628661  6122386299    1938359354  5928828629  2982350611  0526546703  1469697527
5926: 6483548140  9533772638  6795084543  6228520111  1249836332    6554130666  6554355424  8443688636  4450907045  4606205794
5927: 2748156094  7853092757  8759321123  2093443543  6011406746    3170383245  1813478564  9781836592  2849575594  2427581221
5928: 2611275054  2608476635  4342777452  7590361349  6686215904    1805080498  2142667568  0704273130  9066018453  3234411815
5929: 9946447355  4796223423  9686639046  1470305716  5255582736    6822069012  5579380939  0620247529  9073680104  4904965473
5930: 6081349883  6327225409  5294738587  5363055877  6651711252    3431949340  1252572686  8179771406  7341176770  4549904134
5931: 9593009122  8049633146  8029588698  8887286816  8326677017    2600621051  5623600089  1801844329  2905885822  4549931908
5932: 0761462494  2274642437  7412911329  5500568720  2231553128    9968912947  5528526188  5300989335  9278147110  6608809786
5933: 0498827397  6200967502  5859567364  9055501631  4885536856    3778760096  2520320032  3539646380  6194870616  0342506590
5934: 5868032999  3461238666  2390018831  0240946850  6867773901    0414751795  9247405495  9183371155  2672670626  1352208697
5935: 5610171280  2123109223  0794156325  5022308741  8237145783    0903427319  3834886860  6173384574  8698099928  3754285723
5936: 6510994568  6703283165  8552392286  9970915162  6707056036    2577658240  4391608664  0913332700  9470773487  1023963298
5937: 2012019809  3432166118  1173545915  5128954243  5819342119    5869166785  5894252848  5905862229  9745233311  1579101649
5938: 9575825067  6994036664  5468219990  6784188224  9855917274    0634419362  9305998061  1341184592  9293327075  2255466481
5939: 3475115774  2911472445  9319466029  1077909708  4239625365    9196427340  0863603223  2272855945  0442103387  9636671041
5940: 6896158667  5559984103  7045144440  7801037202  1342527117    7284172740  6614755117  5744413309  9680976507  4926874446
5941: 5912020881  7686983323  3018778599  5635322810  9237931110    5586495984  6518371173  8708361396  1515534766  3547898093
5942: 7057874676  1112314624  2977402282  2351487360  0003154807    3839020323  3783391827  1108416244  0422858599  2148772058
5943: 4243731312  6744778505  5006630800  9752618072  1094059646    7590944747  5689295008  3882307051  4382046698  5580710019
5944: 5087942220  7602219870  2326397617  0753707976  8650915539    0907269956  3397664410  3284919722  3126453300  9235510419
5945: 0705700634  8194835363  9666436111  0221726383  1298904721    1670685504  7368843266  2044447654  6558001209  4596447052
5946: 4268700926  4233635541  6531129956  0405205421  3909244899    9213692382  2501700319  3855200984  8873985822  2299195538
5947: 0435480690  8875127746  0119689555  8216093033  2482766015    9102602863  2758365492  1721319442  1550983903  3086089973
5948: 0505028922  0213215593  4531922096  3873623441  8989746321    6049264318  6142122649  3997392529  7888288080  8300107372
5949: 0551935784  7624672917  8541109820  3390724068  2372755421    8340761713  7238089949  9560752618  5013331116  6946602612
```

```
5950: 7760458197  9191424211  6490266567  4588867738  0410878195    2501901321  4747533706  5519390010  7272031315  6021687195
5951: 7010579097  3286547693  6494760775  5662817475  1981600300    8890268593  4434207746  5305503720  0373423561  3842184763
5952: 6732274522  3874021082  3982949749  4793520040  0252957379    2492188868  1380449322  8168920721  2407433920  1519546919
5953: 2128360871  7745687001  4913008073  4795812896  8783313858    8624466254  2148436726  1365478754  8019504690  1958996931
5954: 0113671623  1888829393  7926495960  1037955583  0281473078    6991719483  2706770805  0590313877  5517652441  6239261380
5955: 7295398823  0219610401  9466570521  3037709630  6890499587    2955580249  9703997616  7130733520  8834623040  2746933803
5956: 6451162547  2886314732  1679194196  3736207682  2829940663    1081595599  0523695044  7722044832  3237296565  1963922435
5957: 7661263935  7039321776  7655293579  9223998361  0729529491    8415639390  8223697312  5256913546  5775782315  9023916568
5958: 6801308983  9632242688  5440579219  5797339744  7494832642    3657258596  8863299064  4438469616  4109379846  8886244162
5959: 5663517919  8763695026  3904744459  7170016974  1531242345    0798207166  4384133805  2659443872  3344882958  3739303405
5960: 4950949310  6268635165  4480745226  6521221940  4856358591    6391252989  7369457391  7976934110  1800213846  5898590992
5961: 4792523976  6227765102  5335885289  1121314587  4092479884    8449977085  5794391078  0000193572  4409570972  5643261457
5962: 4308968195  7542564953  9686062211  0208568592  8007654598    2991699533  7876061158  6093059204  1549912170  9435858307
5963: 1186112496  3142027553  5507445651  4290356088  7589078926    9444271005  7201018333  3738536713  6889966847  9540222159
5964: 9619025023  4086984326  8918619136  8359831597  0565762050    4326703818  2759972217  1400658771  1062068201  0485135856
5965: 1763742442  8295242901  0706952939  4564744542  3414696582    2316214612  6368754036  8715150298  5334921662  0357871112
5966: 5570412458  2602063680  2290540820  4478083649  1835257455    3375874196  8525287994  3062203582  0630425400  5203584222
5967: 4711905271  6212818592  0381264907  8403641800  3134323421    1032389272  4597417801  8883023009  1032121268  8523854235
5968: 1474063037  0082060298  4509150566  0526013582  8464667332    8132547231  1143173092  3756871845  6686691035  2986187979
5969: 6594719121  1396229725  8105907030  1380381328  1922560065    4877592314  1461370983  0846861429  2458991273  9745611008
5970: 3831293221  7337277763  5910575845  0445326836  2590407677    2831301411  6370904267  8011095359  8038037014  3788222047
5971: 4646102151  4566647945  2350066824  9635818865  3181889745    1421976717  4599489712  5485906899  5428009751  0832227224
5972: 3744827821  2951781038  0966552981  4271286367  5579486658    2202635182  9163047932  4548847260  6538721372  4562229598
5973: 5562482920  0803524405  5566313703  0588013459  4363507738    5130467266  2120528661  0501318727  9183902841  8566671032
5974: 6089718038  3802797446  3721911381  6655225656  7245149832    6957404619  7109106358  2400690605  7250437856  2943398528
5975: 0729270284  4882700950  1001211733  3513577297  9804057825    3735852944  2342218807  9687561399  4852878456  7707642722
5976: 3050150366  1123745249  7172334760  5174891113  8364000272    9710353366  4781971748  3884991801  6805677985  7187031784
5977: 1823852470  0857630221  1261136930  7813104325  9722261545    9529179372  1464791119  5048624056  3451870717  1436801029
5978: 9494726527  1088065202  1661469862  1488904201  8990591680    8602474433  9246089963  2071812529  1355519352  3258784995
5979: 5724628261  7215523872  3784879429  2621666204  0010113047    3677115133  2071052272  8906190582  5501620830  5879709077
5980: 1390532986  6814873450  9157112909  2671843017  3071135589    9917730891  8256746453  4924478741  2235351174  6093409951
5981: 2093119333  4763024396  4506925870  3983188808  6086703593    9557441977  8398615271  7532922650  1599825438  5895765899
5982: 9140157010  9246911090  7180510296  2679321344  6849531782    6104890810  0588601339  3395596807  6242862447  3859247760
5983: 3146128082  7829517425  5070358790  4244937344  4271051385    1965552402  2640214874  4310610792  3682670786  3570570409
5984: 8607997660  9925835350  1412605681  1392959912  7585651607    7409494485  1531859214  1665857702  8473458575  3955954053
5985: 4805902641  8567597457  6554282878  2482801505  4884105508    7379766195  0691929020  7183288466  9104609570  6522225510
5986: 1782524940  6824739246  4537241358  0316874009  7778826581    3866292401  5347101234  9098690172  1349022003  3663684413
5987: 4254434080  3958263667  6751439485  0953854007  2352336877    7759678810  7572431609  4948205920  7614108800  0577376595
5988: 4681735120  7742492788  7511420015  8104083230  3528448291    4829279200  2167321445  2695220388  3759501459  2325345408
5989: 2154940942  8174273553  1045783003  7853973866  9785202872    0243789848  9978153046  4467193649  0812245578  2570532176
5990: 0496404323  2132821586  3621752621  4739755875  0818810919    2023229566  2252290734  0136877727  7661383226  4477685542
5991: 0691306031  8648100118  8810065572  1056584418  0455188148    5383300712  3933112981  6781403031  6473119348  9249372737
5992: 7377688504  8780061643  8974261001  0944603507  3444959649    7573019292  5821934845  8792699568  4283498085  4009962932
5993: 1595553634  1565139396  9250247472  6891146600  2176158843    4559288490  7453305883  5867153842  9817374599  3796946948
5994: 8737571128  5966380289  2869473219  8647316901  6805450109    2475240601  3447689734  0291233029  8892190470  0844603840
5995: 8968318758  1592540432  8848963979  3481703450  8248836163    1413056303  7232550609  4017175854  6371018234  7062843932
5996: 3850590555  7128804482  2470943093  6735155394  0597745150    4511905701  8567266845  7525115620  5726340894  9637062770
5997: 9723031976  6831939876  2930935718  1743880679  3888299011    9005091380  9207198067  0058963638  9352667632  7574400395
5998: 7825326497  5141931400  5293605905  4698348229  0349013515    0338025173  9046362067  9887126752  6218965377  4742718830
5999: 8230827512  3119648611  9026335167  5914143668  4986547535
```

```
6000:  45851 55391   60132 29414   34751 06063   38427 65779   08245 21774      69980 79928   95733 26829   65071 22710   16240 31108   94449 32728
6001:  24179 77749   11010 66356   91203 47845   68149 61151   62856 05219      14957 63781   45455 68345   93706 13598   91971 08360   65149 24675
6002:  95439 94074   58271 92804   97782 33541   17105 32795   75643 23540      99938 72140   23107 04292   01426 70669   27156 41011   16593 85582
6003:  29160 27574   85353 31511   59155 87279   38100 54556   46472 37311      21135 95996   89301 59378   20745 28029   33377 25497   54665 08614
6004:  57373 73157   17509 05326   16165 97367   41503 65649   46729 32175      21135 95996   89301 59378   20745 28029   33377 25497   54665 08614
6005:  34640 62211   35127 99298   99120 68737   37363 20476   75523 51490      65641 97892   58248 09523   44056 53295   58108 40441   91136 85070
6006:  25649 70383   54512 32467   37716 21937   66837 02091   52504 62959      67731 03075   99502 26580   74328 83344   74140 76666   04825 86895
6007:  32663 46777   64225 95804   46852 10594   70885 42852   17485 49296      57291 35168   62567 96123   88876 19949   54999 85062   83878 70615
6008:  71163 63031   02919 18115   92294 72707   87031 01971   94102 44624      19989 03222   47775 12994   70763 73847   78503 39657   88093 61847
6009:  87621 66985   52542 09363   42055 65331   50842 43729   74031 01509      58368 02777   51787 85884   50274 95759   96929 24700   46224 44778
6010:  39543 01563   39652 06148   09081 25383   16748 55816   93820 02891      59733 32135   69797 09551   69572 46787   34887 72077   74890 47137
6011:  27556 67021   37485 45178   35485 55420   45018 09289   23736 89497      97557 52727   23322 41789   02371 69764   05920 21068   71449 49718
6012:  48060 55790   05505 21689   20722 40868   19085 39910   93246 43830      86362 32191   66373 78147   66432 73323   97350 34570   25513 99411
6013:  49806 02711   27854 38963   28482 72202   54404 20374   85243 10087      56518 48674   59355 95772   67908 70199   68404 70331   33848 31527
6014:  93402 42869   72772 61811   42306 18615   48987 66575   54603 77283      53132 96968   11790 90948   66036 57909   85920 67791   66676 85838
6015:  39681 88798   29283 57525   44979 55472   51980 26103   06690 56669      34338 12674   14705 88245   24898 77734   94150 42103   61015 64249
6016:  95378 50248   50256 04426   09227 20395   21523 15868   79119 31454      40079 91097   80077 45897   15571 98005   08290 28480   44972 50878
6017:  89414 74793   96650 18153   59053 09427   71816 74388   13296 38239      29495 94833   12607 64140   84411 41142   32168 25588   31552 68159
6018:  12794 52801   44813 61715   47311 29723   21754 96786   44487 44027      15861 95123   33691 73717   71820 05171   60546 24167   63409 12086
6019:  44344 75601   56228 99343   63970 47484   77804 71259   53522 95913      29791 74475   36073 24791   54327 68143   56396 64935   66293 94671
6020:  79415 18390   31717 06072   12917 22173   42764 82929   36106 92280      08939 08454   41658 41675   20452 60773   78755 61310   69418 70116
6021:  73250 64380   68257 25751   53987 40212   00685 34372   48687 88686      95089 53504   58788 83732   60344 09240   80300 68625   83382 26715
6022:  29055 73833   21823 18915   74072 03044   02996 91764   86370 07988      87379 79881   20723 69706   13248 56399   44849 96388   95304 49190
6023:  81576 81904   71060 27416   37940 98459   79370 50088   81325 74556      07573 44144   86006 04330   59010 43162   83936 11087   89720 89350
6024:  03636 04449   18364 41691   37700 14286   63861 62066   24834 10910      91264 09253   66401 05429   64034 38064   27429 79274   05722 15425
6025:  16534 52847   64560 81303   18550 72247   67894 33578   29614 47667      14716 98308   00642 42931   11462 23261   47903 74491   10927 87691
6026:  49588 61292   29266 67604   59711 40794   50913 61332   46503 21900      48888 27438   51744 66579   04989 61285   85295 22860   59010 20633
6027:  73215 79867   75221 89056   72814 05354   67699 49363   34845 10114      47420 06194   54583 14418   47348 29270   69581 24265   01026 97288
6028:  54765 65412   94495 94143   33669 82436   25159 74863   13395 20210      67149 10493   23375 05023   11016 63884   77319 85171   94146 37078
6029:  65639 17645   60310 04976   22138 97640   11244 08034   86363 35357      35680 24698   24972 14795   19444 96859   74585 72672   85698 38188
6030:  65146 07312   15699 35237   91232 87241   28649 53368   89082 14830      83897 55557   64715 77831   20466 75400   70808 60025   37657 27467
6031:  77313 80605   76879 65441   55917 89946   06086 27063   33726 98197      43638 45221   68410 00506   43783 05051   14676 48543   17283 36847
6032:  63482 78579   38646 78534   01470 70447   17609 25491   36958 42067      79952 56636   13596 58367   82882 56943   15488 14116   16167 70755
6033:  36178 05811   84952 53328   41881 43101   13519 38739   96738 51940      09679 13778   78022 85407   03567 55264   49448 80648   90450 16917
6034:  53696 08978   71988 49645   64895 63788   14685 02590   79460 59939      26346 95914   27153 73541   96822 71704   83749 34795   92165 32890
6035:  80803 10540   94357 28659   73471 30448   03715 58832   97861 14630      00749 34612   85393 60146   63429 25879   59831 15300   14714 23380
6036:  12208 84007   37765 71307   44680 20880   82593 31672   07122 41720      29809 02194   01966 86945   00004 89564   20598 84565   03230 97716
6037:  32644 03156   73494 00366   30779 79597   55178 71061   09701 59035      46812 04573   22604 29137   91546 11473   38588 98958   23268 03142
6038:  77025 46306   29086 12427   03988 05720   17822 57960   43400 63855      32428 92566   78362 58757   56668 30502   70520 49603   97884 41384
6039:  74513 57781   30288 23518   75123 74071   86666 75906   25037 59294      66237 17205   20220 50518   28851 27743   84413 51266   96543 22911
6040:  90414 52532   43861 52459   13441 22419   95857 12422   58322 34605      81491 92978   53657 92169   46029 05335   09431 34823   49577 46020
6041:  36034 79126   86513 82500   69039 35490   92662 88463   00436 60645      88142 03196   28716 42077   16193 09237   16596 05329   83842 09836
6042:  07201 03662   85031 10555   58846 18103   50893 93664   96500 58840      39143 06362   22894 58088   96664 49025   89385 76740   31768 13500
6043:  74870 11632   84366 59912   04320 94715   80974 17574   26561 52223      76492 41443   97806 99321   27219 36688   12160 35233   66605 97455
6044:  41515 16684   48179 42692   60416 20389   68837 97790   33252 41416      26228 55837   67383 05241   50446 10745   24359 42450   69881 88237
6045:  94989 46107   34907 35888   66218 07898   80948 20166   77520 57974      77096 20393   11595 26219   48712 87693   86952 59651   91259 42860
6046:  22640 08014   83513 03482   82960 66858   63074 36987   72284 82599      64198 07244   87722 07577   68292 76905   28790 65243   07294 10551
6047:  44474 83512   55364 00031   73797 18426   19283 61625   21645 77964      26818 41311   23882 31398   74089 18255   27933 80761   92044 87800
6048:  91040 83155   89026 38647   87980 30130   33767 23011   24949 88900      07673 95408   82690 77529   84759 79509   35629 50613   75236 05914
6049:  76143 15119   50514 59320   85778 88592   52446 05542   76198 42293      43823 72529   30516 25461   26453 43179   90959 13641   72904 49922
```

```
6050:  1619276272  1561262888  7709329917  0146578588  4008078073   0582542095  2329026091  3236249652  9365370282  3766461150
6051:  8964815961  0091098230  4743655177  0014451761  4392032907   0552687462  1563414759  9188919930  9383220234  6979707062
6052:  2225578315  3703871420  6083325975  0029024501  8108868247   4864461535  0692128710  3852330935  9699778662  4534588522
6053:  1133660217  8802352474  3673817422  3034407202  9106189386   5529224530  5398262042  7560317331  3423715785  3849974515
6054:  7119253270  4445852987  9795834719  2487064416  6673742103   8435435641  1895030859  2640067917  3132956196  1606360123
6055:  3143634541  8571867782  6361761635  8446616006  3181471431   2655992262  8262854969  4214161257  6981398675  0051788136
6056:  2835868179  6623668493  1280011851  4834750448  1133415080   2411203545  5024943849  9449530975  7722082200  1264392379
6057:  6134054595  2819776880  0625360826  5202529856  8533935261   1290391420  9420398690  9989901187  4194840046  6539966420
6058:  4496012581  7673033913  0901832691  5714507983  5306590583   3208634936  6814745592  6990492928  9618867533  1376342536
6059:  8397716373  3513075749  6626171765  6996692883  0274629049   0576708742  1984680431  9940773293  7316866391  6926813679
6060:  9468876420  6890916039  2397899802  0331258751  7816567451   3790167205  0923547986  4505942914  1949485575  9134550913
6061:  4348400890  1034293111  8317054182  7391286992  8374476584   7555479042  7961843364  1574307238  2487744473  2030017825
6062:  5277819359  1726853544  5526828854  0035818104  3620411261   7663526487  3805458572  1836943337  9466992274  1056455259
6063:  8217717440  4718804812  4188717092  8698130893  3632538530   5809643375  4590515969  2801155185  9254186884  0384039856
6064:  7636548387  1110021698  4575742870  8972014820  5955481680   2020990752  2376412395  3459919991  1961756275  4044456704
6065:  7764523233  7568150927  8698732569  9543134209  3105310507   6516434322  7527318724  4712768951  0440539014  7454992554
6066:  4197491491  9155775113  4719220166  6617183352  1584112171   7688112113  9606029439  2419693878  4174568895  0697106760
6067:  0433635961  1083518704  6158473979  3952919473  5874293844   6813046230  3563793643  2146991998  9041179374  9852963649
6068:  3406454662  6834907330  2556829642  9811027157  9015182448   2879836079  3577926014  1271919368  0276907256  8094551503
6069:  9900303109  4038994210  2338777119  7070949657  7054033109   6097851520  5063211765  8077131212  8055834170  3144751051
6070:  0231252955  0020415429  0888786558  6391833660  7400131574   8237242966  3336441404  3889756322  6810266768  9977437821
6071:  7218650762  2597004533  9503422212  7644452397  6703262788   0847809963  3335865144  5751828584  7434652512  2040734712
6072:  3592717071  2630919507  4978532267  8143712110  5488009440   4137622833  3982728038  5327748306  3409017377  9362473404
6073:  9924120801  0871215789  4805473836  9650120338  5075915708   9814876674  8436317655  0015726867  2103750159  3554634179
6074:  9411038894  4913090187  3956413542  5031908987  0841574196   5001796985  9200168054  9556254633  0359747378  9764202649
6075:  6481204487  5360277811  5894834475  0015764713  7409389799   4648568494  7617623847  0236568238  2062032406  5429465228
6076:  8693302349  8811672024  2887649571  9242173048  0670508678   6418297778  2023561021  3861742928  8345311148  1465403906
6077:  1315063026  3831992204  3043565722  1296677525  2811568304   5662396677  0862491637  9136013129  5131312328  6991536580
6078:  6308054810  3798430335  6308544965  4510519584  3849324510   6475408107  6448904640  5394488954  0291532004  1765202112
6079:  9243358317  2046772806  7946620065  7622907512  9795747557   7375677771  3788874709  9851102758  9024641930  5793230295
6080:  6892552930  5819808374  5544789314  1611525670  9943517007   9159506000  7218260469  5021763795  9283150628  8929651324
6081:  4634774508  8732668233  5080977212  3474229467  3696077962   7697412399  5335783790  4626368211  3418295202  0913147652
6082:  3588111927  5029020494  9129245651  7921144756  6831200000   6104792320  0732881626  8244292696  2495405966  9449620595
6083:  9617624303  5520315207  2314470861  7542778346  2084973590   6441053859  5625921152  7911804571  5540716991  1966752823
6084:  8320302985  8362768798  4433388445  2700041802  3501736748   5490644312  8115246151  0512994871  3767228024  0793915005
6085:  4964632190  6268395686  2262721712  3291576063  7776085904   9342215882  7950634426  6433619044  0342121015  9304629725
6086:  3347305479  3239438415  0402577240  7660971027  2324897551   3063005838  2209176102  6068156472  6860072274  5247447331
6087:  6143750982  2632910658  5199354638  3875231038  6426700747   1809422413  8133542142  9794388779  9898600674  5781133836
6088:  0199859537  4749290474  5447372705  1930424119  7303123640   0699768970  1416267339  6670716927  6994796785  6760921813
6089:  2816414165  3362886432  4287840835  1939849570  4233297608   5789958826  3457007323  7793845155  1733456833  3288604615
6090:  7458534962  2753906980  9344453308  6309092713  8528988000   0256399770  4244763720  2222848135  7420706769  6265770155
6091:  0617872646  4767534608  9968918556  1147217779  4888512989   8208317089  8014108352  1343255084  5402214307  7018478004
6092:  5911724212  9590891106  0483784132  3058362212  7946927469   9557187661  0319056087  2022852973  2158570388  1512935024
6093:  7235202661  0969178880  6200118397  3357027367  9293677570   3369837042  7676682166  5635169249  4390754293  5474277829
6094:  9439753997  0819856373  0186382895  9886954259  4608808152   4763877920  8272871550  7859241372  6206442719  2345810328
6095:  3651532220  5645888416  6969404165  8405408039  2470925130   7540774559  9982660283  8269244605  6810501167  7886325607
6096:  7410862568  4193421986  6955803647  9064373040  7708080119   6324394791  6565808410  3325073827  9542047141  8873175818
6097:  9659623844  5575380907  0180083474  5693069294  2490106526   4646223275  3760197706  3022397448  9108855981  8154474526
6098:  5390284974  4791523182  5367427329  6792980678  3003027399   6615939230  8906372206  1135786655  3524074456  5664825493
6099:  7152011918  7720574920  0891747340  2628306269  2611174695   4646263869  8641089745  5438913742  2682380673  8658340021
```

```
6100:  63952 28297   73578 75146   68871 55203   69042 80465   61482 63978      55928 84851   71098 94580   65371 06476   82163 63865   46254 15784
6101:  99186 16075   58170 43598   40377 93295   81378 83811   98228 79551      36807 55796   31287 31264   12192 02930   64398 32265   03189 29257
6102:  02395 20545   30772 29328   94384 00751   10312 65597   44400 21280      42885 69232   45363 22184   70960 40913   92740 08525   98402 34686
6103:  11247 90218   80255 68490   27270 10555   05748 15183   35447 91138      15879 25003   13313 60592   55570 01970   62864 69149   25818 20664
6104:  40359 32059   06628 29275   52539 55644   83649 83829   10662 94323      06068 01188   82367 65165   00725 75800   81494 03602   25167 12624
6105:  66803 36152   95667 30743   00884 76362   05845 44837   96193 61745      08434 10404   32384 25756   76852 49987   41336 23159   79842 10702
6106:  81338 38364   55503 24342   05268 42513   27772 46791   77715 90071      63163 71201   37183 94334   61360 23011   86779 43051   39452 20618
6107:  53638 65605   83264 87579   01502 14554   61050 26116   75284 57806      91187 14534   31801 19726   74083 72783   68402 12842   93437 00467
6108:  38295 84701   31706 69787   42737 41968   07493 83000   11929 88435      38153 53860   02403 75070   51283 17138   14034 47033   58589 61524
6109:  57345 71348   27165 40279   72288 55770   29760 64767   13374 38992      04804 11857   02835 61590   80887 42054   63532 94358   77940 69893
6110:  38603 89061   97828 46102   32483 04399   23811 97377   81187 44742      21871 19966   47520 88954   17485 84218   94207 06735   70314 29353
6111:  26507 18297   34222 49001   05376 60817   78196 76872   39890 91397      00109 36970   69542 42329   07322 55805   15480 36016   57473 98212
6112:  34083 66418   86278 56433   93114 58461   29277 84089   94246 31526      80807 42580   63547 89710   18621 36888   22860 89525   67363 78056
6113:  30489 33823   22021 35339   01580 34205   80226 08771   74847 48324      36281 80780   40941 62757   16348 39475   45616 98725   30503 93938
6114:  10474 50493   75391 09016   24877 43491   96536 49615   88947 48514      71262 09228   38405 04101   68369 20795   01141 43097   33621 55642
6115:  61483 38236   87400 69643   10050 27211   63890 06931   92182 38112      48029 90260   75541 48629   30862 05873   06286 87798   29130 15954
6116:  60572 82341   73965 04816   40076 99690   19523 36659   35685 14835      07845 76983   88074 54064   45403 74983   07899 43059   00300 36447
6117:  91505 07084   16436 48838   48264 39071   62228 51802   84658 49198      19014 94232   65929 61152   03987 68996   16079 96784   97147 92788
6118:  28488 94260   94325 89982   68484 85262   67162 89956   67572 61385      32930 29414   56562 22140   63130 47363   75222 79704   76315 57670
6119:  49427 87818   06244 93867   80261 76977   09549 25541   42527 64425      35444 63735   40149 28674   23818 50835   78682 61402   54374 91774
6120:  06060 73400   97050 75100   25346 74424   27627 38998   99056 37317      79857 99939   88715 41756   24381 82847   88029 00223   63418 70902
6121:  07845 49223   32704 59855   09947 18544   18876 24280   27699 54021      36682 25871   11653 74506   43962 04221   35935 71518   37066 00532
6122:  45516 16788   39829 82893   27611 04613   66926 20418   45557 19267      77963 19595   64140 16691   53245 66352   47734 77962   67659 92089
6123:  08235 40351   59477 22458   12054 35700   89991 08916   51985 36192      02119 07559   45412 72785   03293 75285   87735 66934   14220 91404
6124:  80124 46254   01329 81478   33547 80350   23627 97868   22265 03628      71459 82527   07127 64140   78883 82687   80237 44628   81759 74250
6125:  48605 24294   66349 52836   49541 31772   91571 75825   21325 49103      73973 77103   58807 05234   02344 97808   12309 66871   77094 27079
6126:  83696 64005   11504 59660   35868 60973   50105 30687   63463 25769      29814 35089   80014 71157   10361 99059   65224 26491   28041 10406
6127:  40470 38709   14372 03500   73797 80302   95705 49880   93895 77215      02805 13220   61216 58138   39881 28600   80946 40874   80593 66273
6128:  93139 27171   88517 21493   63023 53414   59442 06031   65312 35393      33478 46708   26023 93541   89628 18770   03594 15284   46352 99219
6129:  95030 14784   70141 69576   86270 16393   43846 09136   68358 05541      00948 76655   68607 46903   67057 32493   03674 70034   59514 25142
6130:  98512 88017   63388 67846   75303 27530   91110 10402   54160 60787      10631 70300   70394 05293   12808 25713   37610 16653   12923 12286
6131:  97300 82618   32151 13091   63331 31178   13940 74062   70725 94839      39654 07168   41781 42174   85876 60283   44968 00898   43366 59507
6132:  10445 71026   23523 91003   77866 47093   71572 83336   53103 38594      29494 53719   42194 31409   63607 15456   92011 03182   76453 73243
6133:  56448 38699   71370 35180   46379 77175   69838 88958   04084 55582      81576 43885   52551 79104   87767 14304   10637 54041   56954 23874
6134:  81838 36635   14140 30910   12596 78327   84637 07302   19102 68135      81576 43885   52551 79104   87767 14304   10637 54041   56954 23874
6135:  75322 61066   69314 75852   67495 40830   24409 23308   07078 79295      98384 33630   28773 48464   73402 91627   24840 72530   23524 71070
6136:  24537 18346   33973 47273   30220 09663   43741 25495   30330 24364      26261 25230   69293 93239   50407 35340   47587 22978   56084 20928
6137:  22387 62194   63706 01014   82498 35561   11355 40849   81842 60454      43863 53491   19204 52698   22252 98009   88848 90546   03926 39152
6138:  10100 32653   94838 88452   91011 48087   85854 61558   49339 67256      10001 85296   01216 23725   17158 51345   32158 36719   55485 68142
6139:  51107 49967   42685 72479   12344 22928   08339 67931   64850 17549      73894 87447   98740 17489   03653 10360   54239 28538   28602 30876
6140:  16005 78461   41985 83035   69715 78783   42899 73978   49634 54743      49958 57281   28760 98823   28746 16436   98602 77211   06160 56886
6141:  71032 21994   99207 24488   98775 61480   94347 42224   31830 56790      87793 31902   91889 50686   51551 54826   91148 15496   45816 26373
6142:  94260 22196   74602 49695   07632 66073   40380 80654   83532 88052      91217 35557   94160 68051   24888 25909   00373 85964   18675 53687
6143:  11413 54157   90640 65657   93678 71913   27223 51855   34597 12030      38579 34299   48129 74755   27209 10279   57323 94007   40752 84000
6144:  85828 48681   31212 00116   22813 42957   98819 83254   87829 71946      10295 88497   09613 33721   27027 60314   02923 11726   25026 15074
6145:  57117 24947   24864 47811   32743 09795   09630 38146   13394 60630      47367 69456   85132 04398   74946 41551   77489 99111   92877 34149
6146:  34886 76565   22062 24148   98830 46721   22492 51948   53111 29182      23670 31651   18614 10309   74109 64878   87287 47291   61557 49455
6147:  55684 72550   28257 29798   48876 56536   56885 17112   00830 28267      85316 62390   12140 84064   94135 48748   39325 12852   08307 82221
6148:  49710 33696   31017 59802   55222 32530   48868 42157   82843 02183      03820 34550   27933 98658   72456 98852   72256 82156   27687 09873
6149:  48210 83984   27460 33268   57929 14954   15924 31430   90724 06887      72937 83128   63229 65217   48315 42586   39247 34161   84353 24110
```

```
6150:  1072414250  5173199297  0205593960  0969619143  5304415680    6715503033  1368419335  4454411028  6704746045  0888547464
6151:  6887871771  2723185603  9731225491  3507205931  3773658105    4798805156  2165807873  3702168378  7256401544  2278815988
6152:  1321579052  5100400475  4536718945  9371706126  2832448090    4327833648  4304092025  0473392843  7755602021  5204136664
6153:  4001378814  1681024358  1938542260  1195109365  6698136535    5497410688  6234054699  4757953334  4542513499  7530829069
6154:  3171838945  8106079395  1786805276  4027674435  6413149915    0142812706  3140277513  1566597446  9584960911  9258941534
6155:  8501944179  5057764624  2690440281  6757998800  4522468889    2865953906  2718672363  5144957017  8931203385  5293123928
6156:  2515460331  8172344779  4418659042  8379053696  2999125799    4092924585  8920273147  8933740627  2504161173  0331693301
6157:  8661990611  8277332932  8266102371  4572363769  8607618290    5959610140  2707627300  3142329359  6267261749  4616452541
6158:  2457909909  6771142435  2541235350  0813021490  5163537699    9374904978  8256819326  2939876142  6130686134  9703407518
6159:  9401154210  0247245362  4744188434  4208328743  3728591997    2370640101  8454624088  7901914071  7537664612  3586863551
6160:  9411311580  6125268944  1621398411  0513350768  7164493926    9602603040  1432896519  4023628140  3857950122  9468873404
6161:  0993289127  4004171906  8950813705  5959313356  2109999804    5722185652  3296375064  8440822330  5526774956  0337507021
6162:  0209884391  4079455445  5192649504  9142559313  1110456679    3076069107  2041205849  4152739449  6889977656  0052707256
6163:  3189178606  4142096674  7057048791  9843880565  7339249743    7570792596  7198670016  4805353028  9444195705  3242247107
6164:  4458935588  3289491452  0311765871  6212420949  7560363640    3346827978  7056156987  4561193523  8966482214  3360052145
6165:  4783317902  1794771088  8762652092  0805892342  3733776837    4777652302  2518692382  5046748973  0321319082  7731878530
6166:  9069170649  6301742877  5117310861  7520233220  7777239602    2285261864  0560211905  6121768768  3105126980  9474633432
6167:  0225085395  0511927409  9936237111  6260508730  6630425994    9749438374  4657360765  7182356146  4092162985  5824007712
6168:  1379553382  8282094976  5959740196  1315655536  0635443985    7368390215  4019120425  9536566465  4361549102  8979809619
6169:  3436258950  4891533023  5150899461  7989792196  2396751995    0066982226  5460028201  7522061974  0805364379  3963851843
6170:  8520506411  1525559589  6336440080  1465189178  2607603555    0763831510  8750061680  2183024693  8870174704  8989499962
6171:  0342384614  1520942289  8594370015  9018208606  7574805978    5693528531  1275082052  4001978715  1528101856  0734538157
6172:  5771222464  4577295499  4059127655  3557212714  0750848452    6351023608  0581593496  4602603821  8877958336  8239429672
6173:  0008835299  5713592558  1771280875  3038247396  1156825230    2236059922  4519617738  8245429225  3686667581  7986216122
6174:  8777827485  8718899489  6983953075  7643744966  4188279865    6534301331  3889890814  9616298463  7339643940  6817102575
6175:  7128002173  8024338814  4210012772  1340999490  6928698932    0961252618  1366523904  9514376611  0600460505  8029848104
6176:  1353343370  8529435950  0686355816  1651010330  7017120514    5195976073  7279191910  4027127805  7356920695  5100950616
6177:  5245637494  9736739314  8556345704  8668793790  6044271292    4528007992  5497712574  6076072180  3278091083  4096707797
6178:  7484080812  6172471329  7148021240  3666768389  5404783094    1576051169  0759962961  5572513450  2645784229  4257926606
6179:  2427314204  2808107298  1196900874  3018088422  4102949537    6789555324  0592812192  9620549641  9714819065  6657491467
6180:  7881843908  0381738730  5404486490  1252134026  4102949537    6823463034  7357421130  8313477437  5736367356  4408104313
6181:  0541299536  5840378992  7030572680  5753845063  3387953125    4621473578  1962195565  3854029768  4300939148  9752051511
6182:  0951570789  3215792785  1821331055  4583280400  0269727121    1120907556  5690154463  7424783213  2763239287  4678780949
6183:  1681637898  1190271692  0414500758  8148524828  5078702664    5511485642  3704552406  1668269482  0929779905  5568386038
6184:  0867234513  0762153396  8366254473  3243926239  8642475048    5729207957  6692185679  3666195022  8675591349  4318010591
6185:  8359697820  7514853580  3916685770  1255788856  3743787404    1197209308  1569824885  8162344903  5388939770  5282166883
6186:  3712389570  3402866767  6379394593  0291199507  6196551669    0409588158  1689752942  6722313503  7614880552  3959045845
6187:  2335384506  0315437852  6965054090  3424945506  8959374806    8181524595  1532168308  3796284283  8848875307  7502635972
6188:  4465968642  5261231526  8329719842  5237716186  9271138826    8681315440  0292900097  0804028062  1250020501  8003965099
6189:  8184447407  2238495116  9352791197  7232255644  6413990840    5888404290  5854478361  3770358082  8186225287  4644694818
6190:  9098665803  2796168272  3112578696  5354533910  3375087679    6792989834  5208288674  1093403808  2439595597  9728763634
6191:  6235834330  8160622770  3712423076  6161174275  9605271929    9288867033  0341554096  9886542798  6994243348  4561561397
6192:  3174508798  5655107810  9178636478  2655869316  9581516328    0100228516  8918191102  6701783636  8391653561  4005713593
6193:  1637868754  4087620606  7073482800  4054034977  7755979603    0031782004  4994549153  3253390643  1854402641  7015850962
6194:  8444071139  7655404978  1458469672  4369480824  0790645841    1141526158  6537299230  9258806070  3678137703  2710440751
6195:  5676657453  7548235465  2470651742  0509629376  0318007453    1701708536  2465257299  3178482108  6181650833  6064098546
6196:  3797977500  7480709671  8618521664  3089461890  6394006647    8156924745  1690309550  7646068768  3011668134  8773710353
6197:  4137467713  1430270431  9311923884  7954330085  1136425871    5990116807  3882586993  5138584476  8136160754  1837405579
6198:  7062917579  0714841170  2576080896  8431180408  1953913362    9052472607  4410002692  3945334241  0376829370  3123844062
6199:  2934779646  6927303258  8209739762  0186711480  6961905866
```

```
6200: 96938 27728  86660 80325  90819 36052  53404 18632  18492 08386    56451 72652  39597 32868  12132 51099  50945 72363  05131 36929
6201: 66218 87978  98011 35078  75351 89983  73498 99397  01698 19018    44016 93838  37142 59435  68841 88094  63170 87971  09905 92275
6202: 74662 58496  08340 06051  63274 77192  06242 55495  16747 42373    49978 44648  41862 79094  40101 04221  50840 06757  80815 76295
6203: 58475 66890  48134 09786  16315 78351  86585 33023  27403 98576    99930 50653  44977 28123  15440 86806  65465 97396  01064 21899
6204: 06990 79212  38578 41670  20318 41561  52066 53933  58463 15680    17559 17662  98622 73207  98427 11717  46841 84965  62386 61775
6205: 04690 05204  33100 72672  92550 04750  55732 59773  85315 75430    47986 77053  11361 91452  97304 64601  43514 28518  25098 69440
6206: 18793 84419  95669 96402  13430 11556  51690 87747  91279 12664    42005 39541  14035 46547  37603 41456  29823 76829  82074 27318
6207: 00705 37400  95730 79567  83490 28677  69219 44264  27317 86261    52545 80182  08271 66283  55607 85694  13830 53691  78852 94161
6208: 56282 58175  86508 17374  39234 34440  42461 55303  55578 73551    27729 71888  00157 86543  04480 28636  61588 65456  19546 03844
6209: 76405 44343  72990 57685  18727 18951  19540 95977  70241 18944    06426 69486  07702 22232  50954 28726  33547 95223  68927 99315
6210: 52623 76316  65083 24225  15215 42555  14490 17955  06963 99526    98846 73437  27862 07406  28017 87887  96642 32223  49683 71064
6211: 88553 50377  47977 24248  83935 01171  44661 23773  37493 77051    59791 78993  12216 31768  85464 60243  94443 02585  97739 09096
6212: 83824 20517  16473 75000  21551 47851  04666 04595  57107 18356    62674 90563  73542 19035  07389 68130  02820 96440  67707 50116
6213: 27605 00273  12350 21746  53658 09641  17832 53142  37850 43167    61934 09983  60875 31902  65248 73112  38126 74419  11692 80235
6214: 70761 57238  53970 75187  63308 75639  01963 92290  49280 42971    67371 09112  82612 93770  52577 37282  36204 96598  73524 03355
6215: 55716 74005  84505 88490  47346 90597  74951 36804  31437 63660    23129 45222  94465 68581  76216 83313  97409 77893  61803 68717
6216: 78100 67613  14830 67275  96315 64981  31643 97514  07507 03229    92522 72076  19064 14746  23795 88495  16606 28734  31197 68886
6217: 20524 72731  62421 86790  87387 59266  22418 80142  63993 98405    77013 93784  24809 58518  66249 72199  78443 91243  09081 82350
6218: 71060 67446  53813 10147  48096 75656  31962 45131  35800 73182    02879 97016  33836 08450  21528 22510  19250 71331  48563 03255
6219: 65111 32497  25590 67592  73737 56557  75193 65154  89312 27966    94631 02464  10932 84525  82012 63046  64246 63427  53297 03971
6220: 29812 43928  16828 42351  70541 18475  50487 69646  74700 59345    24510 55507  29865 94820  18018 22274  30344 18751  60837 08316
6221: 14568 64514  53135 57236  76119 68057  78265 82916  82806 74571    87133 63815  56480 01667  01201 19462  23477 25754  74165 71496
6222: 10532 78316  16453 83333  11468 02254  01947 54019  67056 77521    12383 21443  88083 83037  37644 55078  00379 44598  33667 71748
6223: 78128 22121  85553 43019  23286 02411  93340 41194  96774 83119    93780 06638  47309 69505  37882 83679  88974 43427  77171 56500
6224: 98262 75651  11769 27282  40962 50557  55928 24291  67443 44956    15240 19074  75483 12222  26080 14274  27288 15488  51377 80992
6225: 32496 83731  23235 10732  31634 85367  84373 91112  30604 68955    73585 04174  47628 88642  43358 98373  18941 39753  96648 02457
6226: 03018 60585  56297 93258  35355 06763  00888 43355  03835 85674    37073 54400  22833 83413  53081 34943  95963 93883  98712 91436
6227: 13063 20152  21183 13599  33153 94105  39976 74911  59374 90778    62685 86613  30569 26945  79725 08849  48077 41860  82437 77661
6228: 76798 01545  40045 80672  50475 65666  74317 02472  18726 94553    53779 21083  55070 37086  78303 23549  86372 13485  09208 71908
6229: 07927 52482  78540 12464  04961 89255  50907 10092  86895 33974    65400 62804  76521 98177  12997 50564  96524 01301  87110 38484
6230: 88320 05702  84616 97672  53009 46503  52550 91278  03584 38579    65739 40149  69410 69272  09712 88229  48643 50465  11525 99317
6231: 21289 23949  04428 42504  95744 83957  36658 44731  11379 16048    90269 67134  41510 16080  36684 09942  51007 46276  01233 91854
6232: 76341 24897  78723 23768  75533 41905  16208 81325  31711 20240    57555 86477  77152 06170  91865 77029  50775 68727  87286 79978
6233: 19743 98522  40353 49062  39259 45368  80699 22417  85811 02389    08026 02078  07732 57257  82450 43100  65349 17681  60197 58389
6234: 70618 06457  21870 51265  11927 08001  66749 84122  12242 80948    95613 88666  62272 79508  82185 30672  71750 48607  49080 85326
6235: 42776 04148  45629 39980  40654 58079  02795 36343  77667 64286    71678 69784  11836 87852  79086 43920  66763 05922  80035 23929
6236: 37141 40470  15677 36511  21647 34786  33297 93467  73100 72860    47168 09975  69480 70808  93138 96848  43478 72229  48345 43771
6237: 95844 79251  37070 17195  54610 70425  56805 96134  98575 92225    92258 53499  19864 09397  63422 13084  15407 07891  39105 44276
6238: 57546 06832  66525 74215  45181 14413  59623 56719  89944 31398    13914 63936  08336 85794  15603 21318  56835 45792  70642 22235
6239: 21471 80413  49529 31372  16443 57321  34433 78109  33315 96992    54115 85600  06039 40504  55274 82687  92316 36913  01977 93559
6240: 03508 67890  24989 55471  21951 25637  16955 31539  67061 68524    00893 09107  44019 00190  58363 06584  90307 95118  81968 32749
6241: 30275 80404  31292 98637  08880 76967  33684 85363  63989 24242    23148 71004  17243 37929  33978 12315  61650 49919  86164 27017
6242: 79417 61156  01098 31847  15471 91051  03281 14371  13010 29431    71063 17603  32102 48073  66691 68053  87537 49684  19276 62616
6243: 45946 16592  15609 11536  84884 98572  41374 76639  51846 33040    76867 31084  91005 15637  18204 18895  40464 65918  51054 47785
6244: 56461 13246  16907 91862  95961 14230  35297 58172  81464 58440    54576 89938  62675 25919  22840 96209  54250 55845  43232 14553
6245: 89311 53624  06621 29982  15411 17719  15132 31009  35349 01876    80906 64385  77389 74521  53381 15929  72156 92838  11204 94535
6246: 61387 17612  31307 64327  91761 26681  46885 17965  65104 01708    64277 70234  82467 79551  48329 15630  90183 35978  48304 95087
6247: 27232 00321  56339 11412  08353 19233  81862 25387  81835 72251    53760 05757  21240 64553  30632 44346  69995 92021  96481 49258
6248: 81165 75445  15399 90666  68625 87136  86225 70876  93357 16953    38763 93190  41937 64643  70708 13905  35563 77711  00071 56137
6249: 40617 57310  33919 48887  72266 98635  90891 42150  89193 57959    67280 91505  82982 73235  62009 70843  00819 34471  27699 99841
```

```
6250: 80373 76125  55356 23792  56576 88671  49142 26046  63952 34589    20195 95208  47156 81835  62999 65768  05019 46905  04987 93413
6251: 13410 76739  83707 35935  29645 44919  98808 10342  46624 01103    40464 10896  39379 11230  38319 20578  19672 40297  58017 74520
6252: 01714 14677  57409 48215  31163 33999  12348 88460  61385 52194    72733 91877  14548 83406  94201 47240  17941 26966  91299 48135
6253: 09178 99850  58626 03157  84185 33674  15291 28532  29116 01529    16239 21089  10773 09800  46150 49601  33309 94087  09195 26092
6254: 40094 31387  97631 60246  63839 92884  15932 25742  57042 98570    82467 35558  51581 38445  04417 82163  22318 65494  53305 78574
6255: 23473 14529  46491 18783  28437 80917  85983 06271  77751 35104    98488 81284  80283 61967  39593 25729  45613 49524  03148 61933
6256: 15415 00194  96812 04177  27292 90937  93230 64428  81953 06482    40986 75201  91811 46799  42593 94925  36033 15125  04795 91406
6257: 99235 96938  41103 67852  49518 45222  35853 51014  19651 17268    39216 61917  91792 91257  64845 93309  59515 17345  74611 11587
6258: 60632 89720  23665 92181  70316 92440  69080 83926  38064 69241    19157 38817  22230 53917  17023 26545  85465 71386  00957 98572
6259: 11929 18901  97445 89649  38995 71613  09187 99505  21769 05582    17965 99183  00620 28067  02404 69610  34721 37082  35647 29989
6260: 81958 67527  60474 42552  43344 47676  60157 84698  78500 70462    89359 52004  60406 54175  73145 25254  60781 50340  27483 87465
6261: 93563 83804  40412 87773  61843 90191  18517 86274  64947 86042    32666 06754  06486 68657  67010 46726  58363 00077  20154 23709
6262: 93451 63101  15416 77246  06910 07797  30347 02056  55029 38969    84059 11924  63509 90455  36855 67070  36131 74658  75966 80555
6263: 50242 74835  01626 54906  63969 21123  98415 37376  18290 22026    96665 05332  97871 58104  99609 75690  98045 60579  63914 85232
6264: 35596 61092  77852 93364  03494 62009  55392 30801  31412 94047    17263 62460  51501 35106  10262 06260  56718 12902  25793 27909
6265: 46457 08516  58149 52099  38227 59810  07744 13397  94420 99424    09452 51140  95642 27546  72538 15085  54570 11618  20919 79856
6266: 25347 41591  83246 99325  27250 11053  56870 41600  42894 43871    72985 35002  38226 01364  70412 77950  73434 04076  94032 65691
6267: 39884 58873  90705 89739  92471 10720  36016 62116  44534 23451    55576 04097  48674 28769  44550 59271  86436 64129  67204 68461
6268: 51579 69771  76180 25064  90513 62252  98350 17236  31853 25996    88168 70912  07502 23413  96560 88456  83880 59307  99613 85013
6269: 17593 07541  13678 54299  56401 45171  98289 11596  03038 18164    12074 81309  05622 48034  54040 95981  27033 09848  93787 92309
6270: 84468 13295  73666 78771  78656 15247  67268 44935  16845 58792    25119 23434  66365 46855  96336 61405  09158 40605  60277 13153
6271: 12287 19951  85657 86453  86268 89568  37857 39435  90394 34584    49755 43941  66581 34542  20498 56745  45257 35183  93294 62225
6272: 16440 50497  88490 42103  59498 85453  82910 10748  57295 21332    71699 80217  78225 23865  26036 10890  39706 90148  28837 55736
6273: 10371 55004  91010 46845  68982 50365  55736 77495  01081 53126    83227 25601  74066 52482  11276 17801  85757 74848  66850 70780
6274: 36754 23678  69723 75028  11869 75203  17785 54280  33900 14452    27004 83927  60767 55688  96095 44533  12904 17697  86261 59678
6275: 19565 22986  39333 33155  87772 51677  87913 38073  45643 79802    24700 18616  95967 13045  43202 46835  97977 57009  59968 19328
6276: 74753 56057  00578 90126  73702 47184  01719 92741  33981 76491    57852 06820  65153 24938  52629 19000  02641 24185  18485 35541
6277: 26350 56409  20686 22988  60116 34808  39611 12240  63348 27709    77145 76903  00079 99616  95718 06885  93628 75256  17624 88000
6278: 41476 07367  21885 25810  77265 40397  98936 77982  50532 45651    85421 12269  73132 12794  12601 91844  73992 08527  28119 77056
6279: 37170 48000  89918 62785  10867 15491  33245 55544  44748 21838    73044 14409  44641 36082  82920 51297  00492 16887  52489 27681
6280: 67654 72126  77472 38034  24812 19872  09664 49464  82943 82197    47118 16966  26687 12833  85288 32847  42831 06699  89799 79522
6281: 12365 20607  33482 31645  66820 80820  19676 34580  29783 28071    41362 98212  60046 49572  16395 86119  68962 88641  73919 04759
6282: 49769 80145  11393 75068  80923 35264  44354 84537  77442 44885    17955 92381  25697 79251  14919 93324  18268 21581  70746 59409
6283: 87123 67640  34298 24663  04645 64612  32199 18528  16884 61400    72762 95561  30261 82420  48474 76223  18197 17455  59222 79896
6284: 02358 44251  15770 16836  67174 58374  56332 62068  63623 63394    52262 74482  57938 75038  72049 72163  34386 21108  80318 47533
6285: 92935 42439  63068 91035  29984 62855  39717 72174  23102 97288    86361 78670  87943 59377  89139 12250  55632 85344  43001 81600
6286: 39799 85969  84966 59192  44492 58333  26715 85820  20075 06392    37540 82493  67361 68546  90186 25648  05269 57975  66345 01742
6287: 80891 38598  77258 46889  95645 58064  20356 55026  79004 87993    90542 83483  78336 21765  55839 95792  31547 07852  14674 08867
6288: 12426 31158  20230 13842  20942 16005  46264 51082  23717 32261    24307 03780  68312 95403  14438 24355  43476 73895  33664 49741
6289: 78255 82825  73767 32433  10722 34187  72837 76042  20700 55579    96417 84862  42764 11118  19134 82369  34592 88553  32925 90949
6290: 12748 41242  68075 03482  72748 59606  18947 78556  54687 65710    21059 86478  68481 75960  07441 71060  89958 51410  91632 77998
6291: 99169 83044  91088 95802  58838 13409  03412 62042  92008 97741    15408 66923  52852 46349  38474 61733  28508 65050  19458 64559
6292: 62112 06666  69426 16866  61883 85880  87586 94474  94229 49763    66811 26774  78739 09922  23775 60501  50378 20750  27497 49598
6293: 19824 90993  16335 14177  64179 31746  33337 15148  26326 84059    77974 56431  17777 98448  46146 67076  20293 53052  28854 88698
6294: 11561 32179  94937 08617  74443 50141  60561 22573  83528 22908    00691 50937  66934 48970  62607 09505  05657 02646  96150 97641
6295: 26080 16229  57434 14039  13275 56877  30036 82365  64140 73600    00696 16308  04502 53707  59547 96778  56553 06006  12052 42548
6296: 79667 57452  01967 36724  78338 65181  53461 59572  06478 96142    66370 29870  34495 72237  85830 10503  64282 98198  98259 56350
6297: 33130 93099  81925 67436  53711 87422  89133 92370  31884 33170    12777 57827  92866 76931  42268 71141  88551 58587  10980 92196
6298: 65690 43386  63337 47961  21407 00143  23657 39185  72467 27058    42852 55818  11382 36038  81332 16706  54348 82670  64848 59472
6299: 85828 86064  00340 30262  23066 95232  25391 94471  78212 64619    83690 85828  42241 46467  26951 12445  77025 06121  80616 51979
```

```
6300: 35239 41120  21975 54321  87268 45222  98184 50155  11583 73199    08307 39222  79916 74587  62121 89319  62145 86041  53554 41285
6301: 91795 57024  28968 92878  05486 64459  65083 29083  83967 69794    18442 14985  61396 02642  29077 05222  72019 93237  55418 96372
6302: 92973 74331  27104 58993  64409 38833  97812 22675  48861 13463    61882 63176  09312 64307  93871 44117  92828 22447  71338 23915
6303: 96052 76674  83465 08532  75956 14162  52097 64602  94395 93534    24272 64918  54423 31369  21600 45988  53469 78140  80672 64616
6304: 88816 21905  88511 97121  49763 62986  39103 68416  69875 17305    08482 87008  39247 24362  68044 84576  53587 69894  55208 12382
6305: 49097 52917  02379 28708  66955 82435  78296 65824  24020 25357    46936 25884  43320 00271  85528 73902  94876 59858  85239 21111
6306: 83332 50542  82888 44701  90136 11497  41153 75866  67908 81945    43265 50660  11741 50542  97056 51346  07912 45218  43777 82295
6307: 76719 73591  86194 07117  88329 52698  96111 41049  72129 28284    16581 85197  56031 18684  37431 33696  51807 94532  70090 52942
6308: 04722 20879  40711 68419  23008 58289  80123 55239  51790 69632    09005 28681  54488 45309  53822 61191  33871 40318  15603 58133
6309: 35203 78795  01297 10171  90664 79036  00038 53873  47284 79830    44466 00605  89203 73769  37858 83611  55475 92664  61035 13581
6310: 93228 57095  38567 95171  80550 82844  23184 44698  91370 53250    70590 03220  38540 15490  53704 86108  43170 07029  90579 24369
6311: 13978 10646  92838 34390  57859 27817  67540 89213  97281 88863    28680 45073  44807 76036  16896 32804  29256 52154  53756 58152
6312: 42016 89184  37462 23996  15842 63508  13903 84004  92789 43746    80341 45122  53759 67620  90456 44456  60337 94705  43742 09525
6313: 27560 56063  88599 65920  82179 16892  46199 54366  32644 01084    28438 36115  87523 53108  01411 87437  48556 86993  28770 57462
6314: 83375 28766  08820 50494  72644 16414  73911 45181  99273 86492    17056 37822  15463 95050  46402 07791  36281 11426  97570 33485
6315: 05681 69553  24020 47482  25916 90098  97338 29432  25660 61589    84996 02566  76263 43017  57405 06259  49660 13494  18754 20034
6316: 36657 35958  41243 11141  62971 71055  50577 74668  74339 68885    82669 81894  28692 86861  05871 81410  22158 26488  83634 75534
6317: 14442 35302  07935 72495  04307 14071  97253 01014  85593 78739    20220 57327  15584 00090  71397 59145  85998 83421  27201 72291
6318: 02401 90546  00362 47614  33335 29073  97950 69337  68323 56526    26124 95081  12471 11436  22628 65953  58690 28027  46811 59585
6319: 92492 93724  12788 36647  00593 99006  06632 50553  18605 92661    95730 82784  14625 21297  79790 80071  00301 01965  69613 82682
6320: 38663 66324  85989 02265  82763 44334  71524 50851  16295 00498    60682 72163  97495 69506  09658 51195  14141 37625  94687 90995
6321: 93789 23944  52801 67082  80712 42599  65246 75637  29978 10508    92374 49581  90396 08803  18626 40504  85420 38091  75031 51554
6322: 55068 14971  85440 98129  33963 16023  68939 82790  63664 03915    97340 12347  23295 87487  90575 75232  13661 15254  70011 29166
6323: 61183 44281  98053 24543  90535 71670  15158 18423  02699 78231    21099 68293  47151 56652  27170 62791  92046 74057  29068 56658
6324: 70671 27935  96239 87048  98490 52635  54150 41194  05827 53567    45659 14144  32291 64163  78685 42363  60913 24198  80692 53032
6325: 45306 14286  16622 47923  68581 43228  36506 23087  04558 75278    69130 24257  59064 66380  87255 51880  15637 35859  68292 95351
6326: 31225 84353  68387 43447  40149 91491  15964 18241  74277 37931    08542 88052  32376 07751  94146 03629  74485 17445  38944 44839
6327: 11476 60823  70017 30459  26253 53071  32798 41996  71800 91448    33480 92780  64909 58370  51115 62254  96950 31735  53750 58355
6328: 26397 34558  08315 74787  63947 80146  74017 83588  31797 07891    37327 06296  22752 91428  50667 63877  63160 99410  51852 35302
6329: 82522 17670  21145 00747  82775 62951  03280 76527  82666 95400    13249 90734  12125 70803  66401 36969  49405 26996  03651 37973
6330: 67359 31461  21802 52441  83030 16566  95384 55258  87832 32854    87133 20272  41323 34110  08058 73017  97554 59936  49508 10385
6331: 21288 94364  16319 98491  83087 93078  38887 25335  26026 96529    41024 55620  74117 33084  08508 67265  41601 57277  97134 27525
6332: 21742 23168  67765 52412  90316 29220  85961 05386  39151 47307    02618 22119  92775 11867  89427 46527  79468 71385  64480 60364
6333: 62934 70024  84761 14333  59216 41691  86045 04539  26971 60844    90282 67834  52188 76145  49157 02350  24142 78534  16333 26884
6334: 61856 49646  80759 58159  83602 66774  79992 48467  36291 70818    43540 54600  20269 10581  77859 36162  31681 40658  63991 47837
6335: 79039 86539  07536 22077  37534 91692  62153 06180  27735 98540    67630 53418  75153 21136  88957 41321  50321 58616  45314 73462
6336: 49018 04133  59554 52965  74527 44540  91404 28197  65369 79585    96869 33300  32470 39120  63736 86403  85884 54078  30612 02053
6337: 34509 65362  26824 56524  68157 15989  89715 54096  85480 93102    69517 56934  47277 45505  96709 32305  11249 03722  67693 52876
6338: 72654 14568  43731 24133  38616 87562  58271 93782  56846 65294    81723 80585  93035 09919  93386 10478  60505 15372  78143 36964
6339: 03074 33618  98043 90338  88736 90544  50432 08822  20230 64404    79042 37247  87559 90868  14210 25918  28645 46875  48030 35868
6340: 88750 44079  90345 48421  62838 13604  99875 82574  34704 85642    89292 59306  69444 21086  99457 70177  90356 64702  41898 17819
6341: 05661 77907  68916 63514  75350 55630  35743 18006  26806 42583    30224 25693  44568 31369  63196 10638  21760 33336  34439 80882
6342: 11339 75068  71577 19937  75468 86017  00198 62441  68936 37707    76546 75991  66925 30537  37886 03118  43167 48376  73092 60263
6343: 97420 70684  69931 29082  59842 59072  90924 22445  82012 10656    42216 43925  09472 33713  81480 96690  90532 54742  29333 17470
6344: 52808 97493  89148 83886  41002 53247  70332 88112  34771 30981    44418 77208  01443 06616  40910 16254  91044 83006  92306 48078
6345: 77240 75165  57654 15137  23269 15869  49269 73579  70329 82728    34895 97331  94550 54924  13679 81213  67696 14060  45058 15038
6346: 92980 22987  95582 21428  48680 73141  86155 55461  66046 81491    17050 39188  11145 79733  00558 11976  94484 99687  95375 54741
6347: 17389 26936  25537 33683  49348 42658  19117 36308  86984 04931    17098 56656  10826 27603  95121 13989  23650 45099  72941 38448
6348: 88132 63143  63774 33540  17014 66926  95536 33142  77100 71417    63457 09443  94037 44823  48743 57225  64182 20048  69476 24037
6349: 83153 62893  60710 40223  10700 61592  37004 03522  77363 14114    78369 26866  14113 28635  39566 97523  13475 92904  52359 91683
```

```
6350:  0557248663  1903109089  7019156358  4215698530  5413892193    3243319454  7965047954  8233031885  7698027064  2728986594
6351:  0735695523  9725506018  9246970800  2336303115  3416707993    1853309787  3475726629  0355987977  7048127986  3095810505
6352:  9553031013  0429032605  4716922829  0718603364  0948075161    6857214230  3863849161  5814718370  1935723559  3079328597
6353:  8311457948  3628461541  6954730830  6998562301  7002057714    3322271078  6257249664  9555828771  9533298382  3490698516
6354:  4369727316  8212252178  5270982901  1360611669  7828172462    5097840408  3423742677  1788664914  9249326519  5373452653
6355:  3467948905  6076056501  4234413326  2490845298  1975186227    5268534560  0121399932  6831088021  3396382709  8719926812
6356:  3599429455  8501921890  1488569029  7857873196  8012914353    4255018030  9526386946  3204501932  7461433052  7206628832
6357:  2229532968  9765232671  7170135969  8137486873  6985532107    7786917827  5557789960  3208838679  0241073485  4924500574
6358:  4008475934  0390728919  3723778449  9026886010  8964007420    4537812838  9906224790  0179550947  8290370823  8614632985
6359:  1961614658  2489096798  1727971186  6177575689  8523255236    4311245585  9994561072  0356333171  6631503611  2532204019
6360:  6254914238  4180347553  0141176182  1968598120  7013727399    8356163735  4986636630  9808844144  4849834279  5270923466
6361:  8478733778  0638518062  7716971196  7226621615  7129264389    6783081685  6414339237  0467957417  4197300794  4206287567
6362:  7160795453  9066193756  5515976092  7499900847  2851744967    1749468075  2638507209  7563074611  5802764334  5509874346
6363:  7582329365  7521637460  3597568985  7288545787  5381679626    6135251357  1216154613  8647340510  8474118831  1494939408
6364:  2140602455  1313909866  4480546385  8820190116  1211193658    8133677772  7949880328  5188037922  9274696438  3648243377
6365:  5233879621  8204958785  8173207978  9774665945  7422647528    5959902716  4167307742  1779257215  9014085258  3106563811
6366:  8250934602  9167230095  6205589266  9956554785  1785795738    4289453344  9477353669  4423746342  2140563452  4058369564
6367:  4064284828  4769983099  7132907445  7118459461  2093888535    3719539609  1596375333  4787604349  4741061799  7670574735
6368:  8243600006  8387257343  8617450390  8776500356  3784358552    2696560632  0113575205  3625290536  6259473685  9633382914
6369:  4832012350  9475030855  3038981622  3809735434  3277558204    1706430589  4586459872  0165823201  2023458787  3113153216
6370:  9844299377  1258863821  9118470564  5138844605  8883915660    1500162525  2927744958  0284992364  8301382028  3076713734
6371:  1869357403  7211337884  0386491338  8377371985  9317017227    0202432377  7862632073  6504316744  4683604710  2153030063
6372:  6419348535  9572167401  2157074361  6032503953  3946184652    3101625013  9910716584  6111599738  0839613498  9250753683
6373:  2263098618  4237952353  0799681661  6075347588  1072397914    0809279281  1512958552  7611148860  6529951138  9022791142
6374:  4418458404  7864527810  1392929027  2208892769  0158855087    1650368484  2987337082  3925879173  3418777638  2459020430
6375:  7175733015  4298947691  7512683120  8568156149  8229258266    5802823266  1913087965  6289356557  2524689611  4721970453
6376:  3001166202  6695228890  0066406325  4519429105  4758677005    0598155572  9269282214  1883995856  0194112759  6496668808
6377:  2393044417  5533532288  6031368568  3575637122  1559840969    1973582297  0260837671  5899144947  0748293112  9828063753
6378:  1793620191  1532195326  3408194807  8859749056  3068760703    8622412261  6742865475  8423991255  6902919322  9308633023
6379:  4556593078  4466577980  2606424070  2103341246  5647167813    2030034043  8660207387  9908944146  9518104568  3750420063
6380:  7698109236  0903623830  0219882414  6054331297  7358630491    2306650834  7287011963  5164114582  9892078727  0972548560
6381:  9060489887  8377463513  1468410160  1164995869  8898230378    8998335778  4896420365  1804016199  8177550913  6385547921
6382:  4878289563  7296659342  5838836666  4829181671  9303896908    7712676848  0857360957  4746431901  2234017016  2148729215
6383:  8480514422  7348270486  2803281908  6113167920  6221623623    9435798957  5190214011  9906098636  7215752335  4650360472
6384:  3474564407  4486489661  0908521257  5925479218  2218643533    5524718769  7002927902  1869234925  7660219373  8242032766
6385:  1669629598  2890539101  9387839148  9642715207  4713921030    2359075880  3822447900  0442006926  0309674582  2008855921
6386:  9435136398  3419164651  0020318094  2934393363  7259266235    8174832912  1164786990  4041380152  7824358225  5316051279
6387:  7452394457  5024529494  2414516922  7051151351  0780595556    9188082800  9756408764  8455416377  8465480118  4863422732
6388:  3980865961  2895967810  1710794333  4666282482  9013995981    1062818432  9034683166  7250368156  4058863041  5682546985
6389:  2995930135  7893942242  4555256550  7688509489  9733560709    6980541693  5577135326  3519546299  5346101771  3164567660
6390:  6096898320  5057375541  0469163455  7802448052  3491572662    9551156834  1928366702  6528724098  3090018391  1855869211
6391:  7558831356  0024753188  3098013989  0876817536  0653716019    0654882624  3264410901  9403400195  8601889019  2141136197
6392:  7620137636  1778083814  5047211465  9660921275  3626409439    9940342097  9067939451  5171043268  9071605713  7894833460
6393:  4442369724  0189512319  8781324803  9364523322  7576850479    8978520662  4588011025  0252071443  9376618380  5566840093
6394:  7902836471  9060975528  9124023901  6110100938  3115078706    5869915186  5247721419  4407981595  5829028915  5700198656
6395:  8125380921  6886562854  9727763486  7120725234  3821643276    2748082100  9574719949  1245961370  6663457202  9824242703
6396:  8361426667  2727152581  0568407096  3302528164  7518115058    2120729727  4068863307  6073990070  3978422120  0426916739
6397:  4296010567  3071508576  7200164053  0382549980  7098732037    6372503220  2144538686  4827801201  3120925061  8647220221
6398:  8500066320  9648803277  4982275734  2508619176  7596201793    4008578822  9934298495  8276311264  8689989412  5551037691
6399:  1871601378  0190784797  6747526523  6839830173  2750517864    2484713331  0160586385  1895966804  9184280255  9415694044
```

```
6400:  60866 73043  42523 74549  48739 39975  44232 81835  55638 85505    66796 88583  26619 69910  07903 21646  22891 74738  09602 68025
6401:  19202 69806  40956 00597  43178 61058  77481 17160  20689 24270    26675 98032  38605 70079  43711 45225  49579 01613  39294 62401
6402:  18935 69784  62974 66820  30495 26607  50164 32476  40780 03684    27347 31051  90327 13277  77767 92975  23802 76975  88606 25691
6403:  58499 56514  81920 19472  01612 67449  52947 58351  33976 82441    28811 14594  54831 90456  34498 42958  71199 52798  29245 96026
6404:  03788 72339  49673 79319  46402 25482  56634 37872  77401 72224    73916 85995  89943 49359  19975 53216  83627 64612  59310 90979
6405:  92124 47853  50217 87282  20959 92476  94700 54926  10614 36106    12349 69762  52471 61116  07037 35967  69684 38786  60510 09356
6406:  59679 46084  68287 70819  38543 70297  82627 56557  52172 45032    56838 09315  29265 67371  32847 43270  08709 62346  23348 48591
6407:  75754 56052  97128 24590  49696 83885  62466 86710  33884 03280    50352 91636  17335 64548  56285 28428  24850 35114  89093 43307
6408:  52993 69506  19099 73432  83721 44009  51895 02462  66405 75250    60822 70703  58582 82698  78451 25010  63970 14999  95564 15673
6409:  98487 23239  96549 66392  87136 51580  56704 73680  29738 14868    74110 08744  76514 53164  57283 02526  61929 70929  28199 35514
6410:  64535 25542  11186 34122  25028 58709  26397 34499  19222 64452    25641 96600  71113 46530  07381 25646  23047 08904  43701 47300
6411:  93720 62932  19315 87309  06668 14587  39919 72632  46397 63436    00809 16385  46780 20627  72060 78355  49795 67675  89228 29544
6412:  04498 80368  87119 79522  37477 03622  35323 38309  82657 78246    06682 63677  42551 89804  26638 18369  63373 13191  52492 64522
6413:  38897 79302  77069 22099  65593 43727  71243 38427  09948 51702    25254 90388  53670 18102  36782 21109  09730 64310  24076 62206
6414:  97160 12864  72822 73240  55295 82570  11592 28397  36752 21550    68821 16886  43418 29342  84036 10495  48472 53195  70920 94876
6415:  29683 43316  53890 73755  58776 42557  74696 54899  71857 47235    71736 16761  64697 88001  78793 07629  87748 27780  68502 35524
6416:  61331 69379  23046 70916  46870 86242  23363 55453  72530 96276    60381 69244  20521 53024  60761 02694  71605 52161  71096 85835
6417:  04667 56062  87327 05397  08280 25317  12307 43802  68971 58668    86425 02934  06826 74232  60070 96773  98815 67771  10767 21233
6418:  11725 92961  46359 22595  14145 92007  30581 23606  53718 14766    86332 19928  53526 08777  24779 56294  48102 45143  69447 30006
6419:  64858 74212  44371 40701  29605 76015  75792 03364  41641 70177    18755 41663  70982 63231  23301 65860  73170 48250  99077 78397
6420:  28338 38043  74824 73948  68819 96640  43405 45765  27190 04735    76855 87262  88019 70205  03566 83535  74097 04440  89483 62291
6421:  37439 94673  98396 96855  84987 47346  73027 73182  77522 18518    52814 83144  08657 39348  70937 80942  55749 59626  37204 58752
6422:  49141 94477  60977 29305  19431 00334  34928 34715  51065 24173    49775 34579  17053 03749  38029 71745  93343 95530  88503 92828
6423:  75115 23531  40545 24357  76193 14475  95339 50352  00813 06003    11417 37391  22232 49633  45013 92070  87450 03221  71221 44216
6424:  26069 49587  12726 41633  66658 78954  76233 30896  01367 55872    34883 25713  93502 43785  79889 72254  72516 54669  25502 23584
6425:  56454 51833  26952 98279  21028 03308  99372 37628  03120 16203    18916 37282  92318 27487  11237 34267  44052 91440  46535 35961
6426:  24207 34906  38461 16709  87131 06027  44517 09607  79178 80713    98066 86050  23024 28796  50770 55449  82035 73543  37952 25204
6427:  09641 95837  44289 10876  34340 60238  21861 16748  88233 43069    14984 70002  66466 42062  88987 96293  61101 71023  92158 42075
6428:  87689 81314  67946 22426  11554 72509  86904 75573  93233 88503    70190 41546  48824 49253  56802 36854  77279 42143  40729 70604
6429:  25897 15192  90638 13171  76975 89731  99303 66166  55861 01835    26918 19019  13109 18511  61923 94026  18994 31328  95560 44270
6430:  98131 95006  86044 71644  76738 90016  00920 72953  90483 38987    45973 96553  58824 55863  70109 31600  96159 50799  10167 30066
6431:  52756 48642  99731 02315  67129 49790  19151 84164  39233 04921    41591 90316  40180 19907  90097 54277  27586 17699  20427 66954
6432:  91039 46124  55957 33314  11467 11420  68147 76699  48949 29646    97535 94763  72948 35614  60556 48715  93607 13139  97458 90561
6433:  00607 62923  04209 94363  40392 16448  01250 07734  85084 35698    26530 70952  16937 15167  97045 23450  60180 70192  90978 11766
6434:  33662 35857  61769 89481  16308 46104  35562 61622  02319 91189    41868 67148  20749 71210  24846 18136  67526 42277  88534 64356
6435:  34287 49342  37405 12478  78170 15431  62446 79630  79756 98927    88624 77351  19980 09625  22825 35999  85593 33561  62290 14278
6436:  37411 64636  24281 15071  98308 26910  42925 88292  37485 47875    89152 59172  34888 08066  80612 24523  21250 32321  99221 14276
6437:  64922 25089  02383 41473  79673 84299  99985 25872  93147 60163    62310 93604  09517 77882  67709 79795  52061 60702  51094 99603
6438:  24621 94720  54519 88199  01173 83625  93978 85087  12554 80547    61937 58200  17873 45867  71687 75767  22752 43796  05419 65014
6439:  00572 93676  26730 76573  97525 95304  53093 68081  72349 29042    66825 98022  43564 92118  87222 46366  53097 69571  03893 19279
6440:  49891 13659  24651 01651  64920 92699  54350 53991  66744 57583    85350 46113  37130 19719  64299 69080  56597 82213  21446 46102
6441:  42150 31625  28506 23341  62990 68908  69444 97184  39280 46389    66805 28046  27366 49984  45341 21592  77648 55422  48278 15657
6442:  21383 55925  15719 38284  17371 75292  78338 72034  48785 68694    86182 57836  67101 24517  59562 72969  31434 47054  07163 52881
6443:  46244 75158  37028 84735  14399 49239  90860 85576  36827 88191    74576 71980  19621 83670  51310 00390  27354 11688  96822 95303
6444:  47879 80334  28417 14068  27704 00845  20448 35655  29156 62588    25333 93335  70052 16845  47599 50094  54774 37503  71418 20290
6445:  81273 02989  22949 47747  89107 81404  46183 72619  88195 56391    98522 45497  06504 20172  73914 98481  25117 35415  54639 97149
6446:  68903 06033  79528 14680  85684 97640  49816 22593  54270 18835    91211 54356  75435 89876  34169 35812  18413 49716  82163 95805
6447:  20592 27927  80563 16819  76656 21804  70196 60305  93648 08539    34255 44793  38764 68968  59474 64430  14886 16038  27657 63772
6448:  81386 86004  49215 92088  85132 48754  86102 16138  55068 23640    06300 38044  74055 95354  79629 19471  10924 53802  44990 78743
6449:  54445 39338  74175 91601  41907 96146  99767 56907  11426 79409    90785 52865  49047 13914  45041 60168  18397 08640  43637 81703
```

```
6450:  4509886645  7870052995  4632079079  1598138276  1675303324    8578942186  8498817539  2498276066  7451609880  5778900043
6451:  3707496381  0671911815  3263816081  0063546562  5840634493    5284525167  4820648996  9803597513  9345144484  4048874190
6452:  4143658842  3791442484  8884965358  3625787728  4246762030    2009422258  2865036302  5160201609  4214528426  1893178472
6453:  6527297465  2577648053  0991897202  6687792597  3482703304    4444675090  1581792942  6412814792  2116283144  9576396698
6454:  4130525431  1486838235  1204620345  4740855585  1543799513    0497626614  8943748746  8778250692  6586384892  7120398331
6455:  9498718107  4828563064  5619074799  9869897280  0605339478    3128173437  2576761290  9183947367  8729295975  4878540733
6456:  8698169577  8805088715  2888912715  2508929593  8509709029    1399469831  3668296335  5928054758  5992336471  7639345366
6457:  0917491462  5126874538  5985755755  9366722022  1559234619    3536384253  7728840592  1362909292  6068613968  2930947763
6458:  6610201862  4705431423  4280173386  9667323298  7295902554    3078235238  7051358299  3512942199  1563432474  9923840647
6459:  0850974786  3373069894  2044275950  3765512500  0818223991    8544739229  1029714789  5884621827  6628881517  6718512451
6460:  8958634078  9435681347  4670738978  3338917260  9538453866    3412939240  0832787676  4958141424  9436379457  1013620984
6461:  4201279908  2406707554  4664660472  8330127497  5124125194    9605210942  9387923851  7590629489  9824317201  2674137459
6462:  8971946763  0148525368  5284120208  8902387534  2732186170    5089447993  7226799295  3312681056  9105338910  0355176092
6463:  0616797092  2440366882  1504657183  3418619061  3042430054    6539128860  1490437823  4265208362  4657804999  5799879991
6464:  6201347650  7509385375  9368477014  4569214668  8760437703    3243070910  3944272808  3987152075  6436200571  3628361396
6465:  5406360377  4376233913  4955760923  3444311806  4578167749    9432580635  2355699227  9135506784  0719579147  7973173872
6466:  7685613917  9178439773  1318253866  4702034192  5382080075    9711972511  6448143279  7110660678  5912838834  6303316422
6467:  6511516772  4634822396  8391661479  7712509348  6132129532    4903890933  8702280705  1252643103  4499155844  3440995317
6468:  1506291812  8426054523  4530626919  9729790977  0445068379    8010128006  6779963710  9763811357  5767847701  8000056122
6469:  0441967246  7933651181  0072046056  5054881929  9047362947    4830287095  3300025119  0276894635  3457225055  9361742012
6470:  6318263804  0476860296  7323394978  7787312357  4150520716    0408778200  8494705290  1472841879  9877979936  7544286073
6471:  1635706714  2195495953  7355537205  0862636140  1195375741    7924503426  7753187898  3163321964  8630648313  9298055269
6472:  8692485195  3981795753  9712700558  8189518758  1513552628    1948513803  5842549614  9035678023  4514788040  1138956280
6473:  8637572587  1326115448  2433246991  9030644666  5046571138    1811269986  2546069377  4373583101  1292660298  9789341981
6474:  6081307800  9096600016  4526067243  8436246630  8807750632    2820074081  6039423305  2918723098  5046598609  3497663380
6475:  0329668565  5309250655  9011769170  2108505033  2442002913    5270610088  4194405047  6766420899  2865758807  2398869527
6476:  4178130364  6545579747  1078328111  5118979395  6491561333    0898742429  2703798736  0417814859  8326493186  2951042756
6477:  3843269579  7299397442  7542105157  2473634152  7894524082    8766257573  4571245943  8657880624  9352330957  8856181610
6478:  8537557502  2557210575  3316313584  0854293141  3390573925    0217010312  0871909137  6169156029  0025406291  9713697071
6479:  0355079272  3643810356  3167540201  8064671257  2526719704    9525206674  5803961537  4974414952  3744339386  7045660906
6480:  7658811452  4638265689  2374542438  4588218352  6975140159    8888792497  6444316617  5771378684  8121338042  4667272990
6481:  8368452951  8101974082  9395942224  4503044011  6263581928    9887290476  0570205371  4203803902  6709109086  1015828346
6482:  1494116724  4428363889  9499446389  5481189943  8786465557    3844672304  0617106016  4999052692  8995720890  8478663190
6483:  0153307928  9458545177  6178227257  6129859959  2317572091    2695824741  9869647060  2751873876  8927918241  4719112703
6484:  2577150679  6970463699  1739754253  1148237699  5846659662    2491181338  0327415172  2242973486  9123273608  6277704522
6485:  2822228374  8509255736  6110521186  0984365003  6702578229    3291877233  8143163223  8249426116  1212287445  1307799935
6486:  6170821563  4231121800  6473722898  0250958244  6747828592    0319885504  2969700671  5340507584  3040382935  0033127625
6487:  7684075800  7983076426  7707730041  8869485030  5845641499    0429623704  0480277760  8599317175  2538030521  6371778644
6488:  3308778139  3497032827  1814793521  2753644189  8615545077    2715381521  0768562566  8711908684  2994395315  8862259627
6489:  5654418021  9239088065  7693027625  2062522989  0245764335    9248999479  0887905597  1277525578  8373204031  8854784991
6490:  4023892696  5183093291  6954701626  8229993217  0806094280    5265006448  9099012278  0324059437  9990214892  5942666726
6491:  9702609188  1799642022  0540465087  9650997226  4722808707    9572960318  7750391996  6897959459  5809538834  0063586245
6492:  3982932415  0641794917  0231492976  8486515177  4887370974    3484703841  0360776764  5440188829  1563204078  1082610510
6493:  2055602443  5821477490  5987013700  4606388400  0660995591    8707247898  0359285403  2323177431  7728629974  6596056390
6494:  5106205087  5358148985  8124472501  9226892366  1010596669    0623579828  1315385541  1035454062  7849868590  7867405400
6495:  1661215731  8667434028  6010750557  1202576122  0461298061    7723368352  8797410133  6158305803  0877644174  9837378419
6496:  5571194380  5861480526  4823685298  9802654901  9321094164    3706131221  7290808581  2540301090  6994637173  5070505522
6497:  4849233310  2656565647  1132508409  3099645797  2206773215    6176161677  9654841670  8987717324  2991269900  5774326531
6498:  8864549276  9535486922  9216474073  3538490585  0281028852    6349721488  9974307121  1689462201  5751834970  1567517400
6499:  7405427020  9935619985  8071599736  7673585058  2312514213    2618700675  5789226934  3806661006  3001262444  1549536051
```

```
6500: 62715 35915   90274 05874   89117 03707   65334 66427   73538 66473      89821 78766   68730 53966   80261 21709   58064 93229   42358 32122
6501: 51927 14001   52404 76616   65809 00206   56911 48953   32935 09866      21839 73069   85199 81670   28712 42590   62284 93454   25047 60580
6502: 69834 55546   86998 67039   91129 16253   46242 95054   70529 38965      07776 93836   79993 55463   87262 59942   09623 27672   15273 39999
6503: 81195 79571   72824 89170   18963 28841   08778 73877   75904 38087      84473 28997   33412 51821   31311 84363   57281 52685   49366 39667
6504: 99925 92955   60354 25676   95777 45749   43754 00255   37624 67112      66958 22371   53457 83634   43195 29662   85313 68063   97950 68697
6505: 51946 40918   22869 18698   84424 56880   79160 94876   73206 33009      05451 69688   19328 66000   46372 19239   69486 52404   34938 49738
6506: 69423 00584   46638 14768   80974 87601   29044 80029   89732 08386      58277 82652   74989 85101   72206 16678   72476 39259   33828 17356
6507: 17051 51368   08765 97989   94463 98342   52206 77948   21670 37870      25725 57398   19732 35368   32846 26965   06726 88414   31398 72234
6508: 79636 95763   28831 06383   10556 07147   25578 04228   67957 07876      68067 30947   49222 16166   91434 20954   14273 51941   22029 34589
6509: 96597 15845   34495 64586   96911 15256   05629 74408   61348 41709      95412 06838   63982 42729   89890 61072   19452 06630   13208 08343
6510: 74944 12039   08592 33621   53668 95043   11644 37941   04516 16960      67089 29361   86989 85294   39589 05762   57290 67897   52728 10942
6511: 21006 67777   48092 80722   02005 97272   82143 25190   74162 61633      70987 65389   99320 82806   92228 80154   76953 82250   37777 78620
6512: 71812 70232   49067 27125   85565 68597   11424 87775   26871 09286      10689 87762   53237 05868   36726 03037   01160 26650   38010 64993
6513: 15712 88767   20896 15005   53197 22479   04705 21766   15172 14719      99059 98080   21435 33289   86549 02221   87390 73574   32405 93017
6514: 13270 67838   58964 70592   56347 19986   84730 68077   85465 46405      19375 09803   68210 82640   16720 74026   07810 69758   85494 77204
6515: 41105 75105   11220 81226   24182 54051   98088 04167   20199 72352      88806 28247   10318 15948   78641 44276   30749 50458   27920 23593
6516: 01805 57294   92474 50667   69431 11036   67610 43204   02116 61781      76945 24455   75138 65748   65879 94591   34052 80278   15718 20699
6517: 34037 04630   71350 99881   62594 02155   47718 66135   77744 12718      52660 82743   25195 81282   36238 30857   08809 43926   43321 18613
6518: 56085 36969   73097 72006   14411 13562   09626 34678   39739 46147      62410 44594   81314 81852   43237 37493   28527 79704   62766 39448
6519: 40879 73421   34602 62246   77763 91101   05917 11900   59777 87833      46195 61599   34679 55463   59914 89182   02106 43880   45598 93235
6520: 36424 95403   35755 07659   57225 93044   39557 50478   41966 49030      43583 24105   69819 36473   74911 29950   55951 70967   34541 06796
6521: 18313 60290   78021 83396   30170 54998   67679 72670   53347 31086      91013 77802   56396 93523   09176 41586   47264 23340   12415 48724
6522: 51551 15526   29851 20645   20733 79271   67254 32780   82499 40118      26464 30342   26591 01998   53469 38014   15267 93390   19222 67836
6523: 99023 92294   87549 24233   51529 30142   92464 81098   91987 54829      11107 16500   33953 47780   98543 21282   55972 80485   61365 64373
6524: 77732 39298   34673 65527   92487 95009   77909 37351   56664 46597      27056 26903   30113 92015   90644 90917   03802 43779   78498 22800
6525: 46822 98494   18100 62276   43018 71937   84655 97216   29628 62337      11286 71577   00516 06184   49692 40817   34344 79377   37826 43914
6526: 79950 38415   39432 35760   39604 70481   77617 06901   69663 68253      66270 84231   87516 50743   33159 03276   99714 21221   09588 50507
6527: 38310 03951   46354 61230   63152 54768   06618 48116   40421 37326      31879 24543   18402 16000   83133 05042   69378 43263   62774 03698
6528: 05885 75958   55749 03664   37765 26159   92260 26979   72528 19112      44473 70514   57342 17982   45740 82329   80939 93732   29312 54781
6529: 06599 66441   61859 62022   73708 76351   75376 75453   53225 00498      61697 38816   09771 20789   48234 40097   85655 73005   90110 51480
6530: 29268 86787   92784 56519   44849 19161   06725 72530   19533 21993      38315 24374   43407 17693   20634 40245   62730 54744   60822 06572
6531: 49396 59762   38709 08896   66196 76600   63458 01273   33096 29322      70498 79541   00747 04942   68188 70528   78272 53556   65877 12936
6532: 11696 71319   90759 95697   44554 99104   43456 83830   23110 81570      26677 79669   99516 45651   83141 41567   38954 96010   47235 95877
6533: 48888 73207   81866 73415   95350 76246   42114 17683   84371 08349      03810 07842   79241 49630   47643 12692   44317 17906   54247 28776
6534: 47482 85851   47012 44068   49503 38972   31490 71940   51494 52064      06655 50728   06514 76404   07812 98581   61116 23695   22707 25417
6535: 06111 46035   41432 88984   37010 68458   98439 71120   97167 87606      19490 28319   23776 06547   04663 10741   46560 78199   22059 27831
6536: 43116 82574   46832 64855   13994 37254   97875 72045   29530 87150      76081 83572   11906 21453   93173 78206   86226 26146   72033 52289
6537: 82624 21076   11502 80340   71858 38306   13388 78882   30555 52980      79774 75713   93274 64488   04446 11685   09588 79268   66815 34396
6538: 55560 65081   24137 38049   97143 50543   00079 88174   88942 42341      15842 39849   60840 96645   82562 44649   00284 43133   22266 69126
6539: 76816 45818   49149 57129   41537 36228   57184 49764   36305 89868      39582 06505   82598 43194   14082 66129   40630 91133   30907 86522
6540: 97732 64535   02071 18424   14568 68021   37312 88610   37989 57762      44340 27661   35762 75240   10696 27493   80016 78665   71796 74436
6541: 91916 18036   14460 58286   85883 32269   86456 30840   30875 16605      44719 11291   63248 20693   97186 02042   29838 23298   79751 53859
6542: 42694 02444   29755 84074   38113 93365   63389 83502   42672 19859      82043 32554   22984 52618   25808 73482   34628 20599   67768 35537
6543: 70659 02541   46686 93349   59569 57715   21954 49857   62938 33957      36733 18200   07367 50726   66355 06201   63909 75476   18768 06379
6544: 29115 55609   52256 58531   78170 51298   01016 14361   85515 17561      06747 38938   18249 90977   29791 38735   74070 08706   15185 07484
6545: 15846 96122   62800 28243   06447 51723   50965 98543   54568 52337      89880 78183   61628 33439   33414 57493   95834 49207   30526 88084
6546: 90492 90279   31001 46015   71144 24836   88864 79540   30680 88084      91306 17018   17401 91361   32784 40687   50259 27977   65694 13789
6547: 03182 29253   27002 50819   97666 30328   96773 16112   00200 87984      86314 64586   07277 83099   61174 17994   50972 92287   89621 88014
6548: 95488 48754   55581 37196   03914 96060   14189 00360   05283 40405      75949 45941   86219 99236   75419 89381   36469 95979   58724 90013
6549: 20996 65657   29459 43468   76761 72368   81807 74939   36175 97414      76840 98621   28446 96826   93800 96113   81377 86297   36331 19987
```

```
6550: 83757 06328   61405 90647   40564 02549   38004 24203   35094 08370       86520 93515   39928 70196   90124 67281   67963 04511   63852 46020
6551: 85624 59501   99255 66687   65660 07056   06102 03299   13314 19960       60067 22333   58853 85301   39279 13986   42922 76730   44788 73835
6552: 65778 74724   43901 44151   30901 05617   78709 72781   06815 03587       59965 04033   04783 23448   01703 77368   13476 41118   54164 52514
6553: 87007 21923   60253 70331   85324 81971   07220 66355   37057 74622       13051 02717   71958 00246   60574 95425   84227 98963   65040 43048
6554: 16704 94595   23690 07813   43298 86339   91158 51861   20439 54043       94515 81653   22241 36380   53268 60927   42146 20735   06047 88864
6555: 35556 66901   48363 05405   52608 04860   41645 79939   94430 81340       65365 02708   88422 46163   15622 06921   65567 80494   22028 71883
6556: 56551 00743   81966 86919   95170 56417   60878 48480   29804 97579       23303 94729   64795 95409   59889 78783   34851 41886   78416 52159
6557: 99324 65964   07810 18815   92237 88953   79837 38488   73390 62071       03982 70819   45574 94291   32869 98119   10235 58714   00045 07022
6558: 82122 85700   58198 42752   56900 19066   16522 38398   62620 48067       94034 43559   52860 82552   11161 75839   60989 50542   87712 59077
6559: 80804 15173   54184 62126   89686 55364   41291 97903   21113 29848       90590 49577   51933 55187   67888 21296   66037 18066   88095 31341
6560: 19195 40047   28676 69027   64436 86665   33424 49374   67025 00944       82840 92267   11539 08143   66343 68804   35164 19074   39922 56622
6561: 36740 91349   44987 35727   63464 22591   24028 64127   89479 81165       01294 63642   24926 90711   38689 17710   28511 74030   36207 09462
6562: 87531 80828   60050 05406   96522 54769   20160 02931   93993 16645       97638 23303   02750 82206   63051 88772   95341 86653   48816 26727
6563: 62668 14374   05292 27484   38338 98867   03510 86583   13262 45218       91449 99226   35354 38609   78720 43169   43266 81686   44312 59844
6564: 16770 33982   89819 88249   38657 10322   54493 03142   88075 48019       54856 20471   44889 82280   29311 81358   81910 41587   06447 18121
6565: 52936 66113   18186 99936   25430 00325   05731 63940   70566 09803       48215 97311   75027 94093   96612 44232   00946 31278   78668 15548
6566: 29671 44537   17360 73700   42174 61619   04222 04431   09491 48498       04574 42494   12013 44298   24515 87490   48479 49045   10704 38940
6567: 27158 38473   48452 16229   81778 70035   22982 20014   18103 98686       76263 20589   69698 17291   73508 98716   37178 41238   48667 64728
6568: 56918 32153   65507 30283   66524 57592   52410 47317   48580 68179       42490 15381   69931 63711   70788 74522   04554 04981   27567 34963
6569: 54566 36559   37135 55296   53481 02998   66751 87271   54804 70248       82110 14419   72040 92687   00641 91694   49390 83994   11036 34554
6570: 75076 81900   22826 89968   42523 92960   79712 69537   93500 33728       03997 71208   20268 24086   33530 54464   90492 81937   52574 71896
6571: 42782 02580   57057 39532   35008 71855   14827 00973   67550 56379       91484 84563   82141 64356   98262 34534   98614 35508   11185 27220
6572: 47649 24477   08913 14916   56421 92789   67682 94574   86054 10344       48527 60575   66610 63064   09010 94176   17802 45569   20277 11832
6573: 22457 37570   36936 54073   80048 80187   31559 59764   74332 82245       50406 15148   48951 47950   42156 33281   05647 02057   38304 64058
6574: 49139 63242   63849 33750   65004 26188   84334 15621   97281 98316       54135 34365   70897 13760   67282 20819   92522 77539   42651 12807
6575: 78092 96775   52530 87547   97579 78858   49792 47232   27395 11676       69703 20054   68394 93093   34500 36482   22147 68766   42449 27779
6576: 48496 83716   86754 51579   18071 80676   55737 02230   45122 47317       54710 88187   21283 77846   77791 93688   54343 15196   34078 73446
6577: 24785 28015   00448 26953   51310 76313   44919 29632   87755 23042       14023 89814   01456 89017   42783 28519   16511 58019   89439 55824
6578: 23333 54016   10398 74345   89076 74367   01444 04610   20938 34725       57416 98764   02394 03928   47467 85332   99725 25085   45410 56853
6579: 72569 55829   82700 07330   74190 39890   61383 98783   37838 32124       04062 68954   59995 51991   91142 23336   64266 27550   56688 32200
6580: 90964 06492   94089 51130   97646 04201   88203 50862   00559 15711       89778 07701   45049 86223   90467 08562   39983 28912   62536 20193
6581: 13049 43721   22266 39236   50110 43930   40225 85005   61911 15377       94035 76482   61941 92674   05952 34144   93078 57811   92376 24740
6582: 19517 61658   24519 89615   50858 49626   24736 37689   29476 84984       06395 51110   49512 59230   09047 13527   06807 96669   75938 80616
6583: 15547 25171   11481 76541   98648 96892   61113 05665   64955 99192       90132 05218   19433 81588   18607 18211   23711 48891   86178 73357
6584: 29335 20722   90279 68142   38360 78143   70636 67990   34122 43447       75419 99574   38446 78352   44000 08776   02306 71736   42689 17228
6585: 50105 94314   79793 28478   08988 69494   65959 25330   26177 67318       10449 51225   55091 03573   21801 57434   44955 27324   61118 81602
6586: 21202 35623   05038 88927   38115 39829   50008 57992   76005 12663       17416 12397   04539 37123   08276 60459   96532 21080   39137 62104
6587: 99713 54888   65120 03295   52503 42311   93528 29056   42007 00332       84284 70105   57081 57046   32692 22699   63614 70687   68533 32808
6588: 81186 44588   50483 55125   26925 98156   52711 68939   54474 23479       69182 81038   74165 26763   30369 23012   62499 29687   72178 86674
6589: 71867 37098   71796 74763   23902 92841   43548 61667   94791 28566       32892 48900   41454 35395   99849 19923   21094 50726   45551 79508
6590: 87501 84735   17064 90251   56055 77621   53323 62260   31930 52884       38541 11857   40127 33437   83573 56749   57358 24377   33884 36317
6591: 68745 45286   26202 94021   07724 49642   72931 43375   54234 37911       47231 53833   98152 66134   78275 74910   41642 51032   13090 66301
6592: 92835 68090   67780 38475   06942 58872   91365 03715   32832 02540       45162 94590   99375 45856   39577 18705   68909 11422   70901 84296
6593: 52010 52643   93908 14051   30395 28783   67019 78620   00053 36884       88466 54011   69648 03177   06746 46107   80964 48114   72484 45901
6594: 26961 59562   21448 68556   35532 73407   93986 18255   84937 88321       96283 30024   66619 96174   18788 07312   38381 82052   43705 48701
6595: 15846 00341   80520 34371   42351 31785   40751 87288   62666 81806       30295 63694   52520 61481   56078 54047   20202 81202   40256 83918
6596: 78572 54393   63111 76977   44481 50871   63047 32844   26853 68499       43583 10269   18306 89394   09878 57750   06216 89596   99734 51258
6597: 33885 49100   74077 30284   06130 60390   82757 62528   14753 24232       17866 66710   39372 93925   28624 58853   64603 40943   60764 09225
6598: 96428 56645   95049 06057   63461 12954   24104 11916   84208 01043       06088 66882   80720 43545   64578 14423   56678 56720   29130 93141
6599: 07612 74604   69844 99924   87771 58239   09822 89592   83898 09952       05680 43447   43157 10099   61026 67988   20888 41386   28014 32600
```

```
6600: 06546 11398  77626 39399  33787 58277  10342 42713  82619 59517    91788 89659  14882 72771  86442 70078  36614 84778  40142 45069
6601: 13140 11088  38070 89920  28778 29875  46925 14929  05889 55382    92316 87152  89758 77270  29651 50268  68133 46163  00791 64422
6602: 70267 95875  85496 68098  82608 01724  53317 24419  86639 35762    94158 47904  77328 49440  96351 04781  86498 88622  61218 15285
6603: 65033 03350  74732 91166  40089 87370  15438 81809  94671 04777    77993 41003  02528 65000  80384 03449  33723 29037  76357 59683
6604: 94572 37271  11406 35946  87177 26197  47424 69301  97115 99822    54002 50662  16294 09896  24559 10091  70940 56731  60164 92896
6605: 09097 60350  55246 56825  48844 06222  46079 37505  28352 44244    60975 46916  54804 42498  10365 47558  08360 38673  68598 98673
6606: 07592 94149  00441 62570  69282 63603  59845 74371  99482 47831    05742 29523  10744 51277  98806 49999  12126 81877  66817 44458
6607: 45354 52736  34613 66354  17417 52499  13842 60423  52703 62412    02079 65640  14841 25294  01018 98509  14447 63330  38145 77116
6608: 95238 57785  58406 82895  04057 52343  27181 06670  59920 05191    69783 31212  23534 21873  52344 38708  12717 98642  10667 89599
6609: 17057 96655  46449 55507  44648 96132  31165 13018  97047 32563    96853 07881  47434 91470  57811 91526  71519 99787  91796 84368
6610: 35521 89835  94610 57393  89571 45139  71368 44558  79818 68083    16748 65953  13969 81820  84793 32227  62545 69829  28382 64379
6611: 04327 73387  63872 87987  63631 02743  86720 61836  13079 30891    26757 03062  02515 63194  27485 77569  66733 65607  11270 05486
6612: 20046 54791  54417 76846  55126 70007  26076 65477  73597 45632    60387 95140  27164 16470  95530 04979  67789 62026  08744 37372
6613: 49412 32710  85603 22068  08240 84828  76577 67293  66617 28772    11768 16130  70729 29593  34801 40479  06193 47221  61252 28992
6614: 70435 68581  30798 09768  77797 25764  42421 78223  77446 07101    55460 59197  56690 28594  33066 19610  76707 60681  17185 08231
6615: 79868 03922  38229 19441  49161 74568  48644 66503  85766 91215    72081 14044  22389 66967  36283 63317  45657 95086  37796 55602
6616: 98251 03346  63679 10123  01331 19949  56281 51098  84757 81028    12304 99900  08546 94433  74993 09689  40558 54840  25347 46255
6617: 18577 12243  62417 85824  17456 73913  09637 70199  97144 54664    29075 66537  37257 69528  24226 58573  20346 02856  49623 51891
6618: 33013 74964  03577 56721  06654 29100  23981 89989  70224 47621    25503 98244  92999 78234  46310 64526  10506 44816  51562 97656
6619: 21202 98032  20250 86254  44967 75129  82513 13379  45143 93476    04118 09748  56991 67860  60522 22205  09218 21496  65750 52261
6620: 41775 90514  35623 84472  82937 68945  05751 48886  02720 06390    56528 56374  32484 05762  49335 29008  59462 02251  75223 39292
6621: 48698 31180  10941 28675  59110 13778  32168 37286  65678 89649    83621 56922  27744 73050  49770 20216  36604 68117  34665 27750
6622: 27995 96992  67363 58646  75010 16342  90615 57169  38007 98270    61505 98304  01290 86966  19231 61274  05657 35599  39179 37164
6623: 94310 72422  75646 88844  42010 81484  40725 98279  53954 19822    27084 13018  24033 65791  61476 60659  53287 30949  51056 91875
6624: 81675 36793  70753 48191  27829 57733  43910 31289  50008 24264    38169 59454  56063 07658  95686 12432  69820 02722  27870 45200
6625: 44295 33267  44090 62178  56658 25367  97651 54842  17267 79740    00608 33458  28116 39897  73393 00394  21734 40904  26090 61801
6626: 71820 63751  17619 48652  17108 46783  01044 54771  10823 47846    73433 97355  63111 75274  29353 56149  63924 38861  45815 96800
6627: 76085 81237  66198 83663  90421 70865  49725 11177  90937 78665    01166 57631  72187 15122  58724 98886  11852 92635  91800 77344
6628: 48914 23572  66940 28414  04714 43626  30968 19342  81667 28535    88517 68284  17797 12889  31645 71206  32610 93959  97483 02056
6629: 32093 22323  81427 24895  23743 35869  74793 44524  88933 66687    48384 39678  92226 35793  27965 04729  03481 94742  84088 58733
6630: 87313 25540  59341 78734  16681 61611  41757 53128  40969 22490    76207 49518  78831 20641  51446 25237  63521 97799  47530 28416
6631: 24080 49451  39553 24370  50262 31389  05546 90727  19254 19297    68546 78186  32572 74120  65787 03137  34286 69845  38685 51551
6632: 26624 83564  94765 46089  09196 63002  79863 88820  58862 95810    63673 55813  82040 10530  30250 47773  22707 78402  94598 92840
6633: 58677 11774  54001 91385  68342 59580  10878 21082  80172 73286    19481 50635  49715 03460  39526 35056  68719 04746  84884 49908
6634: 22143 87275  60299 00156  38757 37146  12589 05343  97452 91359    55479 55776  69464 38753  42890 61800  57543 95207  05430 01829
6635: 43748 97659  75248 47121  92519 85546  02967 37781  05299 87397    45995 74995  37044 37104  71862 53684  70509 19951  81551 22453
6636: 67197 49022  64481 20993  66959 46694  54632 81554  54848 35893    98260 98843  64309 51882  91922 10915  47892 22539  05664 37918
6637: 92121 34414  11898 38864  07317 32963  81444 65389  63473 70419    49096 51371  82530 93228  70177 55360  96384 79926  10872 14363
6638: 05521 74109  10935 43118  36175 84783  61662 46601  80368 93439    26426 55727  93024 96987  57592 36182  33734 69464  59858 57605
6639: 45337 83073  87736 65585  57953 77562  96332 60690  22877 31433    11050 28447  88504 03694  10984 69043  12958 71887  71775 02785
6640: 31359 21306  39398 35932  89469 38649  75443 07875  19578 19960    05479 85026  86245 88902  79070 52157  91929 38366  60530 91332
6641: 02600 65876  52506 53549  64563 05556  53974 22086  47752 03877    06000 44920  80773 87355  68620 01787  51288 27699  16202 79814
6642: 22627 60270  51577 24806  46850 25200  73968 71278  49926 23687    77524 57671  16244 80994  55350 63807  38127 42767  39035 30912
6643: 38791 19226  66067 03801  87860 19187  06940 92491  20553 04346    54485 88108  40839 00587  30218 25008  91795 70013  35749 27188
6644: 98699 07602  96216 29179  77702 59758  42316 02586  33368 85461    38635 55376  31886 87322  27121 19267  38541 75079  91627 99473
6645: 94166 89845  44938 70927  32621 12998  95333 27802  01727 42299    69349 12875  11565 18965  36475 32618  88633 70830  23439 10383
6646: 34845 60301  44500 98031  71209 80147  24867 66928  41908 18852    70095 31209  17830 21515  06637 27081  29367 39099  86705 53766
6647: 40519 62714  14904 57097  08067 27280  97325 75073  42508 18116    54736 87390  60946 81447  72030 55635  58784 78666  15488 64003
6648: 95683 93619  86584 25607  04815 71462  25998 60959  10350 28459    98387 95371  59233 73414  97352 27045  83620 47337  27120 05723
6649: 41818 40843  41621 27830  69602 83702  30756 93043  43272 83360    27303 18636  20966 10216  38351 50503  43558 77888  18563 95759
```

```
6650:  2992245651  4004770573  8227376338  2753966610  1837222099    3078431049  3310406922  9020938199  2976508177  1570670578
6651:  6959600752  0530826246  1336619841  6946065398  4887375251    7392856184  0870286996  4794275883  5831201044  6136849107
6652:  0185048647  5843338473  4291031318  7372267983  0175440137    9428196307  7718145121  2231379520  1474635395  6450382289
6653:  5699739577  4108529366  1632641775  6608418286  5128223517    2460059973  9870971184  9908792851  1426699565  4083772471
6654:  3882906209  7904639352  2539216784  5674831548  8521723350    1717665892  8075946414  0161657014  5923878374  6994598438
6655:  5835424479  2198884715  2971994032  9493836677  0257687583    0745203209  9882046542  9601577346  6547329954  4412919738
6656:  7153340200  4996261586  8397017980  2544923059  5284704845    0582316432  7692497136  3314790665  6637281865  6658456974
6657:  0179029474  8718076139  2685588597  3201580340  2744250650    4167635774  9979347131  7623640888  2047101767  5999288673
6658:  0845958595  5339818487  6290134587  6474953662  9542971177    6727643982  7270876572  0255759633  7488012581  5368032237
6659:  3830291462  2863330216  8378036643  4584219316  6612087272    6243175642  1992156961  3312577419  0030778758  2560358731
6660:  0767962941  4218579400  3975644481  7713343011  2073197993    7678095127  5434693350  2633489701  7998209297  2954190473
6661:  9714765347  1147536963  9355023931  6803375017  4747591369    6393159569  2760499362  9993259141  1440884718  8088271003
6662:  8654925694  7044979237  5267909652  8214429274  7327363613    9524397697  5356998928  2689608336  3936794438  8670242471
6663:  2613953036  4896652950  4625279636  4972659721  2757071821    6170629466  1248627748  3870771918  7288799675  1174142596
6664:  6355416068  9982014062  8160576314  8908790824  9970161108    3013041451  8145965499  7983812119  0016721960  7684558754
6665:  1776526831  8756911434  3093885331  7972013761  9080724099    6382212712  2606600865  8525168920  1731654742  9806632573
6666:  6687531200  9738820975  4240871627  3935108790  1926968970    2065315305  1896234596  2291774720  5416369507  3807720701
6667:  0604452550  9903743852  0652153287  8682519953  8171672596    0936563559  8838137514  2436347447  1635560640  1951335406
6668:  0009284036  0043055978  3129864864  2201060525  2636824689    7652323756  8862190543  7933825286  4662587543  4749815778
6669:  2656643034  9569752020  3400208466  1579171011  7296061064    4719511207  6632662975  7650273904  6248115837  2489605504
6670:  5858451047  3896033915  4669746287  6935882499  6642178738    0715778912  6811052444  1349747931  1473972231  4546099513
6671:  1832586048  3311511591  3802621212  6176541949  6960810664    1030057627  5982859775  0427583908  7591796952  4429543069
6672:  7896158646  2005833516  0199894114  2105485631  8955226612    3783351168  1836600356  1436284521  9746858885  7578808304
6673:  5203618330  9030904176  4993503323  9255812335  4280695201    4627896810  7812906087  0724116866  0041260689  2636628051
6674:  7998640483  8706379915  4899652421  6915899999  3569260439    0461976792  5965416860  0471980388  4000305725  7055446269
6675:  4336747807  5251540718  6308137871  6537485984  7011085403    8124573526  5371863459  8173329397  1201428343  4253747928
6676:  7940524846  4520178560  6476135727  4144408754  1455209225    7932088700  9655202806  1919317008  9039107943  3564262000
6677:  8345504786  4236779454  3071078137  3434715826  5859353128    6246162659  5003478151  1408673459  1618307148  0738250738
6678:  1201463845  5575518691  8975480585  2266168034  5261711954    3097442268  1419249131  5121977212  2656355628  2584586282
6679:  3868277489  0578571350  3120522891  7708974118  1219659964    3633033099  7813265908  4010340078  6835838579  9628704211
6680:  7595483154  3574235374  4151851373  8225903084  0727240387    2577587079  3069527649  8391800823  8997284119  7788439555
6681:  9607617446  9328925610  6295316810  0012864818  7517348531    1207881986  6984095914  1410409251  7556488054  1727138958
6682:  6861744455  2854658568  9070806917  8981614793  2221996850    0912437719  5988895492  5290930050  0478150994  4911584724
6683:  6187825792  0529992809  2296577830  6672336932  2532515167    1183920814  1531172362  5456039047  3930772303  7183029798
6684:  1561703180  8811836795  6092828592  6285438176  9263580681    3147995832  0166458714  0659703304  6279664922  1075642569
6685:  4722366598  6105236981  1592225127  9972290288  6789753403    6604028376  2931193763  7250630691  8842070339  4469858884
6686:  6138336320  5374605103  8051403452  4978019046  6176569172    7000910904  3556348385  6124323506  2505694111  2198995372
6687:  4302151223  2447438777  4000822886  1041028127  0454877443    6616185668  5689710296  8668327146  7471929269  4562590625
6688:  0920535766  1978882230  0337013906  7676357560  9865727784    4458568583  8097491973  3126573791  1129389254  3168890438
6689:  9939950543  4270272893  7018760590  9689017044  1832288731    6410816830  8474332480  8761349281  8669293851  8977374621
6690:  8271828195  8727579289  0382872110  5606638256  9715327461    4023386515  6329604397  0231570108  4462435804  6279087948
6691:  8535139658  6479571181  8315617595  0623682810  7099821058    9368991486  8355393307  6451071320  5547547848  8075929257
6692:  7158902683  1892222733  3817424425  8568558473  2796178542    0996842775  9776553951  2424616637  4310928304  5442189681
6693:  3934747726  5047571927  5886734737  2315585314  4588607995    5765665195  1438573956  3968187177  8152209935  1103201166
6694:  1489284250  5212314990  0490291283  6390388598  4626520809    9890547144  2041332285  6125565995  2424011500  6204762418
6695:  4403465251  1319858277  1353143223  0421754543  4859665750    8242867723  6358321512  7334413646  2059859600  2223768579
6696:  9880327448  7624514971  0385239512  6956349219  5039968106    4714729072  1929714427  2761393238  5241005667  7856922723
6697:  9597867582  5640458367  6561901210  2977548455  4882351515    8359833702  8483426896  2401147994  2855838516  0947820867
6698:  0551600341  2174054353  2522070047  1419871621  2278035309    4520901124  3288765745  6116067413  0748378957  6442392871
6699:  9766154176  7852647676  2600053290  1747190181  6973988963    3308904760  0989471239  7979040021  7434343575  3690499330
```

```
6700: 68146 28506  63823 70837  73993 44788  19708 35157  26859 02098    60822 77545  52450 42753  91121 21365  03871 24418  42061 24952
6701: 15522 13540  25730 31608  63094 59180  39129 42087  77783 61022    84326 81018  12394 18922  77166 42253  94615 47550  72843 06700
6702: 31420 34412  75201 49627  96740 38614  87887 10411  09149 30884    65741 20437  92640 99942  19271 23778  07263 88551  28266 21262
6703: 72468 48610  84982 37158  92306 72222  16145 99514  43302 57683    79036 86235  03552 78194  41409 80304  75772 91134  14048 72300
6704: 39537 30565  35174 94102  59010 83109  68192 62537  82139 76382    71195 65757  14349 01422  08303 37672  97845 83195  28061 29910
6705: 89262 91438  67013 84820  44155 29760  82511 79791  43864 44053    64091 60017  55846 75836  36429 50196  79590 23456  37384 60271
6706: 38853 74086  61800 79514  32111 53740  56536 95681  44233 40240    32802 15403  99549 78202  07469 24514  24438 39403  59756 87217
6707: 43777 60916  65381 91124  77800 38431  29843 94121  79457 81464    64838 05116  05808 58408  17154 01444  04885 81948  24989 67099
6708: 07400 75347  32711 13707  26431 76993  07150 98036  31820 47695    23922 06976  75250 69737  91851 70562  82558 32473  08213 75379
6709: 34080 81727  51513 77323  01865 31935  52609 41489  82135 11924    88918 03049  79626 34508  52502 69417  85351 93626  36744 72200
6710: 11089 38646  62769 36904  89707 59237  86175 99568  89051 09493    20409 41032  60902 39677  76690 50912  91418 51119  28039 30960
6711: 05328 78937  15664 82482  32120 32686  25086 30898  20876 70529    76441 26247  85856 66033  92209 51914  28190 14443  33621 00277
6712: 09158 21849  15414 92802  84878 25881  05705 69994  44611 15190    38338 81808  62326 85362  58998 59022  47196 35819  37753 16320
6713: 19805 99965  54351 71690  91617 99837  90313 47241  21820 71585    51359 74281  56958 66735  00053 63004  95493 82709  74582 59691
6714: 61470 79188  67767 59741  68254 57560  04911 14952  34199 98884    92838 44176  99215 33998  79053 02400  16426 12859  68597 98062
6715: 82182 99316  72013 10360  07903 89433  95766 66800  14021 92399    45290 33118  20241 02058  88387 54734  75582 85999  15132 92433
6716: 45548 32723  12186 77853  67660 95761  42740 74631  41273 45497    59422 21992  98745 84690  79398 02358  05669 24823  75258 37405
6717: 89598 43922  22946 20920  41612 11845  49754 57621  51829 74023    82921 43750  06229 69267  21457 72791  74010 26253  93937 66158
6718: 11043 25716  55664 59084  89834 67412  69625 87297  92155 20812    04528 84531  47623 41993  23423 26393  21182 23637  39440 46411
6719: 03367 55206  92022 72235  95433 51611  06881 01321  08828 91476    09745 63650  92476 76649  94676 86948  61274 76263  70315 43503
6720: 92733 46890  29569 58787  74489 88195  94447 00821  35365 88698    29727 56508  30450 63319  84567 03003  04238 30703  76238 17118
6721: 27106 55388  67973 58216  44701 22724  69250 04864  98960 58052    74474 39922  54113 09610  41068 20163  90032 52514  38871 84831
6722: 09320 85628  92273 65758  04862 31729  48670 90238  53666 57318    07860 39636  95343 21714  11206 63478  04382 41341  53530 42145
6723: 58121 02439  87679 36373  51469 06217  04286 56968  64932 14593    27164 94719  49824 76683  42419 68914  72288 72244  40827 12714
6724: 67644 40847  90625 12579  81533 18424  76173 93383  58101 71070    24633 79205  40891 12547  54123 12121  37840 15913  01328 37157
6725: 63225 00694  33877 25093  87539 70006  13008 44230  42813 70722    55671 55047  31749 94825  40739 82980  83995 47950  38622 71991
6726: 64243 64657  43039 58264  40071 37245  05612 64918  95777 89926    70913 36506  19194 23640  66617 57209  08893 60507  86178 88108
6727: 46272 25840  79146 28447  38349 42139  84239 18724  74099 24001    21211 65613  44705 97197  34644 46274  65878 28945  64689 61360
6728: 48545 24906  48672 39417  47058 48292  42750 86574  93192 08690    36966 17491  28150 39061  96781 04441  56481 85031  10953 17500
6729: 69139 34116  12312 56370  84292 74533  63856 36820  44797 26879    19841 45489  06063 20422  86594 85665  27286 60840  00223 27389
6730: 53145 68409  10571 04034  62468 48201  83734 28862  74899 91492    85736 54091  44318 84965  79640 87361  38771 54579  49110 45554
6731: 89183 59872  50203 99602  27953 33409  57379 74750  64912 32086    36126 54581  97000 89177  14765 48138  52034 16356  66207 73565
6732: 12750 55418  29866 90815  36722 40958  64975 79448  00434 27837    91007 61257  04295 20754  05405 53875  00623 26560  46578 09050
6733: 22681 74729  39324 23553  83921 17060  72339 11434  65666 78598    72512 40383  27306 08309  66414 98839  04606 03054  03757 85584
6734: 43310 01853  56912 01636  81987 64299  46113 15945  60948 46825    96208 52656  99417 23260  87239 16184  26472 56462  23567 30589
6735: 59959 09627  12778 73976  23183 04279  80737 79737  77401 00466    82470 24644  15795 61232  13084 09314  02497 63443  16463 98851
6736: 50333 52452  42484 16577  82446 79078  93158 02951  68998 52811    54321 32594  07675 58051  71129 23694  82424 93093  77935 98597
6737: 75071 65059  79711 30992  34800 98992  22693 79575  13992 62858    82237 51060  14095 52925  30978 13864  88982 27927  28379 70724
6738: 09436 11253  51365 41043  39113 99213  83834 58490  63963 28157    27735 01952  59540 73788  82998 88910  97000 73308  80049 05134
6739: 28012 65902  89966 56414  74490 38495  17476 79785  17005 29537    39258 06663  13040 37944  61548 93052  45034 07765  64865 92145
6740: 74743 81177  98521 48232  43348 09948  20216 47982  27106 87754    32936 94948  09009 18395  38836 04001  93379 19056  63272 80975
6741: 01914 69980  98117 91668  27453 54877  02863 21587  35688 05474    31480 03836  03179 83573  85574 15897  75541 58591  49709 11087
6742: 24086 06108  58269 95746  00185 19246  65539 87277  77421 23480    99489 99106  85107 02036  76922 02098  06784 11554  70119 87551
6743: 18461 54342  47319 68191  90835 65500  42999 84072  34540 52349    78424 78178  27501 99792  72819 06396  26784 69270  96542 21857
6744: 98680 10891  91753 54153  52548 83146  65578 95484  60804 22697    48096 87219  88053 86453  50865 51297  74930 40320  99913 02485
6745: 83101 81302  10650 35942  63691 51328  08890 99984  19356 46745    55121 01614  36991 16548  27088 03276  31926 06990  04110 38777
6746: 59348 31322  96160 60595  24919 96416  98183 97553  71096 59323    00932 51247  70792 46170  90359 53626  83892 24223  70377 48181
6747: 67404 97172  13066 16285  88087 39955  96555 20633  49135 87243    03399 77782  24976 48952  04062 88601  08267 53175  84316 27909
6748: 63996 26885  40261 69482  95335 98813  31705 03073  41582 25012    74157 99046  37009 30904  75543 86514  41217 67511  50112 07304
6749: 57985 21795  26185 99765  18832 07011  18267 85315  91470 40077    45034 34016  29062 73204  03537 42458  00498 27123  46178 20714
```

```
6750: 0424300720 5780938053 7259458870 8175020666 1521561218   8958167759 0445020543 1688570507 2475033560 0489547639
6751: 1457732767 0901191466 9895334351 0684624870 4396390523   1339302750 7316538748 9192526051 8392083695 3993552485
6752: 2631923335 8447489871 5688527804 8643847528 1549500576   6660852700 8713268300 4779951624 5070707768 3372489261
6753: 8898600584 5609723802 5408741816 8541171287 6634885756   8584017890 7927702350 6751208312 8045243118 6150747544
6754: 2841432326 4738567178 7295801725 2329628892 7365823082   8783197908 4514056470 0012881891 0809915958 5599209875
6755: 4974065091 0952749608 7048307613 9433199360 1796987118   5329596776 1930786006 4772881908 6857591377 5350228366
6756: 7955131089 7376508925 5178971751 5427918868 5836807029   4090844838 1625361272 4455881450 8522844620 2906348697
6757: 5257559995 9862082040 0553117942 1001698081 2875502216   2143893583 5686373584 9274669137 2891254954 7568143140
6758: 6902862431 3732953506 7285184454 3501738087 0108456100   5327840681 4000752819 2258917119 4822552175 8520311316
6759: 8055727768 7705892836 3049940301 5717458641 4388917284   7330037686 4993914500 8943990604 8693768236 8834323063
6760: 0463282135 3397782058 8212426717 4194066015 1281425380   9098769128 1213488183 7415951728 6918864059 8983953905
6761: 7725732290 5438125439 6831938393 4569958052 1798057897   6011140655 5716335057 7006021950 2253331415 4697979817
6762: 3529127008 7440681441 6401112687 2443535481 8376488484   8491784411 2324189216 4371580832 3380450320 5755250975
6763: 8701795020 3302785708 4387793066 7809405823 8224063620   1927774935 2488567776 6691523365 6679984562 6047721591
6764: 1819325385 8016202342 7191062890 4306572584 1821724827   3151405488 8871817158 7331605029 3984872272 6973870491
6765: 7188817500 4726806869 3743936546 0657475609 6859055452   5704603754 9961575322 2550167472 9497552946 3380287765
6766: 8494779516 6450655776 7413209333 3716458870 7024308432   2819651315 9230309011 5958279426 6590245026 9052944688
6767: 0289418272 6293011390 9445712590 1992383052 4506797669   4452633138 1636036878 4128171801 4624420497 5673259279
6768: 1713466508 3758051226 2637224894 5910025386 7652399943   1250186242 9213453226 2417402932 4549751296 0024728601
6769: 2394994819 8543556032 0539246897 5055410039 0409500146   4969641257 0633283237 3827243584 8374558008 3144073662
6770: 9161724878 5891323767 5508809446 5541851862 4876722786   3186832457 6860549171 8508546208 7169168333 2122382909
6771: 5603107173 5869822480 2415561499 1475010049 9603042608   7657906749 3062057763 1980077871 0847915787 1841214880
6772: 6359069963 0288422872 1160306836 3772323369 7421905251   3506014021 9308609737 9739913473 6796370388 1093238824
6773: 1190279263 5161941366 1160424416 3197458363 5032924667   4123509413 1223835736 9093074599 4364441720 6478278389
6774: 9541823339 5335375370 7280764708 9761471911 7179495247   5699649187 9400118948 4850284787 1586656246 1795342334
6775: 1270070553 6575615679 2912345149 1178146559 9336529568   5699649187 3963216439 4449058647 7165181392 1570787218
6776: 7431348559 9248443439 2703184329 2209913199 8895293715   6612486635 0569121333 2305637570 9204362567 3208431533
6777: 7830956822 0207162713 6954765144 3477120370 4798685621   5995757380 2115580201 6112050574 3263933888 5821921008
6778: 1412033976 1567234568 7840619566 5836074582 6193321727   3954403774 3359113481 3538184057 5700845416 5077390352
6779: 1348011951 2860199111 1213722101 8817901028 2337642729   9223833635 3202321364 9305663352 8253810358 1876985823
6780: 2954185868 7946543567 4091829916 1692525174 3834678499   8874617334 1246231578 6068593726 3386166097 9900886998
6781: 0487701754 7206590048 9017692302 0275713374 9073790722   0284818082 5235586932 5120286797 9560439362 4725203141
6782: 2828444105 4383979399 4478716524 9536875240 7984610371   2015787772 1639191315 1710110761 0097386683 8727818834
6783: 5653981515 6991045122 1793011456 9987453456 4592145300   6773394664 3624719371 5274690657 3379857799 1563610605
6784: 9481874485 4899141471 8740029559 2234368456 7135846059   4845884356 4191030830 2808012208 5108866742 4401219516
6785: 9329936449 7854394964 5895564629 5825169192 5747576644   9681829428 1139730232 1423196254 0857562436 5364423499
6786: 9646461121 1503917114 6328158917 9732531417 1664045162   5261895665 3079265352 6892732218 4353202161 2556656715
6787: 8454246280 8697278963 1742966374 1416632781 1215070687   9469290589 7108614692 4644807014 4223439301 1045094020
6788: 3018708143 7883108987 4561162364 8212906282 6695665199   7532690242 4995251619 6787205967 4716548385 5754263039
6789: 0524749252 7929272852 7437159939 1453678172 2028907649   4756011857 2794704829 2130560900 9320873373 4873128220
6790: 4466556952 3943094026 9951965386 9934533376 0866523962   1351382369 6222245747 4517631257 5078375635 1872317075
6791: 0038481834 3216921326 4253107553 9754422784 4637868657   8364809042 0675714651 0943634663 7033332861 8586737642
6792: 7928634513 7334240490 6964312066 4127080418 6979239581   0982247687 2436843234 3442362101 1231416307 7115563464
6793: 8261277998 3957190474 9441390450 1916342938 1388302377   4666240153 4761953550 6295657245 6167710361 5211603260
6794: 8184241016 2745707861 6187786116 3665340878 2315274420   1184295503 4673960149 3260935047 1345373957 9875452065
6795: 9683866860 3596694775 9487559676 5129742897 0427957051   9072005965 5171450281 4874570407 9442653655 2772513459
6796: 7276662427 3448983552 8786133074 8894436663 2964467499   2589746176 8424402406 6850016128 1473390872 6859450630
6797: 0363364961 0029611080 4809813359 5838249973 2375202255   4360808117 7790849524 3637371273 6488281677 6808522726
6798: 1638586633 6785303426 0799773394 1730525015 8364429200   0970584781 8153388000 0458018687 9384549106 5899662934
6799: 7587248470 7001004060 8993060969 6048260054 9230780633   3615288007 2397270786 8067231795 9957614874 7964563768
```

```
6800: 26827 85816   36147 57395   00738 38130   89453 96238   38687 91202      87334 86190   34318 19565   48540 55549   41066 09206   21644 87536
6801: 29172 25824   05536 91527   85010 51479   46800 40479   77209 08542      30090 22703   82826 84223   85538 64237   30765 85992   31015 38963
6802: 64086 50307   23934 44398   11353 78322   36658 66868   13857 05595      06223 70292   38806 60444   71792 03604   37814 69508   41243 99225
6803: 14692 22156   76167 53804   64560 11436   72953 67118   32621 10008      92477 50104   20213 17890   93794 00100   29592 84279   93475 56630
6804: 12479 73278   67046 25277   02352 52646   16066 55072   34388 05641      75751 27138   14740 64403   51956 97092   64786 92947   40967 52301
6805: 18826 47179   39937 33315   21417 33299   21372 32458   96225 54172      95880 81709   80987 83484   12123 06241   31106 67348   73261 18873
6806: 03012 98105   96486 58019   10870 50323   69087 43796   15288 95919      75245 24855   97221 61510   97611 77242   70362 99556   69806 48545
6807: 00736 85867   68985 88210   37841 84772   16072 51866   34573 22986      64485 89454   98020 53428   68064 31704   66793 38340   93775 02105
6808: 95188 30418   29842 07012   04721 72492   37122 31273   31046 97777      97183 80705   23957 38192   79527 53980   68904 07352   79964 91159
6809: 63296 56644   59884 65714   97733 73546   23457 24409   76374 07380      81502 37577   84276 90953   32249 77787   76302 98928   33453 38052
6810: 28538 62219   94232 58769   38846 53972   55689 28521   54535 86859      14291 80415   26241 30984   47821 79090   08029 75023   88305 21175
6811: 62310 76163   58404 27744   07806 05571   45667 50662   30405 43075      74389 64768   17545 80947   49251 50279   79691 72311   29788 36548
6812: 75794 83651   92957 40985   59049 13560   04461 68593   73033 58882      95683 62214   34014 51007   39872 70521   33096 73569   11623 90238
6813: 17972 42393   57667 39628   88805 37877   56964 25884   54729 09974      60199 40612   26931 97271   43506 05377   52211 92962   77836 78555
6814: 54402 99603   07989 40649   58040 67649   26097 16455   52436 76788      26017 70793   38321 67517   96994 68624   41495 01793   82048 32022
6815: 06359 57133   01224 97634   21716 17002   87336 36414   09498 43856      07420 93648   36383 58644   00145 59612   30687 73929   78971 47772
6816: 52651 50509   66493 90419   49587 17491   90841 02230   50499 50623      43307 71911   93706 70868   22375 68341   84528 10861   59302 57886
6817: 48365 31474   94048 77408   25528 63101   75608 85539   07865 80106      63272 62049   26369 82179   27351 54340   35317 67023   83844 89134
6818: 21738 45571   24460 83081   31746 98420   95692 68596   47631 00523      43523 16724   37143 36141   36390 84695   06823 42830   04248 12976
6819: 34741 46890   02001 04754   59745 63406   28408 29947   50075 58421      51141 09678   12399 95445   78518 77022   41937 93269   36357 37869
6820: 88016 65314   10711 75591   97682 89827   91965 30517   30675 33766      45317 21523   64703 66040   28346 01356   69615 15968   31449 36007
6821: 92271 09769   09356 47108   14656 89991   36210 51353   85322 71223      09211 22729   75585 62567   42698 16222   53555 28724   72842 87300
6822: 83805 56133   61748 95251   32488 15944   13906 92308   70708 91572      75710 47904   53864 51734   99165 81100   15578 52729   81054 79683
6823: 22879 87524   88519 25586   00319 32779   23731 09631   11560 64375      97099 36063   19769 52256   06698 06929   57210 03864   42176 74217
6824: 51780 01538   59650 11096   02264 43535   25933 95366   46227 53877      22447 81036   22353 58150   30998 66201   99565 62180   96033 96318
6825: 64966 49619   94102 94707   60606 12599   89515 77307   38171 04402      06425 00447   30210 98693   85398 80383   12309 21232   56877 11507
6826: 10637 51253   41803 81473   38642 89665   64665 09111   71048 93570      23435 07167   12285 11724   85685 44972   88369 70378   37156 36248
6827: 18840 17914   12364 48047   50011 30251   12499 60984   14630 44067      72027 29192   47351 92011   61920 29382   67564 14251   60204 56944
6828: 57508 99408   39123 15316   43822 84011   34125 27666   12765 92928      64221 90750   26064 96095   70559 05443   60179 49421   51942 38753
6829: 37688 15616   87273 06457   49718 05246   51568 59294   54153 14851      29110 62033   40908 56501   61509 90159   42477 97066   72262 67378
6830: 47170 80757   05180 19926   80329 64595   76004 28369   37587 05236      48411 82078   43514 11545   05895 08577   48026 51888   83013 80734
6831: 99248 60770   78587 32363   52537 56309   56372 90940   16571 05617      94076 84107   00403 94197   21902 83482   41465 49454   10649 64714
6832: 21208 20664   60970 47658   90943 05494   19453 93723   00977 59844      73153 89281   68048 76381   36156 31903   59810 05129   14563 95825
6833: 46525 26975   64857 48337   84145 24994   33846 26241   96271 55333      17584 14565   27319 39636   56666 99657   61294 74787   71550 29432
6834: 92428 66766   32579 25013   24530 75357   94220 37503   76847 72730      05611 19816   98941 55671   01899 40409   77731 60197   47902 91785
6835: 17097 76965   29686 31859   69552 15771   31634 18088   50496 58063      63539 47881   96844 63673   13663 22376   61746 39622   34938 90296
6836: 25337 76084   23714 28499   44049 72174   89372 50044   83441 24242      23067 40160   06294 18843   83971 25798   18196 91700   52341 00027
6837: 16213 34936   01517 71321   93967 50169   63305 85835   52569 79647      24682 48924   81496 44148   84458 40666   88964 60288   56119 14569
6838: 67878 65237   41933 73779   18016 00532   02127 09409   37911 70452      26079 11594   18320 49236   55559 55662   98334 00640   38457 09460
6839: 58591 59761   81635 04970   87508 78468   43855 81973   96469 40828      92123 90183   03618 29953   64326 97642   73415 62122   53270 10713
6840: 64286 45928   55013 52548   89976 01972   22438 89991   54163 54517      29287 19863   56227 85247   29773 39304   17320 07505   05069 31380
6841: 14155 66722   58804 03792   80697 80177   82267 85694   31622 04505      72351 02523   56769 00689   37393 48789   46566 71387   95284 22070
6842: 97025 84922   14301 44135   20015 53292   70230 71096   72922 27487      16070 68658   78227 78299   04057 01374   03159 08915   09339 80465
6843: 51014 43224   84699 96315   38629 57641   76809 81071   88910 17742      39124 56275   60125 20802   19138 86865   54098 98570   14236 54794
6844: 38822 65868   18171 54570   41882 73291   56935 06379   03344 95717      12594 43422   38146 44690   76012 69266   23214 93384   04246 83944
6845: 22749 33149   14736 02118   46391 63243   92765 59754   29431 80376      13079 49299   09102 37112   27873 49662   68080 43083   98511 36101
6846: 50257 67675   09502 21055   79141 49012   73715 50095   78714 55173      43567 08167   91353 23032   34546 23517   05609 83787   40053 20951
6847: 84932 02900   96267 44995   37987 08788   39370 51061   68311 93868      71551 93109   07178 61761   88405 99899   62235 85317   97745 96180
6848: 87259 14528   19777 36957   71658 06384   80006 64409   13025 90033      76636 95746   11113 40046   57007 18272   98424 44898   65081 10261
6849: 13609 95870   83876 90956   95725 53786   11211 58594   89166 89915      70470 46988   78896 96986   30989 54444   40449 25300   80568 56787
```

```
6850: 3376505118  0847619995  9187754135  3501731397  1395649418   6961302721  6513579081  6489055627  7623589650  1517169346
6851: 1127382759  0165826777  0647412251  5523033033  4932799647   7617061720  2303197103  3867020966  8159467477  1499528272
6852: 3066212501  7545543415  5987214526  6888236676  2231254917   8839019373  3414589528  9861243175  8130617666  0062223755
6853: 2224576959  1061077556  5789055440  8489693719  5919422608   0696473828  2786467086  2138667206  5019855062  9366872185
6854: 9628014182  9272493140  8394409755  7809839741  7274552632   0988179563  6216019117  5984585713  3130520566  2903415477
6855: 0677040902  2416914084  2403107741  3190482285  9891703798   6192513936  3314433910  0643777280  1265915257  2482376210
6856: 0121058039  0364959198  7274074070  5763843917  9646562839   9544696384  0875089591  9952670287  3685449112  9613842420
6857: 5552409004  1694735590  3530590245  8998589820  7929337572   6484191515  2280839864  5008210881  3384156322  9735507417
6858: 4101361939  2231883596  7757256262  5734234159  0201204976   3409517742  1701958553  9973695853  2737931064  2172251844
6859: 1275980145  4072863806  7566614677  4701623740  5896386101   4645379762  3028293417  9705045070  8844213169  6921943444
6860: 1469792197  5640107743  1038739327  0841638277  8462801970   2048188757  3431043988  2011311230  8396375153  8838634025
6861: 5828889583  9966002948  2846004187  5226173777  6180131119   1198414201  5199216274  2761026811  8102397040  8465913107
6862: 1690857713  5313866228  3667286248  6915906719  2014009704   4296793264  3786935833  9837765850  3076006445  5035299003
6863: 8618252250  9597377519  5735552164  4930695033  6503310136   3996582651  9710800259  8160219076  6715237125  7046783511
6864: 3858456973  7058188220  2551375762  7093914488  0543995407   7102333311  4952619038  3930819544  1939074284  5231109245
6865: 6264414726  3916225841  4922427368  0489478294  8031410649   8647322419  5955484937  4434720686  8948963541  0938228626
6866: 5450457016  1142793090  9431137251  9465581639  1468335124   8114284507  0329089158  6788777272  1435557746  1923218613
6867: 0134613521  6243287764  3262389530  1920348562  9271002929   6548868917  8848429529  9788642604  6519957959  1450382352
6868: 2185410055  9955220468  1417129316  7100320338  0848767236   1875462335  1906402201  6781965115  2381464312  7974926918
6869: 1376005385  6149297755  3203452130  1784245394  1350286154   7047269923  7669482519  1212700650  6283555196  2100204185
6870: 9612957454  1157039010  5675977046  6825205850  8794134400   7968936628  5452849746  6865026234  0079726745  6530594954
6871: 2855123611  5446524759  1858111490  5158203763  5730733931   1394538312  3244721986  0908246545  2378895512  4857212695
6872: 7246711318  8221213695  6200342831  8260741358  0796027771   2269114680  0952725742  9217996044  3572124628  2221339368
6873: 4393903680  2908998105  4365308245  3067208492  1235065800   6575205933  2989136476  6115778883  4850776904  8764110847
6874: 6761333497  7579319281  6508407768  3259808099  0050192522   3437048146  8330894560  8605570694  3927651624  2749326921
6875: 8467166107  2520500023  6207091502  8963537180  4673923132   1452795113  8822784233  0725069352  3112691462  9479857850
6876: 9093570523  8737624388  3701731740  0284501179  7429847123   0237463423  8038900878  5879325705  7210122263  4942160224
6877: 8960869878  9591365244  2260393453  9102235550  2779699660   7093935057  9953327235  4633929217  1046907199  6193299332
6878: 1685847735  8017746590  1001105405  2063743464  9801101907   7134267144  0156029048  6157990705  4988477304  5062022350
6879: 8800661673  2019505856  2636695241  0153238245  2874816587   9127912106  3796974023  7696199233  2411071489  9941978145
6880: 6009542988  8036246104  0745947648  4527451294  3579183460   9996185201  7986516656  4235650314  9138910100  5673820633
6881: 9642704148  1169170861  1362221944  9735442690  0662681079   9220096900  0052410735  8560257699  7809091340  4444309017
6882: 3517419330  9543635686  9380440576  4315795146  9212022037   7642515758  1935835788  8894419910  5947676151  9552172070
6883: 4825244046  9295905453  2627202041  8693512924  3980977429   9558606652  6789557643  6670984787  7219349596  9052267881
6884: 9123677277  6598053134  9834407563  2201230462  3931138726   9328060753  1598729613  5623754365  2314981862  0698053105
6885: 4937514959  3312821455  9935927217  4410913044  3691989381   9366104656  5483775524  0504524080  4876724656  8649107597
6886: 1735271235  8740823676  8917830603  5161748849  8206083037   8063110952  0894024164  4308632055  2108472878  3313216218
6887: 6579639290  2135606654  8952120839  6420858108  8892638736   5045611926  1035064201  5378126861  1904053983  0503938044
6888: 3908567682  9929101818  2966748646  2130296961  7954468059   7497069318  5013552448  3071559218  5344744167  2676523778
6889: 0926819534  6633413803  4038589287  2202192203  0306793834   0668039204  7346513098  0208688633  0962024480  7922725352
6890: 1918366531  0638640860  8929529401  4467342512  8682871764   0367682140  5732333127  6810217415  8088503650  0548562011
6891: 8154598908  5104921233  7872678710  0363777377  6559561275   3766111153  8233840413  0170706071  8072069975  3568374888
6892: 4884082574  2323581581  8049000431  0744392553  0788021905   2286968586  1087285154  5870822227  7323139810  1946633243
6893: 0575387314  4756098481  4707163376  5452230767  9217719686   1364907217  5347958579  4430083072  1584306108  9313546330
6894: 7456114913  3121673755  6249887182  0853513342  3050247559   4479481738  3321226211  6824153990  0924908832  5158369754
6895: 1256162105  4306705044  2252735440  5167652865  5283794450   2646691800  2911963248  2801009286  7303143610  7840039744
6896: 7052728700  0773126952  8526287070  3427636390  2608878562   3236660838  8487829671  8726110992  8175518854  8400229755
6897: 3559293195  3805943465  6425636344  3780993641  0623470715   6642195121  9082484974  7418314539  7458992679  9722092388
6898: 8888685889  2039690678  1125676772  9969746711  1543240475   7706309443  1176330585  8542857269  0355316840  9071307202
6899: 0366306981  0267040316  4161611242  2563513143  5639828614   0953705498  4446382281  2530056802  2229123038  1579405887
```

```
6900: 04882 11004   71142 76408   63248 56485   55029 85025   57893 59576      82864 96854   99907 75275   74018 57470   39606 81437   91416 27774
6901: 45658 32516   36000 08726   45927 66779   11818 80173   24344 95250      54695 74031   83806 01843   42386 06183   10384 42040   53015 83236
6902: 18179 23754   89861 84068   77949 99833   86605 02761   82926 24055      54729 73900   96362 71548   08796 64404   60332 54696   52917 95853
6903: 81842 25244   09946 36216   63600 11804   05205 84671   78832 59342      05346 75515   26811 79303   68622 59703   92440 47195   24654 26111
6904: 85967 65736   52005 53805   74102 72413   67296 71828   61577 03316      67075 54568   95589 64596   67442 06089   70057 20523   72916 73555
6905: 19224 38363   34439 33404   21246 60434   00277 95457   91410 56579      11168 26012   69244 57983   38331 05737   93584 50224   72403 14203
6906: 56903 13438   96447 51367   05447 55045   29806 51308   95695 10329      08475 06530   81412 56714   85849 77380   12269 31757   20125 91232
6907: 75743 00585   17088 72762   41164 51835   95830 46925   39434 59738      10397 44549   55148 30895   84070 07753   25946 91197   77266 88007
6908: 62946 23255   86119 50099   85598 64461   29936 22252   09672 95453      53552 60993   00036 47001   64736 66536   54704 20486   19883 06504
6909: 64271 78876   22375 48560   06390 62455   74752 55911   97788 47951      38415 45178   70963 12524   83996 78256   89506 72057   03769 03606
6910: 49314 39454   29901 20836   84035 34958   01415 30248   77231 23208      45564 00435   77527 92742   41511 98139   32967 32637   17911 90393
6911: 06293 11468   97171 93363   74146 16736   26659 17784   66785 44459      89105 57516   34075 91650   42879 48624   78473 86505   80461 34945
6912: 01969 45408   55268 50817   50860 39721   17880 46410   10449 59930      96636 57224   71183 88990   62123 55068   99118 47354   42453 33636
6913: 80071 10911   04214 07738   18783 65392   66705 10729   18877 12066      96591 50383   41138 96798   00484 73079   93167 19563   22671 64742
6914: 59584 15466   72155 94945   93920 50610   23350 16656   66573 76505      73620 68920   35371 88528   80556 40877   53449 48210   56477 24391
6915: 35116 99610   65147 76786   71462 92155   29792 76962   61882 48112      86777 14269   02570 96322   37119 52708   59626 43631   66113 26427
6916: 34317 40213   30401 17941   17866 74417   44690 89975   84054 06455      46761 69570   47132 97671   15908 53476   87390 43973   46081 48467
6917: 27942 35043   38586 87716   27881 16478   43627 35798   92140 14676      06403 30905   20974 29679   80278 29090   50086 81321   39536 78115
6918: 38850 71095   63442 35220   89289 63978   55937 83179   81209 99064      78494 49478   74225 84618   68599 52881   22415 73811   60945 33288
6919: 22848 14703   36066 67722   59967 24443   86065 33453   21765 61432      68126 42047   08331 26558   64873 74093   94018 83475   41124 58961
6920: 04309 51733   95385 89354   93685 10682   99247 98656   16643 91254      10143 40345   89808 10289   74193 73598   62842 67936   57890 05587
6921: 13276 31812   17348 73230   97761 40447   94706 45570   09818 64277      60701 18248   17770 37984   95681 46363   89534 11004   86845 72593
6922: 30616 10232   01263 14898   14414 00590   47709 34418   66445 61351      18092 31411   52764 28744   71666 14637   06566 82738   39416 36099
6923: 34483 27449   76383 47369   05374 30339   31847 59391   25527 91177      96374 76328   99795 00203   85482 75929   64613 93918   25324 35762
6924: 19867 89739   23946 79953   56666 58710   80863 10564   10518 81965      07793 35563   89961 61080   24136 16938   89655 59706   69962 92117
6925: 65416 29079   53629 77413   08660 29629   11216 04423   91759 26972      18208 31102   31089 90585   08648 11547   63272 61832   82290 44079
6926: 43643 42483   61356 80589   79794 69656   77358 22243   84165 11704      70109 68170   99615 43838   90395 58806   79945 89901   37853 35893
6927: 66445 61104   86072 24768   96932 12976   99245 49121   81429 12658      87242 99054   32313 54398   62457 01507   05637 65313   53883 23500
6928: 07149 30797   90376 27859   95153 47893   75038 82773   79482 02290      59637 86916   94362 45401   23005 47889   69474 53843   71497 86868
6929: 04652 22057   34089 98285   38874 19392   03934 12668   54993 80518      03291 74430   55447 61359   68999 93705   92759 49537   61971 91164
6930: 76401 13385   07144 43650   71006 55298   73708 92774   24949 76326      88166 33812   29189 10771   66199 29122   86820 63305   21012 27586
6931: 81566 99374   20880 48814   55901 45665   42723 37271   43254 86405      83653 23206   39428 82059   37498 47358   93100 26720   97892 93018
6932: 39581 99615   69709 17599   96726 94498   15384 83221   94712 99285      00565 03123   18357 97026   43247 37947   19993 46966   99606 93838
6933: 93018 71959   08999 26557   46791 81572   75702 16989   62049 29343      02488 95644   14660 49197   47200 95097   10810 43217   57878 07613
6934: 42419 41638   20687 06332   34964 12764   96720 30614   84947 52581      19539 09743   99415 54874   54964 82551   03612 89277   14031 03716
6935: 69397 89986   49384 51408   90105 51265   02608 56156   72831 99090      01101 99259   07986 34856   71647 60945   11297 11287   92590 30480
6936: 88293 75640   80532 66433   11264 80389   95172 61562   21370 44509      82120 80997   03359 66867   97669 53021   54299 54034   93606 64285
6937: 99987 86491   37481 25169   45243 93945   41378 17937   55349 34948      59638 50389   92541 02455   53192 82934   36453 39110   65266 68528
6938: 06364 11416   17238 44652   83166 16742   73957 55505   97944 51333      76464 44874   38636 31054   46782 48171   62984 35883   17583 98780
6939: 31695 05894   45047 40277   89884 94631   42108 54330   17817 41849      70897 33804   16223 97094   55613 61007   52380 12450   49010 46629
6940: 60308 43479   10248 23433   41676 39426   43553 79695   05155 61207      99526 41758   86473 72683   06947 76686   73034 90987   22353 04601
6941: 67674 16853   59630 05972   71165 76596   18196 68192   46334 62315      75553 63356   46028 06844   43646 60492   27857 64645   05067 00712
6942: 49668 95653   44889 97563   04521 84618   54802 04287   16265 47347      19239 47314   00205 90004   39848 26904   65903 36266   89753 90631
6943: 81107 44686   88176 53084   25521 09975   80213 90134   02197 33867      53200 80921   64751 18947   31236 10035   60667 70943   72080 37072
6944: 83385 62762   47550 98738   84287 36993   51371 58905   24927 17394      94288 90959   32596 09734   45118 38939   16654 64356   59633 02070
6945: 31544 85483   98123 95772   41447 24751   08786 55307   11599 16240      03093 23836   83916 27294   60585 42845   95474 92253   34843 81442
6946: 47857 17714   13213 50638   58912 32357   41487 40700   88703 41146      43432 36397   75265 38941   04768 61620   28329 07189   30393 47834
6947: 04368 13262   20242 47366   61857 37062   62502 82309   23142 75235      73580 86692   93437 92120   37427 74523   19567 09263   01940 86452
6948: 18088 19828   38382 78198   50400 48097   22184 67738   44991 35139      07522 88704   93648 70699   60245 52258   15239 59696   70655 69865
6949: 85638 56756   27953 25129   31174 39536   71691 34277   81389 14486      04558 59228   74461 68954   69317 09185   05683 99918   82651 36402
```

```
6950: 44836 02566  87616 72362  27332 29329  30362 39907  88865 70814    20561 17876  47741 03244  45723 87818  72201 88254  07277 26450
6951: 22497 80648  32074 12830  25110 59344  41957 24364  58077 56767    12755 36064  03559 62237  35014 93452  93744 68686  47425 29682
6952: 16348 67495  57350 20551  20868 16161  32680 33583  37249 21683    11554 55658  19811 41963  89142 56951  28517 80680  62021 07925
6953: 90992 36516  11541 91086  19349 27594  02814 93842  90034 06150    23238 74354  91298 15388  03735 21085  17822 96702  23422 46003
6954: 41207 20503  23997 84746  90281 44590  41098 85243  54843 90071    56629 66322  82468 09154  03168 34752  98325 30934  88388 30476
6955: 76114 89290  91848 98436  36393 74110  92012 98143  63475 62229    62670 36321  63357 87934  48566 29984  10527 15567  91905 84023
6956: 04567 65114  88986 27849  14090 63076  91488 92837  02504 87150    47380 43841  45250 29063  80634 17108  34789 79019  42165 39482
6957: 07905 64649  92234 43793  37799 27489  21013 98704  93793 61791    16067 56656  56481 68576  52734 02602  65583 21786  24427 80369
6958: 19000 38359  47850 81068  99418 61167  89311 17365  75753 54737    72658 86580  52094 78283  26515 34463  61586 07882  82748 17138
6959: 68399 86033  21979 32924  94408 08755  99710 07209  06197 70417    67873 41112  08662 21743  77781 92097  40114 02063  89545 41179
6960: 90198 83899  91256 26940  16490 11268  97627 61205  50985 73299    76981 87401  32833 40000  67715 42137  49549 26537  17142 50063
6961: 88534 11033  54031 79215  11037 14301  64568 10812  92961 55433    52884 28953  27176 12076  25197 26309  93181 96191  27071 77775
6962: 66218 89580  69460 75455  09047 96123  72190 60486  38405 96240    60002 79355  78697 67396  73808 91183  99165 31813  09015 92662
6963: 53129 06004  25235 63985  50838 63237  40725 61191  37007 41671    20201 76815  80690 07788  73550 86184  99140 40650  70265 71107
6964: 10340 40406  22882 08042  44711 15914  53566 19060  06152 57968    30312 01676  92028 76271  29004 30703  64321 22908  23585 83487
6965: 92465 22765  44916 80811  15731 08451  66773 50422  02509 06707    43248 24091  16940 78983  43056 38641  20879 92996  34622 47913
6966: 60212 77114  06094 25017  99920 25949  85820 61097  57519 32288    39491 89127  14065 39347  19319 91262  57316 85844  53915 65674
6967: 45429 97692  07457 96838  42575 09824  98065 93975  63886 78079    58310 55204  83829 05925  47723 68193  21041 38869  25783 10664
6968: 88425 92615  41158 49827  52631 87234  32362 52975  33507 38787    95289 22929  64105 51319  68156 23585  72653 26794  67015 14498
6969: 41452 13793  68666 04467  61718 50354  04434 17165  99215 50127    96033 43323  71946 07829  35701 33142  87162 89643  13166 17982
6970: 34117 50361  22009 68263  56533 39105  40775 84751  06981 25857    50090 33813  10702 44570  52753 59421  63215 93097  96100 49221
6971: 68706 10503  05319 37861  72556 02312  80756 16533  49877 68381    51100 57764  91066 18136  78576 57635  28330 02411  00833 16716
6972: 08367 24212  01369 08358  26153 53377  69098 20062  63084 30404    83022 20377  45280 05625  88225 34911  66306 37658  02854 53515
6973: 68419 79322  23819 69844  48958 15278  97229 46502  24641 48375    85806 17033  62345 78425  97220 95003  35804 15806  66983 71867
6974: 68301 88423  20531 49127  58275 12401  26901 09619  40725 09058    05922 15511  34651 26938  83374 82244  87407 97941  02835 08058
6975: 44882 52761  82217 58461  07166 29964  99701 33459  18518 11928    23466 18465  26550 93098  83325 32341  35649 35951  57860 82118
6976: 08448 23322  13698 86837  19378 87936  64143 35730  83064 27039    40072 37674  16087 80241  95958 22835  21098 11944  28298 55554
6977: 78380 78525  35637 71567  13279 04892  40529 22481  88808 26716    89728 22222  40150 87664  82448 60305  19867 88390  65931 07423
6978: 13860 16567  02961 82931  70700 42633  10417 12663  07564 31471    72598 60995  11038 92096  90823 83257  59831 47523  77661 56879
6979: 40417 03458  58340 98741  59844 31387  49277 72897  95629 78147    80049 57248  78632 12005  39044 55142  47855 42375  07562 12781
6980: 59006 20363  96021 16945  18021 34483  17677 56024  94460 69284    24158 67594  57165 44338  86732 05371  45609 73449  00049 15567
6981: 67845 52559  76539 69819  02414 80640  22199 99756  46657 37388    15750 65612  55933 25599  17550 26257  03200 47996  48070 32475
6982: 52380 22024  32885 58572  46411 09525  61499 43295  05381 99075    07384 82737  52842 00329  94153 59398  46053 40235  53598 24750
6983: 53981 53630  60864 50157  22755 64228  15484 84051  14903 53026    46457 33655  62571 19415  79549 05678  17335 89708  77414 39757
6984: 60695 07891  83614 22520  16209 47796  11339 27503  96426 82845    92437 78790  39807 98640  61824 86442  41514 87169  23287 22091
6985: 16555 47918  85367 28925  92243 38455  30686 57510  58275 69429    05235 71207  78158 60576  41179 39663  77295 68060  09907 90455
6986: 18432 43270  17893 28965  03145 30386  34608 66433  66284 54126    91228 94984  28373 16623  74423 18701  13348 56044  62128 57573
6987: 01936 01246  50424 10518  69149 66105  21450 28207  10824 15422    47133 95919  82208 26494  48962 39899  76825 53306  22719 20216
6988: 00623 57590  22076 95600  81558 82997  08892 12396  41556 25389    97080 47572  22833 15104  03996 59701  02736 94207  77917 88490
6989: 39193 00424  45989 32161  93900 35881  54146 01314  13902 85735    55320 48541  89488 18308  51924 90150  06416 53297  86926 46771
6990: 54063 30485  01643 59064  08785 22465  55843 43383  44525 10330    53489 26772  16663 55548  20689 95378  30019 94136  34937 51210
6991: 15080 36191  72497 53512  85265 87823  76667 45144  53961 77023    24566 99538  95777 43854  72194 69887  93907 65361  60927 68304
6992: 37125 23075  61031 78981  22107 04475  14296 64743  04927 20337    51318 40288  51702 44390  70818 82586  42489 31136  39730 54054
6993: 51394 03208  37062 21734  83063 72934  56531 17178  88103 13168    33314 23530  19292 75408  53198 82495  97085 04717  43323 40426
6994: 01863 61945  08760 16424  75250 61443  55196 54195  75715 28830    01284 72761  40715 71729  68408 84275  35071 73438  05453 99306
6995: 20168 73260  74434 40019  00009 03182  13467 48089  44567 83410    44400 69284  29760 53025  94473 68908  30100 99468  05610 30518
6996: 63413 96860  35386 43321  86059 60812  87958 02351  35368 38219    39440 90227  93142 19233  05147 32518  16732 02707  76847 79457
6997: 48621 91497  07575 06102  87559 81187  20735 83384  29900 85868    75076 31824  09549 66873  05104 62327  59447 37074  74003 26582
6998: 60151 78021  22343 23005  49821 03495  53809 93167  93573 85645    00938 38765  31072 70358  76688 03133  59890 01336  53302 33667
6999: 15088 14121  10890 39178  84354 74622  71163 07889  76577 00741    69709 01365  36950 99275  06461 64947  66839 78085  66142 94541
```

```
7000: 0896589197  2578481395  1345684671  4658471272  6475844237    4529918431  1308796807  7999434032  8800218817  3533128405
7001: 2463485251  6064274736  7654964986  2399963677  4685986509    8143749035  3769399538  4387497892  0900968573  7598242386
7002: 2996332629  3897191683  1555822494  1957952499  8349443602    1310891084  1582756217  5088617717  5012904243  3618385499
7003: 2300161834  8531313137  0826824214  0809028477  6799787266    6602759416  1336151204  6162396140  8911260226  4256899569
7004: 2809382083  1585745229  8065080456  8884736371  8221919022    9801807377  0583833562  3569286752  2167423280  3816370921
7005: 0524653313  1787294130  4004097529  9114566668  8702294547    4713578681  4526402099  0830637821  6833959693  0554158703
7006: 6429695746  3618460131  2504013323  0327850431  7632366252    1312447523  8138226737  8136674710  5126322664  1607220448
7007: 7857060969  3453176315  4440795448  8228485126  8023736809    0924414561  3911053510  3967904269  8182468275  2117291223
7008: 1207348112  4811776413  6350020551  6411921508  8164215830    4260708096  3125438283  8459603389  2005616262  5662620111
7009: 6913810651  2759184464  7931317302  1100411481  5998986885    8303361741  3745759259  1404170554  7663852950  3604316325
7010: 9041840853  1297286346  1040869679  8604151653  8994149822    7312749467  9046269579  5223427560  6472096761  0299182951
7011: 8922263839  9976196623  8750775120  4892171924  0306319043    4669508252  8350131279  8035287475  1560333374  1770561711
7012: 6379058195  2948626596  5771817169  9571509398  1159164859    8329223227  1993902188  2074453344  6234381531  2381113537
7013: 4542963809  1502164140  8005550260  6423481511  6029559907    2792043990  0814098503  6486218302  2280538831  5380185206
7014: 8068461332  0010729910  5897733717  3861237598  4968574794    0289673393  9501977867  1341263902  3499268860  1230335771
7015: 7980445097  2602064177  1818854641  4284104457  1827052284    6386170154  1531987078  4325771425  2783611127  9388580789
7016: 7488772567  3686288788  5048901906  7199796894  7139825334    4655745818  5882775423  3348130342  9350202377  7860715115
7017: 9301656756  6113584261  7798106474  4953078018  2122042053    9906868456  5208949605  2879580105  2818883743  8674175142
7018: 2834320359  2828431301  6540239367  1962205290  5083926422    8947995474  5466082658  2296137953  0058793548  9647180858
7019: 0031613129  0443853346  5338679305  9321330155  5983607484    6091367011  7259979069  8585674847  4943235452  1729549023
7020: 8231852628  8715965510  9958979182  0303316899  4128263665    4187966845  5630167870  2123844534  0876886322  7074683937
7021: 1211077295  9452676653  3884768406  8586758298  4985473105    2724058745  0384660851  1094200109  7752141918  7596319877
7022: 9658427175  1101626510  4974714392  2229936623  2082823451    9749852097  7760714298  9525961275  3708419165  0357480790
7023: 1407392625  3314482466  4459491602  1339649243  3758628316    4409973998  2232581139  9055726466  4255753462  6556811632
7024: 9066357485  6991431070  1981547373  2559807664  3748427262    7811540729  9672550024  1607767484  3783012522  9323769980
7025: 0160567404  1200517799  9854799622  2901167928  7368825766    2296416692  8041522111  0393115744  9417806353  8928963063
7026: 0489154339  6596709042  0653212199  7152614678  4357859400    5516415572  0247038193  9337019356  8728615572  7455615680
7027: 7898434700  2195809000  5537034742  2151006442  9636286681    2397219504  5534094229  8672718948  1234169942  3354361256
7028: 9064929039  5154099275  8470893686  2419569214  4820503945    3515654542  3248562613  4780690628  8177963342  9908176310
7029: 3367118573  1110010943  4461767534  5995685897  6770332167    2382665972  7890157571  4932882059  8270881697  1243192496
7030: 8683249983  0343800594  6537677354  8966402892  1105396889    7969802818  3679805578  0120425493  9368746662  0889868979
7031: 9460434995  3529437286  0870139699  6672030951  2934835869    5467353579  2043284706  0370071301  9261865957  0499795542
7032: 9485639614  1335870902  8836263716  6232540014  5654350332    0476553919  3199428084  5687465143  7818121326  1774661249
7033: 1088405739  1935761998  1261846965  8243597029  2306111987    7993991259  9356980647  5147122645  0715444554  5809866504
7034: 2588295293  9999072696  4742374652  2581597029  0010135008    0284430338  6484499489  1166103350  0701773394  9121928697
7035: 1292359283  8738494125  6884737359  4324657378  2964217129    8786965621  4070825330  3618869595  1377820976  3748647329
7036: 3955872463  2042847367  0369789751  1022792156  8032052725    4265444978  9390529181  3044882083  8633528880  8085711077
7037: 4652883746  1836913220  4923632453  2386086234  0751681225    7881788761  4658611672  5240183321  2895275463  3764864926
7038: 9388601257  5085055104  3146383697  2806701276  2012262326    4859868767  7009710026  6944096009  6652320279  9958617144
7039: 3907163432  6557117766  8244370658  8726261962  2225711282    9871206176  2502302119  5978993574  6643023796  5863702752
7040: 4640981742  9431698868  3661495918  0424314107  8410719042    6111420015  8414828197  1241479682  7396907416  8959176017
7041: 6583338224  8841518899  9328948444  7564414400  0701169705    8797629747  9433384609  4308713999  0502610299  0020137550
7042: 9895567415  2834737092  3715760915  4651865073  7212818671    1360178443  2511241231  4591449243  0893362826  9667554304
7043: 5521243980  8136123451  5505949770  3512427754  9084473860    8067654696  5827205350  5462867176  8692039544  4668876729
7044: 0406768842  6628667928  2444648918  3670152443  1100017914    5640017290  1767189972  2046241528  9099710981  9128897877
7045: 9583067938  5192116987  0587239660  6145192369  2763491589    9466587683  1305237156  4856107016  1329786338  0435342695
7046: 2103555308  4406824907  5768346949  1200657334  4622323511    3931270011  8101361424  3736565312  2158998422  7754867792
7047: 6050255693  8657073560  3728313659  1746920331  2888846782    0069366877  2847320303  1558744372  7320042103  4667637561
7048: 0688568081  0668121538  2521162267  7053606849  9852449421    0192388872  7188437825  0536248295  8554567274  8746122370
7049: 9721140517  2673529424  4140914753  4750455387  1293535292    1133796790  5868233015  3988531265  4392448664  6792988803
```

```
7050:  46329 36109  26631 57589  69224 07294  35726 34262  40636 64120   8320 435135  66081 46903  75825 44506  76450 21928  81720 30048
7051:  75415 29115  42535 79252  93818 10834  29410 52014  36128 68218   2853 206483  55421 29997  10124 01852  25936 71133  57578 84695
7052:  03038 46907  77141 64176  81830 48272  87690 88463  88693 05148   4886 678718  17213 40623  25557 44652  49751 07116  97275 47891
7053:  45068 71187  16364 57076  10159 60844  34504 66469  16151 34086   10458 42609  36692 50053  76014 89412  63680 76771  05796 20069
7054:  04558 35329  11154 85797  06739 52425  71539 91628  04862 34063   26854 71176  23183 66120  47792 19674  95694 19985  09693 33911
7055:  45202 36956  21548 55487  20341 74378  58481 77228  88854 53316   92152 64955  96093 11659  18029 99924  26697 21957  36057 46644
7056:  82356 15554  17085 87053  71111 72773  46340 69052  26010 25237   15358 09307  05077 76099  77670 82962  53656 66869  48101 50882
7057:  47940 20022  31141 65290  44707 67165  48004 40173  75509 12634   83981 93078  42179 23824  95240 14920  95380 58775  15378 43881
7058:  59851 66018  38307 55413  97601 37158  49817 75626  84648 28151   18834 35023  13257 22163  68492 04210  81457 95905  04477 34816
7059:  34540 95245  30565 63797  27552 52830  10004 27367  45549 31813   77857 09470  08323 15381  84209 47736  59518 42019  33859 75206
7060:  76650 25871  47713 09331  15903 99258  29704 13693  32899 97166   11748 45803  45813 15174  73098 77096  73703 46941  62450 31432
7061:  77663 51904  76319 30287  13558 71172  92286 84588  30914 01386   82696 19176  06712 14045  53931 98960  32063 98593  45988 50387
7062:  12602 02344  73874 67529  11038 10695  17982 03023  55632 66265   70773 21661  00273 30943  20320 63788  09538 98606  75790 23261
7063:  91023 55172  36153 50875  70430 68572  01840 32543  32360 01713   53082 67876  92580 44296  05763 41210  86182 53568  11740 99094
7064:  59087 24003  69632 10620  27122 63048  84238 57360  64666 19592   68410 26042  13211 43196  20061 96068  30648 98177  39743 48400
7065:  86688 80618  59030 73056  38345 76712  66771 48179  43876 62070   12936 24202  59410 44387  69332 96316  48834 74783  17254 36606
7066:  11250 70812  16891 76990  22074 05570  99373 95763  40470 65389   66660 70667  60578 61582  02997 90067  87254 97218  72512 61583
7067:  67685 32992  49855 54615  41237 23600  19493 85684  76726 12597   91665 53311  59586 67436  52722 64084  29181 29510  43225 91585
7068:  05571 86991  82106 86306  33298 40820  69348 05363  99451 54694   09054 51587  07147 44531  82745 12314  62069 37885  60982 78958
7069:  49276 29275  09180 08279  47748 30641  18211 90899  02303 99008   36499 96458  36826 71071  72342 64671  22010 41453  08435 88713
7070:  02180 76361  13788 69086  57497 20270  92181 84055  89263 20951   36559 90657  46708 12225  19918 42687  55224 13698  10276 60028
7071:  54767 47733  06351 24479  51492 77559  81079 17054  95433 18028   04279 36466  23773 47553  36485 05708  64030 24988  42472 77531
7072:  33670 96579  13038 14075  81581 77013  83545 25489  48128 21581   60218 97356  45940 69161  75511 60249  40387 67485  17758 52691
7073:  57523 83537  06424 67152  03265 06419  46169 05058  70338 40675   50090 95217  94078 38084  61477 89114  52460 06851  18181 05131
7074:  20765 06985  41942 66971  77599 16969  63671 90017  42558 94572   57676 70142  05809 84054  84529 84312  43100 18710  49097 04612
7075:  15017 49911  94675 59587  33792 96831  78444 49811  40624 14118   71649 76219  23208 08388  22343 81779  26396 93322  84545 81536
7076:  30102 33262  90493 31934  37566 89086  03796 79145  07287 07542   76998 05836  37415 00917  30869 02100  61202 50336  74721 27886
7077:  54738 69250  85177 62380  81957 10003  89854 93437  24519 03346   70074 97560  96934 18221  91217 85059  47363 96883  06878 64690
7078:  17869 75368  16908 55992  04516 99078  15035 63413  61213 45470   29588 39421  47637 21573  77319 82373  46127 95167  60778 55258
7079:  62307 01251  45078 27999  93679 08381  35384 03485  21403 27520   58779 12909  64699 36774  75740 20872  50330 87139  03705 78577
7080:  99467 23677  14545 32504  93416 13243  37490 33431  73840 56408   08017 79452  09717 57941  55199 10632  32620 55416  76785 80329
7081:  65162 85883  14981 83016  61044 91830  29423 82381  46677 71924   83593 62516  80840 16751  79945 31186  76704 90646  42565 81388
7082:  10739 98500  78827 07499  21386 36878  93234 82934  98410 25432   85157 93223  00227 56118  48414 38928  67256 96043  72352 84390
7083:  75969 11227  16887 88419  59102 43209  69909 40385  36997 01489   98162 86520  88319 07693  26245 24395  25036 07456  30798 17261
7084:  05756 20601  50120 94992  92472 29681  82302 92345  99268 30494   34930 10953  37796 09463  50828 40872  76080 13267  98549 99500
7085:  27793 81846  29535 69912  92080 05132  35665 71475  82611 50564   17221 70481  49383 21690  93851 44452  68125 95928  55127 93776
7086:  72288 55048  65498 63084  45545 25015  55807 17675  77190 27096   13763 12495  71818 89967  25648 54992  74667 56276  36302 55157
7087:  86528 62451  97921 14875  29364 10164  22770 26023  11081 27610   87223 04137  25615 19875  67264 20652  58865 84078  10973 93777
7088:  88658 34064  21833 60365  50468 03856  78233 22159  23638 36476   03077 67226  96964 43584  53714 60205  11600 61155  76631 06944
7089:  29740 40760  28135 24892  93955 33279  06468 81073  79778 84300   96733 55224  98324 12471  33417 56206  94646 84171  26397 88080
7090:  50425 61832  44635 10276  82236 22208  55879 05774  13527 86297   82682 74937  91475 93244  13018 49660  55121 94450  79245 17596
7091:  30008 94935  70690 11938  65923 64365  72090 57548  59411 84535   48741 11963  22489 08352  78830 58527  39550 79339  98920 67939
7092:  44298 59412  81581 40674  82138 89340  65187 95970  97905 61188   07732 85475  66337 40880  07261 29350  09885 80502  06867 06670
7093:  60924 16964  30051 45437  92344 04912  29854 36302  97072 84726   16746 81721  41416 31093  34572 29961  32736 00580  52445 88004
7094:  69308 51498  64947 34348  40031 68329  80666 03197  57973 09820   09669 15786  26459 04526  78299 22614  75365 64794  22733 24071
7095:  02795 44674  97818 80213  00876 39362  63331 05621  61568 10084   43025 09371  06693 99414  29494 06614  13760 23827  36716 02780
7096:  33394 60850  91399 99280  14084 72538  71115 28857  20992 31662   72579 69929  33804 68367  90088 23851  33916 11833  84507 85405
7097:  93780 93493  94440 39028  34175 15463  50586 56599  28888 36759   31733 87058  82562 62373  35574 57374  98798 78869  81632 06426
7098:  63351 07978  51631 19802  19121 18035  26608 72679  76240 52958   92808 53463  33818 25462  72901 55684  41284 72035  67186 90831
7099:  25577 95715  60928 51243  76049 95902  51151 59427  84921 60954   13637 89037  79588 69166  70304 44040  44483 50257  85955 63168
```

```
7100: 81495 26223   44161 93975   87211 25917   30148 68315   16697 66519     69044 40148   86362 20437   09989 03324   52546 89460   35513 76131
7101: 46246 22414   32556 81724   34256 79908   07942 93131   46170 21025     93228 12709   09653 61110   41971 79213   14814 60815   16925 60096
7102: 60446 26530   06138 82044   74976 67689   96226 46055   62175 28688     93327 11989   57753 65520   39077 11305   60006 21411   99724 74970
7103: 47717 59789   97892 69896   31045 48820   00306 28573   29001 80560     47125 71722   04135 33460   22876 67805   51473 01132   81421 63416
7104: 58999 75808   42130 40324   48871 14000   90114 84400   52928 28612     41395 06993   94637 44297   85521 47609   08097 20131   57582 57215
7105: 41625 16426   40833 39463   81317 42530   75912 12693   18585 55352     04858 13835   82356 46466   33876 23912   50140 14216   35364 73290
7106: 99717 97107   84904 18535   88886 05203   84905 15099   88164 66187     90990 22260   73248 24042   20109 15836   93377 74076   03755 34457
7107: 48866 53857   41859 74451   70234 54610   45446 41177   75732 14596     95762 53708   68205 08444   34723 19427   58749 60959   32603 67697
7108: 82239 32021   20350 66015   91384 88846   40182 25504   89191 61266     29530 31877   08622 40331   74243 58670   15261 06045   81278 75787
7109: 50475 17907   69408 55179   80747 09115   38746 21422   77326 92385     98035 22802   55476 51190   04530 76122   73894 48290   45428 29116
7110: 14153 56500   82935 89998   41308 75793   42499 52568   29615 72579     57542 38381   08518 67052   18654 39173   17187 37766   78508 09534
7111: 41910 93059   84687 36432   47603 18184   78351 54169   96205 33154     49768 40755   13858 65083   84712 57316   73800 66551   82643 94645
7112: 61847 07666   45162 42533   46112 18831   55168 37455   19813 04200     14071 44216   59409 04354   48679 07836   75163 95751   36703 24172
7113: 39225 95511   76985 65002   14044 38086   81560 06048   30415 98426     88471 71751   64724 81629   81347 71649   63531 37256   81071 89842
7114: 88944 32035   77797 07604   00553 98507   51454 24710   43667 43802     22992 65058   49357 41776   96914 28846   14008 27027   66184 91180
7115: 07832 67933   22314 34023   63113 93967   43773 13166   89739 28998     73481 31730   12094 96806   13605 57104   77323 18705   33914 00256
7116: 07468 86156   94769 93921   69863 21999   02788 84491   93257 82592     51903 36801   86321 98158   56563 89372   67302 30588   35401 55336
7117: 73856 62862   49389 51010   53566 60603   09370 15912   97625 91057     07820 82346   83880 50441   17587 12877   07634 47241   11294 03102
7118: 47329 22258   60396 00373   11715 27234   51527 58835   64333 03704     14819 20713   55883 85382   25113 57623   05553 69102   44018 35211
7119: 03654 28479   06940 19265   59987 59597   54891 99220   93229 38675     81520 10405   15987 57921   98331 73070   97715 00267   64500 57353
7120: 66771 79930   42873 22081   56052 55232   12658 05569   26320 37945     62084 69917   46790 52983   79032 07896   90286 89067   35228 80719
7121: 97061 80025   39125 03400   25774 75209   53569 54732   10698 99181     35776 17398   74622 17260   79558 56619   35133 77742   10411 69944
7122: 73274 07578   02428 31718   60730 19900   41750 12025   13134 83959     09007 61535   49749 54525   70319 91286   82435 38907   81119 99905
7123: 80925 74261   17786 13535   57825 42728   52454 13090   80813 24144     81819 04154   21768 38871   43273 64361   03438 31733   26627 77616
7124: 33043 07618   49757 64961   23322 10886   14350 17175   37827 76185     79954 05390   26023 11521   09307 92350   33832 18745   86242 73859
7125: 47343 56619   28487 30850   78232 25588   03513 52457   13580 75796     11476 83837   99425 83848   44679 98887   87673 81362   43128 77386
7126: 66272 26354   27466 71319   29548 87780   18680 64639   07114 83008     97520 76155   19208 88811   87198 65011   77997 87627   58677 93431
7127: 86869 02340   82240 01083   00732 90517   12055 79822   30702 78615     02996 35030   52631 36504   15138 75013   73569 24631   17442 23009
7128: 45174 44504   30888 80087   20852 04102   28664 48484   06589 68710     75557 97572   02925 58502   56230 83101   26925 82974   93339 09119
7129: 89908 18758   21892 27808   50052 72982   01638 63451   10761 64960     36417 03465   31575 19658   87924 98569   49478 70961   78645 43875
7130: 37796 81479   86566 63499   65419 96535   18459 83648   99684 30498     87666 99210   31767 06594   58098 80279   43696 47979   89213 72120
7131: 88739 28611   12750 03555   87052 76793   52000 75537   43181 76659     02416 12233   82552 99146   35059 10133   08469 92463   65924 93901
7132: 49503 01583   92017 63031   89739 26071   05970 88601   67190 20124     89208 22014   72684 87836   78648 24572   88990 43639   17788 69570
7133: 31669 65476   88551 34908   96992 37109   52818 38794   38612 60942     42868 27692   86780 41549   34656 46588   82815 31929   30721 98157
7134: 21947 38977   36733 28303   50553 03371   46769 08464   30443 56149     88484 73894   85111 88825   47675 37967   40112 05625   89670 77639
7135: 04676 90436   99020 69849   39570 41775   73737 67660   77946 97482     17283 43458   04266 23205   91224 12163   92229 05765   22848 17543
7136: 11721 14795   20960 02212   57375 92009   96626 57764   67984 42162     95649 81559   76651 74309   33457 50576   32565 21389   62223 16184
7137: 42528 36031   30359 20236   82645 83126   22751 66145   04773 16823     98544 53276   55109 48998   14648 51526   34541 60239   75100 75599
7138: 78452 78306   99861 05098   51581 19237   21452 38308   29125 65234     04753 52657   62346 99279   10371 34312   84871 25681   55165 36670
7139: 92008 28844   27829 61287   80925 13894   85652 66673   43321 93816     94681 79156   06291 36724   59379 48127   39946 06259   38053 34012
7140: 36784 44203   87620 33752   14061 82894   19905 86358   05087 54568     08825 52776   50747 20024   60086 83153   58259 47278   17848 29199
7141: 00123 81420   46636 71840   34883 31502   16213 72947   18702 13442     32559 82763   43381 71483   46510 79171   62693 67865   21800 04302
7142: 80880 78623   21844 93850   08268 23153   02837 84389   12938 41600     72255 53329   44119 63384   71613 94922   14244 25126   54770 33063
7143: 39941 76024   83155 22090   04009 50136   35222 16069   17594 61778     48927 94813   16038 55700   54389 78640   25977 89706   15964 93659
7144: 33368 83820   35409 22639   79313 16663   04872 93614   85897 95434     62153 58457   87079 58781   65820 75634   09807 10775   65305 71153
7145: 91091 96773   10923 27927   22510 90287   25947 99584   75212 25055     51259 70963   71323 91924   10811 78738   64311 92313   26107 70998
7146: 47564 68783   92452 20724   52754 58185   04362 59409   55196 98636     03655 77010   87373 37405   38076 81260   36190 92256   86907 78768
7147: 95365 32236   86000 74289   34494 89395   31039 10273   57564 98010     05210 50903   04814 33940   25750 17272   43990 06105   78625 14698
7148: 17980 44296   65766 15880   11649 06854   01558 55784   82577 53813     00442 46875   48551 76566   34915 29261   59214 57777   52670 81335
7149: 66485 34952   49561 66554   25651 70022   95227 56817   09460 31842     77498 08919   60342 37587   97945 73604   89818 30167   94944 52009
```

```
7150: 27246 96852   82229 25346   17342 31428   09715 45155   77805 09512      12318 49067   54711 83080   57813 63947   12930 65814   33290 13625
7151: 46552 62156   91264 74204   99686 62287   08989 97405   83110 02160      90145 52001   54275 97139   90148 96181   53670 79308   22012 86880
7152: 35725 16848   26427 84681   23123 66307   19640 02213   01749 34456      00399 61951   31005 48375   63683 28519   69451 97595   94809 05781
7153: 75768 35633   40405 66472   68742 59914   78892 08341   68281 62206      48948 50370   42436 55731   27751 31561   48595 55598   05315 45907
7154: 27523 24845   25588 04532   58432 18444   26239 30539   75311 28023      85471 24441   28472 47137   34515 76684   37872 06246   70822 63419
7155: 22459 59230   67683 94450   51105 30797   20936 50040   02689 02037      62989 78580   91443 61724   08073 46760   66266 37097   19408 97956
7156: 00890 22699   38183 53288   19962 31233   25689 99043   01435 29052      58392 67692   12436 20700   87658 24390   32271 87004   97779 01716
7157: 98552 42247   16599 55835   90411 98167   69798 72765   10616 61352      04636 05064   89741 47859   52459 24340   75225 45806   79513 05271
7158: 97778 35326   21279 06770   36562 55237   22165 94397   88296 63233      68591 96866   50112 87123   47179 44040   91550 18371   75022 39006
7159: 83232 42084   65558 61929   96018 11662   51205 88159   71222 16490      69626 21582   89347 91246   81585 70458   50326 31448   25745 28189
7160: 27517 92335   78380 70137   60278 02368   15295 78027   94771 79209      36112 21681   35295 97356   19596 20457   27843 41745   25075 31945
7161: 25855 67613   27873 96556   47782 92113   73289 40147   72259 80092      15205 33328   05674 53003   17724 03233   56898 05473   51432 26817
7162: 53966 29809   39397 92719   59082 67648   34632 25438   64231 17478      38858 39756   28331 62208   56935 31067   15769 93688   05845 48987
7163: 59596 46827   09101 71941   89403 25185   91115 37515   72532 54411      70707 06623   69506 03355   30117 58246   18590 95074   81287 22865
7164: 62411 42948   48986 88502   86255 15270   38729 41157   84636 96076      00249 84054   08060 04858   77440 62893   04428 89068   04415 15472
7165: 03592 95988   59213 26927   91226 15131   60810 79278   39243 68181      27269 07675   60700 24448   51270 98740   98736 06777   77737 49860
7166: 02416 66256   00240 40656   12472 81621   08772 72104   43243 73323      89493 69060   02929 04488   90706 50962   20385 84994   08829 28488
7167: 83731 37171   21530 00764   21915 11492   00990 32818   56874 19315      62699 85995   87072 15245   49399 38063   59798 72604   64429 10235
7168: 81510 86708   22007 58106   63381 55554   55245 67159   05481 83756      21399 20301   49575 31543   06714 40185   82794 86660   71818 15088
7169: 49669 64571   38192 26421   58647 79089   05544 73551   01719 83833      98341 94707   97047 69034   65002 97845   44024 98971   57744 55796
7170: 41978 44416   61958 49417   93540 18578   27362 85965   57057 95570      82209 20364   35214 52767   30376 21778   75456 87201   88614 74217
7171: 79086 11206   59402 07290   58172 47136   71509 49716   47407 96403      26493 52598   57626 95904   74121 47305   29074 59798   19343 16716
7172: 88246 20866   91694 52248   39558 32586   61609 85529   84383 80201      65858 73909   48698 16365   41859 43435   68245 44762   72008 84139
7173: 05718 44692   24471 19583   98426 66643   33711 18529   78450 80787      86371 00332   11592 57690   99399 15698   21775 98517   29161 63203
7174: 84918 06647   22705 84185   48278 41857   36230 00333   67978 25816      44097 59653   88608 71816   68982 32796   60431 70883   14107 20856
7175: 21949 80451   60243 16943   20765 47565   54075 85685   34321 20610      60559 16633   89207 94628   15384 39920   94582 75802   07662 06768
7176: 02376 43358   32835 36815   77348 34687   12206 85948   93838 23075      78575 33764   28923 05163   39784 10348   79249 70859   68253 80310
7177: 29542 08210   11908 39402   28189 68546   13837 81123   10278 36385      32602 99643   98640 71777   86876 49729   00323 73702   86575 65346
7178: 58185 36042   51923 58945   14428 23281   50456 16152   48995 39519      01036 04891   63773 11825   01756 32959   05701 40389   24330 59909
7179: 40027 36260   83809 60942   58545 76936   67361 88435   94183 49295      16463 19946   08092 94375   78028 46603   16813 74435   22875 64851
7180: 65396 02345   96165 61686   62374 89128   74196 29574   25729 52599      68248 68709   80374 75968   78157 15724   94219 57810   88198 69738
7181: 93189 41771   07733 00723   12160 22861   55122 52090   75600 30977      69533 89985   26267 51889   11055 60995   68648 31253   97318 78221
7182: 96692 33132   47625 93893   81388 81480   68553 89502   91040 72194      42162 12339   36280 54085   33275 36537   22772 96430   07690 53355
7183: 68466 35579   61794 38137   69447 92429   46462 70470   66251 30909      53483 78923   14319 63477   85460 13764   38909 60085   80420 83347
7184: 23465 34610   23343 78332   71059 33980   53280 27684   34290 25402      75234 64367   53443 87192   81947 67415   36307 30929   82273 47575
7185: 61565 48746   18148 66484   48360 09031   70154 90863   57666 68621      83241 54375   08981 22087   89672 24806   24778 61360   56976 13559
7186: 17668 30673   49978 16742   90169 86454   36584 32580   19427 30736      62844 45602   92801 72947   24125 81396   21377 53622   76252 05775
7187: 72570 06712   59006 77201   51102 04163   93129 21155   08980 98905      66135 81877   74737 43740   30360 62069   96123 44846   27295 15836
7188: 38388 36045   23758 67504   53943 86059   79800 83102   68388 17880      10782 93618   37107 52429   21437 46841   92330 28308   88559 63919
7189: 36911 19595   05738 76803   47773 70442   37381 88930   41604 69370      37491 18411   84252 69553   22889 41686   78487 66270   96158 77697
7190: 82312 43987   71746 60108   71487 48675   39889 69382   97795 28297      43034 03743   46724 18084   92419 11818   37994 04593   59354 94025
7191: 85997 74430   15994 88095   32390 34647   04395 52437   73046 84924      73756 73395   08183 46380   52190 06258   19444 03552   81161 46603
7192: 99586 23870   18431 58511   01065 60184   96012 58073   20602 41319      77874 77973   36613 45861   11370 48317   59065 65697   73895 06897
7193: 66854 71864   76062 82435   00145 56163   19743 57094   17403 90181      11493 01927   09079 99939   29133 04745   52779 64872   55331 49303 6
7194: 04759 63920   80842 01944   66388 45902   64543 28222   84618 19354      40404 07814   19953 56898   74018 55254   66796 27064   83220 70984
7195: 75973 60820   68438 05445   13633 29769   03125 32413   76495 56363      01442 85598   35778 87156   42570 71702   95612 95851   88781 91637
7196: 37419 71390   93708 27948   56726 97606   09535 89475   30633 51002      34370 03842   33057 74422   65314 43495   22823 69807   98890 47112
7197: 54898 14690   97904 33380   10185 56849   10960 06315   55682 67971      19634 16279   07875 93066   94441 32319   23487 52797   63597 39977
7198: 94724 54455   40045 02665   13607 06320   63843 76344   48240 56902      84566 51100   05027 10348   16850 91534   04977 83856   43066 13579
7199: 79941 10660   36665 68610   07563 00495   85062 29426   69356 70196      56891 72628   25556 32579   08127 80434   27927 13561   21069 42780
```

```
7200: 35597 49572  59207 94788  48991 82380  08598 31986  98553 40989    30613 32274  10365 71407  31169 04059  11041 45669  27489 89592
7201: 75983 56515  86071 97894  53429 85684  78700 22537  17546 72861    13784 86914  52254 45875  19986 23843  29525 44292  36892 95809
7202: 11533 94272  93995 79904  48792 56925  50634 99972  77685 32945    83449 59956  86500 49342  59206 27581  96577 35671  60533 39611
7203: 74746 05516  79834 14970  18587 16360  04159 90562  46206 43181    54784 41805  17643 43440  66042 87133  81451 04615  10246 73660
7204: 00280 92123  27249 31474  30868 86889  05135 66075  99796 89611    57928 49266  63223 98317  73534 94450  53636 42349  34252 27027
7205: 13735 35862  49731 53717  89162 18584  68580 14562  41196 58554    39651 24657  46553 69444  71333 47171  46012 40094  21398 97600
7206: 51529 31882  61437 38368  19733 59834  42143 89748  53639 20588    53013 70372  21884 82069  09174 93810  70148 73065  25009 70452
7207: 14776 61882  83875 16453  32713 91703  26524 24817  41837 78145    03417 53864  59862 73828  21014 15749  82305 11413  93002 99439
7208: 28475 35055  66269 47984  65170 48568  97077 86022  97274 95841    57039 61824  88451 58241  06083 43504  59746 25116  73387 84200
7209: 01600 75994  73583 11718  18622 66068  05314 68929  03459 41552    75041 89089  82869 82262  43224 46655  41459 03105  56982 68305
7210: 80818 25225  05309 48147  92244 97529  31262 59027  89668 82592    65433 04140  04504 77071  93793 47715  12908 19856  96630 28284
7211: 67090 47745  22469 99548  34186 17773  19213 58400  88189 25291    13865 64698  91390 99006  56981 45667  89002 76810  44846 30940
7212: 06190 80754  08834 61373  78260 44840  55910 80747  18756 29597    91314 85514  03402 03658  49384 07588  39536 85180  65291 85018
7213: 26557 52983  23629 64563  50568 69676  86383 57668  27449 84488    84365 43838  21807 69767  91867 86469  08492 78559  66615 87153
7214: 90073 72114  70423 74853  62256 70202  37882 07909  61350 51190    80693 93763  34291 04569  85198 04414  88219 05443  66984 83415
7215: 47528 82360  34921 27193  06363 11124  17808 22823  74170 49203    76388 92458  75547 62702  25848 59737  13661 54490  47426 22963
7216: 57866 25959  26873 89899  87785 35880  77661 81823  76208 26119    92648 44005  84260 54473  89565 92052  53044 52050  15169 79102
7217: 81428 17168  74930 76817  91334 71522  76710 15837  26291 22971    51567 26664  64302 66176  45822 11465  98975 96585  90879 96407
7218: 30588 59728  93676 85643  37370 44710  47291 36017  85113 27921    88006 09491  11899 73401  30543 11177  33187 19524  14058 15633
7219: 02990 80590  02748 66024  33418 21751  11067 94572  18296 56435    04288 21425  23301 46909  51930 68235  05189 61436  72426 09477
7220: 19947 20811  46077 92240  93124 53459  97654 66497  70164 67560    16791 55640  97214 03096  88199 29489  14248 47226  85194 55270
7221: 45097 36503  81517 08052  94635 99223  90342 23486  77356 28308    29631 68018  65827 32530  73252 06900  48399 14180  76001 68270
7222: 49527 76270  54067 09001  52465 15849  66457 03688  87904 33987    79737 80858  57964 76284  15439 97277  34518 81381  90843 93470
7223: 48224 47190  42960 82622  97224 26231  31017 31181  68379 11360    82021 27538  96780 60975  92668 33175  06453 58605  54617 73723
7224: 08526 62249  53155 69115  63288 98361  74865 90492  31153 61818    96402 95568  18227 20618  05743 50783  88963 80309  49358 82232
7225: 11152 37145  85045 29171  80791 02034  21479 19961  12037 14114    83691 54666  80584 73977  86781 47866  12784 64061  05415 57735
7226: 43024 17676  64416 84678  07831 34056  64030 17709  31874 16502    84557 20333  17664 25605  08891 15516  31415 68616  05067 67221
7227: 55351 46581  90658 56866  65340 70518  94404 80503  85598 47488    02478 73609  60164 73167  02134 75469  21357 00329  52565 72175
7228: 02170 75218  24950 58371  26587 61136  34253 49984  02048 88990    23726 99142  24386 10985  10402 82841  05288 68901  15080 77286
7229: 87191 33991  43778 93767  13794 88858  05066 99020  63227 03336    26029 69260  94584 17890  35925 43000  36514 14631  71776 72560
7230: 05375 07038  22223 42656  27213 27008  19696 00595  08058 37122    92334 31417  51017 59547  30721 15329  45680 36967  05429 41980
7231: 85669 14767  25350 76744  65340 36582  22003 65542  76572 54422    20046 24506  20657 29720  36333 00979  51125 70792  81961 06808
7232: 74224 00359  62212 72696  04437 90073  02946 44002  82697 89273    83899 66100  26145 02216  03430 96309  23153 96676  40750 02338
7233: 55805 79364  79193 70021  77698 31954  92022 49058  77148 77891    20360 13388  56577 61863  98813 95662  74668 15869  80565 35965
7234: 40910 76890  00622 80346  14867 96680  02647 72792  75974 46560    83978 83472  49200 00918  22150 50966  73277 85449  17013 56384
7235: 44377 42063  97066 10464  72496 50132  69671 55809  99116 15459    99365 35979  13608 42091  99825 75957  54725 58431  69406 36637
7236: 71404 40629  82552 01287  58530 74916  08236 61381  59779 81981    82941 94558  57528 93008  71139 25426  79933 66816  19525 65884
7237: 22868 06130  05956 85900  52684 04832  94593 70275  47435 80116    14146 11770  96771 70469  01792 97352  76451 93009  42417 51693
7238: 58808 82024  47654 35913  82768 18442  34972 24321  12289 89547    79133 30116  78137 87181  61988 62915  09032 22122  69075 65967
7239: 35863 49186  38624 56968  62284 97769  63320 07608  92459 17264    86291 76393  51338 57876  44850 72426  41826 27106  24447 99588
7240: 59222 72733  47005 28010  12516 75755  34962 48190  86733 39481    75445 98652  12462 34642  53274 22936  43549 64988  50297 42443
7241: 07953 33079  48768 89512  16811 01895  50279 57264  49264 73276    34548 80443  52305 45023  12776 61518  95089 99592  70486 41657
7242: 98913 96972  31027 62765  22448 65687  01639 04569  30592 97558    00363 35911  48512 58618  64490 79439  58421 78769  09741 07567
7243: 76245 33652  31435 58747  14835 70230  93485 45042  33762 69665    58776 14290  70865 27966  82012 96612  60977 28730  57370 50425
7244: 12101 00485  06834 66429  58311 59694  50909 20204  14341 68622    29672 55213  21824 02251  51103 72761  50867 49306  87157 88685
7245: 04531 60038  48279 94395  31614 65830  54523 93441  09021 19575    85733 40519  31650 94230  78472 60827  90040 57578  43078 42867
7246: 08530 38089  49018 63468  77816 92846  41353 48548  01711 23768    56829 29844  24628 37026  59233 96030  69925 64522  60057 94744
7247: 09302 89714  99400 92593  75329 66545  66664 20773  88908 09043    42185 18971  41885 53526  67859 80176  67117 14728  84920 88187
7248: 84632 38341  04831 22746  19730 06730  45393 28115  63161 98654    79070 29433  36035 50013  28821 51300  58930 09670  07308 63381
7249: 30013 78664  62965 62386  37426 90257  86460 30764  30076 98203    94085 60033  96168 46774  15320 65597  61957 70573  75051 14272
```

```
7250: 0408182715  2323658058  8968305441  4990607686  7380834594    1120541552  7506529473  1978398101  5044707158  1799475592
7251: 8883718381  6443813066  2581724255  6392266177  7985288433    6380793798  1262014261  2022178930  9866499212  4600731525
7252: 7611546516  6354601921  6395066782  5807540502  0799015534    3780481781  4016011438  1468535337  3066180879  7601376898
7253: 2365864349  3183583384  7464910850  2661112721  1748861155    3817004289  8725704509  3758165301  4877693476  0809860759
7254: 2221169031  4453327005  4336416708  4011923370  9324977714    6751564771  6130842391  7193872320  4722486869  7167170822
7255: 1595667405  4810230495  1180680696  7148119789  9581342670    5445348971  4239843776  1497377507  7016528223  6564661241
7256: 8479211673  9277029962  3920654496  5645361221  3371454956    0467565879  2466615123  1562534700  8259083951  8738234366
7257: 7771753324  6149434488  0206704636  2963472870  9556328555    3654629417  2385373342  5827761156  9230303172  2941351740
7258: 9175981833  7717704891  6761288059  4775681155  4217376865    5236263959  6038146949  5437390366  9681030542  2984136948
7259: 5714174950  6510434012  1124155985  1863551175  7612044137    0628123235  3889851699  2523650018  7118734279  8481028326
7260: 3584188459  9770399264  9372405661  1295899455  4551600692    3335947176  1060450744  5871366621  5613219200  0088566175
7261: 0654745640  9135140348  8058584901  1225238660  3945612883    6106271526  4802026559  1382597084  2125956027  3021976639
7262: 4733313150  8272026826  8521701088  0859456137  7859234627    2430219920  2973622480  4313634254  3373919052  6317022138
7263: 5774668638  7430060211  5768384435  3736253066  9249315783    8483742350  6089214732  7132748731  2193281959  7656511045
7264: 4597156639  4186056767  8105972628  9255657722  7888884235    1478847553  3169372423  7444452714  9698082761  8534246522
7265: 3870926452  8795951533  3314653434  7069590952  6201009782    5986738280  2608390648  4853089551  2892408194  9538353652
7266: 2215618776  2314081911  3688998330  1705473922  1480516120    9260755295  2641086873  6724086770  3847916061  0852566912
7267: 3008644335  8670176831  5058807434  1140011759  7255465781    0914814843  4406370061  6029175495  8499557374  0453083006
7268: 7589348662  6670869553  9182599369  9530390235  2334052924    8682268273  5125636018  7029646919  7780630124  5504277754
7269: 1607706413  7005770820  3220527108  6136430036  9759854663    3700173143  8861414610  6124448444  8075114750  3951406799
7270: 0452977904  3437190547  2075496314  5945287615  4385048429    9935470559  9668265411  9733369080  6657263969  9025471244
7271: 9406357917  9220657704  9592944873  2622272603  3896698625    1076586683  3401984871  7723641178  3022792098  9223394240
7272: 7516843605  6296775282  9841862972  9287281095  0071178899    7244253636  3443270264  2032245819  7314737534  0244293638
7273: 5233552043  7949840926  7664656121  0875302642  7781447374    3077864585  6652635021  5548928414  2767135009  7515833368
7274: 5549828484  8359328553  7208169977  3926186697  0898199336    1611681399  3548535466  5388411957  8473642141  8356235030
7275: 6205752713  0482100617  6488131803  7664624309  0742397354    6205715571  2939025125  0884658568  6017852534  1644413269
7276: 3952127736  3237092528  4732848843  3863373345  6064646999    8178510097  0301473984  4693128178  7210382655  0296679043
7277: 4102126848  3007334721  4947262840  9916254854  8058962074    6890479074  0937199916  3325852278  4315111563  5258933365
7278: 8515147669  4196136419  6604774403  7606821086  3092286167    7637659770  3412008468  6326134374  7781789402  0955916521
7279: 4026464394  8309017880  9317952563  8008967163  1832907699    1231293386  5270885985  0431338008  5647507515  9893265703
7280: 6718766364  9662938933  6889240087  9908879102  6222472990    3770844970  0615255730  1370063263  2479023130  4659837892
7281: 3337045472  4674452626  1431387963  2866461925  2305113667    1454778593  0025453107  6694696878  2483658790  2333601884
7282: 5818316662  6431397005  2837279536  7930272602  2224143230    2825403508  2598816628  7661346595  8580335686  4761392868
7283: 2796448976  1079156559  3649663632  5841172689  9984986885    1113774038  6836965131  2334076421  3596160944  1404281517
7284: 6566461732  9390960201  3818826275  7591732608  7520171558    5426662364  8394870896  4243319666  4219642694  7706975881
7285: 1014841969  5071131959  4090058209  1485929503  8748930759    5477409329  9376012970  4736309604  3820855352  2208414468
7286: 1286730370  7952676499  5034726626  5745630975  8118229338    1533283912  4587754495  9568492110  1947680400  5829232048
7287: 4363356812  4350587217  0994814496  4253162962  4805328005    4153793735  6432722271  6283793768  8713607868  0234347883
7288: 5439830828  0534750915  6714618837  3864166165  1638575790    0354666065  8815911037  0879867737  5484997388  3263973086
7289: 5017533773  6812253322  1283521266  9477202766  8072980667    6170767219  6843316908  9531352920  3076143982  8114660423
7290: 7699380652  4073511869  7892026899  4888487293  2183520250    3599047031  6712588999  8114510126  9513845635  0425551510
7291: 6811440916  5474070568  7485443320  4159279847  1486182684    9231186383  4940868638  2458751906  6078119347  4226250498
7292: 2097188205  9347265827  0979674805  0779870727  9360490925    1335565098  5074835134  9589347798  7702514891  6541123988
7293: 3682208401  9829027757  4671797467  8605444212  0769823292    3459609548  9212133214  6323570659  2158757616  6443163211
7294: 6170568552  6692531626  6080395633  1174136666  8322403460    8558823637  3367384390  1238403415  6871742625  3570461882
7295: 0053228620  4254661520  6104097914  8320937712  2919405091    9258248310  8118464394  6473238534  4495237051  2978716142
7296: 5817587648  4608430339  0795395480  0061109252  9204063060    3024668103  9529895292  1999382985  4460363378  8442077375
7297: 4588379595  3373764708  7137582076  4842330815  7531928946    3154395695  4381406879  7979106319  6281613183  0777073742
7298: 8481079323  1127445433  2813422067  9698602343  3021344883    0044243823  6807609220  4937751485  4623728919  0170605084
7299: 0599254501  8219137180  4583363557  0286279607  0747467131    9772335155  0619409148  7198716668  2855721859  6845496646
```

```
7300: 17895 37007  27323 63753  30248 62173  59748 44179  37028 01133    85251 14634  71406 21412  30922 94316  03566 61019  89212 60509
7301: 74698 19282  95089 70003  68874 09290  13701 89817  30320 27486    49397 29312  58360 52196  18735 99037  21071 51367  05954 13134
7302: 50205 46515  72897 24100  82759 62392  14324 73916  66906 13014    45138 71980  97501 52574  55178 17988  52916 05860  71128 28879
7303: 99897 68456  31771 35845  35339 23506  79098 34479  05337 01589    93232 71922  88756 17382  93308 35055  24099 39998  10487 99682
7304: 60371 11571  04020 18627  22117 87700  29036 98355  86265 09594    45750 81238  97311 48771  61329 47961  09574 24347  25490 73077
7305: 74018 01949  21653 30306  33587 71867  83667 78700  28969 31370    56812 11001  89049 90624  60711 84228  50716 08005  96636 05979
7306: 28600 66603  66586 00896  96477 89768  75274 94586  59291 19105    28151 65842  71671 99146  85455 75339  22832 46517  68968 17533
7307: 90547 82128  95860 53617  74624 54759  35277 40338  98628 60982    72142 01759  03441 23390  48900 84753  16759 97853  22854 37420
7308: 09348 51379  86134 92098  45594 44680  80652 96283  04995 01191    59616 37177  56207 91874  00666 60857  29311 15976  81490 35057
7309: 48975 71267  11043 17226  43637 19914  30962 05615  67878 93946    92762 87560  41798 94087  02736 68833  78469 07546  55954 03415
7310: 51066 06027  84195 99965  78556 23867  50646 14182  13132 38608    98576 96265  64178 83732  53673 67109  79999 69536  41482 12044
7311: 86805 45159  14589 59851  78654 23610  98713 63430  26438 64486    33174 91943  04726 62459  89771 54514  02106 41749  47697 08809
7312: 47913 88395  08400 69779  85939 67751  83624 33339  46782 82253    91746 46741  38161 31149  39497 10755  36526 00401  09784 98144
7313: 01035 39643  26624 65460  59062 87456  01775 38935  25016 30924    64370 65438  01820 23345  72571 33845  17219 83600  64129 56978
7314: 88649 69309  59095 04899  42172 21233  72393 54350  06731 86314    45826 64944  63672 96335  12412 78895  80413 89274  91029 23645
7315: 07868 96248  17797 66175  88885 02307  51085 50041  95455 78114    50123 27173  95681 08884  82056 07220  64915 62593  41764 69382
7316: 88807 71819  44913 62608  00500 71484  42227 81860  39291 17593    64783 82737  41539 93318  12662 92141  20878 16198  57073 36626
7317: 46063 07239  16474 74513  74604 48762  30893 37742  89982 28054    82506 52573  32062 62651  18733 82770  46582 62253  38587 16670
7318: 09288 08735  90741 98675  53379 17569  40301 30050  83818 65136    35896 70006  96985 00717  41989 67863  89670 31863  97635 58508
7319: 44407 63676  45030 00604  35020 85569  62753 05356  55892 00059    92983 28433  22851 48245  64413 02017  66848 70846  55655 16956
7320: 32766 21624  30704 98777  58323 21582  08318 13826  25306 14694    47617 00433  25103 45649  49239 95892  51372 41547  36744 05275
7321: 29466 78232  06442 13616  11933 92187  13674 00403  40873 58509    55440 09010  08506 79300  88455 46252  49476 49356  52939 75836
7322: 13949 93564  34237 40091  35671 71592  56475 69558  98880 48896    86161 11772  12100 85549  41844 08115  24940 88032  75248 35730
7323: 34545 47610  89467 96719  25453 44386  70646 89607  67404 15455    99761 73844  84258 05774  03299 61652  37647 41922  59920 40892
7324: 04825 55051  98360 64592  27134 06956  13962 27607  48448 12823    22923 60363  96558 37898  76527 56892  02795 69306  27909 74919
7325: 18891 40686  74229 01732  68823 15516  61375 78801  02648 65133    53160 32940  48793 56534  29615 00939  55799 34829  45994 87370
7326: 20032 35023  03620 58980  75395 77479  36383 60818  95092 30604    36235 84865  87961 08294  25983 11387  94325 09495  56380 17090
7327: 53730 39075  40096 86490  60520 58650  19116 79392  21290 50040    43416 51614  57939 60005  57705 05921  49581 39771  49571 80947
7328: 55824 07579  39881 43323  17418 99080  57273 45027  33772 77216    85736 53398  53568 82356  96947 06356  35771 95791  88490 43728
7329: 99791 53262  30522 60398  02706 72111  33555 36961  62037 03972    54654 94282  35652 38396  57212 77353  88613 73055  01712 52979
7330: 44834 11450  13431 72624  25259 04920  29224 81456  08827 82609    69695 30829  90980 91655  92714 49046  92016 92788  65023 90712
7331: 45781 05590  30180 90988  24964 10894  90649 22166  73124 69861    52032 85045  00070 24813  67970 63419  25201 51650  54441 10472
7332: 53223 94700  31001 80559  13259 50060  40153 63227  36857 31666    17343 52289  21433 97621  47739 43270  28743 24503  89655 51684
7333: 18630 69627  07193 55001  90219 37985  41897 46161  90462 63288    36559 76651  35851 64341  87049 03160  01450 21474  49707 60339
7334: 78066 00139  24836 15201  48482 10320  84641 67631  53584 90457    81546 29038  84779 19810  31684 47069  64635 41962  46175 28239
7335: 53498 32434  66040 96625  31982 39320  38506 31263  47281 13463    34344 41816  09882 69604  54966 30219  79982 93146  99892 00753
7336: 87303 66855  27385 84465  78965 04797  58246 27800  36520 57828    91381 90880  02506 00904  63712 75321  37572 77706  57480 64881
7337: 25326 08203  03358 96960  12419 95236  85781 64073  36210 14122    37201 19728  75105 50558  65741 52938  68413 24378  81031 50627
7338: 16586 07128  70937 02447  51658 22273  54250 17110  00297 28587    10671 75021  00401 81161  01465 39711  94140 38365  85951 15446
7339: 64020 40484  00726 55146  63493 42161  68007 25525  87070 14744    97181 57849  16171 22456  41681 23692  52014 20355  92737 89549
7340: 38308 39316  13273 56028  05802 88090  83891 06906  12359 24691    44541 30065  02135 47868  20821 28639  87753 04445  47412 08077
7341: 58051 65257  61826 70138  65076 71833  26747 49109  17716 35233    88920 39066  62378 34981  14543 05254  90747 83702  37536 67098
7342: 69323 78929  64549 39888  54917 71274  11102 78265  61701 53058    67748 74988  06624 09584  27359 68926  94122 20061  56130 35327
7343: 09994 37650  90429 98308  05979 68231  45763 47863  73382 40012    47511 09535  09971 78352  78199 42035  37591 51103  20122 74590
7344: 92276 29905  40495 67837  15205 11719  93234 91460  77764 81467    55466 88208  94969 91278  38676 14593  97498 01605  59052 27918
7345: 72484 99781  59463 00691  11977 01313  25947 82603  27849 73838    25782 30524  16719 02988  72991 79232  53512 43884  75809 40532
7346: 08872 66388  16288 51561  47799 15949  52076 80138  00539 59306    38389 98291  21884 14273  92690 07837  20150 18270  40549 37586
7347: 63999 31689  26710 97073  78030 06246  83952 64223  75728 53514    76888 26810  35005 00847  06181 49491  36008 42753  19727 06740
7348: 66257 76236  03551 30777  77548 06848  74965 20561  34290 78166    91770 66390  70571 53542  19171 08375  12565 61798  32504 37398
7349: 16413 69824  38880 10927  14233 73451  85578 29080  24955 70868    20001 52466  74536 47298  48670 84464  91773 58215  54559 25508
```

```
7350: 0293601748  6833486774  8042632265  4985119813  4291920419    7140569835  7246574869  3710778749  0090153418  5894060053
7351: 7916258468  2334913045  4878056252  5119353943  8375467567    5497619665  4232326953  6148333311  5290125385  1165489664
7352: 7864519726  0842214267  2394275259  5166481460  5133269368    9920906574  8937326601  5222466690  0614374169  0916329407
7353: 3297484473  8489032098  4322471919  1505134903  0749372320    1089279468  2011562505  6784718117  1155225350  1688072043
7354: 5400200091  2908702527  6516914414  1717955921  1092125737    4207212581  8859424430  4127653809  1893311078  3050946674
7355: 3915719129  8146545876  3737181682  0859256754  3292718854    5596880304  1577828324  4432591629  9677383669  7693914322
7356: 0817977325  4604173565  7583760294  4458152430  6907914928    6981251435  5178128449  0668087791  4237062152  0139461947
7357: 4988154987  4834224831  5760397847  0728615788  4519726708    3685727343  2898313853  9176544311  7535808627  6852213780
7358: 7383086809  7013272282  7002421907  9513497193  1097241748    5053842920  3550059489  6960255437  5779664204  2608503184
7359: 0091168074  9900682562  4146340223  8279550646  5562591740    0693169820  5221698933  0849176080  1509349211  3287688881
7360: 0904945298  2677522725  4383930203  3298001056  4877460229    6370602982  0128872715  6548940383  9856563554  8965996475
7361: 4971397587  9027728870  2950978654  1265019690  2371033280    5752323837  2945114719  2577433264  0521631197  0418392566
7362: 8315164173  3296110841  7165783924  6397255313  9633902596    9057584969  9206824946  9884040899  5060721134  9095676761
7363: 5490779675  4732194942  2687216388  4423847275  6345934028    0901848156  6089088779  5700080358  4106171605  9880458579
7364: 8620431253  5047303200  1091514382  0404968634  8973566490    1674387568  8682168327  0451892690  0340362384  5918957236
7365: 3827876447  7643067524  4318062327  5575042505  6915089314    0434078301  6065746437  4937135245  6412841756  2481234832
7366: 4852302257  9151620852  6873957229  0349049886  5239652935    6402544132  0649339512  9557257918  8635968510  3374903018
7367: 8982844074  5772790431  1370233096  0237394350  8063413246    8389904007  6490286016  6158514611  9572595179  3051194069
7368: 3157676045  8105707302  1267281279  5319718477  9203565867    0657214678  4256727837  6418022592  4224977243  9893334354
7369: 0454605447  9961405728  0653935281  5332114303  2807875431    9873806938  7712463015  0923620107  6322034706  5143377737
7370: 2700557729  1173479398  7515407346  1001291143  7266449526    7653525727  4424008901  2162240815  8796469702  4342209610
7371: 0331986529  7884144878  0910559650  1483665873  7577323484    4958956695  2225095776  4099654173  4057422364  6715602902
7372: 9005240560  2517890699  9316725105  2127512295  7507017681    2311153051  0804223271  3304212236  9875999937  2753926453
7373: 9089434346  7995274187  7304621676  2138927349  9749870432    6742112250  0548775953  5056990819  2687091311  1266220888
7374: 0094984483  7419153396  3079578963  9501857598  4581642722    5146069046  7785052650  5391067485  3065049225  1216800329
7375: 4157958156  7407077210  5288676625  2538936901  6290819755    5250082546  1190334009  1836506247  4180774238  5771784409
7376: 6383137244  0111976202  6750940811  5392618169  1454217012    8002019544  4979084660  6346921090  3472878061  0000877162
7377: 7932081445  0627041460  4704810834  9912605669  4394600860    9270836180  9409669367  9639430157  2231672239  1202421261
7378: 1503868306  3855592273  9794399186  6835709942  4271032074    1151583016  9345711817  9310765167  0212527162  2432630233
7379: 6874356144  4811424665  5801667057  0312329939  5651688679    0036926568  1255373271  6536319012  2670737288  3240698191
7380: 9766344373  4878116256  8061205905  5637726300  9833566312    5295468975  6688086130  8722386726  2270507259  2643451541
7381: 4140775314  8627182121  6506635857  7035268442  5661278172    2641845889  7387485532  1451756722  7903262467  5320574775
7382: 3882629958  6165590683  9925430463  9729007257  7135804119    7614120069  4533645432  4400350214  2241744575  3284260121
7383: 1222835745  1587895648  7014899436  3116987905  9859280763    7599313881  7276346153  9582993909  5565993604  0362022417
7384: 3279861132  3781106011  1894015184  6364733223  8427647600    3275188451  6360979934  5176403950  8049064655  5304288116
7385: 2859160246  0404717756  4016446041  2541183755  4085759083    3749901854  6325127321  8680076486  3498721158  8613145227
7386: 7984891006  4738506828  1534950841  1717514949  0366723957    6565248721  8250354802  6884748478  4859221823  2038375987
7387: 1869332420  7066618001  9523417795  2106118997  4950190830    7272527666  0059676943  0258107626  7437023191  8482895689
7388: 0528854095  4756192911  4931285992  9161435870  9429002251    4030622081  9288908374  5944484165  1332056714  2355746331
7389: 5270872758  0127838644  4819759038  0719386487  9943794799    8958976238  2716042334  0256651856  7375631067  2448436140
7390: 7831149605  2821052432  1139632732  7312551575  0568163767    6917081234  2439330354  3761265739  4676001387  3614053067
7391: 8090436038  1131832639  7539826156  2244933105  9050467542    0712909074  1114229689  2178059192  5902600834  8113833896
7392: 1936863007  1278121005  9518342736  0766808403  1299397393    4968745998  1318582953  6341626902  2343638176  2911699694
7393: 3397733141  2933423290  7772820451  8360646192  6205251071    3655860064  6772277921  6364107152  7765143041  7752769421
7394: 9023299607  8013949214  0220497413  6756389704  1358225234    7970910322  0307556801  8892485189  1900544533  0030125293
7395: 1972270677  1353993242  6115284069  4824074286  0529261875    3438323926  9250668886  8227973502  4139610264  1478156695
7396: 7264772948  9334711178  2802178083  2475367716  9519972374    7000018553  2020302015  8677221810  0356051818  7680354617
7397: 4595790778  9178484823  7600392595  5225348018  8153717286    5242079777  2510078291  1794232577  4309383142  7046337590
7398: 4313684175  2406854879  4814252813  9508442456  2629986838    5826041349  1799447938  1649708235  6393914705  7542865414
7399: 4006204480  1253531585  8984960349  2573108801  8713356346    2526534326  5582081056  4777238095  8296940294  4294309111
```

```
7400:  51789 22597   99348 02333   90020 77648   39249 15589   44596 65410      15517 35043   12576 01422   03310 64487   88098 69486   43639 04781
7401:  36198 27360   31594 83563   63327 54712   03270 82783   17842 89220      66207 62720   20718 80009   47567 84333   19672 71105   31682 78360
7402:  54851 57252   78088 57884   69740 37309   55085 12630   30824 69862      63635 09641   29747 53238   17952 11142   16802 38763   54079 45775
7403:  83933 16888   87706 73432   03100 56391   96159 15707   07521 11622      71426 33818   03771 45381   11162 06139   24751 31959   57767 66517
7404:  93737 34543   40858 45672   75829 74010   28631 14125   90389 40533      72115 82347   00184 25021   68207 05729   32828 47152   93375 53730
7405:  41294 72530   11680 68656   70127 11317   41019 69267   99929 61821      43725 64798   40165 68091   09737 03466   33285 54975   95107 26397
7406:  58291 72732   91632 41660   57616 12388   46475 44046   87627 35220      36908 19314   37192 06070   56383 32554   90440 75357   78881 55629
7407:  77879 03491   25192 93374   21966 02998   80557 00008   86050 68129      78885 25408   82039 34382   75366 10239   58960 22987   95439 33851
7408:  42517 25133   83390 91376   65761 63874   90760 56454   30384 94879      65753 37970   78472 37151   74231 77733   59546 70303   09818 77209
7409:  36182 67778   27306 60847   67120 48000   46319 91643   27390 40344      47111 00329   49868 31857   55910 46502   82638 53200   38139 43299
7410:  47026 11937   34399 61446   62824 37666   58520 45871   00825 35770      18794 25466   81727 43365   65217 58015   15028 83638   28455 12910
7411:  14016 11856   85483 96395   65557 98359   40270 73718   53462 16255      77154 39335   06538 94461   38779 01453   01008 57921   37589 64680
7412:  19547 66428   45640 77671   80008 84807   47268 15978   00449 64629      90515 81055   35589 53670   40101 87299   00986 76718   62782 53045
7413:  38733 28678   58121 59404   46042 68278   71334 64798   75895 61592      61773 25963   42792 19932   79641 95410   78557 56030   45721 66745
7414:  99790 04285   20438 31833   83336 79491   10134 65952   33233 67703      45350 71757   46171 05860   26894 42324   39792 88305   71036 81672
7415:  31443 10796   45828 38711   41107 35334   73428 19538   41973 89844      86355 54245   57953 11323   70462 66001   59045 98137   70784 67518
7416:  15361 28003   97290 51309   66286 32670   87854 35062   76110 39204      58835 08531   16228 93146   29299 47023   89824 39293   06403 32087
7417:  11719 16518   13231 10442   64871 73263   37752 60286   78282 16716      53626 80836   99838 72142   51231 74726   79112 80160   72729 88530
7418:  40698 61554   48421 83064   56787 90618   89825 41654   66087 99045      17578 64017   80101 36951   66225 13116   72604 97027   49397 82345
7419:  46075 53239   34041 79402   29168 24462   69535 47130   18526 07519      29570 41156   58563 52692   76659 52311   52299 51036   02835 96512
7420:  82201 15941   28763 90755   58884 13005   30585 81054   71712 17512      95398 86812   20787 74883   38019 31696   04021 21666   62172 48829
7421:  96782 55019   37143 05621   01890 74336   22894 77999   62193 16260      60277 89759   61559 15685   32424 82183   70603 66928   69190 20943
7422:  94821 53616   44375 13709   08793 78620   21855 36981   06785 32123      99881 84306   12017 70253   22748 69591   07183 91357   08129 50232
7423:  80031 52863   33367 25184   16902 24314   24375 60318   30874 92037      41473 72457   30209 24989   84718 05543   53604 25797   90442 40476
7424:  07928 51535   08265 39721   49736 15239   23895 55625   68194 20416      86805 32002   37864 08997   12052 28392   29299 97220   38121 94557
7425:  10474 45247   57801 89202   37155 28150   92328 82093   25472 77681      82267 63165   08628 63344   45282 23604   96395 33357   53039 17779
7426:  99347 68633   25867 79241   23782 03654   46684 02846   57143 14849      49439 10576   47768 09962   29945 12507   56586 10762   94735 74012
7427:  44339 53714   15154 20576   84229 12957   89957 69089   51745 63641      75287 00057   14409 54067   29915 96061   37232 90930   05673 13977
7428:  06029 79688   41457 44927   12538 99420   23681 86802   10078 79979      85005 37336   43103 43309   68723 41189   90940 37958   89229 77284
7429:  92360 43208   73829 20934   75194 91337   01945 74398   62817 28459      80086 72413   67923 08561   81641 51573   21352 61622   07213 53676
7430:  51313 62081   70623 45107   73925 04357   44565 00464   63422 84287      19684 27756   71880 99750   14796 77000   50602 78247   92265 27816
7431:  36312 05611   29189 96451   21268 60544   25041 23427   47655 48678      75928 27038   60229 54668   59300 63863   68698 82381   16847 42064
7432:  76625 35454   33350 93003   87665 43774   43069 29519   64720 32630      13811 54144   76792 93929   26901 14074   42696 94333   80414 52870
7433:  06942 51536   90775 08037   66733 88946   12403 94791   10171 85069      28936 00066   97893 33756   04157 77418   54371 25511   39766 26141
7434:  74348 60274   99913 30258   80342 75514   58962 66387   08069 89424      10302 20346   68539 06455   65897 42746   71746 36344   11703 36343
7435:  04965 55862   46592 72785   67099 08277   92233 98952   33870 30935      11178 31120   73068 27736   40527 32179   53681 00413   81152 78879
7436:  25176 71631   80037 53880   87524 58235   84371 99419   94887 79186      38004 50363   72461 00100   64521 06790   26673 37733   41191 74260
7437:  23450 06774   35658 54852   55671 14276   40598 17512   85332 05663      66548 02141   94245 29934   88101 05337   86515 55914   50901 90401
7438:  80365 84606   60543 34178   49444 80615   27194 22550   89210 51948      34648 46135   64716 82341   89295 54569   13748 12951   76143 65578
7439:  62156 80894   04841 70251   10261 49542   59714 06582   04702 75376      07024 07313   45963 68513   16525 23052   33190 48962   42563 48579
7440:  49614 87862   54616 64506   13811 82548   38829 02426   06369 94201      52008 25604   94420 79464   03705 75724   01676 22860   51849 59225
7441:  61656 44020   19108 09921   59456 19982   33224 18712   38937 81648      16805 06117   10862 82451   61265 83093   83901 36067   05198 63648
7442:  42518 10924   25432 22576   69050 82718   22367 09066   71810 99340      22618 34495   96615 32878   67794 29060   53654 96081   16084 58638
7443:  65093 84866   33130 69220   11836 49015   55535 24910   85293 97153      54411 75349   73775 50712   75282 94517   41307 00416   65509 68448
7444:  44112 33671   75170 04660   69416 96357   37710 44878   00038 31784      63180 04540   22547 49013   50136 54803   15791 44058   02061 86416
7445:  15542 08459   76712 33362   03223 66622   53668 46458   68790 85606      08618 72384   75924 68179   05987 90448   07693 88397   11327 70304
7446:  42612 59426   12467 35429   71758 96134   14954 04616   85871 09920      93010 73483   55268 39859   98110 61536   80571 58664   90233 92837
7447:  40935 31761   46146 19886   57306 19839   93132 69242   19363 93561      01301 54716   12181 58176   03447 37412   21469 90206   69462 89223
7448:  33535 33320   42124 37889   08190 66524   65010 27338   48839 67374      85245 75150   94257 12990   81129 84245   02868 24570   66052 20336
7449:  58455 27075   52076 66809   58146 33779   15181 52461   85116 70204      09344 18476   03243 89376   80383 04069   00701 66793   45172 51883
```

```
7450: 08609 73022  16915 13913  95930 32136  01804 86234  10978 87498    76885 04227  37480 99608  36772 09650  50170 07768  66408 60717
7451: 19652 57533  72437 23790  01071 03771  56209 83245  14244 13130    21808 39642  78971 82021  27483 25697  63738 98111  80144 90145
7452: 18683 19637  60181 64774  72062 77195  95595 11341  37906 79672    01792 23154  01396 90215  94079 90142  21888 64114  83430 99068
7453: 95840 98982  34891 74345  21468 56238  87712 85523  81807 33502    57193 39918  98455 29829  71774 82662  61347 68928  96393 91986
7454: 55507 03147  99737 46902  08609 10049  82501 71260  85027 45018    13195 32443  95403 18561  88239 01387  57797 30582  56770 09100
7455: 26646 15777  74402 38904  09661 40601  56556 59134  76439 44037    62408 70834  28045 54874  29428 80587  41007 34716  67813 07193
7456: 80481 38926  23567 50216  12617 45598  73143 81892  17536 81534    38126 84655  52052 27771  93316 01409  86411 76341  41547 58769
7457: 12692 14566  27140 15212  54491 22165  58410 81900  65400 81474    04297 08139  25834 29359  64096 49726  00807 79668  28672 51284
7458: 30264 95556  97879 90366  17808 47012  97306 26528  07594 48680    38828 24575  14049 51070  75168 29593  30305 01160  90379 14793
7459: 03605 73895  40364 04229  28275 68164  37788 93332  52518 30495    97606 64090  29052 13047  51315 66706  13043 83775  67174 73305
7460: 26974 68977  85613 69297  79728 38493  30997 29816  79351 25927    37011 84276  71660 48519  42935 02845  26558 39836  01802 52980
7461: 42989 48502  13792 13353  98937 68727  81121 20651  41596 59994    60570 71304  09057 17926  59255 93725  60762 71610  89203 20331
7462: 82297 65662  06519 63033  81286 82812  11107 01532  17023 26878    37029 12687  02979 67030  26465 61312  29381 76104  37948 43992
7463: 58356 40370  49074 61371  95399 53780  72775 95252  09208 74455    55292 96676  98438 01247  95968 22892  08546 03795  23359 08800
7464: 16042 51864  33602 21698  55563 74406  50499 46892  79904 17140    68238 03624  49129 96782  78927 72491  54419 50759  52349 66760
7465: 70017 56351  69854 75352  92500 04253  12574 39823  99104 94921    28351 68227  29716 78245  40810 03637  06048 62415  87004 16276
7466: 43899 11149  02005 81679  73227 90818  40509 85994  84751 60079    33544 56721  92080 08971  03681 12721  99477 28422  75554 46266
7467: 75453 29555  15995 23450  03410 04919  51757 74765  49834 73498    43396 84710  22479 63036  62575 30944  97798 56247  06681 92433
7468: 23970 07720  76680 43384  18807 62986  89068 05102  36428 12329    17567 36402  42729 55973  12146 19927  65895 48482  77529 24955
7469: 06620 45722  02771 73460  65926 36045  75425 79689  81435 67679    49774 40523  19703 23006  24690 21496  75111 44416  93699 95478
7470: 64100 64592  03506 05772  51548 66298  35471 81827  96078 97499    39696 56714  93163 53297  07012 34992  46424 48693  22414 22950
7471: 86592 66476  23427 64646  70223 11810  68579 75930  48909 03299    64436 72864  44588 00864  95441 68504  72680 30427  48386 50645
7472: 73833 84878  92321 08200  02776 34495  68730 97119  54785 74717    13066 06386  16328 15261  09936 86880  83582 28142  55667 70853
7473: 67520 90007  90057 87310  91858 09683  36065 35601  27343 42609    04738 98314  44319 89814  10112 09153  75570 76389  48590 82406
7474: 30488 44166  38523 70945  01792 12188  31011 04728  73314 25888    72821 64002  33691 17326  56453 08208  39722 38958  36350 83581
7475: 29821 88050  61381 23821  94257 59945  79650 90384  56663 79333    38308 24366  07505 34428  08883 96647  28381 32594  26252 33277
7476: 08212 14743  78321 15746  44942 12697  16004 82989  72705 92758    93941 48118  87986 94752  99954 25177  56980 43705  18773 86228
7477: 61205 42241  79399 97385  17912 46185  52578 98930  19317 24865    59394 51731  03491 68592  46381 18688  90620 72847  98524 50339
7478: 06123 60220  63388 26526  47048 57284  63317 57698  15786 91771    49825 67955  88566 03864  20113 20238  30459 37160  99994 35842
7479: 54112 85481  36475 75441  60136 98441  27675 57612  44032 25396    54141 25897  72251 00675  75809 94537  81218 30258  57435 24538
7480: 52710 59398  66082 63403  23059 69468  97324 46856  29041 91627    44187 89235  52814 24908  06290 64206  81451 05444  29355 64947
7481: 73352 11586  07656 49605  14763 81535  02342 47532  86161 71512    31295 34405  38194 06134  09712 26687  75469 65201  34914 51817
7482: 58165 32829  65278 60470  18733 03959  79423 58496  43918 88649    21799 44871  47010 49516  99621 09176  69736 11241  70476 61287
7483: 51794 76968  33354 19874  67513 04925  21587 68692  64831 63329    65416 73123  21954 49445  82741 70246  56348 55936  96605 26586
7484: 29543 94311  11121 59838  50856 25641  43185 91706  98795 42615    96221 73788  18362 33732  61154 31735  41452 19946  56014 78673
7485: 31178 16274  31406 65495  65509 05543  24992 96878  45098 70821    76121 73877  11199 72796  09971 16582  20773 73586  10164 07675
7486: 42647 80855  39281 63473  53901 99427  56538 97221  11702 20635    59015 35947  78481 14627  84528 20819  50159 04929  10565 34113
7487: 62167 55124  05147 75703  64049 55851  97992 92547  26173 54647    57353 73191  81629 65233  18666 35171  07577 19141  93555 61517
7488: 54742 78520  41326 48205  38526 01375  10256 74993  61638 93430    94835 90049  52528 92820  49731 24963  13570 40198  54568 96074
7489: 09881 11447  47248 43458  77471 92264  09499 68827  08004 46819    86290 62270  17096 61988  09297 83474  54457 75615  56164 41201
7490: 23411 95861  22496 61476  35167 85936  88414 11664  53200 98738    05158 75515  07204 60889  14661 88553  29932 26700  87943 30268
7491: 23180 85933  51882 07685  29257 18543  15928 06725  15926 72907    86742 94134  50469 51847  25435 67960  18706 37314  87270 80919
7492: 72085 56467  27754 04713  73591 18458  25135 08833  68699 88177    09561 01789  31941 53914  58684 37915  04135 31415  96344 08350
7493: 34814 67036  19733 06859  42407 30346  05595 28789  81107 82298    59235 45914  27464 59885  94740 57191  03285 14643  87032 51430
7494: 76685 06244  69945 77954  00067 72435  48348 87254  94924 76071    45044 61274  15122 98651  88137 38939  32494 05644  96669 76783
7495: 33583 15547  99850 12983  41291 83749  25707 07694  76153 21528    56348 17671  56588 44352  94143 70611  27838 12234  41527 16500
7496: 07501 38655  46531 39877  03838 81886  37893 92334  10534 01301    47684 26668  22301 00355  77575 46660  44655 44224  61431 25242
7497: 17206 99313  62394 49016  55957 44406  76951 04387  78705 23669    56472 43598  99254 63660  19772 16607  85915 90456  61610 80327
7498: 82679 09149  79990 80452  10697 38272  43340 29878  14093 56413    95003 77749  78029 32332  11216 83679  19149 44022  21965 44888
7499: 05613 78255  91539 84025  30081 33271  84765 24557  77990 45102    98260 52463  92093 45135  14482 34006  66971 48887  62681 63891
```

```
7500: 2498 39712  4543 6 31798  4008 0 37668  0508 0 70860  5235 9 36807    2312 5 71619  8515 7 57723  4629 2 53858  4578 6 22763  5814 6 16448
7501: 3593 388062  3135 3 88448  0503 1 83550  8263 9 04681  3934 6 86913    8858 1 55281  1777 8 26433  4314 4 85048  9856 7 74807  9360 2 44731
7502: 0931 0 47098  7993 9 87143  7941 8 78001  0453 5 38496  3936 5 45053    9481 7 26340  4167 8 14321  5809 7 85788  4167 49 9179  0505 9 31245
7503: 3703 9 20685  0106 1 95796  2925 9 41580  4641 5 86162  7619 3 23383    1249 9 18237  5362 2 88585  9499 1 54143  9997 3 30438  1904 6 27668
7504: 9233 6 02040  0494 1 99248  4153 7 38802  1808 9 20084  7286 8 98012    2642 7 43034  5040 4 95192  0330 8 95455  0357 3 47785  4000 1 35799
7505: 6694 0 13468  7665 3 20289  4438 5 87966  7416 0 11577  8250 8 90911    3876 0 02575  4321 2 98528  5528 9 29859  0574 3 30564  8252 0 98056
7506: 4913 4 20502  8730 9 34760  5633 5 94357  3306 2 56380  2476 9 51239    0355 3 61516  0375 0 79391  0926 5 35952  6769 4 31861  6403 5 05306
7507: 6775 1 73345  0160 4 04310  8552 1 52391  5817 1 39372  4061 7 31771    7391 8 55614  0462 0 52062  5795 3 57263  2001 0 70262  9859 7 58263
7508: 9167 2 25961  0209 6 40920  3713 2 93906  1967 5 09565  0251 7 88335    9094 1 15993  9664 4 50622  1381 0 36551  3095 0 00980  5094 6 41105
7509: 9025 0 31189  7581 5 90307  3699 9 40810  0060 8 92630  2624 2 58265    1652 7 63034  8577 2 89422  2775 9 58637  5409 0 85902  2937 3 45542
7510: 8811 3 60561  6292 4 64914  5312 2 09116  4043 6 26728  3983 0 85816    7517 0 69612  8605 3 20005  9793 5 28414  5979 6 56773  6134 4 45799
7511: 6444 7 31924  7879 4 02294  8985 5 93020  5565 4 72319  8146 8 45275    9926 9 34233  4447 9 12940  6428 5 19018  4356 7 48275  9487 0 11815
7512: 0259 3 55356  5400 0 04283  0217 6 74776  8664 5 57099  5652 0 50294    7705 3 16442  0011 7 56453  6086 8 32583  9532 2 73289  7383 0 39325
7513: 9669 1 95690  6785 5 32546  6573 2 56360  7085 4 80885  7065 2 19183    9248 0 35522  4216 5 19302  4317 1 63576  7135 2 19160  0124 2 25100
7514: 1203 0 35373  5404 2 57747  3350 2 84764  1744 4 75668  4673 3 88083    7198 3 28536  1238 6 63903  5388 3 45458  5277 7 40859  3444 7 15875
7515: 7114 5 80872  6230 2 20882  7313 3 12501  1531 6 71018  7252 2 78872    1896 2 21107  8488 2 39774  8676 0 78189  8843 2 07586  1122 5 85838
7516: 3695 4 19697  2710 1 11835  6347 3 16896  5666 8 17484  2978 3 37932    0882 3 56856  1220 2 16229  7216 4 31431  2819 8 61952  8988 7 58510
7517: 9343 1 89302  5534 3 40719  1841 9 58448  7998 9 98389  0459 2 22036    2656 0 84787  1132 5 76320  5913 1 26978  2855 1 84470  5849 4 44650
7518: 5510 5 41720  7576 8 37293  2384 4 78291  4577 9 63708  4080 7 00376    9870 7 82210  0428 7 46157  4239 7 88332  8731 9 01002  0314 9 85353
7519: 3274 7 60558  1956 0 01054  3263 9 58912  5771 9 65327  0406 3 63807    4964 9 38115  0229 6 19792  0460 0 42507  5090 0 07121  0390 9 89612
7520: 5409 9 74951  8658 9 69401  0638 6 85060  8550 1 85423  1697 7 86989    5914 1 71857  8851 5 97259  5707 3 96388  7625 3 88411  6586 5 41682
7521: 7192 0 00068  6774 5 99807  6675 5 90042  9003 2 26538  5550 4 83737    9721 0 23086  2177 5 43210  6307 4 42536  2776 9 86386  9584 4 14267
7522: 5911 2 31960  2311 6 69221  5537 3 56117  8153 8 78777  8894 3 84900    4632 7 99069  1040 9 86147  5979 3 90972  5561 6 72467  4072 9 76093
7523: 1875 8 60032  5713 6 01575  6623 1 45714  3367 8 55617  3123 9 90328    3161 3 00118  6799 8 46398  2624 7 50428  9301 2 02147  4950 8 49139
7524: 0109 9 35892  8134 0 55654  4398 6 16183  1585 7 73157  5633 0 80888    7151 9 39269  2167 6 80356  5017 8 64932  7860 4 66664  8968 2 42966
7525: 8012 391170  8277 3 14065  1547 1 92215  0779 0 95339  0167 3 29626    8260 7 11293  5631 8 47756  1173 7 30660  5987 9 06461  8515 0 89142
7526: 7501 1 15307  6396 0 20239  6377 8 62200  0376 6 62459  4748 6 19260    1155 3 17848  8942 6 85250  5605 3 53819  7122 5 65112  8806 7 23659
7527: 6463 1 25285  2547 4 55378  9429 9 26090  2249 6 97076  3015 0 30226    9766 8 64047  6187 9 16978  5995 2 39141  5726 2 94303  5778 6 35942
7528: 3911 0 32734  2227 1 74005  4492 1 04525  4350 4 31001  4425 3 64072    6755 8 85272  5253 2 17729  3952 5 32484  4561 4 39559  5256 8 11400
7529: 1784 1 89131  6876 2 49802  3806 5 57681  7322 4 59919  3604 6 52839    0790 4 10261  2495 7 81567  2221 8 03351  0602 6 15243  3923 0 80960
7530: 3001 2 44625  2672 0 88132  4431 0 96863  1914 3 44230  3219 3 86806    3373 8 53586  6175 7 33401  3164 3 00559  8765 4 78568  2590 3 38578
7531: 1065 6 23376  0174 3 27164  9921 4 67348  6249 7 09802  3577 2 35952    5284 1 24657  1025 9 25139  5722 8 41880  1732 1 52189  7342 4 75258
7532: 7996 6 83312  6207 9 76860  7492 2 60603  0719 2 69990  5723 4 56242    1422 2 56710  8189 2 71355  3194 3 49930  7622 8 15175  6477 4 26518
7533: 2373 4 52651  7476 1 93379  7208 4 31781  0288 7 05911  1404 8 51413    1831 7 52538  8872 7 56887  6719 7 93655  0273 2 10695  2564 8 92258
7534: 2611 5 67238  3815 7 06129  6673 0 48379  0129 4 06860  3472 6 78572    3871 6 81729  2384 4 44430  6444 1 19090  5860 8 75332  0829 8 55764
7535: 1318 3 80659  3759 0 36826  8089 2 94018  5152 1 70819  3802 4 52176    3466 3 43102  7326 7 27732  2221 1 64544  9939 6 17861  2646 6 61457
7536: 9675 6 26707  7899 6 82174  9475 5 13007  8950 9 60337  3752 7 89931    0334 5 29033  8084 4 48691  3620 5 01567  5818 8 80872  5783 3 14631
7537: 1747 8 01493  2167 4 58389  5982 2 83347  3579 1 76463  9963 6 98331    4130 6 51593  0748 2 93306  1956 0 46971  2507 9 37566  5826 1 60551
7538: 8449 1 02921  6041 9 84655  7274 9 12234  0055 5 29386  5354 3 73885    4004 2 58556  2510 5 95180  3314 3 29310  5808 2 70094  6969 9 89605
7539: 1756 3 77758  7516 4 59801  9241 5 09938  1918 2 76153  9838 3 97833    4632 1 35676  3676 0 69801  0056 7 36934  3973 0 53660  3927 4 26661
7540: 4247 5 32865  2797 7 20640  7314 8 49103  8101 5 26985  3030 3 71821    9284 9 12937  1103 2 57592  6423 9 15071  6749 0 91081  8207 3 72185
7541: 1238 0 43078  7452 3 12199  9263 7 57189  7410 6 56636  7264 5 05625    8020 2 24566  4516 7 13279  3945 2 49883  6369 0 52014  6498 8 44124
7542: 1680 0 56181  2175 0 06788  5596 2 03594  8995 8 86177  8249 2 58772    6862 0 07491  2506 7 60956  3479 3 61889  9454 7 32776  1737 8 47249
7543: 1248 1 51133  9640 3 98040  2595 6 81084  1866 0 01605  2382 7 63591    3024 4 99853  3565 1 30990  7241 4 17717  2210 7 55181  4617 8 28316
7544: 7848 6 54382  2949 4 64015  4191 1 61283  9541 7 79769  3434 4 02436    0833 3 76871  5394 7 81094  0980 4 12368  6756 0 66178  3362 5 84549
7545: 9932 3 64483  1586 7 29000  2719 5 68998  1165 7 18182  8469 5 22203    8187 9 24656  8425 4 38898  1394 7 22665  0686 5 82061  5424 1 35035
7546: 7515 9 04123  0252 7 47578  2557 5 39964  1492 6 18530  7176 5 46079    2405 4 65562  7052 2 00954  8328 2 48923  4651 0 62390  9656 3 52909
7547: 1019 6 29402  2306 4 46362  8413 3 89569  7551 5 41154  3208 7 20775    7857 1 25557  7002 9 21222  9336 2 03310  4723 6 82049  0126 8 44703
7548: 2522 3 53248  9416 2 66155  1342 0 77273  1621 5 43775  2059 2 55226    5030 4 63950  2130 9 45006  9969 2 08416  3475 4 43031  0142 7 61226
7549: 9270 5 05413  1157 9 34695  1458 5 89974  2359 9 90869  4788 2 20049    6131 6 50753  6202 1 11843  4373 1 83605  8263 7 32476  2095 8 87015
```

```
7550: 27230 76892  61854 29860  93607 00092  17688 34309  55498 49859    45230 78064  17068 25803  65733 78562  13861 32916  43455 98609
7551: 05380 45988  25930 87830  52440 17282  27483 99619  81158 90922    10421 53025  09939 97592  41713 47772  42834 32617  65222 34962
7552: 28145 20969  34774 26270  98528 85006  20808 43891  78425 47111    92365 60267  51361 18182  71202 94047  61778 55556  52186 28112
7553: 27147 15848  18372 23443  37125 99212  44865 27946  93536 85703    38754 25270  64566 56161  25211 94839  17468 71887  90209 91630
7554: 20958 86727  94455 59610  54594 00186  46970 12881  03684 24513    80047 46753  32685 81745  93308 93416  33856 92790  71386 53219
7555: 91608 50013  57996 27190  05756 87880  62639 97503  27989 46259    54707 45331  56449 61551  39893 85492  55468 12228  36486 79949
7556: 74528 57975  08793 99534  54219 17325  08776 84671  23556 86701    43690 09941  38862 48203  82266 67513  83877 05477  86002 93466
7557: 67541 21699  19172 74842  91072 20109  65136 12443  44054 17723    59042 92671  91175 55425  22990 33874  14059 11408  88492 91721
7558: 31002 76584  60120 02563  15830 09236  92766 89840  63662 37911    48726 95914  34855 70270  01695 59348  17175 32210  11620 91815
7559: 44917 17907  14921 17690  00943 76980  26864 48162  68030 89515    94681 18901  55326 27945  01248 04538  57529 40926  28483 84140
7560: 13198 97258  86623 69434  40097 48444  75427 35448  61724 44098    67850 78787  66952 78565  20603 22691  73863 12350  93607 56704
7561: 06438 24954  68536 39561  77754 05843  00588 11698  38624 83470    23805 43109  52661 49694  47612 35837  33763 83852  34897 93395
7562: 02260 53786  65810 60621  10800 05436  36834 64489  32744 10817    81028 95173  38011 34941  27633 14151  39117 66567  92329 48926
7563: 36043 22750  25630 98450  84862 40518  31637 10903  56422 69590    49325 86308  75091 13773  54540 75924  14769 53129  80487 12670
7564: 40573 06904  64770 94375  44441 79142  92045 52322  35358 36347    69696 45574  16420 70683  19885 24146  67419 46214  89812 48622
7565: 57238 58811  61758 07048  58990 77888  01000 06427  27217 87104    41047 10420  53275 23543  05353 29952  66915 52961  85779 42641
7566: 23714 53715  41964 41804  33842 78787  78962 34452  01046 34742    99441 80745  00304 35604  59837 10237  54516 53546  78158 33230
7567: 02531 78498  58588 20709  81332 04151  35121 80936  85972 76062    21231 39517  94934 92459  88514 04463  30466 84588  27624 76020
7568: 52182 82245  08742 88851  70733 19071  40804 86887  58742 39377    91271 38711  97721 81497  65208 51138  93684 96969  51835 88929
7569: 31948 21822  34186 47200  38365 92649  93016 43421  17460 91620    36407 09638  99779 48676  93726 50890  92668 12029  52109 99193
7570: 50773 90353  83660 87544  11925 71282  90313 68548  23759 66780    84395 40644  23996 99816  91455 56027  81222 07072  39236 74108
7571: 62064 20734  38550 07626  78842 28400  92135 49500  19528 63035    06721 77968  56932 78520  98422 25909  93671 81939  15604 34207
7572: 20792 45952  06665 39735  49159 66539  56412 50870  82168 77940    64455 69340  76972 40265  90059 25044  55331 40709  06734 18179
7573: 54089 64752  93666 59327  70530 19027  92398 92595  89397 08958    63945 33236  49389 88939  99629 38481  79001 71262  91725 79303
7574: 28785 99484  40538 98286  24746 01750  90580 36343  86641 89268    70832 93005  31603 86645  02206 96629  77105 13205  76510 96813
7575: 84785 94962  80035 21191  13959 70231  12950 72652  46219 23867    52543 47148  10965 64428  18279 63205  78018 46507  12546 33135
7576: 31053 81541  50634 97136  12679 63117  52410 42463  13721 33441    15062 73056  97096 72673  91406 95492  48022 37346  73357 42810
7577: 63402 87657  72262 11240  74916 99956  93840 11098  55318 24559    70467 94835  72676 45263  31823 20306  99047 87177  20085 13803
7578: 15985 82430  55680 05995  34525 78648  26796 16095  72536 98645    49795 09569  74294 59985  00785 36601  45359 02735  93770 65916
7579: 03645 96232  05215 89192  73902 12560  27298 64739  15354 33468    40627 58718  43511 90951  31749 19200  48500 84947  00800 58218
7580: 55699 53019  12837 82616  73162 45246  20464 11602  13389 36564    58621 32047  02995 59057  62288 68128  73964 30921  16277 60020
7581: 80746 28034  57806 70659  22345 03264  86104 33184  56457 00663    54889 20114  86736 65941  65476 70105  13110 70195  91850 92053
7582: 31656 24783  31199 92455  36937 21643  84328 32485  88585 27003    50847 43795  89787 28263  74135 62883  94892 05301  93268 95260
7583: 09400 64988  50765 88991  35256 11060  38750 28160  46866 41653    48548 15205  16749 23725  56166 27724  20503 07206  33784 31405
7584: 75423 31022  42323 15101  59635 80606  36675 76507  58431 30772    80124 49577  37669 63869  72491 50724  40349 91336  51646 04239
7585: 56441 29869  30094 02386  63022 13595  64865 21110  55707 73679    75957 37633  95439 70240  01104 17774  27545 82857  71087 37306
7586: 41361 75609  57508 91747  77241 64392  23383 33626  06331 42552    70865 42529  86392 88870  14639 43636  40115 12337  92390 55131
7587: 94930 73324  82188 81388  08111 44632  75444 89064  19564 82740    96489 71571  56821 69243  85307 67821  40885 44278  69192 89496
7588: 84187 10958  46491 02433  07196 26409  24412 47858  02621 71516    58295 43450  28791 89588  71742 73800  37497 66714  47645 34384
7589: 02451 77022  43600 21200  50039 05692  06033 08649  28867 38431    67579 59247  41733 48878  17560 84212  00688 74932  52262 62713
7590: 46170 91752  92816 43182  03513 26192  01588 95868  34522 07176    66932 65286  79125 31991  47980 66344  70281 44058  98494 82311
7591: 01770 93931  43169 51469  95421 63145  63473 16013  23396 41973    86409 16893  36423 26848  49702 27350  12738 65481  31912 85123
7592: 80920 60657  18751 95970  69718 35614  25960 31711  75358 36743    28178 31455  94325 61740  56730 78237  89579 78345  75674 75777
7593: 07200 71444  35766 31427  96850 45817  20493 15112  34781 32264    46149 18173  93054 65478  54055 74676  81286 03072  94437 97772
7594: 00458 72733  72255 32198  07044 72890  70955 37511  56264 89436    94011 02042  18901 33247  00179 90775  67009 84501  00517 29234
7595: 73629 32870  39172 05006  49629 77102  37588 90636  38372 44155    04500 78544  35545 83701  81590 96418  68985 02691  48520 22443
7596: 80617 56625  02413 99347  31631 08271  24028 14636  85117 46758    11078 35956  16100 49903  19215 21411  68296 16842  81853 58080
7597: 90928 84971  38961 97876  46416 57781  51487 39861  65806 32294    48348 18061  91805 75716  50851 75031  32020 43523  72608 89500
7598: 95883 41308  74936 95891  46173 26141  65084 82167  53696 06605    38096 32551  51733 27741  89359 93752  52954 01616  53828 63750
7599: 55709 81555  44448 82924  24959 83910  19954 30509  34053 17506    67998 17857  42602 67663  25091 80042  55209 91253  65885 45910
```

```
7600:  65265 42855  45411 05164  70106 17930  87657 66264  87841 82346     78849 16006  83932 16691  68451 63801  48752 78751  68991 21723
7601:  76630 61149  79758 82423  74576 42042  46037 62012  19593 15841     87931 36871  75506 20214  40021 58936  46667 12746  74357 64910
7602:  25423 94413  66840 97089  16862 30770  69832 31461  70575 73324     31083 18011  45253 86371  67167 46835  07454 44924  65622 44104
7603:  69588 64835  77804 10512  32672 02628  03327 51739  72868 93958     38975 87774  79770 34284  21163 61718  42257 49371  59612 32058
7604:  77670 08296  97581 72602  51880 78325  44381 12927  87008 43102     64790 30882  42168 14040  79757 64220  90606 40676  93853 57398
7605:  91989 84748  89964 58298  75861 11558  78734 87674  22117 91283     43404 42918  16226 18600  38309 09850  15292 40426  52484 38329
7606:  93278 04920  24065 56893  58036 24086  53420 59034  09827 35362     70739 57130  08164 51748  71402 61747  13976 31053  41235 70280
7607:  83340 78754  03710 06454  63536 08987  28336 92195  96446 41259     17006 14793  73789 42344  58953 94050  95155 48782  60671 31295
7608:  45369 53999  95611 92461  90225 11608  26986 53230  38578 08182     86339 32326  11431 50556  35065 71266  76050 86722  39152 59748
7609:  74481 74961  35845 75717  76882 69826  76366 82384  45137 94459     01074 46406  57274 99669  34804 26089  77113 12925  04741 42765
7610:  29924 50367  86913 69978  87009 42194  87705 59252  73089 99384     01413 42696  99825 81951  02602 58327  66663 11859  98417 53504
7611:  05072 42202  83556 77782  54266 87935  77605 02895  18704 37089     82254 99967  55238 14012  38510 33954  17843 25363  12251 29002
7612:  10146 65043  53321 67465  28920 68673  67840 80554  05394 83925     36994 31045  39388 97333  36887 05630  37806 32282  48957 24042
7613:  11256 77707  98082 66764  95285 78413  51821 97063  99076 28502     17927 42090  38427 60067  11458 96751  05782 84328  35364 29125
7614:  66905 55233  03717 44286  42379 28132  97547 56890  16518 14578     16445 06292  29291 45666  93008 62507  32634 94695  77459 74830
7615:  38219 99813  51683 13862  08440 43185  70889 41973  40616 42481     27063 67651  12420 27414  14564 98993  31164 41357  59777 90396
7616:  40768 82266  56398 75624  67238 98252  88439 96702  22116 59490     56937 03225  26452 19044  34185 64568  52929 93297  25941 02003
7617:  05900 25191  65574 39175  01284 93456  76937 75042  77928 29202     70368 63141  56488 37433  60909 94249  78420 07613  76827 10764
7618:  90219 46666  02115 13356  77010 10082  14366 61686  63848 79886     37561 58208  43393 86311  03992 63645  66719 08404  84436 16632
7619:  57426 55396  59268 21649  32066 06675  29612 95874  64109 10671     86508 29361  12712 13382  48706 55004  58028 65071  93754 93650
7620:  18937 04860  26208 10024  88919 74784  90691 22056  99077 24046     95948 33272  80031 00096  36513 38712  45004 09666  72041 87924
7621:  30918 85562  70632 62894  14867 21103  57488 08409  38618 29388     52169 18619  97685 79255  71783 87538  45405 24965  88595 95714
7622:  23296 80530  97213 29700  95940 85899  72140 90969  10186 19494     73949 39659  84927 16103  25552 63435  28343 37999  08593 05190
7623:  12566 11885  41886 59462  85890 25796  39892 41341  98161 27434     64620 06437  03338 50994  71856 32767  31359 18295  15165 63575
7624:  24993 35360  67081 70315  92089 09298  58734 71409  33760 96445     08576 65252  75334 99020  70632 82258  83960 43785  19960 84871
7625:  62859 60635  50355 56429  61550 04734  71686 39714  34679 08264     46698 31914  72882 33570  41436 08910  10629 67079  32346 61974
7626:  65816 67585  47687 30423  27294 55376  38282 29073  53485 59149     42572 92921  04991 09494  09957 22390  43035 99742  04077 23648
7627:  34674 05201  52139 66219  13071 98019  33738 12055  99814 02856     94918 55185  43116 12064  97595 46448  61306 66395  88912 22723
7628:  71953 66833  43118 29466  98735 81604  22855 27552  15191 41935     20604 38412  19161 51522  44686 79213  11579 63837  44306 12836
7629:  86414 30294  41026 52811  11185 28707  57826 93245  55674 55958     56876 43789  70813 08058  77057 04347  94889 37597  13508 07855
7630:  82206 43508  08138 05411  29820 63116  49470 02872  25468 35681     22238 85636  71067 88069  07185 07405  00325 60231  45703 68140
7631:  38949 51743  35138 50076  09273 29127  45477 71058  00597 89557     70433 67407  10817 75629  96197 88519  02051 65037  90760 69052
7632:  16782 03413  15475 09438  16472 73551  05507 96157  93064 41341     61010 63579  95733 62249  50110 86879  68781 47964  72608 57042
7633:  81442 40728  65086 04722  71234 21406  32570 39855  14670 55787     19793 74049  34056 68207  92527 92556  81027 56264  76442 60752
7634:  24341 73512  58255 97054  65207 99566  39895 35986  38472 75008     91772 93664  26127 14402  67857 11680  29251 37410  00909 55822
7635:  56051 64428  87910 45719  32520 84629  03468 43581  17864 67421     96443 78279  86657 72007  02164 94194  64845 71263  54017 05038
7636:  55264 31038  05567 17316  32450 96814  54594 96305  23484 03644     33717 64726  89139 63100  43243 07057  71214 79133  43870 77911
7637:  31528 72749  22921 12025  73391 41101  11536 20251  84787 93481     83251 11649  30883 04405  21596 33600  56979 18664  17274 40637
7638:  57561 89895  18638 42103  89465 44860  12606 56817  41758 06296     97228 97702  07911 39809  01740 14132  41597 31161  05505 50947
7639:  62425 26528  54721 34725  90046 56579  73018 70326  30572 97709     67788 04458  35477 05837  91381 38195  97642 07534  02702 81480
7640:  84199 34169  02532 71420  78518 44520  58833 22233  53291 79250     77226 81337  46279 54677  58774 38969  58228 64505  02856 97825
7641:  27720 43688  73301 37253  56415 23750  63685 02988  86378 65502     99980 90510  94415 76063  39857 92607  68042 94350  90712 89714
7642:  58502 07929  02401 28162  49591 18708  51731 08975  32827 80556     88654 19445  63860 34721  91785 84418  88589 49835  80267 26219
7643:  33764 70271  16600 76056  15939 60535  44079 68295  30739 90460     16227 24880  34623 07829  04606 25273  87449 81640  93688 65630
7644:  11378 12832  55502 45960  44666 45452  00984 88933  98107 23096     92672 15552  61823 32015  28857 49526  52591 46536  31517 41060
7645:  25469 50413  99613 06781  78668 82448  64429 79643  05779 95544     00046 03945  18397 50369  09223 28369  79756 84715  03706 66415
7646:  99062 32709  72002 34364  19236 01761  01219 48525  36193 49801     99244 10196  21703 06259  72690 68657  75724 76187  95042 33743
7647:  00330 02769  11166 43817  31339 74288  69971 01030  47077 99548     08128 47177  83166 74513  24728 38889  15099 54223  53836 96410
7648:  64764 85254  31377 60883  29719 57218  82668 09032  35020 78838     22089 08757  45473 35246  53979 23601  76834 65993  45026 40149
7649:  29108 54192  45520 40813  00548 88873  25772 63010  54937 05305     59754 56991  09054 86966  38549 62405  12926 78824  44379 85184
```

```
7650: 6514685132  1988807661  1449154508  1017185427  9132769167   0811712043  0797796505  9526100948  0289933854  3868460397
7651: 4851809319  9669347955  7400035526  0756295527  7366161664   3932251389  4516090829  5182164513  6567581331  1199363402
7652: 1809706854  9055867875  2725745263  8070464447  7441522595   2621976278  6265759113  5402073973  2893051354  3082458810
7653: 5537367808  5041293778  4719374448  7680556352  1306374356   3403553095  8538996271  9631227252  9638390815  8794994001
7654: 6548146521  6334361505  8420831037  5729574005  9427452656   7929747852  5110002120  2466033010  4659674457  7733646490
7655: 1230666772  8346061919  2966165972  9488717130  6168392359   0913196793  7603728426  6657530639  4305254905  6721028685
7656: 4528859146  9418300881  4626691972  3014695938  7978717404   6913449738  1571377471  0336528335  8070807379  9946220014
7657: 9911023330  5494110083  6907689544  0573792610  1626954429   6304495880  5988264307  0618794534  4651747773  1521826774
7658: 6738112098  4827078372  0462162440  1461414891  6878607090   8833303798  9136504561  6083204454  1398771702  8093605050
7659: 2693402602  6029446333  4580776282  7687090462  0961142612   8371785974  8217929214  5321375276  6788277853  5144707373
7660: 9553972053  3313611520  4466707108  5661635858  8998946948   6248946034  5046973977  2533682088  9499808715  4444232048
7661: 6034802392  9870067634  2208125071  7008381594  1672179391   2870868113  5795605336  9802622393  1362438241  8303645095
7662: 9480095961  4680599469  0803308570  3956489815  2551780459   3688233977  5873672602  2628271947  8490329335  4253920192
7663: 3620497996  9040305556  8363824817  6985767011  4168092056   4290938998  7441064918  8590047382  6868428066  0398729893
7664: 2117389607  1842130565  9696318588  1228131084  9719297904   6006853301  4828197312  6818705316  9342178182  0632743361
7665: 3208262740  3953178485  2884442327  3310801477  2110443088   2697842722  6334022801  0882463365  0902559186  1186881928
7666: 0256290408  7371794979  9275846105  1104787904  1786710819   2051634910  7933053844  4274744375  8723866656  9313890363
7667: 7625441211  8595934877  9316000211  9729153414  4350055255   6693540708  1797783647  9739663939  5000194478  6187855929
7668: 9874074756  9629283631  1740390150  2947759099  5213267214   5833135282  8741796804  1686563357  0168862984  5266301472
7669: 2311100306  3568826899  0269343273  9299847885  5568833605   4277839468  2609711706  1604776533  7281989577  2797197524
7670: 5178928567  6473034055  7493379171  2394118667  4436492820   5853192391  2328685160  1327960421  9532466330  4488302840
7671: 2010292187  9750080021  2480877948  9722415913  4509570546   3615969098  3379254524  4503691479  3354756894  3389841677
7672: 0145158999  3762064381  4306530670  7918313833  7143074462   7723716819  1477572498  7225329517  9780383273  4266973304
7673: 1628174578  9629871838  5950991343  6502540453  2811001131   3193367201  1532309747  5565102433  4320843090  0038841953
7674: 1552533601  7905968122  8271743263  2999678145  2469296595   4469434972  7793773238  1977796668  1742181594  8821899822
7675: 7879953180  8639522121  0957870583  7921414859  7134615054   7186907869  8329086120  4460311159  5929855872  2551189124
7676: 6080844626  8165659529  4621499149  1949401531  0230383558   7126858127  9221982661  6086081969  5165238813  9192808147
7677: 4901988378  5985047715  0423781870  3067968945  6277183751   5934242435  9515821094  9558442307  3764614777  5616919104
7678: 8864349097  6139397019  5246264207  2644932238  8616181975   7896587621  3249577151  1599701171  0686145079  0619983228
7679: 6333007401  2897547380  7285724803  0960333736  8048713502   2505678487  5747873958  2489981360  8814182011  7632622757
7680: 4425552662  1418322703  3455735734  2865422637  9284782673   7362073742  3718022622  0170941931  5818306785  9217875304
7681: 1271615923  5953758051  5728356442  8617528182  9353286418   2794209470  3209450302  1343612550  9945078735  0656325459
7682: 6153552075  8282641968  6070164228  1274444422  8469069145   4072642825  5680708540  7141917327  4751684833  3035730554
7683: 4802880242  3039285676  2200864685  9119400133  3023188652   9178035976  5448873958  5564548058  9548373247  4469245386
7684: 6258546056  7770111157  1889048226  3202179996  0887068713   6747994839  4373745496  4173585218  8753730713  3948436902
7685: 0056074284  4854749554  2488720894  1928914304  6408845503   6606634523  6177586616  5572461830  9856904645  4317256536
7686: 2265624259  7952051763  9599580196  4024740258  1504852947   9935793877  0041776125  7488127505  6218799099  5028995355
7687: 9787797426  4215513550  5001745585  0132547447  1674003996   9996442100  7426375954  5913093750  8646292508  5286035721
7688: 2669849820  4069681976  1568406530  4658061172  2717311773   6475970204  0655600848  7824360716  6293210201  3138794667
7689: 8171080613  2139390897  3200411362  9999499907  4040002676   1265402373  3863560225  0494003115  4778394611  3272921684
7690: 6742259265  3702982323  8014859772  5684510865  7239438851   2687995699  8799310989  3708930167  4136099023  5962249465
7691: 7006252180  2519872643  7003798319  8015003332  7940719704   2905098366  2711153790  1306683537  7562581170  7004178827
7692: 7082314253  4143495321  2187532358  8572114001  6050245518   4982778983  4209566716  3622209626  7370366150  6425490727
7693: 2564391348  6903585011  0890140868  9208201213  4289375410   1943578341  1757723007  0510276010  0092457231  1988760971
7694: 7342843618  4455338948  6885316370  4019638534  6173568531   7766410979  6363736224  4824826839  8045616616  3413183022
7695: 8289652867  8179556585  2409935211  4701995612  0710725743   5959048618  0508613035  0467582997  9367204388  9159955357
7696: 3288728033  3953987315  0013513431  6865103759  5648130782   5232956526  7618055616  4925544112  0892983796  5119961855
7697: 6999973359  8016858222  6844806726  9827980758  5585412473   2841341253  8759463160  6547038966  9120031379  2237738372
7698: 0167964722  8455648230  1740570792  8247408193  8183975108   6015912021  3898881098  9022472302  6922869126  4530880110
7699: 8991233134  2839019423  4480934061  8715062350  4854736543   3347259119  5864275253  7244471743  0441493520  1947147768
```

```
7700: 0268293524  3224242212  4241481851  7655352229  4706879269   9663535244  5551837188  8450561359  9498603657  7376256436
7701: 4873737608  8005194055  1192258806  7379304572  6128356277   1311991341  3208558836  0525599404  5608020606  9153766885
7702: 2025130145  5846708700  5919290987  0580145798  9091538344   1707880210  4868205075  0584911445  3270388770  9797941373
7703: 1850046675  1444630390  0727340807  9155689065  4308503914   3567473502  0642468109  9858143380  1067921659  6591537204
7704: 0287743409  8900213050  4515212448  6289537310  2989940987   0804653292  7638290019  9399395798  3423276917  5478356828
7705: 3591834747  7572458529  2751178339  6442559522  0932904954   7390679586  6316890142  0519281738  6966859796  2001424849
7706: 8692206041  6302288307  0856725461  2251951748  2767002377   9116317112  2473050319  4800722463  9605767304  7371385262
7707: 3721574988  8143178562  7863458450  3251809250  6529723200   8391394496  4290703176  5843626376  1070504435  7739894361
7708: 2991177497  6102914101  5175012456  5520170092  7750684717   5239220528  9321466297  8635702715  8143898655  1497539155
7709: 5724533237  2915516792  5375656198  6688286163  0387040203   8324915764  3512007309  2772867690  3459644716  2298940912
7710: 9659652994  4857674993  0244732307  6555640567  1602149522   8490762452  0791475072  6398864364  3307821306  4428012591
7711: 1221976208  6731811948  1337171924  4088695091  6427295072   8023967177  0127611321  6388368094  7364470061  1394293956
7712: 9869082609  1629345887  8765123552  6913824663  4997729968   9563516885  6253572417  0814428068  1197371159  5529495347
7713: 5062714628  9690092287  2605961249  4792397748  3643759666   1760360746  9656213052  5506272008  8133639880  7519029106
7714: 2265944495  1477923624  0841171470  0986939949  9550566282   3819308372  4469513594  1389001327  7587286938  5014523685
7715: 5726480424  6192593607  8446141990  3686425182  1944104705   1290009331  9287823670  9426730595  3201119946  7813524453
7716: 2894684747  3780062003  5361424079  2174108039  0213099321   9023640152  3638376748  0726645996  7289839753  3884239476
7717: 6877997592  0172794639  6059032743  7147958369  1326092587   3035686305  4167910099  2740735683  0973723926  5908752041
7718: 1874734174  0208270344  9340220954  7473325648  8286685095   0526120693  7617501570  0815005453  6148034655  8965851437
7719: 4645277156  2039080483  3662972117  2968082833  2043540704   0047709368  1156036513  7160577656  5198726918  6927555312
7720: 9425818810  4828716318  1098226536  5610530789  9139115817   6283020893  4240871515  2530950688  8431128593  2277151097
7721: 0911648781  3365906744  0120618018  5865713966  5431063484   6122703501  3821255346  2203198932  2923402056  3484379216
7722: 6120468919  9970908421  6300001573  3766875830  1029496190   7271653479  7283369396  4453954196  1380766717  6247610939
7723: 6329892366  8808292875  2191112676  7153961746  4341093203   4387066269  1524054954  2961894101  5865695271  4381976455
7724: 4514416777  3734214740  4725670814  7324050670  3643064062   8355864273  1930582237  1250947436  0068813162  2764932140
7725: 9728575710  9164308038  6150852377  5767897543  5127151430   9929240142  9742150851  7465827499  1687550616  1705803470
7726: 6609279754  2341608080  8397615328  8316934750  5439222600   5626861087  6387313535  5747423397  8396729251  3677913760
7727: 5680657917  0978918944  8562429629  5152605958  3467174203   0717502423  3390547073  9232995010  2795120862  4113381719
7728: 6941752144  0616716643  3330610709  8510671118  3778332465   6037975835  0711216623  7855970182  3933865825  2591905789
7729: 8253538299  9510741527  9855026179  8657385216  3672695971   6952601732  1769580017  7760749496  6222622987  9975551210
7730: 4300435539  8525539537  7436328264  8369192140  7732731190   2628674044  4896483286  7263150874  8444471358  4313776325
7731: 5497599958  0294589223  9501503285  6312884193  2649761279   1587274508  5896922168  7657722628  8450161844  1903555374
7732: 1510737758  5827315815  7831156223  9768278857  9528951868   2495441080  2364532697  3146266299  0373698157  7690127282
7733: 1351705339  2482745434  2298804562  1568171367  4487240936   0223029017  5813739020  9012339600  2929619335  2863662417
7734: 4255556288  0700966884  9838182349  7842307906  1502492545   7415085345  4479339437  8478776955  9676057499  2382938580
7735: 6426940367  9360265860  2653599224  9017759907  9401410119   6421325776  5180669933  5773493302  3063779430  4157598189
7736: 6358742130  6664495611  2371567108  9482696180  6048006429   2620428414  3653107758  7780802793  6088901082  3430863567
7737: 1394016018  7835182926  5097309642  6074744430  6930024364   6387662925  2822638916  5063707170  2129512951  4924216384
7738: 4088019387  1906361381  2333228607  1236668739  5075430555   3147932719  8843157170  0721499842  2841087499  7500095152
7739: 2352031869  4777938768  1322451813  1326941084  2292188546   8765708663  5020308160  2087753848  8225916049  0236487645
7740: 5279079965  2235630195  8146577889  6312511691  5204754120   4561050157  8112058130  4114942914  4500992544  9408924459
7741: 3776841509  2917656397  5370549839  2756035510  1446533256   5803259060  6487438335  8314711611  2529518752  5307315887
7742: 8773164059  7437861035  9401201045  5767894717  6696711911   5321764399  3417783655  6748636630  9135369499  1980408760
7743: 0414371904  6258569288  7524735214  4760505600  2151223455   7174472352  0227783544  9081070932  8867826833  3800140432
7744: 2247208523  7582869118  4881113572  8453943086  5474968813   8082623163  9060262206  5383646099  7251830861  7664128051
7745: 7767334187  1409950855  8282754124  4233360638  4751169132   0866992127  6440806702  2307014033  5112092433  4483989033
7746: 6227310347  2030336390  5646727234  0430390392  2638553109   1500039236  4196836639  6672578234  9645526323  4537865384
7747: 0778400156  0524546375  7075308754  6827655276  8033474629   5581219893  8263256210  9136126305  5486676740  2490123647
7748: 9207848292  5688233365  6768324808  2138728209  7470174438   3522741999  0235115712  7290940992  8064874496  7823888419
7749: 0238000008  9307490453  7514166701  6388159173  0274771814
```

```
7750:  79901 52680   99057 36140   59934 85288   94491 23261   11235 66431   06181 58288   27946 34148   52506 32676   12396 57631   82811 45639
7751:  97215 23582   72704 38846   23445 11985   46588 58299   52601 55664   72538 20126   64449 34771   68832 88213   24361 04716   70821 35426
7752:  33566 81074   98717 47460   05750 21135   80647 23558   71843 46043   33867 64531   15665 47735   26973 22754   27345 23519   56286 38992
7753:  18099 82600   15051 61423   13540 60699   83693 41903   17508 91216   06195 44150   47313 02516   23301 70809   81225 41245   10819 49822
7754:  24765 32832   24284 71876   52023 34799   61326 82754   80802 95107   19731 26090   29810 53196   38495 84043   00619 47702   52643 57076
7755:  82124 65068   59069 76904   07406 80666   79948 80280   46429 90844   96524 35341   74443 79729   30583 44219   07433 94583   67883 13251
7756:  72372 60491   26715 55047   13403 04739   36226 52420   81358 66354   87400 14585   14372 07749   47486 87974   54208 76084   86900 06133
7757:  64062 09439   57200 65670   31484 97786   14016 37115   02402 34667   40705 24848   80495 94247   36637 92765   36693 43462   74570 30112
7758:  39229 35338   44052 48348   08829 16320   73817 37857   53066 86905   04603 69970   20061 71346   41340 39861   90615 94341   82112 50463
7759:  97531 29047   58614 73041   45631 79915   50204 35894   22151 90613   08731 67682   72650 75312   27569 69168   71234 67614   15204 22374
7760:  50086 83565   35752 41308   99876 62806   24400 34819   75363 43515   36508 40552   33619 56403   96095 20285   65697 17005   02542 93242
7761:  26774 31297   16628 28267   77022 41275   26773 26211   39256 01233   03464 08442   73090 98513   62199 45775   08490 69962   21606 48425
7762:  39013 64389   56137 98583   38873 16254   46157 73561   85908 92193   22220 22160   86574 57842   10102 49919   88629 14702   56121 49028
7763:  76067 85241   24301 47621   50484 80758   35445 05227   51698 91880   39832 79992   31720 19402   97464 54759   00666 97759   21275 49073
7764:  16566 02082   31149 31341   92901 46515   04417 60624   01089 51328   77815 81460   81210 10566   12963 56403   54116 86726   19007 22025
7765:  79126 92166   68621 97036   54225 89962   73555 14749   14676 19369   31694 89439   42754 78791   49320 03752   08978 86459   91971 17143
7766:  12423 04558   39079 20623   02141 79900   29084 19193   10820 37112   42740 45665   88521 40412   06804 86240   37363 04079   00644 70316
7767:  28271 34594   28187 60565   43945 56027   59277 12696   70122 83996   73294 12157   76932 75728   44148 37118   10767 74650   25471 09128
7768:  09122 34774   00157 69834   36183 08410   43874 46554   60813 91781   66143 54520   84517 83812   79161 45010   74166 70924   06814 35725
7769:  82156 23532   30367 26217   04768 68624   55211 99573   27026 01600   08215 24528   78327 97081   03870 20519   79890 25734   51918 20021
7770:  97238 09002   84003 30839   00581 59671   23542 90045   05432 16841   72944 16090   60727 33223   48979 29768   93799 76563   56120 91973
7771:  62317 44804   07800 02635   24554 01354   63642 51911   76332 34446   92114 78401   70211 59030   99146 30036   81241 42555   59855 93657
7772:  15350 04403   85527 32339   01864 43245   65966 29517   19915 87537   41913 20940   34645 78597   93961 64252   11223 93320   65367 11206
7773:  92123 29934   40996 67169   84275 48697   27016 76894   08160 74231   20821 72690   21494 29115   26142 79773   60312 63547   86106 49662
7774:  81946 44606   61007 90568   25736 58185   26703 98777   59458 00170   79314 12199   53678 39126   67622 70472   23213 14525   41474 83183
7775:  03567 41881   36991 44379   64655 74218   25865 30452   40874 42175   21006 08687   55048 56402   47071 89960   41891 33995   48261 77948
7776:  29508 40164   01472 58050   15243 08269   20822 19798   72417 27776   80814 03382   70912 74260   16271 01214   65761 74025   67143 96643
7777:  34597 46476   70194 91353   03114 51078   42855 15154   59945 50060   13401 88551   82797 55815   08182 61634   83278 74974   05593 86831
7778:  17677 30602   44606 69344   38391 97775   10150 20291   68796 30393   19340 37733   04432 88453   90439 01744   45240 36389   24080 89876
7779:  34573 89553   77036 57817   07572 38069   18179 56529   93518 01128   08098 87288   83402 44466   02587 33281   80868 30664   23281 46944
7780:  50172 09523   75467 70230   16231 20261   62728 33909   04189 18252   42456 33636   31536 21931   34356 44426   51220 47153   49960 11358
7781:  73357 93097   70732 58810   84035 90026   92771 22510   36934 00600   58582 71240   38360 42256   62356 13324   43930 87347   51633 05075
7782:  73998 87835   59840 49622   88707 32615   76135 68735   54165 52309   02607 94931   21758 05543   53641 72449   97278 27515   77574 02529
7783:  39722 39648   25070 60025   19816 64996   29223 87257   29376 09223   13931 12191   04228 27611   42592 80508   01060 63796   53352 37621
7784:  77676 12312   05947 79106   38523 78714   30853 99128   75063 06919   24832 18619   26040 46467   20628 37267   05780 90954   26491 75739
7785:  00785 91472   85582 40455   79953 16079   09061 05661   49203 55696   47846 00018   36195 90988   83451 49172   42565 82439   68737 04483
7786:  76467 27726   18152 73522   21427 29823   05283 12321   81997 81433   53068 98698   83785 36788   45471 94987   57105 35234   18388 68296
7787:  50236 81237   03854 97506   87915 82031   77625 25582   51048 07457   76337 90944   83720 39952   77905 33838   19082 02301   57807 21984
7788:  73748 24445   31959 50803   59866 32788   91190 01120   68346 23745   13495 39472   86122 28591   38469 43604   28144 52451   70888 44069
7789:  42171 99531   95116 43120   41884 04140   87589 05021   90500 06003   10823 92225   71924 46294   33769 82605   18075 95950   46148 24123
7790:  60634 80942   44842 69104   62199 43788   42427 69223   95588 87335   19078 88836   76686 43336   42771 31787   73197 19371   12624 96181
7791:  98195 71294   26914 76833   55271 26311   70968 43202   08957 28445   72793 44720   52496 37443   58744 60657   20516 71206   94905 53376
7792:  87026 08416   92121 21161   01156 31855   23778 29216   91237 02515   74897 10265   20206 06637   74078 39251   59593 25698   24782 91362
7793:  84433 25675   45564 40547   00709 74744   90303 08726   13050 82562   75868 56660   08796 82911   50914 18672   92244 68392   68128 04840
7794:  70569 44272   87738 33805   82395 52162   70656 72832   69866 10114   44764 93816   93005 95066   26117 58829   97606 37256   04290 56805
7795:  60908 67054   23505 85921   99953 00280   73815 93934   22861 19092   67254 38099   96551 01694   31894 26603   75355 44421   86189 47861
7796:  37005 13524   42908 09682   84869 38802   76812 90354   73834 67145   04487 52457   95868 06715   57350 12281   56433 83307   72237 32171
7797:  23992 59739   18600 89197   37628 13945   09487 22199   26789 84555   17827 24159   32125 25008   72942 13903   00448 56245   83001 94420
7798:  99016 81692   39038 01270   01991 07815   16279 05534   88421 46158   01176 64390   89317 33213   99443 28909   08428 89144   78451 45518
7799:  12724 60082   59393 70486   21442 15183   66136 18504   93016 05639   54720 38526   54603 40595   36742 73724   14133 39927   87202 59095
```

```
7800:  30510 48615   12347 59665   91111 58705   63469 84749   01348 82138      42323 89775   19657 29499   32049 20143   92326 04700   86048 80905
7801:  19647 14306   37171 14679   18817 58573   11028 42048   38928 11583      86886 93073   29429 05709   31176 73218   27679 81309   90501 99048
7802:  17997 61764   95089 20807   19324 52142   68931 22251   15571 88909      71330 65414   42562 68416   98788 23531   01841 10822   53048 49074
7803:  10919 68933   55669 78789   73001 57498   14198 95686   55589 18350      37160 72303   64368 82062   48693 55223   92347 48152   76409 62181
7804:  04947 94295   95497 10579   47486 57011   96506 57077   86099 96791      07522 28310   96079 27908   55828 00828   75598 47657   46432 36856
7805:  49116 87208   65752 26775   20120 60721   73752 23433   91300 45305      11813 72083   99340 77698   62543 74033   62749 97353   31467 78041
7806:  54403 58697   29970 61395   75349 52150   93518 91483   31103 77611      08230 11405   68591 12755   62760 38011   45784 55448   08514 05949
7807:  43733 61387   47512 80205   86810 31060   29424 83954   82144 43056      04769 36894   17338 18460   92099 94813   05193 89262   24511 79834
7808:  91845 70461   80250 99953   46513 24140   85773 93245   23315 07805      83294 52354   26825 32006   52649 92049   89979 89178   53054 35515
7809:  77552 02880   28773 51253   97954 66006   78923 77782   97080 98349      33222 41049   58043 11874   58048 91124   35014 10285   09848 50770
7810:  57430 81765   06849 47276   90272 81394   90849 71233   13345 78850      89763 30938   41779 43664   08609 63242   35383 21635   55873 13982
7811:  51718 52584   53031 70739   81699 01600   74523 92481   02855 73856      00810 74754   34622 73900   63187 90424   30902 49603   69056 56419
7812:  19158 24949   65865 49213   46322 27274   98651 91741   41808 60628      99982 34399   74447 94440   96770 13738   34720 71292   97153 31115
7813:  57809 22320   76541 16470   84784 20606   92647 76406   40912 87817      94018 08317   65784 55408   75960 95012   74018 00340   62959 39718
7814:  91552 82297   23384 14373   40715 20642   40668 56847   69863 20655      87754 49736   98865 78053   33058 94674   39305 39701   89564 25488
7815:  52642 39307   77464 07108   56599 24600   05102 40962   40357 48312      30959 01831   17966 67836   61723 35056   59063 00138   31326 12546
7816:  91870 73178   71365 04359   05243 92010   92410 70372   21891 64982      83619 83140   33854 41073   64252 24788   50690 48665   15555 78551
7817:  83205 59599   84527 20572   57602 75261   85855 23927   73222 69280      21432 55104   94930 27578   16656 94641   11534 91349   67903 62171
7818:  09552 11114   94461 49210   30152 35504   86194 04423   66383 82825      26050 94031   47767 61823   25994 62382   91831 73006   91979 94833
7819:  08631 40324   53048 01915   80822 05434   14977 94500   77564 00416      11464 71600   29656 87969   48917 15247   33889 82310   77277 93172
7820:  94711 44865   13787 03174   63236 74588   04814 76538   28300 05809      48216 44654   57835 15482   73278 38239   55797 54651   96416 97765
7821:  84693 10266   94130 81395   75296 57063   27621 77593   48312 52820      08723 96319   65912 55266   19836 11088   51631 07941   28734 28901
7822:  85909 08692   65119 16447   00346 31546   25796 61477   30739 31318      49585 21185   64874 62040   33122 26065   28221 50858   99519 51270
7823:  15034 09818   75382 49944   82555 06204   49249 07980   11130 03145      63086 37196   72792 04243   84396 37443   04236 88675   49385 43969
7824:  57117 21369   75323 87316   16758 12412   07114 84893   62405 27439      59291 60281   90988 74853   81981 46343   27402 11826   87686 42892
7825:  46098 69142   98888 20625   71156 61473   62952 17267   81504 71015      90113 25504   41845 76165   91438 98326   06310 92791   76544 90608
7826:  67877 95128   60392 95079   11300 19341   41178 84480   37189 98623      98014 25336   50494 31444   88463 23087   77247 73913   05045 26584
7827:  87771 79132   66161 28036   99473 27968   82706 35271   33703 80028      69290 89946   19382 72990   83640 36082   29474 16777   53980 18142
7828:  96991 86474   13012 94623   80616 46107   11523 15690   79015 90744      92381 81214   35820 61144   88914 97450   76899 63331   48127 05786
7829:  76787 08662   56894 62200   63807 00738   10113 73414   01012 94500      17355 25347   90528 47342   61144 55767   04708 18220   25984 00336
7830:  38130 93354   21361 25357   49234 49409   10219 39944   07202 40636      79128 64033   68306 04884   66746 39841   66646 90518   18674 73345
7831:  18418 76258   11744 02272   81722 36457   37871 76979   43555 38382      63053 08325   16564 05451   65439 07425   05160 61738   65154 96630
7832:  49987 21937   56582 47085   13372 06514   46040 28384   13242 00520      83363 16750   21771 08039   89756 08158   69393 81130   86476 15352
7833:  62064 16486   10845 63325   21768 91338   80439 34188   69097 42931      71814 07580   85884 09662   87899 75466   79398 17891   14055 65970
7834:  54233 06934   35187 44895   03612 57318   41127 45321   63979 56945      74197 41455   26455 84278   37418 93408   79791 29778   23400 83496
7835:  42745 47355   59841 03013   87751 69393   19939 40293   46646 34228      58168 43573   45641 61571   08492 96382   23980 05140   88185 99460
7836:  07857 29163   08028 92955   67205 19153   46686 06331   45959 77800      03825 39245   75297 71750   73127 38390   72193 62722   72724 68609
7837:  86406 18902   54627 32229   53123 33878   99480 67013   67763 41928      48543 87717   86902 11560   12924 06926   50885 49820   20817 05051
7838:  82723 73896   02833 33494   28696 00614   68595 95737   16685 71322      39904 48072   86851 46979   01488 01647   15910 12158   17709 24816
7839:  38730 47229   34470 21419   42866 01421   80377 83644   62007 78552      90975 29905   99797 91263   47404 25115   44729 94642   55599 08652
7840:  90429 22295   21315 60259   96047 50725   64659 55935   79676 46441      93043 69545   66603 85240   90947 32947   21697 12544   83351 92709
7841:  51034 04427   57802 96999   49640 57694   19574 15393   64961 15110      14614 24636   71680 11349   37158 90254   22060 51463   59461 55426
7842:  98243 17405   49800 08700   29521 62634   11321 78947   21034 41366      01596 72102   30720 04459   19729 29326   84953 23139   30429 22992
7843:  23084 76472   37010 75299   21487 57149   16501 14247   34419 59254      93723 39425   17079 98527   00096 01466   25515 02114   27971 86066
7844:  25094 42843   86808 21387   18497 30151   18887 79643   99190 64442      34790 36418   83512 00982   37507 08332   63970 60166   82646 00610
7845:  82642 11145   63631 97339   49193 82262   59002 12680   69172 83737      42353 70093   11111 81748   02133 52971   28697 96110   33246 19425
7846:  08788 73147   63165 64282   83580 52309   64606 40667   19904 04519      46739 51772   67274 19186   46098 38245   29424 35933   20490 70271
7847:  90965 74431   23178 13839   22470 04308   24428 76221   78863 55854      21069 71648   69952 30357   20590 55234   62296 84517   84093 80425
7848:  00514 86142   73173 59067   44756 14573   09222 21435   30701 48016      13008 78640   06592 38176   80293 07489   71936 28954   89749 30974
7849:  00614 83641   08341 02213   16130 49601   02377 20714   43494 00646      23021 11102   60080 07717   74316 12222   60465 82268   74334 40563
```

```
7850:  07559 41299   49751 65003   10631 93146   92757 24301   36241 21738      07799 68551   50798 69557   08515 13571   17966 45204   82562 24585
7851:  02505 14529   83508 73978   85835 98092   89785 61220   15293 89391      31933 68156   53961 37310   60007 72558   81124 14402   78952 52103
7852:  15267 48905   71383 11999   80461 85552   41103 11480   58448 39289      15499 03238   60382 96930   41825 38438   96155 35066   08844 78440
7853:  89438 19920   40053 91161   27445 82371   71365 13786   15893 31251      73771 91621   54526 09514   49170 41380   23575 55257   65263 88323
7854:  10383 40377   57514 93476   23333 25004   68130 02504   20335 47153      35137 65300   39408 91166   11739 38475   07302 90709   76000 94638
7855:  76345 81983   32049 45668   42801 83817   94413 72243   22128 58424      59697 31988   92677 17241   73318 61299   15404 92976   57582 42968
7856:  06337 12333   35083 99927   32482 20841   80248 31119   44581 81671      19343 93884   99279 54529   67719 29640   96082 51240   99068 94712
7857:  22919 43021   09137 08984   55896 04334   85211 13089   14801 49951      71869 87473   91173 23959   24729 17157   33892 66227   48396 37916
7858:  16167 51306   16029 04467   12876 52678   13811 08476   06833 20894      89289 52774   14777 93530   09986 50923   28874 77880   12489 01514
7859:  79871 31783   05470 51137   81670 83965   73515 72240   90224 01834      40844 03317   18722 12213   97756 43607   50543 68043   43574 32442
7860:  94918 72069   18973 69633   64116 09554   66152 27633   38651 23476      13339 58280   47086 86127   67050 70538   25419 46950   91739 34935
7861:  69907 89775   31567 46228   29888 90712   71624 05408   38426 55458      23001 46971   93488 30192   58409 26873   48209 46481   17363 88190
7862:  80092 04193   48666 61351   19953 21691   34131 14595   10502 15324      21807 38658   37056 53976   61051 48998   99727 55545   27514 47477
7863:  31549 96927   37432 87878   11724 68802   54434 29719   36840 09090      03075 89214   65655 99300   13068 44882   66983 94127   53165 08826
7864:  98617 68172   39497 88244   60411 39650   30351 38124   97862 79544      01873 64798   12344 35739   66898 69982   91483 76058   92880 11672
7865:  66163 70837   09524 43196   12233 61806   58499 68931   29131 58837      29188 40524   78782 41695   65033 88469   79268 36439   37976 24993
7866:  47547 71147   89117 83362   07094 04241   09239 55313   35299 17006      32270 29278   27248 37621   06417 49990   35204 81834   00812 83952
7867:  24002 17509   40432 58794   31356 70592   68114 84312   17225 99845      43055 33949   55994 06777   72242 77299   07355 61844   53256 16415
7868:  82139 40512   81692 30529   27978 43752   95922 93843   70941 72702      08043 08136   16057 31225   83096 77899   29504 48887   12155 91229
7869:  06461 29417   03276 38490   60236 13673   58556 23577   92244 93733      40664 89988   58410 63854   40874 67564   24459 99161   21950 54392
7870:  47101 16599   97381 09808   21015 67208   96844 64356   87021 90120      26021 56180   01496 99070   70134 67343   61191 71264   16725 92772
7871:  32986 18523   42280 45228   11890 84480   56674 66780   48614 39483      72706 96670   98379 15216   32138 44933   70338 72544   74721 86204
7872:  94421 23370   09571 35130   50115 09034   03837 86918   52155 78416      22674 78273   72786 72941   23127 04097   12993 86909   67845 22496
7873:  19269 93440   08068 39045   16547 12991   53028 03136   78184 95481      28722 76251   91406 62822   60271 16205   24297 39297   10938 46930
7874:  50307 04799   82145 72217   78779 86781   24815 34508   34707 25773      22481 21258   45398 01245   67035 49698   31404 74084   67574 18391
7875:  44894 84518   13811 77268   28153 51474   78370 68604   56214 74143      33952 82023   62315 55248   78649 31233   38408 84943   54369 63498
7876:  98215 64563   10138 06036   42316 51111   58324 28890   49750 71588      30067 38593   65917 38599   74470 18132   95783 89517   37564 02256
7877:  54243 56313   12927 44914   90660 95493   37206 49039   19903 06747      67248 62270   64357 58933   87001 12631   93352 93699   12614 95006
7878:  22418 53126   75175 88533   51351 60451   12752 48994   83248 51104      91116 90369   30274 27482   27891 26345   37150 43307   22883 16166
7879:  60349 95057   06797 10508   58351 04675   91248 82084   98236 44124      03075 04577   08863 51740   40524 36128   18977 03976   99548 64110
7880:  84546 43793   05420 27475   98037 76348   89434 67779   16784 88221      79562 03450   71561 60739   82725 51092   89540 42371   94802 49637
7881:  65190 95285   28490 81794   74449 99023   60630 40715   90498 74662      59565 72663   74907 67959   26545 89583   33405 43989   67517 51379
7882:  29276 25599   17293 68767   84671 52502   07876 83614   13916 91510      56550 03384   71711 25185   89829 11286   04320 78435   18462 84637
7883:  53143 08790   18111 14362   29882 91294   99997 37783   57870 66806      91866 60260   35442 36957   38749 91924   57753 41490   54623 35714
7884:  55469 12091   98296 27195   01074 42691   39329 20868   42138 47732      06379 61869   55219 98265   26475 79044   35423 62978   74857 41395
7885:  55062 45539   51699 11833   77867 58866   39457 05536   31759 72759      70935 51904   92190 91710   44921 64021   04104 50862   87096 06250
7886:  63831 81049   66268 39028   05706 44454   05880 01746   94857 65700      29180 66086   85820 47943   29297 81221   92211 72432   70128 04460
7887:  93086 36711   01105 85359   16017 33230   48876 04060   50452 06362      98496 15112   20598 80308   34040 09391   67910 40229   24526 68257
7888:  76657 32225   85822 64849   10654 94581   13542 50304   73823 30097      28731 53192   44226 16562   53292 48011   53256 18009   06436 82762
7889:  29917 78033   69440 94825   40408 62591   66367 52281   67144 11649      38345 85957   48244 92918   40733 27363   69367 75482   12246 65683
7890:  57087 77376   65324 46143   32411 04336   10944 09599   82066 95840      05018 44028   49259 46274   51582 63167   56007 67481   82088 18933
7891:  33758 42689   47923 36894   78000 85884   23149 83989   32382 61948      93772 82282   97027 86370   03537 08370   99720 69353   29798 33669
7892:  58188 37092   90504 90543   77408 18159   93778 86611   51456 77565      99106 37028   76916 95919   31638 94618   21688 01014   68741 42027
7893:  69879 52368   26255 59223   39525 02784   11387 32228   49733 20007      86012 42372   62735 48411   47233 80098   41849 38451   85163 15450
7894:  66193 10600   39866 05495   01398 62050   12301 35719   05936 94710      94916 56471   16248 33049   49786 10609   73057 20831   86753 54250
7895:  88767 38947   16000 58917   91150 05779   62992 45184   64912 25167      10596 43290   35061 48557   55061 36629   23381 24729   58408 84088
7896:  24488 29205   35237 98101   97049 12046   43269 40037   16181 74867      11398 31351   90480 29856   79013 58677   56762 78015   49450 98897
7897:  32864 91428   91336 59744   55975 83019   13026 53889   64885 37621      11398 31351   90480 29856   79013 58677   56762 78015   49450 98897
7898:  69036 57938   54026 65574   32467 92028   70263 82936   88872 31012      91860 33827   48031 70797   87571 81596   24196 92025   45881 73201
7899:  55523 97942   68447 89844   87813 97051   07978 12033   72110 98307      80722 91336   68110 33713   18118 39559   39077 82626   70343 93948
```

```
7900:  41034 97013   57354 03379   25059 12891   92910 62872   37300 40680      99061 25914   02936 53722   66713 01445   58469 13475   58059 94492
7901:  73587 13140   67274 99455   24978 70065   03646 28605   03987 51787      68532 54014   45871 09939   97858 63296   43083 52932   59258 41320
7902:  07859 59553   30901 36877   89330 84361   86292 78663   47164 47235      78817 80022   52955 37693   88652 33214   44741 45438   66977 10501
7903:  02502 08460   07350 41304   82616 49452   68602 83045   91744 68815      65208 25512   78270 06785   91774 29830   51222 43400   67654 09421
7904:  39380 51532   03178 82398   09751 57341   99030 83258   40969 60462      64957 57321   59791 73374   21418 25565   61985 75008   99735 17730
7905:  47670 84334   39340 50349   91514 17525   76298 99819   18781 90930      84546 32559   93179 70689   01526 34705   73348 45730   06052 55353
7906:  44936 14174   53235 41487   31751 37041   98269 09508   92674 98913      65182 24291   35800 25454   93515 84171   11353 79751   70481 29868
7907:  24073 35627   94807 14495   17332 22312   89504 00510   44395 28150      68205 14209   57651 06895   19232 77411   90880 89979   50558 53633
7908:  58564 74887   89044 84919   13431 09519   12378 07641   67226 86258      31791 86258   88227 06665   67025 43824   29199 72298   93714 91380
7909:  52836 77381   41169 95231   59306 81246   64181 79121   65088 33432      83085 00854   10715 99724   66056 81944   19438 32429   09073 96243
7910:  21859 93312   23475 19254   97585 14251   42047 09632   51298 86495      53651 93680   71455 83970   29558 28736   74770 82720   73615 55490
7911:  86107 13634   20592 37265   62748 68563   93365 28829   52846 28220      90392 18214   72202 69688   46474 52441   14500 10700   73295 14085
7912:  62622 70877   84058 27904   84471 04267   41338 56722   11126 36915      04339 80986   74306 19697   79027 70126   74978 42854   66022 77600
7913:  66828 24109   06461 28966   34590 08955   76269 09694   14896 34403      38204 94818   59734 16402   28368 75078   69247 40024   53466 47521
7914:  10726 14970   22892 31348   67149 68765   21238 56636   90165 67308      12111 01296   06663 23483   94825 25823   45109 72015   60456 49553
7915:  88732 96720   08321 68013   81390 71040   06507 21135   49440 50719      89131 10759   30355 48577   22738 07406   23395 28409   86045 97479
7916:  59637 61652   12240 82065   14007 16364   94596 28663   87240 04225      42757 00726   49946 23838   12198 04502   57346 54538   97604 08124
7917:  15477 85547   02177 58539   81330 44393   40009 74579   40509 59251      28061 03724   64713 77010   48125 08464   75091 96367   63303 62995
7918:  05320 46402   46492 13076   79796 46008   22178 30446   41552 80589      98692 10451   43500 14397   15274 72154   53959 90345   04277 77460
7919:  03431 16573   18451 65026   78970 43387   55469 79924   93839 64548      47952 24164   50999 89592   40418 77820   90327 71998   34980 59207
7920:  57624 78466   74549 82286   63707 36542   85408 20525   95134 12693      22794 23381   76739 92173   56405 96066   70174 93998   29289 91154
7921:  29905 77300   85252 43529   08942 69960   36471 55207   30502 08054      76070 92958   07631 09652   32087 84691   96760 99971   57224 84117
7922:  60027 46483   97549 26576   56848 12192   01811 85744   88071 62444      25884 21733   17117 63056   49642 79250   46155 10759   30654 45670
7923:  02823 23322   41222 26033   83405 99968   12942 03638   38395 72133      59657 61376   19829 33548   43865 42630   98352 62693   16670 85472
7924:  25450 13385   27870 40666   22289 65151   12391 86610   73312 46934      06260 24284   49247 23240   26667 63228   91313 21454   65457 76401
7925:  30861 26577   04022 21183   79845 88937   02646 55044   66178 87688      70349 97606   75898 28794   94980 98269   27159 98476   91370 17420
7926:  47551 10897   01283 54237   94052 93017   05088 86421   28169 29209      07463 52754   15036 83569   72960 11739   35500 95761   81952 45908
7927:  89475 32939   39874 49325   79129 48204   46960 93371   99121 15577      49282 23178   75637 00798   51291 91670   84225 29912   46276 44300
7928:  23214 78628   78871 64562   65076 56790   02167 12405   05812 43374      63916 35977   91275 54934   48751 50121   62667 08734   78175 41176
7929:  81695 41309   18239 74155   57818 83115   72747 60387   59798 62603      34063 47325   98860 46303   39399 48811   26646 72435   89331 38536
7930:  98491 14292   38976 00732   32319 36222   29811 92542   46914 17680      47018 25182   96074 63990   94918 02907   47154 74597   25098 47438
7931:  32431 30063   16109 88309   49678 12677   57820 29124   24508 49038      62071 20202   97151 75501   27957 11976   52881 97313   64007 56389
7932:  30426 13105   44786 45775   68869 93242   69921 32536   96618 64121      63559 80007   37415 59665   54399 53041   05464 20569   03621 18918
7933:  41693 35049   71612 19302   68751 15819   83939 28004   06954 18436      96074 67426   22348 80850   78427 37222   63271 54545   03960 49470
7934:  54381 46822   06084 89995   95179 95509   96077 98475   70862 13846      84754 03333   66339 52635   18043 40487   80696 89474   03592 82204
7935:  62043 19647   98871 01240   76462 12615   74652 08960   77545 94042      30238 02081   85593 58952   64209 46599   04940 35498   13902 84538
7936:  16033 29390   84196 52207   51772 66173   21640 14029   51569 58130      46685 87004   51950 07369   67552 32084   66223 88402   65453 10276
7937:  42916 63099   45011 16515   05613 36666   80612 13989   50183 83742      87921 98697   04552 36500   83237 35888   69456 93985   48735 87928
7938:  41581 91291   77381 79115   03211 36415   24084 14503   70278 68664      13578 93112   84465 70755   21552 43392   78287 17694   13281 95627
7939:  50626 81537   58197 59351   93905 60945   83246 74996   44002 08797      49600 12209   16545 54725   98502 28891   75933 22673   34550 49687
7940:  86323 41522   61203 52216   22028 36501   64391 33461   47644 12455      54799 14282   56260 21602   90245 85573   52545 60939   63700 38321
7941:  74287 65711   76003 37975   82477 02737   76244 40077   46235 61631      48248 61367   45826 57158   19135 26178   88373 87854   00184 15685
7942:  38434 02702   44275 03198   17789 03630   68314 32474   40332 05010      12582 27980   79321 55284   98865 29643   92130 14059   63439 86489
7943:  29983 38596   19399 19787   65621 79870   38927 01729   66636 54376      70172 01530   00998 04058   66691 76061   58355 48034   36732 07458
7944:  54682 90994   71176 90282   42132 78786   11834 70860   49220 17464      16063 40246   18502 80024   59952 60480   15935 50031   30558 18702
7945:  55809 82281   92742 62473   48364 81026   16875 81797   95014 60901      75758 97542   78361 57339   37859 06744   35491 59123   28159 27098
7946:  30276 81853   84130 67391   23974 08893   70579 48787   95387 81211      85147 99264   30709 69372   40578 34367   45797 97626   76847 26103
7947:  78485 86523   80927 53443   06771 77581   26706 76014   66689 44717      40521 76554   91404 23461   93619 18696   46880 24525   51054 30602
7948:  92629 59402   19132 33209   44110 49987   06657 81721   58870 75274      76400 24392   69996 60099   47684 64687   98042 28456   27666 74733
7949:  23790 50744   46440 35814   16280 35723   53837 79512   07670 19277      22384 82517   44682 57263   99867 46555   84032 59972   78725 18279
```

```
7950: 88338 89110  51297 69336  04340 19188  88425 48208  45056 91666    69087 94890  40748 44259  82615 20620  05064 58994  51510 13896
7951: 96402 70685  61686 12763  60093 21082  97528 59131  95568 64968    16557 24504  30781 48329  67889 08391  78096 92338  92722 83415
7952: 11309 99796  22596 69128  57183 19583  67869 03455  45826 66166    14428 92840  48985 72649  71911 10928  87208 32442  62940 39537
7953: 46694 80488  84075 96827  70158 89106  09561 83121  91918 37995    99039 20111  80700 54004  74654 72622  17387 10740  84394 17203
7954: 37086 80573  53514 74439  89906 56220  15639 97515  24438 78999    94805 63458  91899 77043  37008 39856  00540 97555  81852 97225
7955: 75449 77326  04174 40159  59264 31625  52003 56734  61561 16688    89515 45039  07498 41352  21781 90552  87188 86512  21347 70313
7956: 89067 27313  50639 12863  04518 03601  00385 86237  15694 13545    64211 30640  98632 08765  84961 01746  16300 37609  05240 99202
7957: 94743 68534  98069 33289  23108 11016  11909 79724  90935 36138    29573 16629  17682 23470  83731 37123  40361 42415  44256 11008
7958: 00809 81161  47937 54992  36806 57811  01686 94558  55248 83158    40826 07164  61346 66427  79123 98625  86665 46296  99644 44189
7959: 93453 58670  71902 46819  03883 97497  32235 78174  93428 09191    43754 18687  70016 00563  96124 64726  16639 71837  93730 72123
7960: 35968 19386  74348 72148  34412 47778  34228 76191  67033 65380    58509 58870  26954 38115  83951 70548  55125 70685  03289 54125
7961: 26036 36377  44925 35032  86863 76086  60623 12072  12027 10447    73521 50154  56187 46585  20632 06978  35432 04114  35643 26520
7962: 66970 14199  51899 89109  20582 25864  17282 95195  85793 38631    04241 61278  33593 01737  21106 97450  90740 13585  04473 29843
7963: 64580 30360  07492 39105  83212 35966  01062 80707  15034 63618    79373 01355  05149 43457  43547 40182  23696 30841  90080 23339
7964: 93425 33159  96408 89709  19859 49361  82718 42337  03676 81487    96473 81078  76680 87767  62188 33053  59774 01911  63873 16168
7965: 93370 12697  45688 11427  81627 04782  21781 06444  71243 52727    54139 68581  59214 60991  09193 67519  14574 57338  09117 96430
7966: 42171 82825  91516 36336  92476 31284  23457 28447  44933 93384    41552 05671  80350 56219  57477 17980  28502 89811  79560 13552
7967: 57646 29973  20134 37045  81344 44815  58220 63737  02746 22509    04220 90960  02266 98100  67663 05760  42790 70135  90143 35330
7968: 78017 40707  74787 23244  25414 20734  98071 80267  63213 86356    41149 08184  32180 58584  88049 23308  87259 28833  63036 74975
7969: 91583 09443  91453 17419  95557 66368  88149 34220  22739 75843    22004 14978  27215 01681  54994 84193  93298 17056  69049 04771
7970: 19691 81412  40225 46806  83344 60929  23219 23122  84354 98309    77252 19472  34191 00702  79594 98434  02302 46325  38681 51432
7971: 20992 41086  05381 73222  43247 65252  96778 30755  14445 52668    27904 41038  06226 90328  23613 98735  08744 68278  96799 71594
7972: 49167 39818  95198 82938  82553 09128  61036 64427  04096 02780    58999 10748  47179 47072  67567 23240  15634 51851  08064 65022
7973: 49698 81642  98035 71707  40388 34290  12543 01692  16153 87127    73469 73083  30349 88858  78629 24281  01162 52530  49803 41000
7974: 17805 60699  25820 39312  65132 28473  40611 80286  90759 13060    35463 02789  31645 56466  36245 73146  67188 10964  75741 16468
7975: 98840 46508  85079 46910  33732 03381  72992 06453  78412 84190    40022 09241  13439 34927  90768 05117  89435 73634  15664 77106
7976: 78475 89961  82317 99603  03515 50772  73619 52610  31048 68137    97753 24955  28125 04543  78960 66060  19405 14713  89664 62357
7977: 81873 20331  54368 26949  31946 88025  32417 35206  54128 65454    60965 20365  61291 72292  64664 46692  63210 90764  76759 79011
7978: 85920 13398  64915 85172  30282 02127  78593 78359  18287 69434    86546 76234  91966 65461  35582 14729  15234 13838  43156 91755
7979: 02300 83383  90903 68589  71012 38442  03358 67492  77691 40380    84505 61135  09815 72321  92904 05824  31830 26803  06616 07664
7980: 01512 37320  67211 19793  85304 13775  55787 66821  32570 34626    32620 14807  25936 79462  94969 55119  02759 02005  53632 38640
7981: 43677 07443  04722 27163  50307 77235  08588 90860  72868 68580    52876 11935  62053 82996  27240 31555  25998 65401  54040 39158
7982: 78242 59775  39821 16683  64544 07518  82509 98900  85695 32513    70280 09383  37670 74248  14914 76410  31748 00235  86496 42941
7983: 62440 20756  31662 37994  98826 12380  48281 51388  35213 50082    00915 07721  68605 95220  80302 12472  12286 10507  93803 73490
7984: 51930 18040  30803 65045  13386 23954  18309 91200  86176 20103    48651 67341  79605 56634  80836 92937  48836 05267  71594 53693
7985: 80946 24333  52642 65843  48393 38733  99188 85416  17603 42613    58161 75886  45793 82723  39913 13329  90089 47269  37210 51476
7986: 27883 20679  18474 52338  66137 53368  36698 19905  65138 25010    81418 12011  72816 63405  16510 70813  57709 99251  96230 50970
7987: 13371 14998  83137 60861  56077 74870  37768 37724  45838 46216    57297 54372  01360 38904  63305 55432  17179 06631  35333 84209
7988: 54614 73084  01812 07555  87921 54643  93850 98336  74993 83615    83116 32111  05749 44602  92368 61924  01672 22012  24317 41050
7989: 83116 32111  05749 44602  92368 61924  01672 22012  24317 41050    53517 33930  23184 31171  58176 87590  31642 26780  33413 53004
7990: 57967 97326  58651 70748  36776 31792  04771 15596  45386 83880    65956 44195  65412 98971  66701 52746  00807 90995  88102 98697
7991: 89972 64615  10085 93268  42463 43041  39601 46331  45083 18473    69842 89514  14444 88605  36878 28802  18870 99682  47423 07508
7992: 62959 61780  95881 58081  15886 22664  24000 61974  45741 07004    70221 60569  24426 35833  54344 23110  81114 68505  75636 73738
7993: 40664 02026  72594 15278  44135 11486  33267 08526  48817 96084    01275 80857  88348 00849  25085 51497  74757 27431  04824 15191
7994: 08698 90665  70929 51380  77698 05110  19809 79319  91799 74802    20463 55221  29580 08511  56652 79029  24879 72893  75405 82752
7995: 13535 84189  22077 99886  70936 01485  68647 53186  55562 00933    63437 40096  23525 80364  31705 89816  32277 70650  87555 08002
7996: 40097 78938  73022 41941  88570 41321  73348 45569  82557 47463    26891 90765  45038 35459  00753 12876  23391 81058  52743 72700
7997: 26072 01138  65717 91832  83214 43035  37596 68361  64585 58568    13957 24980  48492 15345  02045 12393  76925 35437  96838 88517
7998: 01985 74068  93442 99996  58279 37798  70190 63871  31884 36494    96113 67878  92507 45822  44595 57619  73988 19040  38022 23470
7999: 81143 47222  86052 06516  19277 78891  36085 37220  61897 73838    13031 11583  17002 57172  01640 88933  06465 09482  99091 60573
```

```
8000: 18025 75758  24976 98111  56181 30859  01302 20871  17064 44388    13291 40859  29821 21978  78149 44793  12329 82023  28999 86959
8001: 55263 66426  94464 98581  99069 67296  08594 84895  48299 51797    66360 25549  26275 21528  35409 90175  91374 25599  13171 87983
8002: 51419 75611  78675 59721  81951 16907  76129 74757  40357 49792    46392 08600  30814 13841  14642 55264  29433 49415  56810 40626
8003: 88514 72407  16951 56094  02909 50813  42705 73632  44314 17866    97457 44132  55292 70851  36408 31639  57079 66666  08943 33380
8004: 80559 67583  01586 86221  83265 18746  89419 76051  20739 98220    18150 02485  20988 26915  70266 61438  90898 02024  61203 58326
8005: 54744 26508  60305 54674  99021 90635  41129 42777  31968 51067    04993 93946  30548 41011  74294 11535  71761 58552  77149 52055
8006: 89892 45311  79353 84466  73764 39221  01562 14234  06760 77507    81607 14745  99090 91630  79347 18418  27592 32981  30872 28952
8007: 26206 25981  42237 04587  39758 76151  27207 05820  82065 70239    66529 90425  42821 30996  89189 32296  20605 85362  44673 08751
8008: 62227 77235  73351 43472  60088 08383  50768 23567  97570 22722    49577 02820  60306 35504  84031 08851  05892 64688  60403 84732
8009: 07277 09653  30029 05459  26870 73359  98618 95391  37228 81056    88567 89239  24931 83636  13972 78994  66201 52529  91985 87555
8010: 08317 17596  62586 69012  75717 73067  35373 74767  09468 10345    74499 16434  01104 77793  21809 44697  89556 87761  43455 02020
8011: 07032 70290  44519 61105  51995 86145  41778 57900  08944 40586    48181 81459  72481 99851  16911 26907  00562 59783  20545 08919
8012: 03904 23064  94018 48923  49436 80322  69435 92728  55084 53086    00542 59438  46248 69592  33678 10126  51112 36574  15461 97506
8013: 05061 11596  51939 19483  66865 67075  06498 76302  49128 76068    97201 98350  11615 47569  00563 01718  13501 96501  28293 84417
8014: 44069 07212  48757 75837  74751 20270  52132 92470  80806 25253    56389 76632  94793 88785  68952 93186  49206 82581  11541 60642
8015: 25839 33914  21030 91642  19292 60383  38159 05900  00769 92607    62043 26488  73946 54886  25624 64086  59313 75108  19475 94397
8016: 71029 50906  09669 49895  99567 68660  89095 07368  57608 19567    57983 22841  71853 65189  78764 84664  54410 89622  50523 89131
8017: 13294 07884  34024 39630  28662 81991  24633 46090  65506 17504    37286 45119  67905 14131  91049 10850  90568 44562  04703 00776
8018: 98001 21558  21842 57188  72384 30575  92789 74857  45431 19969    33738 89252  45793 34639  64227 53031  76258 33983  20060 74468
8019: 68083 21105  01283 61450  71348 75954  85468 96178  84747 99791    38779 68244  29724 83829  94080 68006  91842 20453  11207 02806
8020: 74434 49690  22589 03552  22108 60710  71229 74866  35489 80491    89286 02995  87442 47181  22304 47969  10487 89204  54096 32960
8021: 86565 96558  90164 27616  19141 73952  85122 56063  36106 60654    78521 91994  05824 22417  65730 79593  36454 64759  60542 13452
8022: 43549 25053  17339 46638  00790 46841  16651 92966  64433 18898    59671 68791  08690 27022  05398 18020  10170 61900  47920 19401
8023: 99553 48849  09050 40821  39871 87618  86564 00415  36448 71121    97401 45674  15218 65727  02471 38157  18434 38171  09638 06315
8024: 99302 93657  90531 04445  93664 60398  98135 58012  32647 51628    99081 81360  02008 26223  37284 58095  26816 21979  04103 78171
8025: 46151 60068  85054 17569  47417 03058  94196 33582  19443 85780    43726 59576  46626 55915  55409 34825  60064 61753  48828 22619
8026: 17933 30940  89955 49332  95137 62120  95494 75765  30470 91943    58004 46917  06311 88123  06045 86568  24977 33536  81334 42203
8027: 84821 37354  48255 72690  37411 45169  05868 08817  42316 42852    30204 83196  73784 40121  64324 65758  46322 77702  10016 63929
8028: 84004 65375  52146 28258  23075 02961  32205 95471  86173 56817    29786 01238  26379 16676  05442 25490  32473 60522  12483 88946
8029: 24657 12119  99151 97146  24499 74563  92842 37298  18682 48520    15525 05468  21271 58507  36186 30060  20537 15147  71634 39891
8030: 98689 01238  81379 49000  36770 56292  43671 60438  75871 59902    33994 64007  45718 00349  44515 94280  33782 29622  82790 65706
8031: 48848 34579  03695 61514  58095 65238  34131 71736  02645 36654    66642 58498  40643 93808  31017 35977  36045 04919  53875 86588
8032: 42238 46369  55435 38885  10693 19909  48678 99265  00418 76876    35033 16148  81363 71963  52053 84040  45959 76028  09484 14964
8033: 29613 67215  41502 83453  04522 26796  20616 94336  61840 13768    32843 17163  24166 56423  98439 83941  96251 79630  27936 88469
8034: 53925 70722  33334 24826  26212 74983  22865 74105  33475 13879    57483 52220  93693 80069  79780 02068  83826 50412  77445 05005
8035: 99447 52180  51792 23506  06814 06389  10321 38112  03228 47888    36914 43305  72456 90335  99296 80386  80476 68842  22482 13763
8036: 13835 99126  53160 53002  51625 29340  08785 81297  71306 61705    12547 85950  40012 22998  71298 42840  14926 97997  90868 64798
8037: 73819 20835  18551 17785  52102 14029  16339 42516  23989 65665    66883 68868  06620 55290  34236 23532  41678 71476  61386 86004
8038: 86299 28353  69151 35300  25930 93448  18482 98817  79623 46507    87101 06008  96444 71661  29893 49609  74185 35033  71827 31752
8039: 80671 74733  02806 48938  49367 79334  86863 75547  60784 29865    56900 61167  63144 21936  96381 69021  99238 92718  63529 23069
8040: 60026 99563  13521 27991  66286 34107  76606 82373  96606 37821    76162 80987  46627 67851  94787 86876  46052 60237  99210 17224
8041: 84577 57903  66821 20271  17179 60505  49101 87510  06685 52370    74608 82238  38819 22053  33982 88241  49823 35289  46307 07650
8042: 52294 27368  96466 25698  29626 61617  11858 27127  54081 54363    33106 38072  36249 46525  36251 69513  27773 39138  55309 68540
8043: 97686 19518  47711 54188  46441 42319  54485 80623  00397 72635    97766 13946  26137 69060  35081 76208  07447 60401  70576 73348
8044: 37414 45133  70175 66989  20666 06677  72468 54595  03446 12763    05348 54751  33707 47227  42932 92616  19276 04701  44099 71693
8045: 74805 56246  52333 02734  64107 70047  73558 01406  26452 69051    61842 42506  30845 49136  42890 25038  76227 41882  29025 19177
8046: 82589 45409  99088 44918  27362 71108  88804 19424  01758 33542    14332 01961  05892 97930  85834 85142  61863 30229  60742 50200
8047: 04750 10850  27589 06906  33652 57975  39752 42064  84568 88386    34840 02464  06142 09659  23046 95393  14591 25291  08117 68935
8048: 57044 21094  88229 33126  44449 29143  18381 63016  91295 19927    19019 62845  03435 95371  75718 96769  15129 10177  77055 22828
8049: 44287 11630  31162 68465  09528 16007  25296 23620  68867 30048    90864 41073  94657 18357  77556 99811  92979 62143  09914 00700
```

```
8050:  9941396257  6903433531  8691571510  9786095247  5085335322    8222690481  0765906508  9590256868  8443792373  8694009239
8051:  2047807244  3692090991  5966278688  9495432902  7698453604    8918742286  8176842980  7738118050  8142095547  8525875747
8052:  7593897799  2335562799  6488463618  9695959897  9780837375    3587031428  5371910235  6867725696  6144485627  9303688839
8053:  5816193574  7506360344  9727673637  9788674349  8288794140    7216409721  0313234178  5297700064  4531623617  8436215186
8054:  8871481241  1459040110  6522498704  6362942827  7557979826    1436566957  0099867814  0497787798  9686984566  2833174192
8055:  2162839372  4880447967  9625446728  3411691120  9093416799    4816244320  0162012893  4754442912  3840859282  5240228247
8056:  8903690678  1074297243  8682618865  9506522903  0127556238    2779403686  7074004917  2092593713  2709703981  4263817653
8057:  6222492201  9119593581  2583572731  2623911386  0091661602    8296896159  7259228339  1210047860  6567970636  2122280656
8058:  2311469160  6673934941  0912744036  8509985986  9304185782    9757846789  8843404064  6187174328  1371624431  0146148770
8059:  7540353158  6898777653  6956085327  3366265466  7851088474    8905984830  7007046322  6386080780  9621923178  8649907161
8060:  1703600506  3052276243  4606234055  6493074631  9870318335    9761459239  5839882430  8764872351  4015532011  8640905552
8061:  7220810711  1169354139  7504908313  8136127905  2558823297    7960820254  5969649696  2268240761  5390520539  3413125478
8062:  3496737649  8101885850  9217268327  5455039631  1387239155    3548614281  5088973745  4807663947  0485723366  8012078003
8063:  1146248267  2100474979  7865755756  6538194078  6440552409    7956293067  7971990366  4328150863  7226742312  9391597395
8064:  4129585015  1592581298  2866452652  2315879778  5151436189    4563345563  1900557708  9561530609  0660813283  0220648328
8065:  1171557229  6381622409  0434764459  5167626056  7226093151    4645402667  6693982880  0492901067  6531879734  5318257046
8066:  0438992420  1135261585  3533075438  9936872451  5708454133    7468084284  8161979299  5823359654  0184909083  7778985810
8067:  4683082826  2016756707  0031051693  9366524637  4347353774    0678203568  2232108856  5459193288  3846485888  9975328843
8068:  1676968649  4525579699  9872635055  1214272277  3352291534    7525797060  3711066801  4749666515  7165429623  5447233575
8069:  9383446709  2088550178  5679933792  7839459894  7085902430    6396205946  8151162000  1174111946  4521853734  9316919420
8070:  2330784760  7705682970  0622362221  0773892436  1778543723    4180691795  4780451549  1197039624  4639560595  4235095122
8071:  6459955518  2761021156  2729255509  1576536768  6686984381    5884943768  9481483349  5486615758  8454267207  5985808937
8072:  6493650882  3127474758  2242483163  5270522257  0786541233    0922242875  8247688003  4341303979  5671717478  8631423503
8073:  4665739229  9924530229  3916414595  4212833105  8380046018    9474682538  0835612562  0279427700  9490851397  4965423409
8074:  8519424030  4642552582  9297068850  1010794210  6771964638    0938387260  8964090678  6024918577  5558761019  0295115062
8075:  4159075406  3288153280  8274726503  2418161291  7356819695    8520265604  2785876944  3660834716  9605403634  7576856501
8076:  5323034880  4256338040  5693093661  5235930722  6250615590    2038514419  7578058611  0568376956  2918545982  6521561075
8077:  8310241465  7269617811  6911548582  4313821032  9792168756    2362733062  1049720975  7044177517  4092804782  4106395100
8078:  1776724605  0906605356  5750935817  0146306601  8403028715    6691954464  4358997572  9864535467  9471022986  0866793640
8079:  8619027123  4307691034  2397622433  8786189661  9934269472    0691085607  7442836949  3977791956  9573234250  6896848807
8080:  4001633944  2514025617  2191829344  2031525699  0965615962    0684892493  7991535193  4468626504  5340951791  4434704471
8081:  3843770643  6125359640  4203152569  0965615962  0642660284    9181645201  3960749980  5277090023  4049630737  6519961657
8082:  7410658185  6167502625  8172397037  5588119436  1044238015    8746553212  8015803088  6853635433  8458032861  4854321805
8083:  1598981885  7839000474  8526092008  4535803153  7389452429    8587638605  8540265231  4633718405  8604352660  3189636343
8084:  2300464660  8428655985  3310960611  2230637291  7774983830    3314074128  4533313142  8816016150  2675293557  8360026707
8085:  9746417689  6391558689  5749711532  8936455806  2626951445    0301826228  6777861310  7399985832  8480666093  3534928256
8086:  6766517754  8556115684  5179644071  1530092799  5255609951    3176612188  3088383799  4352157132  2262813108  9081987558
8087:  2522131710  4319979763  4217322021  4167661406  0852319995    7414263276  0700363709  6067018931  4027340293  9191691272
8088:  5269327928  4346607575  6816819204  4427061981  4414808399    5123499016  0666485225  7180688205  7868938925  1747133484
8089:  1768076012  5121161885  5929781670  8846760191  0161747617    5558576546  2846299307  1858069515  7548088294  7685839363
8090:  0168446387  2716926130  7670480431  5188770836  3819712077    9106564419  9034860646  7100876841  8217270583  9249520971
8091:  7137169163  3289697993  7496887642  6691608533  4181164428    2729835532  0506250500  3748651388  6630322678  5226161172
8092:  1155247525  6893166977  7580579236  9660238713  2616637436    5088529086  3694387297  4418460642  8536701482  3666748482
8093:  6164315453  9187504421  2461311718  8722178603  6411062686    4823064291  1226734475  7363319539  4222679318  2719542462
8094:  2858696530  6259487621  3533347901  1777478522  7137814828    2699953611  0538725829  0141231464  7770267817  2228892418
8095:  5561549463  4817576774  7734584288  4663844795  8489750254    7628409920  3280792409  0263503191  3021066579  1667095331
8096:  9424007733  0847304615  1837886179  8674905029  8494695638    2742024093  6097211934  2438157437  7979680547  2994432958
8097:  3701629374  9870288484  1393603666  8153441496  1337097568    8841079241  1883919842  1444790961  6644802686  1444055395
8098:  1403863589  1021047304  1916629238  1993183381  7193621187    6509449177  0331914921  0053906040  0223539309  9911090178
8099:  2720444854  5908467092  9781127366  2317778256  5510847562    9643385209  9906076368  0180043879  0540764513  8308221007
```

```
8100: 71608 53158   48248 30959   41450 15008   07501 28714   11588 74012    36161 45271   95339 17717   06163 71335   37911 80765   64676 11095
8101: 65409 96723   49309 66289   15861 55814   55544 18934   46822 57494    20500 32748   65500 94445   19281 56783   89306 21074   82427 53323
8102: 82108 66576   69685 61605   63051 84934   39814 00812   68751 71068    40676 71047   15932 76985   39602 19620   91000 10077   09857 68074
8103: 26049 66360   66886 32204   01565 04438   83287 89712   17869 71151    25525 96700   25460 60966   38497 44222   02712 85120   97411 57221
8104: 81218 52266   29663 90984   87564 35938   65316 20078   33049 58505    98436 24817   28419 84144   24621 11180   18776 96282   92668 09578
8105: 55467 91583   64138 55420   56483 62011   61393 20858   05630 69032    19062 43157   75275 87788   02047 78543   29999 61843   57991 56068
8106: 31295 60914   14200 19310   21729 75094   29611 18103   36590 54055    45395 75362   57645 01629   54054 50987   15271 42138   56869 14781
8107: 72151 16266   68405 32900   81400 64564   69587 50182   69197 75827    12027 68937   90536 92475   75147 04019   64694 92193   24689 75436
8108: 02459 05920   63532 54710   62326 27596   21359 85198   91335 45405    77040 42204   20138 29587   81512 71712   55376 43330   12995 88793
8109: 14008 64725   22671 77624   43948 78536   00318 47389   00726 11817    92526 77657   61627 66891   30698 55069   80303 19194   46439 86853
8110: 71281 70373   11623 73372   57666 42723   03290 53228   45165 82588    62274 26373   78375 44838   20603 71535   37091 55713   89476 91639
8111: 92782 06427   76387 24956   77863 19028   29708 50216   24582 66716    78637 22759   52339 77792   69516 68336   21760 68676   03849 98494
8112: 41378 33897   74885 02150   11234 10876   46868 36610   96220 27092    84590 96292   02694 64875   72294 88161   26843 21825   72209 92377
8113: 90233 97701   04035 94545   25343 26228   16169 04230   75418 90911    05137 35370   46958 69786   59082 53065   71614 26338   22488 83348
8114: 92926 76907   56387 76410   25156 50817   61559 94021   92998 06630    12248 65430   33441 68184   25900 32175   05861 70443   08336 60056
8115: 57597 81838   18535 28628   47656 60160   93268 54016   08837 93623    61159 50893   20220 91416   50090 44724   74936 60584   64312 75169
8116: 17214 77648   58338 31872   48964 29577   02124 36401   27970 65397    91775 73681   26087 06013   43768 65915   45805 02953   52075 20609
8117: 95579 60145   20924 32657   59868 24931   17142 11267   88730 34070    93241 65880   61981 08124   71449 56380   91672 19785   96550 64479
8118: 21729 78330   65472 76702   86324 08536   95060 81107   15552 41337    65193 21203   40139 30511   28807 47679   48005 77723   60107 82013
8119: 03466 61499   90341 90686   13864 08963   89295 21671   62286 36896    72620 75017   01688 20244   21298 15443   29005 78838   32232 54475
8120: 17738 65277   69784 30825   89978 98412   64010 44746   84427 44037    10185 12801   38386 80158   32806 08624   15296 92763   93779 94100
8121: 27822 43984   64246 41484   77746 23492   53635 83559   67039 15046    63880 51801   96891 44141   48205 00691   60410 17705   69275 84483
8122: 79751 28618   47359 50810   74183 11081   60814 98200   27836 11067    83887 35715   30575 84241   29807 53336   82999 35330   36705 61179
8123: 23856 28621   01080 86975   36742 00187   18345 71180   70691 83755    58336 54000   89841 76160   08156 06009   62980 06160   10413 50976
8124: 52282 56692   62361 80632   06657 53138   08492 76574   97155 65919    68442 66195   36191 58828   41754 33364   55395 89680   07961 75294
8125: 47102 82891   26159 10211   88597 91689   47122 62315   74562 82519    07490 39093   09655 68164   03244 92127   85422 90403   84215 54167
8126: 98066 26126   00000 91536   38343 37317   98668 96216   12633 82270    92186 78066   76131 38614   75980 02957   79712 61915   95777 34264
8127: 27487 90346   36100 34255   15162 48597   33869 19836   08050 24193    35007 66619   66289 18236   88049 43912   06624 49604   75140 56119
8128: 92241 09428   55540 69258   25211 70005   06435 63677   61336 14966    78208 73848   13941 93419   99143 37607   04198 90873   69637 83566
8129: 98106 61148   62624 43083   78779 76982   72190 04233   16136 87708    90645 43843   05981 63728   44195 14922   46597 89066   95179 59209
8130: 98654 05325   48467 81061   49074 68688   47643 15931   33389 51875    73053 15403   87216 75665   92969 35213   52494 11791   67696 68464
8131: 90755 73927   99213 14767   49395 32301   07112 75068   81796 03746    84324 65438   97402 44735   50451 39650   31258 19824   77403 98441
8132: 83847 06815   23286 38246   00196 41310   71747 68926   48147 24844    83645 57581   40959 03463   78089 36015   70884 54859   19946 64593
8133: 25386 01118   45299 02105   57997 00544   66375 65586   56491 23773    54448 21558   39049 69666   39901 23837   14528 68191   88794 74412
8134: 88684 51719   79845 20052   07287 12802   92161 97434   19632 78114    98737 20476   91614 13333   56769 71675   78995 30095   72511 96506
8135: 26838 74293   19294 69033   92332 47513   85762 15606   93925 98844    87422 39167   25114 87194   41756 60140   50075 04064   95441 21519
8136: 60129 36879   67988 19997   30982 65805   67731 00224   41500 30100    51555 13019   02742 17440   61109 31810   29879 36393   13479 19284
8137: 41918 25735   56511 97188   51343 29062   78137 88353   42204 66552    73823 94970   75065 46464   30163 29423   37934 25442   84238 59353
8138: 79423 81280   79000 14885   40618 54289   05518 55701   05406 91012    13680 52862   86703 24507   18220 70705   94114 26148   21935 52345
8139: 69390 66568   64278 67371   75742 59003   68562 92342   40722 21579    18715 32499   52702 57788   16126 72568   29098 47163   60106 43897
8140: 27452 65819   64903 17022   73833 45145   03694 03051   03407 64462    27450 13313   56874 21463   21570 51682   60254 78633   11705 87014
8141: 01471 24804   10625 28845   73360 10746   78874 57755   43675 48533    45401 21424   49207 95247   34248 12320   20374 64619   20192 66067
8142: 29708 07188   97855 92193   90806 27262   59081 06258   62885 75336    85851 19703   10229 62058   60277 08461   75701 18247   66322 42310
8143: 35528 54640   98797 75806   99243 24783   67610 48733   59455 54449    15984 79193   26935 89375   80560 68833   64847 90131   80494 18032
8144: 77166 04249   04995 04036   55192 48470   89223 68598   00141 87276    03692 03541   74321 34825   53333 17534   54944 40860   87180 50780
8145: 10102 77418   14050 87778   14005 86263   13887 54932   79330 75189    31495 93850   28243 89051   86449 63462   40611 02424   70128 82630
8146: 98637 18940   73707 36692   45090 92279   65355 26396   54484 04284    06573 17769   01646 70350   15445 70669   20012 74921   72960 59684
8147: 28370 34238   06507 45841   13253 37720   93035 68753   33369 22747    55859 34084   71777 56308   61982 74101   32983 42068   16372 13129
8148: 60025 88627   77038 43377   93874 65796   39012 92395   29387 23038    08946 08720   75405 36631   07263 08561   61330 69760   77356 89682
8149: 54671 47028   66222 75900   53995 00414   38133 89093   46485 38076    14142 91794   47226 31482   90781 02642   07596 80400   16923 59259
```

```
8150: 62851 36906   32029 56955   48417 93605   37211 05857   42228 24846      06573 09887   32641 61970   99980 26978   77666 40409   08513 84285
8151: 43167 00018   34873 79199   30653 29547   07826 98359   15152 00832      15751 97162   69403 02192   62171 57196   04273 05051   74663 20019
8152: 78820 65597   67195 21895   85203 93831   25340 70303   37079 12528      91896 12874   15380 71849   92115 71623   77228 65491   03528 09605
8153: 43998 79369   59558 54598   81117 18194   46924 65482   10479 89929      99079 47312   33187 71301   01687 51059   49002 94986   71108 04027
8154: 24690 05152   64780 46441   16399 74946   30595 34862   31233 16623      63285 01981   69805 50348   19975 74135   42599 46595   16104 66647
8155: 83608 40542   77003 95614   52635 79771   66093 54426   83201 86396      86040 80350   33903 09704   52474 64740   18775 84104   67149 68258
8156: 64862 85585   98106 49906   46024 55016   06500 90088   84938 88774      09288 97005   63152 93084   41977 94426   15723 19969   00958 01556
8157: 50783 45645   35927 17307   66886 12657   39570 95330   39788 51466      08101 51372   12305 34881   61644 96074   16308 76182   85564 60364
8158: 32680 73923   84947 36298   56814 22677   35404 51234   74129 84885      00118 92277   72864 01621   20310 41761   13174 53870   09478 22892
8159: 85905 87784   94974 31028   44192 46990   07605 00873   84105 33631      80231 25003   36453 99336   45965 10970   84391 87375   01559 45386
8160: 99087 53102   93988 50201   00924 51689   46974 03930   83034 70806      65577 10217   48803 80553   25094 56809   53501 56085   64708 32075
8161: 42194 91310   51762 07879   55072 76585   49612 09265   66890 51285      50969 95827   56276 97245   10870 20786   71513 90047   92458 09483
8162: 90468 55015   21116 40501   09021 06068   40796 50983   72146 86328      67781 41625   93267 70426   63829 39231   23806 85370   69121 31669
8163: 10372 87266   07498 30022   10757 80899   97827 96100   01534 73214      88594 02847   91821 41479   46980 18364   83468 99114   33410 86486
8164: 78337 16146   94256 40728   28414 45146   44216 95416   44733 42569      47545 10574   72660 35307   82926 11731   12225 12238   12246 22965
8165: 76035 93773   26926 54019   95761 02531   48695 95673   63849 52078      29557 99353   07562 40505   38597 08964   21949 28355   15456 45506
8166: 81261 21184   93308 58361   80449 54590   54747 15088   62642 69384      20177 29009   12690 09646   16463 62685   18995 17381   53790 56254
8167: 91275 04538   01508 26227   13163 40087   22054 75759   51970 86097      75886 78108   77632 36721   21854 00142   75756 70519   27054 71634
8168: 92796 70957   11565 75173   81231 46274   90569 38924   90765 81899      49981 95149   45425 58918   23694 80729   31661 07289   19429 07217
8169: 61790 49653   41348 58108   34249 42725   72687 77666   88816 54443      22530 07890   21592 94942   28342 24395   30654 32320   39845 13140
8170: 47424 74359   81331 45848   24264 41588   19256 74260   91963 92682      57491 75152   51794 13297   88740 58498   41002 96987   16486 75666
8171: 55009 11310   80688 75586   79087 19433   95909 01261   42223 58703      08776 78113   61137 24854   43167 48144   08434 58983   75995 38044
8172: 62300 51591   76679 93442   10968 33695   65983 34235   52106 89931      58083 97314   33759 54216   01830 05559   53317 00222   51152 14880
8173: 36729 10523   18311 68803   00237 24947   34834 35755   69614 25662      27116 18393   32641 11520   17596 43886   77161 92930   00274 15429
8174: 47927 11751   60657 15725   80842 14887   13363 95287   72738 09410      89110 75257   15228 28407   52675 82932   33843 04847   28382 69377
8175: 31462 79937   33580 48968   43350 47495   14645 92639   60832 17339      55170 81280   91587 78984   73072 97529   81919 22442   94156 35089
8176: 53686 50974   63961 72589   17512 21261   89728 60602   52482 09554      46302 90241   24450 91751   74755 43503   10371 43145   29273 79743
8177: 62352 53487   64706 50503   17339 20210   88124 80409   85183 30611      13940 79543   52563 16868   12066 69658   10675 42814   87430 48104
8178: 94448 66714   80512 14369   17362 68644   30111 55567   47081 78513      47564 77787   14708 27187   46184 55174   33239 76095   96018 45387
8179: 70496 32795   86851 85034   64743 09298   19573 51450   96649 63168      13588 32846   98768 17903   06755 35583   28169 27848   19695 78908
8180: 70515 44168   14911 19827   30337 52374   97515 42797   79811 17551      82652 21283   59410 09630   68733 65755   13884 09294   58568 30926
8181: 76405 21427   76410 72741   05455 57784   76308 94575   93933 23368      86298 02116   84985 17513   12519 76988   36602 29784   12577 30788
8182: 75999 65493   77211 50958   16146 67450   57305 13665   08972 62951      39533 30119   33698 45628   01040 02220   41162 29192   85017 45123
8183: 42219 19000   83767 35528   88593 02413   72189 02245   49474 36412      49086 38253   24267 27321   87260 96948   50490 40837   76327 42621
8184: 47406 35536   26086 28665   86891 28003   81620 66416   56903 55908      84307 02693   02914 76050   91960 21856   65439 38436   36874 90584
8185: 17855 39458   35989 24740   86475 78541   54502 39542   93312 19078      09979 56649   16566 01047   50589 53180   58804 64261   36364 68724
8186: 40517 84631   54244 56736   79495 19163   91712 74589   69918 98508      04985 81458   66007 13878   20056 25865   15440 20046   76768 44732
8187: 60910 36595   55991 08209   04483 91822   91880 14146   18960 00485      30007 77182   65314 73254   60205 25502   87789 32379   41200 04471
8188: 35482 06658   80382 64166   46928 30653   63372 18914   70947 49312      81094 60952   35236 05700   17631 95912   18789 93019   22988 34023
8189: 08922 08633   81489 40708   13899 89645   29268 36032   25287 50171      67384 11317   98900 33368   62715 57232   85397 56946   13739 03451
8190: 25151 32653   53567 54486   00019 23261   33248 67385   75774 41052      96699 66666   59249 93157   00528 98049   33572 06107   86567 67043
8191: 43322 57275   91441 30452   09500 68764   51537 29549   79824 76872      66990 47098   50949 32530   64522 76021   44536 15596   27381 94695
8192: 80497 10957   17343 93088   59349 97965   61000 92490   16106 78661      21287 44873   62618 32871   39932 46504   70115 71221   82590 28230
8193: 76675 24775   57686 17972   73360 25353   52704 30928   70973 32230      85584 37782   86791 15837   73082 23429   80873 68977   34104 13196
8194: 08451 46061   97662 25177   46767 94741   22789 34634   82325 25583      51412 22099   36529 82359   86663 46977   61720 62056   71378 14131
8195: 54584 16407   19702 89750   69070 74349   84534 72950   13404 25121      54706 24213   31999 99652   55611 05183   22740 53817   69079 20760
8196: 39464 37312   23463 68513   11597 44989   95485 11565   99257 71883      36269 62140   04456 44743   22396 87287   19949 25019   49006 28844
8197: 75482 93285   77662 16864   44157 54906   81760 57466   75293 26785      20945 94852   37944 96580   34788 05871   66662 53184   00199 07623
8198: 64709 23081   81053 37110   19633 10848   80999 63355   53027 77033      27044 15674   97594 45011   36417 12695   14869 58008   24352 21571
8199: 66231 85435   33898 34188   23826 73470   61842 71302   23953 99705      35011 58493   61391 97495   09436 09934   53864 23520   91528 31560
```

```
8200:  21793 13916   32206 51863   80459 89268   65963 67734   69045 50996      90917 77125   27442 09765   27879 24961   76930 83113   12186 50731
8201:  59811 95676   15902 26502   77919 48577   94970 05540   16000 57400      29805 62263   15573 67197   43672 80643   20721 11043   49245 58914
8202:  87442 92072   12387 53587   85333 81210   33491 57369   76213 20222      37092 13596   86168 90759   65346 49731   01087 28317   04412 75738
8203:  32867 10421   69565 05527   85333 65074   55100 93368   68902 01720      56540 99975   83047 83604   88347 35911   59214 40794   24008 77904
8204:  25804 74330   80913 01431   43216 85532   28670 01740   80145 64136      49533 44635   73930 91539   13468 49911   98999 46903   78043 53612
8205:  57804 14011   10365 13878   79758 04460   16935 71962   79575 48583      60229 72173   97324 27783   00769 95082   52199 47278   38049 88399
8206:  15286 49232   68313 46963   06079 54795   95238 96072   42446 18132      62397 26619   66633 91569   73778 98315   09505 42890   32984 22076
8207:  90343 39987   75124 39492   58323 69564   67874 87759   89330 28095      78282 67507   66998 48914   75500 23840   33113 32815   46632 80489
8208:  05396 17582   55119 82714   56070 44621   97632 59426   56659 89317      20364 70683   00331 93497   86518 44602   44091 77377   61462 93634
8209:  36152 31317   56580 69482   34238 97164   18386 07462   53055 85868      45587 93992   32951 45112   71133 85541   13573 59288   65481 17571
8210:  39601 88467   24457 21599   69111 95184   96141 34459   22162 64435      37899 89642   21957 52631   36974 75501   64891 17225   91925 06208
8211:  85738 84641   93264 45037   52879 13844   14159 82446   29656 12411      64748 50941   91055 09803   99184 78517   12689 85283   25452 58590
8212:  44121 27535   00539 11720   83935 12125   52584 89415   51859 44426      03881 58141   19008 13168   64883 54936   27888 23227   89118 17225
8213:  75973 69249   78553 97658   81194 13180   91389 49496   61664 50488      57199 69366   72069 31450   69113 11586   87715 06403   38368 51630
8214:  68613 45664   51921 31710   63491 64504   96423 15988   15779 00018      82472 63697   54531 70070   50623 19601   94490 39443   19239 07507
8215:  01178 72153   15265 69409   42640 46842   00881 37442   01611 15716      00291 89810   60830 92116   79629 50761   75114 16455   08050 45764
8216:  67998 37505   06071 50755   16374 91904   46797 83852   11325 60045      67411 77921   17542 40952   33551 69120   33570 24066   80077 11715
8217:  57606 88303   44091 14876   34162 73284   62626 80676   89814 89406      62242 26824   66878 84628   51204 05194   45271 01465   33916 78000
8218:  58870 47633   02233 02403   85603 99769   11095 79433   80161 02924      69011 08303   09346 54339   72098 92901   72255 10552   06942 20065
8219:  55677 42311   15965 74788   32240 46689   18730 02124   70984 98106      93252 10676   69971 77311   79001 43866   16566 41645   64057 14041
8220:  72453 18071   53680 83909   44625 67610   95967 43482   89713 08324      11670 80312   28544 33550   87479 28402   23374 84644   19427 36268
8221:  71903 86179   43764 24593   33658 82295   06435 91671   99688 69671      07270 65297   61201 20905   29590 71443   93694 95950   42133 03608
8222:  87220 94455   67064 78140   93082 69722   25925 02772   25010 62779      65239 92535   47499 24442   75552 49276   51301 06854   81345 44034
8223:  14946 27540   28749 27619   67724 68151   48746 07636   79698 21745      95945 84307   99217 45794   92105 96496   40028 47916   56738 29668
8224:  15709 03152   22165 89843   96847 69735   46774 63888   98636 12479      11431 11836   30885 08222   08505 59852   06466 03893   85840 34639
8225:  39608 00272   53577 62325   91044 48438   08038 20176   52764 30389      42959 03779   69539 31189   07843 57639   92246 84132   78386 19242
8226:  54761 13295   35801 97086   64734 33786   20082 03820   85089 78844      88339 06489   16878 60777   06222 12636   05553 94758   90646 03537
8227:  40839 51853   01104 61624   88717 15021   32221 87431   83378 80922      82185 89399   28024 69645   58672 24361   12559 72773   41949 06305
8228:  17015 43302   55265 29491   90691 84966   75437 80393   39698 83786      52400 62366   76582 61436   56613 22322   22799 29640   83193 81027
8229:  50805 89137   01337 34242   06176 03670   69329 41620   54550 72885      12426 10039   74499 75126   48069 26696   39150 42021   77858 99479
8230:  65391 76062   39568 68290   02895 71940   23156 92822   64012 87948      15850 50284   70445 18101   69657 28689   17878 61976   95925 31101
8231:  82807 33860   49933 99696   73117 89700   23278 92278   30337 07990      43501 64506   84045 12034   40159 88018   74867 92418   76877 29990
8232:  02882 72351   60495 05811   09055 96616   13072 59759   12619 09235      81224 75370   84192 08892   62270 88753   00433 55466   92511 32085
8233:  80227 23128   62420 96462   87595 08276   71800 78758   02696 43457      03684 70157   98131 27583   84473 28386   41671 34311   91327 98460
8234:  98156 96468   06342 64654   38537 16999   60912 28984   21733 12130      23956 33986   47523 01337   93804 67032   55265 85568   92876 42657
8235:  11144 44701   80378 79045   99977 90065   92506 38940   06180 38398      94461 13118   36442 44644   46284 76717   93024 49781   85898 05808
8236:  80113 80477   12972 07726   25716 95009   20217 53540   45333 08201      46042 36825   21125 33720   49455 16034   46039 21150   39885 95149
8237:  89512 27929   64987 03388   34538 37059   80015 89758   85855 26325      78382 35078   27572 30737   51417 53306   27076 85906   56714 62398
8238:  75042 72632   70327 99324   75661 62083   90539 41100   40332 67636      79226 14418   30596 22675   14272 17874   09776 80227   14507 50921
8239:  87346 02105   09961 29454   67475 85179   89556 64759   67642 92339      65211 51108   79924 09669   91894 87849   14250 87288   73213 39319
8240:  15867 45875   05936 98287   50887 87120   42315 20076   51822 93645      48714 10167   23261 41194   84873 73095   30271 62414   92695 76399
8241:  76833 59523   87599 63593   10030 06036   80326 44225   56656 71303      75769 60655   52019 94254   31568 90934   09258 13330   06182 96955
8242:  04354 06866   46468 25947   10864 68185   69648 44768   28752 57569      31960 66801   23688 08865   56933 95094   71386 48019   78233 89860
8243:  07369 89448   43429 61649   90370 61945   21930 96321   41567 87349      19577 05568   89782 66123   39216 17615   10024 92949   51156 97856
8244:  88121 36290   01652 66275   28487 35029   43183 15717   42406 23160      01482 72890   47012 92796   28994 67315   83000 89873   59340 00423
8245:  02657 35964   96593 25833   83901 20193   08954 44132   05377 40462      51611 94247   37032 81128   68098 30249   13660 20876   87614 90863
8246:  47812 62359   60695 73955   44958 59463   57498 88007   35173 03408      07075 60749   61806 27784   62216 12906   76171 62567   74849 26906
8247:  36219 22690   73870 95528   65138 05812   59355 83843   00217 14706      24631 26252   91379 89281   12239 64457   46719 93307   25710 28440
8248:  28260 80246   23713 70281   17888 80211   16032 62421   95165 18887      86681 68077   62261 54570   57142 47530   34470 52782   71290 28614
8249:  87788 49693   96778 99475   78184 53441   29487 95731   95751 73520      95045 95982   39588 29861   70399 57101   72058 86559   80294 32588
```

```
8250: 45057 06275   57390 23193   62236 93527   62292 03281   76269 51167    05426 91591   57675 59780   77909 30248   34634 12130   58840 64354
8251: 39665 53896   64458 02704   57553 50741   21106 09251   16191 07843    05165 33588   39108 97077   48429 29674   98978 87076   12955 56950
8252: 70184 69561   63750 91031   82356 69105   54065 15163   71276 54059    95867 53534   48433 10190   20104 69387   55463 09608   72229 95513
8253: 64237 94415   37100 82493   12486 97350   51193 78720   21094 24336    14906 85703   36378 17857   85524 26555   53234 86805   40603 47991
8254: 59989 17783   51146 79133   12709 03027   86753 00308   94822 22550    38252 56538   42312 49221   52220 80466   70272 82425   70825 75822
8255: 84518 08943   25371 12441   37254 79989   17953 47583   01145 63975    68546 50188   40968 83443   15199 46428   78774 04331   02987 58173
8256: 70402 73066   33827 13105   11883 54293   80607 01726   65527 52178    86583 86179   12007 49472   46892 27665   02055 15521   71867 44307
8257: 68084 74276   28267 97045   08361 97199   72014 07305   99370 42416    65473 99709   59372 65410   48649 19424   35768 72346   26353 01331
8258: 51077 44974   85707 00959   45779 41315   79302 92844   49316 18332    53410 87692   91245 11595   32731 09426   52406 68235   13396 29298
8259: 98067 88462   40324 16956   62479 02095   00233 74817   47307 87161    63779 54964   23159 19221   17377 63869   48601 52643   10062 07206
8260: 29704 54211   83385 36092   99963 61591   36003 66224   70615 15644    78050 00698   03096 52636   52498 02695   54958 82171   37947 54743
8261: 56594 66178   89318 93565   10551 95203   60696 01390   39677 36836    07553 03998   29323 68088   83160 86109   74646 93298   47558 72879
8262: 15101 80122   16890 85938   78995 32481   20532 47833   29020 57765    01021 44867   80389 13863   24699 89083   94267 90096   23484 04611
8263: 30293 70697   95950 75181   26858 00029   56022 53193   65438 30097    27632 22713   11643 67185   26138 96972   57767 42473   91135 09604
8264: 67269 27264   58954 20623   08490 36645   54611 42363   42100 72436    97924 13723   57117 77254   24973 73425   23630 28591   56207 80475
8265: 58655 88144   36226 18026   45022 34092   51808 26443   16176 34329    52305 89543   48923 09111   08471 60867   50435 38260   21073 73874
8266: 80972 43248   26230 22817   65320 98834   43247 13450   11033 23957    28517 34737   91413 08984   52309 69714   66902 25671   09166 08291
8267: 32555 31637   86470 04740   55449 97343   06552 05358   18905 72472    38580 82020   33264 23316   30660 53496   14108 67642   71762 78001
8268: 76249 63308   03281 40256   62735 43759   36863 86054   52115 85712    76397 21263   51064 00559   75904 22358   17633 91598   10810 55036
8269: 02195 83822   96363 77188   37016 19371   42782 01990   04580 46511    39285 80473   58567 34504   40940 02024   16054 46904   31082 96264
8270: 45700 64316   71948 31398   69186 29492   29469 11733   47180 15095    42623 85226   22392 86655   24902 88270   93710 67424   35613 55909
8271: 09199 70296   82112 73294   46489 35851   17304 99798   81675 67180    43033 95527   03858 25132   87020 58177   15990 27224   01675 41301
8272: 69310 82300   86476 36301   43826 18189   65008 00648   85437 69939    18847 67993   06612 77201   67404 93016   51776 09567   15776 04892
8273: 07782 91886   47042 51098   84421 68208   18813 25004   37844 59302    46464 45031   43229 46818   76535 28459   34967 62280   79447 21446
8274: 75202 15467   16134 52757   96671 07000   12502 73870   29561 86315    88370 23598   01207 91799   01415 38623   48715 99289   48220 80958
8275: 01141 35834   25127 54762   42299 32228   61023 99397   80974 22208    16278 25310   54274 17770   86684 87875   23262 50986   47499 08982
8276: 67048 58310   78096 10452   53781 30830   54274 72502   75379 62853    24488 25460   79352 75708   68088 12850   84803 99131   23857 08406
8277: 29847 16394   47626 38608   95846 78550   71229 43996   12929 65769    63731 66318   91152 47099   18780 49087   33054 64882   77729 66425
8278: 50211 36415   50070 13979   45001 97233   56744 15946   84481 75004    42180 38925   45451 52797   02694 42364   29442 65149   19704 71719
8279: 71963 06410   99308 08055   41288 18007   95790 67531   17291 07301    71272 81453   17021 61946   19594 91760   89814 75668   26773 64858
8280: 45849 58047   53380 62393   52596 85820   44855 94854   03455 01916    95118 80246   85977 91842   62890 47381   08064 57151   07459 71211
8281: 56920 16374   83977 47980   91877 45664   55367 71672   92557 84143    65476 42675   17083 44408   59922 50434   41663 29415   31210 58110
8282: 23684 87711   41250 02214   77429 45644   17842 40265   12229 09736    21835 16954   01622 61008   88489 12887   34401 07406   39119 08307
8283: 79660 52133   59447 67145   86072 84331   33543 65782   29215 39138    20940 45055   40072 88103   62399 55791   75953 35542   37609 48553
8284: 55669 22885   58255 64238   62327 91796   03517 47859   80576 73417    86265 94949   64323 71472   79419 87720   72321 63894   72602 31930
8285: 82585 32318   67156 09396   86086 48658   76189 80625   14102 89165    68193 89864   66978 58038   33274 95838   23807 23702   06774 79598
8286: 37790 95646   03944 40017   88681 35393   76796 80654   50938 98890    60148 91004   03715 70084   26896 73466   54138 13104   76172 34355
8287: 60825 79118   16963 39316   71353 67566   65840 75513   30409 11947    99853 81203   03793 06741   35737 54656   48069 14190   51089 53088
8288: 20159 03421   77916 26896   24844 12355   84815 91376   65807 65672    96587 18578   41208 17441   88644 82777   16144 03667   00120 33246
8289: 13808 97486   71552 02222   09676 25379   75017 29138   34100 26171    75024 71712   08081 41935   69146 01766   10104 39690   71401 00103
8290: 19857 28495   42744 49164   51712 60746   96013 56292   38274 47542    44514 40799   31509 60560   76238 69731   30970 64098   20861 79428
8291: 60444 25940   09978 72926   10033 09980   98676 39144   29247 55928    32678 92723   16110 70627   74717 05443   20213 46886   73586 07511
8292: 49941 20129   72598 09449   99295 74167   52358 79460   00343 43613    05838 90618   86906 94316   50375 22427   34567 23209   03744 44716
8293: 11487 18136   55829 09595   27892 38426   83687 78633   56785 50753    75615 51921   62652 86502   62708 92592   19206 46086   17764 60330
8294: 01281 94739   66139 25962   13424 47396   84080 84396   61657 84571    57754 10202   84891 75334   11264 24225   08217 47063   58396 85131
8295: 11716 91749   43221 12060   74166 77218   01443 06625   82323 38818    13806 12990   74286 43415   57182 27196   51656 71883   72624 88550
8296: 40222 50755   11822 42300   70422 15999   05687 85352   55557 61567    37119 86978   13892 52940   04553 28577   67777 63161   82226 62716
8297: 47872 55404   80652 13226   21938 07356   23367 35858   55891 34511    72095 52863   64814 80400   10553 85199   09864 88023   44088 79133
8298: 23279 19714   94667 08603   72888 13634   29689 02229   34293 66379    73990 97195   18072 75443   38154 37566   61201 38697   91810 34352
8299: 16145 81978   34615 22477   34357 05412   23986 44085   32805 83542
```

```
8300: 76235 30815   04879 36949   32887 40959   46593 54238   87257 51799     84741 78466   62922 58140   76401 38787   50955 98953   67269 81440
8301: 23006 69378   81321 19457   47185 43706   79597 88526   19615 35897     94840 71502   13178 74429   52225 80944   90435 37433   86374 69925
8302: 90115 61733   14025 89435   38108 87196   13682 37242   28401 06001     80027 40887   48766 27600   25297 35364   40789 48863   06082 27806
8303: 21196 37782   41917 75524   64141 59012   65128 52485   01898 36239     47573 40898   44355 54005   35021 05691   91392 24602   73391 94398
8304: 96524 70670   00654 19082   05997 76356   58466 92700   70956 17307     44433 20061   40318 12364   93254 44030   72654 62944   82676 46412
8305: 82211 80827   18502 76443   84426 34141   77375 05429   97390 15723     24745 25677   46597 81266   07718 50429   13839 93262   17547 28019
8306: 27855 69212   09015 11960   69730 98175   15218 89637   86301 99823     77006 41722   10294 57998   48207 86110   00979 81678   84069 87601
8307: 67791 92013   22897 61580   66376 12050   63203 26844   52218 64405     88606 43104   33157 66278   63522 71633   90973 51234   49013 82437
8308: 40098 54117   25785 15519   16215 57609   15308 91360   11473 57386     48700 01215   79305 47902   61309 79036   19170 47069   33550 14035
8309: 29518 94464   89564 70153   46575 77857   64025 17652   13991 44944     14190 69263   42900 82628   05753 17204   28171 69224   47386 26635
8310: 29294 93272   07058 94369   34126 90744   39997 22153   96934 97079     47235 10700   79771 96397   22877 62210   10753 56706   07316 44554
8311: 20330 08284   76603 40817   63880 60314   15068 68304   13318 12875     05527 44679   19045 08511   14034 95127   73419 87824   49184 65552
8312: 96568 31738   01842 94934   22378 60212   35653 13981   64703 46301     05409 79617   40307 79899   11449 18237   87980 45146   42764 56776
8313: 85588 24244   74738 29189   48597 35297   70093 81203   27331 72092     11927 84619   61029 98434   99199 87404   80393 77294   57975 66306
8314: 00507 29352   94076 19612   31060 29536   00227 88628   92319 01265     30376 32682   33782 96443   38428 36160   87647 50476   49602 84542
8315: 91221 95807   19582 77577   02144 15079   00698 91012   19049 80461     54871 62045   82582 92586   17594 16540   84605 57964   70569 41173
8316: 48774 92633   70357 57926   32951 25480   44999 30390   16593 04285     22367 71554   28651 02037   41419 12306   22952 45308   51598 03562
8317: 13121 31528   43340 36868   39818 49515   58271 97550   53414 26037     36025 91797   72314 47383   29727 42279   13498 59880   43232 95641
8318: 00250 22148   04812 88374   66065 06256   03335 88411   37221 86641     14060 36975   81045 96970   13801 35583   37781 19869   10430 94047
8319: 12341 14801   13888 83772   51470 45460   96410 20636   95747 90107     77953 68835   84999 00371   60443 65778   19048 94813   17181 94752
8320: 05767 26079   40337 89011   99835 18597   37087 37957   79263 26642     17187 55554   65720 82043   00224 13001   27162 26688   90318 79098
8321: 90901 51578   72453 91733   35309 39020   16094 23596   88997 28442     08309 75724   83119 52082   07876 44130   39802 64874   96260 18782
8322: 74574 18728   16726 40854   98469 46058   90316 66045   88772 03423     77233 83872   62607 79107   54247 05045   99613 18027   04058 49885
8323: 13988 09882   97537 60012   34584 72746   63144 56493   22790 47955     81846 23115   87219 08469   78533 25665   13743 14691   90896 69980
8324: 50042 81574   44683 74689   18935 75363   19921 30215   06394 37449     75677 13576   73750 05003   10178 37897   98872 79065   22814 92514
8325: 37011 57133   39062 58039   49012 76585   79555 80949   91870 83375     51121 64366   13388 65089   64236 21836   47740 87533   09601 04757
8326: 40610 63069   28421 29311   92458 91378   69108 36613   58914 73293     57305 61914   35997 83051   36886 82279   62480 95630   06455 37904
8327: 04205 52325   34025 00519   85044 05660   31939 88002   83752 49011     07920 71704   78136 66135   71780 94882   05703 21029   09831 31695
8328: 45897 34403   31990 80202   30485 76762   97834 47657   27954 02874     72786 59657   12243 92037   82270 96909   74053 18779   41860 55377
8329: 19213 07301   12685 45163   89779 83916   87860 02528   00765 14820     82458 70310   11846 19157   24145 15020   68674 33136   50949 91022
8330: 05123 61947   78373 67873   64977 41872   07755 92026   99313 36741     16517 93409   71791 59799   56512 70502   83453 97936   17401 38523
8331: 39084 90703   03438 63293   92301 00405   12911 64232   95379 13588     70931 60284   60273 89570   87468 05408   74128 28854   12506 70668
8332: 16800 47181   03698 93608   61868 22672   73821 29986   22413 54687     75148 76608   23734 98981   22741 42731   05692 40706   14680 63253
8333: 34031 25848   52829 22146   11722 15696   04387 75787   94791 12826     16959 01559   79661 74209   09517 95951   73022 55301   25483 70066
8334: 88278 64477   95675 60685   04667 75553   58608 33732   62071 08482     06847 55645   80492 79147   18279 15430   99654 75320   84281 12576
8335: 99648 37008   06484 81946   69317 96369   82987 68193   47443 68331     79055 71021   79393 23218   02279 77241   23496 97523   00188 91702
8336: 46013 47515   38743 43504   59262 70742   05780 43685   25546 12846     63087 55951   31690 05141   37999 52202   49817 39208   17152 00191
8337: 93554 16548   31817 43041   93663 18317   83005 63938   99896 41752     28981 87461   00080 23476   75208 87094   31997 69828   35889 72101
8338: 48821 13721   23646 84278   10644 45506   80998 73488   35252 79795     16542 08682   87562 18531   33549 10057   34077 59727   94186 05688
8339: 71128 29864   66963 31128   96786 89272   59211 53173   91148 74799     99871 55821   36641 75544   90345 82618   33286 40993   33459 16891
8340: 20754 87815   19414 29582   60329 68501   92385 31221   75710 78592     94727 15529   01341 33119   77817 49520   94476 65055   61027 89282
8341: 85727 75959   82269 06227   01591 25759   45217 06053   42947 13137     35497 76079   73811 60002   24416 36504   52924 46124   45417 95067
8342: 32878 16876   91150 11252   13170 84200   34633 39934   18616 70662     52568 85221   25824 09633   67903 83126   64067 12875   03925 38339
8343: 91887 86199   25427 95437   52417 79764   60615 50177   04857 17188     89693 82982   64297 70209   71003 17651   28374 77536   17097 44644
8344: 34865 32221   81960 11263   96807 15527   54616 49513   43348 29635     94153 89072   14498 11285   35317 02493   39921 94089   13606 59699
8345: 76830 38059   00839 48389   19938 47632   13877 81197   40062 93991     00864 00882   36073 26961   07573 16455   77402 45841   74744 25072
8346: 69201 67763   85844 98290   58172 31513   92137 45641   78357 10465     35866 41426   12672 09189   81167 95632   26133 45594   63061 93965
8347: 85168 87711   84570 53570   32484 47863   33119 26432   23544 72897     29396 21619   32975 88472   85542 59201   61718 39449   78809 21112
8348: 09386 28323   84217 26109   43184 35647   97147 06923   96225 89429     23872 32535   33251 77434   94484 67522   05825 76105   25576 56505
8349: 78836 32245   40266 19701   92892 55443   01818 92959   43681 02845     80848 97916   40141 62149   56121 65286   99217 68620   87001 23366
```

```
8350: 2465199889  9665417374  0658491328  3537005706  6415730237    5640144711  6580887208  4312610478  0110708390  4793192767
8351: 9628272367  2071766896  4276430065  0014950217  2054066782    1725806243  2652083227  8669171679  2167856054  0633099605
8352: 2960483845  6171892538  6590399082  0272596908  6204769075    3243560982  9423258180  3591635540  7286907843  5307675340
8353: 5122550104  3299450746  5838174355  6889043852  1658386636    8800766186  0929891162  1684435513  8333434173  8926457027
8354: 4849060857  0257661771  5086736697  6698680848  3144839666    6211537965  6541571621  9180956393  0342266279  4515915931
8355: 0454859797  8015739201  7562807794  7410386557  3637224641    7305419014  9673005691  3234973784  1164574895  6243445078
8356: 8057353255  7052061824  8321581698  4946145092  5160480091    0772359484  5692939588  7559610491  0502206514  0957874567
8357: 7887905289  1262342413  5779074417  6228040580  9652405289    7062199376  3650069763  0132076208  9351838772  9340625374
8358: 1731012119  8211234484  2031696924  9681019897  7129781800    4041121702  7821471162  8907329987  5073476061  6928365297
8359: 3431110575  7811693388  0186795105  8309025809  1748905001    2961002666  9968276832  3574347488  8066494928  1758970362
8360: 6687259450  0277470669  4573827244  3704835566  5870320212    8048302422  9928062905  4503580077  2316549382  0917461444
8361: 2206725430  7515266773  4005785524  2363958442  1180183826    9978091347  5085946534  3950804894  8518684660  6419327725
8362: 2258148056  9907792552  0678360737  8377940073  1657319345    2951295813  7049663173  2104123384  6730889441  2955270943
8363: 5687747183  4803963659  0268493456  1236598441  2284132583    3524800066  6410475317  2687718911  4656986004  2820751767
8364: 0366458371  6870830651  9159289268  2193119402  6907985744    2700315716  2013442840  5986346499  2507088100  5571672287
8365: 7132502560  2666359972  8742669330  9995278602  8538700619    6128913310  4887655199  4607577038  5006397815  9577336123
8366: 6014882567  7578944173  0837277621  2583046288  5597436543    8170801095  9186953889  4913606335  1625443031  9152818725
8367: 1427036156  6878075387  3272918172  0769785178  4554888035    0535295003  0370319927  0595441454  6671930039  8278982705
8368: 0721644139  9682388229  8103840172  1482611118  3874183092    8392699810  3228075479  4847535379  9866111331  6747863909
8369: 1115106549  5051774176  3844070561  0687588038  7378379523    5223131313  6366262888  8189774416  9736838919  1411758811
8370: 9182844716  8268551691  5336621660  3327735855  9863348225    7032109672  5662203307  0894162654  1393898570  8966953086
8371: 4531241148  8314722820  4421986786  5354219434  6943144268    7388406230  8824839127  0828156790  1061382759  6137766491
8372: 7636861767  0204828921  0607367331  1425705859  4751395636    9623070195  3105720301  9631291755  2184187851  8590073139
8373: 2191389597  0969279238  2188426443  6167813822  7779414320    5072278942  7836655201  3818410458  1951102223  5741157902
8374: 3013888882  0684061343  2800233305  3456865350  6982641570    0081496820  9343604115  5187484805  1949617859  3036024140
8375: 9866056618  8510912721  0272601328  7426582407  6354705551    0952338886  7470214692  7940479977  3729937941  7338636020
8376: 3939204219  7461625513  3968453861  2635828089  2930093232    2790806140  2523641075  2241480047  2854649489  6101729989
8377: 1695617352  6676255785  0654101468  3932871640  5276284319    3285077965  4197778167  7901053129  5899014092  4900760637
8378: 5804818057  6361518238  3922264749  7576214096  0519485635    0685490696  5521283286  1526172781  7518355299  0652205202
8379: 1127081757  7213409270  8438776684  4322348638  2668110773    5670276306  0975191157  5298297190  3278121900  7672074094
8380: 6492215780  8991059962  9219443824  8065008607  1483198841    3684994689  6296068706  2284810050  7967623681  5618441952
8381: 9945582063  0234725565  5441596344  7061150716  5745431044    1623816685  6000583750  8671533784  2031725137  7152914650
8382: 2542068879  8628952340  6667606895  0820019584  8598326974    1206195321  1654134381  5952767413  8152614759  4598690716
8383: 2283927263  0623947210  9118583253  6315268466  2491284591    7039016448  4520589467  9784310679  3497928391  2095132970
8384: 8710412599  7286362722  8252297693  3399748804  2516066443    8359485630  9202422078  6965766795  9908989184  0909162244
8385: 3126968310  9371202920  9650768584  8541734853  9370654809    9726432229  4049307920  8241480882  5013702586  2449580009
8386: 7851895052  0158391310  7350093768  3326787202  7301215856    9116742253  7264361765  9090033141  1555448935  5927020945
8387: 3347972233  1836096367  9688443310  4952227796  3793750166    8861050405  8932843164  5286827579  9757037452  1841630116
8388: 3849620385  6949379346  6869718756  1933155198  4471630571    5500205240  4713352831  2498942758  4025785926  9835712366
8389: 8513587289  3568432639  7728881745  1692062510  6331363892    8268490900  8246172885  0046741277  9840551513  2671995712
8390: 2740687595  4036722834  4059260792  6140068157  2478897465    5112846696  6096550173  5536064079  3036096756  3764227565
8391: 1720659413  5303414329  1276423514  2079492760  4661954288    1262165924  3748327106  4033384854  3222574400  2192751150
8392: 2384166651  3206615135  0877028262  4100650963  1147294066    3723417089  1588793406  7242745180  4011424784  5431127456
8393: 4318878058  5049802758  9925528554  5117558409  8224154136    0759000100  9102817864  0879989270  4798714346  2675669608
8394: 8665049421  7130143074  8537485245  6193192605  2940874832    1690580338  2152344523  9771990000  6130474226  7614917101
8395: 3058283716  5777742441  5546500767  8315383845  0414068170    2358510506  3210195417  5795004223  3727621444  6692660602
8396: 8045228872  0730514273  8530447219  0478304543  3540262428    4420803044  8833857224  3633597979  5807553486  2211667137
8397: 9749671937  0646960940  1858651700  7122438623  1477923928    7076765354  8598244264  3803818979  3073783094  5517412760
8398: 4975807702  7321743719  3243397238  5794021208  6005461983    9043302645  9427415431  9154830831  0101829189  6141137337
8399: 1050387155  3303634935  2865326950  3737341885  2820279375    6937503334  5834717618  2724356765  8959307790  5750844244
```

```
8400: 23115 03587  29149 93821  02890 21101  31958 44910  42493 68517    86743 21269  91568 18747  05895 37102  07691 26346  46328 42174
8401: 87153 51275  75228 52841  08066 69929  36426 06451  25564 88841    61073 46132  34999 59337  89522 74892  67249 21876  07300 22033
8402: 85966 69516  25210 13573  00109 31810  84145 23661  54875 68027    37249 96563  77812 82033  40309 88906  03042 41787  13267 03140
8403: 98610 07851  18278 70108  96487 23770  57045 46755  61496 50624    02055 88547  60373 66323  02636 20776  99000 04436  94658 63707
8404: 30713 64818  14892 61716  39034 25600  69696 49560  49030 98679    17601 17680  72544 90325  15109 19293  79331 04113  57332 67687
8405: 73075 60195  36428 01658  77503 73808  49734 68028  13484 34202    67883 68690  84445 63421  03165 35359  67731 86366  29552 40176
8406: 84277 16263  77508 32163  89759 51224  64929 87627  94108 89914    94492 22989  78863 24747  08642 11158  57024 10187  85022 88391
8407: 84750 84867  63733 62348  11818 28844  50508 15716  38063 16132    12622 53408  67938 22225  60218 05899  22909 34637  64729 95273
8408: 15638 28317  04471 83334  06818 04581  02367 79538  52924 16682    25902 65456  03923 89627  27090 68491  00752 46159  29110 47739
8409: 18773 49615  49810 53540  99622 15851  53358 73081  62766 43871    16119 76094  56425 76050  76404 31661  67473 57781  90548 73590
8410: 08568 19899  30678 83026  99906 37659  74068 20007  30043 70556    21602 87308  11891 31792  41121 92884  13781 72920  90693 21539
8411: 51974 88139  35691 51530  40353 17686  62846 43894  32453 85583    22605 73813  83967 82076  11641 26528  85951 75431  12910 95221
8412: 59659 43943  29451 62670  99639 61360  18292 84329  91906 02310    15493 21980  65455 21039  57216 12390  72424 71711  16206 98183
8413: 81977 48000  50563 23375  93829 95981  73623 89031  84752 52651    58878 39488  73233 25905  89542 00464  19923 04989  88344 83983
8414: 08827 08251  56934 49951  88343 16361  82851 45634  65297 58081    47454 42173  53604 05981  94426 75772  47247 40633  43468 52300
8415: 10719 58193  80159 41269  16780 38155  07366 57600  82683 53759    91111 22151  01093 61204  56367 50037  59633 67272  61585 73229
8416: 54215 88669  50380 49359  49794 64410  28880 95426  58569 64481    67781 26825  11695 06693  40394 12047  59963 16914  06525 08806
8417: 07635 82241  90058 72941  36198 46597  87011 47121  70359 48689    16447 90778  99393 68557  36801 43708  63079 93449  27881 03179
8418: 15499 48109  61912 79753  89815 06178  45673 66997  87549 29057    62728 49660  96548 74056  77543 26354  31592 01072  56777 16425
8419: 93957 89786  44574 82002  53960 15869  39410 55187  75512 04787    98384 47622  27801 32918  89779 21065  92105 83946  52762 81054
8420: 13553 62889  09413 02211  12448 50007  26533 40023  47665 15050    35694 17870  17666 07551  80390 47527  49764 12394  43408 33787
8421: 13064 10263  09734 92231  05531 20528  28635 37386  63182 69900    30755 37435  23801 43837  64504 65960  07060 01387  01250 12397
8422: 61168 53599  32926 24114  00864 76382  94981 08041  28407 15021    46438 73415  16254 34004  07094 29148  76955 19700  14173 66189
8423: 84901 03112  48179 41996  60864 40580  36638 50832  91234 54510    09852 01712  57924 41669  60869 23945  15337 56170  51674 05093
8424: 46769 79393  94657 03867  34194 24883  10638 41624  38830 78536    95019 12637  71849 90742  94615 96497  74633 41758  11083 22289
8425: 90041 89740  58229 85581  04139 63914  05304 07085  13172 82242    84357 71693  97613 08516  03814 65520  71301 49227  71285 62950
8426: 67331 14901  65230 03601  02520 95382  79256 71211  56309 84751    03189 86283  09025 92253  44138 87192  66818 48155  25959 56807
8427: 80986 37042  64570 26370  54648 68053  84460 44983  53359 64095    24278 84932  16988 09845  06000 28347  54273 45637  99192 76289
8428: 40014 90461  52648 85583  64650 41865  30338 87368  91050 28830    16611 77718  09132 99984  11881 33213  57791 89459  76348 35724
8429: 32128 02735  30585 43022  15438 87370  00456 82441  36188 89450    09978 69801  66615 41457  55501 28175  49106 23694  98563 93605
8430: 85409 27813  24370 37279  21305 13083  05869 39227  47321 22542    35533 83163  13131 07712  46506 31698  93806 27985  63082 15300
8431: 84544 52482  70934 83312  35974 42325  46518 81658  96553 71208    63731 28202  49768 56803  87060 58744  84949 47294  77330 71664
8432: 44326 39722  46030 09595  52164 88378  47211 08254  46734 72305    52773 26352  41851 21927  42936 85813  57728 31724  89155 61066
8433: 28832 20434  05696 00737  54290 16308  65746 78152  29595 04750    52912 14472  77671 80647  80774 81610  01544 53114  43657 24614
8434: 02006 03960  26422 47616  00393 37170  96617 20375  81121 71950    08545 60613  98050 78397  08852 06766  34075 08081  03285 92538
8435: 04597 47097  39944 36299  05298 97392  27710 39088  95600 25742    42354 83170  50157 70386  82192 72138  93867 99031  92065 44112
8436: 28021 49271  38590 29913  77846 99376  97641 43426  76651 90420    62852 85273  62924 22094  54771 93311  50556 07804  09612 88874
8437: 57133 66967  10552 66006  95613 78349  41621 62339  88140 28528    59690 00468  57319 48357  13062 74360  39196 96941  27457 25379
8438: 85633 81971  02146 73052  77166 52004  88960 21936  91006 53324    05905 37781  26616 60042  44138 37393  17558 97988  21967 10525
8439: 46836 28318  60893 34831  48813 89797  88251 29898  81857 67115    07535 44145  55449 04139  04095 87563  91576 72064  52539 33014
8440: 57897 98440  80477 67505  20073 71002  29142 29335  77985 77018    02201 08272  30518 75585  88761 69155  87748 00548  50537 37295
8441: 26073 62022  28544 18623  31711 30844  24708 25471  25986 30021    44293 14812  77688 57205  00392 80395  25251 38588  78141 89205
8442: 21993 16095  18950 60180  53998 20604  49538 90313  03073 44783    89235 42681  92506 22390  58525 28395  00534 14185  65967 28204
8443: 51240 57619  59730 76091  73599 49195  05789 78935  91304 90867    33240 92880  00602 43446  71582 25726  75993 87171  44926 07569
8444: 97770 66823  42730 34332  82579 65252  68647 65592  40745 43445    77715 90955  16448 29945  81413 11272  04646 33447  94468 66334
8445: 84611 56637  14327 54731  26829 31179  10934 41600  02545 27481    02649 47199  95946 76727  17825 04827  74455 43114  84508 23284
8446: 10693 85221  73195 66754  54413 61789  80203 43091  30490 13200    85060 59009  87716 20983  60164 35711  77617 53378  42684 63792
8447: 63360 00375  31727 08044  15525 69017  63698 50883  09823 70008    06295 98589  71625 35340  92859 87368  25725 94445  79998 22373
8448: 47861 02632  65882 24323  67189 96593  49505 15480  35945 32543    54873 91241  63041 68204  61862 54918  80842 78673  92764 82531
8449: 05715 41665  37745 04249  09710 47417  81768 03125  24961 17314    40231 25962  39081 68253  88063 54532  88046 05770  03936 07263
```

```
8450:  36028 83999  55293 52181  35100 65393  40596 32004  48736 41513    66247 12089  17624 70632  79876 92928  31776 16647  08036 70693
8451:  28227 85621  48144 29842  90086 96513  12896 66084  78566 19120    09540 74906  51087 20245  18464 53089  53831 65915  94382 47122
8452:  89951 40684  67566 66580  89825 07882  21071 60669  68840 98326    57453 51386  42407 23658  45404 04234  18745 48001  66033 18022
8453:  22469 20399  33625 42408  08120 88869  26169 20066  27225 06528    75325 33056  93508 31095  11442 51432  67190 05123  09232 41203
8454:  03052 12598  07860 28955  50497 49841  54856 83800  67994 01817    07626 84448  05467 82040  11542 22160  54426 82113  71833 03987
8455:  81260 81257  52746 02561  58396 78152  05172 36196  43558 22561    69561 64770  91878 46364  24939 13522  41984 62393  86432 37798
8456:  80656 05038  67693 79676  53719 52583  52159 92094  23191 06891    35246 08740  76323 95243  54841 88192  46604 00271  24541 25397
8457:  99522 27923  56115 30087  18969 57566  76114 05087  94818 34884    71316 84801  81419 79968  21725 92637  94615 43219  08142 83169
8458:  85217 73737  60174 88921  54971 47391  76637 00547  41190 22475    62660 82030  56497 39923  62900 13426  43377 23553  52676 66952
8459:  42386 34749  18504 36339  30319 36645  00693 27643  13373 10863    82047 34805  75687 27767  78632 36406  82532 82785  77017 25507
8460:  08499 47347  86745 26443  20123 93241  21102 09655  85164 09418    94057 53348  88476 60270  05497 27968  99635 69330  60824 80520
8461:  75737 73026  81283 60311  09761 85478  38558 33983  75118 26860    72382 70700  99066 71187  88574 09799  84781 58189  83621 36530
8462:  83214 67553  50809 94046  84219 88796  64993 04079  60648 99648    44869 09892  74961 42010  66630 36084  53180 72675  58075 86545
8463:  89939 50474  12205 95661  11386 76002  52287 17556  10047 30472    67330 37396  17160 78832  21033 28372  34951 57132  80985 69725
8464:  57253 00394  34182 47359  64158 32797  38331 83777  93528 26683    45704 29088  52460 75769  42484 55104  17575 19708  35671 36363
8465:  74483 96287  05681 24070  66580 31167  17545 85464  22977 79929    28452 75232  55862 43965  75435 82170  27606 91405  94797 25738
8466:  34194 51762  78367 96333  80391 26505  36690 60469  22522 07055    55580 79590  70384 33339  01704 55724  94067 55991  60711 11083
8467:  95666 46526  08683 18428  76315 98593  62528 60302  54926 79876    19287 99522  44818 88099  86859 98048  25081 66934  89254 15787
8468:  11878 55904  39473 21349  55264 81948  32278 01184  06368 59295    98755 09042  77303 53163  80575 51698  17422 18803  34878 14372
8469:  46795 05436  10848 61736  55952 65452  99305 61423  52271 67720    92640 47077  89238 17095  58831 26438  56937 60894  08164 76344
8470:  48389 13281  06578 30411  88600 99215  60514 71033  29161 43094    34551 13988  07312 73630  35688 27294  82496 02631  60486 10529
8471:  90372 26497  38955 76687  37150 20058  43355 61621  88398 86537    54530 41621  08870 96784  63618 56055  89255 87079  97684 40159
8472:  34025 01800  16466 69137  92615 30483  22417 29145  84960 57085    84371 14200  64137 72014  57223 50053  57505 21190  20878 13420
8473:  07957 77494  67833 23730  22842 81633  21485 44100  59194 16651    73702 96135  35987 12835  03398 55535  69362 26319  64419 54833
8474:  86667 85145  35105 29360  06586 06185  82596 51056  14862 76846    98585 37287  10615 98596  48911 64404  55295 86792  11047 75599
8475:  85398 14717  51147 68506  24202 02777  59043 36768  63922 98082    95993 23324  16169 11620  08306 34798  46984 20136  69840 58374
8476:  26137 25728  47494 69707  72897 68011  70242 57879  15541 69251    39811 67310  26601 26569  90984 49616  14815 33181  09455 79671
8477:  24380 75226  31315 89121  90851 18989  11073 65570  60345 54693    37285 66588  49709 26118  95959 48572  40433 77316  13617 76799
8478:  09492 86055  70139 04019  86798 94781  06918 72759  53056 17221    02313 02942  68651 51929  70642 70731  98447 46553  31978 24829
8479:  07954 22898  24430 50451  34775 44457  92876 31065  74243 68486    29354 00569  23825 31701  60776 07987  65386 66771  53185 05556
8480:  72879 45351  12642 42565  42742 72938  53620 01832  26242 58478    75156 80882  66455 52106  99726 57076  30946 71484  43852 05114
8481:  83168 44398  78541 97953  79420 40971  96661 89968  24896 46506    55463 52214  45798 77484  35137 01316  37182 75817  57272 49372
8482:  53544 77285  59209 03900  84858 21261  13684 06203  52231 36284    36660 68692  56123 97395  34939 41367  87217 18025  76817 92871
8483:  15519 05990  84858 22725  72926 89181  32148 11933  74194 81457    44164 62907  69003 84320  46663 13927  31802 64202  10155 83896
8484:  76711 74140  42640 97629  28733 34157  28394 81887  95622 18420    83739 60662  95229 38620  12184 36767  60537 73714  56681 18960
8485:  95568 44939  69226 96942  59343 55006  08201 88442  61537 06318    73636 33579  40435 19458  57599 32646  01578 68281  49409 62367
8486:  39448 76796  40653 17437  24083 91037  20045 90358  58680 18708    48459 28416  70563 01096  67743 26390  00790 31340  77868 43868
8487:  62221 72041  43959 18048  30913 19854  13737 79025  62174 13865    24695 88894  24703 50030  10500 25899  09738 08397  16209 71100
8488:  99556 24909  18501 50347  47904 09111  33490 44463  30688 23238    39111 98467  61322 61613  97554 16814  19862 44499  45315 75186
8489:  99742 05813  87903 03116  52113 25435  55436 85549  86142 73520    29562 34838  65579 60107  94566 82076  61016 15837  62021 82738
8490:  82658 19525  88261 45671  54812 08880  71520 03858  51395 08292    72843 16806  89560 28905  81789 83821  37889 83311  00642 51086
8491:  37882 46614  24141 04421  37344 07302  95427 26916  61539 76103    53559 43015  77977 44643  99751 81188  04559 48178  83339 16999
8492:  95904 68784  80013 36611  19418 13606  98003 57843  61261 97750    37373 73762  84147 84451  49048 88112  47872 58896  79024 04358
8493:  02419 61690  73215 23103  30175 40359  86390 66557  31540 72714    54579 58053  63164 46661  29062 94684  09774 92025  46439 74404
8494:  18865 79377  01844 37964  93847 10207  44117 32668  04734 98876    16381 01461  20202 86252  87519 46239  58323 62987  70830 69401
8495:  13200 36775  73969 58512  74912 96365  70297 43829  19626 55587    57301 43945  96780 11234  93380 47884  00938 98688  32819 46551
8496:  60643 49658  84327 22858  55456 59434  84267 21516  84393 21632    19731 78177  27589 42906  30315 70599  36586 30654  51682 38408
8497:  57335 08482  82093 75191  88495 60364  08538 51218  32756 25171    53094 75743  50567 99034  50142 35126  06355 58219  37734 98737
8498:  45746 06688  10257 11802  87715 45073  01522 06613  68076 14867    94079 68533  72671 07634  28610 56455  75396 89748  29608 21783
8499:  12468 61504  13085 59102  11812 20833  11876 91970  57287 39221    46724 50615  54043 51846  13359 31909  13315 91984  22216 96785
```

```
8500: 97198 40295  45526 02848  0046 487048  80945 70757  36808 05435    79305 31934  27352 68048  79658 37866  17174 34699  6955 576051
8501: 44063 55087  84775 11078  40599 17644  76708 88822  81527 43424    9380 472585  0323 763666  3890 441878  27272 50965  7743 471306
8502: 09735 59106  83130 28446  53131 19615  27296 79191  64387 17717    13219 41946  98780 23240  45075 30271  97630 92267  64689 44838
8503: 44863 54570  41199 79387  91042 86355  86655 75095  75361 32802    71785 96998  57372 17170  66957 93743  08413 55783  70136 64509
8504: 15087 17307  41955 64798  78391 38390  58403 73104  58457 66113    23092 35725  9114 375950  29421 01556  41030 78046  35938 36318
8505: 65060 73168  44496 58649  96769 85591  20610 68266  66494 35367    58060 75420  91237 82624  48984 67681  99212 71191  68245 44821
8506: 40411 81794  45565 44386  15400 63574  64655 44108  88163 37114    0604 379869  94659 88258  90249 44714  96541 99859  82280 09193
8507: 60211 70359  89980 28454  32148 24917  56653 46510  65278 86008    94289 04288  08040 40810  78843 68011  04307 63662  31238 74628
8508: 82008 39787  19097 89454  62262 98777  47704 37524  35054 69522    62738 57701  84989 71935  99971 80969  63016 63557  66911 98992
8509: 59473 39742  03072 86648  58597 79083  38795 37797  64463 22027    64419 92961  87706 12137  00456 53520  49234 49864  43661 83265
8510: 41176 54797  66907 24317  34228 16406  92331 91535  88427 66519    51111 09438  13852 26496  83975 82311  94665 55863  80277 85524
8511: 35147 68823  77775 82133  12744 74840  98254 06405  55022 24758    40882 54084  96887 40796  94043 91667  57984 39111  12222 41370
8512: 99000 14539  91994 08630  86335 48880  50996 22475  62974 45628    39989 65072  00784 36463  00375 41783  00209 75208  78332 38571
8513: 36545 12716  22976 10430  82640 94460  31456 78576  72373 40211    88135 25350  44421 81100  01873 54690  71731 16183  80104 51594
8514: 29881 62817  08657 68239  03742 01429  88249 31545  32819 24435    79982 79131  99577 10768  22737 63916  24282 73663  03601 14051
8515: 73628 22251  10200 89237  51187 80264  61787 87092  36041 29675    93723 27234  91031 23945  07303 87442  10993 78932  70164 11764
8516: 23865 74902  17023 38722  14261 54401  57011 75097  54736 32324    00272 47420  77869 85176  74464 27164  97475 98175  96587 43593
8517: 32289 42721  76485 21287  08516 41542  75658 33431  32070 43035    23581 13147  64709 95024  43828 36716  46008 38242  34605 64659
8518: 72568 64611  96362 42434  02710 45834  07153 30063  84007 56951    48485 09584  21626 67556  10580 19716  99058 53338  21149 77739
8519: 91628 72477  18750 13089  68150 11203  38455 34722  43329 19204    68620 83579  99256 96834  34191 12238  72218 99670  56822 70156
8520: 70772 42701  61073 59992  84204 67603  92663 37780  61737 35763    39748 22117  31402 37320  35786 10816  46402 51716  16328 81506
8521: 04253 34657  73628 67880  92632 96977  28058 89641  21782 53970    96307 27206  17542 87024  97664 17833  94502 18581  98704 48354
8522: 28791 70670  31733 21354  71840 81822  08442 00853  18800 69312    24192 37198  50196 23864  33884 22283  68410 95387  98864 77961
8523: 46609 10836  25504 31941  29722 77514  92206 61980  06848 22810    21515 99786  74018 41685  49023 02931  21153 63418  23572 49158
8524: 51970 46588  88370 57581  07135 47011  39097 60615  40844 87405    66362 22539  56346 92483  37434 49098  35062 32827  93766 02548
8525: 72011 63362  12467 72844  20275 94220  83237 48186  81487 16894    03308 50475  88848 61315  36461 52434  08396 74501  51862 96066
8526: 10321 00968  16452 45277  47123 49399  89173 53890  82249 87514    53655 71605  48636 23281  48629 87365  31222 07679  76927 49958
8527: 11338 51286  03563 38007  62406 27296  56322 09109  82240 95766    40507 30566  15023 92375  80617 78455  11819 65506  65867 70158
8528: 87073 77727  63317 51180  20471 37226  78816 33574  24031 01594    55780 67497  27481 12039  31234 26805  91289 63414  17419 37077
8529: 83938 81979  00979 51793  13591 87611  06181 89692  88959 18991    73053 34249  91603 84488  43715 44951  87322 02151  81165 07389
8530: 97639 47630  53989 55942  24208 85488  13509 00702  12494 01787    75856 29500  80429 99173  55713 72218  81511 82270  93691 77250
8531: 64778 30381  22574 47130  49255 56985  05934 61088  10637 19468    90728 31374  64504 27047  23688 60067  47090 91353  49431 43295
8532: 81265 91036  03251 95286  22634 88266  28517 79773  75885 00926    18456 86868  03627 87063  41391 06260  69042 09836  73028 23284
8533: 89192 69150  43021 23941  91325 08780  04392 62443  79164 94423    92866 58994  45501 28854  57465 01734  27124 58424  78336 72801
8534: 31661 07389  79802 94778  10597 81011  27923 86407  13322 40770    23053 71040  37680 02920  03878 80412  70662 67920  95933 07253
8535: 78941 21299  77951 41387  78093 50664  95077 22658  48091 05391    00560 37423  31318 88401  54157 34861  53205 25605  18923 90680
8536: 07640 18873  64456 46096  36811 82213  51241 93485  86111 77076    51891 62143  95196 71349  68910 11271  88309 76772  45282 63448
8537: 93522 95423  91281 33830  92376 03139  28326 36508  87969 38491    72183 76123  79591 04791  24269 51812  03121 48501  21885 71766
8538: 63598 00480  25504 59892  44023 08151  38123 94388  28583 28354    04098 58144  91570 66565  22215 32047  56481 92625  40776 33632
8539: 21178 50977  06153 31448  42889 05898  11517 02207  28936 02294    75338 05401  90732 65047  43689 50343  73291 27324  77620 46587
8540: 69802 41071  15096 29998  44184 63221  15678 69730  25797 73355    44463 82958  82060 48858  44688 21326  11504 96306  19309 04287
8541: 56838 68040  04299 97583  76723 36043  71166 65775  13518 46912    54612 43807  29698 84228  28125 90654  81677 88306  43178 31952
8542: 32511 83165  79440 05572  06010 10036  49265 49313  30189 38164    34116 38570  08632 76914  93029 06593  40791 41912  86097 31325
8543: 32247 14570  51305 52012  01946 54831  81796 61734  28416 14917    23845 42600  35947 18064  18655 38934  91519 74443  68407 44735
8544: 91547 79686  14650 12517  89501 15408  31198 24484  60505 41029    34011 75912  66096 17880  13494 97372  76617 98226  32672 26902
8545: 19112 35225  29451 14080  09747 90321  32419 34675  58442 90652    00358 94566  26475 81073  07137 90089  16953 61668  59271 04425
8546: 73337 52509  86335 53605  93742 22574  26439 75128  77054 86757    43283 19603  91732 75817  57463 49132  71162 99426  61937 71329
8547: 27003 78471  57034 98221  50619 05942  95919 74382  83192 24318    46810 46833  42472 72013  02948 17785  97166 16340  83786 75468
8548: 73364 34332  26681 28460  11924 59686  23238 86957  74719 31764    55028 19246  68093 86609  45929 13395  91719 67472  77297 38924
8549: 18508 88725  73477 86269  61881 91067  12715 06014  96807 24653    19449 99407  06328 55222  96394 37896  29470 72317  52870 16026
```

8550:	42917 09090	92629 83452	72358 34905	91745 46600	08480 57656	47208 65098	05927 73349	10933 49181	45120 92854	73781 84942
8551:	16189 33918	02521 47176	98619 39010	39760 15650	50106 34591	25991 06634	83291 04496	28031 62003	57427 67301	91327 00991
8552:	35679 76090	62820 52668	13450 16367	66873 27332	69959 49242	14648 16385	32314 96614	33746 90067	05004 34177	24902 33637
8553:	46171 62917	51615 90195	95208 32070	11659 93904	28174 38428	31276 86182	23259 06995	49692 78841	44010 79541	80725 74815
8554:	64242 21037	31135 09569	39211 55645	32148 23984	09504 90004	85891 15425	42365 58960	23521 15434	47975 90062	57816 17913
8555:	05739 19418	71416 88364	85648 56307	27496 87479	98224 70857	50717 40219	52483 56868	78213 74146	62913 65352	19728 32687
8556:	86447 67367	26668 75918	69700 56867	76633 61604	72706 24472	63000 10611	52867 58285	74287 85202	41994 45462	62356 90886
8557:	36009 15970	06274 35925	36981 38088	47609 88208	90886 16853	69353 35319	84527 45142	77305 83724	91700 26845	39160 27613
8558:	46692 92447	71335 84115	02910 87940	42523 86853	37591 60138	20030 59132	31760 12242	14857 36740	16514 70602	16310 10086
8559:	37666 01110	04621 69981	27311 81047	26649 02417	59864 90456	80489 41338	74195 50954	21350 19372	05712 46333	15962 32366
8560:	77878 06717	48424 64381	23047 95598	80717 08614	70242 76626	69285 36922	50540 97990	76906 05903	73708 57553	46973 64263
8561:	83514 14648	12461 54888	19592 37313	06623 31609	37877 29958	96019 66057	64324 40787	04429 16658	43483 56876	58590 49433
8562:	30286 56840	36248 85187	29424 95376	91491 37298	34420 22122	85989 79728	76312 00884	95303 78862	61032 54022	33893 37258
8563:	71270 22500	16423 22077	02243 22171	64369 23497	45238 73560	07893 77993	14955 11850	26999 04972	72361 67228	78524 13946
8564:	70385 46966	63618 89623	94832 81301	65285 67791	40262 38475	72666 10667	53183 57457	62561 72675	03377 82412	72747 40819
8565:	99765 10850	68522 31210	61804 82484	15650 32271	05815 33730	63742 71098	65523 11350	84299 85330	40307 59181	40498 66478
8566:	62263 29860	14206 96289	47784 51533	88563 47181	77267 14447	17906 96059	38143 81850	09493 13737	73398 34830	67235 11660
8567:	52378 41191	65399 70291	45452 26402	08160 74154	48327 12291	95614 92175	61171 06495	03595 02895	40010 96998	16913 42758
8568:	79954 85815	69285 31687	72541 75217	18147 37296	01987 14804	36047 21670	97463 09601	08723 91619	23815 65299	63719 44465
8569:	32722 76014	76210 55038	27926 61529	93162 93665	11913 66845	97533 64684	78587 00833	26693 36264	96854 70674	17240 17764
8570:	01048 77866	82907 42412	82491 43897	49642 08952	37005 48752	43272 82189	84006 90946	88076 02110	59518 84349	54961 48685
8571:	66000 38813	24969 30855	02159 02050	47396 87667	64383 99711	20274 75776	26698 35053	98056 89046	46820 79130	91703 62294
8572:	18676 36225	24980 36484	61732 64898	50169 91908	43937 59095	26389 55646	81617 70072	28865 09087	36594 05510	39974 31463
8573:	76481 28480	59993 53359	48558 19826	16418 15101	39221 75028	79982 18098	62959 54619	51490 30236	92667 85377	58653 57844
8574:	29058 04859	68409 81777	77639 54485	18343 33748	69573 85327	82600 01879	86893 60941	27446 84707	02901 59549	80361 54109
8575:	43645 56418	72414 42066	18794 92757	95125 45613	22226 70474	42511 01672	09546 53415	85663 29340	83381 68463	56275 46895
8576:	61134 16395	98027 23946	70778 92940	21208 96384	74679 93856	37155 02705	58272 28337	95846 57398	64676 02430	33955 38082
8577:	94838 63886	70177 24346	39222 30227	05805 56625	40224 73041	74587 76434	62150 75625	75986 23519	17729 37096	72281 34941
8578:	00336 54852	60381 13001	18874 43963	56790 64121	45815 14303	91443 88035	95532 12244	69458 79278	26585 18446	67940 16254
8579:	64922 33017	62852 29683	02253 13987	52531 43160	00250 90897	56169 92289	39372 45255	27097 95815	91864 29146	50131 88536
8580:	98593 23302	87588 91028	28101 91153	42713 72290	97866 32263	35659 13859	43688 58034	06323 96647	20787 44875	01529 58319
8581:	57364 70972	42108 39851	72046 74173	37437 37539	66013 52609	71173 04543	69121 86275	70921 99446	17603 67313	80251 49399
8582:	83681 76969	97843 91517	15968 16208	27496 99569	75656 77311	11592 27956	34647 55204	12471 34923	71344 13072	67030 54346
8583:	16772 97461	72563 10832	81387 04452	48371 12761	11882 98209	73291 69362	15932 33895	09512 99707	00141 75017	24098 99935
8584:	42241 04459	45031 17391	37756 42775	19220 78321	01858 33881	86277 63676	99706 91721	61308 71247	00277 18716	36699 88726
8585:	12567 95246	50512 42698	00571 82989	44418 40099	17681 06640	75671 22867	20455 51169	68273 00461	29776 43829	20860 29743
8586:	69603 91453	75999 34718	19698 86882	84450 03196	52230 88556	74670 08208	48865 57691	65606 37817	08387 69147	67205 09997
8587:	95225 02310	18467 17443	15241 55707	82641 31497	55905 17908	15036 86448	65409 48257	22228 63454	03276 24194	56229 78103
8588:	73104 60142	65821 72172	86044 28500	67068 57843	53381 88780	77963 86571	53944 09427	00262 32553	64013 18432	39192 52443
8589:	95794 85863	32006 30287	58915 97646	05808 99884	92009 57718	70155 02925	71812 09758	04286 24665	90384 40708	21495 51237
8590:	70155 02925	71812 09758	04286 24665	90384 40708	21495 51237	98187 44670	97255 13848	59796 49480	69259 34643	47161 66814
8591:	90783 82262	92047 33162	81665 12779	92164 98712	70096 92200	75952 14821	18377 55556	54942 82992	25675 48836	63972 53765
8592:	37142 09262	69126 19948	83240 24496	90334 54497	60154 32834	50084 89853	73306 63914	25229 85668	58518 80474	99564 16055
8593:	52349 44905	20210 42693	66576 02503	39581 88566	60583 67841	15265 67506	40383 01021	07608 67434	18395 74052	75467 19844
8594:	74573 44946	20742 95707	79391 91724	25736 96606	12706 76428	79865 27749	14142 14158	14565 18142	68773 76719	83845 59074
8595:	15140 12780	37696 66002	83559 00383	34219 89894	21575 82315	05900 46539	35104 31544	15520 43289	38092 30512	73147 14137
8596:	48681 49995	38917 90579	79761 13354	44182 03826	24155 82259	91879 47641	00183 47825	70578 95508	72595 37431	16695 34143
8597:	91669 95965	75415 36189	79961 28110	21056 67534	63132 56960	36878 04086	64108 99249	27092 35743	80363 91521	35269 82118
8598:	70874 93455	91876 51123	22367 35184	06957 18770	81060 26408	96299 28942	33663 73459	41504 96393	69705 46604	73989 93986
8599:	20885 25254	37545 94445	26035 38476	32741 11696	15154 61304	48578 34878	04550 36086	02910 69143	65034 00953	60291 86978

```
8600: 95900 32974   99911 42482   98474 27822   29064 79577   94502 62468      90374 09741   68782 27073   71814 93120   19843 38637   11705 81791
8601: 08471 04419   26985 37045   20075 18819   81487 69253   92995 38355      95661 94310   39440 60210   65226 81900   42007 75021   75159 95274
8602: 14178 59083   91123 34045   08889 01294   73562 30490   49921 42714      51803 61903   92734 06395   42007 75021   90381 99079   35685 05116
8603: 67331 90020   87547 24082   69263 59935   92364 08764   99312 46300      61042 83449   63124 88887   02331 85592   63947 42752   15859 12215
8604: 37050 18881   06503 72101   18834 81215   07030 44658   20805 30013      64597 52753   62459 05641   55987 43450   63092 96995   43529 53468
8605: 27296 04988   46128 95374   82654 28224   57497 35728   56983 44959      92268 89198   31451 40884   14959 12499   25660 43051   22015 35391
8606: 58060 17085   82333 74269   19107 24547   67799 20405   10726 55235      72793 92680   90026 06839   28485 20911   35972 20094   11738 18737
8607: 41655 67246   65493 41740   17655 84448   54440 89779   87609 18750      66187 47362   56868 49689   29517 39013   07730 04229   52851 15730
8608: 95211 34828   14655 80986   43396 97949   23151 74534   93237 47355      60731 77810   53682 16109   36604 95522   81143 70302   08247 02890
8609: 47551 32573   51066 69099   22072 54838   61869 45247   67182 81474      87766 15894   83951 91276   25908 57278   22230 78250   66395 68802
8610: 21878 51825   09326 51616   89320 79468   69060 97081   76388 95773      00864 02701   79760 89427   27359 34956   60863 87688   73375 54831
8611: 04856 55753   70585 91245   91814 77459   69751 56663   04739 06626      99289 27016   42011 65335   78893 34148   48398 00170   32629 90166
8612: 72061 47235   82501 44826   60493 13896   81668 57484   74325 73563      28003 04847   17751 13258   72557 41327   16384 06339   86655 86352
8613: 87313 79399   19141 70653   25374 52754   42155 34304   79582 69691      77228 81731   64334 22276   31283 04674   19127 39162   63107 43869
8614: 98425 58781   31414 14661   63940 80872   97410 46212   08454 59250      38083 67413   44862 63322   12112 91427   47982 52916   24178 48800
8615: 14238 34292   60759 40896   37756 46572   68814 10571   48307 19400      63513 72078   68718 66441   54616 21012   00877 46680   53244 00343
8616: 04911 76469   98280 14382   63769 60338   66527 64283   51437 42056      18406 63198   04074 21926   89949 28727   48323 42977   58530 22170
8617: 51242 18949   72227 56481   69720 63698   33394 62418   73313 57709      12010 88541   76772 99744   31109 49324   55590 31604   96200 99232
8618: 07846 00569   73431 56753   45409 13669   02164 06475   70808 33029      95900 18169   12580 04631   60350 85965   72158 91616   69979 30622
8619: 79801 74115   99462 51842   95724 07057   45383 71883   03362 75921      95346 40584   78910 98270   46060 91563   71595 72418   38369 19959
8620: 64612 31230   11060 33589   26905 35252   24839 62699   73055 08118      16548 19906   26013 19786   37778 98960   62035 52906   56907 96978
8621: 78456 81166   85275 53883   40440 54902   19696 72267   49433 01639      94686 36762   94566 00805   76138 84165   30624 75534   98645 98233
8622: 94141 57044   07716 51386   62048 00519   22376 52028   33550 09107      11807 81786   86006 02572   65614 70399   15656 30042   30574 09277
8623: 39911 99717   73416 83311   41740 18336   27296 43708   17921 76627      58596 17050   06978 27768   07044 45844   75352 15010   37632 74340
8624: 64087 22942   23967 31613   73463 91598   06616 70288   38731 50359      96381 13046   82718 87056   42806 93355   41091 60505   69257 26705
8625: 87712 16201   23540 21096   05095 02608   49677 34582   23336 99958      08236 06488   21632 17942   50746 83224   23143 78830   52723 04908
8626: 62491 12513   21275 71132   89734 20100   86314 76997   85583 63931      51954 72691   21523 56952   35426 88403   66435 72160   16868 30986
8627: 62543 04970   41470 12454   53660 98465   96634 79198   04566 24829      72061 75343   14860 74276   58305 67545   86475 84429   84671 59972
8628: 52744 45857   88764 34952   52920 35501   00693 79438   90751 33201      43420 56589   25967 96287   55891 55586   44714 99353   51921 24420
8629: 01305 17418   64729 24198   65846 66479   73228 48350   30566 15997      19215 42488   12802 37799   23954 49982   90074 46526   81771 95651
8630: 99507 05808   10834 89189   31372 42255   78039 44299   76300 75329      70476 41835   63578 40149   46695 19533   90814 04647   45414 29053
8631: 98535 84387   58877 58829   19211 58346   94630 16556   98881 27404      74096 53255   45332 20622   41614 72081   53267 51255   86200 26023
8632: 73155 17742   30446 79763   24228 90224   49377 18814   03568 63545      63257 62965   13619 65984   00767 74499   80485 56171   84698 06240
8633: 72432 87163   79361 18303   80380 75917   95915 20826   75019 34824      87401 17390   36104 86366   04378 90538   96746 87983   54374 57610
8634: 72910 96381   50904 38934   94484 97310   69631 98018   46356 86917      70619 72222   56273 89997   88629 07169   18664 66906   99463 60056
8635: 32588 93080   00944 02733   79794 16538   66907 30875   56223 59987      99510 66421   32313 95905   75702 05038   20621 36060   06535 19626
8636: 24815 20085   93944 12792   79807 17866   11214 62707   35276 56849      29662 60177   85788 64106   32148 88481   85766 98152   29730 26829
8637: 69939 87252   48368 15458   03030 81335   18116 45072   38257 91378      96755 69389   99491 18613   60081 20816   58167 99029   31156 89370
8638: 99070 84710   93747 59560   57306 11230   49108 11237   70351 58981      39853 17182   37936 09012   86071 52498   65453 76946   19381 54146
8639: 86321 36110   96820 49474   89726 30049   61433 52476   24980 53429      57154 64805   53463 93421   64607 07736   81336 58319   59357 30449
8640: 38806 51642   44157 65661   31779 30789   77801 61138   61565 93023      22919 34800   08060 94737   78251 30411   20566 81139   69332 29275
8641: 33270 60402   08516 36686   60295 11934   10768 05831   31627 07793      88116 74991   91712 70216   73164 25490   95080 51173   58393 94393
8642: 60261 90018   27363 67101   55231 90142   91610 41511   01553 70536      81351 09711   64141 34481   87187 05027   42778 82361   87575 32092
8643: 61309 63240   14408 40241   14458 01511   16186 20664   26733 64681      65230 17881   21291 21965   83656 06129   97338 18800   81526 92434
8644: 08346 83267   82064 31914   69740 86222   95071 01458   62505 23357      64274 74151   76742 07835   53892 00228   07984 21097   19793 16781
8645: 88239 60423   13013 04256   44818 14072   83782 07100   92341 01404      52823 36025   98579 19279   32227 69871   18195 51543   10221 74735
8646: 18027 26903   70803 53754   04751 78749   19515 83230   33509 54627      86335 40528   29493 22141   83050 36761   95947 82356   71790 82023
8647: 46638 41327   75896 93296   35987 90062   87859 78063   43508 69634      99849 61920   93717 33174   17218 78665   12519 57241   29895 90859
8648: 22469 02039   99329 07614   41118 04573   08147 73499   76879 98565
8649: 00507 01608   36803 45886   35940 41497   03165 85424   14138 38969
```

```
8650: 75256 56082  82246 18432  95528 41786  48220 53244  83698 18163   28911 46035  34367 6979   56970 22116  49843 04835  52812 07061
8651: 16277 05169  37282 60481  98511 54318  78343 94814  72664 66353   05934 43262  71956 68847  60101 44474  95600 81011  13500 62059
8652: 02938 11523  86305 91748  32247 00076  66499 63254  26619 38567   98134 02010  77562 87250  42727 35643  72660 09840  67218 07911
8653: 31536 35863  29183 42567  85863 17336  28096 40739  01468 62235   17285 49085  62631 95969  89672 38848  35056 73561  25287 16889
8654: 34701 69692  91270 55899  83271 05295  64013 95525  79503 10054   33961 47194  66652 50865  35562 74121  26811 31413  21512 55484
8655: 71264 46968  58769 77510  11736 41194  00345 40932  93484 82040   21833 31597  47390 13892  72345 86465  64100 95702  03001 63613
8656: 23148 91835  65058 62724  00495 01107  16891 06923  93499 25827   81059 83861  76392 02908  64501 03756  30438 57942  25094 24928
8657: 55580 97097  80346 85115  79826 25745  33161 99402  80956 04575   45753 57874  24034 83870  53721 84648  56907 14777  97988 80131
8658: 94528 31730  09966 19631  38743 81663  47996 82919  38772 02280   97140 97857  16395 83012  46390 32676  52451 89435  68152 59365
8659: 30619 51687  80247 01569  85227 58464  93682 50496  39381 59845   41679 77862  93930 65614  14239 48976  05875 78669  93015 03002
8660: 69033 42922  69584 23498  51485 68261  58952 03064  13000 34905   96805 84773  89859 34362  47795 13807  99739 52613  13188 72061
8661: 85454 58590  46187 62354  77080 30572  27019 86363  90107 11852   62117 23777  23069 52376  03619 57240  49355 89582  12950 91812
8662: 50193 07281  64493 58177  42557 27661  20576 00714  36515 91261   81686 79826  50885 66639  49564 20646  34646 29472  77158 30086
8663: 81333 45766  87322 03615  83393 83199  04855 30566  80247 33158   85057 19671  74172 32454  02154 98585  08004 95128  36726 69083
8664: 72915 90712  13850 36439  24930 28907  76833 95520  86782 72958   27728 67675  42118 83977  36648 31943  82568 09187  33524 99847
8665: 52418 59217  80646 59844  96394 63044  30909 65101  02895 82613   46222 54177  55584 35976  59686 64466  67841 18621  49883 36267
8666: 40819 53100  25994 70398  94530 33221  58693 67778  94922 05765   88309 24238  81467 75863  38029 62803  93224 16982  21456 58790
8667: 03415 90950  19148 34085  80609 55225  18217 79476  40968 90526   26109 02376  36383 22793  24642 02026  55208 06709  46870 45688
8668: 74546 02575  64734 36997  85680 60100  97309 86631  39332 25732   99065 12492  61747 72640  15720 22694  24750 68040  12888 05076
8669: 08173 55751  15788 76610  83058 13772  54182 24524  29640 21284   13319 45727  81105 63273  02325 18485  94882 11234  72289 94635
8670: 73716 30765  72214 87860  25687 66333  21236 65260  63816 68830   57201 95800  88057 48208  20042 36268  12884 69694  38336 51376
8671: 49993 63930  21102 97887  31849 97243  49520 18856  54330 59882   99294 56877  59134 86033  56585 11841  53185 01662  20929 50032
8672: 05084 64884  95955 16455  32516 39160  01147 17097  79103 79465   94954 11133  18028 15107  16886 74416  88363 52332  84783 86308
8673: 98724 72218  25783 35189  65464 12259  60035 62027  52254 57745   24299 00956  04554 19298  72982 85639  22936 18518  17005 33275
8674: 26153 42159  93190 01414  42115 21403  51589 69708  02893 36482   03754 29174  12053 18590  86253 67154  55749 97137  84556 44981
8675: 09479 45767  52378 10011  15383 22828  82403 45836  23888 70400   19198 14471  24676 79511  58159 86804  45599 50815  14600 22203
8676: 87081 24364  30639 94491  20004 80419  58753 82547  86898 67727   32302 41881  28403 37577  44791 58033  53491 16012  84562 52830
8677: 62992 20022  60468 61269  32663 01452  92031 48151  97302 01500   88662 86308  89468 60789  96079 36180  87290 72757  88135 25565
8678: 98461 31743  18767 15216  59421 06157  19506 38354  52555 42903   54113 38893  40827 89622  15272 18865  14455 20646  17819 08977
8679: 36158 97019  23610 73283  44409 90623  64037 32587  43838 29848   16224 32339  62136 36295  34954 52680  96371 76812  70076 79336
8680: 62202 69420  18561 26456  03988 18152  96481 16772  69645 95351   07343 25129  23715 05752  96914 87342  52634 19411  45448 42720
8681: 98379 65152  84219 81251  33965 68402  23989 53782  24447 04873   19060 87912  44410 19079  58517 69429  35470 05720  15416 43229
8682: 25836 51577  22376 82168  13280 21618  42724 71489  84193 51794   56727 10769  75436 63989  95441 28671  08974 09424  54623 08729
8683: 58976 94737  38296 34737  23158 33953  52601 59301  48922 03778   36052 73322  89375 21822  72613 65724  82302 60078  98031 10515
8684: 61499 75839  21218 31307  46519 91165  75390 86320  99673 04037   22273 53461  57614 63363  70921 54816  13050 07590  33226 17276
8685: 68670 85814  38527 87961  64600 79027  27813 07683  36585 41616   61379 41593  24279 07289  47400 80075  04417 89212  61518 59177
8686: 94942 64896  16315 58834  01721 99380  53323 79131  90043 17897   83114 76270  93836 12667  43356 30068  17879 40578  48001 74485
8687: 33790 08236  42059 25241  25065 95131  05073 11372  54781 08610   31913 84192  07424 74680  15692 26939  62839 61055  54563 31909
8688: 70262 17785  88743 16677  03358 22726  51661 34076  30686 79596   42214 09468  71221 14835  48216 99385  68905 13272  61571 71688
8689: 64723 14269  81047 05456  06054 31014  39502 02947  96586 02090   14119 07551  67051 12064  82517 32034  95327 15276  64940 52503
8690: 15987 53894  12032 83530  29778 91142  69798 32009  01146 59146   63292 79008  84556 02328  73949 46863  35920 10473  80234 00679
8691: 98265 81870  94645 06098  41911 62242  66584 96305  65795 61864   32737 57671  30041 19166  38190 77921  68473 57516  94897 97208
8692: 15016 24599  30404 75854  92377 71910  36724 84665  03333 17739   00716 50030  13680 71488  11018 29590  86675 90414  98029 61112
8693: 78955 90512  23454 31816  79568 56470  29279 10984  21989 50694   90671 13416  78995 89902  04201 75573  28305 34109  37781 61572
8694: 91082 46406  94444 01465  34315 48445  28923 40029  09339 94950   13459 03020  75331 37853  17604 24882  30715 80963  78648 24945
8695: 00485 13042  81032 46747  77521 09775  97090 65530  98567 74205   55766 10500  75690 37269  80882 71570  51713 41736  98469 61551
8696: 66913 87645  19882 00139  51524 94380  65158 95776  93632 30721   04682 45148  05101 17865  38250 16805  19925 66125  37945 02180
8697: 02813 12912  52246 03933  68242 93816  65060 80206  82685 00843   55851 58558  07759 24254  88471 56748  80307 66880  94531 65469
8698: 04084 10108  15276 46603  62497 28952  70556 70210  69638 42498   32317 97531  45185 09921  69247 26194  04117 74950  17067 12589
8699: 47837 86886  58193 38774  02231 07858  07599 62081  08239 85081   46139 46066  23033 03917  65202 21645  88991 50612  98658 84968
```

```
8700: 60948 25976  53630 11468  20560 19238  51362 20676  03366 07495    06225 56944  84643 03395  47278 83420  28276 39868  12356 67850
8701: 26454 13776  84188 10500  38406 07271  10530 91656  96096 68989    85149 57707  58535 23733  40309 91182  89095 80820  98523 62536
8702: 86365 61991  85800 16339  59371 85845  14728 24490  35706 30615    92588 07590  53457 33263  99145 94898  65217 46754  77102 54767
8703: 74942 00276  27801 92868  18286 59609  13402 23961  36473 00958    22562 98027  02432 55319  09895 39657  75773 35220  17880 24585
8704: 20642 65020  51939 00380  49264 74026  38204 38051  80354 45673    48919 92375  83114 80032  20790 43288  74365 30223  83818 35488
8705: 39586 53380  41915 63067  28303 51690  35816 58742  99570 39697    72337 16309  48740 19247  37979 95852  24264 02762  52704 96253
8706: 59715 19532  94926 60460  72887 08068  89203 81897  35194 48913    47548 64677  91154 61567  84448 77350  34234 48789  41226 23783
8707: 08499 15308  21603 91975  03215 14531  09279 50884  34206 17163    14845 08356  88517 45391  46841 58658  55562 86628  57632 84669
8708: 46455 52271  78431 50589  36142 55186  38783 45337  76811 34000    65222 81100  87580 95879  95659 78445  32910 55273  57100 72084
8709: 61263 57238  34647 82967  20614 53037  56962 87911  37362 38662    27114 02765  88018 91611  09905 32843  80209 66099  99086 13931
8710: 40945 15906  53693 54972  69989 99167  36703 43044  05008 35298    55003 66567  44862 97983  04759 00905  13947 07898  69107 77465
8711: 44814 57238  52715 27156  50216 91140  00621 39933  70627 53764    47729 27340  81698 48139  90478 02926  48396 37157  22375 54106
8712: 04791 46372  13798 47438  89060 53195  04231 17172  86857 68978    12567 79958  95369 74403  23435 19363  33378 89420  28321 83071
8713: 08083 18790  68635 69280  58164 61309  30882 85566  30951 01268    95354 76932  33402 00744  71790 95692  02605 75503  36592 57312
8714: 96288 99393  38235 51404  62299 82992  80004 97045  64928 39441    77670 68642  84241 19457  66701 73530  85105 23164  00254 22718
8715: 48816 77150  26179 94016  86181 41095  34017 76813  74212 71170    26501 76878  37054 90571  82845 97245  11572 67642  69126 60066
8716: 06242 17055  28734 89649  33516 94831  82432 13999  44414 71134    78538 42755  65081 08707  72180 16397  68612 06854  44010 04373
8717: 61930 95882  62638 62021  29804 67729  50916 77868  34677 63570    47504 24435  05704 98322  82651 81798  83120 48050  82113 32460
8718: 73427 23899  78983 65225  06850 38665  78396 66321  74961 35167    46341 56507  25549 30055  36842 71370  17876 46334  52851 74820
8719: 05206 15129  90718 98893  72533 02014  44081 77824  86939 33564    28516 95326  86644 80008  97646 54769  92275 93011  29898 84166
8720: 76966 42929  44535 15788  14567 14006  03889 52168  35439 45944    38691 20016  87135 90174  80598 06137  32509 37194  71187 27295
8721: 63271 56959  77858 84484  80167 97747  99586 05574  14169 45935    59777 15874  85825 54975  15594 12420  81216 22194  35087 45298
8722: 96628 55203  82825 81822  38515 14145  86555 19438  45605 52231    49251 39618  65403 88347  08634 69256  52189 60866  36216 53601
8723: 42986 94901  21564 72477  88340 25698  22786 32748  01570 99080    23888 04791  38722 37079  75322 18321  55590 73248  03163 61661
8724: 89594 87916  71346 78269  45209 72212  62844 46939  60293 33314    41410 16965  31015 21926  43231 45645  08866 05857  63578 59230
8725: 78391 34255  89330 71976  21953 83603  94632 27015  87891 31876    10625 49267  95167 05392  13591 21257  27641 59245  87184 92698
8726: 25481 03842  95711 71881  82590 63910  13659 01007  85067 27447    85166 50583  68282 37077  55991 28994  82257 91820  39543 16366
8727: 77169 25407  68687 67094  83834 38938  78914 53305  83064 64613    30200 39361  09261 32528  48265 66924  40234 42855  75573 19597
8728: 37232 44764  24513 65872  47533 13247  22303 17173  09505 17331    69700 09212  00546 08371  77965 66431  01414 60182  99157 07970
8729: 76819 92495  36430 47532  40621 65083  17828 38414  17033 94642    94470 69449  43237 21364  91197 97087  33477 95770  36179 91532
8730: 44500 50758  21764 69161  75607 65279  70923 31480  58142 75868    96999 72702  33879 68847  52841 49658  44419 71697  54689 85503
8731: 18542 42058  93805 43067  45077 83336  29839 76727  21765 49779    69474 60458  74294 04977  40432 78318  39398 92210  78136 95670
8732: 92800 57204  93868 32882  84738 20339  83680 42385  79576 03290    57752 50742  65194 61476  19412 93787  18319 43285  95045 67700
8733: 63638 69040  15053 90779  24803 16803  35571 44857  35363 50530    54363 31818  78829 21288  34448 85168  34056 11244  52039 69981
8734: 47351 28587  14509 63239  82394 34218  41777 92514  17248 01882    82799 75929  36589 69460  57857 66000  37076 54926  52641 11989
8735: 22731 51406  31171 30686  71449 95600  63936 54557  74190 85139    90970 99922  05698 70682  66006 85033  40043 26227  65093 19046
8736: 58555 51372  09386 24231  62621 65061  27323 48539  80523 89221    14459 65205  55354 36932  99772 93838  18703 12767  33161 64911
8737: 29963 22707  94077 98510  28570 51435  62500 45656  86422 75496    33744 33641  15517 86822  18234 73065  18098 21938  59983 62361
8738: 00509 56000  13544 22755  80372 10147  62095 05180  39782 57372    51115 37353  30140 44699  67296 03532  19764 15532  62040 22530
8739: 75062 62630  00682 98975  72250 22571  63056 91140  89879 24283    60928 66494  47897 21514  04185 26703  57021 92279  31132 77582
8740: 37598 17551  67483 46218  03167 29298  74646 59609  04836 23859    66305 05406  31892 77071  92742 82768  33878 98491  74777 76410
8741: 06507 96253  20369 53870  95783 70606  05618 55994  97742 49401    52912 61153  03607 20165  00232 57192  12475 52887  91315 50706
8742: 56786 16096  91404 57988  90068 56973  98426 46025  43987 49425    47481 49793  99917 10293  37488 43856  04563 83881  17090 09272
8743: 86133 04899  66916 36886  07714 56835  98524 01128  90850 11792    64308 41851  73623 37767  84597 88336  79668 27741  34074 59412
8744: 29711 09166  02029 44435  64994 82571  22791 98590  59999 88257    78661 16801  43901 65622  98006 02105  30067 43942  92861 36510
8745: 85882 64548  32853 90987  94851 00270  59413 96817  39261 99202    27980 24834  79619 18135  30636 21019  53693 17600  94461 60494
8746: 58899 68031  89888 85602  90526 35089  82617 82326  36998 58458    94531 26172  61220 41948  99165 73991  05396 62397  23215 55816
8747: 78137 23906  96966 86067  28751 89742  86224 37966  13790 06783    58391 63991  02603 83549  70967 43598  42973 45735  88357 49885
8748: 12277 17269  38358 49280  72032 80826  32077 91487  69248 02565    03201 91410  40515 03641  88169 01942  68855 85078  85322 63054
8749: 21753 92520  10146 46185  28974 79046  79072 10571  20346 18474    76358 32535  97838 42689  29876 04619  08031 73551  76568 13729
```

8750:	90600 42994	44590 59187	81945 21315	90369 00618	47888 04344	31776 62521	57326 36705	86905 15254	74445 32861	30914 98535
8751:	30518 69980	75591 03805	39016 93397	31756 33239	54788 09150	11593 30304	72986 35881	66780 39693	75915 04273	41413 39713
8752:	27639 57448	95729 93696	78321 39029	18515 77096	22331 46709	50261 83567	83737 03990	86132 88062	50855 33098	05602 95156
8753:	77334 03921	27858 94115	96891 73939	74819 87999	44380 96726	49729 76580	66626 03483	15362 74081	00437 14251	95510 47100
8754:	52371 36353	97858 65735	10816 69776	32566 58535	65240 97419	97648 71712	67427 33882	95069 75416	47578 42703	97733 85528
8755:	63515 19279	47346 85990	45054 83162	41419 35519	95422 99894	11568 60159	14027 10240	77918 15755	40659 54019	13846 31989
8756:	97496 53511	95500 58770	46510 39966	94795 75457	52887 93123	23381 53923	53496 18388	69555 74367	82888 52201	54114 14564
8757:	66739 10390	42236 10961	19774 08452	32466 30582	15868 89407	78934 30501	82431 19507	55639 86173	05266 58843	07330 63740
8758:	76924 83797	53219 96200	14143 85553	09891 47149	61375 58811	24808 60420	67644 18914	23411 20291	92453 91144	90690 76030
8759:	19567 50891	25719 91079	12611 77246	19488 28810	79492 70644	51608 69923	08273 50774	28643 32664	69983 65515	05425 75955
8760:	13221 60789	71721 40044	89311 97751	25347 50846	50557 86623	44969 52120	71999 92316	67695 40057	52910 73372	11783 54086
8761:	58687 39403	25078 91362	95589 55119	26637 76872	06707 95388	41267 30948	07669 24089	43663 07431	52378 76964	66424 77182
8762:	51461 20917	78241 63665	83806 90525	55549 59327	72261 50709	41117 21594	44670 06488	08144 58142	97849 03030	07857 31798
8763:	41547 48260	78872 09735	86868 08123	64364 61617	28950 05012	83003 46272	05031 74378	68778 15528	41570 31917	70191 38994
8764:	05607 27046	25912 39801	96605 61180	66287 00373	25837 44261	94877 05849	61713 25041	88783 57749	51594 58595	42757 73067
8765:	72815 10964	07848 28376	55637 61184	97329 61791	57743 92071	03378 22834	24051 81562	59724 99321	56339 01371	59338 25346
8766:	79255 73060	35040 26975	31720 14852	83957 26612	07634 33466	88657 84944	93065 92176	71258 57998	81962 23764	41979 01629
8767:	02815 82177	99813 45700	43129 42812	72839 60315	04583 12787	22635 58210	55793 57720	53958 41933	05952 74557	26020 11484
8768:	99836 36666	78743 17687	15916 89729	69980 67728	08457 70305	04039 18518	10263 18799	69225 90972	25335 12067	25475 05088
8769:	71978 45024	17358 25732	65930 14993	34248 77057	90836 14840	51233 98211	89076 95814	85579 67252	65935 75178	41027 48331
8770:	43766 89363	76122 24368	11583 86506	53546 13805	19536 90366	18618 49215	50661 76882	15038 06598	00687 00661	17630 03860
8771:	51429 98277	15185 79364	77281 29684	41702 27147	10577 91556	04289 88360	17279 62583	03070 16718	54408 66996	27646 38430
8772:	75369 87923	12558 18203	08367 04980	30383 00384	45310 52172	81150 24312	49837 84850	31539 34957	07010 45298	40467 50642
8773:	16806 44756	46338 29004	84901 30326	89529 71187	14696 39374	44810 31215	34305 20239	67410 75151	28026 72547	67515 36506
8774:	70342 35275	41600 80555	88527 48808	63794 69024	06017 44593	18468 46406	30957 14938	34871 90693	40606 79446	92740 42698
8775:	32704 71923	00010 07293	52000 12064	90535 30271	74528 94099	99379 45460	19080 62712	40778 82545	15609 29602	99269 31723
8776:	57388 16966	84146 06186	81896 84064	33961 54248	19510 71875	03822 66541	71174 29272	84342 05692	95487 60849	51271 96508
8777:	54098 60533	35501 59437	21121 90023	70942 08891	20598 01184	30492 01325	52309 24522	35621 70729	74558 34509	97104 66797
8778:	76170 80820	18215 55692	90074 78088	62625 47525	77422 03109	60709 61921	27322 82946	41033 81677	85258 82941	38308 45621
8779:	87585 61105	37139 98529	27602 67119	79362 58116	43298 96843	80030 52762	06237 55469	12727 86499	35150 52265	80485 39671
8780:	77518 76283	19657 07328	09738 54375	48623 55227	84257 63931	13406 06574	33463 47789	82461 91911	85712 59750	90678 08408
8781:	10990 76849	76048 14056	98207 26512	66996 32896	35123 39663	33199 69980	17266 98060	24721 84085	26558 79690	84961 67009
8782:	31942 52048	47771 17923	30626 17995	57383 10223	37605 96989	97922 86088	83177 92783	84216 82110	67093 62197	33193 72939
8783:	20002 27156	98024 17150	03714 81354	37873 37306	87344 62429	59231 93502	81688 97231	04381 83439	84021 01013	42677 31686
8784:	10878 17567	44591 65630	25332 19121	80918 33961	86318 09235	87464 40068	32693 88200	11465 45120	16650 01895	92965 82291
8785:	73166 86041	55362 86050	66713 15248	74582 44861	12349 22878	76142 10085	90202 44050	90229 21513	28725 65806	53256 19728
8786:	08282 17747	71148 21132	85044 02896	67234 51811	81107 10450	13730 26207	30066 53124	68107 90476	31770 86766	15397 37149
8787:	58476 61063	66668 67478	67433 18098	82989 02188	34720 41986	15984 76156	41167 73357	73421 18567	68729 75141	40668 63650
8788:	79687 65297	16198 89750	89042 26637	96617 12150	91066 12354	29060 66194	42120 72128	42886 08995	05445 47347	33374 03163
8789:	44718 31871	17424 29301	56458 88340	20629 28847	13187 49307	95894 31756	14604 11934	65647 60402	35410 49783	53663 76725
8790:	23567 70111	64903 50948	53254 34525	40364 53181	62194 60452	27521 39057	84357 47202	98144 44530	15283 20643	20030 31546
8791:	98601 36739	81561 73269	78131 78074	69198 11048	28371 45662	90767 54129	03216 82417	53799 39493	45294 93370	37075 04730
8792:	56363 75545	70237 17601	27518 49609	92263 92431	23655 35613	28925 97070	41935 40405	75083 93975	29777 22154	90550 25051
8793:	15259 37473	41388 88397	40137 05713	24817 22678	93386 84710	34603 89004	17595 50790	85748 04183	09786 22306	28352 95084
8794:	56996 46200	61985 47047	20197 49576	69333 49167	80753 40425	58430 07987	62549 57113	45826 04735	70787 33425	90389 46533
8795:	95791 20787	84177 79693	05438 33613	08794 33131	63141 97185	15842 32106	38270 61242	66328 23845	94305 98672	00458 95609
8796:	36988 41919	83331 76517	87192 78947	07961 47354	38354 10879	41055 28033	23578 41976	22999 73892	97323 88587	94048 13667
8797:	88394 83184	83753 60213	13531 26316	79050 86541	78763 91803	24800 04181	56567 63949	92575 52424	83237 82224	50321 16405
8798:	89645 15323	90005 92524	01693 90874	43153 62730	29993 81243	96974 19715	27750 43057	75141 44122	09214 63924	31974 70198
8799:	87961 37213	07498 62625	87040 62136	65116 39526	78747 66461					

8799:	87961 37213	07498 62625	87040 62136	65116 39526	78747 66461	96974 19715	27750 43057	75141 44122	09214 63924	31974 70198

```
8800: 24959 47058   41567 63978   46463 54990   64353 51837   29011 41468     84944 74468   50650 44629   42764 15344   76035 30705   90569 99721
8801: 58802 43412   08095 11903   89692 56477   87424 61891   82043 69836     75754 12288   36923 75523   49960 70229   61518 21225   01941 46447
8802: 50107 65301   43755 30348   41556 37304   93498 61247   31190 72621     05801 42950   81130 86070   30743 86156   06887 13514   46287 45740
8803: 92267 37331   16125 36440   67705 81302   26257 82030   58326 46806     24281 98689   40070 47396   92260 64196   42490 09513   98302 68292
8804: 18906 81711   19778 77257   57148 75588   04463 14798   20361 01397     53526 30935   18555 71731   41240 16728   14630 63907   52855 08120
8805: 59197 11306   23581 08381   25504 61216   55239 80689   75738 07653     17367 54867   69300 59325   56932 80567   99735 47873   97688 84050
8806: 26627 49451   35885 27413   64339 42142   34780 67043   12980 53943     66810 93332   79831 70894   56354 69187   81899 32036   92764 33941
8807: 75665 45259   29420 87719   67039 40982   65199 97539   60834 93077     41064 18887   55695 61316   35480 24679   27856 96452   74887 84584
8808: 41146 29843   24289 95765   88844 04603   86584 94638   89426 60215     74154 25175   03869 56364   38592 49732   58688 70644   89301 48214
8809: 48150 85495   98794 48563   75239 83706   78095 27037   93412 22211     34069 85033   42789 12063   72225 07660   27198 40503   36683 49756
8810: 29057 98092   20354 30646   28311 04550   04776 48914   98314 15984     02791 95665   61468 93925   43609 44152   82954 92046   51040 65330
8811: 95555 88220   70521 30327   13690 61698   98809 61590   03855 63596     93833 25027   31958 29169   07496 04733   48561 04025   33279 18418
8812: 46930 04095   11817 22797   31344 43111   79164 15294   14570 40436     95590 74201   25580 18290   69092 54054   46264 67309   03979 36768
8813: 13255 03367   90159 04721   12947 73042   65036 27026   09690 96581     09098 02084   95335 04138   27229 49194   08777 79182   27747 04530
8814: 95131 53738   28471 61381   41528 34146   34817 99203   67017 28646     32064 86937   83270 20631   07278 06979   72782 82153   51345 64673
8815: 98909 23409   05872 40550   46863 13423   97104 46361   36966 41228     18623 76019   01266 15132   01226 41795   39943 64230   19405 82895
8816: 27158 83676   23487 47568   02917 49470   53744 20009   05638 54979     60328 01399   95012 81429   62343 77599   21420 92906   87488 70094
8817: 28933 30007   63638 65133   15670 57438   52620 40196   15492 21307     29496 65304   79373 18480   63581 75279   85265 82294   67956 45400
8818: 44539 67577   28467 85761   74020 09411   17091 34802   09112 51511     23092 79774   38786 90380   76896 23483   70473 39810   30654 16017
8819: 49578 41947   57657 37971   44955 83185   36550 26257   22088 99842     46498 03024   12070 10203   56558 21695   48080 38587   91047 97489
8820: 50032 86384   83569 53725   05740 14826   11214 15465   40265 50839     82638 39483   66212 43863   05223 12367   14176 48705   36047 27988
8821: 74810 18977   04820 31766   50161 87660   70407 37446   17320 15772     17014 42155   17217 04596   20517 22220   03757 36048   55603 41824
8822: 08942 27045   91630 75179   69253 38840   41737 54403   57183 74819     31557 22729   91354 03569   35042 67143   55153 42687   12494 86070
8823: 92540 33300   83711 03984   00431 71029   66001 04454   26426 95452     97776 93815   69074 72917   23432 04940   78527 63383   97252 44561
8824: 80498 23532   71149 23341   39302 57654   50548 88419   95625 73436     92333 78921   99284 25181   57425 53000   96468 27786   83004 30579
8825: 02489 59160   58391 29651   01285 04850   67381 21071   66033 50396     45548 09290   75346 34710   86585 90375   11629 11844   78896 77310
8826: 94078 90172   65248 96602   70848 99739   25021 98534   53383 03873     54119 82468   43837 50793   70754 17985   42747 03508   57213 32552
8827: 75176 18532   56466 18889   67317 21700   69223 36067   48517 00189     52765 90302   28815 54887   77766 60546   02008 61703   29448 15011
8828: 51786 60010   81715 71734   01957 35659   20264 08123   03464 38069     04268 64029   55721 50826   88323 84157   12889 77471   80373 46709
8829: 38400 77265   96214 35992   20939 51765   43299 24623   59910 75283     21159 07920   64966 18048   65428 73280   30592 26082   95209 45849
8830: 54242 39400   63742 42128   19287 59821   87236 28309   10466 57914     15070 85410   71395 07843   35638 26166   36330 15101   09578 63001
8831: 16195 32480   13275 58081   39764 56750   50816 01740   50773 13485     23756 84166   56485 48224   17142 95040   27024 72427   62975 09949
8832: 71078 10216   02249 08170   13565 64888   27889 62451   36147 38626     25848 45410   67180 32814   10653 67715   36752 87503   36669 35394
8833: 34513 88305   23791 16606   88582 23008   52899 44397   09626 95451     23922 06727   73679 19414   25603 44385   46589 53517   77153 68754
8834: 97436 91341   91053 57314   24368 15197   33182 54651   36182 27277     43692 28832   26581 42583   96963 40875   27923 13967   61674 49285
8835: 45705 28461   80158 10627   09625 93079   88219 61372   44718 81535     46776 91109   81538 89399   15442 93072   80475 79592   50910 63545
8836: 02088 43736   78550 91001   07220 07015   58371 97781   06000 64987     28704 97620   46129 98595   59424 04386   00558 15904   31359 66943
8837: 93763 25861   79317 31834   43617 82574   54606 69747   00319 56220     92325 85902   15924 53835   01925 50308   47404 88560   88188 76075
8838: 60038 46767   86059 13029   41576 81627   13338 03046   39142 06073     45508 43369   60974 26894   57835 11807   51693 96295   59190 95279
8839: 59209 24879   84539 56301   91124 94688   20812 46084   06471 27001     33401 04470   28194 81065   66362 70881   83527 06614   80375 69197
8840: 95647 73774   52883 33755   72637 90746   97964 78985   55895 70061     28657 47459   01643 87749   04679 44678   05673 71517   97949 84363
8841: 71997 21681   66458 35597   02994 61441   19898 56478   75602 72915     67630 88914   61402 80255   65599 76424   92618 94882   64253 24289
8842: 95984 62390   78966 52043   72210 41222   56320 31162   52871 99797     42379 63003   98132 69946   02779 22783   38416 31588   06810 88338
8843: 65110 13574   26751 04773   96054 86159   96625 92676   77390 67847     15668 64564   48737 96538   12820 76151   90831 23752   90647 13470
8844: 06779 78933   90348 39539   21749 62247   05877 51209   49833 54510     62478 05867   45279 50978   35618 24242   72500 09086   60578 92774
8845: 64543 79261   68785 57508   65734 57149   82470 28708   08526 62315     01973 64623   30606 75092   43036 11394   46112 44775   61538 61341
8846: 60518 21646   08317 40428   42014 19345   55089 52467   55712 22916     82044 95800   64221 61927   42116 84832   88244 80430   33315 77548
8847: 26635 63967   04795 18593   42576 51515   04521 69887   90890 09224     44615 87797   08528 44995   83873 69584   07553 07956   64108 93866
8848: 53043 32380   42134 75021   04935 39620   75609 92768   83349 28019     85860 79187   35163 30234   01248 08098   11748 38667   90875 32841
8849: 77805 43690   26686 08996   33363 31337   15865 08624   12925 18499     27453 99333   63499 38202   41259 80261   15206 02630   17953 87481
```

```
8850:  83459 92479   23506 97061   32190 00099   08163 30754   51589 12910      03894 96169   86080 84255   05945 36786   20251 04296   69767 33719
8851:  19101 75200   89112 60657   13524 92203   16751 14473   33365 02183      23190 82408   83042 90553   32662 38199   29732 02692   80717 55780
8852:  39099 74930   28584 03827   14598 57294   25359 44239   67325 64933      70826 35910   39630 72086   28035 69717   63136 42891   79222 15750
8853:  09786 73540   35562 55480   71399 19814   93489 46169   96476 35830      36543 59344   84568 59970   72884 64565   87456 58258   47831 93632
8854:  65929 56319   70735 56386   98843 58073   13265 13437   02468 77476      48676 24331   29533 48513   68752 78301   11155 47419   27874 29706
8855:  20130 55607   56098 73785   50096 37860   31345 71091   15823 79101      23697 70576   06990 34420   34925 37177   59508 40901   81135 63457
8856:  59987 29376   66982 01909   82703 47386   09044 44579   90978 06426      71936 71329   65971 09236   15218 46646   12108 33622   47811 60059
8857:  32730 31557   06581 99288   04207 18048   78032 27607   28649 64615      90986 46135   16611 93864   93005 78782   27078 93438   34654 28205
8858:  46668 54305   89273 39848   80412 37995   37110 47692   38176 37251      92924 52574   51650 16015   05552 23067   59968 48786   59250 23475
8859:  73819 33110   12853 20433   95830 22457   02299 24108   79783 23935      20809 57537   06901 94004   89268 11568   29705 97410   90777 19439
8860:  66921 90901   82778 58744   23216 44997   12729 29572   43243 11218      57685 95521   53534 26077   43315 04584   53574 21844   96657 88122
8861:  54846 72221   97760 81268   15438 73634   91848 32291   01849 63837      49813 46167   33193 57247   22080 46109   88850 15011   49765 92652
8862:  95572 49615   07462 89557   48088 07565   84212 58647   94069 19838      80138 49567   41391 89776   77082 50358   73174 33383   56964 95984
8863:  05499 15150   60817 31313   22748 71034   80682 75088   11443 87032      64378 58767   08410 02908   29089 55675   64187 44835   49361 63814
8864:  40820 16690   20486 20074   25621 30109   39792 79462   47817 75069      20282 82675   89094 50737   61285 16477   02739 37655   09355 59934
8865:  76029 44629   17595 23717   09362 72832   79632 83312   47543 09562      08239 02417   69544 53873   00515 38424   19965 43233   19008 25146
8866:  02775 04659   77941 72412   86559 07500   61520 89352   30204 65788      01164 68839   91421 83408   30639 95277   31062 81879   13586 54450
8867:  31207 75681   05932 76659   17603 00251   06270 41028   11928 54358      85481 83885   33295 30424   98016 52146   88381 98566   12924 17650
8868:  39904 59532   41275 57761   54359 11661   78750 01197   79919 89757      40244 38048   57688 50401   51538 27126   99038 63905   68714 48290
8869:  51770 27116   94758 58955   34340 79760   48067 20041   11416 57196      90632 58121   17531 35038   20549 90495   94347 92191   78801 60302
8870:  32257 80776   80632 09287   93629 74890   06854 66075   30328 56085      78384 79789   24005 55273   62662 79542   18535 30678   38168 92981
8871:  52901 95235   79826 33483   70466 95797   94307 67365   26184 28249      68007 90682   14635 36869   09586 91193   49794 06178   05155 30459
8872:  52914 64008   11205 42439   29980 74623   68242 80659   84020 07945      24265 27927   00989 48314   95578 98743   42638 85441   86921 32791
8873:  11110 55479   01737 29545   94596 37337   82036 79666   80148 46326      08474 35255   57657 60142   73971 76099   04883 62798   80833 76800
8874:  64179 42526   08294 39903   39488 89700   46967 85580   51842 80999      13487 47387   41850 57926   70090 19618   80678 37773   96341 93641
8875:  90371 63631   13799 25474   76687 32981   23805 39636   67723 06775      22521 68455   93129 80754   22382 29229   73249 53476   12618 55294
8876:  04849 35095   19587 46678   10190 41285   40727 73155   20513 76732      96354 78952   69875 23747   04969 25524   18461 94098   05375 62222
8877:  84872 42101   62570 17977   66622 98076   79113 85588   62794 75404      25718 50010   97643 87061   49016 87128   02549 07839   46241 98009
8878:  07682 51333   27616 21257   56656 39789   62046 27009   45778 77069      72991 68270   44035 86718   02700 37973   55037 80036   65187 55654
8879:  30797 76956   30182 41632   19679 74484   76965 80271   84159 14248      50149 27409   27449 28216   66535 97333   18315 38556   51648 26420
8880:  87200 68160   80037 14224   40630 27980   14605 62806   08919 77281      84565 63698   11448 76670   56988 87180   51607 95490   24241 48515
8881:  80546 20022   68584 34238   96203 15377   45872 65104   27032 86935      95932 98823   89925 29979   72058 15647   94762 47863   35450 07264
8882:  56840 97267   43741 01986   00550 84343   64069 64677   10118 04138      50136 09283   28672 60550   15182 13283   30153 72211   79076 68991
8883:  68959 74127   38527 96169   98253 01155   64069 64677   10118 04138      04952 32083   83887 75006   64731 90996   92295 57684   39387 88817
8884:  06690 16991   95015 16234   30221 85914   87175 63960   13769 08640      04952 32083   83887 75006   64731 90996   92295 57684   39387 88817
8885:  73505 65054   21052 17378   15300 39787   18376 03986   47271 72483      71135 09278   54872 87547   44058 83608   38621 94331   08438 43825
8886:  06513 62173   15960 89586   70492 70825   33654 89587   03239 43325      01551 87876   52253 33854   65461 78768   88053 84535   68555 31975
8887:  05016 61791   85443 05022   56737 91349   49551 42251   70949 90463      15836 83861   39019 16724   58953 65984   71495 24028   99307 85340
8888:  71316 57016   54162 79912   64600 70469   49379 66437   48213 83576      65025 90740   05815 08872   50701 62663   10095 19664   02265 15909
8889:  44604 13866   42076 95493   58506 57519   68410 68651   39477 27144      89451 77674   86257 74683   99611 66650   93273 23898   27804 29251
8890:  77913 86100   22170 29216   42052 75926   51247 48756   47963 07938      91279 04150   39928 27666   10013 16979   68338 76197   49554 11281
8891:  99893 84196   13335 11883   85653 68424   75558 07111   08401 11960      14693 33460   77450 05567   84070 41275   31039 24076   32332 27966
8892:  03429 24265   73591 08510   72421 81378   34577 55724   22687 57704      46367 83597   55221 40930   22166 41318   83799 39932   23834 20069
8893:  25441 57933   61237 13247   12756 81016   17443 89789   83152 76156      41097 92967   09471 93109   42615 30167   08793 41958   27177 37835
8894:  95988 68715   17225 74321   92138 40528   09505 45862   68496 99599      73906 74889   01870 68713   80496 25372   86231 97302   36599 48932
8895:  72791 51482   21840 04496   61018 09761   06765 03938   90452 04919      96913 89460   50611 08083   56041 74101   73465 42832   74716 35696
8896:  78467 29923   57797 01344   88316 53405   15665 09688   91478 20506      25279 03757   69106 25403   97666 43277   88831 50356   68928 14390
8897:  40465 22475   49788 97905   69598 00592   04810 15945   57791 92951      81994 61560   18265 09890   11497 84214   65586 63248   98141 02356
8898:  30785 19001   80949 08099   38540 50001   17457 29092   07199 68206      56716 34638   40962 49616   53146 82242   21990 25342   85294 44089
8899:  43442 68649   04939 12504   96617 31628   20775 36794   23848 36945      71346 34041   39254 22951   04955 60949   63922 09640   62461 36270
```

```
8900:  64043 22209   74106 62287   38781 51224   61430 57536   42003 87419     87317 02439   48528 85420   26702 46115   01709 85281   83402 83958
8901:  96418 76560   07211 23376   95552 64156   48517 69371   60235 38498     53231 75114   89190 22234   98329 47091   58598 87058   75448 02005
8902:  67447 40458   69463 08771   51959 86495   51186 80663   76335 07567     91920 33705   68941 69934   88370 63871   30738 94228   50553 40014
8903:  67331 54889   56871 77127   99479 10129   72744 59169   73200 20316     19343 30830   23622 65082   73842 29513   28998 02134   71566 71587
8904:  94417 57399   10516 52482   79637 89653   61003 40716   76400 18257     90397 75377   68810 33103   31245 86815   12017 37427   89992 84565
8905:  08168 73368   63965 15723   54865 45706   77851 39931   31477 88605     50163 90887   45257 32094   32552 61464   75807 88732   09590 99257
8906:  28409 52573   65073 39884   18696 06699   61521 73212   35205 06414     56801 28730   05678 38399   48036 04871   83216 01430   63096 24092
8907:  33335 98403   69343 74779   88471 23731   18549 61243   52262 63403     38603 12879   87436 52871   81237 47487   59571 91146   99392 27434
8908:  19637 06046   19025 95753   78006 88618   45599 28435   70898 07627     28999 14706   03212 61612   50670 32602   17499 09717   26870 08527
8909:  85423 23545   65581 33432   48767 33719   39153 97074   79588 04210     66208 77103   90758 33814   73750 68398   31952 64100   35166 94789
8910:  26363 82743   13245 68724   72474 32404   44457 79931   12496 56203     17378 36655   72015 04246   06486 95851   98188 51369   00201 70600
8911:  54811 54369   54321 06099   57186 36609   57265 76199   78712 80919     28731 23459   87974 17132   22640 65038   67128 22308   97149 54133
8912:  75465 91898   81731 24038   29137 68041   56582 11559   07499 73559     24404 06831   99292 07689   60207 55857   32776 21256   31211 77253
8913:  13214 44815   69371 80202   93082 72971   21382 79598   69478 92328     12369 23267   82700 75596   66705 05734   13277 68623   88652 77742
8914:  09016 30288   26485 75179   46145 20899   73864 02968   31218 60088     57698 33766   76533 64062   57009 29699   72138 14924   49200 56970
8915:  18608 58188   21356 79847   32447 05069   20590 22277   04888 02984     51404 29458   98632 09765   41252 41295   15543 76535   36007 91280
8916:  37964 16431   69841 78827   45747 73538   11770 08929   93668 32187     13141 53698   23059 09165   14874 95783   74487 20719   50615 09797
8917:  45616 69223   02008 89928   27756 88683   00610 86593   65984 43352     03375 94074   31407 62946   72935 77057   89349 93325   32557 46794
8918:  80089 12259   51893 71382   12125 23946   29567 76060   91945 09365     22861 89316   25331 94281   55578 71840   72853 68018   84237 39584
8919:  35269 83298   38839 18088   03117 97586   56513 38596   46414 95226     28767 35612   76648 86969   95601 72172   43021 29953   54232 13530
8920:  53467 59512   09501 89533   37776 75100   58538 85486   77996 48874     04229 38208   48507 32811   50133 66176   32762 51345   17318 89749
8921:  57212 31015   93961 93829   92783 49207   05945 89324   70498 66521     24512 02549   58006 25299   56675 69677   12016 81075   60607 56271
8922:  88055 03243   12768 27642   58400 26142   27367 78327   17861 45141     03555 28859   75529 22329   93667 97480   15907 48291   86853 75317
8923:  28663 79623   19818 64967   57829 83590   37969 63656   66996 33778     93434 58766   18421 79431   19245 57379   87107 39339   08017 57839
8924:  60399 01128   77620 71097   47932 54096   34196 49637   06281 03110     43739 81431   53059 40371   04163 69122   26699 01159   30891 49737
8925:  55070 86990   11944 42883   42923 61300   68778 53300   50164 10468     28484 14246   66645 61935   32797 82591   76616 59243   69138 72785
8926:  01329 65271   95530 04782   97213 75163   14234 87781   65467 24478     94765 31495   03758 54889   19370 48802   16646 22186   33447 58121
8927:  71866 95739   51102 57084   62725 57387   37799 27516   16795 06428     34397 50349   03554 85854   41983 80262   29933 96240   87013 08982
8928:  36648 13181   46417 92219   00704 94636   82905 65909   54989 47377     94438 76624   67537 61020   35455 15172   21508 52456   61878 24341
8929:  14219 94661   58015 48965   95574 67308   55106 24319   58309 84685     25264 84303   75952 62201   01868 10144   65826 90680   13288 98940
8930:  22596 62369   41106 35950   85794 67603   46518 76783   48503 25021     94923 70351   73694 75620   90612 84296   83477 82536   93763 07635
8931:  41384 12196   99174 58334   67866 94275   99455 69946   70915 42128     42427 95113   89253 00663   26764 05167   29868 26233   20193 55534
8932:  49214 31544   72591 85764   35499 98801   36701 82304   00643 74601     98252 00303   66576 00442   76619 54149   65617 48473   54626 21888
8933:  32843 06729   79430 67311   17110 53894   35187 19385   74994 72993     20986 59286   13561 23615   22361 29922   69645 87761   27348 73161
8934:  03567 36524   81953 67510   72999 18855   21052 99556   56768 11642     46861 24613   15672 04003   14845 30271   52668 07399   42224 63175
8935:  22945 02699   96493 86406   23493 83591   24544 55447   22816 60618     09926 05338   41243 25778   74357 54359   04452 81234   37204 36781
8936:  06793 10996   80355 26991   54663 12428   29548 10045   34920 93086     04012 34744   42739 95211   90754 07378   28412 22325   29892 10954
8937:  14362 28256   33926 85135   76426 22441   15892 06413   61151 93116     82009 88844   11507 03518   62174 76598   47336 13119   58035 76957
8938:  07062 61594   82790 34161   17304 86939   68037 85526   54547 82544     83389 45692   35821 61070   39981 50316   37796 93739   70528 08188
8939:  17934 71144   33984 09418   54944 35293   70643 71555   02655 14170     91583 45481   22801 76425   43074 99207   66827 57986   55055 74931
8940:  93076 35146   64040 20766   27593 47677   82507 80597   95626 48450     68561 66942   95711 26287   64388 13492   71500 20865   12764 19468
8941:  57713 53768   03563 83539   57621 66131   59915 12932   10299 09763     62379 31929   05553 47126   49293 09988   78168 87446   01874 16460
8942:  65921 11008   05006 30329   34444 40704   14215 19799   79569 36660     68329 06580   54831 19764   84892 86649   77624 64152   81931 86375
8943:  19507 63921   54272 04649   50779 54643   95402 15428   70902 79481     96058 97554   44993 18364   80679 21868   29636 38483   72716 65043
8944:  31024 62973   00720 50527   44566 57575   78689 22327   72061 14673     15300 62054   11649 38344   68314 93319   93444 03239   04149 01970
8945:  77958 22996   45689 75130   70771 41768   61277 93253   51889 94581     17306 60178   00550 35285   90357 81935   73549 93612   57936 75846
8946:  89618 83836   24195 70420   52436 28957   49781 32056   67045 45767     72141 55777   75870 36734   42357 09318   02194 44447   84848 77544
8947:  61397 92809   27241 20414   44866 99273   68531 80186   81429 25904     33102 58409   02132 43783   75780 64429   44645 68860   99375 69124
8948:  98414 66136   59083 86926   89900 01359   28383 86978   59164 39435     67336 46497   37168 14400   83336 15105   40952 46989   47428 30574
8949:  28255 06241   46122 93335   20576 76783   85932 27580   89257 29869     92619 53508   20061 73858   69207 32368   27751 01817   08685 66434
```

```
8950: 84546 50180  20866 12807  08638 40618  67217 86208  67264 75454    74065 30831  91310 19209  04285 18486  25515 39887  53210 42504
8951: 56615 96081  43760 88750  36117 26349  63912 51089  80325 04975    08357 54334  69323 30068  22701 76867  02575 19112  87405 70619
8952: 27294 23741  95472 72638  07996 42609  11885 85812  06684 23013    59799 28898  46777 49297  25860 94398  43712 13986  38126 61346
8953: 13203 11795  42983 13890  73439 37729  66502 23242  83465 87803    45845 36940  98289 07378  81183 11090  57385 81135  46450 23311
8954: 65682 32333  81952 29188  57610 58942  79674 20436  06264 74865    14666 52785  33127 49988  88766 88843  24300 90128  02254 57443
8955: 10037 69727  90345 38460  14034 55775  61236 17561  35512 46988    30503 58660  35897 33803  23552 15825  44202 18409  25990 43664
8956: 48888 63240  46250 61472  94648 77542  97032 72225  09713 49621    30624 31065  61218 24373  05949 86026  48843 51339  74269 65959
8957: 52112 31269  88552 81414  17435 12167  58013 03182  92390 60203    43689 63760  47554 88024  87485 98655  00165 50924  09733 94741
8958: 52243 86690  25482 92635  97018 15064  70859 54789  89110 78610    44187 10739  42611 35520  75119 50208  50488 31893  44402 28159
8959: 58113 15290  78419 68410  57966 97677  84965 75048  30588 59841    13991 44942  74388 64302  74909 26541  51747 40055  97198 17545
8960: 76397 50264  48647 24444  20516 18746  92522 67186  27232 32607    80904 32812  07053 08405  68034 18285  67512 80271  51996 82749
8961: 70497 11382  17232 30370  33077 21767  39171 72478  80641 45706    41723 97543  11309 00847  52768 36130  20882 86796  69081 51364
8962: 46226 34757  97421 13426  45130 71160  27184 96649  39027 88475    54434 73539  36862 53096  85876 87232  79191 21040  43904 11784
8963: 25402 40896  49954 70331  53553 65256  90784 54564  65622 47419    42673 95816  49296 23062  85734 92098  81566 72159  24531 41696
8964: 89749 32381  82086 60336  86472 83278  23736 89317  61299 66812    81286 58804  94104 18247  42304 32840  68949 09202  93446 85579
8965: 11890 50240  69725 26924  11973 96462  56468 30776  25850 63633    86886 78852  92949 15070  96929 17195  05388 67951  41883 22218
8966: 09864 60620  61699 73719  81234 20081  53962 64176  26698 86497    38944 88607  71779 46209  97498 82474  09962 01083  19033 34560
8967: 32671 45879  44091 66054  77288 92523  74778 66566  78834 11772    63547 41229  03743 60742  64060 89715  69818 15353  83547 03689
8968: 59802 88444  22045 48969  20788 45126  51558 39766  43390 69994    35823 59112  17294 02228  12784 38203  84737 79519  91408 69750
8969: 76582 96295  35259 50415  03165 49735  43822 45641  46560 07538    44518 88176  18910 96141  54741 86366  39458 45872  68739 27251
8970: 97175 65206  60356 72590  23991 54165  53874 02503  10349 71718    16675 22031  01264 42576  85972 78612  70434 96811  02313 26693
8971: 68784 89219  61323 72181  40556 34052  76089 47960  92067 27063    34342 69593  44737 81363  60746 23008  01109 68065  94451 53129
8972: 86138 30317  46289 13018  12196 80682  24424 62273  16807 96033    06635 10912  84464 76161  17135 64544  64803 07153  23091 13478
8973: 60752 38621  60763 30418  91183 06059  33345 44696  20738 61007    40662 14619  31139 08243  95401 53669  23571 63347  73913 76123
8974: 58936 09044  67996 53803  80184 38018  67389 28669  97858 44122    87379 67079  04163 18595  30852 83393  74873 15003  21583 90097
8975: 90098 00557  45835 21376  00013 37173  28042 97564  98156 62748    20661 69194  49384 38404  04584 73847  89828 68003  90894 79191
8976: 47275 23496  55837 85316  94046 48431  35726 48856  43545 04994    22058 27525  71422 66143  96514 22125  30364 35861  16529 50062
8977: 42388 12097  90272 52795  44260 56615  50794 62361  14096 90613    36956 29603  24051 59696  48104 99224  30449 67368  78175 35092
8978: 50735 26311  89285 69171  75079 60659  59741 25551  48997 04157    05669 78465  67087 69205  97078 26849  57532 35646  08165 65408
8979: 11936 21746  66046 05401  35786 14311  76854 70622  32637 96854    30071 16569  12533 25240  55755 66861  02601 07900  61288 67151
8980: 19597 32857  32777 79900  67709 97895  08937 80362  15401 13842    84411 38913  83403 27412  96318 36067  11627 62727  10843 81988
8981: 99489 01927  00094 57889  20677 65053  62927 14275  31177 30268    36590 28524  17587 90078  26309 26101  44448 02092  51336 91622
8982: 99528 62590  20754 34167  10495 52024  84261 67237  90506 90064    37016 29257  73563 87654  17058 37815  02677 31415  08983 99733
8983: 43767 76922  67777 07672  31830 06709  49912 77568  86884 01803    16552 15528  32456 86821  31590 55684  81353 19819  33729 75010
8984: 11193 46089  48979 47700  28918 47104  30341 24476  82956 14663    42954 58351  95453 58264  51664 97860  13980 78044  31148 83664
8985: 58558 32792  25034 67017  00575 87907  16283 93800  02417 04671    86471 61306  17874 96522  59436 27027  51507 57855  62424 67988
8986: 16747 19899  21740 40687  75276 90320  89954 59567  97484 46120    32333 85216  84904 35150  92306 11828  30002 68695  48120 35612
8987: 73599 44436  00829 28456  06921 25025  68343 58499  22857 15116    81070 81039  35371 06854  09310 59673  22082 08367  00768 52588
8988: 78505 32825  46115 72707  61053 10475  29420 58931  10717 50840    94003 32152  56405 98112  10068 66425  76756 46957  40655 44978
8989: 69827 05584  57211 18653  25210 68396  88551 72875  67463 14886    44736 07264  42369 27554  19562 32968  49353 55094  55274 32271
8990: 46638 09443  33327 97935  92079 79589  13825 23267  31000 42977    19067 91820  88228 81805  76905 87611  62268 18988  46793 97780
8991: 00655 75220  49152 35105  50874 60316  64586 95067  76656 13041    84451 76128  43041 97880  37459 83998  01859 15013  39772 83857
8992: 86818 96717  02043 94449  92486 70527  37012 31504  69294 16280    91645 83328  55653 66681  27590 95422  68332 31333  82598 79404
8993: 85403 15028  94153 12471  37525 15254  26324 36734  88487 52378    87828 44882  46357 31404  47685 05748  45509 14898  99656 34221
8994: 56782 78967  37040 06361  19759 68615  74501 52021  41784 73261    24235 68821  92116 29844  27627 05877  90776 24540  31466 54553
8995: 57547 88146  21997 79990  36730 97617  14614 73713  18625 54775    63525 82875  65734 23623  20320 63281  82992 45309  45470 90274
8996: 20963 24308  55893 93228  11220 64884  34928 81389  98660 95631    78745 54942  53171 70635  52110 02363  70039 18397  22712 75047
8997: 14993 57417  79989 90980  82108 00757  63908 99101  83865 88569    04955 42334  72856 96955  19367 14837  05414 58976  61573 24733
8998: 07388 88187  15964 51868  96630 14534  29287 76403  69216 36713    92275 90550  88668 16888  18634 79745  29498 25232  50769 92171
8999: 06223 52047  76646 35310  16635 23022  01029 45749  67971 56127    44748 22132  86203 81641  94231 42699  70716 72232  83548 56673
```

```
9000: 62847 09669   78225 33774   01000 79045   25936 58297   94130 07432      97515 64849   56679 17807   12629 32977   95134 62989   60143 08114
9001: 85759 81115   00868 86160   54984 63596   09897 67467   20552 06800      14987 08579   51513 53388   92093 98600   74065 13227   64055 43271
9002: 14370 04689   26853 91073   59039 50019   29037 25586   99940 23746      37651 30627   56775 85343   75494 13752   41207 55185   78439 33882
9003: 15634 88603   56550 21814   64689 91829   79775 94383   28156 90721      99244 66662   54254 72800   32722 66617   06167 03891   15639 91223
9004: 85797 88391   62955 83865   57721 06977   08947 69156   37347 63785      03609 70536   29723 32012   70661 07711   56200 04945   69883 28044
9005: 73515 26383   47906 34765   45335 33700   38982 42078   96673 26797      54660 69517   44498 82855   44923 78065   23574 48665   90692 48459
9006: 68617 47402   42237 90385   39105 05146   72471 06790   35322 71349      07500 65030   33899 92790   50877 98879   61646 48020   37969 68227
9007: 09287 47683   93422 96103   05053 84973   51511 14035   10597 73135      77933 97740   26390 65849   15325 39977   17323 61656   74979 30746
9008: 93712 45835   57463 85498   51165 64270   48754 53809   67798 05841      61375 92462   13625 51936   45705 76773   49693 29680   94485 76035
9009: 05579 37860   72785 62006   79569 00049   32835 66773   41040 60910      57430 93968   78404 43816   49156 49316   89358 12102   41083 38001
9010: 91221 07822   70339 63621   73531 52670   19617 74840   84524 09636      02707 60071   69320 34766   47904 55509   46371 00087   37448 38916
9011: 86086 72967   40958 44763   05737 46614   25636 09346   08951 11613      82546 10642   05221 58079   58984 10674   60314 28994   88975 35399
9012: 90063 61543   84698 12870   49214 54517   62940 50768   61681 91075      39380 43640   84553 39508   20900 03400   87553 33748   17045 35795
9013: 85811 91901   21320 50435   15249 97458   86317 46060   81094 27359      54898 71916   97101 89593   77934 25138   26208 89809   52764 91955
9014: 76479 22540   00203 11163   20080 01451   06781 26126   47959 23335      01487 11251   31855 93270   08328 42531   45053 77219   72702 21687
9015: 28952 08814   73225 53558   23250 47147   71224 94445   70269 93919      71910 76280   34814 10046   38711 48655   17251 54594   79205 84890
9016: 03293 60756   09445 41226   34750 13086   88040 31358   15926 63393      28745 44101   34851 15215   80560 26009   10072 45665   99382 05789
9017: 60205 61775   49603 09063   42234 18225   05881 20305   02701 64285      83329 20842   77845 04083   96990 15987   85640 24038   84549 66139
9018: 62710 43768   93218 32420   72516 88967   58509 93421   38561 31144      30990 04476   32763 11101   23380 57841   26592 08260   11713 03817
9019: 28950 19807   02774 43146   87673 07169   05790 73641   88494 40416      55989 80042   08416 47569   93159 80270   64443 63091   37361 66159
9020: 73925 47579   61538 68327   75341 82379   60306 60044   60606 79013      88394 16086   45867 88641   67255 43627   77208 76430   51107 54325
9021: 66619 32783   81480 30658   10794 85616   84081 59721   32021 72578      91441 28135   15643 31618   17991 32678   36530 45082   91972 67397
9022: 45855 94025   30485 04814   19425 57995   27671 46571   43947 93839      35032 77295   06837 40198   10064 52128   16016 10523   94741 56143
9023: 93826 17630   01607 90682   59023 32568   11721 54755   80837 54288      15641 58452   45607 67639   90274 48732   39644 35332   11663 28362
9024: 75700 21998   73402 59698   17826 98519   48277 63917   40588 66349      24019 31765   04068 17095   04712 48913   52721 41615   98169 57857
9025: 90840 72939   36408 29742   36392 57089   90506 51213   41710 64355      03456 62753   75150 88347   95414 31039   89878 80984   78906 23344
9026: 76091 54909   59826 40039   46631 06160   22068 26681   93326 53996      89411 77506   02079 75975   53743 10983   64394 72893   95108 54314
9027: 68078 06366   94815 56700   38413 26147   36445 96018   58513 43845      78578 22486   49670 92756   61507 52227   49531 31584   56117 81736
9028: 66587 55739   06858 87126   63638 36718   96846 64329   85667 55579      18608 32376   14636 83849   68051 67833   10140 50707   66967 25907
9029: 65819 84447   31634 23680   65182 05967   54725 46989   75334 22904      71786 99735   52296 97033   04187 21541   89772 87867   02784 54109
9030: 00986 12969   18158 66484   37480 36454   30370 87370   47409 25139      19430 60783   99428 15645   29884 96438   87856 58533   63487 15507
9031: 62084 62790   06243 49685   27684 42039   25039 51634   59229 97353      38468 35039   90083 61904   36235 82968   67994 91597   76423 96144
9032: 24688 70956   11343 43265   16458 38763   32939 67165   63596 14838      66743 33964   35736 08122   61403 81837   58638 35828   65940 56719
9033: 15247 41320   48316 88388   40711 45906   41421 07932   98308 81446      57527 04453   23204 27900   80748 31229   46164 63424   99262 24585
9034: 39036 42430   79182 78698   71072 72247   44731 35285   95377 27160      25834 04501   18202 65906   77640 88663   62226 70772   01100 11177
9035: 90933 53635   34836 46597   07794 91617   19509 67057   61406 04950      92365 36965   16475 70188   72584 94609   24259 18023   66813 09014
9036: 80662 99714   08817 31462   96666 83149   41423 49742   26748 46913      63095 10884   58341 55962   05015 78710   42588 59890   40523 61477
9037: 37030 91795   05519 66270   98611 73677   68102 07573   44753 64697      11332 22994   72394 03968   51716 69451   24004 34406   29315 79224
9038: 12860 03378   16824 57536   75791 82313   50242 53297   31541 41021      80385 17279   17870 16203   68437 50073   76510 51433   18960 13012
9039: 70970 82156   69838 23936   99339 95839   20090 45126   05471 17168      96543 34781   88775 63247   11932 42614   45910 80634   22498 67108
9040: 53986 62642   51577 30308   54812 21175   43341 61719   66964 69471      95764 64813   81687 28807   54583 77263   76770 16646   55229 53201
9041: 60891 06820   95191 69719   17266 24517   82246 98104   65575 16096      19284 06756   65391 85219   39114 18161   53003 07418   83036 94625
9042: 21379 62512   59964 25004   96085 27391   40249 82750   13879 05325      94269 97568   61450 68349   81046 00052   13725 34977   02209 62538
9043: 38277 55967   28560 17347   21735 34922   84789 49951   92312 09208      98979 58726   18188 51373   00231 72631   57156 18562   48370 61933
9044: 65875 34597   49836 94759   58400 68101   86675 55228   98162 36511      16546 22215   38182 38128   88468 72758   01987 86266   53219 68358
9045: 98256 34287   34774 24632   85462 19919   22322 17606   93122 54436      59612 17096   81882 84673   93458 89074   77861 09508   81823 01808
9046: 97497 89763   46436 39228   36781 20138   30390 08569   90956 24347      77658 76274   85867 79864   55452 81841   88218 03436   15099 44054
9047: 05516 18902   97387 47643   28110 76651   83796 98888   41751 78343      76800 10255   16995 45011   33698 53181   58255 91957   87631 40759
9048: 28984 60759   84658 68718   46560 56930   06129 18665   00186 75282      89485 99226   33350 57308   77775 39584   43284 46386   64556 93186
9049: 89532 56889   30827 81861   44732 90835   82694 13642   54382 75188      20428 68273   83583 91806   89983 74125   17991 87560   78545 18635
```

```
9050: 15969 72254  15652 47644  47330 10853  94780 38068  37571 44395    92908 14017  69754 23978  06670 70680  24489 78321  00917 79045
9051: 75288 06651  63873 49760  22531 90646  04722 60859  59821 60915    34775 16438  39909 36365  58113 70482  49631 03331  07422 94586
9052: 48822 00524  56235 38615  08762 33651  86814 92523  48307 23302    07212 45492  88431 02744  78102 66096  13056 16481  54849 62960
9053: 27339 38587  83683 11279  97716 49169  34480 63002  30215 61052    32541 80864  83304 54120  10778 45529  81136 74195  18181 93335
9054: 41037 69155  80255 33871  09928 31027  74407 80745  14469 65125    00828 33883  43298 70939  61596 65840  75411 44494  80875 17788
9055: 75209 19893  12089 50533  84462 52407  92382 91975  40242 43835    11678 07962  34058 97924  33641 17948  51748 13393  46257 41543
9056: 02644 11186  65640 74738  57307 95504  74078 49714  30439 46614    37707 67665  14840 24535  25507 78009  86305 08798  31695 18210
9057: 77775 97862  65505 50289  61713 15771  90970 97083  75781 04319    55974 00403  45095 84347  97759 76341  69214 85633  67102 99731
9058: 30587 18457  62186 32792  03823 64961  35163 85661  22370 54675    14651 02455  58379 94590  54910 93533  17501 90850  21882 97408
9059: 15623 27415  39128 32147  05612 12264  54013 61843  85772 17542    40868 87575  35714 56783  19360 90431  79725 33520  58954 83083
9060: 32803 15777  83428 02666  90443 82366  57736 43143  49716 87125    97935 22187  90769 47459  25969 76424  54976 87320  26463 93315
9061: 02189 20504  04391 10689  18475 13076  11927 62934  35094 28096    26737 57493  30183 57912  18238 88950  38408 02314  84198 33349
9062: 59935 18223  41992 70899  36695 86211  85522 06411  47877 06829    32370 77173  79249 84152  55313 91373  23442 82323  92128 29535
9063: 72590 05470  27601 56196  78711 10406  03521 64622  65667 09300    04491 75023  60377 45485  13831 28229  06824 38891  16030 08419
9064: 06589 69979  99786 34494  43759 99627  10254 90504  87368 05048    18193 95037  09658 06826  79510 45634  25925 09894  83304 12145
9065: 18469 65781  70264 15397  34329 15235  75600 86159  34448 71979    79873 83568  11927 86945  52414 41809  75810 06519  04617 39434
9066: 61580 90703  84515 96408  41161 12527  40749 51751  33166 49299    70143 69611  50074 92581  13211 43088  03646 07575  11204 34630
9067: 23442 40167  17740 63781  78916 73278  02316 76887  56450 96854    19657 47247  29320 60330  51121 96418  04469 71823  01762 43410
9068: 48414 75072  10101 36008  89051 17518  98090 49694  08304 34706    63491 77603  37595 21915  68790 91586  57917 02306  08945 85947
9069: 14149 09631  35314 59277  72536 96285  18164 88097  48848 03595    80680 51600  80932 42055  22118 90424  97757 07011  16284 90675
9070: 51193 62104  85134 35093  43824 88759  50697 97687  13843 01563    19136 98015  26336 29222  91650 91308  98056 98828  30020 05396
9071: 73343 04857  67788 38473  92423 78910  83458 53087  64875 56955    06562 93521  00790 29324  97552 41121  30772 26124  08034 92141
9072: 20751 69443  31115 23258  62021 97329  41947 79827  76830 47720    29862 38184  63358 50335  41422 57078  31607 39228  60368 18703
9073: 73543 59142  30407 39037  36086 02021  73992 91111  17842 61951    80990 43449  58217 41206  04536 49566  22049 04788  56666 77498
9074: 66617 70247  33974 86663  85268 25597  81003 99386  19145 82792    89786 84507  52113 85987  39371 98289  10005 29031  48455 73095
9075: 01845 31487  77364 81267  02416 40840  87245 26982  94007 40141    90829 49428  06852 93818  83976 58845  12140 90864  64626 16242
9076: 19814 48336  24959 73121  36056 16220  55162 74952  79783 98348    84410 74375  37592 14907  25349 60520  95127 93841  74534 01088
9077: 24113 95294  14152 43943  65603 95274  18598 64420  58620 58678    75326 51741  89645 18470  90115 74091  11976 15622  74417 07876
9078: 00226 29702  97095 75691  73241 88676  01489 93798  64924 24725    58730 08603  03460 24975  12166 28955  03685 81224  93826 96440
9079: 94202 02876  76410 51298  31602 58433  46829 31485  95002 72910    59831 14412  65504 47655  11339 25827  73970 45619  38674 70630
9080: 21701 11273  14553 81945  58611 08674  26061 14229  66611 26578    15409 71066  25285 82464  76009 09387  30539 73642  49858 80209
9081: 17919 77803  28896 55880  83878 72818  44243 96560  21998 07081    42990 45135  74383 67763  42602 05725  12828 52721  87043 24730
9082: 72681 27064  24408 81664  56603 15317  81991 84451  29359 17586    01349 21989  53153 74480  96995 47003  29435 02401  09836 97813
9083: 42797 76937  43942 88978  98505 26542  75676 60757  06471 42805    86307 82893  94052 85725  06919 12831  57872 45824  36368 94222
9084: 23472 89913  73837 40897  61971 47287  76079 76209  18087 04507    58926 73058  68097 40955  35677 80595  77934 62970  74662 50846
9085: 98208 72637  79835 51065  37945 28673  42916 78763  30221 23520    39037 23891  19550 51380  62336 07408  75021 22102  19359 56838
9086: 24997 22574  28523 31892  75688 31681  19802 30432  13310 85306    52972 39319  91717 40505  83134 99323  52921 43838  59762 36532
9087: 26427 00358  52929 91709  35340 15280  32512 41742  37939 00283    46318 42015  30609 54911  67272 58839  95155 42865  29978 14766
9088: 41383 66652  40942 57828  65759 13992  41799 22146  53404 27719    47275 27736  18009 72880  45110 23066  99860 45828  45291 51262
9089: 51457 90196  51423 20472  13612 46092  06768 79107  80410 79434    36692 64316  16627 87489  23078 17693  25891 77071  05846 47433
9090: 44385 97316  18521 45036  87854 93508  88827 25540  18774 22131    67046 16158  45506 08388  66519 76004  07269 46318  04638 83422
9091: 62768 72544  41263 20412  94580 44160  52378 86311  16017 10717    56165 69746  12360 65357  55467 97906  88739 82470  84430 09127
9092: 36801 00758  07407 88832  71899 42974  80800 76625  51867 11218    21985 08902  67457 07017  32650 45755  86692 12879  14698 09662
9093: 02134 75669  68438 50242  69739 45267  52617 69334  36660 08156    28269 13955  42300 90917  07169 91919  79113 51052  51372 18786
9094: 55839 49523  70492 45742  48559 88850  55771 17260  75268 29573    42259 64595  19355 24446  77043 31999  28781 04567  29876 80863
9095: 04743 22835  47899 93338  72297 47915  63993 06979  99906 79337    37865 59535  04131 99769  93497 40998  39398 22854  94646 92652
9096: 54938 48139  21017 98703  71280 43806  54654 13867  44578 96449    09151 61483  71828 18201  64579 92414  46921 07792  04179 55153
9097: 50153 22883  65494 46534  05872 97442  80101 45514  37200 56645    37789 19814  88082 21331  33987 29241  88383 55832  20670 55425
9098: 54272 00438  99713 02325  36943 40445  15770 84437  40438 32367    16844 47011  10046 26690  37080 24307  78502 97766  87067 09151
9099: 95482 31552  45462 68691  40703 38327  00625 89048  73746 41258    49899 87606  71369 57246  20397 08972  54923 68047  53282 69297
```

```
9100:  32717 67359  51450 21899  47073 01927  69627 01720  45887 73249    97648 79374  49062 69659  73947 08804  44572 40457  90070 92041
9101:  77598 14118  01605 24766  83869 35552  35799 19658  14053 69730    77336 27439  23769 59700  41299 09450  93889 93000  15717 58217
9102:  35703 47522  11571 25714  13103 70287  98763 75488  12781 27134    50648 95821  64918 56137  04320 56469  55844 72048  71452 94522
9103:  02648 36796  15997 88452  51047 15173  57114 66938  83681 36607    67615 05846  62639 27167  87387 84470  37886 88137  33181 12950
9104:  57614 49547  08575 18907  39927 64748  94309 03520  50483 84465    16287 18901  52245 84894  58401 28429  25946 20789  24828 49950
9105:  56171 99599  00995 53302  91169 77241  55160 95299  90775 07702    42598 72476  53574 97478  42665 72683  01620 62699  60084 77991
9106:  74825 07628  35183 40000  31424 14788  57348 89281  59525 99911    44680 90703  78364 27351  96027 40751  35237 08634  68015 20512
9107:  04246 35683  51887 47269  15168 66743  32937 13003  32406 58930    68121 01900  32210 37480  01415 25520  34446 00870  02861 00092
9108:  35406 81966  72139 29058  77295 94179  28937 22476  12745 75989    72171 37537  85013 80571  37267 31549  17167 84544  17930 60999
9109:  97629 15086  94137 82901  70702 42508  95066 87648  35593 36549    91473 89704  90615 52141  60215 17047  85456 07478  73552 29302
9110:  76492 76466  92798 60454  79032 56671  95324 81520  30455 67755    59859 11875  78373 73842  52360 49442  53599 32681  72943 74751
9111:  42376 73635  69334 45217  08912 36395  68331 65254  50146 98122    35188 78678  67929 04250  59909 96931  32051 14933  79043 11102
9112:  13834 60711  04658 30588  71405 51863  59109 77704  76639 31787    44545 46938  07019 71330  68416 35654  82388 84370  06589 25319
9113:  15285 59280  91056 95626  77791 75819  46291 01188  49635 72616    98001 39106  64292 87750  93554 27463  67441 82355  99383 47581
9114:  90876 90276  67525 05606  47406 28751  09269 40100  93511 47919    73639 27115  15431 83768  85302 66204  54878 88180  74371 34162
9115:  35203 14792  73828 46135  21893 72521  51943 14528  32519 62095    45912 68273  51020 55209  25149 76950  07104 27089  28416 58723
9116:  51608 43624  28040 42304  07399 92484  13220 49386  44304 32103    27312 86966  97507 31924  17754 07700  63826 35298  04426 81851
9117:  61815 96124  20878 10560  38973 77168  34398 77105  46054 63821    55914 54794  95862 21164  11130 93302  55595 27986  34426 80004
9118:  83512 53714  79439 18913  18206 44636  62326 41262  50116 29618    07849 98649  80734 35875  01461 21072  95781 90113  45230 54311
9119:  38923 32958  90940 51250  97731 65433  42689 56249  79426 42931    90008 88028  46737 37226  70762 56251  59134 13292  69087 97936
9120:  86030 55704  77984 89199  74360 71502  75117 22132  87273 52466    81656 21223  14298 46665  29403 59043  37750 39110  08866 98453
9121:  70722 76336  54907 74475  19778 59658  98013 47201  78916 40130    13430 85145  91103 52349  34553 31760  22016 65226  67143 05422
9122:  87968 84024  26417 35174  14553 96544  56061 54327  67840 32119    79137 68975  81032 65406  76692 03464  68635 22755  92603 89804
9123:  82999 17854  93826 70249  12450 52226  12394 25719  94438 72281    75913 09574  99382 12383  11442 90207  12782 48727  63115 43921
9124:  92992 97668  84137 10733  58878 28542  48667 08575  54579 78492    08380 22275  86617 49894  94423 78033  38445 83940  95121 19448
9125:  03974 74678  92921 63884  01758 18462  75090 43180  01119 21687    84238 55146  62945 17458  21308 71473  86456 82793  01902 82659
9126:  21205 93007  76943 23985  79223 49785  60992 36854  38961 33892    35954 87046  15464 39154  61351 01283  45732 09756  85250 94819
9127:  43170 91173  64302 97143  03960 04645  52044 52188  57217 34291    73298 47922  10287 49690  97075 67980  77082 27995  78010 29903
9128:  95305 81810  31304 24685  84317 36108  04078 23895  64126 81793    55883 52915  23067 46906  68989 48053  28403 05338  36285 71183
9129:  15246 51636  65648 55575  12013 28383  91015 00963  99036 54600    17833 93279  16241 46626  09365 96111  15728 40535  94338 36173
9130:  09541 43458  75095 97991  37270 19163  29175 73849  85335 17833    65924 99892  90201 40422  63756 02432  67601 52769  13868 73152
9131:  90686 29430  44776 52971  95937 14318  13841 55769  73388 22951    06195 89207  33674 70350  39787 84171  52843 37434  18836 27784
9132:  25614 68039  47629 27148  09818 71812  02870 15443  70717 23933    59725 81730  93032 56235  04705 52462  75946 90070  97778 89310
9133:  66229 31751  56731 95329  26971 05146  01467 17631  03333 59635    83522 28988  21076 43428  18209 79737  25726 83509  89089 71852
9134:  56830 83607  99421 97361  00631 62234  65685 19022  90989 77916    45721 91961  52859 92887  36301 57443  90860 25793  26709 82352
9135:  85548 40304  64473 55564  52894 77288  12142 16181  28592 13965    65025 07049  84529 63051  04884 24328  57661 07813  42591 83022
9136:  35999 33110  93386 96976  25126 15954  47450 75025  42506 32875    39105 16828  42387 00201  47193 32876  11579 93416  68274 73110
9137:  83942 54188  04619 14810  24563 83774  07038 58363  02055 48315    12165 84865  58210 93925  68577 28296  50815 69805  09040 10408
9138:  06381 85924  50829 94019  94739 11554  01221 73146  16453 65194    92479 54022  68430 51036  56983 12231  38830 30752  46437 72291
9139:  11791 91772  96369 04457  55699 94253  24752 98252  32373 24963    16503 75261  66711 48367  12825 38465  87813 24680  64786 91157
9140:  29254 27768  67713 15822  42931 80628  38044 71677  30855 65551    41288 88669  71739 93533  90454 12050  45654 12364  01268 82484
9141:  63054 07487  95485 08262  77659 38444  89016 03180  50229 70502    59835 57693  58412 30455  77000 14738  74534 14412  39788 81808
9142:  60464 60879  28073 74680  17626 05342  04869 27799  88353 92450    24402 68058  92744 53368  10098 12261  31412 82687  61779 92007
9143:  17451 64302  24770 44111  93011 77255  91744 61641  84041 07174    36729 27964  86013 86967  11415 77687  50032 73926  06098 42739
9144:  55871 58433  17500 54351  25087 86379  55270 97733  07880 81037    54785 25779  22896 14249  79923 14775  91775 12522  74816 36558
9145:  48670 01333  17233 60800  38772 01296  93797 27774  29374 00809    90485 19196  30026 14776  57553 66655  25306 84275  90417 70452
9146:  17552 73546  59469 61857  95422 57688  93977 34993  32929 81578    53665 66013  94307 39784  72578 68516  23109 20475  63294 23462
9147:  75529 34630  18648 30400  86625 37503  88093 92911  44148 82011    00408 73614  09257 92850  01637 23181  39352 88622  25256 51420
9148:  57000 92919  23056 46383  71709 34248  19081 55839  33838 37489    61927 73008  74192 23896  55587 75512  24974 87580  35368 91012
9149:  16101 72959  53955 19866  28977 74375  12549 96540  39662 17547    78328 59235  89288 09702  44044 30976  25764 27168  07463 74493
```

```
9150: 95405 58428  96954 59533  76995 53625  52123 85828  41704 79688   30685 57620  53227 95822  55344 67785  06436 27312  05184 73663
9151: 57970 26733  43451 65688  49994 69391  21606 36951  78173 03000   87687 21998  85300 69095  99106 78443  74348 11242  08687 54120
9152: 47655 29761  47506 74737  75727 08024  25055 24555  72981 57312   82465 40054  68889 08846  53024 18983  35664 13064  11412 39050
9153: 06353 16798  65130 86329  29425 31254  46544 17350  93214 97202   35655 07940  03332 78774  51600 43249  32220 49944  17777 53148
9154: 30264 00438  22532 45470  36612 08345  42391 63657  25191 49450   49350 01189  89868 15375  67592 39238  79035 54547  70395 79658
9155: 89887 22941  91405 53065  15466 81352  98141 18494  86193 61453   44852 50608  76713 00024  91298 72167  85864 29561  01454 63792
9156: 58676 39672  34035 29151  22333 81081  64617 70204  26152 23464   52429 02960  26364 00117  17765 61508  71418 80173  85293 20325
9157: 86808 77483  42431 56118  82509 27703  73042 38454  08776 33919   62642 35841  84820 75821  27650 93331  44032 54120  70140 81896
9158: 33519 51482  79178 37461  03849 12869  25546 47095  96015 62838   41865 21220  10324 61725  16383 51814  39247 67021  89742 20690
9159: 01471 52497  46365 01813  93616 41703  85996 58358  44339 89583   41055 42884  59545 96413  35713 10313  88839 45119  96380 45203
9160: 92919 87967  17218 20563  52046 08898  49879 19387  79321 77541   01421 10997  69108 45987  67775 76309  00906 04590  42386 99476
9161: 81307 61263  31718 79979  10942 29617  17139 04967  86543 77349   70182 50140  81143 81556  73126 34988  55965 48652  16756 91951
9162: 65898 19896  18120 86109  41967 88800  69194 24171  47274 28152   89411 82050  16266 37738  52949 45768  93469 66629  02085 02839
9163: 75188 85158  58058 15751  66161 87373  54908 18407  79531 88657   26492 41199  69357 28415  57133 89956  99001 37806  74928 40796
9164: 39456 15628  20141 86811  96185 95919  38212 55947  97320 26608   67128 32805  00145 91919  74380 65224  18627 61523  36091 74398
9165: 48242 06483  51203 47513  50281 86183  26379 38084  60747 46407   15736 82503  36610 84377  02411 11071  02344 83756  94363 56858
9166: 62190 32049  08327 29107  61096 80929  95353 85927  31180 47721   44737 22516  68558 93403  07249 54140  63670 78448  65980 48655
9167: 66234 45829  17907 91110  06995 51627  42616 26720  61372 47935   50430 10060  41970 30738  26250 04644  42104 14132  77064 98720
9168: 65917 99915  80392 84034  07756 70706  57593 60259  28764 56822   73872 36260  31675 11677  14185 98475  26274 64523  66336 56105
9169: 33314 59451  50519 63004  19700 91424  82051 27667  26096 25605   93886 62430  04579 00287  80361 04997  03371 10228  23043 48368
9170: 88134 18456  35558 66628  58550 03371  59632 88673  05291 10496   56798 17192  19097 69298  66548 86247  42011 46807  86956 53270
9171: 83358 41903  53181 59752  56635 81390  51762 22743  97229 93807   03205 78340  97336 97522  38288 31361  38173 98602  86753 33663
9172: 27806 86302  92849 40190  78990 25727  01836 42674  54554 54026   35142 48191  27689 46346  95614 80575  46459 48395  49134 13303
9173: 89468 23981  20844 71395  25500 04324  97951 93994  70074 83538   42438 61446  65872 26755  44095 08945  74509 43189  95799 74306
9174: 44894 42653  36199 98278  85493 13164  54940 74884  60319 12524   92875 09402  00760 56433  68098 89549  72896 76096  43486 23326
9175: 33726 31306  16365 71456  52215 83344  21750 73668  01388 03018   44051 87995  80900 28332  61968 34869  16770 46106  20697 77437
9176: 90463 43645  14579 34227  01800 95894  01328 04909  13580 85813   00463 20609  61154 64956  86681 56635  46470 73942  18152 73866
9177: 99965 71687  73283 90055  21687 58894  44247 67339  59994 15395   31566 75851  56485 59715  00553 88342  51780 01610  14014 76792
9178: 23582 57989  11369 21106  32364 62117  75869 81825  75967 75432   26523 95753  46414 25257  75898 63293  11557 83452  93935 60801
9179: 67509 43596  52225 30815  36185 49081  93741 53310  49657 58895   27418 38206  15490 41757  54860 93553  91269 84583  31731 46268
9180: 52728 41825  76982 96688  04925 97521  48759 29053  25252 23010   90387 31838  02899 58299  89946 23981  30699 44426  76975 45583
9181: 44668 43728  02789 81501  91626 52282  77672 51629  66322 96168   48526 22849  11994 08354  16226 14158  85233 10716  60310 58500
9182: 99513 79211  41482 28191  14931 41381  89610 58308  29628 35726   63339 56559  96458 24292  05649 60798  40677 82938  19020 25234
9183: 12697 17177  92588 34541  47261 27043  10545 91517  96996 23816   19544 77428  73336 86067  16028 07235  99271 12272  28477 78876
9184: 63349 72046  44970 82534  56343 06815  16018 53004  28203 06567   65562 66861  78788 67776  21647 45084  44137 08530  35227 01136
9185: 83615 15392  24736 73922  77114 75504  62168 69299  83153 86062   30780 57183  18673 54508  04797 52273  30231 54803  78206 39700
9186: 34460 44190  40876 83270  95387 36052  21712 62434  32835 43308   58203 44452  86307 64135  71373 49624  12843 66216  39978 37003
9187: 47163 54800  69536 38301  37729 93914  36540 98021  04442 12815   24573 66412  55290 03668  33785 29830  72943 88363  38052 40141
9188: 26301 26748  10632 38941  48616 94230  79429 67622  31222 42330   95467 41838  24220 29631  94149 75389  24675 03937  97528 72945
9189: 80904 81383  39211 68586  16869 69817  37379 89219  46057 25801   13665 20148  25897 59410  35832 77850  38940 91965  60303 92100
9190: 53429 25518  47948 14672  83120 95548  75905 61758  17869 91589   96629 43723  44891 52867  06305 95266  88447 88930  75023 57310
9191: 68480 01273  43467 88538  18060 71164  56481 09843  45283 59654   82090 84675  56257 95098  50161 68351  24791 30404  55193 09973
9192: 31770 12819  80097 42110  82851 15830  05118 85436  59095 43072   36826 96374  36539 15761  29106 98373  66846 63952  26868 17069
9193: 88540 80133  82624 28257  66984 85634  99250 97826  02580 90584   20865 49296  24568 98025  21088 36458  66769 37978  32934 80722
9194: 57874 26738  71986 44501  09019 32611  80687 23263  54260 56832   17780 64180  76275 82233  28593 71592  87171 24522  27079 97300
9195: 64709 58121  11472 73641  02007 03658  46884 45310  45704 37437   62531 16021  92503 08855  22768 50379  87746 60375  62275 16350
9196: 80592 50819  61407 90162  51018 83042  04713 35669  72546 63414   48060 67883  44513 31586  58997 37849  17952 31863  65719 59926
9197: 58276 73053  35405 96559  94872 44157  36771 09890  44298 11560   75746 71361  52594 56559  08151 22454  24732 50840  46458 48261
9198: 29759 19304  80630 16543  01178 85091  46238 68180  79714 08186   43858 71585  84721 12160  85923 60862  88978 02001  17575 20889
9199: 14349 23969  23157 10331  57643 97150  28019 33453  61798 95204   59482 34924  19730 01116  69292 32611  84001 38461  91646 62562
```

```
9200:  77105 32833   47482 91946   23593 13725   83481 87181   58193 29979      08221 58053   15548 52847   89571 32106   80666 48650   04227 06453
9201:  21907 40257   51434 23538   91054 71168   98363 77821   69387 81497      67291 58980   07965 39538   48481 73920   54027 60494   37335 83581
9202:  60146 18085   39529 28818   60642 65813   71305 71915   12625 55011      82246 22466   38486 51056   08136 78530   47301 75280   83180 11939
9203:  00250 87391   91113 09494   67073 47831   08744 82178   77691 45391      07585 01997   23739 72299   45398 90947   31556 77309   37693 43080
9204:  31003 83153   17741 19211   97157 67548   43768 91638   60131 01314      40234 18099   02601 92417   32350 49487   83674 42380   03499 33862
9205:  62167 88555   32961 26702   26644 73039   86796 93662   29024 52373      56107 70123   97335 70806   95351 03149   26858 59943   76774 58215
9206:  79705 07214   19393 00022   07852 80470   16369 32003   21630 15519      70690 64359   85183 65346   48632 04514   11708 48950   38109 45544
9207:  84324 86441   65540 75484   82380 66374   25765 64581   61761 01285      09766 93587   57339 35549   45661 62766   62333 28939   35326 92785
9208:  58823 01279   46324 63441   06840 38921   37127 62703   43151 07588      71735 33567   60912 64243   34655 57731   57148 11148   70021 37515
9209:  96646 14097   24568 49914   75406 73788   75091 34275   75310 09472      88250 06536   88417 35323   41865 83457   45556 74782   86270 75631
9210:  62435 95145   32674 81249   16989 95532   75769 80193   41022 39792      71564 89239   70601 66978   15886 29597   32809 86433   37719 52393
9211:  44334 74005   64879 78309   82066 55750   49262 88380   79379 12913      53723 23097   63125 82848   60884 34543   39459 29684   85978 62868
9212:  75456 70935   57669 42647   87725 73063   11954 03552   49163 17444      44546 24385   55322 80216   21531 21780   49818 96800   02907 82107
9213:  93989 04907   49657 19992   48040 53931   02409 56303   13306 10138      18794 36736   78763 55690   11921 77091   73990 21896   65745 00485
9214:  20522 66908   42810 34078   28207 95098   39763 29369   87990 37414      49813 53429   33493 77505   88127 45262   64312 32756   68120 37363
9215:  80232 22406   73560 34516   84519 85186   89816 91253   83689 42981      18485 89178   52051 54185   74665 34426   43048 91277   70664 60056
9216:  77133 96815   83925 94711   00133 36753   95180 74176   73588 54663      16805 59385   54679 38086   56449 12284   65693 33413   25411 49935
9217:  87983 79375   29830 80736   28618 28324   86310 49492   03760 79986      96359 28828   15748 13023   11946 19555   01464 76691   64510 08343
9218:  44881 52399   92440 75631   60978 45693   92766 32150   10540 81909      16848 12587   66270 47356   73104 09071   66162 74714   23409 65662
9219:  03875 90840   66134 48297   19969 17154   50027 49151   91553 31659      92163 17731   03809 99140   56226 25482   77680 59466   71643 42546
9220:  76524 57138   98883 19993   90379 15517   05087 92941   94632 26160      44076 34558   28894 20110   93817 62969   75230 59715   29543 86941
9221:  95211 37030   48492 77251   65352 60021   80891 65902   71211 45131      59941 28327   18406 69848   91811 78002   79944 50423   30944 66338
9222:  54083 20902   49861 96313   08128 06796   90907 36779   05021 21430      92291 81697   11809 31063   30237 47566   40769 88347   14887 44345
9223:  60742 26101   19484 28233   34744 20032   45304 35721   49047 80760      71273 21526   77295 73872   77421 35126   78195 57459   78173 28628
9224:  15166 25253   05140 26707   76373 52247   28319 26356   95808 16816      66046 90125   73771 30281   07816 98254   90117 32846   16472 30895
9225:  07619 12191   21644 10663   45975 18220   05029 16732   84717 11005      70137 22257   55911 67701   71259 42191   39371 39004   92639 00204
9226:  72224 83674   31602 51542   11016 54831   98805 87605   51074 05496      29471 10664   36797 28128   99768 50133   44511 42485   79386 59166
9227:  08789 15426   22129 23672   78440 36398   04349 13620   08525 93323      45149 49398   89179 11575   20025 31394   43814 85976   81266 74564
9228:  84603 65100   20223 75786   61269 92733   23751 08447   97112 08490      69332 41721   28758 35212   68402 64602   73951 94488   50387 59989
9229:  68598 79474   04212 81713   49463 43490   03529 55676   98497 64641      39779 43044   89238 71382   06024 37275   54273 34122   87072 80083
9230:  56298 76352   47193 10684   17964 98225   09127 06610   31008 43824      54687 99171   72259 89954   63374 64852   28208 89923   86973 57177
9231:  76797 71642   93025 04992   96045 80448   93429 07880   85570 57331      47739 17411   39513 41233   39942 76959   31575 61490   87939 48838
9232:  65220 63097   29862 25055   04281 15787   42435 64169   88502 13844      83520 73810   73997 57948   02481 35793   13375 68308   21718 01732
9233:  63545 47592   22819 61896   71978 17060   07302 21578   45527 15599      21604 31215   83287 88024   87671 50766   81952 47120   65953 38813
9234:  79200 67468   20468 01515   99674 62080   36894 81609   99154 28394      84608 80082   19791 93027   05224 37482   22997 67179   47514 76730
9235:  22856 94819   85037 97867   08435 22394   80740 10233   11095 94593      17166 45018   92883 61680   56018 32298   66830 39817   17177 23935
9236:  68597 47408   52570 66764   19143 55963   65190 93455   81019 62786      36449 90333   63526 36265   87516 34784   69455 98207   90103 08648
9237:  57716 50033   59092 40024   31733 57579   48462 12143   19505 05800      20022 32730   70676 80786   93035 44157   79849 01547   23617 98007
9238:  56418 91324   56940 13735   78630 14624   38955 82003   99042 44337      95012 90763   52139 28502   79456 48697   44528 40885   95885 09146
9239:  31761 46871   02240 89997   71578 51403   88160 85895   73027 53949      09973 08705   67152 00837   05084 55457   71934 85714   50266 07366
9240:  63210 71506   73633 82974   35652 34911   36687 70069   81386 36801      80680 44757   29819 36040   28112 61298   84632 87111   50824 06520
9241:  05294 59985   24775 91894   49830 76061   40151 28255   89434 21162      57014 47273   67665 71785   49881 25727   47393 25027   45556 17931
9242:  70327 91026   59600 13769   57595 24311   47759 11750   35231 14217      94401 26916   78140 57452   00260 01039   61923 93302   99663 28651
9243:  03648 46646   19607 29363   28052 46946   46696 52196   33458 97705      55116 95629   52391 24826   03670 34448   57598 57228   48305 47371
9244:  88990 35554   23371 96332   70491 51931   17725 27916   57592 79755      07365 45507   11294 57233   68352 29937   10850 21315   40419 52638
9245:  12975 48242   33634 05015   94535 57951   75743 84822   03021 13602      04104 74292   24017 63832   48402 95720   18201 87637   88504 08725
9246:  81013 62007   72862 46682   18674 74524   50739 13161   15438 23431      35884 66545   06241 00159   27194 41229   67737 07609   19475 58077
9247:  76918 60605   98070 35245   02833 40457   29436 03336   27967 94253      85274 87496   22299 12503   83891 62582   35389 03890   72679 05250
9248:  03400 49578   74937 42904   78189 67414   14596 66854   06977 49348      00842 70131   99815 57067   25763 16501   29980 65328   15985 38724
9249:  94884 01003   38925 31879   48713 40647   26264 50124   63272 09624      30154 57410   27046 54104   14010 94115   70175 31975   47523 80475
```

```
9250:  7919078226  7944609982  9334189635  3516965708  5005707075    0780337950  0045750142  2281571181  8519508474  9276113963
9251:  8934110342  7069817134  1238480800  6543208126  5276328383    4344789141  9109823575  4702001724  9415249010  0090150305
9252:  3578812451  9449170926  9084918148  7516244266  3457602648    1098363155  9769707206  9076423029  9747441812  5629992539
9253:  0328362087  6959829507  8665101664  5391857426  5549512638    7361158763  4179387806  5967409335  8687849684  1997823119
9254:  9176701162  0353517333  4439568086  1096181125  5126096444    6651418211  7289402419  4231118499  1984993349  9956283150
9255:  9564582091  4961724485  6686316155  7054532577  0086998235    8471899613  6264589371  8504509776  8572035684  6247711655
9256:  2414278596  2477226006  2173912262  2992427863  4718902905    7411039255  7666335815  1211339360  4996962950  0962812318
9257:  9493205814  6109862670  0896184257  4199784081  5981864230    2451436877  4605379146  5720438282  0499545154  9011061370
9258:  5545462787  9016715751  5894044618  8740564697  5119752115    7592359696  9184968505  6747953248  5662148453  6134156298
9259:  3251719107  4009482837  2138674562  0573582959  4994154662    8900322538  0806578829  9860700029  7979366472  7697844040
9260:  2339475705  5803958454  6703404357  4084208263  2436797985    1904471028  5270066936  6813675388  2906459324  0473749051
9261:  6099020803  7726981048  4736238297  6912526172  1601793747    9321473850  8455737971  0618627932  9603942313  2125698883
9262:  2048492270  5771822271  6602843374  3921676332  8341800357    4402171566  6081594091  8860861742  6494210664  2790532256
9263:  2529764147  1167632178  5663325709  0675182139  4810339554    9021417519  1224371391  1576321110  0981955799  8706762443
9264:  0147042780  4013897433  8409413779  9355406367  2186459491    5641354401  4065996248  0301612184  8669650255  7244026993
9265:  2012783050  0634994203  4612027893  1059929485  8400872571    3037099553  3773866006  3293444065  3054580337  1290644144
9266:  3109921958  7586812183  6632423062  7871655939  3608516754    6855002728  7269356390  1221062471  0245926347  3181920505
9267:  3142149867  2802649556  1203562128  9787903960  1075399469    9977780150  2013478240  0388393581  9204813388  2843509489
9268:  5034320325  0985800701  7700696710  0681131451  9678997426    8935950969  3709331756  2569151862  1128204122  6945777233
9269:  8377985559  0819551958  6040997381  3793783768  3726228348    5934499084  6014862687  5353478211  5601558224  6228446361
9270:  3299739112  5536247885  7259134258  2948996365  3294811507    3609915811  9322777875  4623766868  1681775693  5462633416
9271:  2829683866  6079877078  5744766956  3992689224  5529421636    5185166456  7170397213  0155921893  5446316405  8532171312
9272:  0938079571  1348383277  3804212671  0735538905  3196220139    2838061333  3349668421  2917565698  6189828645  7445981020
9273:  4949445445  5319338460  1772310026  8639655931  9983289926    1567869858  2704695216  1355579386  9718713030  8039415368
9274:  8027354765  9714005854  5048153670  1546743224  4539688251    5002141336  2954315922  6685991847  1318090158  4137469328
9275:  6300624234  9105043746  0600259122  0731165186  5489094591    5175462847  1088533369  1858079610  2336192245  9349852685
9276:  7502165156  7419966212  4187732002  3449069553  7063271346    6817658717  8304333306  4314933957  4433600150  1296518320
9277:  6050603248  9323241012  7420810274  5504127894  8518160577    1149487483  7272444812  4685062295  2964336032  5737413883
9278:  2869177062  6535751315  5882942865  7873173172  1802113072    7323515059  4330315374  4081781726  6440146146  3847548679
9279:  7399894265  1713174563  2505605556  7540072092  3755106536    1935554098  9045460142  6740151631  6633289533  3495434543
9280:  7151413002  7092670951  4743151506  9798583434  6158713597    5310815775  7716510451  4937859987  6473728773  6408388380
9281:  1479389157  6440445167  4187732002  3449069553  7063271346    6817658717  8304333306  4314933957  4433600150  1296518320
9282:  4142840537  8970351740  3213423207  3561905066  4175366255    9413833309  8076513139  3453615630  6927820611  5444052045
9283:  1913855013  7244137483  2156382092  9679372520  3709960697    6811816006  9592329268  6481831756  4031713157  0286656482
9284:  1072447800  1093846778  0546662181  3241131636  0200993959    0322774882  6684101581  7499089771  7773590197  6900530987
9285:  4632016420  1036303593  8295909203  6280141386  1549390568    5589079590  8519767104  9527192968  0856631974  6762740658
9286:  8670903378  6191319350  1053755667  4539901746  1619515427    0635393885  2471018949  1983726895  4029098969  4626728885
9287:  4098580096  9961268369  1316255965  4219066635  6310654096    4942839396  1985914754  3762205019  1919105033  7842695693
9288:  8262876869  7688320763  4368525005  3849582742  7938295574    6942115605  3516036031  3783057297  6800402118  8944600632
9289:  5881757695  5218445100  2947311592  7050526140  4248531297    7287131368  9702544125  3237755493  3746376618  0210962947
9290:  3933911034  9785080060  2748490327  9699442627  1858555481    4926821821  4321324311  3144623883  1648786448  9880525106
9291:  4737794741  2962650491  8608224295  2344713445  1854677110    9141651214  6300518602  4903155661  6194236571  1431301664
9292:  3602558709  3234818309  1132469686  0117489496  2710973333    6066981940  4757428948  0451331427  1967332096  1230123875
9293:  0869437841  1823300302  8426951814  2811835148  7497215648    8836476099  5829109527  2916038448  6741135642  9413205977
9294:  1953144688  9800998516  3534021546  0529154346  0368599163    7270308527  5885753346  2466611785  7880399010  9893874514
9295:  4511957103  0294428666  4169633868  7228061087  1530137806    8047114449  5738446747  7357219763  2976108253  4511661968
9296:  4731531239  1377041095  4161142237  6996059125  8102997145    6706686556  8964743130  2962166434  9622767655  0055257787
9297:  8543113005  2316456417  9865073913  8176870810  1061635176    0235062771  3630228891  5761392431  3199713735  3730327919
9298:  2336357491  3363563287  2882435160  4359248424  2727232975    9666512959  3228410541  4148291047  0490770618  4428727034
9299:  4534006881  7711452274  2000604630  2792578246  8911944077    9525669636  1982298646  0503126935  4208413748  2660854402
```

```
9300:  16571 92686   46059 48997   79026 71695   39100 38400   06798 23409    21667 26236   81343 85662   68944 68315   26588 88003   80375 19890
9301:  73354 66524   32110 90276   15998 53001   93392 71263   96815 11104    52064 97974   32701 15483   20656 84450   26757 56599   41441 48358
9302:  00683 35182   31065 16639   06087 06638   34103 63412   21551 87706    39387 58608   23316 76142   17225 19705   81656 67283   66681 14364
9303:  59679 53438   68215 15942   03625 45297   68403 95010   74970 95147    02760 33913   85713 88970   60850 83956   81656 67283   66681 14364
9304:  28292 48080   58928 10042   65860 90393   56548 12634   63795 69436    34210 24282   15624 25903   92785 34571   21051 92743   66933 80071
9305:  97030 26136   70245 84962   27742 03046   44915 65764   81036 95953    62267 13046   71512 79349   73339 96693   29166 95305   17721 73348
9306:  01316 55334   40108 64500   86295 51687   52649 48481   75049 57900    94605 12916   97922 54247   53760 14260   20948 81886   16078 77441
9307:  36887 63027   57968 36845   36972 32733   36595 89261   21565 49879    43025 24942   76847 17771   20203 12067   33078 10590   49512 77140
9308:  57994 88689   32634 81972   64906 44623   39441 87406   51864 84116    82763 30573   12473 13888   70616 42665   13022 41355   87520 90417
9309:  86746 33458   24352 84530   06006 58492   83635 97939   93256 04615    41411 58653   80447 32152   23255 13191   88690 51024   14899 53601
9310:  18582 96864   46099 37929   87867 63106   52721 75800   14623 81834    12727 06780   31326 45505   53050 88281   99463 51700   33057 48269
9311:  82515 15178   90554 79017   16374 59820   42652 59427   79645 97500    89239 86810   00958 60837   02732 72416   28368 63082   00028 09542
9312:  69692 73038   25147 64803   11611 36706   04324 26572   13476 75967    10808 51963   69662 58449   85954 43094   93698 12468   75213 90251
9313:  22061 01187   63392 46189   39897 34142   48572 96905   84226 54966    46234 92889   85299 71198   08389 96306   61322 96646   19773 17072
9314:  66570 71103   10200 41132   20129 78972   08380 75971   76743 44927    32332 52816   36076 51755   84234 40374   82270 01945   66801 73347
9315:  78175 89942   28785 80019   89539 58705   49659 69144   04278 52190    42113 18857   98473 29414   88474 98165   62074 29562   57690 56667
9316:  96538 13515   83079 14351   18853 58654   28733 69799   75776 25728    77565 94500   21741 36614   46282 97134   49877 26126   66337 92751
9317:  15573 15434   40812 89257   99493 29713   14196 62035   89622 06863    40482 39115   05936 40364   98572 25398   02191 97808   91123 61973
9318:  72221 28025   76882 43251   63557 34475   54184 80355   63236 98759    40832 58527   97226 62263   94046 71075   64107 18604   06577 74185
9319:  77161 97326   47944 43841   57599 49064   47716 73629   86602 73879    00176 01458   03190 04746   73257 15404   55316 35078   77998 78914
9320:  66251 00079   51764 50646   47228 78108   23766 00751   74394 68723    74454 98198   76947 44851   38891 63473   91436 10652   43552 74722
9321:  71706 20049   78936 59635   35223 68738   86050 40338   25679 81453    84785 92409   24738 04220   87805 55991   81942 18383   63538 52732
9322:  37166 63167   93006 87379   82665 11242   68597 70877   60310 02053    27030 36066   12995 58007   36173 87384   92653 07139   77983 69244
9323:  30049 80925   67945 33514   72104 05585   49093 67793   11544 37844    84476 01363   66702 49499   79112 05943   70282 96860   64962 94335
9324:  62061 27544   18360 39243   54500 35680   20652 95988   64201 56442    08988 02108   65224 07788   43749 52922   47340 59623   82958 27147
9325:  68495 28279   71640 42580   33492 03656   47650 45514   36451 82173    66696 67051   38574 79287   18343 70203   65539 34057   64926 68624
9326:  80726 03563   21718 81543   17753 70126   88319 10912   77822 22658    74591 46666   78638 13170   29243 53411   00791 23349   93394 36915
9327:  54680 88636   17714 65744   53449 12304   91829 82721   10303 00910    56646 62479   45829 55292   03098 59394   50795 36694   76408 39471
9328:  04425 61836   52716 81380   81487 71821   53581 47188   02014 25830    96381 93734   14367 69740   75205 98229   75690 05512   70263 42181
9329:  38285 65717   26268 56785   90200 44219   20856 23978   67741 47509    84501 93703   65213 07890   40441 83912   30099 25125   80005 95144
9330:  72423 61734   62000 02116   51377 60825   25263 17733   24062 16142    44387 27118   82323 06626   27345 05063   95303 67849   11597 90807
9331:  98879 96399   09346 55044   19058 06435   26290 19288   73183 10442    60494 17242   26537 33947   34310 72766   17681 27952   44127 74660
9332:  11309 31996   90059 10228   39189 55897   19603 08709   37461 17783    04627 81708   90805 27008   62613 09680   24658 79865   23403 15104
9333:  21434 13323   94844 26082   82411 47615   92729 98992   04217 51833    79617 12948   85135 65494   22235 56697   26604 58714   41259 70423
9334:  23027 76303   98635 89114   69412 03689   63146 02514   55767 47033    02520 32589   69346 03894   06543 09073   68818 77130   32212 81135
9335:  09313 96699   17419 61720   84189 88403   01226 92538   58195 10929    36337 78513   72491 06019   54203 65116   14726 94501   91347 76502
9336:  93028 41804   61456 29475   14540 67325   05983 70097   43636 35906    51151 47755   35551 45455   99622 17772   53684 63545   14550 97067
9337:  76034 86041   83647 43684   61892 82062   02373 57313   78297 40775    01718 97693   16247 39765   02431 12501   84639 17309   55831 86188
9338:  70221 14674   14743 37351   54686 16052   79818 69906   06638 53158    77005 18583   95864 49372   09822 96010   81018 06854   52505 27426
9339:  52578 03849   19809 64498   49986 55619   58684 06898   83034 96377    29102 07538   35232 29932   18182 38572   36220 39078   71843 44784
9340:  39313 01416   82739 56341   00229 07811   47012 60066   22422 51784    91569 55494   01076 48340   13292 33475   57424 84583   91365 23497
9341:  36632 16838   03072 36297   34369 84519   50280 68081   22670 94857    40053 37714   87209 86024   15396 48800   57901 91252   55576 28264
9342:  65479 21478   78377 01578   80855 99773   20598 75518   32946 10073    10828 31612   06719 33757   70508 83656   32211 93962   99132 87837
9343:  14480 02694   23247 08943   04894 04170   66027 43202   16962 61426    72547 20656   69747 94914   63883 40502   23295 40589   96303 62040
9344:  70147 16854   76943 17827   61709 99795   15028 69472   58789 94866    03403 45644   86123 98201   13972 77688   85250 77969   23210 12588
9345:  75154 16766   52205 15996   65705 18749   05854 27109   66000 67164    14305 85435   46227 73835   94210 93383   07296 28421   38036 10820
9346:  33968 18927   47757 74565   81490 16779   86508 46418   62395 27383    19183 65380   21806 32047   55873 58634   81489 66379   69395 93700
9347:  73967 82132   24039 12490   59150 81077   78091 10173   78387 09587    04022 65902   86183 48863   43984 30962   43877 99464   23971 22405
9348:  52870 65545   82745 47445   09867 80593   77267 82488   75321 47513    72212 67502   88821 06497   53037 31704   18319 91320   87704 31307
9349:  72952 38559   91234 80658   38791 31582   44403 51017   94299 99592    99864 70160   32318 73042   94707 89651   11796 93250   49432 43399
```

```
9350: 8338669287  1945557064  9395573176  4907015873  1326926064    7072250910  6694631736  0650146541  0553366302  9094168475
9351: 9906997905  5031953121  5838937112  9743037666  0531201616    0742576003  7294541397  1549738651  3059916728  0483852627
9352: 4738193005  8711820153  5363409988  3336123078  7119857892    7029955839  5505608211  0592586747  7737121968  1869277345
9353: 8313673896  9367012231  9452772029  4285589438  4539411383    1158379020  0331452521  8001104423  0357554101  1704717470
9354: 4923273866  6685416426  0735408588  9929625103  5552233860    9436023235  0956226965  6720639196  7105221227  5792155091
9355: 3997201656  6986275270  3360671358  8066380821  5835705126    6810923127  2273955001  7528185237  3248917134  4690446987
9356: 4347788656  7749645295  8552326260  9155289394  8988616134    1739910205  7916889517  5330519255  2980324627  3384051367
9357: 8024515295  3763032556  1479311843  5315571343  4310948184    7910973398  7501953200  2719522610  6260764383  0146295630
9358: 3020274611  9489625058  9878004684  1484342005  0087293718    4356569550  7204669182  0357683907  2364058403  7065686101
9359: 6530631661  7160842350  3839860818  1526726846  6622337747    3069699371  9301116887  0830111575  4862093872  0529651852
9360: 5552980940  8157771689  0205775941  7185094856  9704179604    8333803514  3932324687  0117985358  1879154081  1224690248
9361: 5563731814  1571124642  2990139776  4705829044  2735684312    5092224892  9273673807  7172803948  4035233342  6896743296
9362: 5837374862  1033693977  1618759258  8486943955  5094019149    1816923929  1067271955  8515590749  4315740213  2665799270
9363: 3028159138  8093932693  2367245226  7913674763  0182981460    1855118380  1011773820  4802666735  0264017219  2333766811
9364: 7383432637  5259085171  8866518305  4553971181  4505732476    1916333997  5818702868  0622115770  6314984826  6244364111
9365: 2169386444  6111510821  4373714577  8456894433  0558326844    0540169719  7432236898  0723276308  5645990717  9679132081
9366: 2225751764  6575530415  0465595932  9902878661  5497248060    0020712763  4276342410  4528233773  5758959953  8626459020
9367: 7545423577  8357863410  4948415431  0943178503  0435398644    5227275018  6327093030  4586348475  4770314407  1001165399
9368: 4874220221  5306895812  2138466701  3498414572  9925649440    5135025431  7334684826  2945443759  9476388229  0771209404
9369: 5104864041  6628646831  5482177606  2972285662  3612393373    0792453822  3716924130  2316648418  9724064031  9237872681
9370: 0188997904  6217162769  8256996515  3392805593  1730062102    5824651013  7064853509  9554724319  3402025973  4911043960
9371: 2784996358  3870718831  0380795954  8736366718  5053803281    2243242734  2139181631  2790767497  0186280885  8240972280
9372: 0046763557  7853080694  6388357340  1331704015  6577664003    4825178204  3196475371  5785406833  4890152147  4353217054
9373: 1797951827  1607791727  0354772480  7978827628  6976679471    8574071913  2381431688  8651864282  9038310320  2332660242
9374: 1886940432  3583845908  6988864854  4845197009  9953543391    1401754795  6395391155  1152738823  9855420240  5247497169
9375: 1191717700  4447872888  0643064811  9713868594  5523847928    0923043315  4600659429  8099239380  8165730411  1595441903
9376: 3714915107  8479490080  6622843595  4701172837  4037248769    8550202930  0820206679  8130390720  5486148980  6268343607
9377: 3648241471  5458926892  2061229098  9145303485  7969034593    0266425296  4882738435  5792596169  0023543897  2545906329
9378: 6831161718  0545808882  4390440087  3765977590  1280913974    0547490822  1933169113  3450002014  0065125664  0025555999
9379: 1612907682  8467078298  8286879509  0427470745  8858504259    9040397806  1148961217  9232321053  8653873063  2563134245
9380: 7385516985  5373119102  1907515705  6359021794  8209018898    4874411801  9766466564  0822057552  8586466987  3868863181
9381: 4376097202  3892217237  6137208382  0606191170  8854809426    7495709906  9504759791  1761638496  4832701516  1373038873
9382: 9964188010  6405360215  9502919537  3643902398  7109110619    9232873873  6110236963  5235473395  7897088887  6823465917
9383: 1384191156  7921803797  0255783925  4131516565  7383251775    5345421700  8010366659  6877800087  5471627285  2475861012
9384: 8723225718  0235711414  5839604214  6599400957  5025198069    0764025520  2688589926  8934257390  6542299797  0449885698
9385: 6715571267  6064567715  5570971244  3855473933  8455185363    6121319975  9007161310  1991455634  5940657435  1149612029
9386: 4593894944  8602283154  3444698576  9659643520  1364282051    4632022233  3873910109  4939495150  2039627017  5206586827
9387: 3510677197  3612124375  6531552552  9495095941  7047793840    3048681410  6526194147  9908114908  0759879836  1425580213
9388: 6698137424  8574496680  9327453102  2867242787  9755245190    2601424096  3854412529  9875647850  1636621415  1993460523
9389: 0345437370  2624609112  2833652252  5336214222  9951466872    3483161995  3453417415  9488827872  4494852352  0726100405
9390: 5198432084  3214999706  4227091446  8502023636  3073812314    2251747409  1925444700  0013887510  3336579816  4363740427
9391: 7081867610  0330740564  4015862099  5992200903  5486502822    2429609810  0653427380  5987010208  2053061235  0178293493
9392: 0776307293  6517170364  8115885025  4235263595  8192424400    3073166228  0142739007  2927270528  0709680253  4003566355
9393: 5864156233  3949195125  5634134024  0040424205  6712298576    8680522884  9130532754  1098254872  7497159018  5569780119
9394: 3101058104  3842173652  1398549237  9551212460  0175351113    3968832323  8901140970  3640859642  8172962652  9230562273
9395: 0741641511  6465398399  5288707969  9188061634  4828595879    1018539987  8810052933  2684120612  0538805930  3170603854
9396: 2380262744  8221451476  3666241062  0716912164  3229027801    8191152469  0680375997  9029313356  3982702872  4079269733
9397: 5045968189  0607020554  0855791898  4701692127  6351785902    0298407519  4516530320  5934698254  8442593633  8812928558
9398: 8401898389  9961665554  9251573272  2822355504  5469116579    8871455238  2924953605  3139449730  6004731266  3106383902
9399: 4396800030  5158833751  3309490942  4072643412  5311820827
```

```
9400: 23361 87573   13400 15326   13937 64345   31549 02935   37134 80987     76616 00519   35092 45959   17745 46957   47830 82781   90730 64604
9401: 68530 05107   78212 22219   48213 88731   10201 84841   14150 00900     04248 36176   08388 12163   24011 87090   49442 22798   87207 51190
9402: 23891 90278   56893 64886   66692 54068   26547 90681   55357 44232     82650 32778   61283 64607   00467 45332   45043 73454   77221 91419
9403: 93714 72962   03012 82114   08872 30718   27413 88689   71692 19320     49931 57202   71942 15481   21363 53603   07209 26298   46319 36787
9404: 38000 11793   27908 23677   69600 09888   97332 34318   82448 91986     55575 43802   89364 16196   13028 86042   35157 23049   04386 32075
9405: 12486 86604   92450 06430   41486 68093   86490 98479   64984 63403     48967 96172   33511 13988   33832 07177   49967 20934   80837 56836
9406: 40313 39514   86924 31452   20621 50860   00710 97266   67482 35759     89818 31998   85358 54991   89432 11867   84226 56158   08348 00213
9407: 34607 75677   05116 60470   06635 02164   35318 11489   84316 09664     59422 61739   55860 74218   21454 97781   08546 90044   27175 06900
9408: 17727 53545   37168 03895   76184 96804   01096 64230   53956 18982     14965 17668   21536 79239   87906 13618   75604 34526   50799 39471
9409: 19650 60074   24475 67445   34998 48289   40137 63164   43964 77558     53472 45478   36855 80043   46937 16625   34400 82633   37339 71699
9410: 52157 47740   44301 61735   73943 16896   78275 59676   41464 55050     86823 00143   89482 39335   45984 18382   54360 72242   96083 63281
9411: 22091 50492   53813 68116   46656 03822   14400 27095   70335 47911     81874 27789   72377 96240   93446 55999   51466 86869   69726 10713
9412: 38998 22616   89545 97573   82984 97871   92129 25039   80006 84133     90497 46536   57998 47275   48079 12178   83816 33701   58244 61546
9413: 16534 75150   56380 09149   63432 04593   35752 91558   05342 21684     57249 88594   71763 95248   90922 81026   43625 54170   12108 78484
9414: 39657 13640   26198 73700   96230 96304   98759 97940   10994 39541     20985 47054   45950 30140   84713 02581   45183 55528   28699 51840
9415: 21246 08932   06637 35136   81296 64278   13731 16224   03090 62867     78749 81003   76425 47795   05080 28186   68207 67019   81925 85645
9416: 74449 03255   09865 37799   72030 58169   03999 37081   85953 04564     76134 79258   09801 78505   59320 27980   46450 88981   84508 45709
9417: 79385 07361   32657 75253   23815 85805   48690 94623   71292 77091     78179 45710   67471 53303   85512 01029   49560 50594   20236 05287
9418: 25546 86564   70755 29557   55649 94106   23479 20523   65853 01264     43863 26385   64334 39362   90136 14755   22890 65340   62351 63088
9419: 73154 15658   41393 58784   75550 71545   89290 62462   90290 39108     31218 15755   89754 89101   84762 51359   83493 34137   92325 01405
9420: 38767 21894   81588 19048   93302 46437   64602 93758   13816 26102     48981 02730   74926 12543   53792 00554   83846 53732   03178 76579
9421: 38644 26985   14223 67872   52564 78506   39348 47797   26131 55555     57304 63412   17252 41182   96018 43319   57445 82807   73025 27600
9422: 86380 56520   16833 91629   47344 52851   61168 31364   31943 31258     17560 95155   58598 66325   98378 37015   17144 29319   26251 01514
9423: 75352 11462   17048 49175   25638 16480   20381 90153   00823 97308     57865 53665   12250 55474   82570 90766   89250 08075   71450 34789
9424: 69049 22942   48071 81265   84931 66278   00556 84870   10164 53331     81033 82884   22488 08012   18181 07258   08837 12647   07952 77365
9425: 08050 67613   17279 84855   84895 06335   02433 01593   15920 20598     95151 35053   80657 22651   30992 83850   58614 36222   24911 48413
9426: 85394 70523   97297 29639   12050 93585   60569 62127   75425 89946     80079 83873   79986 32209   13144 46968   18472 46999   51412 90066
9427: 79400 40527   09264 69179   55357 84663   47296 90752   53529 04250     67457 53298   49445 48310   43412 80484   10826 72150   35933 18214
9428: 34698 94344   87943 90073   74985 59368   23800 27193   90224 70433     98545 04624   97279 00340   19285 37120   04611 01322   94246 07553
9429: 74467 12378   81066 00012   07166 74462   10616 69948   17392 99663     80650 85027   81240 92730   43378 97124   12669 07625   40094 43365
9430: 88169 98209   26090 26676   29235 10947   67601 99401   81278 69010     04170 14912   22601 31897   16106 55785   71849 45321   07262 15278
9431: 80648 99799   68364 18053   19800 29230   36543 28176   23771 66054     66068 55603   29371 73216   10270 75492   18406 35169   30377 94580
9432: 16875 70886   10451 07776   17672 08492   32303 03899   45995 63083     93408 63343   38773 59417   70008 11532   55767 24293   86574 72725
9433: 15350 08119   83666 52412   70420 61919   25048 89783   20308 26885     85899 77514   47337 02641   17378 85405   06136 10752   02045 99677
9434: 09308 92752   13540 08124   91787 14935   96626 16440   96272 01911     20404 94040   24250 17620   13882 45693   94511 64659   88640 82921
9435: 58514 28822   14553 06006   79136 44591   53765 74179   15452 44539     11484 47434   58445 55261   52114 65642   80022 58706   09190 40556
9436: 56557 40342   46113 99588   08845 74603   59145 88374   61137 30605     74356 92676   50969 22257   69956 03513   47183 46125   15441 74741
9437: 77946 53849   41782 76723   21840 70963   14897 57801   02844 51736     85557 56214   48643 84226   98255 65992   18741 06024   70646 31877
9438: 89612 79111   99899 83042   19204 42138   14349 28236   84467 29692     96870 41195   29377 25370   95869 94339   09528 20132   18790 35257
9439: 56279 93959   98773 52637   51082 22201   76290 79348   89747 07145     40029 05776   12645 67412   22436 34199   27288 78268   15365 44499
9440: 47838 97449   27401 72823   65621 91083   16751 95250   65950 55728     46697 94885   00797 45766   53700 72540   97650 96643   55630 90642
9441: 54935 56174   95994 85974   30072 23584   70857 48689   06888 87719     34719 39646   22739 62933   72511 95807   46627 78825   17523 44750
9442: 68209 32034   10175 53586   56174 39368   03713 27007   13036 74033     06615 02775   53206 68635   74028 31477   16904 20715   66194 81530
9443: 41537 74412   42141 10112   04968 99143   04534 34279   71253 02894     61946 90755   87186 43208   34579 84520   53023 95883   90228 95006
9444: 08095 46721   89826 27059   45261 20439   47769 86141   80860 16021     23193 45480   11175 73792   54947 45053   32798 29571   07157 07817
9445: 78777 33580   25039 98450   36807 97863   89393 22578   60388 45556     68641 04305   17110 61414   56459 28090   84102 80631   31586 44831
9446: 04633 36363   03617 58492   61652 53184   09485 87553   22468 50855     94116 67483   30728 83596   43253 61765   55004 12390   23180 27589
9447: 71339 61670   45263 43010   53768 27827   72367 02735   03282 27465     73418 01763   08682 60406   37285 09445   45744 79224   48206 62249
9448: 03793 44925   12942 20496   51755 02141   18546 55216   46033 95134     14422 05432   53144 99516   81735 45672   94710 40718   55999 22186
9449: 51275 42414   28618 19815   89715 23220   18739 47330   58176 54717     91494 23198   95673 63741   84968 18760   42123 95238   05033 77536
```

```
9450:  52352 36977   37406 82217   51950 25443   15058 98731   11566 15545   30490 65148   26298 04267   13891 52417   73937 32612   83695 20790
9451:  42592 44659   67075 78005   28462 88672   04335 69926   25766 37758   48853 54112   58247 68240   89852 38369   35134 39370   89024 29019
9452:  60654 42473   27221 61423   69678 20267   50660 68427   40235 02751   21909 93008   32390 30917   16171 65469   53896 48058   33274 54007
9453:  31052 62760   74451 15079   50481 00600   77271 46227   98457 17628   18571 43710   70235 36153   85014 86383   27639 13615   48135 87419
9454:  81497 36613   72652 85837   24936 58878   97846 44078   58242 40305   97566 08283   70758 41595   49892 18039   10772 13848   66592 21239
9455:  37920 14410   93668 61371   57437 92820   32221 24229   23988 67921   01835 21330   07480 49484   86747 86336   13665 33281   55371 70353
9456:  74299 08306   78308 71125   91758 45790   84419 38191   83114 94013   24352 99125   73471 46084   40271 23885   25048 18370   05796 75800
9457:  39495 70790   16150 49005   12649 67166   72218 42585   04423 10757   58469 85817   32321 51415   90685 24869   75057 21996   71522 76054
9458:  83713 57439   98616 58554   43795 13080   52249 16671   79879 26946   63816 71303   66178 47343   22044 04801   23654 65012   62224 22390
9459:  13606 17975   11544 81645   07749 16429   76325 78366   97061 07103   19892 09453   78460 33152   10088 31110   41353 32500   75383 99324
9460:  17195 57631   17593 63850   32478 53033   48020 04907   57963 14834   73107 80629   35119 01250   21976 22642   10663 71433   20100 93771
9461:  32905 09187   79686 71354   52359 39742   18236 67828   76920 07063   66112 88529   25699 30842   92435 94389   39839 75460   48747 07938
9462:  94522 38368   79003 22390   90956 34855   03966 22101   84866 90556   53287 12645   50625 56682   41177 28000   42205 27210   84959 02984
9463:  79318 98334   90388 25628   24982 71333   24654 75321   19880 41366   06522 96843   51395 74670   14120 96809   90400 51313   94250 26284
9464:  83135 57771   64126 99806   28133 98763   59366 78117   11542 73852   97992 92381   81128 31738   25148 11833   74340 01010   17800 13309
9465:  58787 77413   98819 48635   17303 21531   91709 72170   02204 12905   47007 94400   66869 86577   30482 60576   99347 51114   78428 67300
9466:  69534 42054   58819 97915   36814 89056   86736 87593   34236 29847   45771 60067   48220 06103   07171 00935   49236 10652   98890 23599
9467:  37638 92644   04969 82916   97332 29581   80958 61672   52842 80069   79802 59758   74216 89755   19355 58130   24775 41196   67411 78689
9468:  81731 79956   91679 50780   92036 08280   44123 91573   87735 03311   74570 58688   36591 55304   12129 99345   85128 38096   30676 35924
9469:  37818 26769   45950 79689   96539 04957   07097 24114   77109 64774   86357 35871   52107 13511   91041 10656   94190 11583   66679 16716
9470:  28673 27389   36131 30566   56351 71890   45872 70030   49042 67403   51374 60895   90257 77779   52771 53740   70576 21530   36379 87841
9471:  76768 15035   12912 51843   67127 86115   83896 66035   81028 72860   78154 62054   05988 90957   36095 38239   13270 31600   93031 89150
9472:  43766 25503   55430 53390   55689 35398   98661 69858   80020 60433   89301 58835   56450 28825   43194 40186   48369 11990   09490 09706
9473:  60239 73010   67065 34247   54175 38950   14892 08201   76960 58474   67731 41522   53034 94742   42398 07967   43099 79895   28002 44867
9474:  81134 15380   09632 90911   38968 45311   26492 24752   93624 32669   40003 65496   31179 29830   99175 82037   16494 55685   23919 85859
9475:  39292 79408   44118 29965   23384 39012   07140 94768   90612 72444   67733 02166   46821 39750   21060 71238   29308 92545   11303 34137
9476:  22478 67532   91595 69925   41404 75194   66644 50031   83037 86122   07765 84482   69080 69099   77519 66519   72588 83994   76624 58804
9477:  01878 22786   23830 55543   60603 04715   07959 54058   31913 04282   46105 49302   97992 06789   64645 01288   14538 66966   59821 23840
9478:  43252 65879   59629 28953   29933 94323   86137 18222   07738 35169   79542 82009   65909 16444   69243 98347   54834 82431   94899 20007
9479:  14244 10839   17202 62394   62994 77295   41508 40970   59365 40099   15138 08204   58474 76755   80370 28861   15949 11074   28125 41321
9480:  62684 29157   92644 62356   37345 12249   71315 05320   33567 71867   76038 31889   41360 26646   89080 34813   13046 08016   11149 31403
9481:  55528 63893   98950 21988   37117 35140   74679 15985   51370 56935   69402 23425   24541 16667   92582 09076   32513 39827   37165 34913
9482:  40390 49613   90820 56034   95265 78173   86521 50490   58450 09148   83330 77672   12859 89902   74834 01241   88523 88944   49249 89775
9483:  30427 39784   68987 58521   97013 89759   47085 10114   46138 02379   16496 51908   63350 63488   73696 37174   45840 09476   20003 11263
9484:  62054 48035   97757 24470   94734 92319   39130 18418   15300 58292   94753 67370   43176 40178   26163 02275   02476 43905   97674 86450
9485:  89140 17968   63251 77590   17658 69448   51917 06013   27666 67330   64599 80080   49916 52130   58778 84789   43714 61616   67731 88467
9486:  12585 85391   89507 65873   30540 80080   41916 39396   90316 30489   37022 60972   81534 03135   16508 54333   38289 39637   57275 96859
9487:  87973 66536   74048 39060   29943 16569   70601 53492   03538 40750   49683 62585   06826 46194   83228 64802   98379 44714   74554 52021
9488:  64511 51847   30329 36927   04240 19570   89785 15189   97384 61191   02169 69846   74393 28680   17676 50654   61339 57521   77061 34749
9489:  16875 39644   27026 77342   55999 35439   20398 76452   72181 27372   56392 03530   05704 35619   50127 06032   98644 59389   34018 22855
9490:  49907 98569   92247 15809   35117 62041   64353 44039   02607 63597   75542 97376   17244 40375   48700 55748   87863 62963   82309 53271
9491:  16743 84165   49584 65158   10824 89140   73823 62109   13735 27187   28702 19328   15115 07177   51944 00803   29937 36255   01935 80829
9492:  19799 42459   28427 61107   13331 47693   08340 36103   70231 52268   53159 48412   47502 70333   59730 04794   77412 66180   78714 02255
9493:  80664 30215   99862 52101   61525 60267   45536 23764   52832 89238   73466 96093   89269 34376   03660 82624   30585 86144   62388 26895
9494:  00081 42594   78727 29422   07007 92013   04513 59224   08654 69321   30666 21038   06973 39145   69094 17736   74738 65367   33028 24156
9495:  07282 51114   27932 75667   04028 19036   48841 03479   30025 23781   08877 46668   27950 76478   57214 96003   63936 92512   58158 07900
9496:  92193 67322   67057 84355   32803 52932   57162 07513   37986 54575   74932 94492   98907 67657   43596 26524   78853 72286   66233 94027
9497:  62314 41775   00043 91208   97645 88864   39928 98078   26420 40357   01938 77524   22703 88116   24274 14598   85517 57550   68109 62003
9498:  99204 29811   41430 88929   69334 90384   58400 90485   60630 81820   21472 74780   35950 38223   99673 14696   13840 78955   42422 72813
9499:  98782 12402   39248 71101   67034 25338   09265 41859   75627 90706   11255 44010   30000 92820   54679 76073   68973 13318   34943 99043
```

```
9500: 13594 75529  00981 64102  25701 07662  79921 23614  54985 70038    36431 00360  00663 62770  01127 90978  10355 87572  00017 29842
9501: 98217 88794  99318 10359  14122 90346  31356 85850  10848 32330    51099 77756  04620 39797  49450 27283  59452 63263  02603 22778
9502: 18665 21740  44389 21486  09290 21624  28437 50524  01955 70244    49878 08678  08568 03398  60540 63218  51444 21431  30297 68668
9503: 64735 98479  40779 29764  42587 47523  32952 06249  66289 27090    07076 38873  84864 87699  93278 16249  35019 59610  20398 89272
9504: 37354 53474  01598 09621  75028 51790  55164 05419  13479 97836    97484 66159  37268 55480  40701 15535  85796 79193  14729 56136
9505: 27948 42117  72415 45342  39520 45553  89677 50242  96651 14706    41021 18269  33422 36099  57262 74769  78593 73651  46647 92682
9506: 51784 45493  99088 72868  91873 58563  02872 31664  76086 23326    92694 04696  58722 48677  15013 43839  78223 56193  54449 77306
9507: 29108 02441  30580 90554  95595 14251  65297 42778  63544 44437    28514 34386  21727 51343  35910 48340  07487 68844  32326 06443
9508: 54012 22344  85823 75944  83118 20735  84679 16322  99522 20134    53440 46472  19739 98030  09956 49257  24471 99986  40584 03097
9509: 32411 14478  95093 12741  78665 77644  96645 05024  29110 02736    59685 82503  34294 18910  85033 04207  44081 19127  48365 15905
9510: 73306 77186  53377 27201  96316 11492  71103 70367  68308 43069    53205 39378  68133 81042  70691 49697  92227 34393  36967 91156
9511: 70650 64799  14590 08314  58686 01729  84182 14359  15449 24978    98977 16974  53096 23860  92966 37463  49243 58209  14269 69229
9512: 59717 57543  60483 84812  94406 92482  23780 31852  00345 13149    88195 81386  03266 60134  84795 59075  97204 47089  66512 56876
9513: 35313 16019  87155 26393  86874 26978  82600 26349  82097 77250    41684 10938  69764 74674  62616 18246  52958 68785  03419 23252
9514: 98848 49423  55201 49529  82395 48914  62367 86484  11066 36016    46030 12824  07865 22776  57193 04259  29889 14173  68662 74732
9515: 36693 40339  04986 34634  44870 55665  66767 59406  08401 25918    86100 49336  17265 69082  41449 82175  92282 16165  37823 57948
9516: 34661 30209  53850 32921  82568 23516  10658 31144  24159 14486    39346 88523  80956 48972  90269 22793  57021 68902  82779 41275
9517: 34149 13667  50863 26280  83606 52040  40105 75756  68888 96091    41272 30111  54630 14840  65125 87381  68229 05547  05880 41478
9518: 83416 02471  28603 19974  03314 99381  25186 83853  72277 83417    10405 14287  20753 58909  61966 50012  62384 42715  34976 14575
9519: 00281 46298  64379 82632  25489 74698  26729 68491  58629 70351    46943 49208  56902 78792  06600 92574  69199 83456  95368 81454
9520: 67841 97273  90161 41309  33486 32727  88733 20060  44447 24385    03317 06262  50710 63244  25708 64412  97430 11195  39732 14232
9521: 40806 39064  88993 08634  79114 24443  15448 62361  09964 35250    27136 90576  86577 13626  02125 62471  63007 58693  48655 63503
9522: 44437 04622  17041 95477  64766 78088  97909 04941  46323 08336    10393 47959  29845 02823  65469 99651  83081 36572  83698 80787
9523: 46169 61282  80145 67694  46364 07428  98596 75983  81425 61964    90517 67193  08864 00960  35914 49370  87934 80402  44731 72460
9524: 83570 68858  53613 81429  46108 50511  49342 92299  99115 04337    35117 97727  26004 14656  63971 53196  84250 62200  07316 22723
9525: 90619 39083  80970 54361  40367 35253  66200 77958  49044 24750    28393 58473  18168 58552  85196 06895  22253 29623  47671 51509
9526: 66483 23681  13325 35669  28644 70336  26445 47446  78699 02651    71850 90385  53961 49693  81311 40407  06327 09796  13231 53830
9527: 20275 84784  22294 26803  91652 50277  39257 83816  75222 06254    60756 09393  85800 54628  61701 76067  69651 32861  36898 02257
9528: 45475 85480  22905 65752  59293 52340  45169 20680  78794 78390    91179 95687  24540 98096  47576 48514  22518 09054  42253 13037
9529: 49970 02800  19092 07348  65682 25045  80025 89863  01724 06544    37670 92391  02387 54062  96161 87423  89562 05248  32444 67185
9530: 07511 48968  82058 90283  06480 94013  12751 00588  43952 42941    21081 68528  71184 66137  74799 84061  27750 08711  27263 37864
9531: 93305 57761  33234 37085  34887 40504  65825 75743  42251 42659    11112 91584  67495 21914  47992 40163  03419 20521  11994 36477
9532: 17581 49990  37968 44964  42631 34819  26468 72575  95717 67018    70913 62155  34392 57232  39949 62334  28291 07435  99054 77165
9533: 74936 09150  04501 14361  23619 93318  65454 44181  20761 40851    88176 34661  99906 78435  15293 20050  84627 89502  41264 91903
9534: 51249 96692  74407 32787  05821 40898  18221 92080  83908 33482    92365 95422  68075 08075  50000 82502  78107 60062  59704 37340
9535: 72836 23967  35922 89340  14944 33862  64960 49136  07075 07728    28142 83753  82323 74134  41657 49292  63580 36419  42517 48992
9536: 34547 90036  25403 93329  57628 63684  76110 32177  09906 73195    01880 26181  41720 90994  85049 98298  70529 61901  92016 76178
9537: 57690 72203  50892 37542  62677 66298  11354 37853  72118 23902    12877 07897  49201 49929  85738 36171  40643 56174  40643 69320
9538: 74064 70621  01087 25951  12848 69713  07049 50970  00357 65911    10089 17508  06802 51937  13476 74849  66903 52724  10330 04255
9539: 62422 50374  47604 30248  85000 03773  66179 37904  42082 98441    82982 45125  86664 45548  90953 63084  57303 27323  22419 89453
9540: 14486 18138  34215 53349  96579 81810  28640 40475  47568 20498    20214 72138  37532 60530  28418 28636  31908 58942  37140 56746
9541: 10591 87112  75075 22899  37267 15679  27780 37435  52152 89264    91551 94627  39721 59175  20724 25372  03006 82149  63255 00517
9542: 42255 08566  19234 16413  67389 44578  33087 12684  19171 30488    75378 33803  19291 74283  15768 73566  73872 10936  43097 17636
9543: 26649 35378  06045 57711  69362 92618  72355 19383  35233 14443    47656 02604  46587 00866  48181 40579  61619 10413  24964 95070
9544: 23633 78492  42390 63634  41563 92993  19166 08431  26535 27229    53682 00173  23232 21023  11261 83803  96997 72099  13117 70222
9545: 17039 78650  79429 59800  84839 23936  12880 03405  67920 33755    39571 77187  70712 64493  79897 29334  88878 40433  59368 84581
9546: 07159 72713  34643 22213  03389 29910  11308 55682  87383 25588    56400 94653  77341 00239  41949 20369  79812 18092  67440 84602
9547: 15354 23528  57281 30110  96523 25100  30731 38092  02810 20373    29100 17093  15963 94597  84022 87780  45988 64362  44993 18043
9548: 81563 92598  40163 42409  20207 24116  74364 34437  58808 60534    92892 51392  46192 63065  49967 11247  58052 00951  30549 16045
9549: 62611 76507  90496 37721  58082 72637  92320 96323  78258 47518    53909 32751  21263 24042  46999 27377  97315 79529  91081 62721
```

```
9550:  95751 48956  36484 92309  38139 02355  79462 46974  81081 65658   73767 20568  26118 63651  26388 37831  99282 53467  49330 79827
9551:  28962 93721  40463 25475  05418 62604  16069 60412  65287 01815   81272 59232  01120 44893  74667 65838  69125 64045  28446 89960
9552:  74188 46015  54877 89645  93937 66975  04190 41285  89333 66910   06385 18968  24149 19622  10085 17993  85830 38901  15370 06126
9553:  50989 80076  56483 84455  70095 57183  82331 65220  30640 89799   86561 96534  72257 92136  08005 64212  17005 55065  86266 04488
9554:  49639 04587  40243 98319  30230 89086  81339 78329  65008 08068   86725 08131  62427 06330  18079 39406  03934 41043  01318 68113
9555:  25205 63065  16625 12775  85303 37602  83332 75628  44121 23120   36810 97022  52265 92880  05666 34474  75244 69122  35931 23989
9556:  26386 98991  72986 65538  38658 12709  62294 97649  53134 07679   80119 52845  14051 82171  24744 38181  56073 82584  18782 88990
9557:  07905 50764  88429 02130  01945 17209  62769 57486  84715 01620   02810 98495  59571 11200  22014 51482  39057 90820  20422 99033
9558:  69598 83923  17390 16906  51587 18860  05778 55232  30833 99100   04994 80458  86561 55763  38240 42823  22745 71765  00680 65807
9559:  63555 84089  54954 50085  84712 09971  82123 37745  31409 91729   68454 61015  31527 21306  33270 71667  62654 60615  61403 87457
9560:  68493 35402  12715 29986  73767 93262  57332 22867  01689 82218   89405 44662  95543 76253  76252 66221  67696 50151  77118 14724
9561:  15894 36316  43896 41407  70541 45970  55080 30916  96206 60551   70781 23917  08708 26681  05265 76886  00824 71940  15698 28890
9562:  29099 87110  81581 32220  72397 76836  42205 95910  74590 95956   93153 20594  70035 14860  15885 32129  68592 97763  63986 22596
9563:  92950 85436  42462 04293  55249 11492  46467 57701  89249 31073   99033 47067  98044 69491  79667 65776  34944 26985  50701 76503
9564:  52354 76008  31479 92072  98038 55331  58767 04765  68572 61590   68293 85896  72571 36081  40309 44371  17720 62852  26169 22574
9565:  66786 68769  99312 00987  55671 53282  52732 17451  10466 60449   16140 36594  78384 32242  87003 02295  24400 56671  82450 08144
9566:  80437 40391  02641 63555  91879 37752  04535 41146  16728 75642   28860 61201  29559 57489  21661 65028  98523 79315  28112 79679
9567:  06061 28386  13849 53219  77144 58245  86367 30487  60554 42752   50593 77640  59371 42773  32724 58338  94922 39563  43913 32404
9568:  42170 82782  05947 51509  93510 89565  91347 64291  46081 43134   82770 79359  19506 92900  31750 88364  90820 50337  00270 42124
9569:  41905 57227  80116 06252  34483 53147  24925 44012  02969 31550   97917 96755  24162 62095  54247 81887  52316 77200  10777 00916
9570:  56960 29322  67849 88479  62585 00035  19990 72254  37415 21634   49837 89616  60752 87271  31404 29431  66631 04907  13612 31478
9571:  87130 95750  42419 08439  30978 82755  62419 88532  99226 76817   54614 83621  62308 59354  69745 60818  53497 50118  69387 05236
9572:  02431 35118  79025 67338  46972 22038  12278 7230⁵  84415 12655   70203 60715  21727 77622  77320 30298  30284 34329  46960 88440
9573:  57223 02988  19021 29685  51623 26639  36509 17743  50996 91605   36426 77932  64695 92681  32348 83871  32291 38178  73914 84412
9574:  53690 64213  04174 09199  47834 17704  94411 56950  02953 09548   24467 43436  24180 17872  46716 58200  26406 99052  01665 18826
9575:  92648 99995  63321 42129  96419 07023  73491 18583  88468 87859   94145 76886  23021 58829  78922 44173  40218 06476  82636 33178
9576:  31906 89054  81296 29265  39304 93401  35718 66314  99864 41177   06196 00574  21026 72854  73668 40072  34141 46151  00905 35443
9577:  75683 58014  01089 64131  24737 22787  20019 15696  42692 87653   75650 24789  08753 55892  88256 64226  21444 21519  27246 60898
9578:  90337 89405  20220 13759  94002 28812  74200 78125  96151 22879   06924 98547  27662 07807  79702 01447  27538 93192  06582 67268
9579:  63080 42726  95348 34220  42754 00549  31952 42003  27371 45449   08300 86808  33882 03591  23517 41082  60643 77593  92217 15939
9580:  16917 42736  98398 41198  82687 56090  43283 88432  76525 96979   17410 56735  29803 44707  60502 35038  82566 46616  87861 73724
9581:  29643 59099  83975 47217  76016 89967  28777 95211  09878 80352   95779 20896  72229 90151  92029 63927  50063 55750  37935 45218
9582:  90715 19081  17178 42675  23514 46490  08913 54161  19346 24441   26113 80663  15439 18864  57812 92253  16002 43758  47960 81921
9583:  68108 51005  95651 46583  14760 32026  97422 41165  95341 90478   37227 04522  34093 36410  36926 41229  84499 17252  53936 17729
9584:  16568 12980  45568 39136  71789 54121  02599 07709  74379 28860   36354 33611  64462 72401  95681 91492  71029 89336  73206 27674
9585:  13853 16101  49837 69754  76978 40448  32385 74891  57329 38704   47791 05611  30608 61190  91555 31279  78527 94060  74723 98606
9586:  98238 36554  97572 29605  54186 18445  47351 70079  40393 65483   52555 35119  75759 51605  51311 61804  95116 44005  73814 56336
9587:  21870 92435  43670 25136  18592 29213  75479 24059  31909 94163   01435 40529  39741 42515  05244 18853  97309 94181  32598 19366
9588:  00638 04007  31200 00886  24074 43158  99232 09668  60019 86913   69960 25805  21579 38539  96635 47348  94147 67244  20659 73489
9589:  04419 72486  80499 62128  21010 88347  49527 00161  62440 16794   68947 17945  11180 15059  67356 55666  11062 42364  25804 24854
9590:  92237 26030  02777 51949  78198 54943  59357 89014  39181 65660   44226 27579  04136 49050  54195 99146  24634 16623  74122 65169
9591:  52537 43031  20831 85455  79200 21013  95803 60894  33101 59525   82729 00464  91808 53775  37114 54529  06612 02661  84850 83168
9592:  27079 28574  62913 88780  51555 81856  12590 40012  24124 35905   84691 62622  74572 67994  86543 79669  40075 48919  18676 14119
9593:  58738 61052  67989 66689  75765 19569  59135 24797  93240 48714   37104 32129  86860 46648  38122 87192  22201 06045  32639 75930
9594:  85280 07006  51219 75493  22270 76501  15910 72263  53886 96593   11392 71730  28933 99121  90671 36609  22244 23598  41007 81763
9595:  57664 14373  67816 63498  29910 41613  91210 78114  11797 94184   90390 39148  40000 28261  67602 04567  76988 91511  58197 93054
9596:  60432 21955  27760 20836  25387 35399  86288 97391  70561 67155   00564 56177  67690 19977  77468 98075  14604 59130  70430 25642
9597:  13838 07278  37531 63916  87413 23537  79660 72181  98101 52476   43986 48312  05293 18819  11117 57104  65786 94975  65186 60319
9598:  52677 24264  71225 33606  51411 91222  99957 17843  61008 25285   73202 17429  84600 34305  74579 60348  23554 12628  66195 28092
9599:  96530 09118  21418 09755  66850 24065  19347 72526  24439 93239   46398 86505  67836 31965  88772 36494  22245 21407  48858 67466
```

```
9600: 56807 00372  05217 81144  20906 42007  37848 81693  64373 30470    41726 08320  73398 35974  90692 69500  30354 71012  28453 50932
9601: 22970 97823  17920 99635  52491 05189  20206 99291  24561 83952    68187 47232  02315 21793  32706 27319  36561 56946  26969 11878
9602: 49652 08470  81296 84541  89512 38524  84363 04760  09781 45256    54373 03636  35430 19059  28921 94750  28081 10377  50097 12904
9603: 21754 32987  09825 37097  78172 79741  83389 60591  52754 88038    14737 47027  29917 56146  61184 09073  15688 50891  30905 95917
9604: 46335 54429  84817 42572  48261 50586  34930 21376  48994 99367    86326 71188  04257 79856  05986 01737  00855 30078  51807 56781
9605: 59905 39272  75665 46883  46654 18006  06389 53937  06370 81075    28989 98036  35271 67017  77897 25351  28162 37669  16261 17826
9606: 96316 05597  56456 65191  41431 62545  92658 58150  84433 35467    64221 72871  35043 53235  27452 38989  05500 15755  82436 91042
9607: 97840 35684  30671 78996  91426 85209  26243 50081  90746 80916    09933 72038  83371 16884  06648 64106  96056 51953  77426 82608
9608: 19479 34287  25140 30373  07035 46706  37259 75926  65490 17075    87574 50341  89552 25209  99498 83993  97078 36877  83534 22549
9609: 08471 15488  88994 21764  87030 44666  08037 40902  46133 49006    64837 88574  49866 73012  14110 82184  95811 61975  65816 57877
9610: 95290 51664  80816 77007  41356 58566  40668 31000  19774 55790    87279 96341  06910 04685  23361 00481  80189 82289  50024 76971
9611: 24718 51485  80185 69949  45804 00656  28722 98490  13980 49052    09100 33669  98925 85444  15440 67414  17718 94758  93945 82971
9612: 97444 43808  19169 28140  00677 25769  82764 49438  98895 65804    80196 64153  82310 92760  01022 04189  22903 85687  88954 23904
9613: 12239 70027  78962 43390  14334 52080  66170 60620  45313 35334    89574 53282  28501 06882  00759 53385  40876 89549  07315 30688
9614: 15937 16536  91780 48159  27148 11849  53843 12648  47210 21774    69611 91897  43561 26035  46943 23712  40991 31507  48160 81911
9615: 42762 47021  42985 99050  18601 10616  18032 32399  97498 51046    45550 24086  87426 74891  80247 74852  87809 89339  01291 85819
9616: 02744 14633  70856 76280  67176 59564  42415 65606  40466 67314    10483 16122  04963 14193  25867 96319  17079 58893  96532 71387
9617: 34400 67022  13689 09716  26880 14692  07544 67157  65843 70928    45636 21298  20051 93007  17989 87643  68763 19444  11982 63785
9618: 92327 44839  03709 65629  46927 01778  50722 24617  45979 61020    93733 15298  93939 57514  16963 52501  50919 65469  54127 52956
9619: 12698 95017  68258 21055  03203 34478  55181 57871  60054 32473    62645 26459  34527 03911  35004 55790  88896 98572  80003 44157
9620: 24609 02306  49013 97591  91055 95696  10887 03705  86172 93043    28144 36313  55288 58866  91467 98735  59331 62188  96055 33159
9621: 71683 97910  56329 22347  58865 99276  00112 70062  63344 15217    32705 73018  38160 72176  85094 14449  05395 66700  53641 05130
9622: 84425 31592  01352 21958  13915 36787  41764 30487  10215 73077    86323 77021  14770 52320  62720 31463  62130 97226  05566 34415
9623: 13255 63631  80084 63625  29336 12716  05066 95328  41327 11521    61433 73694  54358 09743  46555 08653  74157 79307  83525 69563
9624: 88513 45620  37253 97371  55262 82985  65027 50275  05961 19727    77755 68421  80911 30199  59887 74423  83595 27476  22370 79129
9625: 20678 22847  37228 48218  43715 36420  40524 70594  00494 54753    29437 07920  82893 55265  87969 09120  21329 08690  21935 55672
9626: 36884 31712  56042 37714  48399 03798  89255 38087  54142 12450    95637 62125  96519 21502  02274 10473  54887 95589  06271 85371
9627: 60470 66874  41502 04109  51525 24115  58576 66360  22477 15253    88509 47646  56603 25585  58279 29445  48839 68653  02688 90583
9628: 37003 33115  64647 28313  71828 08221  95568 57773  81452 48186    27347 76366  64314 86629  51671 48738  83521 97056  28890 84456
9629: 02647 31654  57390 87165  81649 67567  33859 82314  69270 28995    54999 80117  04100 82859  67637 24374  53177 51947  15493 81603
9630: 35155 97834  60867 50075  05115 41428  57489 77709  68524 75658    26264 46694  89583 40153  66097 37314  07418 54781  31818 10366
9631: 47733 86193  65188 88759  40323 32318  31300 34636  19701 23460    49372 35836  78167 33977  45521 22643  56484 32956  18671 66960
9632: 01051 03502  78424 11507  66525 34272  00855 48096  98857 97537    30310 90627  26778 54025  10673 62092  81047 92825  25332 02862
9633: 00021 26636  28780 60569  26861 96159  04246 85586  87327 09316    06384 95380  93036 50661  42528 94639  65237 53832  52899 66918
9634: 54363 92891  08485 80260  53741 34691  58606 77508  37695 00540    00692 11908  22894 15980  49804 03047  80182 05994  04589 10556
9635: 59142 90139  69564 04351  84446 63806  81952 92087  42435 08492    97992 02153  31242 61781  78911 88064  21509 00204  07860 30077
9636: 12830 48381  52354 33772  09820 78886  12941 38719  34568 35505    48471 14921  10044 92954  99114 58094  59499 19429  32716 54573
9637: 22765 25430  84247 71336  09683 04916  71405 47386  90589 80784    04332 56375  95682 16710  17305 22963  02366 88235  48881 78774
9638: 90203 46637  34783 05561  49748 87493  04887 28498  34435 26621    79113 80330  77391 96593  63528 38766  25750 04558  56997 70903
9639: 64076 62117  02513 11495  47009 52110  52189 35261  69292 60129    83777 54882  04358 34841  62586 07162  86264 82917  85210 52855
9640: 12123 91407  27958 20674  97865 64018  14993 22505  95550 09805    38939 19600  94327 27348  97027 92674  60977 30204  78399 16036
9641: 27886 64009  84951 78865  64518 15438  84053 99717  70416 29576    86576 57852  38489 16035  26577 10419  09434 61390  42790 12333
9642: 19220 95102  81444 55753  53041 84566  83646 54441  31895 21216    71645 43606  94223 10071  51095 81575  40388 85694  17641 39042
9643: 43052 86686  65838 46542  56500 34965  72840 00277  48219 21940    13838 30263  55115 22857  18127 38808  95589 17909  16885 52236
9644: 51498 22719  50297 86635  01282 06830  64284 63935  14369 95118    29847 40453  18630 11285  89227 73212  65557 36357  84977 68118
9645: 74096 44965  36106 32501  36887 01227  07553 44728  26082 07266    47746 81859  84550 39070  87308 03738  21127 30953  32876 61540
9646: 88163 39092  12673 51066  08664 16468  45189 20269  69905 20592    27085 85901  22216 34101  10649 30324  25522 63400  65728 43653
9647: 51045 29108  70185 18396  99235 33189  55155 49538  73869 59741    80795 45263  12890 89063  79163 99981  47116 82310  10697 05207
9648: 59040 04304  89732 16362  12613 76450  44215 21693  84161 68074    03143 04331  40934 04772  81439 36289  10967 10164  09026 91050
9649: 68093 69840  42242 56639  95670 81613  15301 99360  53495 26810    29665 59549  03133 69023  60936 62622  42173 56855  47944 55152
```

```
9650: 99456 12751  28399 35585  39879 73398  48013 43733  69764 45115    18940 62457  06700 01647  84875 84139  19269 55706  53306 39369
9651: 22806 97505  02696 72602  20421 76584  58936 82554  41561 24098    35066 29146  01864 12246  50144 07160  50473 07309  34071 39110
9652: 69213 15923  18567 92896  89928 20599  56699 07851  59377 90662    96399 99125  06427 77166  69047 90219  88617 90580  95930 13621
9653: 75410 91325  33042 49973  09847 96446  16248 31072  68804 97665    75862 35190  27761 26399  87725 16002  44527 60905  53083 70662
9654: 70332 44776  47706 29844  69742 32986  09491 65484  25313 09352    69131 54987  73716 34881  97327 34653  87999 89922  49692 01125
9655: 49830 17947  90169 65308  80377 76944  63594 36071  24094 25730    00075 87382  60158 41649  36472 95792  69286 65301  37654 13279
9656: 38951 75586  44139 30169  65443 33123  05377 40677  71295 52191    87709 37552  69445 72991  68170 26252  83611 60836  27846 79148
9657: 46871 29953  72442 40470  01177 34806  51883 77836  25578 43162    08877 78183  85168 87230  88133 81393  02045 64773  01326 45234
9658: 23831 65315  35656 69046  88359 14404  10807 44944  98886 42855    56066 28344  15555 35355  54712 70876  35038 92174  15675 78954
9659: 82561 43531  32453 18008  54140 09644  72477 37294  99608 41100    22946 10388  98455 50450  10850 42747  64497 04436  62132 50462
9660: 90489 69009  87328 47081  09588 58296  99556 47404  70094 86912    03675 09706  17297 67636  95015 23719  07869 81258  86138 79593
9661: 51246 62072  69083 13092  27161 26185  19014 31937  06238 31097    19473 25213  21319 55169  13689 64547  40812 49914  48804 46068
9662: 18792 17622  27921 06555  71901 29586  60636 72254  67571 52428    04838 76722  71274 40202  69660 03413  12120 52378  11650 85084
9663: 12187 51772  30961 92303  59755 19677  54108 91680  57485 72716    26400 16355  46614 00113  91824 21194  47836 13719  99572 44501
9664: 42785 57879  93821 19502  46920 51118  31786 32672  73521 09960    56060 90350  91610 15099  36285 27805  65219 34958  85904 14908
9665: 41433 21761  57502 61758  79009 53193  69813 10824  03335 54829    86104 66066  00756 40521  20973 81049  33456 31756  03919 26173
9666: 19432 26975  14855 81355  33768 26207  80139 28093  87138 07671    77177 98646  44607 11191  57422 11993  96334 16595  93729 74375
9667: 14543 34493  64226 01106  90017 19694  60983 40843  39357 60715    90857 01579  74047 79985  18719 61652  76058 65040  42471 91492
9668: 76626 77130  85339 65123  03858 14905  36900 34207  62781 46525    84099 42736  28785 35380  11542 00336  10553 18469  19699 01885
9669: 73519 92081  99573 31157  88733 79706  28728 93887  38103 08585    32922 42956  96031 64747  54806 29897  68917 36981  17786 73470
9670: 91339 95673  10436 02074  98011 79560  02006 89207  59610 86906    66975 42911  16445 54103  68139 89965  55005 92763  65879 16393
9671: 96871 38512  36629 51428  80704 03031  01599 51186  73706 91416    46863 49500  84795 76542  18722 95982  62010 56427  84103 78518
9672: 81889 17091  61993 23004  93872 71401  44811 92727  44968 06817    43521 16219  96377 53631  55173 53792  92874 90859  67822 69895
9673: 99705 68981  61571 83958  90342 90475  69741 40195  43199 93579    88750 40231  20780 85477  04696 63108  55618 95165  59253 33650
9674: 93721 28715  13957 70680  35641 14936  08114 15627  65459 91013    27375 17888  22599 76479  03349 00463  39052 79851  84589 73893
9675: 24669 14565  86809 46025  95942 00738  34275 04761  17113 63051    23140 02095  08610 47290  31591 67172  36764 77389  87937 22033
9676: 71097 10683  06464 78583  19246 22458  10908 85526  89081 18927    95906 21447  02746 94376  29148 33821  47918 96270  12622 99489
9677: 28633 12000  34042 03233  71903 82742  18039 28304  44388 63269    59652 97062  05516 60033  21471 40065  85354 01138  41055 66584
9678: 08728 57751  00769 59097  46012 43720  03441 01140  01510 57313    15033 31960  59151 87879  77237 04701  69022 38916  35538 53742
9679: 51211 92026  72518 43073  93263 86237  23517 90607  13859 62048    50145 39232  48592 63079  30295 91150  67822 98937  41510 19324
9680: 47208 80693  98542 75745  99669 22189  91994 75081  68572 67142    42324 01804  45182 70392  93160 00454  66747 98717  54793 88083
9681: 13408 85754  91876 64629  60245 72135  05180 28416  11469 29370    75444 73929  76771 97329  63802 30992  22781 14533  00438 73693
9682: 69127 13065  37396 66162  17010 09128  86993 22432  77197 20760    30153 09081  19214 13413  64090 40886  30029 58437  19369 31631
9683: 28011 47037  68116 55547  86978 04500  81137 98388  24688 68778    42023 80192  50824 44078  66462 26982  18232 25556  02706 20765
9684: 31787 83624  90147 25849  95247 36693  72408 11496  36259 29147    29291 34056  61035 81090  05094 47721  43484 04099  70971 40790
9685: 09565 17633  34013 82563  95242 76698  47905 43183  86359 26598    78605 32868  72419 10133  10710 67970  27706 04466  82063 67443
9686: 73527 03169  79869 55665  27942 70270  27627 74149  76787 00764    60691 18921  95833 51426  19613 33649  41842 48245  27229 83099
9687: 31706 38067  98634 57129  45939 62428  31890 72267  98539 81141    75098 46499  90124 42801  35956 62276  79691 98135  85076 30313
9688: 57990 27161  07853 41558  21457 89816  88100 78432  92012 46882    86538 74455  07151 43172  02598 41477  47495 26246  36354 74751
9689: 10368 13329  57444 42026  52218 71620  15276 61207  53215 03558    13207 41920  19479 54158  68755 15572  58593 12232  24902 97674
9690: 90320 26310  69899 69696  81969 65189  81536 70351  97474 02037    72430 73395  94244 55628  23539 85216  01716 81501  54620 17095
9691: 03546 61571  05582 38239  39398 93839  37187 79647  98924 65066    65869 58350  99648 12580  98908 22437  03621 99510  76871 76103
9692: 04691 39891  72253 65802  79650 32601  41693 89752  43931 17153    22604 15598  49983 36254  95570 88776  08950 19613  24803 73697
9693: 50239 09120  53112 95180  99773 92405  70130 30715  53027 26298    49585 85941  39254 93894  82755 81192  64966 20315  10966 86509
9694: 25058 70925  32948 24852  95887 87594  37070 45153  69791 44045    27906 10638  24991 97802  90872 95575  98201 12323  47839 69690
9695: 14956 32809  24047 11869  37045 16677  07387 30489  22789 31454    38792 49845  57893 36029  48955 54052  94368 19547  09419 19150
9696: 33176 01472  76099 99133  38594 15864  60247 21661  26350 14412    74006 26483  75903 34262  53395 54997  46723 91931  11937 60534
9697: 84111 71165  22555 22910  54460 63189  89565 68065  75282 20581    69191 86394  70358 91052  50510 16474  78556 50330  82980 53643
9698: 44327 35774  47528 62130  58610 71289  38920 60950  79560 60456    12532 88545  23030 92288  01777 31584  87381 12326  54925 95158
9699: 67834 14225  57628 85312  70866 15874  83132 38033  98186 41687    47117 33792  52842 02917  26269 54698  98030 41841  50215 10089
```

```
9700: 6999092773  88936 33093  42427 51039  12140 53964  7428679058    85855 87567  8612361074  1314292128  0172939046  0490059926
9701: 4284316206  9923273821  3720004604  6771098882  5961982059    30674 87470  44039 17847  7354497035  3298178122  0486400469
9702: 4025672458  1492724343  7311098296  7162409336  4119 99005    42884 35944  25973 06586  9137755004  8483397664  4008406317
9703: 8006522890  2096557042  0739296454  9732211267  7707548933    29303 63413  7040681498  4331004534  6303400817  3108707466
9704: 8084806744  4512751883  9259433176  1933120376  7345149979    24116 49545  0170373507  0590944208  6184818643  5664302388
9705: 6561973987  3011119261  2716707398  3880656418  9806354657    31566 14895  1625300859  4632264052  6121124231  5029437855
9706: 4396358005  7387770409  4067422760  9743721051  8895778138    98920 68477  3344325981  6476943594  7591998009  3636167704
9707: 1800827800  1117789206  5583026581  2933936304  5059233547    29398 27100  0137558259  5501683220  4195218460  5299130694
9708: 0753940369  1864121139  7297401388  9836516933  3190452130    61198 09360  2458507026  8649847068  2526316678  9053457351
9709: 1761458929  1066143974  9019600313  3003458828  1761526646    42217 25518  6901113064  9570427218  3907635623  3075695927
9710: 5047629724  4784711563  8372953969  0048767963  2795572839    69177 06648  5862085192  6443051479  1946838587  2589344516
9711: 6318733556  6946854054  9293630901  0964906130  6292674319    82207 92605  7334555470  2701246557  1253370254  0589786326
9712: 7877164599  8727301794  1153680870  3945577303  9116868542    57621 45160  1160055191  2899497938  3079411072  3233414089
9713: 0364728366  8528750882  1110113688  6458101404  2552163209    28235 95835  8998317529  6547144972  5314805118  3240246744
9714: 2754932850  3491989273  7384765368  1094835312  9684282907    66371 02609  1020127813  5217755312  2697029621  4028579978
9715: 7353504985  5222794320  8000667571  2896940198  9024633885    28782 92219  1516843081  2897346808  2177673203  4866265322
9716: 4917914530  9201011756  4863490350  4323009442  7133773035    08562 38345  0038793616  7472644817  1094018529  0322048152
9717: 7138315940  6609763060  8699489280  9864462512  2351761222    57280 56779  2084470986  5950920863  2744577053  3295044195
9718: 2470177392  6895084415  5399711323  7299091242  6296946926    35334 94587  6992100941  3228882042  2026178773  7777455512
9719: 7840536099  8189332267  8733877692  7655085870  4752974800    02984 46127  4156130785  9044542233  7076820583  3823768636
9720: 1278002726  8354280074  4703352276  7102059119  5811866983    54962 85780  8084653760  1711321130  4475277485  4722397104
9721: 6555760830  1389899596  8023884093  5834448293  5736690406    50780 34093  7791085182  7689638817  9045349803  5112874014
9722: 0766055217  5398488462  4977764228  6212532997  2910768742    15110 97450  9947467923  0041082594  7442698495  4003763094
9723: 9890998456  1011169709  4217915797  6451100702  5144763968    84406 14602  4862478774  1308442467  0189344683  1041148452
9724: 7505428925  5474812879  7274335209  5406495538  1962075624    17072 77380  0028642403  2472497410  9888148471  7751647740
9725: 3352799189  5497814184  8135731125  0039756097  7393896908    46313 84985  6634898655  4031874129  5837603243  7939000927
9726: 8451647889  5201430034  4228411893  6976785113  4072667452    50107 80278  9886000277  1868273648  8907043050  8101884865
9727: 7856350815  9067252076  9011068960  9386996150  8059936213    28661 84494  0228542811  0806661249  7749002899  8641709107
9728: 2553119495  4909212421  4765606657  6714269619  4065585562    27946 27649  4983833056  4210046213  6936468625  5228967931
9729: 0158408148  2103335136  4391869920  2517997416  8585251315    53597 67411  0473181547  8338851392  5466279437  0517141434
9730: 6460817282  0270418569  2474838969  3933646944  8615954059    50461 38621  4281569382  5686395211  1527299785  1470022740
9731: 4758248787  8053692325  9916041363  9473466566  5339194222    56402 89841  6127045057  2028871182  6470686385  6644911773
9732: 9781885111  5829662998  2424022215  0915712841  1407045454    23334 49619  5220126690  8997643320  5417245497  2789469871
9733: 3783121560  6103161424  5957325383  7732463398  8557539974    95521 99176  5619461576  9804841853  4044240700  6799264947
9734: 3803595724  5508178742  7630938363  3195293267  8287289194    08961 62972  0402059169  5494044969  0564384853  6673420306
9735: 9275269578  1186541282  7356864764  1482484432  3229871163    09642 90870  7072753386  7749445725  5751617991  7088225436
9736: 8246598260  7891390616  2987546926  4145221916  8578046084    67107 00260  9914982882  3572837085  5429106631  4032689425
9737: 3332128997  1059740523  9767077754  4724670005  3237000579    27828 85145  2550917878  1519938837  2385279728  4250267279
9738: 8970075643  3620308605  7141577085  1922630566  9944610205    98090 04126  2706118012  5320836030  0507077700  6077206177
9739: 2605983609  0688530778  5396454368  6498196499  2110830965    97851 98764  7980397398  1582214372  9345696467  2355350018
9740: 7803809743  1490654639  4175595548  6317481581  4163519484    99867 56116  9316867430  7307988997  6105605121  4860735637
9741: 9643675536  1537247116  4219136557  5349822363  2405677139    18395 42688  1968804792  8393047927  3461260658  5000093397
9742: 9733881437  6191453469  1916546504  2962601229  7190622931    93042 99029  4041919217  6596904580  3988044223  5131880014
9743: 1839964974  5979471132  2246972404  7471385772  9971275241    29632 06401  3984821125  7400287680  4932772394  8557291623
9744: 5615875749  9743523887  5385465788  0388984877  6830882128    92817 18288  7775089928  8117079999  6226823486  6935890000
9745: 7576652642  0104899324  9396854021  2652369064  8504482634    43860 73350  9461403314  2962603872  1647697238  4003056950
9746: 9235854554  2884822051  8258009951  7527385033  8765052904    38992 26316  0523768417  7721062306  6149933677  1480239610
9747: 5654642315  7344775430  1115702150  6444500114  3268251575    87578 95500  9854133993  9077385000  3391761464  1298242471
9748: 5207525589  6954896479  8716601941  0814672742  8489250471    96961 60899  4566290827  2572297553  4230043065  2208222711
9749: 5508543541  9414925881  6721306857  4489826871  8535139250    52508 14174  8906626070  3019910969  4024685623  1808590663
```

```
9750: 07206 24150   83265 00264   77456 85885   31376 54851   89402 92003     82801 36216   26509 83812   43503 68032   77560 41897   25916 99362
9751: 11342 74011   49210 79221   37570 65725   18521 37224   92117 62735     97942 82455   16872 67156   99821 28397   00072 53406   30119 21517
9752: 60327 33684   36459 96591   96083 05731   98904 79616   24693 79761     12665 37884   94645 00473   04874 03468   70237 74347   30281 88672
9753: 46951 88051   45442 22256   79312 23894   86132 92371   42580 25725     72888 39692   15312 60215   25623 55108   82324 62026   84675 53893
9754: 41582 04855   81919 49028   35193 25912   39353 97487   62451 85492     73768 05902   93591 30608   44785 90658   48185 51688   02186 51749
9755: 25759 12999   88317 77182   95372 61159   89335 12323   34556 55434     78024 03643   77675 53398   65128 30974   18624 66816   42223 69426
9756: 97326 19105   06209 36072   49347 97908   92453 77431   70812 68378     99308 84354   42478 95059   38866 03298   51478 20443   22696 80616
9757: 92543 67216   85201 31592   03347 78524   01324 31196   02964 90138     46838 29284   21432 27616   26226 00886   32300 58830   61487 44390
9758: 01464 38005   01645 47898   86086 79059   45627 91922   63634 17409     94253 29748   88463 75640   90557 39505   37281 80908   77256 09782
9759: 24103 88876   55869 39721   57986 22639   38341 20487   86316 90692     76235 95279   06449 01879   89238 01504   26409 93552   12008 27469
9760: 49392 54433   91697 48187   99948 92663   15991 11294   91064 92533     77536 94244   91856 21735   73409 76728   23258 06370   50409 35256
9761: 47650 73289   25487 53650   66221 03981   13057 88145   59660 92992     03549 23729   55306 72652   37614 42577   23880 22318   48760 04314
9762: 52370 89353   04856 67480   46217 44205   64142 16386   37301 64867     41762 14021   98564 61418   91839 52820   85574 33725   51123 20517
9763: 81470 13187   29765 06256   51879 01488   01188 52257   47944 91525     58981 10723   95637 56495   52074 10011   92141 29073   42540 35714
9764: 93260 30587   58138 45303   42392 21261   10952 92074   66769 51451     93069 05411   94178 53339   55899 42338   32390 25075   36984 30048
9765: 27988 60322   20059 15876   85565 51864   06468 33354   70684 51879     23433 74061   23644 67863   85826 43192   50662 82306   05289 76963
9766: 77426 37527   34318 34227   13454 52414   81560 86316   16723 67071     15608 85708   21411 54372   67041 39418   95200 35025   91837 93109
9767: 68426 56941   88611 12209   48947 14490   48631 30983   51748 43434     34204 58301   74702 76198   52200 11737   98834 54577   85101 36729
9768: 41778 89407   36429 15751   91645 12984   93469 20935   32730 76494     07469 72756   24472 57435   87933 26347   98253 02456   64110 28949
9769: 92358 48208   08775 79244   62292 53703   45107 59116   39425 57575     13072 57863   17514 47436   04867 53094   99609 68053   49066 50046
9770: 24567 50627   18061 67802   88925 70889   04268 78299   75322 69066     29630 92766   93590 18004   98781 59845   14557 10622   43567 97591
9771: 11298 51833   87764 78482   26808 86090   09672 65170   51837 00007     75062 33805   17624 95403   79550 14718   26256 08078   86188 33133
9772: 80919 34791   17478 47065   30898 38830   56922 55812   28989 15515     48490 01683   68058 97555   55903 49162   84353 57042   26238 73712
9773: 80394 09812   09535 02719   39178 29500   76760 19707   10123 20099     83446 46467   57733 89996   76376 96959   66397 43889   49606 00927
9774: 87241 73606   27054 69172   13864 53104   52704 96748   09045 18998     75781 80787   58557 88235   87736 00385   09073 42900   17359 46524
9775: 20214 19093   29711 14505   99639 86485   08460 24852   33581 34807     81427 86581   85796 08019   47597 39872   24604 14584   42099 20375
9776: 24095 95369   87897 02218   18129 44894   62114 08747   34548 67952     52235 15290   20267 05082   36009 48040   02181 34417   57908 27044
9777: 32973 02987   14730 41024   53888 93987   96858 23932   79207 68685     60737 13120   51223 66216   83223 29917   91727 90007   93119 97995
9778: 12535 57308   98747 77178   59641 05114   99002 36839   11003 37943     69073 11117   11001 02783   76414 38672   22267 72890   95827 75977
9779: 29307 80338   22983 77821   62512 65103   94737 06720   90621 50855     16418 40698   89830 83060   59038 47306   90460 28660   02852 94250
9780: 21744 49510   44056 52239   33910 57984   60755 70486   43392 84669     31469 58438   39674 68689   66182 42471   25532 21758   20701 49835
9781: 15910 46373   90382 38862   06692 69459   78345 61340   11906 16576     59824 62123   04370 51254   06220 51447   32247 61120   74261 13160
9782: 71194 84983   03230 51567   92282 81039   35881 60547   48759 22720     27389 20751   35773 99234   86867 54604   33983 18698   08955 75759
9783: 84258 39493   80369 99961   94592 76184   24617 34312   73660 66472     06987 36502   10493 40153   44502 71189   28879 81571   01404 70830
9784: 51020 85849   54381 84358   35459 32176   07204 26162   41364 46843     14618 36723   46096 85566   78257 43842   56613 36170   19367 83508
9785: 90354 22414   99678 67833   42058 17913   50131 78409   27823 11837     75553 62863   11918 41241   73330 97682   66545 97308   01991 77994
9786: 17490 51746   68122 76355   44092 80325   55430 53816   46399 82813     34792 98856   84473 53918   27540 34291   97293 24461   26447 55846
9787: 97536 74860   22278 22957   59828 74853   63826 13584   05593 39744     33601 37614   62311 09032   59896 05839   84318 57377   03240 08746
9788: 44528 36721   34598 37116   05656 35378   23499 39988   38898 31955     55452 49697   55076 30777   82140 44178   87797 05226   73200 33746
9789: 24543 83424   63344 55155   92296 50076   55175 35097   72257 14659     20532 32438   28786 28033   77487 24535   13362 06566   52753 95710
9790: 86557 77812   57247 37491   15905 95472   04030 41842   11855 78370     72425 78567   79917 00726   52817 55792   93546 66661   14856 51485
9791: 14932 55716   36218 55752   30657 18668   01197 20474   05855 80051     57967 06295   25648 30634   58637 51530   73772 09182   06357 19603
9792: 55476 13812   97715 28085   56045 01490   37831 23485   44471 18993     78673 97977   67235 86685   67903 49088   03695 11464   34729 03826
9793: 72128 44344   28657 30621   13479 99229   08043 70017   53323 56713     92678 77394   77208 29242   97677 14108   51156 99332   22580 97788
9794: 88754 91217   86565 26598   48686 17026   46394 41584   11947 90957     30143 31547   90878 33027   68216 86979   74324 65535   30835 80960
9795: 93926 45264   76050 83468   41165 28199   43550 96726   28317 01438     62643 41776   14786 44183   68471 31546   91310 61412   40760 88624
9796: 98605 81911   98594 26909   59081 36629   05732 38277   46217 11537     76238 09560   81441 02427   90477 60587   02615 58250   03217 82357
9797: 06551 46483   80667 98931   55826 21974   95284 10281   96644 70869     76238 09560   81441 02427   90477 60587   02615 58250   03217 82357
9798: 99204 10285   48835 12575   78793 75680   63017 91686   38726 38499     52151 75502   77703 51397   69889 47788   03662 51682   70437 06745
9799: 00758 25822   41125 07567   32991 36153   70059 44658   21825 81908     56660 73276   39347 40374   89506 35167   67076 14372   25990 94547
```

9800:	67612 95666	20384 31143	70498 15455	19724 19760	76059 85133	60880 54157	90620 52513	61931 39457	32445 50271	81366 03557
9801:	85661 26172	22120 90056	10905 36268	92246 58889	76417 69855	51856 90440	44604 88094	93688 27703	82889 69159	83068 96906
9802:	76330 17238	38853 57662	51015 18769	76756 35298	70557 72666	91348 90534	29722 42381	71818 65552	46147 20866	99354 43892
9803:	75648 82826	67092 05528	15602 72575	46636 60504	53237 42091	64300 51567	96421 17109	74896 53254	16431 84164	25842 96424
9804:	79348 83279	90778 89618	56491 51358	59290 44944	72012 64048	40707 02174	04984 19700	01767 13551	81414 00219	86262 14561
9805:	73639 82952	79056 29063	07594 65430	53139 86879	77971 65950	05334 50839	12609 10468	29393 24052	33162 94079	44903 21274
9806:	69318 46832	66620 79756	13797 77471	67764 54227	14776 45411	31000 87403	94050 61724	36555 29744	78151 51294	06167 67557
9807:	95185 31957	35408 69907	11196 86150	61034 97789	96866 33172	34790 00233	03457 61101	26289 41354	67534 52483	97444 79263
9808:	62553 95973	03685 10850	26088 74340	45448 20122	63313 05130	83801 63323	38491 84439	11027 21791	09149 32057	04173 73141
9809:	96153 31351	74574 56561	05212 28928	41051 14848	84998 30218	35629 94681	38428 98342	62554 05066	73354 78606	16366 33092
9810:	10228 94275	59141 34247	67612 86285	57736 53975	97793 27590	08475 65768	26279 47030	90097 12180	02264 42232	72524 72251
9811:	89124 91346	45681 40024	46005 31588	60326 96076	60859 12055	38042 13790	45995 07363	42871 36907	97581 50735	67713 57572
9812:	17336 43309	79954 54947	61812 04052	41432 80803	86713 72483	62546 56511	93992 79315	36795 34426	57559 64874	54587 16774
9813:	01800 80766	16739 85290	19877 48214	14432 46322	92119 60174	79631 46343	85316 20618	55352 65574	18365 07600	97894 82800
9814:	14042 28193	02361 71131	17577 83852	34222 63732	50491 67276	17373 26329	61389 05256	37537 15914	13799 75687	10198 47219
9815:	79935 66647	83552 43788	13908 91402	30922 42326	23596 98367	98046 68167	43810 40342	26433 91118	43628 53717	72564 18710
9816:	69470 47850	43796 78906	23813 37492	59785 83481	87407 47814	53050 05830	44441 03266	09870 43242	10580 74271	49626 91239
9817:	07762 53580	79142 61143	90169 54751	30403 33366	91109 86345	60073 66017	55318 28282	47593 05002	95045 97690	29779 61688
9818:	33186 53891	73381 33626	71254 78372	25193 40404	74733 58443	27443 02199	06933 02574	06462 19466	86918 74490	44061 53863
9819:	57196 87107	51163 00301	71407 73728	21162 31041	50264 49081	55735 34645	51814 26724	05030 95174	80426 05539	08719 19745
9820:	16032 79711	38839 73453	59723 38958	24450 40653	95503 21245	74521 97090	47513 58356	03730 84066	79825 90283	38822 24606
9821:	02106 01230	04194 89571	84876 98828	72002 86926	53966 56824	24749 89102	75487 01139	27034 98952	85027 41773	33052 66490
9822:	01616 51448	86140 02164	08767 57398	78682 00506	01515 78739	32121 86524	44731 98966	14306 63336	89565 30759	61321 99117
9823:	75157 58703	45872 99735	76660 60297	83526 87955	00185 22079	52966 42258	78117 57035	09275 61120	27710 78843	63381 35787
9824:	84574 93878	39060 83196	56420 62578	76458 99541	04094 78591	27567 27161	99389 28746	90665 27847	50540 96277	98961 33440
9825:	56394 95037	55615 93123	28202 22506	02419 50939	61735 92940	55044 76648	85282 10462	59688 95365	61538 24155	96957 51992
9826:	91077 11754	98541 15708	03658 26576	73357 42046	80253 74621	25237 79451	26899 54254	13770 14702	98906 05818	75624 04254
9827:	72564 73844	10213 20837	43743 57524	85136 59149	60137 79190	08796 17908	34130 50788	33635 85754	48742 56462	27287 58449
9828:	31275 61186	77724 48586	74639 31304	94158 55988	34688 09711	28279 40745	41526 09582	20466 98181	13853 81197	95935 96732
9829:	76549 50566	37991 69617	65085 52191	35281 06590	02242 96911	45498 84370	49199 87574	54230 36849	02758 79788	99582 51929
9830:	08683 87982	15835 75878	91736 77927	67189 88525	45924 83241	67209 43144	55627 51319	85949 54228	05415 47764	42900 94000
9831:	06082 43734	79961 76872	25042 26193	10956 30958	88722 36373	41477 52024	04981 07767	76657 33925	56504 48465	65559 07084
9832:	44587 28178	29693 51133	00204 96694	23842 77533	89009 63140	66708 42171	88902 21524	71858 31492	53090 77559	40146 77698
9833:	06865 73398	07337 99786	60257 48551	16469 14464	90515 42978	13056 45220	70770 61674	10698 69142	96906 02216	79119 50208
9834:	25115 39129	22520 13282	14683 27470	51327 17346	00592 32881	89724 08584	49814 75766	14028 65450	91322 30576	98694 47398
9835:	49037 15286	82470 65782	93961 73396	28711 30972	82106 67361	73860 60853	34849 34068	62055 36568	28597 11700	29388 30275
9836:	81622 00025	23009 90081	62843 63538	09643 81968	57920 06197	24190 04731	97145 65228	97437 01554	21589 97930	39983 42143
9837:	02785 94047	73371 69165	75210 41212	83464 23368	44178 31065	68311 62302	59068 10113	35253 53829	36507 54329	93312 12240
9838:	22488 93471	80383 56796	80161 60650	90755 69132	34644 32038	54358 60931	79458 70467	98803 01908	73306 39087	46446 36170
9839:	78183 61870	64603 03761	05948 42041	95217 33484	99134 28829	71188 47649	90089 75459	02261 83479	48172 59713	72064 88631
9840:	67994 09819	99097 51826	03120 50951	07502 57261	00085 74828	75843 39798	62557 81901	35493 36608	89905 73584	65775 32881
9841:	72717 80126	22792 58139	12903 36106	24962 45175	47602 81869	04805 93462	51230 44470	34344 55315	61194 23582	94832 79902
9842:	33699 08571	33369 38897	03644 86981	96431 84622	86036 28056	25150 76327	52972 97497	96035 46759	88683 99222	92614 99730
9843:	71888 73727	86596 98864	95640 32180	10888 32567	43331 41477	82928 69577	34087 15653	53315 60809	65127 24462	08330 18928
9844:	98218 17520	95909 91415	18796 81536	27024 11411	93700 68832	08815 62038	62725 89910	93737 46879	40820 87835	30485 58840
9845:	17831 76904	62958 88876	57856 20024	59106 97134	83136 07877	64363 06975	75905 14996	68041 42784	28241 00124	56790 91689
9846:	24735 52344	96663 52162	50967 97158	56420 99949	08852 82106	79149 23691	56915 23265	69733 77448	02082 31621	75706 35936
9847:	06016 80833	25593 17227	50539 35787	66751 97206	69556 97792	77486 39663	89811 44280	49171 46200	50991 29480	91899 03753
9848:	25648 49572	19766 89766	16183 89368	22350 60674	75308 54853	71238 48243	57405 54766	85209 53022	14210 91738	25141 27729
9849:	48162 49553	04582 01801	58622 29069	18452 69346	72955 76612	86370 35368	70591 87430	78000 59470	72997 45280	66076 93425

```
9850:  02737 31779  44476 12940  80654 58936  51677 09548  75886 69883    12715 88424  54520 76944  91204 08465  35431 32314  53292 22349
9851:  38965 48008  16258 10014  93042 56210  30924 72822  78536 59332    93921 21477  03666 66066  03418 54126  63749 81683  43547 16499
9852:  20220 20257  99278 79420  89297 61170  33330 32456  14215 13236    96794 08861  45667 07774  43535 02081  23820 52519  65755 28990
9853:  53353 24171  65521 16149  63106 88156  45134 35888  59741 57898    43659 45695  31924 64085  01249 96474  73818 92812  72426 46043
9854:  59811 08331  99450 66705  06600 95736  32317 75610  45729 97267    14208 42480  82510 95946  40312 05984  23130 23412  43607 48791
9855:  21132 21562  72684 03077  39175 55948  76192 43608  52251 43768    59817 76563  71026 05283  82494 44082  67629 89120  63088 51944
9856:  05393 68574  10081 67338  99464 57470  97028 41574  44076 34377    53784 33425  41739 39087  47389 28142  49489 95785  03411 11901
9857:  88843 35021  16758 74868  29756 20365  53180 27431  50088 49555    50009 10853  68801 59289  21951 89439  31251 11850  25697 01074
9858:  91711 46263  31637 03657  90642 11646  77973 13181  18685 17338    71081 39640  46639 91100  58985 83089  77793 06127  94512 51989
9859:  90404 26051  73859 64511  18114 66547  24620 18238  88300 66348    62002 80524  56773 74998  80185 71399  26522 51922  00774 02914
9860:  37836 27994  07082 27249  38074 96712  21608 31196  27107 87370    93810 51066  27664 26840  88337 68679  08230 12952  38694 36787
9861:  97140 18712  17655 86030  44773 15755  50255 08239  83579 76461    10212 60796  79165 90581  07542 91027  17776 60569  78935 76467
9862:  42964 83510  55876 51451  58520 98899  84695 17480  65413 21910    79794 25784  29165 96470  26614 94363  13763 88457  95386 31641
9863:  82191 46891  81031 27554  24591 59585  08253 21168  41594 60579    81407 29230  08930 39843  04828 26377  65926 45139  21088 76545
9864:  81577 58990  59357 54128  85697 69193  08173 75443  29503 44299    69630 53763  90838 02872  85180 93452  81276 97096  21732 12027
9865:  53977 47662  04335 83996  48196 18531  64361 78129  28329 30385    50985 29828  11241 16537  65751 61702  61075 58052  03013 84370
9866:  32504 52057  93465 20908  84273 69944  07170 84661  21313 17113    89633 75997  90340 84835  60705 09350  53219 01260  50668 30686
9867:  92680 32816  01813 04192  72608 15585  69596 81696  51059 82939    96378 90867  34062 59007  38487 04152  47593 12671  89960 51253
9868:  30426 76470  84000 08790  17531 18855  95028 04352  91501 22773    45270 25822  35978 29564  07907 88954  31887 62004  20864 15474
9869:  38825 76416  19687 15124  03841 15609  20249 79042  25751 59649    76799 77678  82063 77419  63067 88629  27058 19185  32638 09326
9870:  15767 86359  52681 10336  63536 36514  74658 23278  17114 87445    82056 27730  50926 77174  66525 10924  69483 24672  79447 51543
9871:  33297 84222  43173 09487  59521 13150  45458 48231  84113 77417    87449 72796  12472 42064  98487 14820  66153 46303  16406 92396
9872:  55653 05655  31330 66908  98421 88352  67109 28838  14548 46837    41581 76989  92232 98488  35291 86270  16639 15179  36777 38193
9873:  75362 64638  53105 33711  77079 79609  22638 47272  71411 57910    79609 90366  38874 70822  76566 88406  74575 09314  44762 43876
9874:  87389 20163  30638 62647  65061 71412  18397 78988  07147 79348    25681 71226  97048 06755  11537 35061  47746 71662  97036 01327
9875:  62252 06841  75704 36499  18211 18364  94743 79010  00408 15190    33423 37141  95297 35034  72704 90105  20500 76430  97362 05271
9876:  29422 96486  17620 57935  73576 06161  78943 72819  59295 94610    96110 56300  26781 53382  06871 94950  32531 87707  79160 29713
9877:  21432 09984  73385 65820  07420 80989  22723 94938  62331 57389    68146 33688  85222 15258  79739 84740  10483 25053  33505 46737
9878:  47457 73955  65911 47196  18043 50236  02451 84184  80318 02312    13346 02908  69041 94024  75643 83344  95273 73378  62829 44990
9879:  01904 78419  51521 53733  96537 81371  09286 28443  06146 12849    27582 73788  79190 08431  53999 62582  62440 48132  78037 23689
9880:  29636 78558  77881 71353  73063 21277  96984 01036  41712 09811    83370 20238  35392 22341  67195 19264  75153 20592  03500 05854
9881:  98491 79894  39237 79373  66464 68567  07942 46639  86029 84547    42631 94931  55649 10386  86914 33722  01161 79260  53969 83233
9882:  52771 43404  96152 14202  74664 57011  83338 78325  25394 84117    79645 32305  71107 20273  18225 25959  92396 54819  59958 67307
9883:  90516 11605  20604 49699  47711 94257  09681 84941  37388 24110    61658 49864  56350 44541  05319 25715  72881 08597  24597 60073
9884:  49899 43101  62369 25785  64125 27019  74210 91085  55035 65951    07706 16048  99298 20981  00486 77325  60948 37362  68238 65697
9885:  00130 13972  29455 71609  15081 52430  99159 67047  19852 06571    07431 11868  77088 41877  19160 99163  69444 87235  42831 54414
9886:  53819 75707  76042 74572  62078 68209  05907 41660  29516 89625    61503 08603  81219 87734  07770 24178  03412 88583  32679 11181
9887:  88113 03788  67154 30584  85299 33093  65471 16687  29516 89625    76229 97974  87655 71452  54890 13802  00761 10028  66940 45359
9888:  03527 35215  96231 05571  51576 68389  46484 63441  23014 35471    25754 81487  54478 55685  01152 69486  56327 06222  07400 07712
9889:  44939 40864  05665 82127  41203 91232  28422 63202  84553 04741    21987 86763  71293 28744  58160 65626  64398 87714  06820 74629
9890:  99445 31637  53713 29293  06533 34393  46635 58487  73881 93019    13421 15966  12233 86693  14849 78732  05796 28319  02548 52960
9891:  54248 35449  25527 48366  40646 82118  35492 81137  24655 60500    33011 11672  36209 16966  65186 73791  89910 94286  51934 69268
9892:  38511 50540  44600 94845  80660 66162  73990 77071  13766 25655    73834 90642  02614 85523  61810 01806  80811 71886  44791 32397
9893:  35332 82876  73894 12383  94444 42475  96833 01284  09704 62032    34810 98115  03044 63351  82951 62684  02619 60037  09015 66656
9894:  68741 31715  25896 77202  36806 73853  65368 39134  23263 55227    99009 44180  32885 77161  49390 46072  38680 32859  82261 60152
9895:  64883 94558  15317 96157  92375 82385  25900 80141  94240 34251    86798 69327  76508 74190  58441 71056  87113 31401  03952 89898
9896:  81252 97474  44875 23230  69615 81022  05421 92085  47557 29529    44704 33229  70798 49245  70847 59633  19494 87159  81078 88257
9897:  08335 81883  23535 87204  02463 27255  51016 20569  05740 60645    29853 34217  85722 31447  09200 37417  30733 99257  67350 74353
9898:  12609 74655  87680 10524  46226 33216  39849 76692  50685 11445    44553 32553  54553 06415  67699 94913  66567 42289  70995 43737
9899:  12108 83753  95137 93934  88038 16053  50695 17767  75599 10269
```

```
9900: 78953 24342  10342 89152  98196 70134  97306 16012  01192 89720    93156 85330  63146 95063  82848 16508  18525 47253  07261 85288
9901: 24582 40276  36563 93278  08873 23685  09306 45109  06032 11017    62183 48870  22555 91284  73916 91269  33155 42141  53051 34124
9902: 52408 17496  54318 77236  39332 20936  05459 52979  58528 54656    59361 64549  79887 22745  55633 40369  23179 87019  42196 19417
9903: 57131 82790  09548 42107  14828 72710  83778 02893  48658 14649    57804 14260  17814 96555  42273 56976  72457 85197  19288 27328
9904: 40013 54325  94447 23280  68218 00987  06131 46481  37991 82944    12343 84077  69992 32052  12668 31351  06892 68670  87443 99366
9905: 01727 10237  58770 60887  25683 07489  49934 55007  99277 71388    05602 05595  18513 23123  63663 12166  37899 51729  63061 24239
9906: 63192 00815  19231 66060  67945 37956  23655 88817  46361 87668    30814 90443  18313 93130  01689 78339  48839 05073  10048 52471
9907: 17722 15904  40193 09285  65550 62080  44227 11489  18150 81305    19014 42568  16601 40328  38632 84690  42734 67665  37219 37464
9908: 20748 10413  49983 38785  40657 56488  98793 37310  71487 37479    78187 86259  52937 24169  61508 34047  93838 57136  74754 72333
9909: 19880 37397  16970 26989  66312 79129  28309 53359  05341 44769    91312 47065  61685 70763  06340 02785  91143 44424  91179 55736
9910: 64315 34807  45070 18740  87682 65189  15925 79780  28908 65140    73587 65407  96262 16393  17759 68860  08006 07554  27920 05882
9911: 85037 32777  64069 06937  59687 00206  38095 81682  70177 55100    99725 66914  54286 37910  32559 65657  08225 77294  90828 91501
9912: 64110 29460  09093 47587  91920 09797  17042 19642  00514 35926    94629 42801  24765 59206  45026 89288  07661 80106  79728 28251
9913: 19826 07493  86899 14392  06506 54252  19540 50699  11880 47308    99182 80011  99352 62097  90035 91483  69919 84300  27514 96824
9914: 34983 51586  69212 43399  31630 68243  99635 73535  25546 33899    84267 03849  13052 80404  49813 72243  55625 94711  64553 86063
9915: 09516 33805  72319 24126  76379 01407  74356 97812  01219 41578    30626 45983  07098 40922  32915 89700  18871 94434  41551 20756
9916: 04592 73861  29644 32447  23037 76597  77817 15931  60467 27926    08908 38397  88932 65296  50341 44592  62395 04670  14916 26921
9917: 15242 99327  20539 76889  24416 04488  07882 44239  04618 17122    40597 08132  39605 53868  13437 94327  32538 46635  62226 78324
9918: 64257 27104  44103 86540  29632 40805  81788 89245  80869 05238    02391 28582  99177 87156  50895 60562  37067 62557  14109 42056
9919: 08475 56983  66950 49024  59303 15920  60043 80350  22619 36004    17763 87223  05625 05247  87072 84493  22048 19923  98132 05755
9920: 49021 38754  55249 29612  65501 34983  23451 54411  11684 61234    18649 11272  46953 52916  57075 91273  42155 32233  28003 88121
9921: 67227 24420  05933 00652  17676 37061  55661 10693  01714 85946    01746 74156  74527 97628  97467 77096  12845 01836  64690 10694
9922: 18750 40110  08576 18627  50048 31154  78520 26011  00615 38646    26856 20515  58937 19465  58862 02077  60023 42107  88411 52735
9923: 44358 99365  00533 50219  90677 11317  26574 03039  33714 26920    64892 99357  23608 62255  54674 96307  52045 03907  58204 00667
9924: 60120 20806  19643 87880  20191 34573  85931 00610  55928 75209    66985 78223  64271 55883  41657 88688  18345 66951  17548 80175
9925: 57038 22544  38924 98424  52603 68387  67187 84586  36246 03731    38611 15874  34919 24820  08061 96483  34884 61815  62182 63645
9926: 04287 79172  88681 74958  30426 61474  99010 03106  03805 20777    17482 04457  82202 73012  33435 49952  84222 44962  16291 45746
9927: 58412 59149  76363 63210  18893 96534  76874 03515  56198 46440    25146 44778  83907 67028  60898 21420  45354 01369  10788 15017
9928: 30272 84559  22957 19038  61446 32096  59612 82809  89242 81524    00505 86577  01350 06617  75872 85476  39133 23041  61649 12934
9929: 10451 27320  44046 85190  31534 15966  53828 44292  47698 38964    83070 34267  29612 84840  91493 54245  55334 95744  54569 02504
9930: 75591 96668  97417 02780  25900 80780  41060 88469  52451 41368    51624 16479  42922 27332  04846 46603  92851 98869  84253 76275
9931: 73052 86831  05808 71931  46109 99352  69820 64478  55496 98357    61767 12075  07240 73777  36171 51536  61467 21567  71212 72199
9932: 28123 55524  39599 61703  99487 23910  19215 07518  57113 78847    26464 45104  10185 61917  83227 49563  58586 24049  83456 69179
9933: 06339 02846  64332 50450  76344 72738  53269 31029  59143 34730    12073 91731  77203 32239  91373 26780  71971 71426  26582 36039
9934: 41699 73431  00323 08273  93172 90580  64084 66847  69553 28806    20549 58266  42326 43664  53867 72198  86970 70881  92312 46184
9935: 74242 75250  63389 19630  28190 94307  47391 43145  47116 71596    87858 16862  37076 44733  80048 93362  48342 93152  86323 99946
9936: 98601 79293  14339 32450  60059 83180  08044 80426  44313 96942    27649 04382  88387 84028  20497 54885  24356 69184  42375 59327
9937: 15317 65165  43079 10216  42363 00163  12273 08477  06198 99478    00979 37100  42007 64439  46451 95081  86534 02226  33307 09666
9938: 65663 50943  25383 34024  71977 80164  05730 85056  54424 39182    05594 80050  78831 54840  46233 83907  93229 99451  00306 09908
9939: 06580 86904  88017 10450  50555 02893  55066 26938  34682 17965    54133 80207  98293 01392  89185 54299  54043 40883  75906 28815
9940: 98629 96844  08052 54897  71920 28879  61378 66659  06791 66884    42806 12593  91551 84297  75754 41361  36479 85891  47490 16016
9941: 52674 05220  34315 93820  25393 51656  56183 84985  60854 87860    15893 15485  75822 15283  95477 44639  22421 06259  66157 51299
9942: 34743 30635  25440 57102  42241 15986  89895 93184  41226 38380    40787 01400  00805 95144  81599 00666  62480 49692  13728 05753
9943: 72122 70551  70966 84492  10305 22608  22643 17115  67996 50571    18298 04238  53322 01500  30862 38866  25706 36352  93915 60757
9944: 32159 19358  98609 64571  76407 73742  82803 03485  64075 31042    01498 47171  93715 50703  13556 58856  10246 30818  47838 39252
9945: 73730 55117  82672 24263  33917 15945  71312 98225  41971 74123    62316 42413  06304 26495  14227 73940  12890 91724  99120 67032
9946: 91614 59582  79542 47009  96193 12237  99280 72778  71138 21471    76920 58715  39878 69973  91588 13874  55726 50032  76008 49401
9947: 45864 61266  26277 62854  47656 92065  46554 74333  47697 55677    52474 99516  26212 86213  65151 85475  54982 92775  47530 48183
9948: 23815 91972  13761 54376  72829 25410  03653 66946  21305 75614    29670 04615  75983 30828  24951 41333  05597 43269  11668 53328
9949: 90165 64629  99129 98222  67687 72597  70763 96483  33111 56934    60938 26856  33044 96224  90509 40505  33661 33817  32286 55089
```

```
9950: 12038 66428  94875 32208  37713 05026  72144 35922  07587 47311     39558 12853  64792 21805  58241 45631  74819 59570  93682 87559
9951: 77192 70606  56646 67354  35480 14667  44439 37081  20806 57410     84587 87970  08749 67713  87181 34548  97976 29474  49110 22717
9952: 59340 53611  65054 91858  62871 68289  15840 63343  05869 08924     98807 88346  96318 46796  52638 88516  89381 62584  43703 76967
9953: 51020 05004  09673 09129  49895 68339  31521 13051  58501 31319     14499 00848  36666 89081  28638 45342  66056 91404  88946 91165
9954: 32581 89620  81362 54582  77559 79860  13241 41874  47395 88312     25806 65432  70892 68673  26893 41957  56550 20841  33840 71596
9955: 09846 58306  71036 17445  73825 33741  11378 29241  13162 70601     15476 50672  54404 00649  01999 77179  39833 15669  94015 09959
9956: 15195 79229  83248 96389  80152 45518  76603 05490  31276 09844     55924 48280  85982 97043  89247 06595  54924 58681  41117 74311
9957: 72690 81112  03342 88451  64863 15957  02611 29162  15673 31981     51587 88312  61448 40535  57641 41197  18801 29648  62555 16494
9958: 05020 72344  75479 90999  65262 85534  07598 06892  24059 79167     03771 30174  32617 11895  63538 27034  67017 64662  21347 50570
9959: 06564 74436  85837 29931  93336 71342  35722 70853  63355 91422     74052 63659  64105 58254  00238 37853  14285 01053  92985 31611
9960: 39814 78687  08748 18824  75847 51079  79516 78533  42374 15697     50078 18207  38153 96793  23453 93955  09543 46686  27814 60422
9961: 18726 73557  07721 46751  03276 11818  20239 46973  27239 18340     50078 18207  38153 96793  23453 93955  09543 46686  27814 60422
9962: 67338 49795  28406 01516  29565 73268  55103 45239  73582 10144     49027 78361  22373 01573  85090 26824  27458 09272  40825 72871
9963: 42040 63416  46523 68891  78441 12709  03494 76083  92357 60829     24296 77237  26701 56914  73662 40561  55941 67283  93350 47981
9964: 34379 64162  26395 59957  94673 09681  12108 15260  09710 98715     66013 99373  71258 31641  31126 84174  09066 18489  24890 55496
9965: 92329 96071  13858 08852  24042 20812  98782 08942  01275 08008     53507 73994  40937 22470  26070 62358  35059 61637  98122 84648
9966: 91494 21941  52326 85119  91788 32006  20166 65958  66710 51920     11205 16432  33887 63334  10539 62654  40126 29308  30816 37353
9967: 23831 51972  03845 11769  43951 64802  65368 14629  51835 75310     07898 37730  84756 47599  99002 27332  26946 21335  58494 79114
9968: 26002 10233  68084 85890  20985 57438  37649 85442  08837 62114     71483 91650  57858 43315  04863 12443  87377 88479  59516 32988
9969: 12357 19818  44162 96558  44467 28902  52491 69441  98089 44167     81721 93865  66928 59909  13893 71618  70927 19650  21127 78796
9970: 41272 93652  55555 60861  36435 95742  12445 76242  56104 80371     39193 55336  18613 66139  56224 30318  34060 98980  74339 97033
9971: 66189 14218  76142 05502  47554 21656  52555 90401  65914 20720     51951 70391  87220 99726  21271 40502  25732 19020  19832 31721
9972: 99065 02689  56926 98312  26893 91482  66685 39189  69079 29546     83566 79663  68410 29004  64288 37305  58467 70564  22483 40054
9973: 58666 10866  07836 58030  80532 72391  14173 62661  84017 19966     93655 72479  39912 79025  54244 03477  25841 51685  47200 52210
9974: 23932 46820  30686 20216  99945 38062  16672 91991  15568 49696     04596 91699  80809 41049  90246 89514  01056 06394  37336 48669
9975: 61065 24911  20510 13064  74614 57201  40829 63004  57548 21306     50991 50334  50179 66081  13009 42912  13395 78299  25626 12007
9976: 11342 59868  23217 51258  96777 65532  25652 49114  31431 70965     56639 94367  52100 15379  18666 85678  18640 34673  92308 19822
9977: 91400 87382  67187 05565  52861 86889  21954 31612  14099 08603     80757 48520  77240 27565  72833 90699  54265 52793  50111 03110
9978: 57899 70576  83135 87551  52264 18922  62580 34484  87658 93576     03625 53977  32278 39101  86237 19006  88300 14226  85712 97301
9979: 94014 09923  11385 59667  76627 25575  25310 89909  34222 39121     38111 39861  02058 81659  05084 41959  07778 32471  35135 66582
9980: 75628 94889  04650 04454  49837 01032  01114 18159  84830 34119     80576 12490  94750 35461  39917 08260  66637 94753  90882 09342
9981: 77951 61881  93867 88611  33264 86789  87439 50705  15628 80237     68139 09552  73643 42391  37105 66241  44777 42069  70923 04403
9982: 83678 35593  67430 83779  57203 15111  65266 16308  96058 42310     52455 03671  43515 07630  10198 27882  18510 04172  81645 19387
9983: 04201 01892  99055 50655  32674 93394  92092 98076  57728 65168     49313 47398  76485 59199  75985 45230  09715 51790  86242 79492
9984: 06666 78421  17130 17524  68645 14508  91092 76904  95517 26007     98548 82617  84481 85729  17178 28897  30371 55938  82488 16311
9985: 26257 00908  08727 36360  15093 28011  79171 29886  41207 32134     61400 58227  86289 34951  27967 16532  41870 41865  44206 83648
9986: 48484 13924  77352 61020  70873 71631  68966 40240  73600 46957     24532 03964  31973 63651  23264 75195  30102 49667  75351 88376
9987: 42323 25591  98848 60786  03065 51701  07471 56257  64600 94002     59297 08214  49565 03242  96457 74980  67029 74430  86466 69736
9988: 67147 20399  63420 90498  44304 12079  44678 40209  05818 10696     12521 23425  30663 85986  28153 73539  86934 72360  75571 94984
9989: 11969 84242  35593 84898  70675 10537  34068 37890  09326 48159     63090 15011  50140 70000  45052 40592  23271 44504  68225 14055
9990: 59848 60023  19896 57393  28031 56531  02478 30663  81783 57677     70253 37621  29803 56560  99156 93945  37555 25782  62886 70307
9991: 59028 42648  24606 15959  83739 84212  96955 12749  00448 23751     48310 84638  59058 02932  96607 24893  40702 98920  07953 24130
9992: 45823 80171  93008 13983  98077 17044  29928 23010  70831 39180     04740 94823  63047 20841  80367 06935  50507 98902  32814 93815
9993: 85148 96765  85728 35244  11491 39998  19662 66816  78867 63665     91646 49234  10576 49091  96597 39359  17849 92143  80333 75635
9994: 69084 99060  73694 42978  44429 41797  77979 20195  45072 26482     72840 46245  90816 09169  29314 51978  40611 56143  71062 31990
9995: 97587 12335  09317 99139  28293 94331  50256 34562  19371 36102     18821 22659  29901 52759  32653 47461  54380 34745  50690 22074
9996: 53506 43891  41823 96348  32163 84673  42388 96489  93367 79889     93907 50536  44783 71958  80507 29654  25900 24250  22150 06348
9997: 56641 80892  64944 99201  87584 01815  23053 81652  93364 79889     53683 83679  75565 86672  97585 18654  53321 21636  66623 54382
9998: 90333 47701  04688 77080  71700 83708  03517 59840  01218 54515     65791 99607  10027 77576  66514 71445  12254 14373  62331 01660
9999: 55739 20359  68769 20360  07938 90885  23734 86970  10132 51836     95518 00323  36537 24450  82229 58164  74145 35489  25764 39998
```

One Random Digit

0

CPSIA information can be obtained
at www.ICGtesting.com
Printed in the USA
BVHW020832080223
658123BV00005B/209

9 781949 815030